MATERIALS SCIENCE RESEARCH
Volume 18

DEFORMATION OF CERAMIC MATERIALS II

MATERIALS SCIENCE RESEARCH

Recent volumes in the series:

Volume 8	**CERAMIC ENGINEERING AND SCIENCE: Emerging Priorities** **Edited by V. D. Fréchette, L. D. Pye, and J. S. Reed**
Volume 9	**MASS TRANSPORT PHENOMENA IN CERAMICS** **Edited by A. R. Cooper and A. H. Heuer**
Volume 10	**SINTERING AND CATALYSIS** **Edited by G. C. Kuczynski**
Volume 11	**PROCESSING OF CRYSTALLINE CERAMICS** **Edited by Hayne Palmour III, R. F. Davis, T. M. Hare**
Volume 12	**BORATE GLASSES: Structure, Properties, Applications** **Edited by L. D. Pye, V. D. Fréchette and N. J. Kreidl**
Volume 13	**SINTERING PROCESSES** **Edited by G. C. Kuczynski**
Volume 14	**SURFACES AND INTERFACES IN CERAMIC AND CERAMIC–METAL SYSTEMS** **Edited by Joseph Pask and Anthony Evans**
Volume 15	**ADVANCES IN MATERIALS CHARACTERIZATION** **Edited by David R. Rossington, Robert A. Condrate, and Robert L. Snyder**
Volume 16	**SINTERING AND HETEROGENEOUS CATALYSIS** **Edited by G. C. Kuczynski, Albert E. Miller, and Gordon A. Sargent**
Volume 17	**EMERGENT PROCESS METHODS FOR HIGH-TECHNOLOGY CERAMICS** **Edited by Robert F. Davis, Hayne Palmour III, and Richard L. Porter**
Volume 18	**DEFORMATION OF CERAMIC MATERIALS II** **Edited by Richard E. Tressler and Richard C. Bradt**

A Continuation Order Plan is available for this series. A continuation order will bring delivery of each new volume immediately upon publication. Volumes are billed only upon actual shipment. For further information please contact the publisher.

MATERIALS SCIENCE RESEARCH • Volume 18

DEFORMATION OF CERAMIC MATERIALS II

Edited by

Richard E. Tressler

The Pennsylvania State University
University Park, Pennsylvania

and

Richard C. Bradt

University of Washington
Seattle, Washington

PLENUM PRESS • NEW YORK AND LONDON

Library of Congress Cataloging in Publication Data

International Symposium on Plastic Deformation of Ceramic Materials (1983: Pennsylvania State University)
Deformation of ceramic materials II.

(Materials science research; v. 18)
"Proceedings of an International Symposium on Plastic Deformation of Ceramic Materials, held July 20-22, 1983, at the Pennsylvania State University, University Park, Pennsylvania"—Verso t.p.
Includes bilbiographical references and index.
1. Ceramic materials—Congresses. 2. Deformations (Mechanics)—Congresses. I. Tressler, Richard E. II. Bradt, R. C. (Richard Carl), 1938- . III. Title. IV. Series.
TA455.C43I58 1983 620.1'40423 84-9938
ISBN 978-1-4615-6804-9 ISBN 978-1-4615-6802-5 (eBook)
DOI 10.1007/978-1-4615-6802-5

Proceedings of an International Symposium on Plastic Deformation of Ceramic Materials, held July 20-22, 1983, at the Pennsylvania State University, University Park, Pennsylvania

Softcover reprint of the hardcover 1st edition 1984
A Division of Plenum Publishing Corporation
233 Spring Street, New York, N.Y. 10013

PREFACE

This volume "Deformation of Ceramic Materials II" constitutes the proceedings of an international symposium held at The Pennsylvania State University, University Park, PA on July 20, 21, and 22, 1983. It includes studies of semiconductors and minerals which are closely related to ceramic materials. The initial conference on this topic was held in 1974 at Penn State and the proceedings were published in the volume entitled "Deformation of Ceramic Materials."

This conference emphasized the deformation behavior of crystals and polycrystalline and polyphase ceramics with internationally recognized authorities as keynote lecturers on the major subtopics. Several papers dealing with cavity nucleation and creep crack growth represent a major new research thrust in ceramics since the first conference. This collection of papers represents the state-of-the art of our understanding of the plastic deformation behavior of ceramics and the crystals of which they are composed.

We are grateful for the suggestions of our International Advisory Committee in recommending experts in their respective countries to participate. We are particularly grateful that the organizers of the previous Dislocation-Point Defect Interaction Workshops agreed to participate in the Penn State Symposium as an alternative at the suggestion of Prof. A. H. Heuer. We acknowledge the financial support of the National Science Foundation for this conference.

We thank Mr. Ron Avillion for assistance in planning the three day meeting and for coordinating all of the events, and Ms. Linda Rose for her expert assistance to the registrants. Thanks are also due to Ms. Judy Bell, Ms. Linda Decker and Ms. Marian Reed for indispensible secretarial assistance in organizing the program and bringing the volume to press.

July 1983

Richard E. Tressler
University Park, PA
Richard C. Bradt
Seattle, WA

CONTENTS

DEFORMATION OF COVALENT CRYSTALS

DEFORMATION OF OXIDE CRYSTALS

DEFORMATION OF SILICATE MINERALS

CREEP OF POLYCRYSTALLINE CERAMICS

CREEP CRACK GROWTH AND FAILURE IN CERAMICS

ELECTRONIC EFFECTS ON PLASTICITY OF COVALENTLY BONDED MATERIALS

P. B. Hirsch

Department of Metallurgy and Science of Materials
University of Oxford, Parks Road, Oxford OX1 3PH, UK

INTRODUCTION

The plastic properties of insulators and semiconductors can be affected by electronic mechanisms under different conditions:-

1) At "high" temperatures, when plasticity is controlled by creep and climb mechanisms involving diffusion, grain boundary sliding

2) At "low" temperatures, when the dislocation velocity is controlled by the Peierls stress

3) By the injection of charge carriers, as for example in photoplasticity.

In the first and second case the charge carrier concentration is that corresponding to thermodynamic equilibrium. The phenomena at high and low temperature have as a common feature the movement of a lattice defect, i.e. vacancy, interstitial, dislocation in the bulk or in a grain boundary. In a covalently bonded crystal the structure of such defects involves strained or broken bonds, and these are associated with electron energy levels in the band gap. This leads to the possibility of charged as well as neutral defects, and the relative concentration of these in the thermodynamic equilibrium with the electron gas depends on the position of the Fermi level, i.e. on the doping level. (Similar considerations apply to ionic crystals.) The motion of the defects involves the breaking of bonds, and the saddle point of the process at which the bonds are broken will be associated with deep levels in the band gap. The saddle point energy and therefore the activation energy of migration of the defect is likely to depend on the charged state of the defect.

The effect of charged defects on diffusion and creep is well established (e.g. Blanchin, Castaing, Fontaine, Heuer, Hobbs and Mitchell 1981). Recently enhanced grain growth has been observed in doped polycrystalline silicon (Wada and Nishimatsu 1978, Mei, Rivier, Kwart and Dutton 1981, Smith, Tan and Fontaine 1982), and this effect could well be due to a mechanism involving a charged defect in the grain boundary. This paper will review the present state of the theory of the effect of doping on the dislocation velocity in semiconductors, under conditions when the generation and motion of kinks are rate controlling, and discuss a method of obtaining information about the effect from indentation experiments (Roberts, Pirouz and Hirsch 1983).

THEORY OF EFFECT OF DOPING ON DISLOCATION VELOCITY

A number of theories have been proposed to explain the doping effect on dislocation velocity (Frisch and Patel 1967, Haasen 1975, Kulkarni and Williams 1976). More recently, however, the effect in the Peierls stress regime has been associated with the generation and movement of kinks (Hirsch 1979, 1981a, b, 1983, Jones 1980, Heggie and Jones 1982, 1983), and this theory will be reviewed here. The experimental effect was first established by Patel and Chaudhuri (1966) for Ge, and since then the doping effect has been observed in Ge, Si and in some III-V compounds (Patel, Testardi and Freeland 1976, Erofeev and Nikitenko 1971, George and Champier 1979, Erofeeva and Ossipyan 1973). Figure 1 shows some recent unpublished results by Pirouz and Freeland on Ge, using a wide range of doping levels

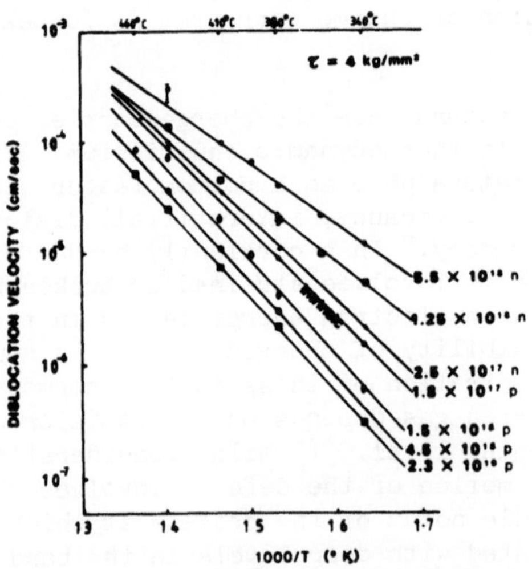

Fig. 1. Log dislocation velocity as a function of reciprocal temperature in Ge specimens various doped with As, Ga (Pirouz, Freeland unpublished).

and the single stress pulse dislocation etching technique. These results extend the original results of Patel and Chaudhuri (1966) to show that the dislocation velocity is greater for n type and smaller for p type, than for intrinsic material. For Si n and p type doping tends to increase dislocation velocity, although n-type doping is more effective than p-type (George and Champier 1979). It should be noted that for Ge at low temperatures (∿ 340°C), the dislocation velocity for highly doped n type Ge is almost two orders of magnitude greater than for highly doped p type.

After deformation at relatively low temperatures, dislocations in Si and Ge tend to lie along [110] crystallographic directions; the basic types of dislocations in the diamond lattice are 60° and screw dislocations, and these are generally found to be dissociated into partials. There is good evidence for this for Si, Ge and III-V compounds; recently dissociation has also been established in diamond (Pirouz, Cockayne, Sumida, Hirsch and Lang 1983). The basic partial dislocation types are 30° and 90° partials. Dislocation motion is assumed to be controlled by the Peierls mechanism, and the dislocations move by the generation and motion of kinks (fig. 2). The theory of this mechanism has been developed for two limiting regimes (Hirth and Lothe 1968). At relatively high temperatures and low stresses the drift model applies in which the velocity is proportional to the concentration of kinks ($\exp - F_k/kT$, F_k is the kink formation energy), the velocity of kinks ($\alpha \exp - W_m/kT$, W_m is the activation energy for kink migration along the dislocation line assumed to be controlled by the secondary Peierls potential), and the driving force, τ^n (τ is resolved shear stress, and n is a constant $\sim 1 \rightarrow 2$). Hirsch (1979) developed a model for the doping effect within the framework of the drift model, in which doping affects the concentration of charged kinks, and the activation energy of migration of charged kinks.

The basic ideas are as follows:- The cores of the 30° and 90° partials are considered to be reconstructed; reconstruction for the 30° and 90° partials involves bonding between adjacent pairs of atoms along the core and across the core respectively. Calculations due to Marklund (1979), Jones (1979), Altmann et al (1983)

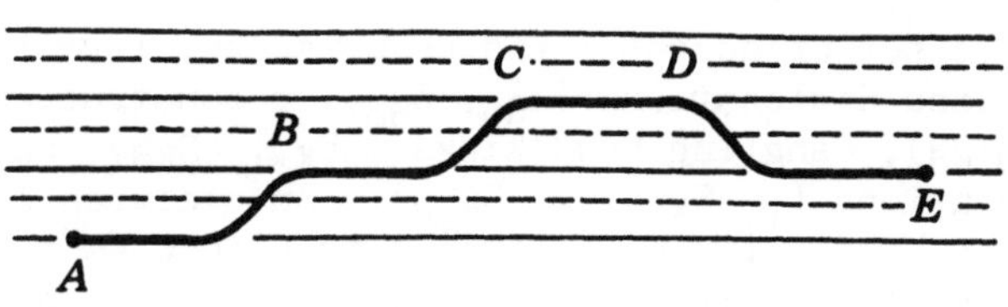

Fig. 2. Movement of dislocations by generation and motion of kinks.

suggest that for Si reconstruction for 30° partials is likely, but the conclusion for 90° partials is more uncertain. Reconstruction of this type doubles the periodicity along the core, and for both partials any particular atom has the choice of bonding with one of two neighbours in equivalent positions, leading to different relative displacements. Where the bonding changes from one set of neighbours to the other along the core, an antiphase defect (APD) occurs, associated with a dangling bond (Hirsch 1980). This defect has been called a soliton by Heggie and Jones (1982). Kinks in the partials can be either reconstructed or are associated with dangling bonds (Hirsch 1980, 1981a, Jones 1980). An APD can react with a reconstructed kink to form a dangling bond kink, and conversely APDs can be generated by decomposition of a dangling bond kink.

Figure 3 shows schematically a set of energy levels which might be expected from reconstructed dislocations and kinks. Acceptor and donor levels for reconstructed dislocations (E_{da}, E_{dd}) and kinks (${}_1E_{Ka}$, ${}_1E_{Kd}$) are likely to be relatively shallow levels; dangling bond levels associated with dangling bond kinks (${}_2E_{Ka}$, ${}_2E_{Kd}$), and the saddle point configuration for the two types of kink (where bonds are broken, ${}_{1,2}E_{Ksa}$, ${}_{1,2}E_{Ksd}$) are expected to be associated with deeper levels.

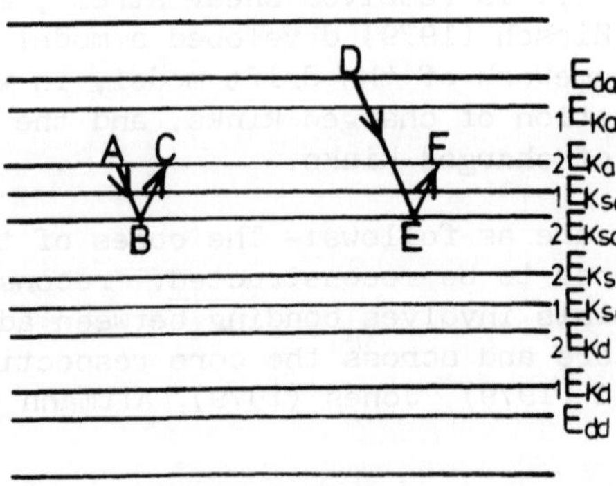

Fig. 3. Schematic set of acceptor and donor energy levels for reconstructed dislocations (E_{da}, E_{dd}), reconstructed kinks (${}_1E_{Ka}$, ${}_1E_{Kd}$), dangling bond kinks (${}_2E_{Ka}$, ${}_2E_{Kd}$), saddle point configurations (${}_{1,2}E_{Ksa}$, ${}_{1,2}E_{Ksd}$). An electron on a charged (dangling bond) kink moves from A to B to C as the kink moves one atomic distance (see fig. 4). For a double kink nucleated at a negative charge on a dislocation, an electron moves from D to the saddle point energy level E, to the kink level F.

A reconstructed kink is expected to have a relatively low energy, but the activation energy for migration will be relatively large, because at the saddle point configuration, two bonds must be broken. A dangling bond kink is likely to have a higher energy, but the migration energy should be less because the movement of such a kink can be effected by switching of bonds without creating additional dangling bonds. Dangling bond kinks or APDs may be stabilised by the presence of charge.

On the drift model, the theory follows closely that for the doping effect on diffusion, kinks replacing point defects. If we neglect the concentration of positively charged kinks, the concentration of negatively charged kinks, ${}_{-}c_k$, relative to that of neutral kinks, ${}_{o}c_k$, is given by Fermi-Dirac statistics as

$$ {}_{-}c_k / {}_{o}c_k = \exp\,(E_F - {}_{i}E_{Ka} - eV)/kT \tag{1} $$

where E_F is the Fermi energy, and eV is the electrostatic energy of the charged dislocation, i = 1,2 depending on the type of kink; there is a similar expression for positively charged kinks in thermodynamic equilibrium with the electron gas, through E_F. The concentration of neutral kinks, ${}_{o}c_k$, is constant at a given temperature, independent of the doping level, provided the interaction between charged and neutral kinks is neglected and the concentrations of all types of kinks are small. The velocity of the dislocations controlled by negatively charged kinks (assuming ${}_{+}c_k$ to be negligible) is given by

$$ \frac{v_-}{v_o} = \exp\,(E_F - {}_{i}E_{Ka} - eV - \Delta W_m)/kT \tag{2} $$

where $\Delta W_m = W_m' - W_m$ is the difference between the migration energy for charged (W_m') and uncharged (W_m) kinks. Figure 4 shows the energy as a function of displacement for a charged and uncharged kink along the dislocation line, the energy barrier being due to the second Peierls potential. Assuming that the Born-Oppenheimer approximation holds, the electron will follow the motion of the ions and move from the energy level ${}_{i}E_{Ka}$ to ${}_{i}E_{Ksa}$ and back to ${}_{i}E_{Ka}$ during the activated process (see fig. 3). Thus

$$ \Delta W_m = {}_{i}E_{Ksa} - {}_{i}E_{Ka} + \Delta W_m' \tag{3} $$

where $\Delta W_m'$ is the contribution to W_m from changes in positions of the ions due to the charge. Thus the energy term in (2) becomes

$$ \Delta U = (E_F - {}_{i}E_{Ksa} - eV - \Delta W_m') \tag{4} $$

This expression is similar to that obtained by Jones (1980) who assumed that equilibrium with the electron gas is established during the activation at the saddle point. The present treatment avoids

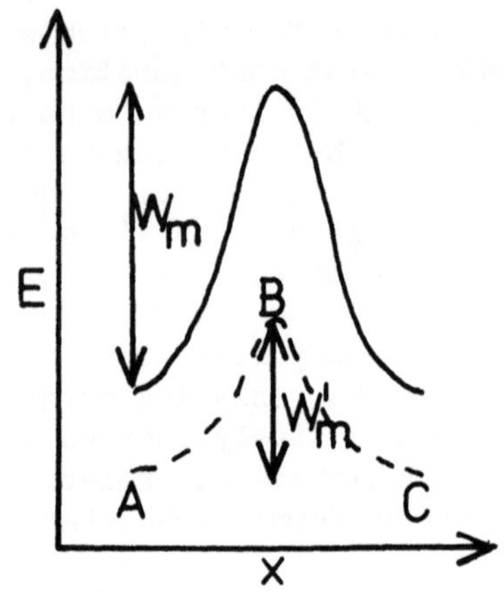

Fig. 4. Total energy (E) versus displacement (x) of a charged (dashed line) and uncharged (full line) kink along the dislocation line.

this difficulty (Hirsch 1981b), in that equilibrium with the electron gas is already established in the ground state of the kink. However, as pointed out by Jones (1980), the electron energy level important in the doping effect is that corresponding to the saddle point configuration. This energy level is not expected to be the same as that found from DLTS, Hall conductivity or optical data obtained from specimens containing static dislocation distributions.

The question arises as to whether the mobile kinks are of the reconstructed (${}_1E_{Ka}$, ${}_1E_{Ksa}$) or dangling bond type (${}_2E_{Ka}$, ${}_2E_{Ksa}$). The activation energy for kink migration is likely to be smaller for dangling bond kinks, because motion occurs by dangling bond exchange, rather than by the formation of two dangling bonds which occurs when reconstructed kinks move. Thus, it seems likely that charged dangling bond kinks are the most mobile. Heggie and Jones (1982) suggest that these would not be stable, that they would tend to dissociate into reconstructed kinks and mobile antiphase defects (APDs), and that a kink would move only if an APD would be temporarily bound to it. This theory is however effectively identical to the treatment given above since both depend essentially on the concentration of charged dangling bond kinks.

Direct measurements of migration energy have been reported for intrinsic Si (Hirsch, Ourmazd and Pirouz 1981, Louchet 1981). The method consists of finding the length of dislocation segment below which the dislocation velocity becomes dependent on length of the segment. For the critical length L, the kink velocity, v_K, is related to the dislocation velocity, v, by the expression

$v_k = \frac{L}{2h} v$, where h is the height of the kink. By in situ electron microscope observations the regime in which the dislocation velocity appears to be dependent on segment length is identified, and the results suggest a high kink migration energy (∿ 1.2 ev) for Si, and it follows from the activation energy of dislocation velocity that the kink formation energy is relatively small (F_k ∿ 0.4 ev). This suggests that the kinks controlling dislocation motion in intrinsic Si are reconstructed (Heggie and Jones 1982). It would be very desirable to have further data relevant to kink migration energy, since the electron microscope evidence so far is rather limited.

At higher stress and lower temperatures, the dislocation velocity is controlled by the nucleation of double kinks (Hirth and Lothe 1968). For the experimental conditions under which the doping effect on dislocation velocity has been determined the dislocation velocity v is independent of the length of the dislocation segment, and is given by

$$v = 2h \sqrt{J}\, v_K \tag{5}$$

where h is the distance between neighbouring Peierls valleys, J is the rate of nucleation of double kinks per unit length of dislocation lines, and v_K is the kink velocity (Hirth and Lothe 1968). Hirsch (1981b) has suggested a) that preferential nucleation may occur at charged points along the dislocation line, in thermodynamical equilibrium with the electron gas, or b) that equilibrium is established with the electron gas in a metastable position of the double kink before the saddle point position is reached. In the former case the electrons (or holes) move from their dislocation levels, E_{da}, to the saddle point level (for simplicity this is shown as $_2E_{Ksa}$ in fig. 3), and beyond the saddle point to a charged kink level $_2E_{Ka}$. Several alternative reaction paths are possible. Preferential nucleation may occur at a single charge, resulting in two kinks, one charged, the other not. The former is likely to be more mobile than the latter. Under these conditions (for a singly negatively charged double kink)

$$J_- \alpha\ f\, n_T \exp\,(-\,F_{DK}/kT)\,\exp\,(-\,W'_m/kT)\,\exp(E_{da} - {_2E_{Ka}})/kT \tag{6}$$

where F_{DK} is the formation energy of a double kink consisting of two dangling bond kinks, f is the occupation probability of the original charged dislocation level, E_{da}, n_T is total concentration of possible sites. With $f\, n_T = n_o f/(1-f) = n_o \exp-(E_{da} - E_F + eV)/kT$, where n_o is concentration of uncharged sites

$$J_- \alpha\ n_o \exp(-\,F_{DK}/kT)\,\exp(-(W_m + \Delta W_m)/kT)\,\exp(E_F - {_2E_{Ka}} - eV)/kT$$

and from (3)

$$= J_o \exp\,(E_F - {_2E_{Ksa}} - eV - \Delta W'_m)/kT \tag{7}$$

where J_o is the nucleation rate for a similar but uncharged double kink. After nucleation equilibrium is established with the electron gas, and the kink velocity controlled by negatively charged kinks will be

$$_{-}v_K = {}_o v_K \exp\,(E_F - {}_2E_{Ka} - eV)/kT \exp\,({}_2E_{Ka} - {}_2E_{Ksa} - \Delta W_m')/kT \quad (8)$$

the first exponential being the fraction of negatively charged kinks, and the second due to the change in activation energy for migration associated with the charge. Thus, it follows from (5), (7), (8) that

$$v_- = v_o \exp\,(-\Delta U/kT) \quad (9)$$

where ΔU is given by (4) as in the drift model.

Several other alternative paths are possible, and some of these have been considered by Hirsch (1983). Heggie and Jones (1982, 1983) have suggested that preferential nucleation may occur at an antiphase defect (APD). In the model both nucleation and migration of a kink depend on the presence of an APD. It then follows from (5) that the dependence of dislocation velocity on E_F is again the same as in the drift model (eqn. (4)).

Figure 5 shows the results of an analysis carried out by Schröter, using the above treatment (eqn. (9), (4)), and assuming that the dislocation velocity can be approximated by $v = v_o + v_+ + v_-$, and eV can be neglected, for dislocation velocity data on 60° and screw dislocations in Si, obtained by George and Champier (1979). The best fit is obtained assuming an acceptor level $E_a = 0.67 \pm 0.04$ ev, with $\partial E_a/\partial T = (-1.6 \pm 0.5) \times 10^{-4}$ ev K^{-1}, and a donor level $E_d = 0.28 \pm 0.17$ ev with $\partial E_d/\partial E_T = (-2.0 \pm 2.1) \times 10^{-4}$ K^{-1}, all measured relative to the valence band. Heggie and Jones (1982) have reanalysed the data, making different assumptions about the temperature dependence of the gap energy, and making some allowance for eV, and quote $E_a = 0.55$ ev and $E_d = 0.34$ ev respectively. These energy levels correspond to the acceptor level ${}_iE_{Ksa} + \Delta W_m'$ in eqn (4), and a corresponding donor level. The precise interpretation of these energy levels is in doubt. Even if $\Delta W_m'$ can be neglected, it is possible that the velocity in strongly doped material is controlled by charged dangling bond kinks, while that in intrinsic material is controlled by neutral reconstructed kinks. However, the results are consistent with a simple theoretical model in which the saddle point configuration for kink migration is associated with deep electronic levels.

The data for Ge in fig. 1 have not yet been fully analysed, but at present the data do not seem to be consistent with the model if the eV term is neglected. One possibility is that the eV term has to be taken into account for the n or p type material, or that the

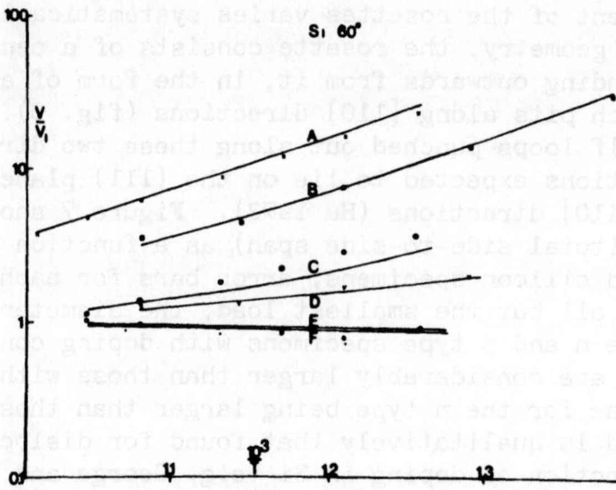

Fig. 5. Ratio of velocities v of 60° dislocations in Si to that of intrinsic material, v_i, as a function of reciprocal temperature for various doping concentrations; donors:- A = 14×10^{18} cm^{-3}, B = 6×10^{18} cm^{-3}, C = 1.2×10^{18} cm^{-3}; acceptors:- D = 80×10^{17} cm^{-3}, E = 6.8×10^{17} cm^{-3}, F = 2.7×10^{17} cm^{-3}. Measurements of George and Champier (1979); fitting by Schröter (Hirsch 1981a).

rate controlling kinks are different for n and p type material. The interpretation is continuing.

EFFECT OF DOPING ON INDENTATION ROSETTES

The effect of doping on dislocation velocity will be reflected in mechanical properties. For many materials it is not always possible to obtain single crystals of a size suitable for velocity measurements. Maeda, Ueda, Murayama and Sakamoto (1977) have used the extent of dislocation rosettes on indented surfaces of GaP as a qualitative indication of the effect of doping on dislocation velocity. We have used a similar technique, using a Vickers indenter, on variously doped Si specimens (Roberts, Pirouz and Hirsch 1983), and more recently on Ge specimens. For Si, after indentation of (100) faces at 400°C, the specimens were annealed for 30 minutes at that temperature. They were then etched at room temperature in a reagent consisting of one part of 0.25M potassium dichromate and four parts of 40%HF (Secco d'Aragona 1972). The rosettes were examined optically.

The results showed that for Si the indentation diagonal, and thus the microhardness, are effectively independent of doping.

However, the extent of the rosettes varies systematically with doping. In this geometry, the rosette consists of a central plastic region, and extending outwards from it, in the form of a cross two arms of dense etch pits along [110] directions (fig. 6). These arms are formed by half loops punched out along these two directions, with the dislocations expected to lie on the (111) planes intersecting in the [110] directions (Hu 1973). Figure 7 shows the size of the rosettes (total side-to-side span) as a function of load for differently doped silicon specimens; error bars for each load are also shown. For all but the smallest load, the diameters of the rosettes for the n and p type specimens with doping concentrations of 2×10^{18} cm^{-3} are considerably larger than those with low doping levels, with those for the n type being larger than those for the p type. This trend is qualitatively that found for dislocation velocity as a function of doping in Si (e.g. George and Champier 1979). In addition the rosette for n type float zone Si is larger than that for Czochralski Si, showing the same trend as that found for dislocation velocity (Harada and Sumino 1982), and which is attributed to oxygen.

The results have been interpreted on the basis that the dislocations punched out from the central plastic zone move until the resultant stress on any dislocation due to the indentation stress field and the interactions with the other dislocations along the arm is insufficient to overcome an effective lattice friction stress

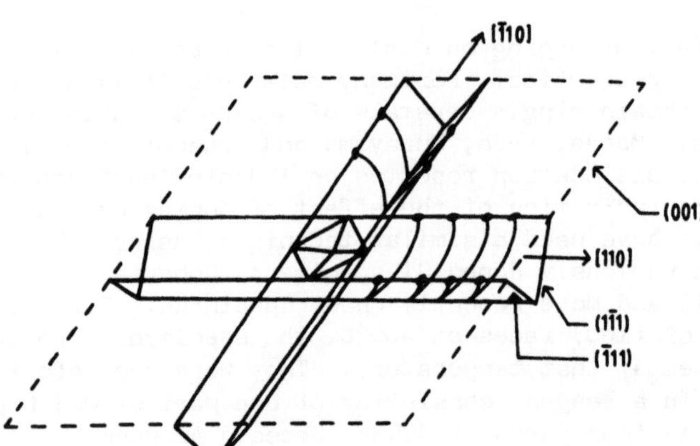

Fig. 6. Form of indentation rosette on a {100} silicon surface (Hu 1973).

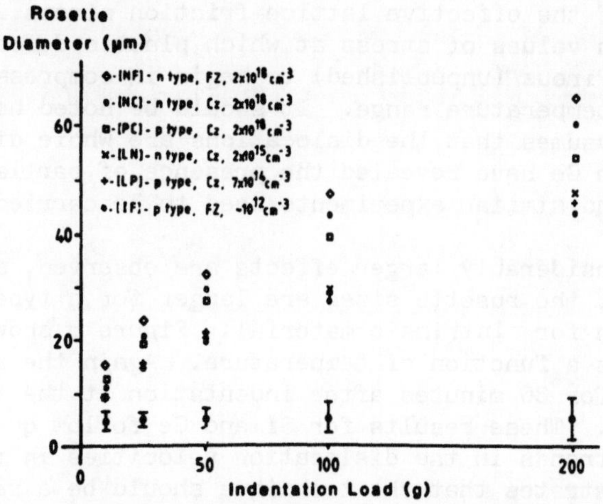

Fig. 7. Size of rosettes (total side-to-side span) as a function of load for differently doped silicon specimens (Roberts, Pirouz and Hirsch 1983).

controlling the dislocation motion; the indentation stress field is assumed to be of the type proposed by Yoffe (1982), which consists of a compressive radial stress and a tensile tangential stress, both decreasing to x^{-3}, where x is the distance from the indentation centre. Table 1 gives the mean effective friction stresses (averaged over a set of values obtained for each specimen from rosette sizes for different loads) for Si with various doping concentrations.

Table 1

Values of effective lattice friction stress in Si at 400°C, as a function of doping

Si type and doping (cm^{-3}) (Fz ≡ float zone; Cz = Czochralski)	Effective lattice friction stress (MPa)
Fz, p, 10^{12}	102.7
Cz, p, 7 x 10^{14}	98.7
Cz, n, 2 x 10^{15}	98.6
Cz, p, 2 x 10^{18}	83.0
Cz, n, 2 x 10^{18}	77.5
Fz, n, 2 x 10^{18}	73.1

These values of the effective lattice friction stress are reasonably consistent with values of stress at which plastic deformation has been found by Pirouz (unpublished) to begin in compression experiments in this temperature range. It should be noted however that the analysis assumes that the dislocations are whole dislocations; observations on Ge have revealed the presence of partial dislocations (see below), and simiiar experiments need to be carried out on Si.

For Ge considerably larger effects are observed, and in contrast to Si, the rosette sizes are larger for n type but smaller for p type than for intrinsic material. Figure 8 shows the rosette sizes for Ge as a function of temperature. Again the specimens were annealed for 30 minutes after indentation at the temperature of indentation. These results for Si and Ge follow qualitatively the different trends in the dislocation velocities in the two cases, and this demonstrates that the technique should be a reliable qualitative indicator for doping effects.

Quantitative interpetation and correlation with dislocation velocity measurements is however difficult. The stress field around an indenter is very inhomogeneous, and the dislocation arrangement complex. Figure 9 shows transmission micrographs of a section parallel to the surface through one of the rosette arms in Ge,

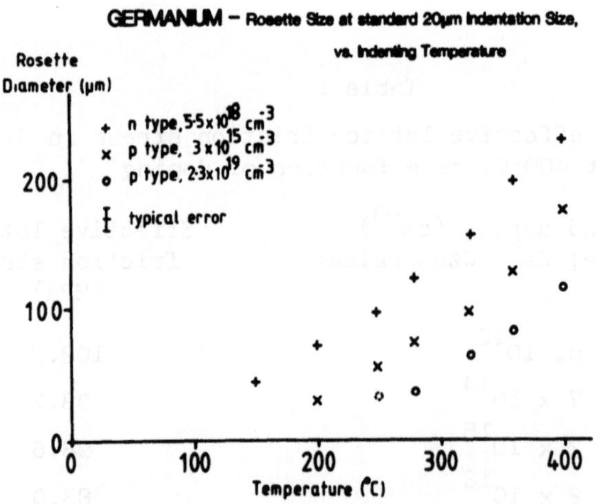

Fig. 8. Rosette size (total side-to-side span) at standard 20 μm indentation size for differently doped Ge specimens as a function of temperature (Roberts, unpublished).

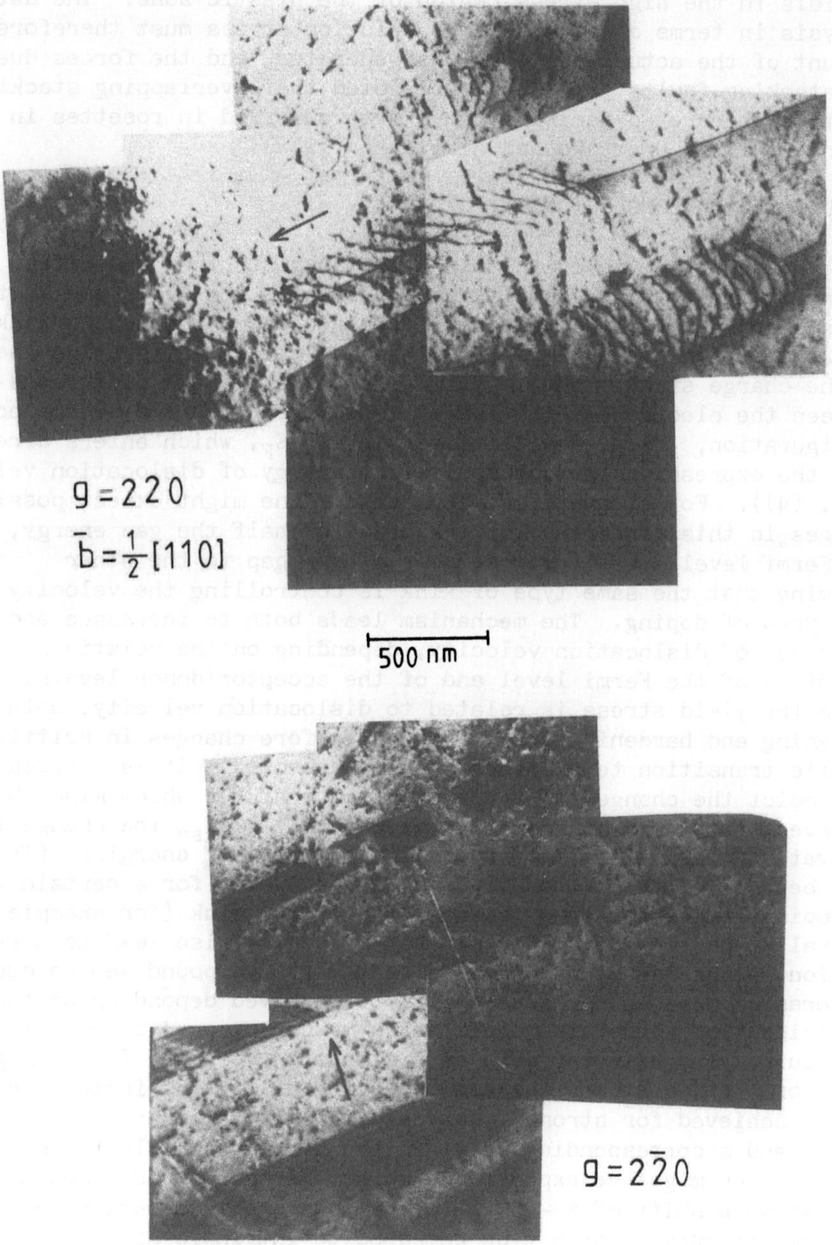

Fig. 9. Transmission electron micrographs of dislocation arrays along one arm of a rosette in Ge indented at 280°C, taken in two reflections, and showing the presence of partial dislocations in the image taken in $2\bar{2}0$ (Pirouz and Roberts, unpublished).

indented at 280°C. It appears that many of the dislocations are partials, and this may be due to the relative ease of nucleation of partials in the high stress region of the plastic zone. The detailed analysis in terms of an effective friction stress must therefore take account of the actual dislocations generated, and the forces due to the stacking faults. It should be noted that overlapping stacking faults, interpreted as twins, have been observed in rosettes in indented Si, by Eremenko and Nikitenko (1972).

DISCUSSION AND CONCLUSIONS

The theory reviewed in this paper suggests that the effect of doping on dislocation velocity in the Peierls regime is due to the dependence of the concentration of charged kinks and double kinks on the position of the Fermi level and of the kink migration energy on the charge state. The important parameter is the difference between the electron energy levels associated with the saddle point configuration, ${}_iE_{Ksa}$, and the Fermi level, E_F, which enters directly into the expression for the activation energy of dislocation velocity (eqn. (4)). For elemental semiconductors one might expect possible changes in this difference of the order of half the gap energy, as the Fermi level moves from one side of the gap to the other assuming that the same type of kink is controlling the velocity for all types of doping. The mechanism leads both to increases and decreases of dislocation velocity, depending on the relative positions of the Fermi level and of the acceptor/donor levels. Since the yield stress is related to dislocation velocity, both softening and hardening effects and therefore changes in brittle-ductile transition temperatures can be expected. It is difficult to predict the changes in transition temperatures which might be achieved; for a given change in value of $E_F - {}_iE_{Ksa}$ the change in activation energy is likely to be less because of charging effects, i.e. because of the eV term in eqn. (4). Again, for a certain range of doping levels the type of rate controlling kink (for example neutral kinks) may be different, and this will also lead to a reduction in the magnitude of the effect. For compound semiconductors or ceramics more energy levels will be involved depending on the type of dislocation, and this complication makes any prediction even more difficult. The existing data on dislocation velocity for Si suggest that for a given stress the same velocity as for intrinsic material can be achieved for strongly n-type material at temperatures ∿ 50°C lower, and a corresponding shift in the brittle-ductile transition temperatures might be expected. Rabier, Veyssière and Demenet (1983) have shown a shift of ∿ - 50°C in the lower yield stress versus temperature curves for n type compared to intrinsic Si. For Ge the velocity data suggest a shift of ∿ 70°C in the curves of yield stress versus temperature for strongly n and p type material respectively. Substantially larger shifts should occur in the wide band gap material, e.g. in diamond.

The same mechanism should occur at the surface if the relative positions of the Fermi level and electronic levels can be changed by band bending due to charge effects at the surface.

It should be noted that the processes discussed above are a consequence of thermodynamic equilibrium. Effects on mechanical properties resulting from the injection of charge carriers (as mentioned in the introduction), e.g. by light or electron irradiation, are associated with deviations from thermodynamic equilibrium, and the electronic mechanisms are generally different. Thus the photoplastic effect in Si (Küsters and Alexander 1983) and the electron beam induced glide in GaAs (Maeda and Takeuchi 1983) have been attributed to non-radiative recombination at dislocations, which can enhance the mobility of kinks, e.g. by a multiphonon process. Similar mechanisms have been invoked to explain stimulated point defect migration (Kimerling 1978).

REFERENCES

Altmann, S.L., Lapiccirella, A., and Lodge, K.W., 1983, Int. J. Quantum Chem. 23:1057.
Blanchin, M.G., Castaing, J., Fontaine, G., Heuer, A.H., Hobbs, L.W., and Mitchell, T.E., 1981, Journ. de Physique, Colloque 3, C3-1.
Eremenko, V.G., and Nikitenko, V.I., 1972, Phys. Stat. Sol. (a), 14:317.
Erofeev, V.N., and Nikitenko, V.I., 1971, Sov. Phys. Solid State, 13:116.
Erofeeva, S.A., and Ossipyan, Yu.A., 1973, Sov. Phys. Solid State, 15:538.
Frisch, H.L., and Patel, J.R., 1967, Phys. Rev. Lett., 18:784.
George, A., and Champier, G., 1979, Phys. Stat. Sol. (a), 53:529.
Haasen, P., 1975, Phys. Stat. Sol. (a), 28:145.
Harada, H., and Sumino, K., 1982, J. Appl. Phys., 53:4838.
Heggie, M., and Jones, R., 1982, Journ. de Physique, C1, 43:C1-45.
Heggie, M., and Jones, R., 1983, Conference on Microscopy of Semiconducting Materials, Oxford 1983, Institute of Physics - in press.
Hirsch, P.B., 1979, Journ. de Physique, C6, 40:C6-117.
Hirsch, P.B., 1980, J. Microscopy, 118:3.
Hirsch, P.B., 1981a, Proc. Materials Research Soc. Symp. on Defects in Semiconductors, J. Narayan and T.Y. Tan, eds., North Holland, New York, pp 257-271.
Hirsch, P.B., 1981b, Journ. de Physique, C3, 42:C3-149.
Hirsch, P.B., 1983, Conference on Microscopy of Semiconducting Materials, Oxford 1983, Institute of Physics - in press.
Hirsch, P.B., Ourmazd, A., and Pirouz, P., 1981, Inst. Phys. Conf. Ser. 60:29.
Hirth, J.P., and Lothe, J., 1968, Theory of Dislocations, McGraw Hill, New York, pp 484-497.
Hu, S.M., 1973, J. Appl. Phys. 46:1470.

Jones, R., 1980, Phil. Mag. 42B:213.
Jones, R., 1979, Journ. de Physique, C6, 40:C6-33.
Kulkarni, S.B., and Williams, W.S., 1976, J. Appl. Phys., 47:4318.
Küsters, K.H., and Alexander, H., 1982, Physica, 116B:594.
Louchet, F., 1981, Inst. Phys. Conf. Ser., 60:35.
Maeda, K., and Takeuchi, S., 1983, J. de Physique Colloque, Properties and Structure of Dislocations in Semiconductors, Aussois 1983 - in press.
Maeda, K., Ueda, O., Murayama, Y., and Sakamoto, K., 1977, J. Phys. Chem. Solids, 38:1173.
Marklund, D., 1979, Phys. Stat. Sol. (b), 92:83.
Mei, L., Rivier, M., Kwart, Y., and Dutton, R.W., 1981, Proc. 4th Int. Symp. on Silicon, Materials Science and Technology, H.R. Huff, R.J. Kriegler and Y. Takeishi, eds., Electrochemical Soc., pp 1007-1019.
Patel, J.R., and Chaudhuri, A.R., 1966, Phys. Rev.. 143:601.
Patel, J.R., Testardi, L.R., and Freeland, P.E., 1976, Phys. Rev., B13:3548.
Pirouz, P., Cockayne, D.J.H., Sumida, N., Hirsch, P.B., and Lang, A.R., 1983, Proc. Roy. Soc. A, 386:241.
Rabier, J., Veyssière, P., and Dement, J.L., 1983, J. de Physique Colloque, Properties and Structure of Dislocations in Semiconductors, Aussois 1983 - in press.
Roberts, S.G., Pirouz, P., and Hirsch, P.B., 1983, J. de Physique Colloque, Properties and Structure of Dislocations in Semiconductors, Aussois 1983 - in press.
Secco d'Arragona, F., 1972, J. Electrochem. Soc., 199:948.
Smith, D.A., Tan, T.Y. and Fontaine, C., 1982, IBM Research Report RC9361 (no. 41211).
Wada, Y., and Nishimatsu, S., 1978, J. Electrochem. Soc., 125:1499.
Yoffe, E.H., 1982, Phil. Mag., 46A:617.

DISLOCATION DYNAMICS IN SILICON UNDER HIGH STRESS

Helmut Alexander

Abteilung für Metallphysik im II.Phys. Inst.
Universität Köln, FRG

INTRODUCTION

Dislocation motion in pure silicon is an interesting object to investigate since this material represents covalent bonding in its purest form, diamond being difficult to work with. From the strongly localized bonding it follows that dislocation cores should be narrow and kinks should be sharp. Because breaking and reestablishing covalent bonds is necessary to move a dislocation by any mechanism, the Peierls potential (P_p) will be prominent as well in the glide direction as along the dislocation line (P_p of the second kind). However, things are complicated by the fact that dislocations are dissociated into two partial dislocations separated by a stacking fault ribbon by the same splitting reaction as in fcc metals.

Three experimental findings are fundamental for understanding of what happens on an atomic scale when a dislocation moves in a crystal with diamond structure:

a) The velocity v of isolated dislocations depends exponentially on temperature, while the stress exponent $m = d \ln v / d \ln \tau$ varies between one and two [1] (τ : resolved shear stress)

$$(1) \qquad v = v_o \left(\frac{\tau}{\tau_o}\right)^m \exp\left(-\frac{Q}{kT}\right)$$

(τ_o : standard stress for dimensional correctness)

b) Dislocations are dissociated into partial dislocations [2] also when moving [3].

c) Changes of the dissociation width d under high stress show that the two partials bound together have to overcome different friction stresses, even in the case of screw dislocations where both partials are of the same character [4] .

Summing up these results one concludes that dislocations move by the thermally activated nucleation and spreading of double kinks (DK) in both partials.

MEASUREMENT OF DISLOCATION VELOCITIES UNDER HIGH STRESS

In the course of the work which leads to the observation described under c) above, we found it attractive to expand the investigation of dislocation motion especially in the regime of high stresses and low temperature (HS-LT) for several reasons: under those conditions the dislocation arrangement appears very much simplified when compared to "normal" deformation conditions ($T \geq .5T_m, \tau \leq 50$ MPa): dislocations are perfectly straight, parallel to the Peierls valleys <110> in the {111} glide plane and belong all to the primary glide system (defined by maximum Schmid factor). (We could show that the radii of curvature of dislocations which are not extremely wide split are as small as possible, taking into account the local line tension and local stress [5] ; in cases where the two partials exhibit clearly different curvatures the statement holds for the partial leading in the direction of motion, not for the trailing one [6] .) Therefore, in crystals deformed under HS and cooled under load one finds only three types of dislocations: screws and two kinds of 60° dislocations, distinguished by the order of partials: 30/90 and 90/30 dislocations, respectively. Another advantage of LT deformation is a diminishing influence of impurity diffusion and climb. One has to bear in mind, however, that plastic deformation near the limits of plasticity needs some predeformation at higher temperature because of the lack of dislocations in undeformed crystals. Concerning intrinsic point defects EPR spectroscopy has revealed the production of a lot of those defects during LT deformation [7] . Also from a more fundamental point of view dislocation motion in the HS regime should be interesting: for shear stress above 10 MPa (Si, 60° disl.) the formation of DK in the two partials should be uncorrelated processes [8, 9].

The only measurements of dislocation velocities below 500°C in the literature are due to Ba Tu and Yizhen [10] (400°C) but the stress range being very limited ($\tau \leq 50$ MPa). So we started a series of experiments at 422 °C using uniaxial loading and the double etching technique. By etching layer by layer Kisielowski [11] has found that only 60° dislocations form a 120° bend beneath the surface. As the two kinds of 60° dislocations may move with different velocity into the crystal the depth of the knee beneath the surface has to be determined additionally to the displacement of the etch pits to get reliable values of v for the 60° segments of dislocation half loops.

Concerning first the dislocations of the primary glide system we found the phenomenological relation (1) useful in the whole stress range investigated so far (fig. 1). The velocity stress relation at such a low temperature is far from being linear ($m_{60} \approx 1.85$). Since screw segments of primary loops do not penetrate the $(11\bar{1})$ surface which can be etched we tried to use asymetrical loops belonging to secondary glide systems for measuring the velocity of screws. Here we were puzzled by the fact that the stress exponent m (and also the prefactor B (422°C) = $v_o \exp(-Q/k\ 695K)$ of 30/90 dislocations turned out different for different glide systems (fig. 1). This observation means that Schmid's law does not apply in case of silicon. Kisielowski [11] extended now the measurements to four glide systems, whereby hardness indentations proved to be more useful as dislocation sources than scratches. For crystallographic reasons 30/90 dislocations could be measured in all systems, 90/30 and screws each in two systems. The resulting values of m and B (422°C) are compiled in table 1.

In order to uncover the systematics behind the different exponents m we supposed an influence of that part of the (elastic) Peach-Koehler force which acts perpendicular to the glide plane ("normal force" F_n).

$$\underline{F}_n = \sigma(\underline{\hat{d}}\ \underline{b})(\underline{\hat{t}} \times \underline{\hat{d}})\ \underline{\hat{n}} \qquad (\sigma: \text{compression stress},$$

$\underline{\hat{d}}$, $\underline{\hat{t}}$, $\underline{\hat{n}}$: unit vectors parallel to the compression axis, dislocation line and normal to the glide plane, respectively) the influence of which is denied by Schmid's law. In fact we succeeded to separate m into two parts [11]

$$m = m_o + m_1$$

m_o being proportional to $F_n/b\sigma$ and m_1 determined only by the type of dislocation being under investigation. Table 1 also contains m_o and m_1 for eight types of dislocations which have been investigated.

From the very beginning of research on dislocation velocity in semiconductors [1] it became clear that the operational parameters m and Q are not constants of a given material. Instead Q decreases with increasing stress and m is found the larger the lower the temperature is. To complete the program described so far we have to search for the temperature dependence of v under HS conditions. This can be done only to a limited degree, on the one hand in view of the technique applied, but on the other hand also because we are not sure if the mechanism of dislocation motion is unchanged from LT to high temperatures. Nevertheless we are doing such experiments. Their results will be published elsewhere [12].

GLIDE RESISTANCE OF PARTIAL DISLOCATIONS

Wessel and Alexander [4] detected that the modification of the dissociation width d of straight, undisturbed dislocations by large shear stresses is not in accordance with what one calculates from the different glide forces F_1, F_2 acting on the leading (1) and trailing (2) partials, respectively. That means: the two partials do not get over lattice friction forces of the same size. This apparently does not only apply to partials of different character (and, therefore, core structure) but also to the two 30° partials constituting the screw dislocation, since by far most of the screw dislocations are wider split under high shear stress in spite of the driving force F_2 (on the trailing partial) being larger than F_1.

To find out if the normal forces $F_{n,i}$ on the partials are responsible for the friction we conducted the same experiments as done in [4] with dislocations of secondary glide systems, so varying the normal forces at constant driving forces $F_i = \tau_p b_p$. This could be managed by W. Weiss [13] who predeformed the crystals along an axis different from the axis of the HS compression. So an optional secondary glide system could be nearly exclusively activated during the HS deformation making this glide system the primary system during the predeformation. Careful analysis of the dissociation width measured under the same resolved shear stress (and T) in different systems made sure - in spite of a remarkable scatter of the measurements - that the character of the partials and their order relative to the stacking fault do not fully determine the width d, i.e. the ratio of lattice friction forces R_1/R_2.

Heister [14] extended the experiments of Weiss to further stress states $\underline{Q}$ (orientation of the compression axis) and was able to give a unified description of all results (v, d) obtained at 422°C for different stress states $\underline{Q}$. Heister in view of the non-linear velocity stress relation generalizes the conception of mobility ($\mu_i = v_i/F_i$) used by Wessel and Alexander and tries the equation:

(2) $$v_i = v_{oi} \left(\frac{R_i}{R_{oi}}\right)^{m_i}$$

(v_i: velocity of the partial i of a given total dislocation,
R_i: friction stress felt by partial i, being equal to the sum of all forces acting on i parallel to its glide plane

(2a) $$R_i = F_i \pm \gamma \mp \frac{A}{d}$$

v_{oi} and R_{oi} constants depending only on the character of the partial and its position to the stacking fault.)

Clearly

(3) $$R_1 + R_2 = \tau b = \left(\frac{v}{B(T)}\right)^{1/m} \tau_o b$$

Steady state motion means

$$(4) \qquad v_1 = v_2 = v$$

Equations (2) to (4) result in

$$(5) \qquad R_{o1} \left(\frac{v}{v_{o1}}\right)^{1/m_1} + R_{o2} \left(\frac{v}{v_{o2}}\right)^{1/m_2} = \tau_o b \left(\frac{v}{B(T)}\right)^{1/m}$$

v on both sides of eq. (5) must have equal exponents, so the equation (2) makes sense only if (2a): $m_1 = m_2 = m$. The soundness of this model is to be examined by its ability to describe our results. The experimental material consisted of the (averaged) splitting width of 15 dislocation types (5 stress orientations, for each of them screw and 30/90 and 90/30 dislocations). From the splitting width d the corresponding ratio of friction stresses $R = R_1/R_2$ can be calculated using the relation [4]:

$$d = d_o \left(1 + \left(f + \frac{r-1}{r+1}\right) \frac{b\tau}{2\gamma}\right)^{-1}$$

(d_o : splitting width without stress [15], f : geometrical factor depending on stress axis [4], γ : stacking fault energy [15]). r is known if one knows two parameters for the particular dislocation: R_{o1}/R_{o2} and v_{o2}/v_{o1} (compare equation (2) and the exponent m for the given stress state. The latter quantity was taken from Kisielowski's work [11]. So for each dislocation type (e.g. screw) widths d for two different stress states were used to determine R_{o1}/R_{o2} and v_{o2}/v_{o1}, and with these fit parameters in fact it was possible to calculate the width d for three further stress states, with only one exception (table 2). To separate in a next step R_{o1} from R_{o2} (and v_{o1} from v_{o2}) one can use eq. (5) if B(T) is known for at least two stress states (and, therefore, m-values). Table 3 contains these quantities which describe now characteristic properties of particular partials. Here the quantities in the case of the 30/90 dislocation have particular weight, since B(T) was measured for two more cases than needed for the separation of R_{o1} from R_{o2} and correspondingly v_{o1} from v_{o2}.

CONCLUSIONS

1) The success of the audacious assumption $m_1 = m_2 = m$ in my mind shows that stress influences the dislocation as an entity, even if it is dissociated by 100 A or more.

2) The properties of a particular partial are reflected by its parameters R_o and v_o. Looking at table 3 one notices that three partials are rather similar in this respect: the 30° partials belonging to 60° dislocations and the <u>leading</u> 30° partial of screws.

TABLE 1. Stress exponent m and prefactor B (at 422°C) (compression axis [213])

Glide system	Disloc. type	m meas.	m calc.	$= (m_o$	$+ \; m_1)$	$B(10^{-10}\frac{cm}{s})$
$(1\bar{1}1)$	90/30	1.82	1.86	-.176	2.04	5.1
[011]	30/90	1.87	1.92	-.25	2.17	3.4
(111)	30/90	1.81	1.82	-.35	2.17	4.2
$[0\bar{1}1]$	30/30	1.50	1.54	0	1.54	8.4
$(\bar{1}11)$	90/30	2.08	2.03	-.014	2.04	3.1
[101]	30/90	2.20	2.16	-.014	2.17	1.9
$(1\bar{1}1)$	30/90	1.63	1.58	-.59	2.17	4.6
[110]	30/30	1.19	1.12	-.42	1.54	10

TABLE 2. Dissociation width d[Å] of dislocations differently oriented relative to the compression axis (τ = 250 MPa, T = 422°C). For calculation the parameters from table 3 and m from table 1 were used. Not measured m were calculated [11].

Glide system	Compr. axis	Dislocation type 30/30		30/90		90/30	
		d_{meas}	d_{calc}	d_{meas}	d_{calc}	d_{meas}	d_{calc}
$(1\bar{1}1)$	[328]	43	47	37	40	96	96
[011]	[213]	45	42	36	38	71	78
(111) $[0\bar{1}1]$	[213]	48	47	39	42	95	91
$(\bar{1}11)$ [101]	[213]	63	44	43	38	100	91
$(1\bar{1}1)$ [110]	[213]	35	35	37	35	56	55

TABLE 3. Characteristic parameters of the partial dislocations constituting a hexagonal loop in silicon

Disloc. type	Leading partial R_{o1}/b_p (MPa)	v_{o1} (cm/s)	Trailing partial R_{o2}/b_p (MPa)	v_{o2} (cm/s)
30/30	71.5	$3.5 \; 10^{-8}$	14.3	$1.4 \; 10^{-9}$
30/90	214	$6.25 \; 10^{-8}$	1.1	$2.1 \; 10^{-11}$
90/30	17.8	$6.1 \; 10^{-9}$	177.5	$6.5 \; 10^{-8}$

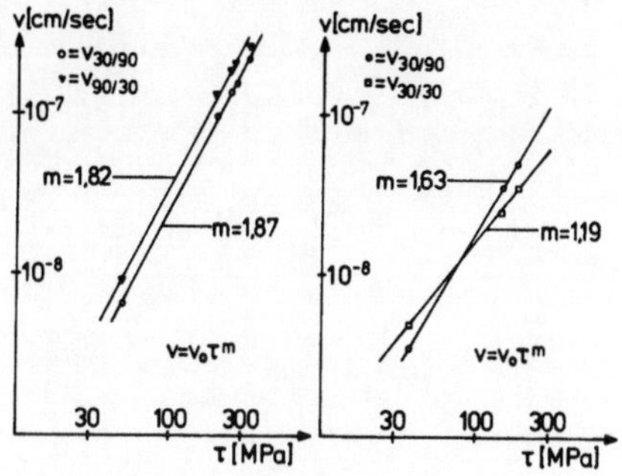

Fig. 1. Stress dependence of dislocation velocity at 422°C.
Left side: primary glide system $(1\bar{1}1)$ [011].
Right side: glide system $(1\bar{1}1)$ [110]

3) Most interest deserves the dependence of the dislocation velocity on other parameters of stress orientation than those which determine the resolved glide shear stress. The reason for this effect might well be a modification of the core structure (of double kinks, solitons? [16]) by non-glide components of the stress in analogy to what is found for bcc metals [17]. In this report it is absolutely necessary to find out if the activation energy Q depends also on non-glide stress components.

Acknowledgments: The author gratefully acknowledges the delightful cooperation of the members of his group. He also wants to thank Drs. M.S. Duesbery, A. George, R. Jones and F. Louchet for unpublished material and valuable discussions. The Deutsche Forschungsgemeinschaft provided financial support.

REFERENCES

1. A.R. Chaudhuri, J.R. Patel and L.G. Rubin, J. Appl. Phys. 33 (1962) 2736 and 34 (1963) 240
2. I.L.F. Ray and D.J.H. Cockayne, Phil. Mag. 22 (197o) 853
3. R. Meingast and H. Alexander, phys. stat. sol. (a) 17 (1973) 229
4. K. Wessel and H. Alexander, Phil. Mag. 35 (1977) 1523
5. A. Tönnessen, Dipl. thesis Univ. Köln 1982
6. H. Gottschalk, J. de Physique in press
7. E. Weber and H. Alexander, Inst. Phys. Conf. Ser. 31 (1977) 266
8. H.-J. Möller, Acta Metall. 26 (1978) 963
9. H. Alexander, Dislocations in Covalent Crystals in: Dislocations in Solids, F.R.N. Nabarro Ed. to be publ.
1o. Ba Tu and He Yi-Zhen, Acta Phys. Sin. 29 (198o) 698
11. H. Alexander, Ch. Kisielowski-Kemmerich and E.R. Weber, Physica 116B (1983) 583
12. A. Swalski, Diploma work in progress
13. W. Weiss, Diploma thesis Univ. Köln 198o
14. E. Heister, Diploma thesis Univ. Köln 1982
15. H. Gottschalk, J. de Physique C-6 (1979) 127
16. M. Heggie and R. Jones, J. de Physique in press
17. M.S. Duesbery, Acta Metall. 31 (1983) 429

KINK MOBILITY AND DISLOCATION VELOCITY IN COVALENT CRYSTALS

H.J. Möller* and U. Jendrich[+]

*TU Hamburg-Harburg, Halbleitertechnik
Eißendorfer Str. 38, 2100 Hamburg 90, F.R.G.
[+]Institut für Metallphysik
Hospitalstr. 12, 34 Göttingen, F.R.G.

INTRODUCTION

The electrical and mobility properties as well as the atomic structure of dislocations in covalent crystals have been thoroughly investigated in germanium [1-8], silicon [9-12] and some compound semiconductors [13-15]. It is well established that the dislocation velocity is exponentially dependent on temperature and approximately linear on stress for stresses $\tau \geqslant 20$ MPa. From theoretical calculations [16-19] follows that this behavior can be explained under the assumption that the dislocations move over a high Peierls potential by the nucleation and annihilation of double kinks. Though a qualitative agreement between experiment and theory can be easily achieved no model at present is able to yield a satisfactory quantitative description of the experimental results. Some of the difficulties encountered shall be summarized briefly.

a) For low stresses $\tau \leqslant 20$ MPa the activation energy Q of the dislocation motion increases both in germanium and silicon (60° dislocations only). Originally this was qualitatively explained by the existence of obstacles on the dislocation line which interact with the expanding double kinks and can only be overcome at low stresses with the help of thermal activation [16, 17]. Weak-beam TEM investigations of deformed germanium crystals have shown that dislocations are dissociated into partials and constricted in mean temperature dependent intervals $\ell \approx 0.2 - 1.5$ µm. The number of constrictions observed was, however, too low to explain the necessary high density of obstacles $\ell \approx 0.04$ µm derived from velocity measurements. Recent TEM studies in germanium and silicon with the lattice row imaging technique [20] seem to indicate that the cores of the partial dislocations may be contaminated by oxygen, however, it is doubtful whether these impurities can explain the observed activation energy increase.

b) Theoretical calculations of the double kink nucleation process on dissociated dislocations have shown [19] that below a critical stress $\tau_c = 3\gamma/4d_o$ (γ stacking fault energy, d_o equilibrium width of the dissociated partials) double kinks must be created simultaneously (correlated double kink nucleation) on each partial because of their elastic interaction. Numerical estimates of τ_c fall in the range where the activation energy Q increases, therefore it was proposed that the increase is due to the beginning of correlated double kink nucleation [19]. Fitting of the derived equations to the experimental results in germanium, silicon and gallium arsenide yielded reasonable parameters. The model was based on the assumption of a low secondary Peierls potential or a high kink mobility. Recent theoretical and experimental (HVEM) results [21,22] seem to indicate, however, that the kink mobility should be low.

c) The mobility of single (geometrical) kinks can be investigated by internal friction measurements of deformed crystals. A great number of experimental data is available, unfortunately, no clear picture could be obtained yet. Recent studies in germanium and silicon showed either no internal friction maximum at all [23,24] or only very small signals due to single kink migration: two peaks ($Q^{-1} < 3 \cdot 10^{-6}$) at low temperatures $T < 100$ K. From these results it was concluded [24,25] that the secondary Peierls potential or the migration energy E_{KM} for kink motion along the dislocation line is approximately 0.1 eV in accordance with conclusions drawn from velocity measurements, however, the density of obstacles, which reduces the height of an internal friction peak, should be high ($l < 400$ Å in germanium). Recent computer calculations [26,27] of the core energy of dislocations in germanium and silicon allow to estimate the secondary Peierls potential. Large values of $E_{KM} > 0.5$ eV and a small kink energy E_K are expected in the case where the dangling bonds in the core of the partial dislocations and the kinks try to rebond as much as possible. In the development of the theory of the dislocation velocity the impact of such a high migration energy has never been taken into account adequately. Because of the importance of this question it seemed necessary to reinvestigate the kink migration by internal friction measurements.

A comprehensive discussion of earlier measurements has been given elsewhere [24] and shall not be repeated here. The investigations show, however, that reliable results can only be expected from measurements with a low background damping and with specimens which were prepared under carefully controlled conditions. The present work pays special attention to these points and furthermore extends the investigated range to higher temperatures T > 300 K, where relaxation maxima with higher activation energies should occur.

EXPERIMENTAL PROCEDURE

The internal friction measurements were carried out with a composite oscillator technique which has been developed by Schwarz [28]. The system consists of a rod-like oscillator (specimen and piezoelectric quartz crystal) hanging between two thin wires and excited to longitudinal vibrations by an electrical feedback circuit. The internal friction Q^{-1} is calculated from the measured amplification which is necessary to keep the vibration at a constant amplitude. The background damping is mainly determined from the friction between driver and the wires and from the glue. A variety of glues (for different temperature ranges) have been tested. Best results were obtained with Devcon "5 Minute" Epoxy* for temperatures T below RT (room temperature) and with a mixture of water glass, $BaSO_4$ and Na_2SiF_6 for T > RT up to 550°C. The cross sections of the specimens were calculated according to Schwarz [28] in order to minimize modulations of the amplitude ε_v of the vibration by matching the acoustic impedance to that of the quartz driver. Measurements were usually carried out in the lowest vibration mode ($\lambda/2 = l$, length of the quartz) at about 90 kc, an amplitude $\varepsilon_v = 2.5 \times 10^{-6}$ and in a helium gas atmosphere at a pressure of 10 Torr. This ensured a good heat exchange between specimen and the furnace or the cryostat chamber (for temperatures below RT). Temperatures were measured at a dummy specimen parallel to the oscillator and controlled with an accuracy of about ± 2 K.

All measurements were done with germanium single crystals which were grown by the Czochralski method. The specimens were dislocation-free with an electrically active impurity content of less than 10^{15} cm^{-3}. Rod-like specimens with square cross-sections and a long axis parallel to the <110> direction were cut out. The mechanically damaged surfaces were carefully removed first by mechanical and then by chemical polishing. The specimens were dynamically deformed in a Instron machine under compression at 600°C and $\dot{\varepsilon} = 5 \cdot 10^{-5}$ sec^{-1} by 0.5 - 2.5% of their length. Experimental details concerning the deformation procedure were already described elsewhere [29]. All samples were cooled to a temperature T = 350°C before the applied load was removed. After deformation they were cut to their final shape (2.7x2.7x28mm^3), polished again and glued with one of their end faces to the rod-like quartz driver (4x4x 28 mm^3).

---

* Devcon Corporation, Danvers, Mass. 01923, USA

EXPERIMENTAL RESULTS

The dislocation arrangement of germanium samples deformed under the same conditions has been studied elsewhere by transmission electron microscopy [24,30]. According to the <110> orientation of the specimens two glide systems operate during the deformation. The dislocations are approximately of the 60° type and deviate from the <110> directions of the Peierls potential valleys within a range $\gamma \approx \pm 15°$ [24]. Their density N varies between $N = 0.2 - 1 \times 10^{8}\ cm^{-2}$ according to the degree of deformation. Weak-beam studies [31] have shown that all dislocations are dissociated and contain constrictions (which are connected with jogs) at a mean distance of about 1000 Å for our deformation conditions. Furthermore it is a typical feature of the dislocation structure in germanium crystals cooled under load that many bowed-out segments can be observed, probably moving dislocations which are frozen in at lower temperatures. Given that the geometrical kinks introduced through the curvature are mobile even at room temperatures the bowed-out segments are probably still prevented from relaxing into their equilibrium position after the load is removed below $T < 350°C$ because the mean free path of a moving kink is certainly limited by the constriction which should act as strong obstacles.

One can try to estimate the relaxation strength Δ_M of a kink migration peak. The existing models (for a summary see [24]) yield more or less similar expressions, namely

$$\Delta_M = \frac{\Omega}{12} \frac{\mu b^2}{E_o} NL^2 \qquad (1)$$

for $\sin |\gamma| > KT/E_o a$ (Ω geometry factor, μ shear modulus, b Burgers vector, E_o prelogarithmic factor of the dislocation line energy). If L is determined by the mean distance of constrictions, we approximately obtain $\Delta_M \approx 10^{-4} - 10^{-3}$. Given that only a fraction of all dislocations contribute to the internal friction, namely the mobile (bowed-out) segments, we still can expect $\Delta_M > 10^{-5}$. A relaxation peak with $Q^{-1}_{max} = \Delta_M/2$ would be well above the resolution limit of our measurements. Therefore one can be sure that the conditions under which our specimens have been deformed ensures a dislocation configuration which should give rise to a detectable internal friction maximum due to single kink motion if it occurs.

Fig. 1 shows a typical result, the damping Q^{-1} of both a deformed and undeformed germanium crystal between 80 K and 800 K. Below 300°C the curves are identical within the resolution of our measurements and no internal friction peaks can be observed. The variation of the amplitude ε_V of the vibration

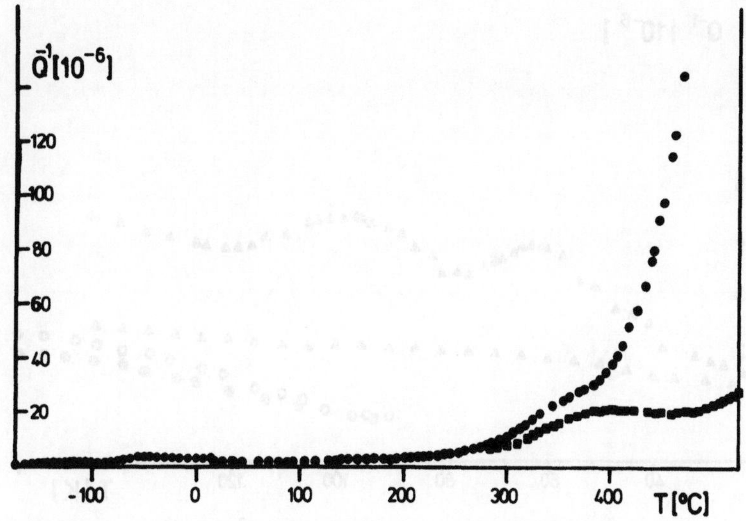

Fig. 1 Internal friction Q^{-1} versus temperature: (circles) deformed, (squares) undeformed germanium.

($\varepsilon_v = 2.5 \times 10^{-7} - 2.5 \times 10^{-5}$) and of the degree of deformation had no influence on the qualitative behavior of the specimens. The results disagree with a number of studies of other authors who have found internal friction peaks [32-35].

More recent measurements with a better resolution showed, however, in accordance with our results that earlier studies must have found peaks which were definitely not connected to single kink relaxation. Haworth et al. [35] investigated the temperature range from 80 K to 650 K at 2 kHz. A small peak occured which could be removed by chemically polishing the specimen surface. This is in agreement with other results which showed that surface damage can introduce small peaks below room temperature. Measurements of Möller and Buchholz [24] with the torsion pendulum at 15 Hz of both germanium and silicon also did not show any internal friction peaks in the temperature range 20 K to 400 K. The background damping of this method was, however, by an order of magnitude larger compared to the present investigation.

In contrast to our results are comparable measurements by Ohori and Sumino [25]. They investigated the low temperature region at 70 kHz and found two small internal friction peaks which they attributed to single kink migration with activation energies of H = 0.08 eV and 0.13 eV, respectively. Their measurement has been considered a proof for a low kink migration energy, although the

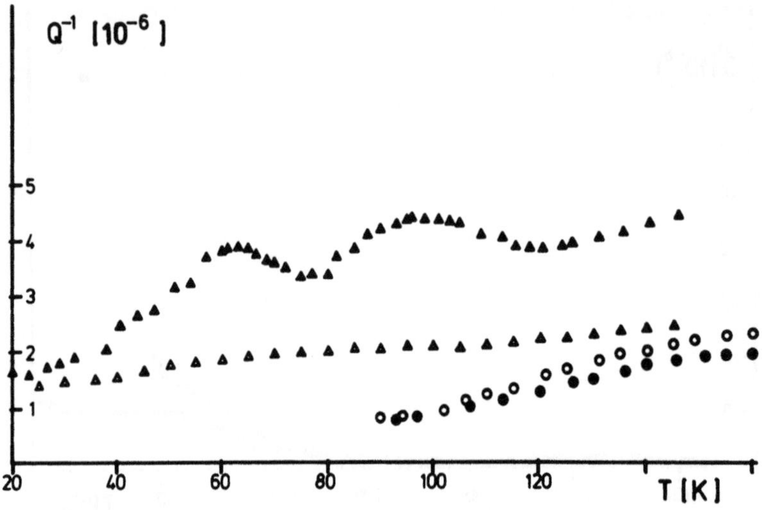

Fig. 2 Low-temperature internal friction Q^{-1} vs. T., ▲ Ohori et al. [14], ● our measurements, open symbols = undeformed specimens.

smallness of the peaks could only be explained by assuming a high density of obstacles (l < 100 Å) on the dislocation line. Because of their results we paid special attention to the same temperature range. Fig. 2 shows our low-temperature measurements in comparison with Ohori's and Sumino's. Our temperature range does not permit us to see the peak at 64 K but in none of our measurements we were able to detect the second peak.

Measurements above room temperature between 50 and 150 kHz by Gerk and Williams [36] again agree with our results. At about 350°C they observe a small peak which also appears in our measurements both in deformed and undeformed specimens and which has been explained as an acousto-electric effect.

Above 400°C the damping increases [Fig. 3]. A logarithmic plot of Q^{-1} shows a linear decrease with the inverse temperature. For the slope of the curves a spectrum of activation energies from 0.9 - 1.1 eV is found dependent on the degree of deformation. Higher deformed samples tend to give lower activation energies. The increase of the damping of deformed germanium crystals above 400°C has also been observed by other authors [23,36] though this spread of activation energies has not always been found. At 400°C dislocations become already mobile under stresses comparable to those which are applied in the internal friction measurements. It is therefore reasonable to assume that dislocation motion is

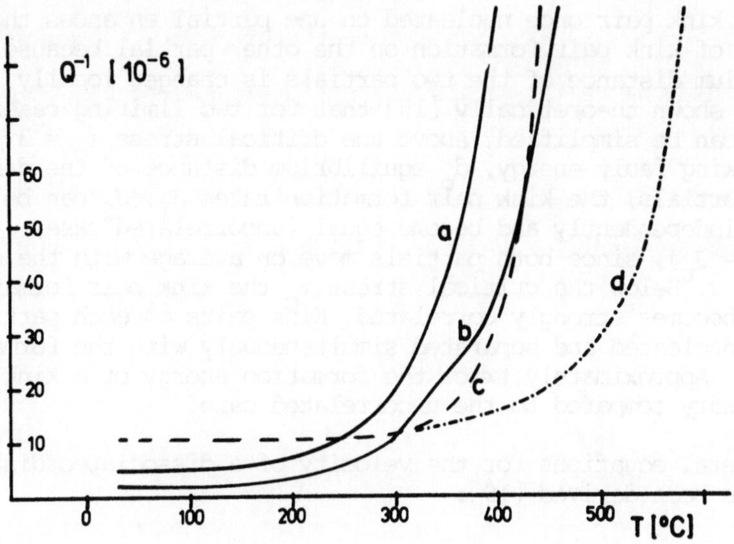

Fig. 3 Q^{-1} vs. T : our measurement: (a) deformed by 2%; (b) deformed by 1%; (c) Welsch e al. [23] tensile deformation at 630°C; Gerk et al. [36] four-point bending at 600°C.

responsible for the damping increase and that not only the kinks have become mobile but that double kink nucleation can occur also. Summarizing the discussion one can be rather confident now that no internal friction peak due to single kink migration is possible below temperatures where double kink nucleation begins. That conclusion allows to estimate a lower limit for the kink migration energy. From the condition that at the position of a kink migration peak $2\pi\nu\tau_o \exp(-E_{KM}/KT) = 1$ is fulfilled (ν frequency of the measurement, $\tau_o \approx 10^{-13}$ sec.), One yields $E_{KM} > 1$ eV. Some consequences of such a high kink migration energy for the theory of the dislocation velocity shall be discussed now.

DISCUSSION

The dislocation velocity v of a straight dissociated dislocation proceeding perpendicular to the direction of its Peierls valleys by successive formation of kink pairs depends on the kink pair formation rates J_1, J_2 for each partial dislocation, the kink height a and the mean distance by which the two members of the kink pairs separate before they cease to move. This may either be the case when they meet obstacles on the dislocation lines that are impenetrable to kinks, or when they annihilate with other kinks. Due to the eleastic interaction of the partial dislocations the nucleation rates J_1, J_2 are not independent of each other. A kink pair once nucleated on one partial enhances the probability of kink pair formation on the other partial because the equilibrium distance of the two partials is changed locally. It has been shown theoretically [19] that for two limiting cases the problem can be simplified. Above the critical stress $\tau_c = 3\gamma/4\, d_o$ (γ stacking fault energy, d_o equilibrium distance of the dissociated partials) the kink pair formation rates J_1, J_2 can be treated independently and become equal (uncorrelated case, $J_1 = J_2 = J_u$), since both partials move on average with the same velocity v. Below the critical stress τ_c the kink pair formation process becomes strongly correlated. Kink pairs on each partial can only be nucleated and separated simultaneously with the formation rate J_c. Approximately twice the formation energy of a kink pair is necessary compared to the uncorrelated case.

General equations for the velocity of a dissociated dislocation have been derived [19]:

$$v = \frac{1}{2}\ aL/\ (J_c + J_u) \qquad (2)$$

for the case that the mean separation width X_{KP} of the kink pairs is limited by impenetrable obstacles at a mean distance L, and

$$v = \frac{2a}{\pi^{1/2}}\ [\ v_K\ (J_c + J_u)\ ]^{1/2} \qquad (3)$$

if kink annihilation with other kinks determines X_{KP}. Experiments [21] have shown that equation 2 describes the velocity of short dislocation segments ($\lesssim$ 0.4 μm), whereas eq. 3 should be valid under experimental conditions mostly encountered. Therefore we restrict the following discussion to this case.

The kink velocity v_K for the motion along the dislocation line has been calculated earlier [17]:

$$v_K = \nu_D \ell e^{-E/KT}\ (1 - e^{-\tau ab\ell/KT}) \qquad (4)$$

ν_D Debye frequency, ℓ mean distance of the thermally activated jumps over the potential barriers of height E. In view of the high value of the secondary Peierls potential E_{KM} and the fact that the density of observed strong obstacles (with $E > E_{KM}$) is not very high it is reasonable to assume that $E = E_{KM}$ and $\ell = a/\sqrt{2}$.

A high secondary Peierls potential not only affects the kink mobility but also the kink pair formation rate J. The most comprehensive calculation of J (in the case of a large kink migration barrier) has been given by Seeger and coworkers [37]. The model is based on the physical picture that in addition to the elastic forces on the kink pairs (external force and kink-kink interaction) the kinks are subject to Brownian forces due to their interaction with lattice vibrations. This effect is described in terms of a kink diffusivity D_K which is related to the kink velocity by the Einstein relationship

$$v_K = D_K \tau ab/KT \tag{5}$$

Under the assumption that the saddle point energy of a separating kink pair on a complete dislocation is given by

$$E_{KP} = 2E_K - a^* \tau^{1/2} \tag{6}$$

with $a^* = (2a^3 bE_o)^{1/2}$ and E_K the energy of a single kink, the following nucleation rates J in the limiting cases of high and low stresses have been derived ($a^* \tau^{1/2} > KT$)

$$J = J_h e^{-(E_{KP} + E_{KM})/KT} \tag{7}$$

with $J_h = [2\pi\nu_D (a^3 b\tau)^{3/4} E_K] / [(\pi KT)^{1/2} (a^2 E_o/2)^{1/4} W_K{}^2 KT]$ and for low stresses ($a^* \tau^{1/2} < KT$)

$$J = J_l e^{-(2E_K + E_{KM})KT} \tag{8}$$

with $J_l = [2\pi\nu_D a^3 b\tau E_K] / [W_K^2 (KT)^2]$ and W_K the kink width, ν_D Debye frequency. These equations are valid for an undissociated dislocation, however one may assume that they can also be derived for a partial dislocation whose interaction with the second partial can be neglected. This is the case for $\tau > \tau_c$ (uncorrelated kink pair formation) and since an estimation of $(KT/a^*)^2$ yields values approximately in the same stress range as τ_c equation 7 should well describe the high stress region, that means $J = J_u$. For lower stresses $\tau < \tau_c$ correlated kink pair formation has to be considered. This case has not been treated correctly yet, one may assume, however, that a similar expression to equation (8) with twice the activation energy can be used for the correlated kink pair formation rate J_c. Inserting J_u (eq. 7), J_c (modified eq. 8) and the kink velocity v_K (eq. 4, with twice the kink migration energy for the correlated kink pair migration) into equation 3 yields the dislocation velocity at high and low stresses. As has been shown earlier [19] v can be approximated in the high stress region by the experimentally well established relation

$$v = A \tau^m e^{-(E_K + E_{KM})/KT} \qquad (9)$$

with m ≈ 1 and slightly depending on stress and temperature. For $\tau < \tau_c$ one obtains

$$v = B \tau e^{-2(E_K + E_{KM})/KT} \qquad (10)$$

with A, B prefactors including all constants. Equation 10 is experimentally not so well founded as the high stress case. Velocity measurements in germanium at low stresses [6] show Q = 2.8 eV (screws), Q = 3.0 eV (60°) compared to Q = 1.55 eV and 1.5 eV at high stresses. This agrees quite well, however, no such results are available for other covalent materials. In silicon an increase of the activation energy is observed for 60°, but not for the screw dislocation. Since τ_c is smaller for silicon it may be that the low stress region has not been reached experimentally.

Summarizing the discussion one must admit that the low stress regime has neither been investigated adequately by theory nor by experiments. Nevertheless one can draw the conclusion that at high stresses $\tau > \tau_c$ the activation energy Q is given by

$$Q = E_K + E_{KM} \qquad (11)$$

Since we estimate from the internal friction measurements $E_{KM} > 1$ eV one obtains $E_K < 0.5$ eV in germanium. This is in good agreement with recent theoretical calculations [27] in germanium: $E_K = 0.4$ eV for the reconstructed dislocation.

REFERENCES

1) A.R. Chaudhari, J.R. Patel and L.G. Rubin, J. appl. Phys. 33, 2736 (1962)
2) M.N. Kabler, Phys. Rev. 131, 54 (1963)
3) O.W. Johnson, J. appl. Phys. 36, 3247 (1965)
4) J.R. Patel and P.E. Freeland, Phys. Rev. Lett. 18, 833 (1967)

5) J.R. Patel and P.E. Freeland, J. appl. Phys. 42, 3298 (1971)
6) H. Schaumburg, Phil. Mag. 25, 1429 (1972)
7) I.E. Bondarenko, V.I. Erofeev and V.I. Nikitenko, Sov. Phys. JETP 37, 1109 (1973)
8) B. Ya. Farber, I.E. Bondarenko and V.I. Nikitenko, Sov. Phys. Solid State 23, 1285 (1981)
9) T. Suzuki and H. Kojima, Acta metall. 14, 913 (1966)
10) V.N. Erofeev, V.I. Nikitenko and V.B. Osvenskii, Phys. Stat. Sol. 35, 79 (1969)
11) V.N. Erofeev and V.I. Nikitenko, Sov. Phys., Solid State 13, 116 (1971)
12) A. George, C. Escaravage, G. Champier and W. Schröter, Phys. Stat. Sol. (B) 53, 483 (1972)
13) S.K. Choi and M. Mikara, J. phys. Soc. Japan 32, 1154 (1972)
14) S.A. Erofeeva and Y.A. Osipyan, Sov. Phys. Solid State 15, 538 (1973)
15) S.K. Choi, M. Mikara and T. Ninomiya, Japan J. appl. Phys. 16, 737 (1977)
16) V. Celli, M. Kabler, T. Ninomiya and R. Thompson, Phys. Rev. 131, 58 (1963)
17) V.V. Rybin and A.N. Orl, Sov. Phys. Sol. State 11, 2635 (1970)
18) T. Ninomiya, R. Thompson and F. Garcia-Moliner, J. appl. Phys. 35, 3607 (1964)
19) H.J. Möller, Acta metall. 26, 963 (1978)
20) A. Bourret, J. Desseaux and C. D'Anterroches, "Microscopy of Semiconducting Materials", A.G. Cullis and D.C. Joy, ed., The Institute of Physics, Bristol and London, p. 9 (1981)
21) F. Louchet, Ibid., p. 35 (1981)
22) P.B. Hirsch, A. Ourmazd and P. Pironz, Ibid., p. 29 (1981)
23) G. Welsch, T.E. Mitchell and R. Gibala, Phys. Stat. Sol. (a) 15, 225 (1973)
24) H.J. Möller, J. Buchholz, Phys. Stat. Sol. (a) 20, 545 (1973)
25) K. Ohori and K. Sumino, Phys. Stat. Sol. (a) 14, 483 (1972)
26) R. Jones, Phil. Mag. B 42, 213 (1980)
27) H. Veth, Doctor's Thesis, Göttingen (1983)
28) R. Schwarz, Rev. Sci. Instr. 48, 111 (1977)
29) S. Schäfer, H. Alexander and P. Haasen, Phys. Stat. Sol. 5, 247 (1964)
30) K. Kojima and K. Sumino, J. Phys. Soc. Jap. 26, 1213 (1969)
31) G. Packeiser and P. Haasen, Phil. Mag. 35, 821 (1977)
32) P.F. Kromer and L.P. Khiznichenko, Phys. Stat. Sol. 21, 811 (1967)
33) B.M. Mecs and A.S. Nowick, Appl. Phys. Letters 8, 75 (1966)
34) K. Ohori and K. Sumino, Phys. Stat. Stat. Sol. (a) 9, 151 (1972)
35) W.L. Haworth, R. Mattas, H.K. Birnbaum, J. appl. Phys. 43, 3658 (1972)
36) A.P. Gerk, W.S. Williams, J. appl. Phys. 53, 3585 (1982)
37) A. Seeger, J. de Physique 42, C5, 201 (1981)

PLASTIC DEFORMATION OF INTRINSIC AND EXTRINSIC SILICON BELOW 500°C

P. Veyssière, J. Rabier, J.L. Demenet and J. Castaing†

Laboratoire de Métallurgie Physique
40 Avenue du Recteur Pineau
86022 Poitiers - France

INTRODUCTION

In this paper we report an analysis of the plasticity of silicon in a range of low temperatures where it usually responds to a standard uniaxial compression by brittle fracture. Indeed, it has been previously shown that silicon is able to undergo macroscopic permanent strain in compression provided cracks are prevented from propagating, by superimposing an hydrostatic pressure onto the sample[1 2]. This is but imperfectly attained in microindentation tests where the stress and pressure gradients are not quantitatively known resulting in ambiguous information. Concurrently, a modified version of the Griggs device[3], mostly used by geologists, has been used recently with success for spinel[4], sapphire[5] and silicon[2]. It has been shown to offer sufficient accuracy to allow acceptable standard measurements of the yield stress and of the activation parameters of the deformation. Therefore, it was interesting to extend our knowledge of the plasticity of brittle materials to : (i) low temperatures ; i.e. temperatures at which the point defects (dopants, impurities, thermal point defects) are no longer mobile. Indeed, aliovalent impurities in ionic crystals as well as electrically active dopants in semiconductors have proved to affect drastically the velocity of dislocations. It is therefore important to carry out experiments at temperatures where the mobility of the point defects is minimized. Furthermore, the core structure of dislocations in silicon is matter of active investigation since, in order to explain several electrical and mechanical properties, it has been repeatedly postulated that it should be associated with rows of point defects[6].

† Laboratoire de Physique des Matériaux, CNRS, 92190 Meudon France

(ii) high stresses; i.e. stresses comparable to or larger than γ/b for example. When the stress applied to one partial dislocation is larger than this limit, the partial may overcome the spring force exerted by the stacking fault and would then move independent from the other partial. This has been already observed in silicon[2], it was shown that at stresses of the order of 300 MPa, the dislocation microstructure in samples deformed along <001>, is mostly composed of extended stacking faults and twins : features which suggest that the individual rather than the correlated motion of partials has been operating during the deformation.

The macroscopic yield stress under a confining pressure for two dopant concentrations, several temperatures and strain-rates are discussed in this paper. Attempts at relating the deformation microstructure to the deformation conditions are made. This work is part of a more extensive study of the properties of dislocations in silicon where the effect of a varied climb force, should help clarify the present knowledge of their core structure. Indeed, the theories which have been proposed so far to account for the doping effect on the dislocation velocity have in common the general mechanism of motion by the formation and the migration of kinks. These have been empirically correlated with the climb force acting on each partial[7]: varying the hydrostatic component of the stress tensor acting on dislocations, under controlled experimental conditions, may discriminate between some of the theoretical arguments.

EXPERIMENTAL

Silicon single crystals oriented along <123> for single slip, <110> and <100> were deformed in compression, under a confining pressure of 1.5 GPa, at two different strain-rates $\dot{\varepsilon}_2 = 2.10^{-5}\ s^{-1}$ and $\dot{\varepsilon}_1 = \dot{\varepsilon}_2/10$, below 500°C. Two grades of floating zone grown single crystal were selected for these experiments, such that each would keep the same electrical activity over a wider range of temperature than that investigated in the present study ; they will be referred to as "intrinsic" (10^{13} atoms / cm^3) and "extrinsic"($10^{19}\ cm^{-3}$, atoms of P, n-type). The deformation samples were prestrained at a temperature of 400 - 450 °C in order to minimize their yield behaviour. This procedure is particularly well adapted to the case of silicon where the samples may undergo stresses larger than their fracture limit, along the transient yielding regime (Fig. 1).

THE PLASTICITY OF INTRINSIC SILICON UNDER A CONFINING PRESSURE

The engineering stresses at the lower yield point (σ_{LY}) are plotted in Fig. 2 as a function of the temperature of the test. For each strain-rate, the engineering stresses fall roughly in line for both orientations, i.e. it is apparently orientation independent. When trying to derive the yield stress resolved on the more stressed

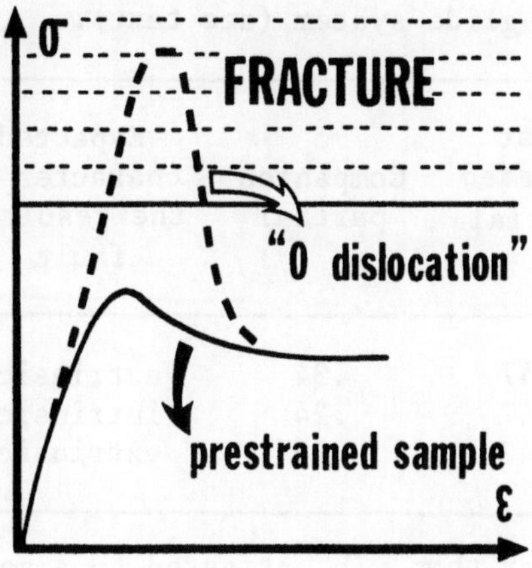

Fig. 1. Difference in yield behaviour according to the mechanical history of a sample.

glide systems, it can be seen on Table 1 that the best fit is obtaitained provided that the rationalizing Schmid factors on the most stressed partial are considered rather than this acting upon the perfect parent dislocation. This suggests that the slip systems which are "seen" at the microscopical scale of the deformation experiments might be of the <112>{111} type rather than <110>{111} as usually postulated. This hypothesis has already been invoked in a previous study[8], where a large density of extended stacking fault and twins, associated with the fact that the resolved shear stress applied to

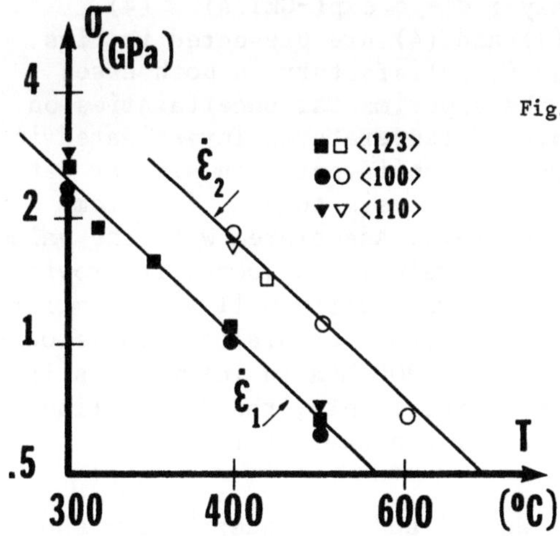

Fig. 2. $\sigma_{LY}(T)$ for different orientations of the applied stress and for two strain-rates ($\dot{\varepsilon}_1 = 2.10^{-6}\ s^{-1}$, $\dot{\varepsilon}_2 = 2.10^{-5}\ s^{-1}$).

Table 1. Comparison between the resolved flow stress at the lower yield point for the three compression axes, depending upon the choice of the potential glide system (see text).

Compression axis	Schmid factor on <110>{111}	σ_{LY} (MPa)	Most stressed partial	Companion partial	Expected character of the resulting fault
<123>	.47	691	.47	.34	extrinsic
<100>	.41	646	.47	.24	intrinsic
<110>	.41	713	.47	.24	extrinsic

the sample during the tests was larger than γ/b, attested to a mechanism of deformation by uncorrelated partials at low temperatures.

In this limited range of temperatures, both the phenomenological equation :

$$\dot{\gamma} = B\ \tau^{n}\exp(-Q/kT) \qquad (1)$$

and the usual Arrhenius law :

$$\dot{\gamma} = \dot{\gamma}_o \exp(-\Delta G(\tau)/kT) \qquad (2)$$

where $\Delta G(\tau) = \Delta H_o - v\tau$, can be fitted to the present experimental results. Indeed, the activation volume v shows a variation with the reciprocal resolved shear stress (τ^{-1}), the free energy is thus given by[2] :

$$\Delta G(\tau) = -\int_{\tau_o}^{\tau} v\, d\tau = A \ln(\tau_o/\tau) \qquad (3)$$

where τ_o is the flow stress at 0 K. Since $\Delta G(\tau)/kT$ is approximately constant ($C \simeq 36$), combining eqns. (2) and (3) yields :

$$\tau = \tau_o \exp(-CkT/A) \text{ or equivalently : } \sigma = \sigma_o \exp(-CkT/A) \qquad (4)$$

The curves associated with eqns. (1) and (4) are presented in Figs. 3 and 2 respectively. The agreement is satisfactory in both cases : this ambiguity is probably due to the experimental uncertainties on σ_{LY} and to the very restricted range of temperatures investigated in this work. In order to get a discussion consistent with similar data by other groups, we shall restrict our analysis to eqn. (1) from which a value of $Q/n = 0.4$ eV can be drawn. Associated with the value of $n \simeq 4$ determined experimentally by strain-rates jumps, it provides a value of $Q \simeq 1.8$ eV. This result compares fairly well with similar data derived independently from determinations of dislocation velocities between 325 °C and 420 °C, with $\tau \simeq 300$ MPa in intrinsic silicon prestrained at 650 °C [9]. In these experiments, the dislocation velocity was assumed to follow a power law behaviour :

$$V = V_o^{n-2} \exp(-q/kT) \qquad (5)$$

resulting in $n = 4$ and $Q \simeq 1.6$ eV. Wessel and Alexander[10] have shown that the main features observable by TEM after such experiments were the increasing density of dislocations more widely dissociated than

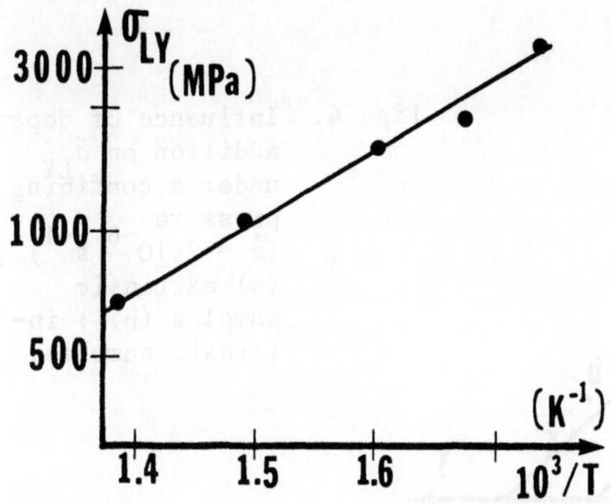

Fig. 3.
Plot of the engineering stress vs. 1/T : Compare with fig.2.

after a deformation at larger temperature, together with the apparition of "nodes" of stacking faults. On the other hand, undirect dislocation velocity measurements at high temperatures [11] yielded values of $n \simeq 3$ and of $Q \simeq 2.4$ eV. If these parameters characterize the deformation mechanism, as usually postulated, the observed consistency at low temperature between velocity measurements and our results, together with the discrepancy between velocity measurements at high and low temperatures, reinforce the idea of different processes controlling the deformation, according to the temperature at which the test has been carried out.

THE INFLUENCE OF DOPANT

The strong influence of a dopant addition on the dislocation velocity in silicon is well known : the mobility of dislocations is enhanced for n-type samples compared to intrinsic ones, but it is roughly unaffected by p-type additions. These properties have been examined at low temperatures, the results are plotted in Fig. 4. Contrary to what was expected, the ratio between σ_{LY} for the two grades of single crystals, which were deformed in single slip, is temperature independent within the precision of our measurements. We do not have any satisfactory explanation for this observation which suggests that the ratio of dislocation velocities should also be temperature independent, in contradiction with what was observed by George et al. [12] at slightly larger temperatures.

TRANSMISSION ELECTRON MICROSCOPE OBSERVATIONS

Several thin sections have been ion-thinned from the deformation samples. As previously indicated, when the deformation has occurred along <001> the microstructure is mainly composed of twins and sta-

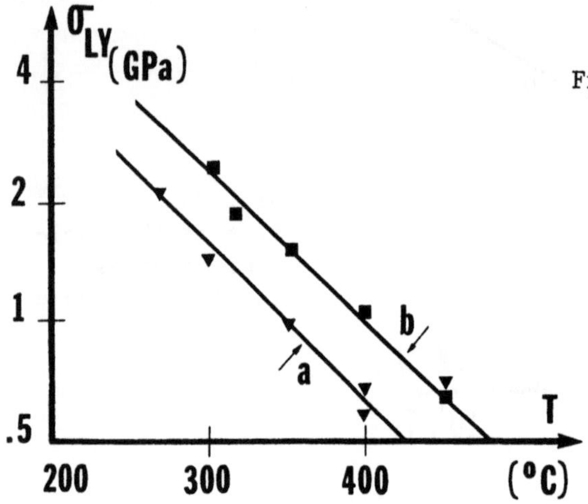

Fig. 4. Influence of dopant addition on σ_{LY} under a confining pressure ($\dot{\varepsilon} = 2.10^{-6}\ s^{-1}$) ; (a) extrinsic samples (b) : intrinsic samples.

cking faults, which attests to the individual motion of partial dislocations. We have carried out a TEM study of the microstructure present in the samples during the deformation at the higher temperatures (this is also the microstructure after prestraining in single slip) : an intrinsic <123> sample has been deformed at 425°C, the deformation being stopped just after the upper yield point ($\sigma \sim 1750$ MPa). The foils have been cut parallel to the most stressed {111} glide plane which will referred as the (111) plane. They have been observed in a JEOL 200 CX operating at 200 kV with a double tilt goniometer. The microstructure is typical of that observed systematically in high strength materials deformed at relatively low temperatures : the deformation is localized inside of bands containing a very high density of dislocations. Out of the deformation bands the dislocations are totally absent. In the present example the density is so high that it is almost impossible to carry out any satisfactory weak beam study (Fig. 5). Several less typical features have been detected : presence of sparse deformation bands inclined to the primary glide plane indicating that secondary slip systems have been activated but to a small extent, absence of extensive twinning but groups of stacking faults inclined to the foil are occasionally visible. In the areas where the edges of the deformation bands intersect the foil surfaces, it is possible to derive some qualitative information on the fine structure of these bands since the dislocations are seen in a lesser density (Figs. 6 and 7). In Fig. 7 the dislocations have generally the $1/2\,[0\bar{1}1]$ Burgers vector, they glide on (111) and they exhibit a segmented shape along the 3 dense <011> direction of this plane as usually observed after deformation at moderate temperatures. However, unexpected was the

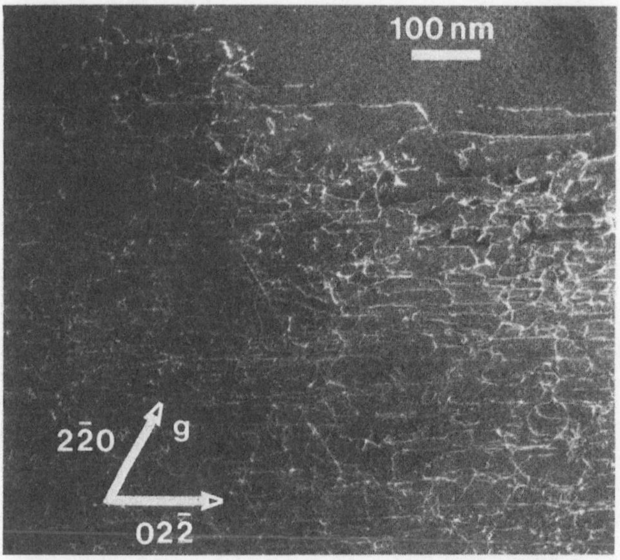

Fig. 5. Weak beam view of the very large density of dislocations lying in the deformation bands. The foil is cut nearly parallel to the (111) primary glide plane. This area is close to the edge of the thin foil, its thickness is less than 500 nm.

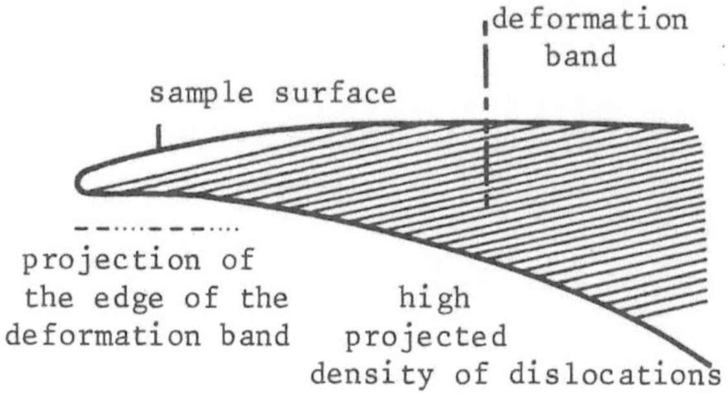

Fig. 6. Schematical section of the geometrical position of the deformation band with respect to the foil surfaces.

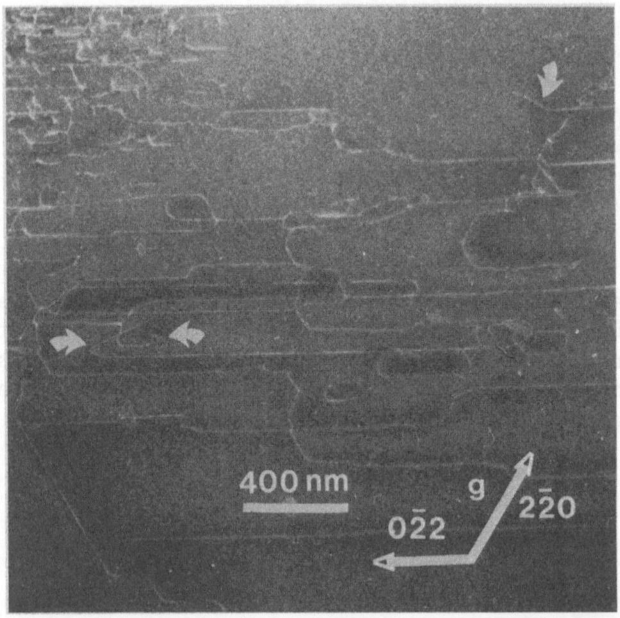

Fig. 7. Weak-beam image of the edge of the deformation band (see Fig.6). The curved arrows indicate segments of dislocations which belong to the cross slip plane.

observation of very long straight screw dislocations, appearing in a much larger density than the mixed segments. We suppose that this can be ascribed to the stabilizing effect cross slip may have on the screw orientation. Indeed, indicated by curved arrows in Fig. 7, one can observe dislocation segments which make an angle of ± 30° instead of ± 60° with the screw orientation. Appropriate tilt experiments from (111) to (011) and (101) sections under weak beam conditions have led to the conclusion that these segments are oriented along <110> rows in the $(\bar{1}11)$ cross slip plane for these screw segments. The absence of extensive twinning is not understood. However, extended stacking faults have been seen once again. This is obvious from Figs. 8 and 9, where two families of faults with drastically different spacings are observable : one with very narrow spacing is probably

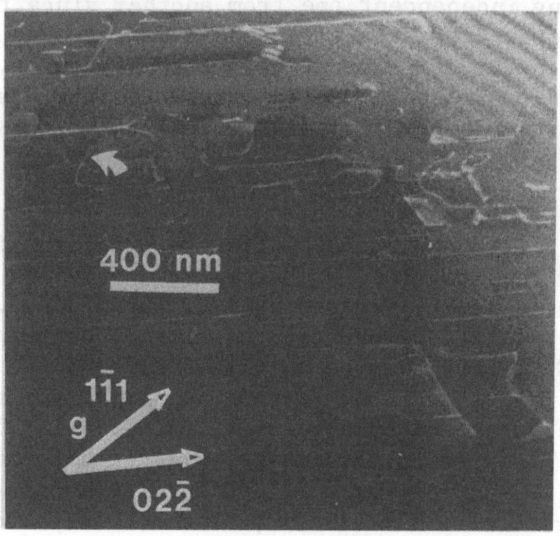

Fig. 8. Weak-beam micrograph with $\vec{g} = [1\bar{1}1]$, showing the fault contrast. The fault with large spacing, slightly inclined to (111), is indicated by an open arrow. The small curved has the same meaning as in Fig. 7.

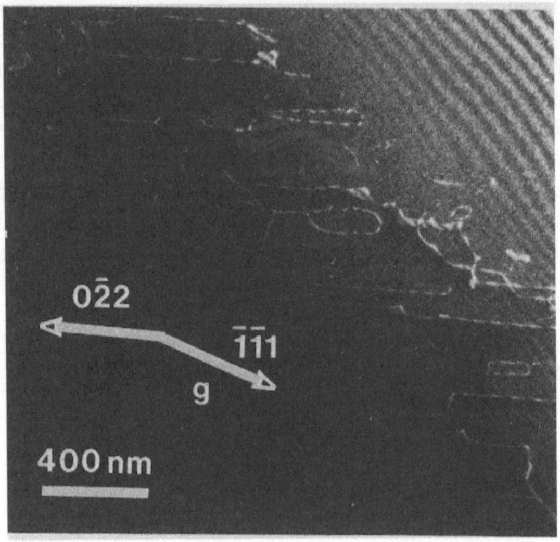

Fig. 9. Same area as in Fig. 8. The other {111} reflection of the (011) foil section is operating.

associated with a dislocation which has cross slipped, the other shows a very large spacing due to its small inclination to the foil plane. These features indicate clearly that the partials related to the mixed segments which belong to the primary glide plane are moving independent one from another since the stacking faults extend over distances often one order of magnitude larger than the equilibrium distance at larger temperatures. Further TEM experiments on the dislocation microstructure resulting from tests at temperatures lower than 425°C are in progress.

REFERENCES

1. M.J. Hill and D.J. Rowcliffe, Deformation of silicon at low temperatures, J. Mater. Sci. 9 : 1569 (1974).
2. J. Castaing, P. Veyssière, L.P. Kubin and J. Rabier, The plastic deformation of silicon between 300°C and 600°C, Phil. Mag. A 44 : 1407 (1981).
3. D.T. Griggs and G.C. Kennedy, A simple apparatus for high pressures and temperatures, Am. J. Sci. 254 : 722 (1956).
4. S.H. Kirby and P. Veyssière, Plastic deformation of MgO (Al_2O_3) spinel at 0.28 Tm : preliminary results, Phil. Mag. A 41 : 129 (1980) ; also : P. Veyssière, S.H. Kirby and J. Rabier, Plastic deformation of MgO : n Al_2O_3 spinels at temperatures below 1000°C (0.5 Tm), J. Physique Coll. 41 : C6 -175 (1980).
5. J.Castaing, J. Cadoz and S.H. Kirby, Prismatic slip of Al_2O_3 single crystals below 1000°C in compression under hydrostatic pressure, J. Amer. Ceram. Soc. 64 : 504 (1981).

6. P.B. Hirsch (1981) Electronic and mechanical properties of dislocations in semiconductors, in : "Defects in semiconductors", Narayan and Tan, eds. North-Holland, Amsterdam.
7. H. Alexander, C. Kisielowski-Kemmerich and E.R. Weber, Investigations of well defined dislocations in silicon, ICDS-12 Amsterdam (1982), to be published in Physica B.
8. J. Rabier, P. Veyssière and J.L. Demenet, Plastic deformation of silicon at low temperature and the influence of doping, J. Physique Coll., to be published.
9. M. Omri, Thèse 3ème cycle, ENSMIN, Nancy (1981).
10. K. Wessel and H. Alexander, On the mobility of partial dislocations in silicon, Phil. Mag. 35 : 1523 (1977).
11. I. Yonenaga and K. Sumino, Dislocation dynamics in the plastic deformation of silicon crystals, Phys. Stat. Sol. (a) 50 : 685 (1978).
12. A. George and G. Champier, Velocities of screw and 60° dislocations in n-and p-type silicon, Phys. Stat. Sol. (a) 53 : 529 (1979).

6. P.B. Hirsch (1981) Electronic and mechanical properties of dislocations in semiconductors, in : "Defects in semiconductors", Narayan and Tan, eds. North-Holland, Amsterdam.
7. H. Alexander, C. Kisielowski-Kemmerich and E.R. Weber, Investigations of well defined dislocations in silicon, ICDS-12 Amsterdam (1982), to be published in Physica B.
8. J. Rabier, P. Veyssière and J.L. Demenet, Plastic deformation of silicon at low temperature and the influence of doping, J. Physique Coll., to be published.
9. M. Omri, Thèse 3ème cycle, ENSMIN Nancy (1981).
10. K. Wessel and H. Alexander, On the mobility of partial dislocations in silicon, Phil. Mag. 35 : 1523 (1977).
11. I. Yonenaga and K. Sumino, Dislocation dynamics in the plastic deformation of silicon crystals, Phys. Stat. Sol. (a) 50 : 685 (1978).
12. A. George and G. Champier, Velocities of screw and 60° dislocations in n and p-type silicon, Phys. Stat. Sol. (a) 53 : 529 (1979).

PLASTIC DEFORMATION OF TRANSITION METAL CARBIDES

David J. Rowcliffe

SRI International
333 Ravenswood Avenue
Menlo Park, California 94025

INTRODUCTION

The transition metal carbides possess a combination of properties[1] that places them in a unique position in any classification of compounds. Both the elastic moduli and the melting points are extremely high which are primary characteristics of covalent solids. On the other hand, some of the electrical properties of the carbides are similar to those of metals although some important differences have been described that reflect other contributions to the bonding.[2] Close to room temperature slip and cleavage occur on the same planes as in ionic solids with the same crystal structure, but at high temperatures the details of slip show great similarity to those of metal crystals. This multifaceted character makes it particularly difficult to predict and understand plastic deformation processes in the transition metal carbides. Individual observations can be explained but the details of deformation are not as well understood, at the fundamental level, as they are in a material such as silicon.

Much of the information on plastic deformation of carbides is derived from experimental work in which many of the important variables could not be controlled completely. This is due in part to the scarcity of suitable single crystals and also to the difficulty of carrying out experiments that will cause observable amounts of plastic flow. The lack of large single crystals has been particularly acute in the case of WC, because this carbide decomposes before it melts and single crystals must be grown from solution.[3] Consequently, few large crystals of WC have been available for deformation studies and most information has been obtained from indentation experiments on small crystals. Single crystals of the monocarbides of the Group IV and V transition metals can be readily grown by zone melting and there has been a considerble amount of work on strength, brittle to ductile transition and deformation mechanisms. Two types of investigation have been made on single crystals. In the first, high temperature compression or bend tests have established brittle to ductile transition temperatures, operating slip systems for bulk plastic flow, the variation of yield strength with temperature and to some extent the influence of stoichiometry on these. The second type of investigation has been concerned with local slip at hardness indentations made over a range of temperatures and has led to the identification of slip planes and to a better understanding of the brittle to ductile transition.

This paper concentrates on plastic flow and the brittle to ductile transition in the monocarbides of the Groups IV and V transition metals and tungsten carbide. The determination of slip planes, models for slip and information on the character of dislocations are reviewed. Finally some current work is described that aims to apply information derived from single crystal WC to explain some aspects of deformation in WC-Co. Details of the mechanical properties of the carbides have been avoided because these are well covered elsewhere.[1,2,4]

CRYSTAL STRUCTURE AND COMPOSITION

Table 1 lists the carbides discussed in this review. The Group IV and V carbides all exist over a substantial range of carbon sub-stoichiometry. The Group IV carbides retain the same rock salt structure over the whole stoichiometry range with a slight decrease in lattice parameter.[1] On the other hand, a number of ordered structures form at particular carbon contents in the Group V carbides,[4] and ordering influences not only properties such as strength and

Table 1. Some Properties of Transition Metal Carbides

Carbide	Structure	Stoichiometry Range*	Maximum Melting Point, K
TiC	B1	0.65-0.98	3330
ZrC	B1	0.65-0.97	3690
HfC	B1	0.65-0.99	4200
VC	B1	0.75-0.88	2920
NbC	B1	0.75-0.99	3870
TaC	B1	0.75-0.98	4250
WC	Hexagonal	1	2870

*Expressed as carbon/metal.

hardness, but also affects the brittle to ductile transition. None of these carbides appears to achieve full stoichiometry. Most reach a maximum carbon content corresponding to a carbon/metal ratio of about 0.98, but in VC the maximum is reached[5,6] at a ratio of 0.88 corresponding to the compound V_8C_7. One explanation for this can be found in band structure models that indicate complete filling of the d-band bonding states before full stoichiometry is reached.[4]

The other carbide of interest is WC. This Group VI carbide is a stoichiometric compound and has an unusual hexagonal structure in which the c/a ratio is just less than 1.

BULK PLASTIC FLOW

At room temperature the amount of plastic flow that occurs in bending or compression is extremely small in most of the carbides. However, at temperatures around 1000°C the rock salt structure carbides undergo a brittle to ductile transition and deform readily on the {111} planes in <1$\bar{1}$0> directions. Table 2 summarizes the brittle to ductile transition temperature ($T_{B/D}$) for a number of single and polycrystalline carbides.

Considering first the single crystals of the cubic carbides, $T_{B/D}$ is < 0.5 of the absolute melting temperature. The brittle/ductile transition temperature for the Group IV carbides TiC and ZrC is relatively insensitive to carbon content. On the other hand VC shows a discontinuous change of $T_{B/D}$ with stoichiometry[4] with a maximum value around $VC_{0.84}$ (V_6C_5). This $T_{B/D}$ corresponds to the disordering temperature of this compound, and this is also the temperature of the maximum melting point in the V-C system. Comparable data are not available for NbC and TaC but qualitatively similar behavior can be expected at those compositions where similar ordering occurs.

From the limited information that is available, it appears that $T_{B/D}$ for pore-free polycrystals is only slightly higher than for single crystals. This is not an unexpected result because {111} <1$\bar{1}$0> slip provides sufficient systems for polycrystalline ductility. The data in the literature[1,4,10,11] suggest that grain size, porosity and carbon content have a strong effect on $T_{B/D}$ for TiC. The low $T_{B/D}$ for stoichiometric polycrystalline TaC is also worthy of note. Martin[16] has also found evidence for plastic flow in TaC fibers in directionally solidified Co-TaC alloys, tested in tension at room temperature. In addition, indentation in stoichiometric TaC at room temperature[17] causes plastic flow on a similar scale to that seen in MgO or LiF. It is therefore likely that the brittle/ductile transition for single crystal stoichiometric TaC might be below 700°C.

Table 2. Brittle to Ductile Transition Temperatures for Carbides

Material	B/D Transition $T_{B/D}$ °C	$T_{B/D}/T_M$	Remarks and Reference
$TiC_{0.79-0.97}$	800 - 900	0.32 - 0.35	Single crystal in compression[7-9]
TiC	800	0.32	Hot-pressed polycrystalline in compression[10,11]
$ZrC_{0.875}$	1100	0.37	Single crystal in compression[7]
$ZrC_{0.95}$	1100	0.37	Single crystal in compression[12]
$ZrC_{0.94}$	1200	0.40	Fused polycrystalline[13] in compression
$VC_{0.88}$	1000	0.48	
$VC_{0.84}$	1200	0.50	Single crystal in compression[4]
$VC_{0.75}$	800	0.37	
$NbC_{0.76}$	1100	0.36	Single crystal in compression[7]
$TaC_{0.98}$	800	0.28	Carburized polycrystalline[14] in bending
WC	20	< 0.01	Single crystal in bending[15]

Information on the brittle to ductile transition for single crystal WC is incomplete. Very little work has been done on high temperature bending or compression tests on single crystals of WC and most information is derived from observations at indentations or from polycrystalline material bonded with Co. Schenck et al.[18] compressed single crystals of WC along the c axis and found no evidence for plastic flow for stresses up to 450 MPa and temperatures up to 1520°C. Luyckx[15] carried out bending experiments on very small single crystals of WC at room temperature and by careful observation found small groups of slip lines close to the fracture surface. Luyckx identified the primary slip plane as $\{10\bar{1}0\}$ and made very convincing arguments for a $\langle 11\bar{2}3\rangle$ slip direction which has subsequently been verified by several workers.[19-22] Tungsten carbide cannot be called ductile at room temperature, but some detectable slip does precede fracture. In this regard, Chatterjee and co-workers[9] have shown that extremely small amounts of slip on $\{110\}$ precede fracture in stoichiometric TiC, between room temperature and the bulk brittle to ductile transition of 800°C. In this case they have identified slip on $\{110\}$ up to about 700°C. At 600°C slip takes place simultaneously on $\{111\}$ then above 800°C only $\{111\}$ slip occurs. Thus there appears to be a transition in the primary slip system in TiC just before the bulk brittle to ductile transition.

INDENTATION DEFORMATION IN CUBIC CARBIDES

Much of the information on deformation behavior of carbides has been obtained by indentation experiments. The advantages of this method of deformation are that it can be applied to very small crystals, and the large hydrostatic component of stress tends to promote slip while restricting fracture. The main disadvantage is that the stresses and distributions are not easily measured.

Crystals such as TiC show no evidence of slip-line formation, when indented at low temperatures. However, the hardness of TiC and other carbides indented on a given crystal face, depends on the relative orientation of the indenter to directions on the crystal. In 1971, Brookes and coworkers[23] improved existing analyses of hardness anisotropy and showed that it could be used to determine operative slip systems. Figure 1 shows the change in effective resolved shear stress with indenter direction for Knoop impressions made on the $\{100\}$ face for the cases of $\{111\}$ $\langle 1\bar{1}0\rangle$, and $\{110\}$ $\langle 1\bar{1}0\rangle$. slip. The change of effective resolved shear stress with Knoop orientation for $\{100\}$ $\langle 1\bar{1}0\rangle$ slip is the same as for $\{111\}$ $\langle 1\bar{1}0\rangle$ slip and the effective stress on the slip planes is a minimum along [010] and a maximum along $[1\bar{1}0]$. Brookes et al. further showed that the hardness is a maximum when the effective resolved shear stress on the slip plane is minimized. The model is open to criticism because it is not specific about the sign or the magnitude of the stresses and some caution needs to be exercised in the interpretation. Nonetheless, hardness anisotropy has been a remarkably successful tool in identifying slip planes in carbides and various ionic compounds.

Figure 2 shows anistropy curves for TaC and HfC, from which the slip planes were identified as $\{111\}$ and $\{110\}$, respectively.[24] The $\{111\}$ determination is not unambiguous because this form of curve could also indicate $\{100\}$ slip. However, dislocations move unusually freely in TaC at room temperature[17] and etch pitting confirms the interpretation of $\{111\}$ as illustrated in Fig. 3. This figure also shows four pairs of $\{100\}$ cleavage cracks generated at each indentation. These are not the radial-median cracks commonly found passing through the corners of indentations in ceramics or other carbides. Instead, each pair is located in the center of the slip band and the cracks are thought to arise as a result of dislocation interactions on intersecting slip planes.

Indentations in HfC are similar in appearance to those in TiC, ZrC, and substoichiometric Group V carbides. Figure 4 suggests little

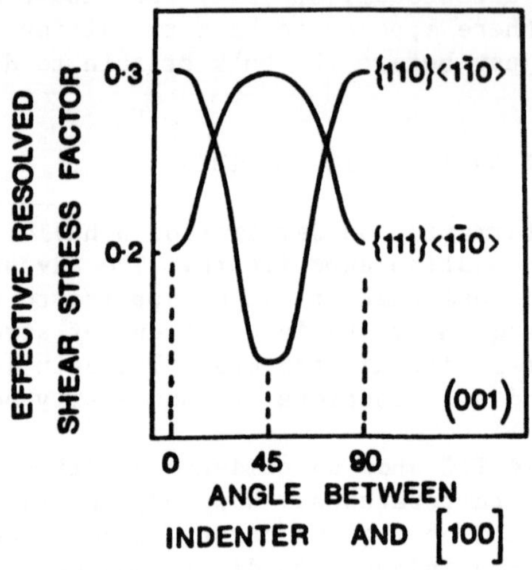

Fig. 1. Variation of the effective resolved shear stress for {110} <1$\bar{1}$0> and {111} <1$\bar{1}$0> slip induced at Knoop indentations made on the cube face plane.

evidence for plastic flow other than the formation of the impression. However observations of this indentation by transmission electron microscopy (TEM) confirm the existence of high densities of dislocations, with the interpretation that plastic flow has occurred. This is shown in Fig. 5. Conventional g.b analysis and the use of large tilts confirms not only slip on {110} but also the <1$\bar{1}$0> Burgers vector. At the same time, it must be pointed out that the dislocations generated by indentation do not necessarily have the same characteristics as glide dislocations produced in more conventional deformation experiments. Although some of the dislocations shown in Fig. 5 were shown to be mixed glide loops. The loops at A are prismatic since the Burgers vector is normal to the loop plane. It is likely that these loops arise as a result of the very high local stresses that cannot be relieved by the simple shear processes that are associated with unconstrained slip. Instead accommodation dislocations might be generated to allow complete deformation to take place as has been suggested for silicon.[25] Even though the interpretation is not straightforward, this example illustrates the type of information that can be obtained from very hard solids using a combination of hardness anisotropy and TEM.

Hardness anisotropy measurements made at high temperatures, have been particularly informative regarding slip planes and the brittle to

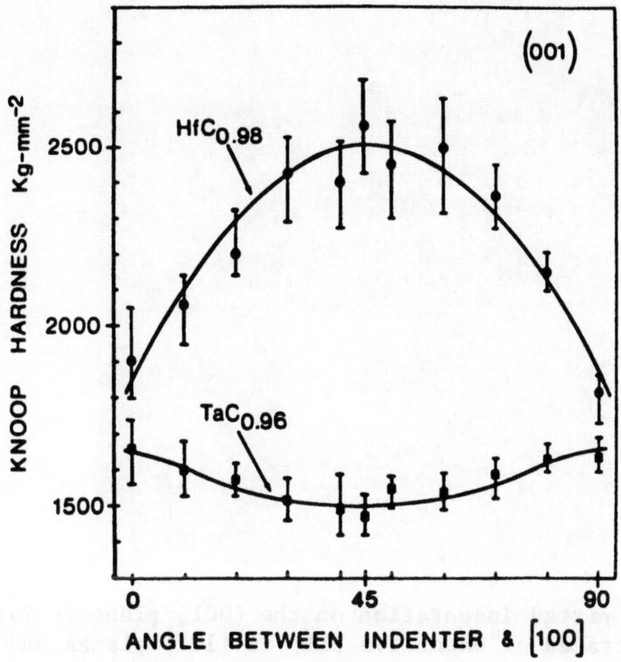

Fig. 2. Hardness anisotropy curves for HfC and TaC.

ductile transition in carbides. Figure 6 illustrates Knoop hardness anisotropy as a function of temperature[26] for TiC. The significant feature is that as the temperature is raised, there is a dramatic change in the curves which denotes a change in slip plane from {110} to {111}, the usual slip plane of rock salt structure carbides at high temperatures. This type of curve indicating a change of slip system at 250°C in TiC has been seen in several substoichiometric carbides at room temperature,[27-29] including $TiC_{0.8}$, $VC_{0.84}$, $NbC_{0.83}$, and most recently in a (W, Ti)$C_{0.8}$ solid solution, as shown in Fig. 7. In these examples the hardness rising to a maximum is interpreted to result from slip on {110}, whereas the fall in hardness to the minimum indicates slip on {111} or {100}. Similar curves were found[28] for $NbC_{0.83}$ and TEM observations confirmed both {110} and {111}. However in recent TEM work, Shinde[29] has identified slip on all three primary planes of (W, Ti)$C_{0.8}$ solid solution indented at room temperature.

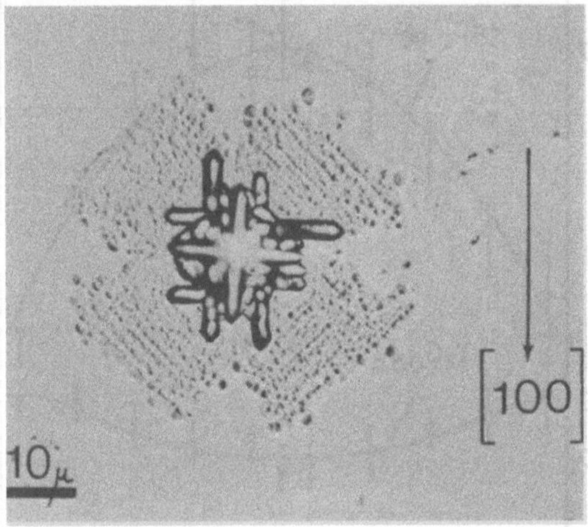

Fig. 3. Etch pitted indentation on the (001) plane in TaC showing the traces of extensive slip on {111} planes and cracks in ⟨100⟩ directions.

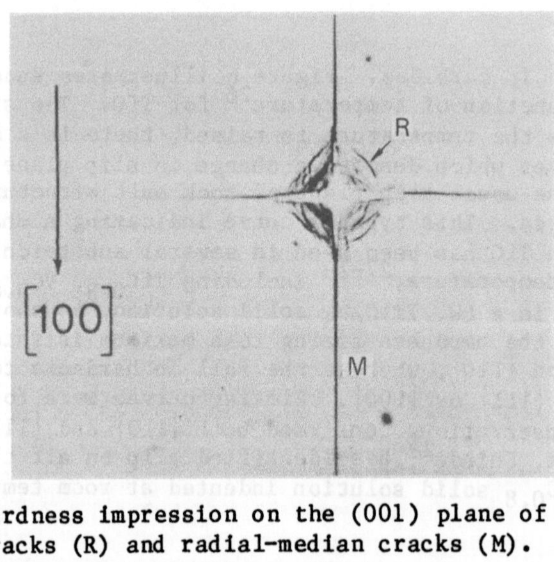

Fig. 4. Hardness impression on the (001) plane of HfC showing ring cracks (R) and radial-median cracks (M).

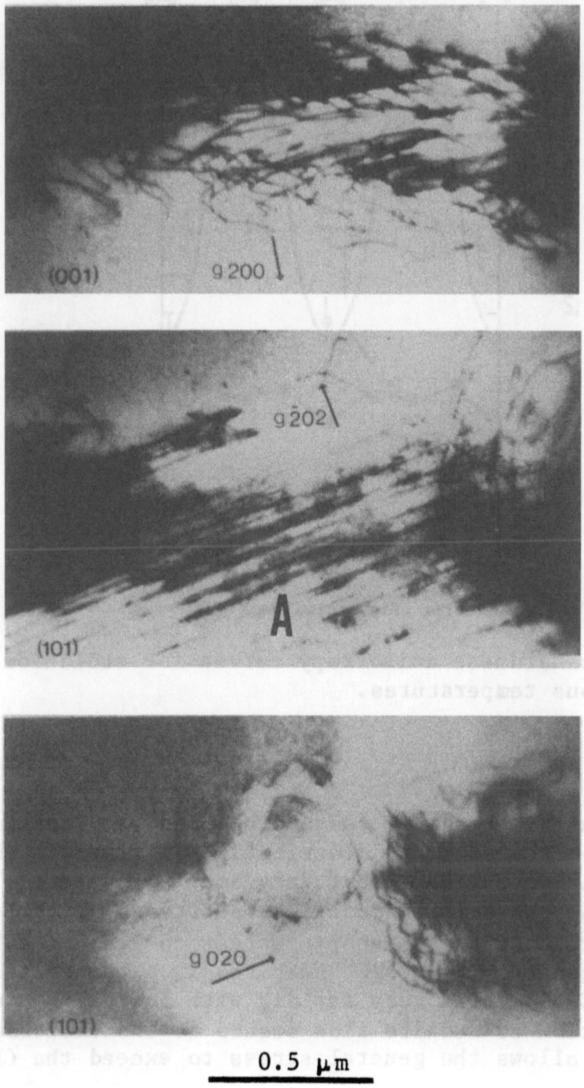

Fig. 5. TEM pictures of indented HfC.

THE BRITTLE TO DUCTILE TRANSITION IN CUBIC CARBIDES

Table 3 summarizes information on the principal slip planes in the carbides. In the cubic materials it appears that one requirement for ductility is the operation of slip on the {111} planes. However this clearly is not the only requirement since {111} is the primary slip plane in stoichiometric TaC and in several substoichiometric carbides even at room temperature. The brittle to ductile transition

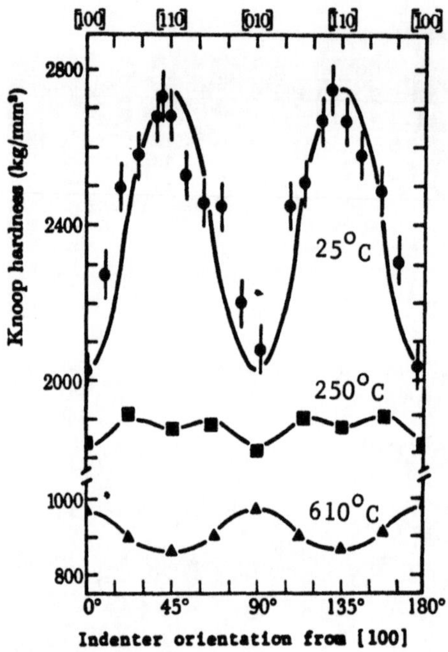

Fig. 6. Knoop hardness anisotropy curves for stoichiometric TiC at various temperatures.

occurs over a narrow range of temperature and the location of this range depends on the size and distribution of flaws in the crystal. Current information suggests that very small amounts of local slip occur well below the brittle/ductile transition, but this is not sufficient to raise the macroscopic stress to reach the high levels necessary for general yielding. Near the $T_{B/D}$ the critical resolved shear stress (CRSS) falls very rapidly with increasing temperature,[7] and the transition to ductile flow occurs at the temperature where local yielding allows the general stress to exceed the CRSS.

The question as to why {111} is the preferred slip plane at high temperatures has not been answered in a completely satisfactory way. On the one hand, for a given Burgers vector, $a/2[1\bar{1}1]$, the energy of the dislocation is independent of the plane on which it lies and a particular slip plane such as {111} will be favored if dissociation and a reduction in energy can occur. Looking at this from the viewpoint of bonding, Hannink et al.[26] suggest that the preference for {110} slip at low temperatures is a result of strong carbon to metal bonds that prohibit slip on {111}. As the temperature is raised they suggest that these bonds become less directional as electrons pass into metallic states, and slip becomes more similar to FCC

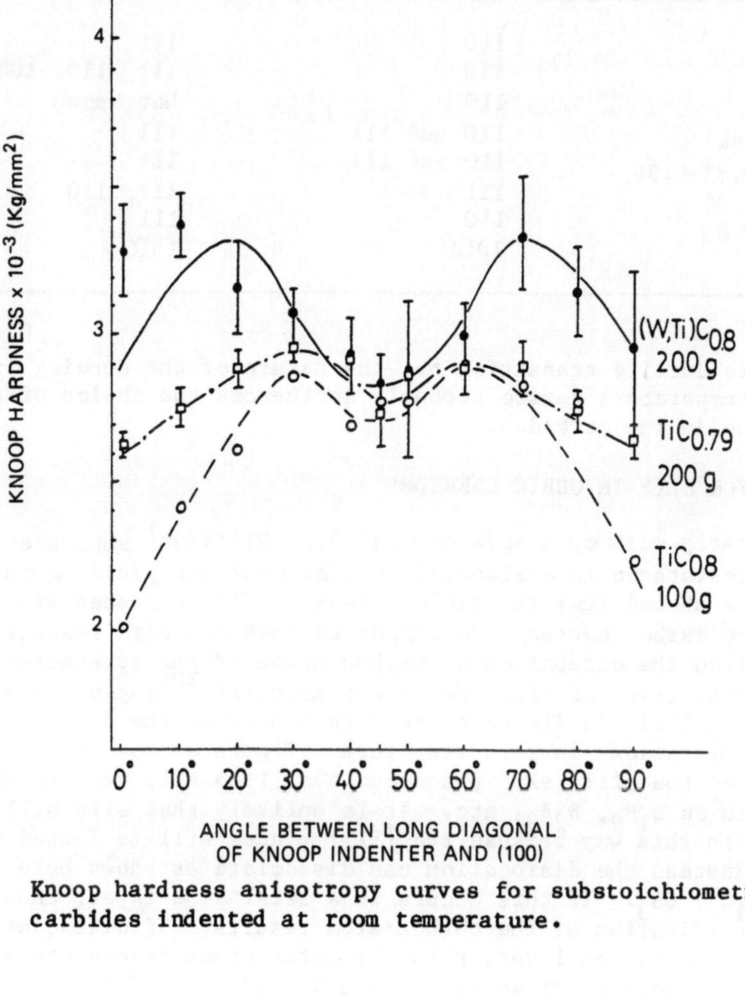

Fig. 7. Knoop hardness anisotropy curves for substoichiometric carbides indented at room temperature.

metals, namely {111} <1$\bar{1}$0>. Similar implications are made in simple hard sphere models,[27] which suggest that close packed structures favor {110} and as the structure becomes more open at high temperatures so {111} becomes favored. On the other hand it must be realized that changes in bonding need not be invoked to explain the

Table 3 Slip Planes in the Transition Metal Carbides

Carbide	Slip Plane: Room Temperature	Slip Plane: High Temperature
TiC	110	111
ZrC	110	111, 110, 100
HfC	110	Not known
$VC_{0.84}$	110 and 111	111
$NbC_{0.83-0.96}$	110 and 111	111
$TaC_{0.96}$	111	111, 110
$TaC_{0.83}$	110	111
WC	$10\bar{1}0$	$10\bar{1}0$

brittle to ductile transition but the nature of the bonding in a particular temperature regime probably influences the choice of principal slip plane in the carbides.

A MODEL FOR SLIP IN CUBIC CARBIDES

In early work on single crystal TiC, Williams[7] suggested that lattice resistance to dislocation motion controls yielding in carbides. He showed that the yield stress in TiC increases with increasing carbon content and suggested that the high Peierls stress results from the occupation by carbon atoms of the octahedral sites in the FCC sublattice of TiC. Kelly and Rowcliffe[30] suggested a model for slip on $\{111\}$ in TiC that takes into account the difficulties of moving dislocations in this structure. Figure 8 shows the arrangement of atoms on the $\{111\}$ slip plane and the $\langle 1\bar{1}0\rangle$ slip vectors are directions such as B_1B_2, B_2B_3, etc. It is unlikely that slip will proceed directly in this way because the $\{111\}$ planes will be forced widely apart. Instead the dislocation can dissociate as shown here, that is $B_1B_3 \rightarrow B_1C + CB_3$. If this occurs in a metal atom layer, then tetrahedral coordination of the carbon atom results. If dissociation occurs on the carbon layer, then the metal atoms become stacked over each other. Both of these can be avoided if each of the Shockley partials itself splits according to the scheme shown in Fig. 8. In this case the dislocation core extends normal to the slip plane. The inference of this model is that high temperature deformation will be governed by carbon diffusion near the dislocations.

Evidence in support of this mechanism has been obtained from a number of studies in which strain rate sensitivity was measured for several carbides. Martin et al.[14,31] suggested the following equation to describe the deformation of TaC.

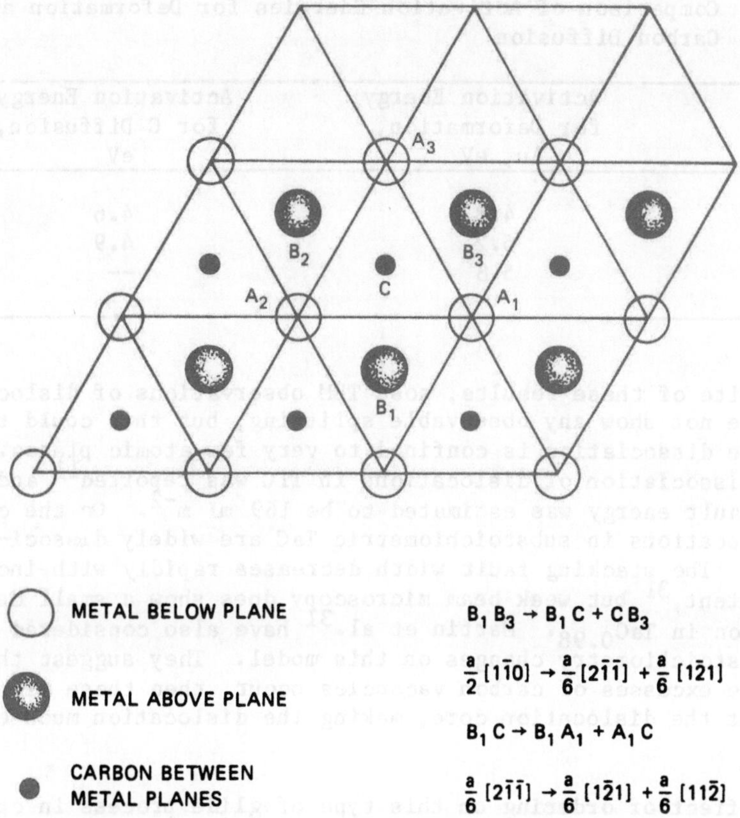

Fig. 8. Schematic diagram of the {111} plane in FCC carbides showing possible slip paths.

$$\dot{\varepsilon} = A(\sigma - \sigma_i)^{\alpha} \exp\left[(\sigma - \sigma_i)\frac{V}{kT}\right] \exp\left(-\frac{\Delta u}{kT}\right)$$

In this equation, $\dot{\varepsilon}$ is the strain rate, σ and σ_i are stresses, α and A are constants, V is an activation volume and Δu is an activation energy. Table 4 compares values of Δu with values of the activation energy for carbon diffusion. The value for TiC is calculated from Williams data,[7] and the value for ZrC is derived from the same equation by Darolia and Archbald from compression experiments on polycrystalline ZrC.[13] A similar activation energy was obtained by Lee and Haggerty for creep in near stoichiometric ZrC single crystals.[12]

Table 4. Comparison of Activation Energies for Deformation and Carbon Diffusion

	Activation Energy for Deformation, Δu, eV	Activation Energy for C Diffusion, eV
TiC	4.8	4.6
ZrC	5.2	4.9
TaC	5.8	--

In spite of these results, most TEM observations of dislocations in TiC have not show any observable splitting, but this could be because the dissociation is confined to very few atomic planes. Recently dissociation of dislocations in TiC was reported[11] and the stacking fault energy was estimated to be 169 mJ m^{-2}. On the other hand, dislocations in substoichiometric TaC are widely dissociated[31,32]. The stacking fault width decreases rapidly with increasing carbon content,[31] but weak beam microscopy does show a small degree of dissociation in $TaC_{0.98}$. Martin et al.[31] have also considered the effect of stoichiometry changes on this model. They suggest that where large excesses of carbon vacancies occur, then these might concentrate at the dislocation core, making the dislocation much easier to move.

The effect of ordering on this type of glide process in crystals such as substoichiometric VC has not been considered in detail. High yield stresses can be readily explained in terms of creation of antiphase boundaries,[4] but these crystals are disordered at temperatures where substantial plastic flow occurs. On the other hand, some short range ordering of carbon vacancies might persist that could exert a strong influence on dislocation motion.

High resolution microscopy and lattice imaging techniques could be usefully applied to provide more information in this area.

INDENTATION DEFORMATION IN TUNGSTEN CARBIDE

Most of the information on deformation of WC single crystals has been obtained from indentation studies.[20-22,33-37] Figure 9 shows an indentation on the basal plane of WC. This beautiful slip pattern arises from simultaneous slip on the three pairs of $\{10\bar{1}0\}$ prism planes. Using TEM, Hibbs et al.[21] found that most dislocations are widely dissociated into partials of the $1/6[11\bar{2}3]$ type, but undissociated dislocations of the [0001], $[11\bar{2}0]$, and $[11\bar{2}3]$ type were also seen. The core structures of these dislocations have been discussed by Hibbs,[37] and a model has been put forward that suggests a possible

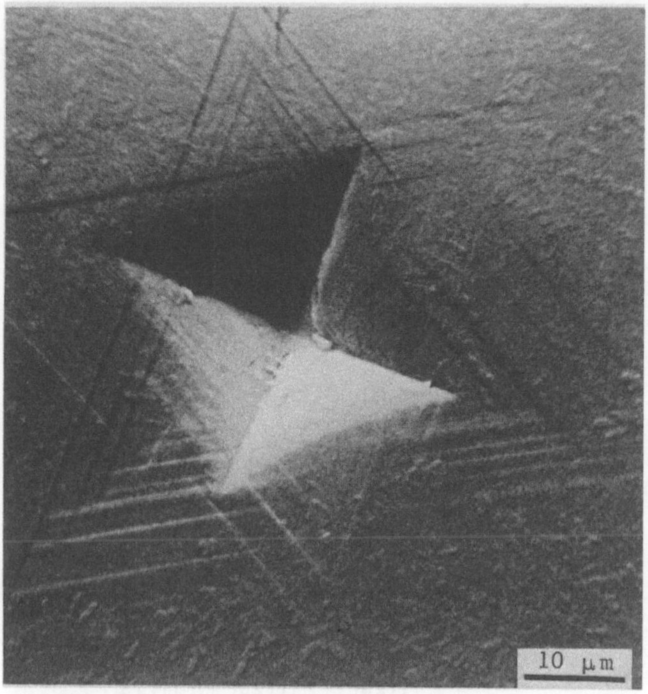

Fig. 9. Vickers indentation on the basal plane of WC at room temperature.

configuration of a dissociated dislocation.[20,21] Figure 10 shows a projection of the [10$\bar{1}$0] slip plane. This model considers Burgers vectors of the [0001], [11$\bar{2}$3], and [11$\bar{2}$0] types. If slip is to occur in any of these directions, it is likely to do so by following the valley direction AC to avoid wide separation of the slip planes. At point C there are three possible routes which would give partials of the 1/6[11$\bar{2}$3] type but of different signs. The energy difference between these alternatives is negligibly small, and slip proceeds by the motion of a second identical partial along the route CA_3. Figure 11 illustrates the configuration in the saddle position. The trigonal prismatic coordination of the carbon atoms is essentially maintained and the faulted structure is simply rotated 90° from the perfect crystal. This suggests a relatively low energy fault and is consistent with TEM observations of widely dissociated dislocations.[21,36]

Various types of cracks form at indentations made with high loads. The crack shown in Fig. 12 appears to form at the intersection of major slip bands, and somewhat unusual in that it is sharp at both ends. It is thought that these cracks arise during the loading

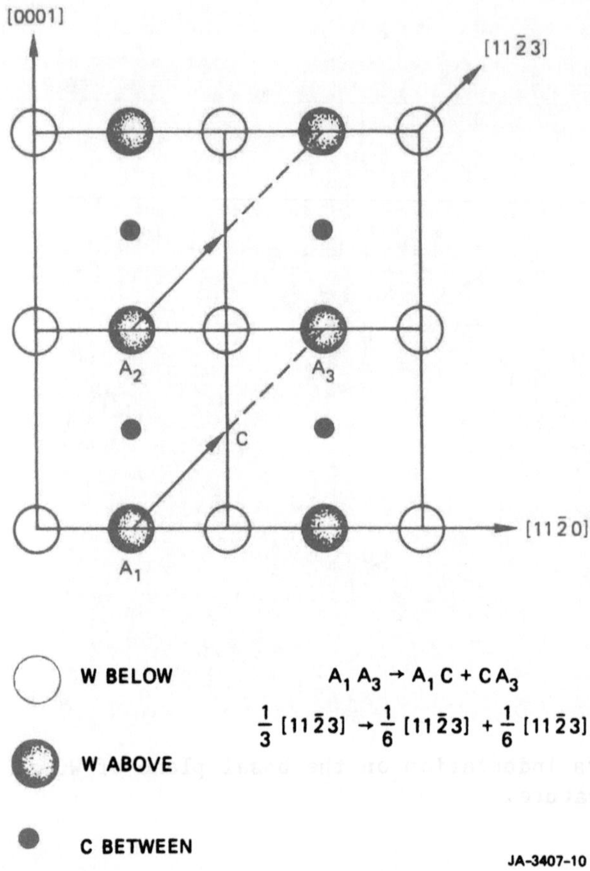

Fig. 10. Schematic diagram of the $\{10\bar{1}0\}$ plane in WC showing possible slip paths.

cycle due to pile-ups of dislocations. On removal of the load the sign of the residual stress is reversed which drives the crack backwards towards the indentation. The nucleation event for this crack is illustrated schematically in Fig. 13. This reaction is similar to that proposed by Cottrell[37] to account for cracks in BCC metals. Observations of small cracks by TEM confirm[20,21] that these cracks are produced by dislocation interactions, but partial dislocations, of the type $1/6\langle 11\bar{2}3]$, rather than total dislocations are involved.

Indentation above room temperature[35] results in an increase in the amount of plastic flow and a decrease in cracking. This is

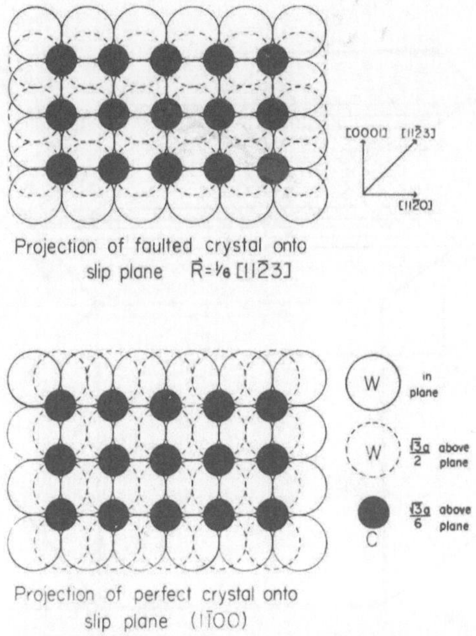

Fig. 11. Schematic diagram of the $\{10\bar{1}0\}$ plane for perfect and faulted crystals

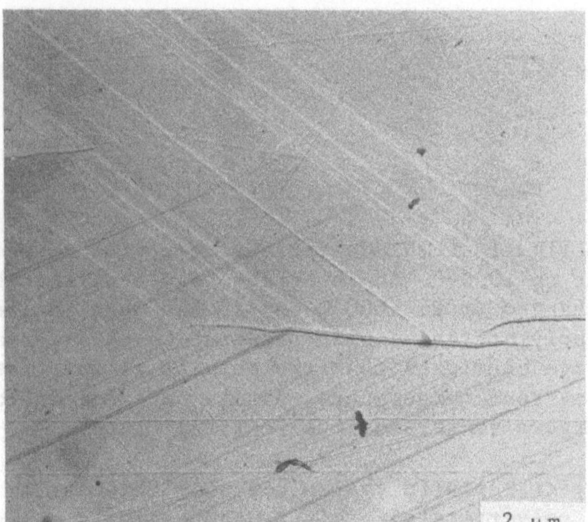

Fig. 12. Replica electron micrograph of a crack in WC formed at intersecting slip bands.

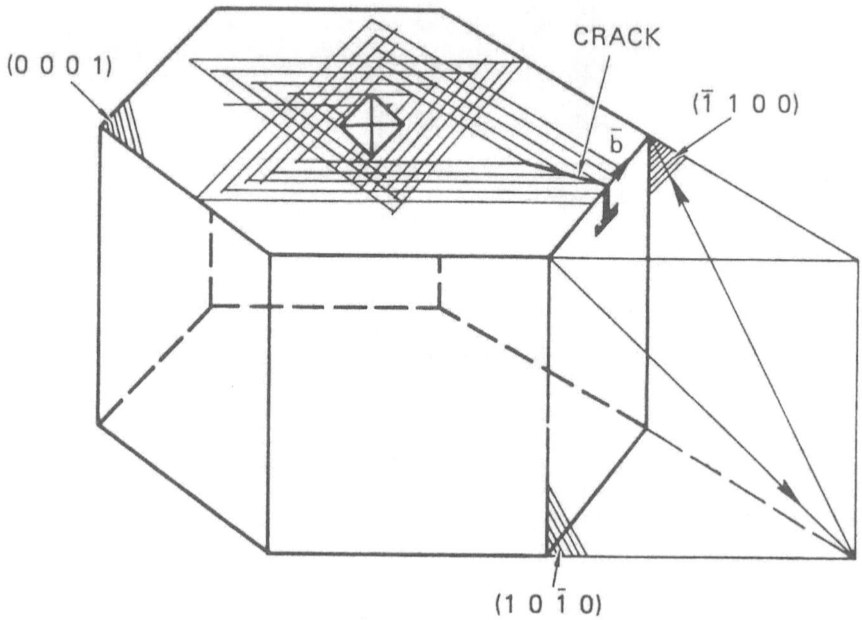

Fig. 13. Illustration of possible dislocation interaction to nucleate the crack shown in Fig. 12.

illustrated in Fig. 14 for indentations made with a sphere at various temperatures up to 1000°C An important result is that the slip trace pattern is unchanged up to 1000°C and it has been concluded that $\{10\bar{1}0\}$ is the only operative slip plane up to this temperature. TEM studies[21,36] have confirmed that the details of deformation are the same as those at room temperature, namely extended dislocations on $\{10\bar{1}0\}$ with Burgers vectors of the $1/6[11\bar{2}3]$ type.

The $\{1\bar{1}00\}$ $\langle 11\bar{2}3\rangle$ slip system has some interesting implications with regard to polycrystals. This system can only generate four independent strains. In particular, it can be readily seen that WC cannot sustain elongation in the c direction, and therefore limited ductility is expected in polycrystalline WC, or even in WC-Co at certain types of contiguous boundaries.[38] This rationale is the basis of on-going work that uses the information obtained from studies of single crystals to explain some of the effects at indentations in WC-Co.

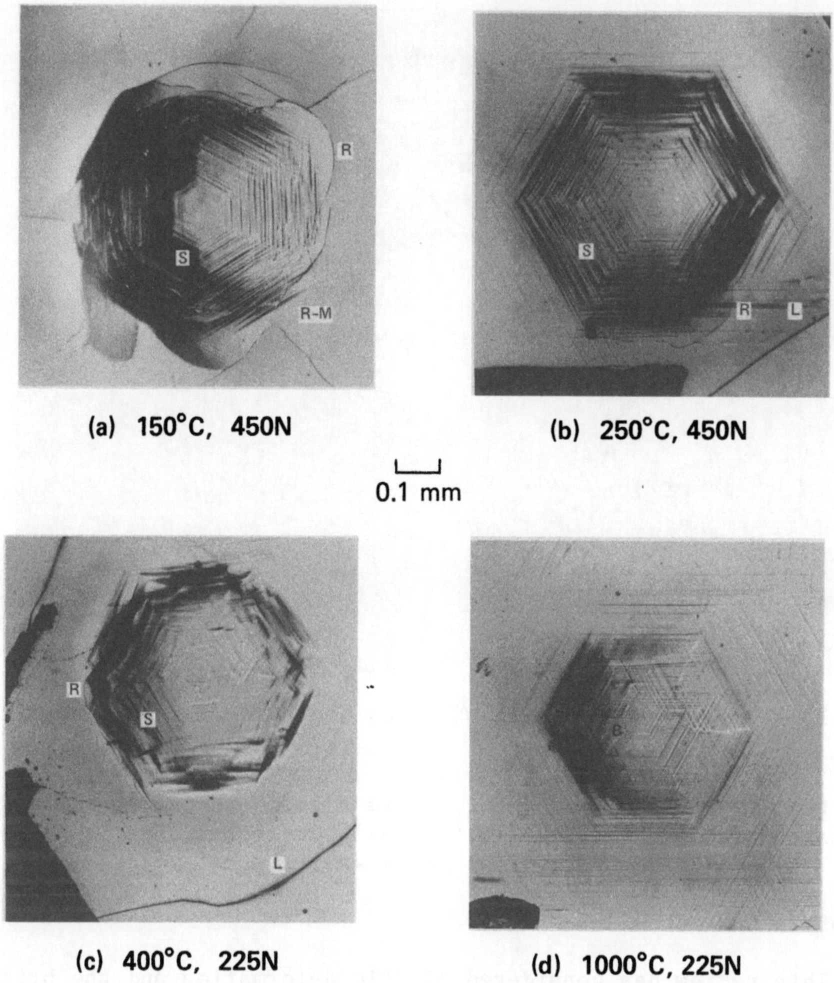

Fig. 14. Impressions made with a spherical indenter on the basal plane of WC showing slip bands (S), lateral cracks (L), ring cracks (R), and radial-median cracks (R-M).

Figure 15 shows a cracked boundary between two WC grains near an indentation in WC-6%Co. Not all carbide/carbide boundaries crack, and two conditions are required for cracking.[38] The first is that the c axes of the two grains must be at a relatively large angle to each other; in this case, slip cannot be transmitted fully over the boundary. Secondly the boundary must also lie normal to a c axis. When these conditions are fulfilled, as in Fig. 15, then the boundary is likely to crak. This appears to be a particularly fruitful approach to understanding the deformation behavior of technologically

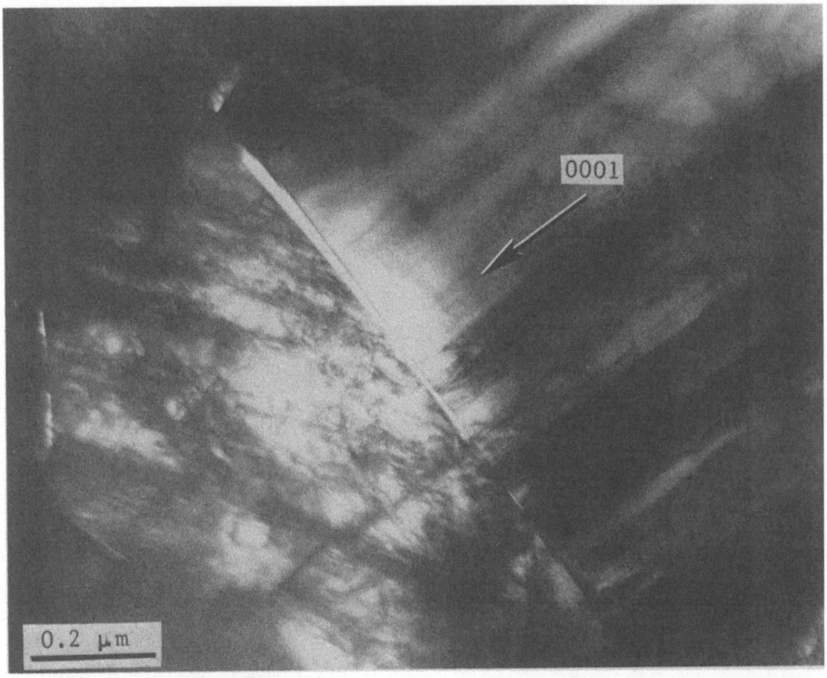

Fig. 15. Cracked boundary in WC-6% Co indented at room temperature.

important materials like WC-Co, and is a good example of the usefulness of fundamental studies in single crystals. Studies of other types of boundaries in deformed WC-Co are in progress.

SUMMARY AND CONCLUSIONS

This review has considered plastic deformation and the brittle to ductile transition in various carbides with respect to the slip systems that can operate at low and high temperatures under different deformation conditions. Measurements of hardness anisotropy, TEM observations of dislocations and direct observation of slip lines at indentations provide much information on local plastic deformation at temperatures below the bulk brittle to ductile transition. At low temperatures slip has been observed on {111} {110} and {100} planes in the <11$\bar{0}$> directions in the Group IV and V transition metal monocarbides. The choice of slip system depends on the carbide, carbon content, and temperature. Above the brittle to ductile transition of about 1000°C, {111} is the predominant slip plane, but small amounts of slip can occur on other planes just below the transition. Strain rate sensitivity studies generally support a model for slip on{111} that requires dissociation of dislocations and carbon diffusion.

In WC, indentation experiments indicate that the slip plane is $\{10\bar{1}0\}$ at least up to 1000°C. Dislocations produced by deformation are generally dissociated with Burgers vector $a/6\langle 11\bar{2}3\rangle$. The $\{1\bar{1}00\}$ $\langle 11\bar{2}3\rangle$ slip systems can generate only four independent strains and the inability of WC to deform when stressed along the c axis leads to grain boundary failure in WC-Co in certain cases.

ACKNOWLEDGEMENTS

The author is grateful to his coworkers at Stanford University R. Sinclair, V. Jayaram, and S. Shinde for many valuable discussions. The work is supported by the National Science Foundation.

REFERENCES

1. L. E. Toth, "Transition Metal Carbides and Nitirdes," Academic Press New York (1971).
2. W. S. Williams, Transition Metal Carbides, in "Progress in Solid State Chemistry," H. Reiss and J. O. McCaldin, eds., Vol 6, p. 57 (1971.
3. A. P. Gerk and J. Gilman, The Growth of Tungsten Carbide Single Crystals, J. Appl. Phys. 39:4497 (1968).
4. G. E. Hollox, Microstructure and Mechanical Behavior of Carbides, Mat. Sci. Eng. 3:121 (1968/69).
5. C. H. de Novion, R. Lorenzelli, and P. Costa, Existence d'une Surstructure dans le Carbure de Vanadium VC_{1-x}, Compt. Rend. 263:775 (1966).
6. C. Froidevaux and D. Rossier, N.M.R. Investigation of the Atomic and Electronic Structure of Vanadium and Niobium Carbides, J. Phys. Chem. Solids 28:1197 (1967).
7. W. S. Williams, Influence of Temperature, Strain Rate, Surface Conditions, and Composition on the Plasticity of Transition-Metal Carbides, J. Appl. Phys. 35:1329 (1964).
8. G. E. Hollox and R. E. Smallman, Plastic Behavior of Titanium Carbide, J. Appl. Phys. 37:818 (1966).
9. D. K. Chatterjee, M. G. Mendiratta, and H. A. Lipsitt, Deformation Behavior of Single Crystals of Titanium Carbide, J Mater. Sci. 14:2151 (1979).
10. K. Hara, H. Yoshinga, and S. Morozumi, High Temperature Mechanical Properties of Titanium Carbide, Nippon Kinzokin Gakkaishi 42:1039 (1978).
11. G. Das, K. S. Mazdiyasni, and H. A. Lipsitt, Deformation of Polycrystalline TiC, J. Amer. Ceram. Soc. 65:104 (1982).
12. D. W. Lee and J. S. Hagerty, Plasticity and Creep in Single Crystals of Zirconium Carbide, J. Amer. Ceram. Soc. 52:641 (1969).
13. R. Darolia and T. F. Archbold, Plastic Deformation of Polycrystalline Zirconium Carbide, J. Mater. Sci. 11:283 (1976).

14. J.-L. Martin, P. Lacour-Gayet, and P. Costa, Variation de la Contrainte avec la Vitesse de Deformation et la Temperature pour le Carbure de Tantale entre 1200 et 2200°C, C. R. Acad. Sc. Paris t. 272:2127 (1971).
15. S. B. Luyckx, Slip Systems of Tungsten Carbide Crystals at Room Temperature Acta Met. 18:233 (1970).
16. J. L. Martin, private communication.
17. D. J. Rowcliffe and W. J. Warren, Structure of TaC Crystals, J. Mater. Sci. 5:345 (1970).
18. R. Schenck, R. J. Gottschall, and W. S. Williams, Deformation of Cemented Carbides: High Temperature Stress Relaxation and Strain Rate Dependence, Mat. Sci. Eng. 32:229 (1978).
19. S. Hagege, J. Vincens, G. Nouet, and P. Delavignette, Analysis of Structure in Tungsten Carbide, phys. stat sol (a) 61:675 (1980).
20. M. K. Hibbs and R. Sinclair, Room Temperature Deformation Mechanisms and the Defect Structure of Tungsten Carbide, Acta Met. 29:1645 (1981).
21. M. K. Hibbs, R. Sinclair, and D. J. Rowcliffe, Defect Structyure of WC Deformed at Room and High Temperatures, in: "Science of Hard Materials, R. K. Viswanadham, D. J. Rowcliffe, and J. Gurland, eds., Plenum Press, New York (1983).
22. J. D. Bolton and M. Redington, Plastic Deformation Mechanisms in Tungsten Carbide, J. Mater. Sci. 15:3150 (1980).
23. C. A. Brookes, J. B. O'Neill, and B.A.W. Redfern, Proc. Roy. Soc. A 332:73 (1971).
24. D. J. Rowcliffe and G. E. Hollox, Plastic Flow and Fracture of Tantalum Carbide and Hafnium Carbide at Low Temperatures, J. Mater. Sci. 6:1261 (1971).
25. M. J. Hill and D. J. Rowcliffe, Deformation of Silicon at Low Temperatures, J. Mater. Sci. 9:1569 (1974).
26. R.H.J. Hannink, D. C. Kohlstedt, and M. Murray, Slip System Determination in Cubic Carbides by Harndess Anisotropy, Proc, Roy. Soc. A326:409 (1972)
27. D. J. Rowcliffe and G. E. Hollox, Hardness Anisotropy, Deformation Mechanisms, and Brittle-to-Ductile Transition in Carbides, J. Mater. Sci. 6:1270 (1971).
28. G. Morgan and M. H. Lewis, Hardness Anisotropy in Niobium Carbide, J. Mater. Sci. 9:349 (1974).
29. S. L. Shinde, to be published.
30. A. Kelly and D. J. Rowcliffe, Slip in Titanium Carbide, phys. stat sol. 14:K29 (1966).
31. J. L. Martin, P. Lacour-Gayet, and P. Costa, Plastic Deformation of Tantalum Carbide up to 2200°C, "Electron Microscopy and Structure of Materials," G. Thomas, ed. University of California Press, Berekeley, CA (1972).
32. D. J. Rowcliffe and G. Thomas, Structure of Substoichiometric TaC, Mater. Sci. Eng. 18:231 (1975).
33. T. Takahashi and E. J. Friese, Determination of Slip Systems in Single Crystals of Tungsten Monocarbide, Phil. Mag. 12:1 (1965).

34. D. N. French and D. A. Thomas, Hardness and Slip in WC Crystals, Trans. Met. Soc.AIME 233:950 (1965).
35. D. J. Rowcliffe, Indentation Damage in Tungsten Carbide and Tungsten Titanium Carbide, in "Science of Hard Materials," R. K.Viswanadham, D. J. Rowcliffe, and J. Gurland, eds, Plenum Press, New York (1983).
36. M. K. Hibbs, R. Sinclair, and D. J. Rowcliffe, Defect Interactions in Deformed WC, Acta Met. in press (1983).
37. M. E. Hibbs, Ph.D. Thesis, Stanford University, (1981).
38. V. Jayaram, R. Sinclair, and D. J. Rowcliffe, Integranular Cracking in WC-6% Co: An Application of the Von Mises Criterion, Acta Met. 31:373 (1983).

DISLOCATION STRUCTURES IN POLYCRYSTALLINE TUNGSTEN HEMICARBIDE W_2C DEFORMED AT HIGH TEMPERATURES

T. Epicier, C. Esnouf, J. Dubois and G. Fantozzi

Groupe d'Etudes de Métallurgie Physique et de Physique des Matériaux, L.A. C.N.R.S.n°341, I.N.S.A.,Bat. 502
69621 Villeurbanne Cedex, France

INTRODUCTION

Preliminary remark

The mechanical behavior of the hexagonal tungsten hemicarbide W_2C has been recently studied in four-point bending from room temperature up to 2200°C[1]. The main features of the results based upon the thermodynamical analysis of the thermally activated plastic deformation have already been presented[2], and an apparent coherence was found through a comparison firstly with the results obtained on various transition metal carbides and issued from the literature (see review in ref.[3]) and secondly with Transmission Electron Microscopy (T.E.M.) observations of the dislocations structures in deformed samples [2,4]. Recent refinings [5,6] in both areas of the mechanical analysis and the T.E.M. study show that difficulties arise to get a strict consistence between all the obtained results, and this leads us to think that the identification and characterization of the deformation processes were not so cbvious as we believed - and unfortunately as it could be expected - in a first approach.

This paper is so concerned with T.E.M. analysis of dislocations in the material of interest, especially in samples deformed at temperatures where a transition of deformation mechanisms seemed to take place on the view of the initial mechanical tests analysis ; this - hasty - interpretation is briefly reminded in the next paragraph.

Previous analysis of the mechanical behavior of W_2C

The figure 1 shows the evolution of the apparent activation enthalpy of deformation, ΔH_a, as measured by different methods[2].

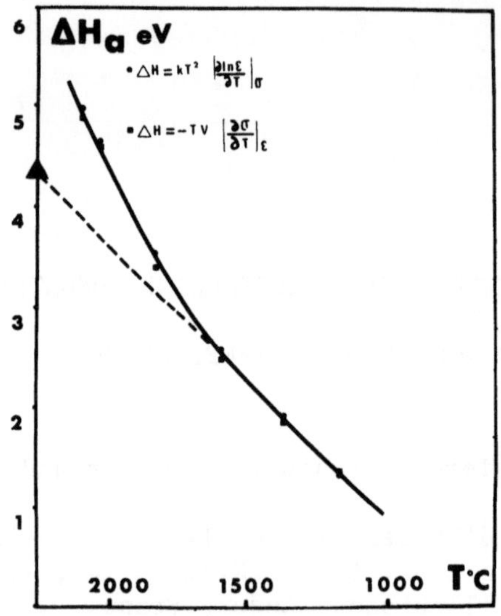

Figure 1 : apparent activation enthalpy versus temperature (with the assumption $\dot{\varepsilon}_o$ = cst[t]).

On the basis of the classical assumption that the pre-exponential term $\dot{\varepsilon}_o$ of the constitutive law :

$$\dot{\varepsilon} = \dot{\varepsilon}_o \exp - \frac{\Delta G}{kT} \qquad (1)$$

(where $\dot{\varepsilon}$ is the strain rate, and ΔG the activation free energy depending on effective stress σ^* and temperature T) was independent of σ and T, it has been deduced that the parameter ΔH_a was strictly equal to the true activation enthalpy ΔH (see ref.[7] for a detailed analysis). This led us to propose two temperature ranges - below and above 1750°C - where two different mechanisms were supposed to control the plastic deformation of W_2C.
Up to 1750°C, a "shear-diffusion" mechanism* of dislocations glide** was assumed[9], which involves carbon diffusion in the core of the dislocations narrowly splitted into Shockley partials following the classical relation :

$$\frac{\vec{a}}{3} <11\bar{2}0> \rightarrow \frac{\vec{a}}{3} <10\bar{1}0> + \frac{\vec{a}}{3} <01\bar{1}0> \qquad (2)$$

The good correlation between the activation enthalpy extrapolated at $\sigma^* = 0$ (from the experimental points in figure 1) and the activation enthalpy of carbon diffusion into the material (i.e. ≃ 4 eV) supported well this hypothesis[2]. Moreover, the expected dissociation of

---
* derived from Kelly and Rowcliffe's on consideration on TiC[8].
** the basal slip system <11$\bar{2}$0>(0001) was the only one activated in our experimental conditions up to 2200°C.

perfect dislocations has been clearly observed by T.E.M. in weak-beam imaging[3].
Such an analysis has been suggested in most of the cubic transition metal monocarbides [10,11] and effectively applied in many of them (see review in ref.[3]).
Beyond 1750°, the increase of ΔH_a could suggest that an other mechanism (involving diffusion of the metallic atoms) is operating, probably the dislocations climb as it has been invoked in various metal monocarbides (see review in ref.[3]). T.E.M. evidence of dislocation climb in W_2C has been obtained, see figure 2.
Extensive investigation in electron microscopy has been performed to verify the occurence of climb, and to specify the core structure of the climbing dislocations : as the gliding ones are dissociated, the question comes to know if the faulted ribbons subsist during climb.
Typical results are presented in the next section, and they will be discussed in relation with a fully consistent analysis of the mechanical tests.

DISLOCATIONS ANALYSIS IN DEFORMED W_2C

Experimental procedures

Samples preparation : thin sheets of polycrystalline W_2C with the predominant ε- Fe_2N type ordered crystallographic structure[4], were bended to $\varepsilon = 5.10^{-3} - 10^{-2}$ at various temperatures between 1750°C and 2200°C. After the effective deformation tests, the samples were cooled down under stresses (C.U.S.), in order to "freeze in" as best as possible the dislocations arrays in their testing temperature microstructure*.
As the deformation level was quite low, and the strain confined in the external fibers of the samples, it was of the greatest importance to prepar the thin foils for T.E.M. as closed as possible of the surfaces. This was achieved by carefull electrolytic polishing (details of the electrolytic solution in ref.[9]), sometimes followed by ion-beam thinning at 15° during few hours.

Transmission Electron Microscopy observations : T.E.M. observations were made at 100 and 200 kV (Jeol 100 C and CX respectively) and 1 MV (E.M. Cameca - C.E.N.Grenoble, France). It should be noted that the use of a 200 kV microscope seems to be particulary suitable for observations of materials such as metal carbides especially for weak beam images, which were obtained relatively easily (dissociations of 30-50 Å were sometimes directly resolved on the screen during observations at magnifications of 50 to 150.000 with the X10 binocular system).

* the applied load was generally removed around 1000-1300°C to avoid undesirable cracking of the specimens.

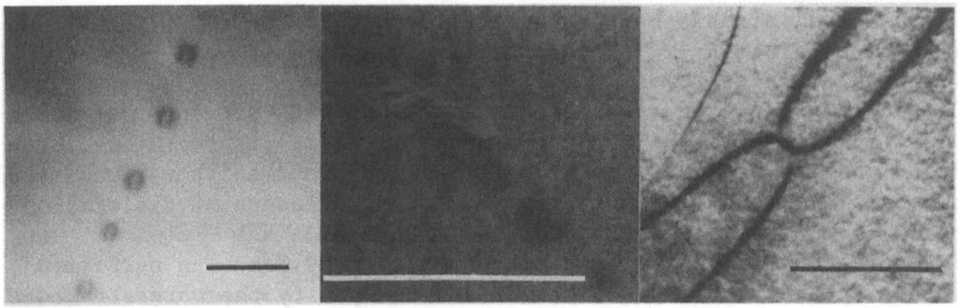

Figure 2 : dipoles and loops formation as observed in W_2C deformed respectively at 1600°C, 1800°C and 2100°C in the presented pictures; bar = 0,5 μm (see ref.[3,4] for a complete indexation).

Concerning the technique of dark field imaging under weak beam conditions, multiple beams such as $\{22\bar{4}0\}$,$\{33\bar{6}0\}$ and $\{20\bar{2}0\}$, $\{3030\}$ $\{50\bar{5}0\}$ or $\{60\bar{6}0\}$ were often used, with high values of the $|w| = |s\ \xi_g|$ parameter (generally 10-30); it has been checked on summary calculations of two-beam dynamical theory that these conditions must be valid to allow a direct interpretation of the dislocation contrast[12]. Anyway, an "a posteriori" confirmation of this was indirectly obtained from the sharpness and intensity of the peak (s) observed on the micrographs.

Evidence for dislocation climb

More than ten dislocations arrangements have been fully examined. They clearly show that the climb of dislocations segments (and not only of very local parts of dislocations as those needed for the formation by pipe-diffusion along dipoles of the loops shown in figure 2) has occured in the studied temperature range.
The figure 3 shows a dislocation node where the intersecting defects appear to be out of the basal plane (angles of deviation varying from 11 to 16°)* ; the weak beam dark field (WBDF) image presented in this figure shows no evidence of dissociation for these dislocations, at least for partials separation greater than 15 Å - 20 Å.
The figures 4a → d show an interesting configuration observed in a W_2C sample deformed at 2000°C. Before to comment upon these figures it is necessary to specify that the electron beam of the microscope can cause irradiation damage in metal carbides ; especially, carbon atoms displacements can be induced by inelastic electron-atom interaction[13].
In W_2C, when no motion of the metallic atoms seems possible at 100 or 200 KV, a spectacular spreading of the faulted ribbons has been

---

* the determinations of lines vectors were achieved by the classical stereographic analysis.

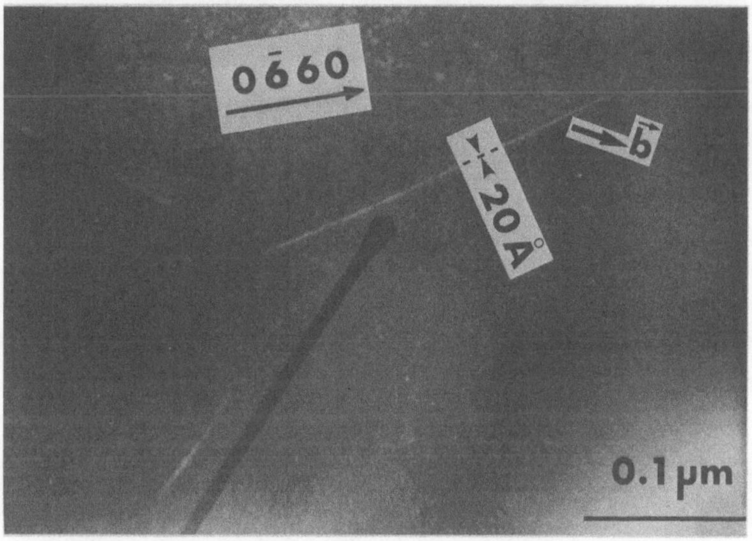

Figure 3: Intersection of dislocations which are lying out of the basal plane (T.E.M. 100 kV, W_2C deformed at 1750°C). For more details, see caption of figure 10 in ref.[6].

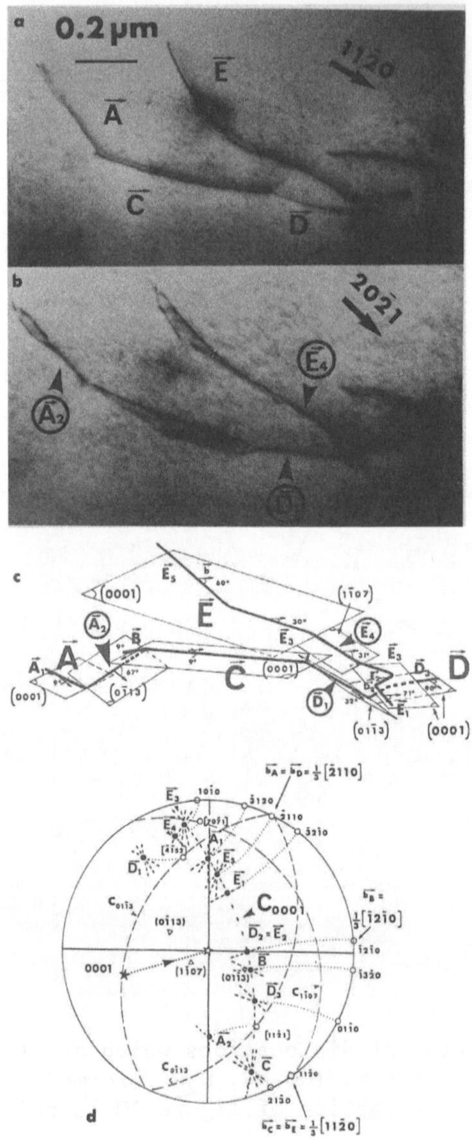

Figure 4: Evidence for climb (C.U.S. sample deformed at 2000°C).
a) bright field electron micrograph (200 kV). Azimut [1$\bar{1}$02].
b) idem. Azimut [$\bar{2}$3$\bar{1}$2](see text for details).
c) schematic representation of the dislocations array; the planes formed by the Burgers vector and the line of the dislocations are indicated.
d) stereographic projection (the full symbols are referred to the indexation in the reference's system of the microscope, when the open ones are the corresponding pôles in the (0001) standard projection).

observed when the incident direction of the electrons is closed to a direction perpendicular to the $\vec{c}$ - axis of the hexagonal cell of W_2C* ; this unexpected behavior has been attributed to a local decarburization of the sample, and an associated decrease of the stacking fault energy because of the deviation from stoichiometry [3,4].
The drawing of Figure 4c represents a schematical "3-D" view of the configuration formed by the 4 main dislocations A, C, D, E ; the encircled index (i.e. A_2, D_1, E_4) corresponds to those segments of dislocations which lie out of the basal plane ; the analysis is shown on the sterographic projection of the Figure 4d . An attractive interaction has led to an alignment of the parallel segments D_2 and E_2, the common direction of which is easily explained by the fact that the edge contributions of the Burgers vectors of these two segments are colinear as in a classical dipole.
The Figure 4a is a bright field electron micrograph taken after 1/2 h of normal observation ; the micrograph on Figure 4b has been taken after 1/2 h of irradiation in the azimut [0001] . A distinct spreading of basal dislocations occured, and no similar evolution was noted for the deviated segments A_2, D_1 and E_4.
The immediate conclusion of such observation (repeated on several dislocations configurations in different foils) is that climbing dislocations seem to be recombined when the remaining basal dislocations are effectively dissociated ; this point is supported by the weak beam observations presented in the next paragraph.

Dissociation of the basal dislocations at high temperatures ($\gtrsim$ 1750°C)

Figure 5 shows a dislocation arrangement in a sample deformed at 2200°C; although the contrast of the micrograph is somewhat diffuse because of important thickness of the foil, the dissociation of some dislocations is observable. Some others, like the one at the upper region, appear to be undissociated. A correct analysis of the dislocation line shows that this undissociated dislocation lies in a $(1\bar{1}03)$ plane; it was verified that the faulted ribbons lyied in the basal plane. An irradiation experiment resulted in a similar behavior of these defects, as the one reported in Figure 4b.
The micrographs of the Figure 6 show a W.B.D.F. analysis of a network observed in a W_2C sample deformed at 2100°C.
In this network, the three families of dislocations lie in the basal plane, with a nearly screw orientation, and they appear to be dissociated. The " $\vec{g}\cdot\vec{b} = 0$ " analysis of the partial dislocations at one node is shown ; it confirms the expected dissociation according to the relation (2). Moreover, white arrows on the last micrographs (each corresponding to an extinction of one partial dislocation) point out an inversion in the partial sequence; such constrictions, which are probably jogs (see discussion), occur on many of the dislocation seg-

* this operation was done by removing the condenser lens aperture, and by focussing the beam on the area of interest.

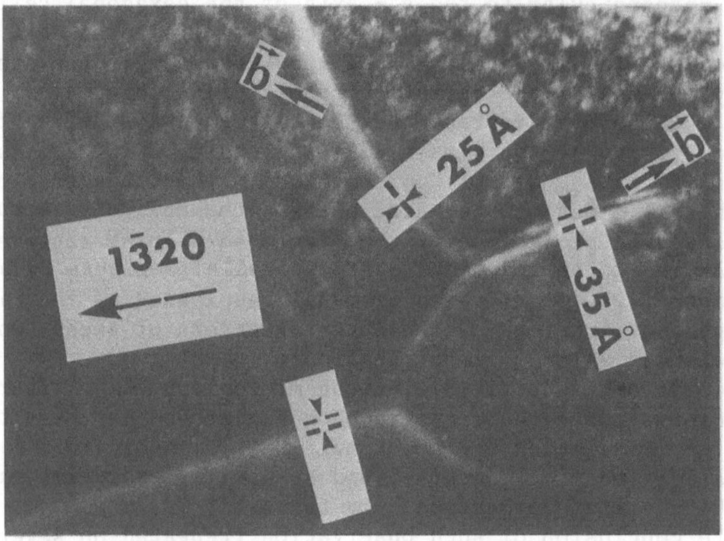

Figure 5: Non dissociation of climbing dislocations (C.U.S. sample after deformation at 2200°C) W.B.D.F., T.E.M 200 kV. The non-basal dislocation in the upper part of the micrograph lies in the (1̄103) plane and appears undissociated; basal dislocations are dissociated.

ments of the network, as it can be seen in the general view of Figure 6.

DISCUSSION

Deformation microstructure observed in "Free-cooled" (F.C.) specimens

A great number of observations have been made on F.C. samples after deformation at 1750, 1950, 2000°C. In these specimens, more or less well developed tangles or networks of dislocations were observed (see for example Figures 3 and 4 in ref[9]), but none of the configurations analysed point out that dislocations are present on non-basal planes; moreover, dislocations gliding on the basal plane appear to be dissociated into partials (all of the dislocations analysed by the weak beam technique -see for example Figure 5 in ref.[9] -).
In view of these remarks it seemed to be preferable to analyse C.U.S. samples to obtain evidence of dislocation climb in a more direct

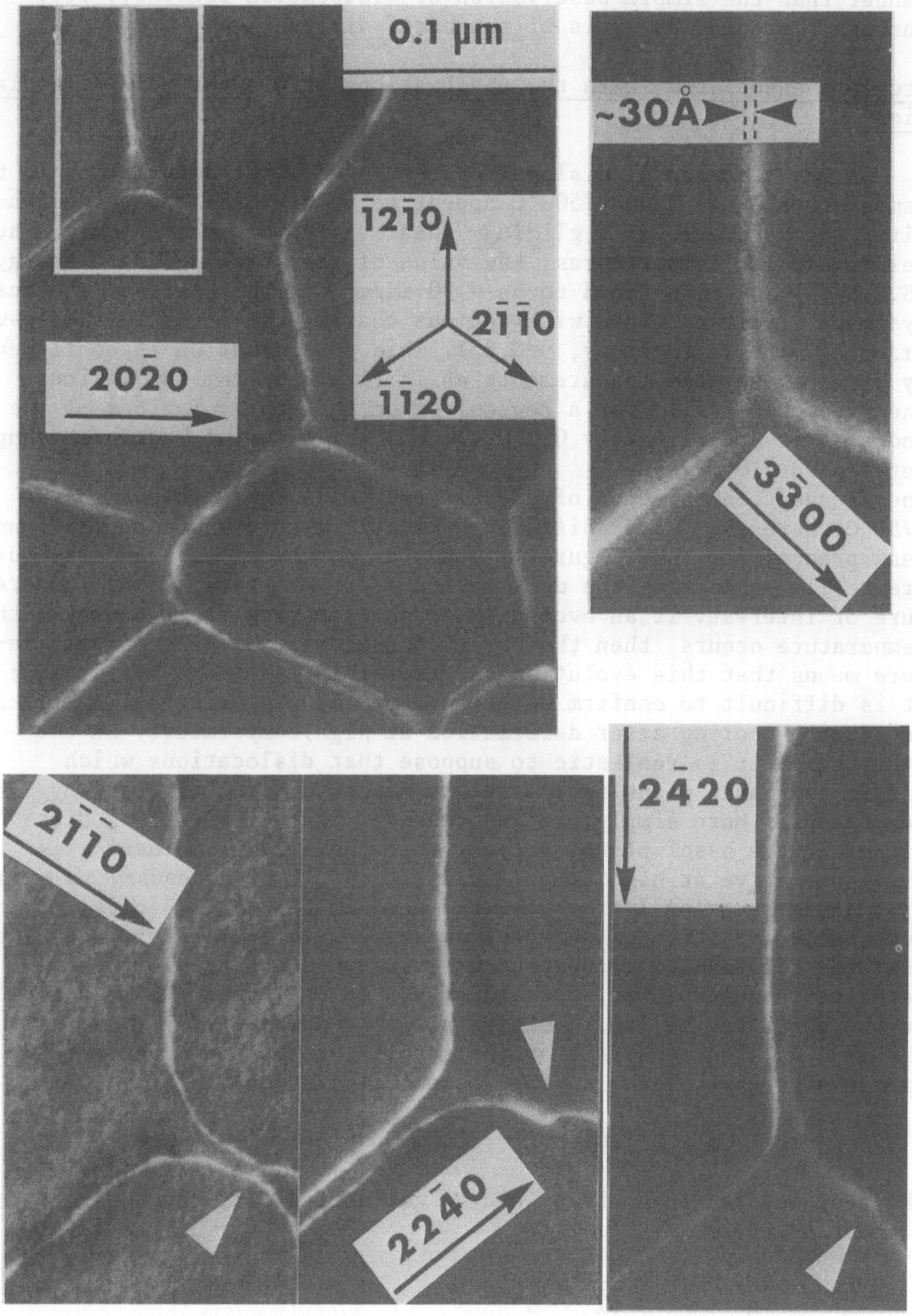

Figure 6: (On the next page) $\vec{g}.\vec{b}$ analysis of an extended node of a network observed in a C.U.S. sample deformed at 2100°C (T.E.M 200 kV). The white arrows on the micrographs indicate an inversion of the partial sequence.

manner than the simple observation of dislocation recovery, i.e. through a precise analysis of non-basal dislocations, or loops.

Problems associated with the observations of dislocation dissociation.

A great number of dislocations analysed after deformation in the temperature range 1200→1500°C appear to be dissociated, which confirms the occurence of a gliding "shear diffusion" mechanism at these relatively low temperatures; the value of the stacking fault energy (S.F.E.) γ has been found to be $\simeq$ 90 mJ/m^2* on the basis of the analysis of 13 dislocations with various characters[4,12] (with the assumption of elastic isotropy, ν=0,25). This determination is confirmed by dislocation node measurements as in Figure 6 from the various theories (see ref.[14] for a review): from the configurations of the nodes, ν is approximately 0,2-0,25, and the value of γ deduced ranges between 90-110 mJ/m^2.
The frequent observation of dissociated basal dislocations above 1750°C where the "shear diffusion" mechanism should become less and less predominant (see Figure 1) leads us to wonder if such structures are really reflecting the deformation microstructure at the temperature of interest. If an evolution of the stacking fault energy with temperature occurs, then the observed dissociation at room temperature means that this evolution is reversible (see next paragraph); it is difficult to confirm wether or not the dissociation is "frozen in" during cooling after deformation at high temperature. On the other hand, it is realistic to suppose that dislocations which lie in the basal plane have a net tendency to dissociate at any temperature. More simply, we can consider that dislocations dissociated in the basal plane must exist if the gliding mechanism is still operative at high temperatures; then, it is necessary to reanalyse thermodynamically the results shown in Figure 1. Because of the detectable evolution of internal stresses with temperature we believe that the deformation microstructure was not constant over the temperature range explored; consequently, the pre-exponential factor $\dot{\varepsilon}_0$ in the constitutive law should vary with temperature.
A self-consistent analysis of the results, taking into account the previous remarks, leads to the following relationship[1,5]:

$$\dot{\varepsilon}_0 = cs^t \left(\frac{\sigma^*}{E}\right)^n \quad \text{(with E = Young Modulus)} \qquad (3)$$

The true activation enthalpy can be calculated from the apparent parameter ΔH_a from:

$$\Delta H = \Delta H_a - kT^2 \left(\frac{\delta \ln \dot{\varepsilon}_0}{\delta T}\right)_{\sigma^*} \qquad (4)$$

*The value reported in ref.[9] -i.e. 50 mJ/m^2- is in error and must be corrected to $\simeq$ 100 mJ/m^2 for the chosen example.

which is easily transformed into[7] :

$$\Delta H = \Delta H_a + n \frac{kT}{E} \frac{dE}{dT} \qquad (5)$$

Data corrected in this way are described by the constitutive law:

$$\dot{\varepsilon} = cs^t \left(\frac{\sigma^*}{E}\right)^n \exp\left(\frac{\Delta H_\circ - \sigma^* V_\circ}{kT}\right) \qquad (6)$$

with $n = 2{,}9$ $\Delta H_\circ = 2{,}1$ eV, $V_\circ = 0{,}6.\ 10^{-27} m^3$ (activation volume)

over the whole temperature range studied, except perhaps for the measurements at 2200°C which apparently give an incorrect point[5]. The surprisingly low value of $\Delta H_\circ$ (compared with previously reported values for W_2C and most other carbides reviewed in ref.[3]) is discussed elsewhere, and is compared with the energy of the carbon diffusion in the grain boundaries[5,6].
It is probable that the difficulties encountered in this analysis are caused by the weak plastic deformation which can be considered to be in a so-called "pre-plastic" or "microplastic" stage.
Nevertheless, it can be concluded from the previous considerations that the glide process predominates over the whole temperature range except for some influence of another mechanism -most probably the dislocation climb- above 2000°C.
From the above it is not surprising that dislocation glide in our specimens, even if dissociation cannot be correlated with the actual deformation state because of possible spreading of the very narrow faulted ribbons at high temperature,when they are cooled and observed at room temperature (see next paragraph).

Dislocation climb

As mentionned already, manifestations of dislocation climb have been obtained through observations of hexagonal networks and rows of loops. New evidence of this phenomenon is deduced here from the analysis of non-basal dislocations (mixed in character); the examples presented in the previous section show segments of dislocations lying in various planes ((1$\bar{2}$12), at 58° from the basal plane, (10$\bar{1}$7) at 8°, (10$\bar{1}$3) at 31° -this last one has been found several times-) that are not glide planes for dislocations with $\vec{b} = \frac{\vec{a}}{3}$ ⟨11$\bar{2}$0⟩. Such manifestations are not seen in F.C. specimens, probably because of the easy rearrangement or recovery of climbing dislocation structures on the close-packed basal planes.
Irradiation experiments or direct weak beam analysis in T.E.M. have shown that these climbing dislocations are not dissociated to the extent deduced from the determined S.F.E.
One can argue that the S.F.E. increases with temperature, so that deviation or climb out of the gliding basal plane should be energetically favourable above 2000°C (nominally); very narrow separation of the partials could lead to recombination and subsequent climb if

the local stress is favourable. Moreover it is possible that carbon segregation in the dislocation can locally modify the width of the ribbons and initiate the constrictions necessary for further climb.

Discussion of the constrictions and inversions of the partials observed in the network of the Figure 6

The network shown in the Figure 6 exhibits number of crossings of the partial dislocations, leading to the hypothesis that not all of the fault lies in the same basal plane. It is well known that Shockley partial dislocations of the $\vec{a}/3 <10\bar{1}0>$ - type in the close-packed hexagonal structure are the boundaries of an intrinsic fault in the basal plane; because of the sequence a b a b of such planes, the dissociation (2) can be written (with triangular bipyramidal symbols*) in the form:

$$AB \rightarrow A\sigma + \sigma B \quad \text{for 'a'-type plane} \qquad (7)$$

and in the form:

$$AB \rightarrow \sigma B + A\sigma \quad \text{for 'b'-type plane} \qquad (8)$$

If such a fault jogs from an 'a'-layer to a 'b'-layer, an inversion of the partial sequence of the ribbon has to occur as shown in the figure 7a[14,15].
This seems to be the case at points such as that arrowed in the micrographs of Figure 6, i.e. a change of fault plane from an 'a'-type metallic basal plane to a 'b'-type (see the schematic representation of the Figure 7b).

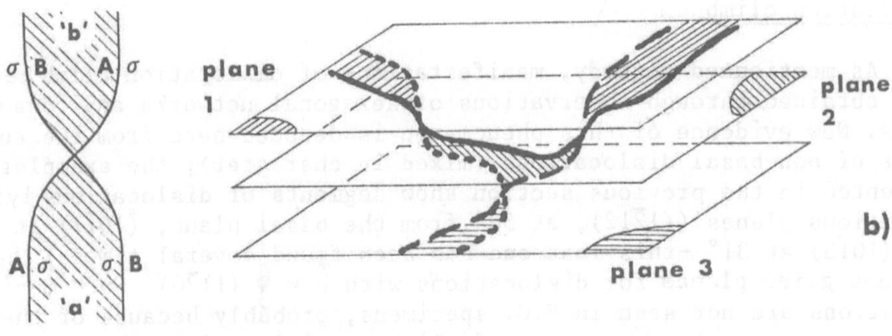

Figure 7: interpretation of the constricted jogs of the dislocations from the Figure 6.
a) inversion of the sequence of partials accompanying the change of plane of a dissociated dislocation.
b) schematic representation of what might be a node in Figure 6 from the interpretation given in the text.

* equivalent to the Thompson's notations in cubic structures.

A well developed network with such jogs may have been formed by cross-slip of the initial screw dislocations[16] or by climb; in every case, the observed network should be in a "just-before-equilibrium" configuration since the nodes are not in the same basal plane. Since cross-slip is not thought to be the predominant deformation mechanism in W_2C at high temperatures, we believe that the network has formed by a climb process, the dislocations having the tendency to redissociate in the basal plane; the jogs would be then the last step to overcome for perfect climb on the same basal plane; this stage could only have been attained by diffusion processes in a time longer than the test time. Such an analysis would be interesting for highly deformed samples, where the polygonised dislocations structures are expected to be fully developed.

CONCLUSION

The work presented was undertaken to complete an initial study of the mechanical behavior of the hexagonal tungsten hemicarbide W_2C. Although the plastic deformation of the samples remains very limited, a coherent constitutive law has been proposed, and the "shear diffusion" mechanism, involving glide of narrow faulted ribbons and simultaneous-diffusion of carbon atoms into the core of dislocations, is thought to be the predominant process controlling the plastic deformation in the temperature and stress range which has been explored.

T.E.M. observations have confirmed the dissociation of basal dislocations into Shockley partials, and climb of dislocations has also been clearly identified; the occurrence of climb at high temperature (i.e. between 1800 and 2200°C) may explain a slight shift of activation enthalpy determined at the upper limit of the temperature range; if there is a "hesitation" between the two mechanisms, the analysis at that point is not meaningful. It would be necessary to go to a greater plastic deformation (in compression) and creep experiments to determine precisely the activation enthalpy at these ultimate temperatures.

REFERENCES

1. J.Dubois, thesis (in progress), Lyon.
2. J.Dubois, T.Epicier, C.Esnouf and G.Fantozzi, 1981."Mechanical behavior and Electron Microscopy analysis of W_2C", I.C.H.M, Jackson Lake Loadge, USA, to be published.
3. T.Epicier, 1982, "Observation de l'hémicarbure de tungstène par microscopie électronique en transmission...", thesis, Lyon.
4. T. Epicier and C.Esnouf, "Observations de l'hémicarbure de tungstène W_2C par microscopie électronique en transmission", submitted to J. of Microsc. Spectrosc. Electr.

5. J.Dubois, G.Fantozzi, T.Epicier, C.Esnouf, 1983, "Determination expérimentale d'une loi de comportement dans le cas de la déformation plastique à hautes températures de l'hémicarbure de tungtène", Science of Ceramics, St Vincent, Italy, to be published.
6. T.Epicier, J.Dubois, C.Esnouf, G.Fantozzi, 1982, "Mécanismes de déformation des carbures métalliques. Exemple de l'hémicarbure de tungstène", Rev. Int. hautes Tempér. Refract, 19, p 345.
7. T.Surek, L.G.Kuon, M.J.Luton and J.J.Jonas, 1975,"Linear elastic obstacles: analysis of the experimental results in the case of stress dependent pre-exponentials", in "Rate Processes in Plastic Deformation of Materials, Proc. J.E.Dorn Symp, Cleveland, USA.
8. A.Kelly and D.J.Rowcliffe, 1966, "Slip in titanium carbide", Phys. Stat. Sol., 14, K 29.
9. T.Epicier, C.Esnouf, J.Dubois, G.Fantozzi, 1981, "Observations of dislocations in tungsten hemicarbide deformed at high temperatures", Scripta Met, vol.15, 1279.
10. G.E.Hollox, 1968/69,"Microstructure and mechanical behavior of carbides", Mat. Sci. Eng., 3,121.
11. L.E.Toth, 1971, "Transition Metal Carbides and Nitrides", Academic Press.
12. T.Epicier, C.Esnouf and J.Dubois, 1983,"Dissociation des dislocations dans l'hémicarbure de tungstène W_2C", communication at the Colloq. Soc. Fr. et Belg. de Microscopie Electronique,Liège, Belgique (unpublished).
13. J.D.Venables, R.G.Lye, 1969, "Radiation damage of ordered V_6C_5 by Electron Microscope Beam Bombardment", Phil. Mag., 19, 565.
14. S.Amelinckx, 1979, "Dislocations in particular structures" in Dislocations in Solids, Vol.2, North Holland Publish. Comp.
15. J.P.Hirth and J.Lothe, 1968, "Theory of dislocations", Mac Graw Hill, New-York.
16. J.Friedel, 1977, "Sur le fluage par déviation", Rev. Phys. Appl, T 12, 1649.

PLASTICITY OF WC MATERIALS BY T.E.M. INVESTIGATIONS

Sabine Lay, Jean-Louis Chermant and Jean Vicens
Equipe Matériaux-Microstructure
Laboratoire de Cristallographie, Chimie et Physique des Solides, LA 251 - ISMRa - Université
14032 CAEN Cedex, France

INTRODUCTION

Structure defects of tungsten carbide crystals in sintered WC-Co alloys were previously investigated (1). Dislocation networks with extended nodes and stacking faults on prismatic planes were frequently observed in the as-sintered material. The Burgers vector 1/6[11$\bar{2}$3] was found to be the displacement vector of the fault enclosed in the extended nodes and in stacking faults on the {1$\bar{1}$00} planes.

Recent T.E.M. studies on WC single crystals (2) and also on WC-Co composites (3) deformed at room temperature were published. They confirm the dissociation of the perfect dislocation 1/3[11$\bar{2}$3] → 1/6[11$\bar{2}$3] + 1/6[11$\bar{2}$3]. Another dissociation of the perfect dislocation 1/3[11$\bar{2}$3] was also proposed (4) (5), which has not been confirmed by the other T.E.M. studies (1, 2 and 3).

The following results are related to structure defects of the WC phase deformed at room temperature by compression of WC-Co specimens. Different values of stacking fault energy in WC have been published (2) (3) and (6). Measurements of stacking fault energy have been performed and are reported in the first part of this work.

Pure WC samples were deformed by compression at 1450° C. Defects induced by deformation were studied and are presented in the second part of this work. These results are the first observations on defects induced during well defined steady state creep conditions of WC crystals in WC-Co materials.

EXPERIMENTAL PROCEDURE

Specimens (5x5x5 mm^3) were spark-cut from sintered WC-Co alloys containing 20 to 25 wt%Co. The samples were deformed by compression at room temperature at a strain rate of about 10^{-5} s^{-1}. Pure WC samples (5x5x5 mm^3), annealed at 2000°C, have been deformed by compression at 1450°C and cooled under load (7). Thin slices were spark machined up to a thickness of 80 μm and 3 mm in diameter. These discs were then ion milled for electron microscopy investigations. A Jeol 100 CX microscope operating at 120 KV was used.

T.E.M. RESULTS AND DISCUSSION

Structure defects in tungsten carbide deformed at room temperature

Dissociated dislocations are common in the deformed WC-Co composite. An example is shown in Fig. 1. The defect denoted A, in Fig. 1(a), is a faulted ribbon. The fault has a displacement vector 1/6[$11\bar{2}\bar{3}$] and is bound by two partial dislocations with the same Burgers vector 1/6[$11\bar{2}\bar{3}$]. The fault contrast is visible in Fig. 1(a) and out of contrast as well as the partial dislocations in Fig. 1(b). In fig. 1(c) only the two partial dislocations are in contrast. A weak beam image with g = $10\bar{1}\bar{1}$, Sg = +1.4 10^{-2} $Å^{-1}$

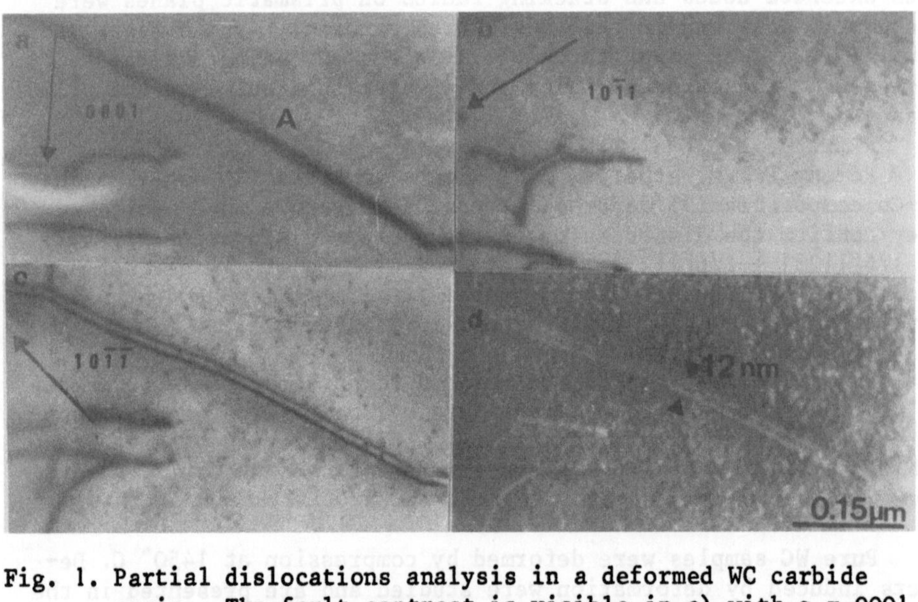

Fig. 1. Partial dislocations analysis in a deformed WC carbide grain. The fault contrast is visible in a) with g = 0001, out of contrast in b) with g = $10\bar{1}1$; the fault is out of contrast in c) with g = $10\bar{1}\bar{1}$, only the two partial dislocations with b = 1/6[$11\bar{2}\bar{3}$] appear ; a weak beam image is given in d) with g = $10\bar{1}1$.

$(W = |Sg.\xi g| = 6.7)$ is shown in Fig. 1(d). The glide plane of these dislocations is the $(1\bar{1}00)$ prismatic plane.

The separation d between the two partial dislocations is equal to 12 nm. This measurement allows an estimation of the stacking fault energy γ given by the relation in the isotropic approximation:

$$\gamma = \frac{\mu}{2\pi d}\left(b_{1s}\, b_{2s} + \frac{b_{1e}\, b_{2e}}{1-\nu}\right) \qquad [1]$$

where the indices s and e refer to screw and edge component respectively of the first and the second dislocation (8), μ is the shear modulus, ν the poisson coefficient. In the present case the angle between the Burgers vector and the line of the dislocation is equal to 30° and the relation [1] becomes :

$$\gamma = \frac{\mu\, b^2}{8\pi d}\left(\frac{4-3\nu}{1-\nu}\right)$$

It has been found $\gamma = 170\ mJ.m^{-2}$ for $\nu = 0.2$ and $\mu = 292300$ MPa (9).

These measurements of the width d as well as others performed on ribbons, evidently presenting a constant width, give a mean value of stacking fault energy γ of 170 $mJ.m^{-2}$ with an accuracy of 10%. In some cases, the dislocations are widely dissociated and their width is non constant. They are of course non equilibrium configurations.

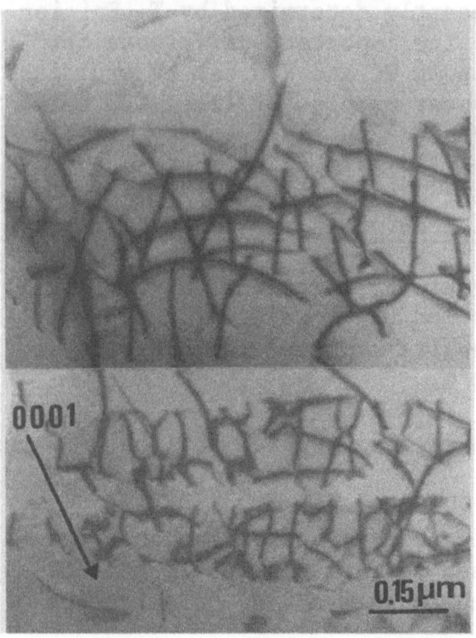

Fig. 2. Dislocation tangles in parallel $(1\bar{1}00)$ planes imaged with g = 0001. Partial dislocations with different Burgers vector and perfect dislocations are present.

Dislocations tangles are frequently observed in deformed materials. An example is shown in Fig. 2. The dislocations lie in planar walls of a thickness of the order of 10 nm which are parallel to the glide plane $(1\bar{1}00)$.

T.E.M. analysis of WC -Co composites deformed at room temperature show that the glide planes are the prismatic $\{1\bar{1}00\}$ planes and the glissile dislocations are the partial $1/6[11\bar{2}3]$ dislocations.

In the undeformed WC-Co alloys partial dislocations with $1/6 [11\bar{2}3]$ Burgers vector have been analyzed in a previous work (1). The presence of faulted ribbons in the undeformed samples can be explained by the consequence of stresses introduced in the material during the sintering process.

Different values have been reported about the stacking fault energy in WC. We have found a value (170 $mJ.m^{-2}$) close to the first estimation of stacking fault energy in WC (200 $mJ.m^{-2}$) (6). This first estimation was performed in an extended node configuration. These two values are higher than the two more recent estimations 50 $mJ.m^{-2}$ (2) and 50-100 $mJ.m^{-2}$ (3).

A variation of the separation between partials can occur in various ways for example as a result of stress effects (10) or of variation of composition (11) which is unlikely in this highly stoichiometric carbide. Stress effects explain more probably the formation of irregularly spaced partials which have been observed in dislocation pile-ups (3) or in dislocation tangles (Fig. 2). The stacking fault energy in W_2C was recently determined by T. Epicier (12) : it has been found 90 $mJ.m^{-2}$. The important difference between the two stacking fault energies in WC and W_2C can be explained by the difference between the two shear modulus of these two carbides. A value of the same order for stacking fault energy has been published for TaC (13) and TiC (14) (170 $mJ.m^{-2}$).

Structure defects in tungsten carbide deformed at 1450°C

Dislocation structure and subgrain peculiarities have been studied in detail in specimens deformed under well defined steady state creep conditions (7). The results reported in this part are related to samples deformed at 1450°C.

Dislocation networks and one array dislocation walls have been analyzed.

A complete characterization of subgrain boundaries observed in tungsten carbide crystals has been performed by using an accurate method presented by E.K. Polychroniadis and P. Delavignette (15).

The small angle rotation axis R of subgrains is determined in this method by the different positions of the Kikuchi lines when two selected area diffraction patterns are taken on each side of the subgrain. Two different positions of the Kikuchi lines give a shift vector : two vectors, for two different orientations, allow the determination of the rotation axis, their lengths giving the rotation angle R. The boundary plane, the directions of the dislocation lines and the equidistances between dislocations are determined from their projected images.

Six subgrains have been analyzed and the different elements characterizing these subgrains are listed on table 1. From contrast experiments it is deduced that three types of subgrains exist. They can be formed by one array of three types of dislocations : 1/3[11$\bar{2}$3] (n°1 - 2), 1/3[11$\bar{2}$0] (n°3 - 4), [0001] (n°5 - 6). The direction of dislocation lines L and the boundary plane are also given for each subgrain with the rotation axis and the rotation angle. The equidistances between dislocations measured on the micrographs are also listed with the calculated equidistances.

Table 1. Sub-boundaries analysis of the six subgrains denoted 1 to 6.

N°	Burgers vector $\vec{b}$	Directions of dislocations	Boundary plane	Directions of rotation axis	Rotation angle
1	1/3 [11$\bar{2}$3]	[01$\bar{1}$0]	(0001)	[01$\bar{1}$0]	0.47°
2	1/3 [$\bar{1}$2$\bar{1}$3]	[01$\bar{1}$0]	(0001)	[01$\bar{1}$0]	0.65°
3	1/3 [1$\bar{2}$10]	[41$\bar{5}$3]	(1$\bar{3}$20)	[41$\bar{5}$3]	0.64°
4	1/3 [1$\bar{2}$10]	[10$\bar{1}$1]	(1$\bar{3}$20)	[10$\bar{1}$1]	0.75°
5	[0001]	[11$\bar{2}$0]	(0001)	[11$\bar{2}$0]	0.57°
6	[0001]	[11$\bar{2}$0]	(0001)	[11$\bar{2}$0]	0.95°

N°	d_c calculated equidistance between dislocations, nm	d_m equidistance between dislocations deduced from micrograph, nm
1	34.5	45
2	25	28.5
3	26.5	26
4	22	18
5	28.5	31
6	17	17

The micrographs of the six subgrain boundaries are presented in Fig. 3 (n°1, 2) Fig. 4 (n°3, 4) and Fig. 5 (n°5, 6). The corresponding stereographic projections are schematically represented for each subgrains (n°1 to 6) in Fig. 3, 4 and 5. The directions L of the dislocation lines, the rotation axis R, the Burgers vector b of the dislocations and the boundary plane P are plotted on each projection. The boundary plane is parallel to the electron beam for the subgrain boundaries n°1, 2 and 6 (Fig. 3 - 5). The dislocation structure appears clearly for the subgrains boundaries n°3, 4 and 5, the boundary plane being, in this case, perpendicular to the electron beam (Fig. 3 - 5).

The character of the six subgrain boundaries analyzed is tilt with a rotation along an axis ∿ parallel to the boundary plane. The directions of the dislocation lines L and the rotation axis R are found ∿ parallel. The deviation for the pure tilt character is of the order of 10°, a maximum deviation (∿ 20°) is found for the subgrain n°5.

The four tilt sub-boundary n°3, 4, 5 and 6 are characterized by a single set of edge dislocations with a Burgers vector b = 1/3 [11$\bar{2}$0] (n°3, 4) and b = [0001] (n°5, 6) perpendicular to the boundary plane. The Frank formula, $d_c = b/2 \sin(R/2)$, was used to calculate the equidistances d_c between dislocations. d_c values are in agreement with the equidistance d_m between dislocations deduced from the micrographs for the subgrains n°3, 6. These subgrains have no long range stress field and they are in equilibrium. A same deviation between d_c and d_m is found for subgrain boundaries 4 and 5. It is of order of 20% (table 1).

The two subgrain boundaries n°1 and 2 are characterized by a single set of dislocations with a Burgers vector b = 1/3[11$\bar{2}$3]. The dislocations are not dissociated and they have a mixed character (∿ 45°). Only their components which are perpendicular to the boundary plane (0001) will be considered for their participation in the tilt character of the rotation axis. This component is b' = [0001]. The other component b" = 1/3[11$\bar{2}$0] is parallel to the boundary plane and therefore gives no rotation, but gives rise to long range stress field. A maximum deviation between d_c and d_m values is found for the sub-boundary n°1.

From the analysis of subgrain boundaries structure some conclusions can be drawn :

- Three types of sub-boundaries exist : they are tilt and are formed by one set of the three possible perfect dislocations in WC : 1/3<11$\bar{2}$3>, 1/3<11$\bar{2}$0> and <0001>. Some twist boundaries have been observed but they are not yet analyzed.
- Some sub-boundaries satisfy the Frank formula and are in equilibrium. We have also found sub-boundaries with a long

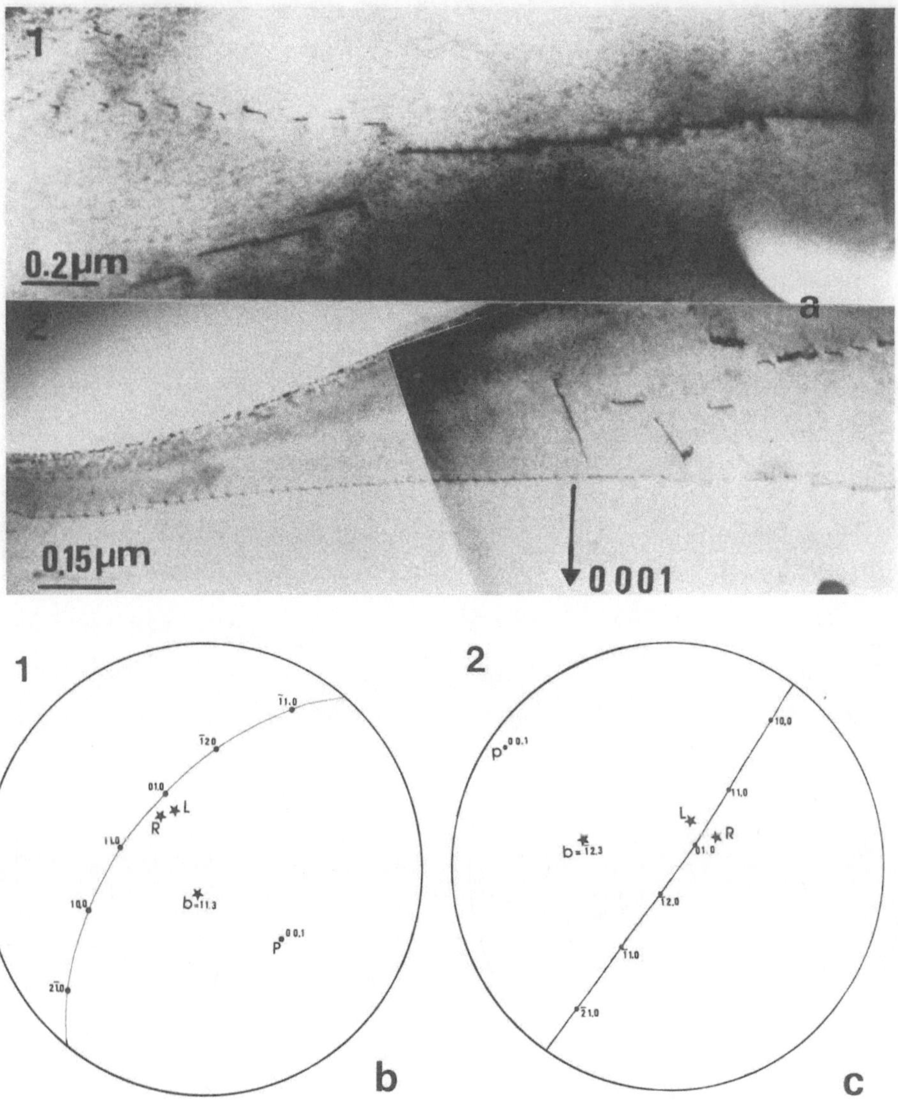

Fig. 3. a) tilt boundaries with dislocations of Burgers vectors b = 1/3<1123> b) and c) stereographic projections of the sub-boundaries n°1 and 2. Tilt boundaries observed in pure WC sample.

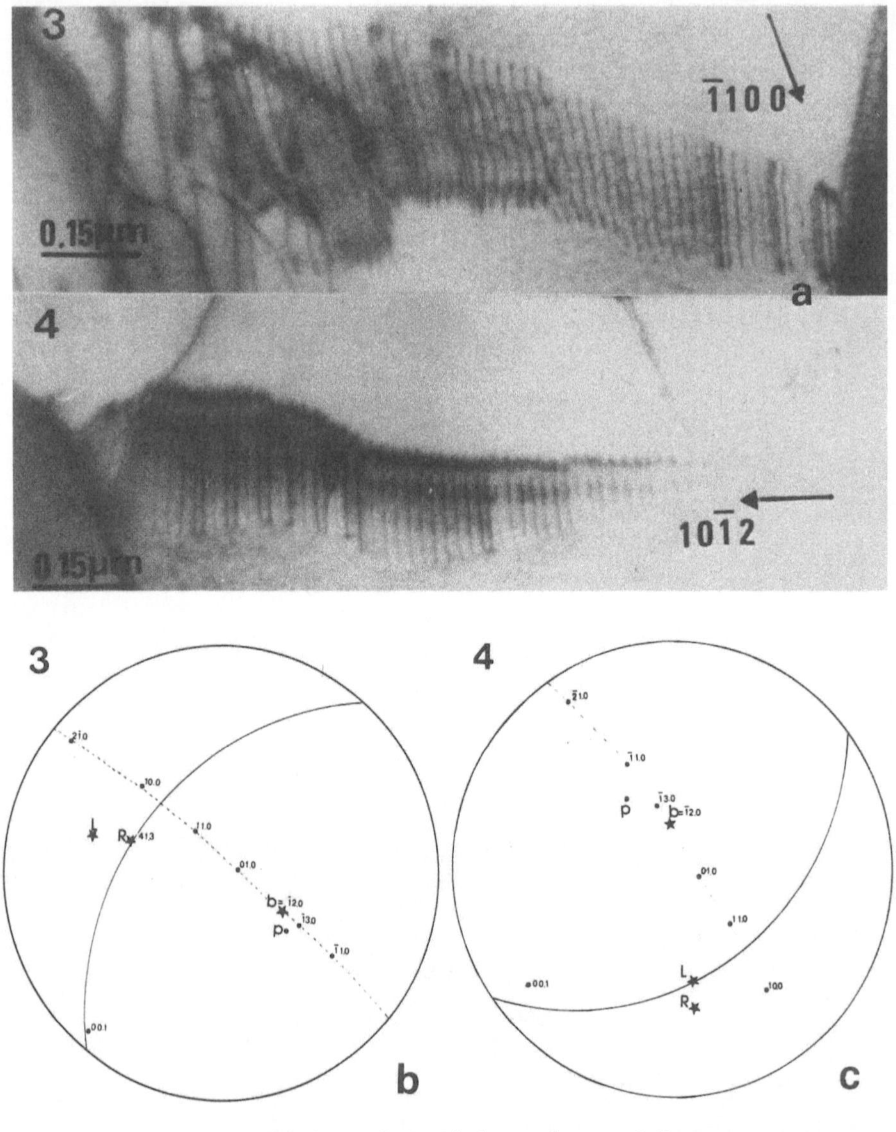

Fig. 4. a) tilt boundaries with dislocations of Burgers vectors b = 1/3[11$\bar{2}$0], b) and c) stereographic projections of the sub-boundaries n°3 and 4. Tilt boundaries observed in pure WC sample.

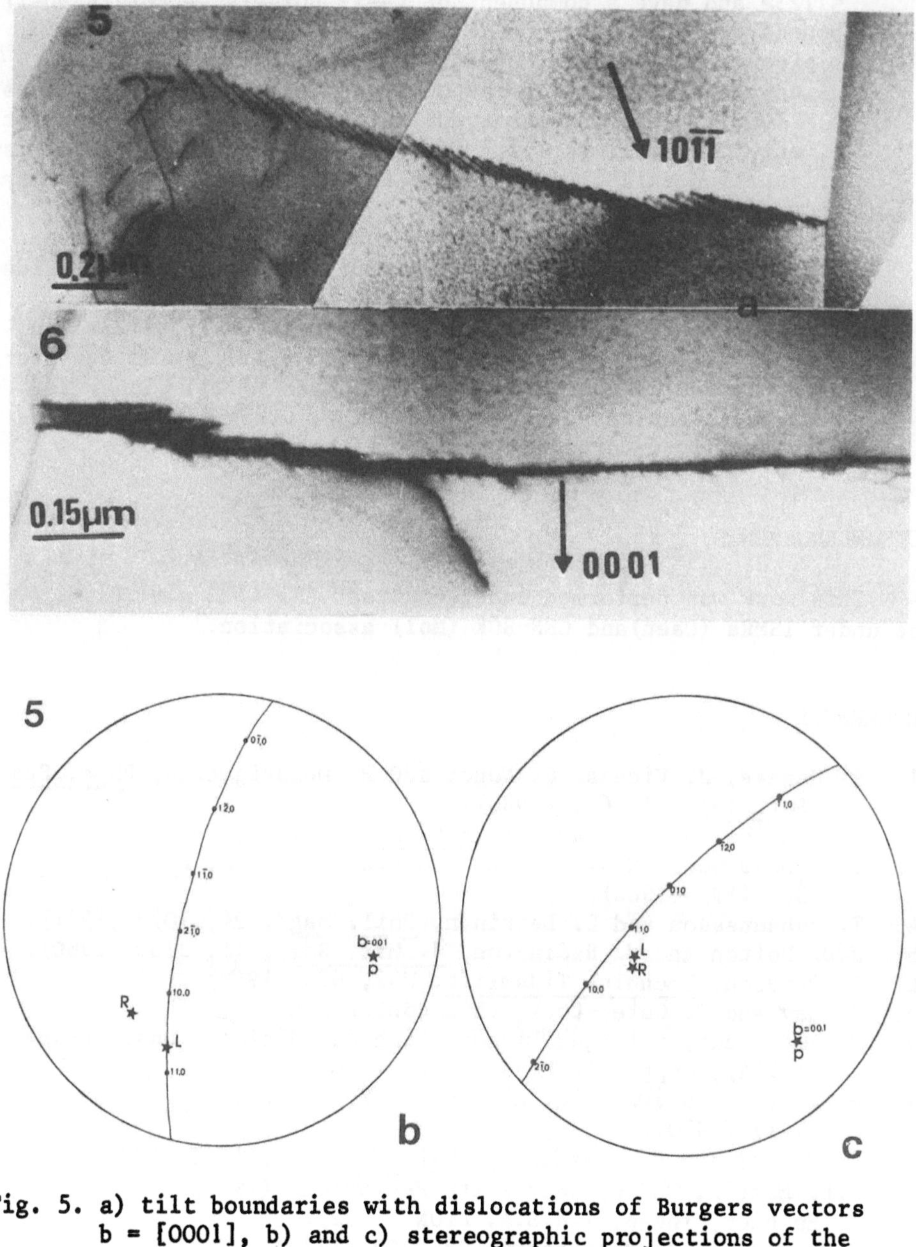

Fig. 5. a) tilt boundaries with dislocations of Burgers vectors b = [0001], b) and c) stereographic projections of the sub-boundaries n° 5 and 6. Tilt boundaries observed in pure WC sample.

range stress field : they consist of perfect dislocations 1/3 $<11\bar{2}3>$ and have a component of their Burgers vector in their boundary plane.

- Dislocation segments inside the sub-boundaries are close to glide planes $\{10\bar{1}0\}$ for the subgrain built with dislocations b = <0001>. They are observed in {0001} planes. They consist of edge dislocations lying in {1010} glide planes. Their misorientation axis is parallel to the direction of the dislocations $<11\bar{2}0>$.
- Dislocation segments inside the sub-boundaries built with edge dislocations with b = $1/3<11\bar{2}0>$ are not in common glide planes and they can climb.
- The sub-boundaries with perfect dislocations $1/3<11\bar{2}3>$ can move by glide of their dislocations on $\{11\bar{2}1\}$ pyramidal planes. Cross clip can exist from $\{10\bar{1}0\}$ to {1121} planes for screw dislocations with b = $1/3<11\bar{2}3>$. Other investigations are actually undertaken to confirm these results.

ACKNOWLEDGEMENTS

This work was performed under contract ATP-CNRS n°91.81.80 and under ISMRa (Caen)and CEN-SCK (Mol) association.

REFERENCES

1. S. Hagege, J. Vicens, G. Nouet and P. Delavignette, Phys. Stat. Sol. (a), 61, 675 (1980).
2. M.K. Hibbs and R. Sinclair, Acta Metall., 29, 1645 (1981).
3. R.M. Greenwood, M.H. Loretto and R.E. Smallman, Acta Metall., 30, 1193 (1982).
4. T. Johannesson and B. Lehtinen, Phil. Mag., 24, 1079 (1971).
5. J.D. Bolton and M. Redington, J. Mat. Sci., 15, 3150 (1980).
6. G. Persson, Technisk Tidskrift, H32, 683 (1969).
7. S. Lay and F. Osterstock, this conference.
8. S. Amelinckx, Colloque"Dissociation des dislocations" Beaune (1974), Suppl. J. Phys. fasc. 12 C-7.
9. H. Doi, Y. Fujiwara, K. Miyake and Y. Oosawa, Met. Trans., 1, 1417 (1970).
10. K. Wessel and H. Alexander, Phil. Mag., 35, 1523 (1977).
11. J.L. Martin, Thèse, Université PARIS-NORD (1972).
12. T. Epicier, Thèse, I.N.S.A. LYON (1982).
13. J.L. Martin, J. Microscopy, 98, 209 (1973).
14. G. Das, J. Less Com. Met., 83, L7 (1982).
15. E.K. Polychroniadis and P. Delavignette, Phys. Stat. Sol. (a), 77, ... (1983).

THE OCCURRENCE AND BEHAVIOR OF DISLOCATIONS DURING PLASTIC DEFORMATION OF SELECTED TRANSITION METAL AND SILICON CARBIDES

R. F. Davis, C.H. Carter, Jr., S. Chevacharoenkul and J. Bentley*
Materials Engineering Department, North Carolina State University, Raleigh, NC 27650
*Metals and Ceramics Division, Oak Ridge National Laboratory, Oak Ridge, TN 37830

INTRODUCTION

The Group IV and Group V transition metal monocarbides (TMMC) possess a combination of some of the highest melting points, hardness values and mechanical strengths of any known materials. Because of these extremes in properties, interest in these materials continues, especially for structural and coating applications for utilization at temperatures where most materials would be molten or would seriously degrade and for employment as abrasives and cutting tools. However, to date, extreme notch-sensitivity, difficulty of fabrication and poor oxidation resistance have limited their acceptance. Several fundamental studies have provided significant information concerning the mechanisms of plastic flow induced by diamond indentation at reduced temperatures and by standard deformation equipment above the ductile-brittle transition temperature. A review of this research as it pertains to single crystals of these materials is presented elsewhere in these Proceedings.[1] The principal thrust in this paper is a review of the results and a discussion of the possible mechanisms of high temperature deformation and steady state creep in both single and polycrystalline TMMC, including recent results obtained at NCSU on single crystals of NbC_x. This will allow a comparison of the derived results and proposed deformation processes using different materials and applied stress parameters.

Silicon Carbide (SiC) also possesses many similar characteristics to that of the TMMC and is a competitive material for employment in many of the same applications. Its low notch sensitivity relative to the TMMC and its resistance to oxidation and deformation at high temperatures also makes it a primary candidate material for use in systems for the production and conversion of energy. A review of prior deformation research on the various forms of SiC as well as recent results determined at NCSU on the reaction-bonded and CVD forms will be presented herein.

TRANSITION METAL CARBIDES

Background

The structure of the TMMC is B_1, i.e., the NaCl-type arrangement where the C and metal atoms exist in two inter-penetrating face-centered cubic lattices. The structure thus consists of identically stacked, close packed {111} planes of the metal atoms alternating with {111} planes of the C atoms. The C atoms occupy the octahedral interstices and are essentially isolated from each other with the C-C distance being more than twice that found in diamond or graphite. This suggests that little C-C bonding exists. Furthermore the metal-metal distance is only slightly greater than in the pure metal structure--about 3% for Group IV carbides and about 9% for Group V carbides. Each of the TMMC exist over a wide range of C/metal ratios; the metal sites are considered to be filled while the C sites may be depleted by as much as 50%. The main impurities normally associated with these materials are oxygen and nitrogen which occupy C sites and may prevent the maximum possible stoichiometry from being attained. They are essentially covalent solids. This is manifested in a high hardness/shear modulus ratio at low temperatures. As the temperature is increased beyond the brittle-ductile transition point (1073-1473K for single crystals,[2] 1473-1973K for polycrystalline materials[2]) their deformation behavior is similar to that of fcc metals as discussed in the following paragraphs. Previous general and specific reviews of the TMMC have been presented by several authors.[3-9]

Group IV Carbides

Williams and Schaal[10] were one of the first to investigate the high temperature mechanical properties of TiC_x single crystals as a function of temperature, stoichiometry and strain rate. Plastic deformation produced in bending was found to occur readily on {111} planes above 1073-1173K over a wide range of C/Ti ratios. The Peierls lattice resistance stress barrier was suggested as the mechanism controlling the temperature of the onset of yield and the dislocation kinetics in contrast to impurity pinning or defect formation.

Williams[11] subsequently conducted additional research on the plastic deformation of single cyrstals of TiC_x (x=0.794, 0.829, 0.900, 0.930 and 0.951) using uniaxial compression over a temperature range of 1073-2073K. (Crystals of $ZrC_{0.875}$ and $NbC_{0.761}$ were

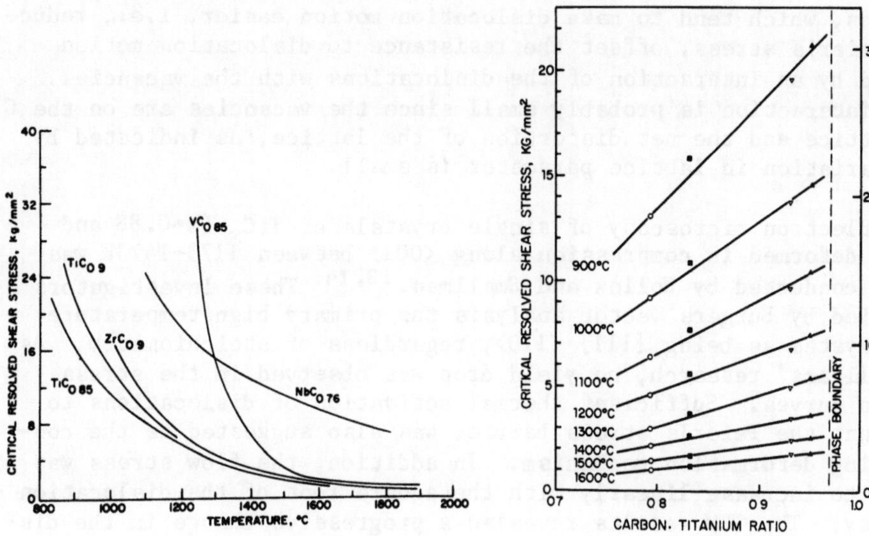

(A) (B)

Figure 1. (A) CRSS versus temperature for several single crystal TMMC (from Ref. 7). (B) CRSS for slip in TiC_x as a function of C content (from Ref. 5b).

also similarly studied but to a much smaller extent.) The critical resolved shear stress (CRSS) was found to decrease exponentially with increasing temperature for all materials (Fig. 1-A), increase linearly with increasing C content in TiC (Fig. 1-B) and increase as the logarithm of the strain rate in TiC. Of interest is the increase in the slope of the curves of CRSS vs 1/T above a temperature which is dependent on the C content; this break occurs at 1423K for $TiC_{0.83}$, 1578K for $TiC_{0.95}$, etc. This indicates a change in the deformation mechanism. Finally, none of the more than 200 samples examined revealed a yield drop in the stress strain curves.

In contrast to nonstoichiometric oxides such as TiO_2 which are stronger than the stoichiometric compositions because of the interactions between dislocations and vacancies,[12,13] the strength in TiC_x and other TMMC decreases as the concentration of vacancies increases because of the decrease in the contribution of the C atoms to cohesion in these materials. Since the number of C-metal bonds which must be broken is reduced, dislocation motion is made easier. The strength of the metal-metal bond is also diminished in the formation of vacancies on the C sublattice. Both contributions, which tend to make dislocation motion easier, i.e., reduce the Peierls stress, offset the resistance to dislocation motion caused by an interaction of the dislocations with the vacancies. This interaction is probably small since the vacancies are on the C sublattice and the net distortion of the lattice, as indicated by the variation in lattice parameter is small.

Electron microscopy of single crystals of TiC_x (x=0.88 and 0.97) deformed in compression along <001> between 1173-1473K was first conducted by Hollox and Smallman.[13,14] These investigators verified by Burgers vector analysis the primary high temperature slip system as being {111} <1$\bar{1}$0>, regardless of stoichiometry. As in Williams' research, no yield drop was observed in the stress strain curves. Sufficient thermal activation of dislocations to surmount the Peierls stress barrier was also suggested as the controlling deformation mechanism. In addition, the flow stress was found to increase linearly with the square root of the dislocation density. The TEM results revealed a progressive change in the dislocation substructure from elongated loops and loop trails to an increasing density of tangles and cell structures containing some dislocation loops and loop trails as the amount of deformation increased from 3% to 9% in the temperature range of study. The formation of these microstructures was attributed in part to the process of cross slip--a process made easier by the high stacking fault energy in TiC (estimated to be 169 mJ/m^2 (= ergs/cm^2)[15]). Finally, neither fringe contrast nor dissociation of dislocations into partials was observed.

More recently, Chatterjee et al. have reaffirmed the TEM results of Hollox and Smallman in TiC single crystal deformed in compression from room temperature to 1373K along <001>. Dislocation tangles, vacancy loops and cell structures were present in the samples deformed at the higher temperatures. From approximately 873-1073K the primary slip system undergoes a gradual change from {110} <1$\bar{1}$0> to {111} <1$\bar{1}$0>; this was interpreted to be the mechanism governing the brittle-ductile transition in TiC.

Finally, Kurishita et al.[17] have observed yield drops or "work softening" in single crystals of $TiC_{0.95}$ deformed in compression from 1280K up to ≈1840K. Following this phenomena, work hardening is observed. This represents one of the most complete deformation studies on single crystals of this material in regards to the ranges of temperature, stress and strain rate employed. Stress exponents of 10 and 5.3 as well as activation energies of 2.49 eV/atom and 4.87 eV/atom were determined at the yield stress for the temperatures below 1510K and above 1840K, respectively. The deformation rate was postulated to be controlled by the Peierls mechanism below the transition temperature of ≈1650K and by C diffusion above this temperature; however, no TEM evidence was obtained to substantiate these conjectures.

Since the slip systems in the TMMC are {111} [110], the deformation of polycrystalline compositions have also been of considerable interest in the search for ceramics which may be ductile at low temperatures or whose high temperature ductility can be controlled by suitable alloying. In this regard, deformation research in bending and compression has also been conducted by Williams[18] and Kelly and Rowcliffe[19] on polycrystalline TiC. In experiments similar to these described in Ref. 11, Williams found that the yield stress was higher at all temperatures than in the single crystal samples. No cracks were formed at the intersection of glide bands and high angle boundaries, i.e., slip was readily propagated across the grain boundaries. Additional studies on polycrystalline $TiC_{0.88}$ and $TiC_{0.93}$ in three point bending and compression, respectively, by Chermant et al.[20] allowed the determination of an activation energy for deformation of 5.0 (3.95) eV/atom for bending (compression) at temperatures below 2023K and 7.55 (6.75) eV/atom above 2023K. As the lower temperature value found in bending is close to the measured activation energy for C diffusion[21a] in TiC, these authors propose a mechanism whereby thermally activated kinks are formed at C vacancies which propagate under the applied stress and the whole dislocation segment moves forward. The higher temperature value is close to the activation energy for diffusion of Ti in TiC (8 eV)[21b]; thus, the deformation process in this temperature regime was assumed to be controlled by the diffusion of Ti vacancies. No microstructural confirmation of these mechanisms was reported.

In a recent more extensive study of deformation in polycrystalline $TiC_{.97}$, Das et al.[15b] using four point bending and compression found that the ductile-brittle transition temperature (≈1073K) depended on grain size. A yield point drop was observed in specimens deformed in compression at 1273K; however, this was not observed at higher temperatures where the stress-strain curve exhibited almost parabolic behavior. It was also determined that the yield stress was a function of grain size and could be described by a Hall-Petch relationship, i.e., yield stress α (grain size)$^{-1/2}$. In lightly deformed (<2% plastic strain) materials, the substructure consisted of elongated loops with some initial tangling (1373K) similar to that found by Hollox and Smallman.[14a] As the temperature of deformation is increased, the number of small loops decrease with corresponding increase in dipole formation (1473K). A small amount of climb also occurs. Following deformation at 1573K, the microstructure consists mainly of long glide dislocations with a few loops and dipoles being present. At higher temperatures, evidence for polygonization becomes increasingly apparent indicating a progressive increase in the amount of climb and the formation of a recovered structure.

The activation energy determined between 1323K and 1773K are of the order of 3.2 to 3.6 eV. Although these numbers are not close to the activation energy for C or Ti in TiC_x, these authors have postulated that the controlling mechanism is that originally postulated by Kelly and Rowcliffe[22] whereby the total a/2 <110> dislocation splits into two sets of Shockley partials and the movement of these partials on {111} planes is controlled by the diffusion of the C atoms. A review of this mechanism and its possible relationship to deformation in TMMC is presented in the Discussion subsection.

In studies on other TMMC, Williams[11] has also shown that single crystals of $ZrC_{0.88}$ are stronger than all TiC compositions between $TiC_{0.79}$ and $TiC_{0.95}$. This has been confirmed in $ZrC_{0.945}$ by Lee and Haggerty.[23] These latter investigators also measured slip on $\{111\}\langle 1\bar{1}0\rangle$, $\{110\}\langle 1\bar{1}0\rangle$ and $\{001\}\langle 1\bar{1}0\rangle$ systems when the crystal orientation was chosen such that the Schmid factor favored slip on these systems. One observation which has not yet been explained is that the critical resolved shear stress for slip on $\{111\}\langle 1\bar{1}0\rangle$ appears to be slightly lower than that for slip on $\{111\}\langle 1\bar{1}0\rangle$. The absence of both stacking fault fringes and extended and contracted nodes indicates that the stacking fault energy in this TMMC is also high. As such, no evidence of dissociation into partials was reported. However, at the relatively low magnifications employed by these authors under bright field conditions, very closely spaced partials such as might occur in materials with high stacking fault energy would be difficult to observe. Additional TEM of samples which had undergone steady state creep in the stress and temperature ranges of 1673-2273K and 39.2-294 MPa revealed the presence of well defined cell walls similar to that found in TiC by Hollox and Smallman[14] and Chatterjee et al.[16] Some of the cell walls were simple tilt boundaries or hexagonal networks. This information coupled with a determined activation energy value of 4.8 eV and a stress exponent of 5 prompted these investigators to postulate that creep is controlled by the recovery process of dislocation climb. A similar opinion concerning dislocation recovery mechanisms has been expressed by Chatterjee et al.[16] based on TEM observations of the clearing of the cell structure in heavily deformed TiC at 973K. These results also correlate with the earlier annealing and TEM research of Hollox and Smallman[14a] in lightly deformed (3%) $TiC_{0.97}$. In this case, dislocation loops were found to diffuse in the sample with an activation energy of 5.25 eV/atom. Subsequent studies by the same authors[14b] showed that this energy and annealing temperature decrease as the C concentration decreases.

In additional ZrC creep research, Miloserdin et al.[24] obtained an activation energy value of 5.0 ± 0.78 eV/atom. These values from both groups of investigators are in excellent agreement with the activation energy for ^{14}C diffusion in $ZrC_{0.94}$ determined by Sarian.[25]

Darolia and Archbold[26] also found the compressive yield strength of arc melted and homogenized polycrystalline $ZrC_{0.94}$ to decrease dramatically from 755 MPa to 186 MPa with an increase in temperature from 1473-2073K. In addition, yield drops similar to those noted in TiC by Kurishita et al.[17] appeared for strain rates greater than $3x10^{-3}$/s. The activation energy for deformation at the yield stress was calculated to be 5.2 $\pm$ eV/atom. The proximity of this value to that for ^{14}C diffusion prompted these investigators to conclude that the Kelly-Rowcliffe mechanism noted above is also the controlling process in ZrC. Again no microscopy was conducted to substantiate this conjecture. Finally, it is interesting to note that the three processes of yield stress deformation, steady state creep and recovery in TiC and ZrC have an activation energy very close to that for the self-diffusion of C. This possibility is explored more fully in the Discussion subsection below.

Plastic deformation in single crystals of other carbides have been examined to a lesser extent. Hollox[5,27] has reported the effect of ordering and carbon-to-metal ratio on the CRSS in single crystals VC_x. Single phase VC may be more correctly described as a series of ordered compounds. The primary slip system is again {111} <1$\bar{1}$0>; however, in this material the yield strength passes through a maximum at $VC_{0.84}$ (V_6C_5) as the C content is increased which is also the composition having the highest order/disorder temperature. The strength appears to be governed by one thermally activated process; however, two processes control the deformation behavior of $VC_{0.75}$.

Williams[11] has shown that single crystals of $NbC_{0.76}$ exhibit a greater CRSS than either $ZrC_{0.88}$ or $TiC_{0.95}$ at all temperatures employed. Creep research on monocrystalline $NbC_{0.89}$ by Dement'yev et al.[57] between 2073-2973K using stress levels of 11.7-68.7 MPa. At temperatures <2473K, a stress exponent of 3.3 and activation energy of 3.7 eV/atom were determined for the steady-state region; above this temperature, these values were 4.8 and 5.2 eV/atom. Based on this evidence and some topographic analyses, the Peierls stress was assumed to be controlling dislocation glide at the lower temperatures while at high temperatures the controlling process was considered to be dislocation climb.

The major creep research on NbC_x single crystals has been conducted by Chevacharoenkul on three different single crystals of NbC_x grown by the float zone technique and having the values of x=0.775, 0.789, 0.804, 0.842 and 0.856, as determined from lattice parameter measurements. Chemical analyses revealed the amounts of O and N to be each less than 50 ppm. The dislocation densities of the as-received crystals were $\approx 10^9/m^2$. The temperature range of these investigations was between 0.35-0.53Tm (Tm≈3873K).

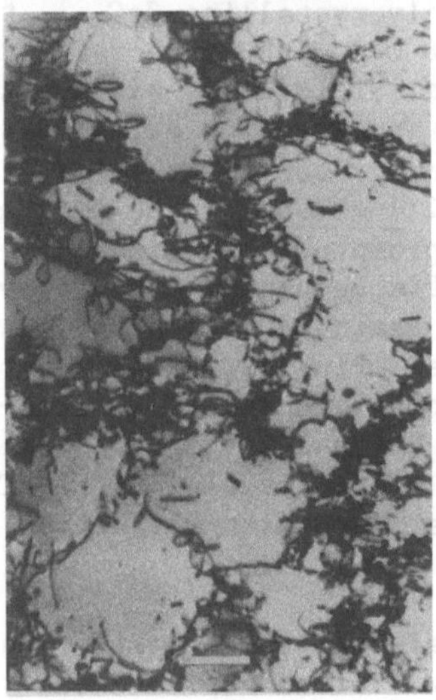

(A)

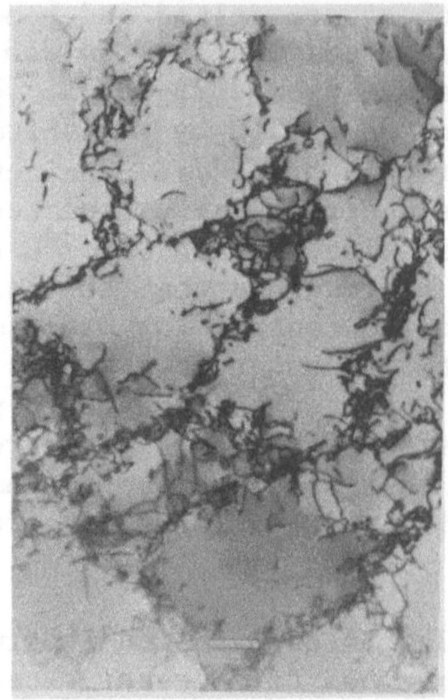

(B)

Figure 2. TEM micrographs of (A) $NbC_{0.775}$ crept at 44.4 MPa resolved shear stress at 1373K. The total resolved shear strain equals 2.7%. The foil normal was near $<\bar{1}17>$; $g=\bar{2}00$. (B) $NbC_{0.804}$ crept at 15.8 MPa resolved shear stress at a final temperature of 1873K. Total shear strain equals 3.55%. Bar in both figures equals 2 μm.

In the low temperature range of 1373-1473K, the creep rate decreases gradually regardless of the stress level, yielding "logarithmic" creep. This type of deformation is characterized by a continuously decreasing slope in the strain vs. time curve without an apparent steady state region. Furthermore, the strain drops to zero after an extended period of time. These effects are caused by the progressive entanglement of dislocations and cell wall formation without sufficient thermal energy for recovery processes to be important and or sufficiently applied stress to allow the dislocations to overcome the obstacles hindering their movement. A typical microstructure developed in this temperature range is shown in Fig. 2-A for $NbC_{0.775}$ crept to 2.7% total resolved shear strain. In all samples crept at this and higher temperatures, Burgers vector analyses of the TEM micrographs showed the slip system to be $\{111\}$ $<1\bar{1}0>$.

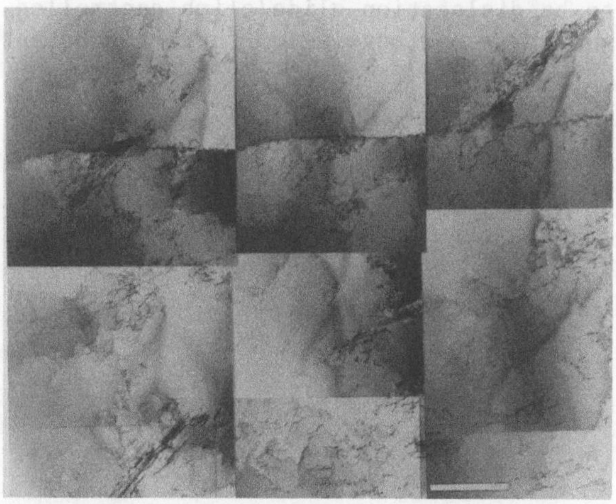

Figure 3. TEM micrograph of $NbC_{0.804}$ crept at 17.3 MPa resolved shear stress at 1923K. The total resolved shear strain equals 5.2%. Bar equals 10 μm.

For samples crept between 1473-1873K, steady state creep is achieved. The stress exponent and activation energy were calculated to be 3.0 and 3.2 eV, respectively. These numbers would indicate a dislocation glide mechanism controlled by the Peierls stress, as the activation energy is considerably less than that for C or Nb self-diffusion.[38] The TEM microstructure shown in Fig. 2-B provides additional evidence for this reasoning and is similar to that found at 1373K; however, the lower stress level has allowed cells with larger diameters to form and the higher temperature and associated recovery process have reduced the density of the dislocations in the cell walls. The structures shown in these two figures are characteristically very similar to that found in a deformed fcc metal at intermediate temperatures. Micrographs of samples crept at 1923K such as Fig. 3 reveal a transition period wherein the dislocation tangles and cell walls are collapsing and glide bands are emerging. A very small degree of polygonization also occurs which indicates that dislocation climb may be a small part of the overall deformation process.

Samples crept between 1973K and 2073K revealed a stress exponent of 6.2 and an activation energy of 4.7 eV/atom which is very close to the activation energy for self-diffusion of C in NbC.[38b] The microstructures of the crept samples reveal prominent glide bands as well as the formation of low angle sub-boundaries as shown in Fig. 4, the latter structure is direct evidence of the occurrence of an extensive amount of climb. Thus operative creep process appears to be dislocation glide/climb controlled by climb. Additional research is now ongoing to further substantiate these ideas.

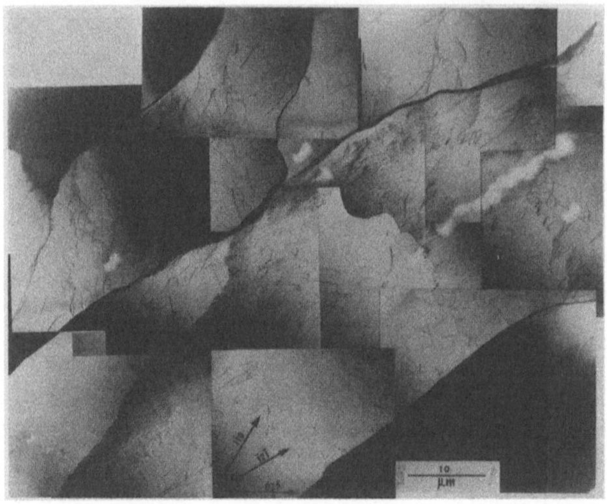

Figure 4. TEM micrograph of $NbC_{0.789}$ crept at 11.1 MPa resolved shear stress and 2023K as the final conditions. The total resolved shear strain equals 22.75%. The foil normal is near the (111) slip plane. The slip band is approximately parallel to $[\bar{1}2\bar{1}]$.

Finally, considerable weak beam dark field TEM has been conducted on all compositions; however, no dissociation of dislocations or extended nodes were discerned at a resolution of 30Å. Calculations using the recently determined elastic constants for NbC_x single crystals, showed that for the disociation to be observable, the stacking fault energy must be as low as 73 ergs/cm^2.

As for polycrystalline NbC_x, Kelly and Rowcliffe[19] have shown that hot pressed $NbC_{0.95}$ is stronger than $NbC_{0.88}$ of similar density indicating an increasing strength with C content over this composition range between 1773K and 2273K. For small amounts of plastic flow, very low work-hardening rates and some recovery during deformation were present in all materials at all times.

The most extensive and informative deformation (4 point bending) studies in a non TiC polycrystalline carbide have been those of Martin and coworkers[25-28] on thin sheets TaC_x (x=0.78, 0.87,

0.90 and 0.96). TEM research of the undeformed and deformed samples revealed stacking fault fringes bounded by Shockley partials whose separation diminished with an increase in C content. Only weak beam dark field techniques[32] were able to resolve these partials in $TaC_{0.96}$. A discussion of this substructure in this material and its possible occurrence in other TMMC is presented in the next subsection.

TaC_x was also observed to undergo extensive plastic deformation above 1773K.[31] Annealing effects were revealed only at very high temperatures and low strain rates: 2173K for $\dot{\varepsilon} = 10^{-7}$/s, 2273K for $\dot{\varepsilon} = 10^{-6}$, and were considered to be caused by dislocation climb. The controlling mechanism for deformation, however, was considered to be a Kelly-Rowcliffe type of process but with only a single splitting of perfect a/2<110> dislocation into partials accomodated by a low C concentration in the core of the dislocations. The process was considered to still be governed by the diffusion of C atoms near the dislocation.

DISCUSSION

The primary experimental observation derived from the mechanical property studies noted above is that the movement of dislocations in the pure TMMC requires the breaking of very strong, predominantly covalent bonds at dislocation cores and the surmounting of electrostatic forces--effects which involve a large Peierls stress. This mobility is markedly increased by thermal fluctuations, leading to a strongly temperature dependent yield stress. Also, as the experimental temperature and/or dislocation density increases, one or more additional processes including dislocation intersection, cross slip, dipole formation, dissociation and climb may become active in the deformation process of a particular TMMC.

Among the several deformation studies conducted on the TMMC, only those by Williams,[11] Hollox and Smallman,[14a] Das et al.[15b] and Darolia and Archbold[26] have provided graphical information on stress-strain relationships analogous to that normally given for metals. Whether or not multistage hardening normally seen for fcc metals may be observed in TMMCs is not known; however, the curves of Hollox and Smallman[14a] indicate a similarity to the curves for the first stage of work hardening. Thus essentially all the information concerning the mechanisms involved in work hardening and annealing in TMMC must be obtained by careful observation of the few available TEM micrographs of the dislocation substructure of progressively deformed or crept samples. As might be expected, these observations have generated various interpretations regarding the mechanisms which are operative and controlling in these materials.

Electron micrographs of samples of various compositions of TiC_x, ZrC_x and NbC_x reveal that in the early stages of deformation (<3%) and within the temperature range of the brittle-ductile transition and approximately 0.6 Tm, numerous dipoles form apparently as a result of cross slip of (and possibly jogs on) the screw components of dislocations gliding on {111} <1$\bar{1}$0> systems. (Samples deformed at higher temperatures have not been examined by TEM.) Hollox and Smallman[14] have shown that in TiC, the dipoles subsequently "pinch off" to become elongated loops which in turn decompose into trails of much smaller loops in order to lower the energy of the configuration. Similar structures have also been found in lightly deformed (<1%) NbC_x by Chevacharoenkul.[33] As noted above, continued deformation (>3%) of these three materials causes the formation of an increasing density of dislocation tangles consisting of irregularly curved lengths of dislocation lines, kinked at irregular intervals with usually no discernable obstacle at the pinning points. These tangles gradually become arrayed in walls of cells whose size and dislocation density are a function of the temperature of deformation, stress, amount of strain and strain rate. Generally, the lower the temperature of deformation and greater the amount of strain, the smaller the cell size. The microstructures are very similar to those found in fcc metals (although tangles and cells are by no means limited to these substances). The most typical characteristic of dislocation tangles aside from their irregularity is that they represent three dimensional dislocation arrays spread vertically over a large number of closely spaced slip planes. The tangles in the TMMC crystals are commonly associated and interspersed with small dislocation loops.

Several mechanisms have been proposed to explain the work hardening process and thus the origin of the tangles and cell walls in fcc metals. It is very likely that one or a combination of these reasons also describes similar phenomena in TiC, ZrC and NbC (and possibly in the other three TMMC as well). Generally speaking, these theories postulate that initial work hardening is caused either by dislocation-dislocation interaction through their elastic stress fields[34-36] or direct dislocation-point defect interaction.[37] In the former process, dislocations on secondary glide planes subsequently become active and interact with the primary dislocations forming either (a) grids[34,36] parallel to the primary glide plane and containing a number of sessile Lomer-Cottrell locks or (b) complex carpets of high dislocation density.[35] Each of these regions can in turn act as a new obstacle to further slip. Since each slip line is stopped at another, the dislocations tend to cluster into local regions forming cell structures. If cross slip and annihilation of screw dislocations of opposite sign are also prominent features in the work hardening process, the edge dislocations remaining on the cross slip planes may also form tilt boundaries in the cell structure. In the latter theory[37] the dislocation vacancy interaction leads to superjogs or prismatic loops elongated normal to the Burgers vector of the screw dislocation. The heavily jogged dislocations will find it difficult to move, and some parts of them may be pinned so effectively that they act as dislocation sources. Also, if the jogs themselves are sufficiently long, they too may act as sources on different slip planes. These two sources will interfere with each other and possibly induce further cross slipped segments that will again act as sources. The expanding loops continue to interact with the defect concentration and thus the process "mushrooms" until the stress is too low to cause the loops to propagate, i.e., when the dislocation density in the tangled cell walls is too large or the defect concentration is too low. The growth in number of the tangled walls eventually leads to the formation of a cell structure.

Depending on the metal and the extent to which it is deformed, evidence for all of the aforementioned dislocation processes have been observed by various investigators. Analogous microstructures in TiC_x, ZrC_x and NbC_x point to the similarity in deformation processes for the two types of materials. However, understanding of the exact mechanisms operative to produce the stress-strain relationships in these TMMC is at the moment as conjectural as it is for the fcc metals.

Another aspect of mechanical behavior is that of recovery. Although the extent of this process may be small in a standard stress-strain experiment, it appears to be controlling above 0.5 Tm in steady-state creep phenomena such as that investigated in ZrC_x[23] and NbC_x.[33] In this regard, Hollox and Smallman[13] have shown by TEM that the annealing process at 1423K and 1623K in lightly (<3%) deformed single crystals of TiC is accompanied by the coalescence of the "pinched off" dislocation loops which are vacancy in nature. Heating the samples to 1673K for 15 minutes caused the loops to anneal out to form the general dislocation network of the crystal. Using the much lower temperature of 973K, Chatterjee et al.[16] observed a considerable reduction of the dislocation population as well as clearing of the cell structure in deformed $TiC_{0.93}$. A climb process involving the lattice diffusion of vacancies was suggested to be operative in both cases. Hollox and Smallman[13] calculated the activation energy for the annealing process to be 5.25 eV/atom for $TiC_{0.97}$ and 4.88 eV/atom for $TiC_{0.88}$. A comparison of the latter value with the activation energy for self-diffusion of C in TiC[21] in Table 1 shows very good agreement.

As reported above, the related high temperature creep research of Chermant et al.[20] on $TiC_{0.88}$ and of Chevacharoenkul[33] on NbC_x have determined the activation energies for this process to be 5.5 eV/atom (T=1673-2000K) and 4.7 eV/atom (T=1973 - 2073K), respectively. Furthermore the activation energies for creep in $ZrC_{0.945}$

Table 1. Activation Energies for the Diffusion of the Carbon and the Metal species in $TiC_{0.891}$, $ZrC_{0.96}$ and $NbC_{0.860}$

Carbide	Q(eV/atom)-^{14}C	Q(eV/atom)-metal	T(K)	Refs.
$TiC_{0.89}$	4.68	7.7	1673 - 2553	21
$ZrC_{0.96}$	4.9	7.5	1853 - 2273	25
$NbC_{0.868}$	4.4	6.1	1900 - 2315	38a,b

(1673°-2273K),[23] ZrC_x (2448-2823K)[24] and $ZrC_{0.94}$ (1473-2073K)[26] have also been reported to be 4.8, 5.0±0.78 and 5.2±0.2 eV/atom, respectively. All of these creep results also indicate that the lattice diffusion of C is controlling deformation in the TMMC at high temperatures.

A review of the foregoing results indicates that deformation at the CRSS, steady state creep and separate annealing processes are all controlled by the diffusion of C. However, the exact mechanism(s) by which this control is exerted and the influence of the C/metal ratio on the occurrence and nature of the C-control process is one of considerable interest and debate as described below.

Both dislocation climb and a synchro-shear process have been proposed as the C-assisted rate-controlling deformation mechanism in TiC and ZrC. Although the latter process developed by Kelly and Rowcliffe[22] is the most completely described and popular explanation for the mechanism of deformation in the TMMC, it has been difficult to prove experimentally. This theory, originally proposed for TiC_x, has as its basic premise the double dissociation of dislocations into partials. Furthermore, it is assumed that no rearrangement of atoms occurs during slip wherein the C atom becomes in tetrahedral coordination.

The stacking sequence of {111} planes in a stoichiometric fcc TMC is shown in Fig. 5 and may be represented by the sequence

$$A_\gamma B_\alpha C_\beta A_\gamma B_\alpha C_\beta \qquad (1)$$

where italic letters represent metal atoms and Greek letters represent C atoms in the octahedral interstices. Slip by the movement of perfect dislocations in directions such as B_1B_3 and B_2B_3 place the C atom in a distorted tetrahedral arrangement and would appear to require a large lattice expansion normal to the slip plane. The normal dislocation dissociation found in low stacking fault energy metals (and proposed for high stacking fault energy materials such as Al and Ni) is of the type

$$a/2[1\bar{1}0] \rightarrow a/6[2\bar{1}\bar{1}] + a/6[1\bar{2}1]$$

which in Fig. 5 is equivalent to

$$B_1B_3 \rightarrow B_1C + CB_3.$$

If this should occur only on close-packed sheets of metal atoms, the sequence $A_\gamma C$ is produced which corresponds to a tetrahedral coordination of C within the stacking fault. If, however, each of the two partials noted above is further dissociated into Shockley partials along successive C and metal layers, e.g.,

$$a/6[2\bar{1}\bar{1}] \rightarrow a/6[1\bar{2}1] + a/6[11\bar{2}]$$

$$B_1C \rightarrow B_1A_1 + A_1C$$

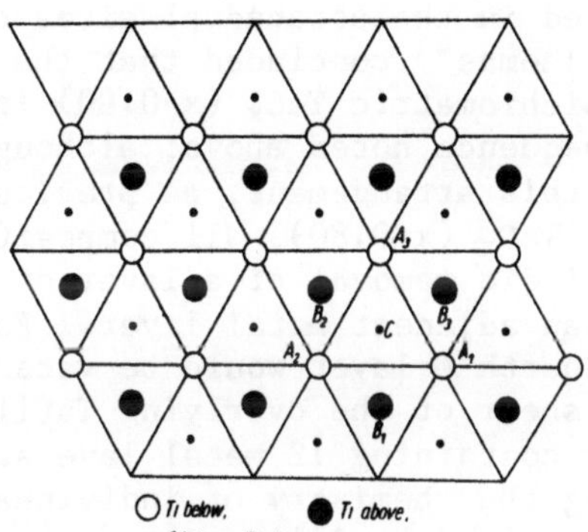

Figure 5. Plane view of Ti and C atoms on respective {111} planes (from Ref. 22).

an intrinsic stacking fault can result having the stacking sequence

$$A_\gamma B_\alpha C_\beta A_\beta C_\beta A_\gamma B_\alpha \tag{2}$$

In other words, dislocations would be produced with their cores extended normal to the slip plane. Although all the C atoms are octahedrally coordinated, this arrangement with the three layers directly above one another is expected to contribute to a high stacking fault energy because the closer C layers increase the energy as a result of repulsion. However, this energy could be lowered if the central β layer of C were absent. The basic nature of the intrinsic fault would be retained. Yvon and Parthe[39] originally proposed such a stacking sequence for ordered phases of TaC_x such as Ta_2C and Ta_4C_3. More recently stacking faults have actually been reported on {111} in VC_x (x=0.75-0.8)[40] and TaC_x (x=0.70-0.80)[41,42] crystals grown from the melt and in carburized Ta sheets (x=0.75, 0.83, 0.90 and 0.95).[28-32]

In the single crystals of TaC_x (x=0.7-0.8), Rowcliffe and Thomas[41] found that the {111} faults were indeed shear faults, i.e., they have bounding Shockley partials of the type a/6 <112>. The fault density also increased rapidly with a small decrease in the C content. A maximum is reached at Ta_4C_3 (the composition of the zeta phase polytype) with the formation of regular {111} faulting which produces sheets of h.c.p. and f.c.c. layers in the material which, as a result of compositional inhomogeneity, also coexist with areas of random faults and a fault-free matrix. Diffraction effects from this regular faulting indicated that C vacancies in this defect are ordered, i.e., complete C layers are missing, as postulated by Yvon and Parthe,[39] rather than the vacancies being randomly situated in the octahedral sites of the metal lattice. Rowcliffe and Thomas[41] concluded that the stacking fault configuration for stoichiometric TaC_x ($x>0.80$) is represented by the second stacking sequence noted above; although problems in C-C repulsion may attend this arrangement, as previously described. For substoichiometric TaC_x ($x<0.80$), all compositional changes can be accomodated by periodic removal of a layer of C atoms followed by a single shear of an adjacent metal layer. For example, in an ideal $TaC_{0.75}$ every fourth C layer would be vacant which, if accompanied by subsequent shear of the overlying Ta(111) plane, would result in a unit cell containing 12 metal layers. Additional information concerning the chemistry of individual stacking faults in $TaC_{0.78}$ has been provided by Allison et al.[42] using electron energy loss spectroscopy. Their analyses revealed a significantly lower value for the C concentration in the widely disassociated faults than in the matrix. If the removal of the C layer precedes dislocation glide, stacking fault formation can be accomplished by the mechanism for compositional changes in TaC_x detailed in the previous paragraph, namely, the shearing of single a/6<112> partials without the need for the additional dissociation postulated by Kelly and Rowcliffe.[22] Not only can this mechanism describe the changes in stoichiometry in TaC_x in the range x=0.7-0.8 but also plastic deformation. A model for dislocation motion developed by Allison et al.[42] requires that the C atoms in the central layer of the dislocation diffuse out to allow simple glide of the metal atoms. Such diffusion-limited motion of dissociated dislocations is analogous to Suzuki hardening which is based on the relation between the equilibrium concentration of alloying atoms in the faulted region of the crystal (C in this case) and the average concentration in the matrix. The important point to note is that this theory of plastic deformation has been developed for a narrow composition range of C/Ta ratios where the zeta phase occurs as a precipitate. If it is found to be operative in the deformation process, the theory will likely be applicable to NbC_x and VC_x which also have a range of compositions where a zeta phase occurs. This phase has not been found in any Group IV TMMC. It is therefore of considerable interest to determine (1) if the Shockley partial dislocations at the ends of the stacking faults of the zeta phase are mobile and therefore allow deformation to occur as a result of an applied stress; (2) if dislocation dissociation occurs outside the composition regions where the zeta phase occurs and is active during deformation; (3) the applicability of dissociation to Group IV TMMC and its role (if any) in the deformation of these materials.

As a possible answer to the second point of concern, Martin and coworkers[28-32] have reported dissociation over a much wider compositional range in polycrystalline TaC_x (x=0.75-0.95) produced by carburization of Ta and subsequently deformed to approximately 1% strain in four-point bending between 1473K and 2473K. The width of the dissociation was reported to decrease as the C content increased until at x>0.90 weak beam dark-field techniques were required to observe the splitting. However, it is very difficult to discern this disassociation in the TEM micrographs presented by these investigators of the higher C content samples. These authors have also postulated that the C concentration in the stacking faults is very low which thus allows for single 1/6<112> partial dislocations. Deformation would then occur via the motion of these dissociated dislocations which would, in turn, be controlled by the diffusion of C atoms away from these dislocations in order to maintain the deficit of C atoms in the stacking fault. This process is similar to that proposed by Allison et al.[42]

In the research by Martin and colleagues just described, no evidence for cross slip, climb, dipole formation, tangles or cell wall formation-processes observed as a result of deformation in TiC_x, ZrC_x and NbC_x--was reported in the deformed samples of TaC_x. Only in relatively long term relaxation experiments was there a modest indication of climb. Thus, there would seem to be marked disagreement concerning the mechanisms of deformation among the various TMMC. However, we believe there are several factors which must be considered in regards to these materials in general and the experiments conducted by the various investigators.

In a comparison of the TEM micrographs of the entire range of TaC_x compositions reported by Martin et al., the microstructures of the undeformed and deformed samples appear to be similar in regards to dislocation substructure and density regardless of the mechanical state of the samples. This is not surprising if one considers that the maximum amount of strain imposed by these investigators did not exceed ≈1%. The dislocation density for the deformed samples was noted to be $10^8/cm^2$. Unfortunately a comparison of densities for the undeformed and deformed samples is not possible. Nevertheless, this value is considerably lower than the 10^{12}-$10^{14}/cm^2$ value reported in deformed TiC_x and NbC_x single crystals which originally contained a density of $\approx 10^5/cm^2$. Thus Martin's observed extended dislocations could have existed as a product of the fabrication process rather than being produced during deformation. Their participation during deformation is not proven. Their reported existence of dislocation dissociation above $\approx TaC_{0.8}$ is countered by the results of Rowcliffe and Thomas[41] and by similar research on VC_x and NbC_x by Billingham and Lewis[40] and Chevacharoenkul,[33] respectively, although all groups possessed the means to make such observations. The lack of any evidence for similar dislocation dissociation in the NbC_x materials can be attributed to the fact that no composition investigated was near the C/Nb = 0.72 ratio which is the boundary between the pure NbC and the field of occurrence of the zeta polytype. Thus, any presence of dislocation dissociation in the pure monocarbide region must be explained by other reasons.

The possibility exists that the stacking fault energy for TaC_x $(x>0.8)$ is measurably lower than that for the other TMC in the same range of C/metal ratios; although, from a consideration Seeger's amount of d band filling and the very large values of the elastic constants, such a reduced value of this energy would seem unlikely. Another possibility is that the 1000 ppm oxygen and other impurities in the TaC_x $(x>0.80)$ examined by Martin et al.[28-32] exceeded a critical concentration and thus acted in a manner analogous to the effect of <100 ppm B in $TiC_{0.9}$ which caused the formation of very thin partially coherent platelets of TiB_2[43] which precipitated at widely extended dislocation nodes bounded by Shockley partial dislocations.[44] These defects so closely resemble stacking faults that they were initially identified as such by Venables.[45] In this case, the precipitates pinned the dislocations[43] thereby increasing the CRSS by a factor of 6 at 1873K; thus, it is unlikely that the associated Shockley partials would be active during deformation.

Based on the foregoing discussions, it is believed that had Martin et al. more extensively deformed the TaC_x samples, dislocation tangles, cell walls, glide bands, etc., would also have been observed as they have been in TiC_x, ZrC_x and NbC_x. The occurrence of dislocation dissociation in substoichiometric TaC_x and VC_x as an inherent equilibrium structure is proven. This effect probably also occurs in the lower C/metal ratios in NbC_x. The participation of these Shockley partials in deformation has not been sufficiently substantiated. Furthermore the occurrence of these partials at high C/metal ratios in Group V compounds apart from the effect of impurities or other extrinsic phenomena remains a point of debate. It is further recognized that all TMMC could deform by the movement of Shockley partials, separate from those already identified and whose dissociation is less than the width of the resolved dislocation image (≈25Å). Indeed this has been postulated for the high stacking fault fcc metals such as Al and Ni. However, in the case of the high C/metal TMMC, the creation of tetrahedrally coordinated C or the stacking of 3 layers of C over one another and the considerable bond angle distortion in these predominantly covalently bonded materials may cause an increase in the internal energy which is greater than the decrease in the self-energy of the dislocation as a result of dissociation.

The zeta polytype does not occur in the Group IV TMMC; thus, dislocation dissociation similar to that noted for TaC[41,42] has not been discerned. However, as noted in the previous section Das et al.[15] has reported the observation of occasional dissociation of dislocations (probably screw in character) in $TiC_{0.93}$ having a width of 25Å which have occurred during deformation at 1773K (see ref. 15b for an account of the deformation studies). Most of the dislocations observed by this investigator were found to lie on {111} planes with a Burgers vector of a/2<110>, i.e., total dislocations such as those determined by Hollox and Smallman.[14] Thus whether or not the deformation proceeded by movement of the observed partials, unobservable partials or total dislocations is still a matter of conjecture for this material.

All investigations to date on TMMC indicate that while all deformation above the brittle ductile transition temperature occurs by dislocation processes, the controlling factor is either the Peierls lattice friction stress or the diffusivity of C. The occurrence of these phenomena may also, in turn, be a function of temperature, C/metal ratio, stress and strain rate. As noted above, in those investigations where an activation energy has been determined, the agreement with the activation energy for C diffusion is very good. Kelly and Rowcliffe,[22] Martin et al.[28-32] and Allison et al.[42] have offered alternative mechanisms for deformation in TMMC and in TaC_x in particular which would be controlled by C diffusion. The operation of one of these processes in the group V materials within the C/metal ratios of the appearance of the zeta phase is possible. Little evidence for dissociation AND the participation of the resulting partials in deformation exists at this writing for Group IV and for compositions outside the zeta phase region of the Group V materials. It is therefore very possible, especially in steady-state creep at high temperatures, that deformation is controlled by dislocation climb which is, in these materials apparently controlled by C diffusion. The agreement in activation energies noted above may also be fortuitous.

Finally, it is also recognized that for fcc metals such as Al, deformation and creep at intermediate temperatures (≈298K) may be controlled by cross slip of screw dislocations. This has also been reported to occur in TiC[14,16] and NbC.[33] If disassociation does occur and the partials are active in deformation, cross slip and therefore deformation should become easier as the reported[31] separation of the partials decreases at the higher C/metal ratios. By contrast, however, the CRSS increases markedly with an increasing C content. This latter phenomenon is better explained by the increase in the bond energies with increasing C content which, in turn, increases the Peierls stress. Thus it would appear that dislocation motion is controlled by the Peierls stress; the overall deformation process, especially steady state creep at high temperatures appears to be controlled by C diffusion.

Summary

Considerable information and theories exist concerning the kinetics and mechanisms, respectively, of deformation in the TMMC. deformation studies within the C/metal ratios of the occurrence of the zeta polytype should be conducted in order to discern if the existing partials are active in the deformation and if this deformation is controlled by the diffusion of C. Very careful steady state creep or stress-relaxation studies coupled with extensive TEM should be conducted in Group IV and Group V single crystals in order to determine the controlling deformation processes as a function of temperature, C/metal ratio and stress.

SILICON CARBIDE

Background

As a pure material, SiC is the only compound species that exists in the solid state in the Si-C system and can occur in the cubic (C), hexagonal (H) or rhombohedral (R) structures. It is also classified as existing in the beta and alpha modifications. The beta or cubic form crystallizes in the zinc-blende or sphalerite structure; whereas, a large number (approximately 140) of the alpha occur in the hexagonal or rhombohedral forms known as polytypes. Silicon carbide is also a generic name for materials which are produced and fabricated by a number of processing routes. The resulting material may be multiphase or single phase, depending on the chosen procedure, as, for example, in the reaction-bonded or chemically vapor deposited (CVD) SiC materials discussed in this manuscript. As such, the kinetics and mechanisms of deformation in each of these materials may also be measurably different.

Reaction-bonded SiC

In previous research on reaction-bonded SiC, Rumsey and Roberts[46] have employed four-point bending at 1473K and 141 MPa. The resulting creep rate of 2.7×10^{-10}/s was attributed to the deformation of free Si on the SiC grain boundaries. Subsequent and similar research by Marshall and Jones[47] revealed only transient creep behavior even after 10^3 hrs at temperatures from 1273K to 1473K and stresses from 207 to 414 MPa. Although an activation energy of 2.38 ± 0.82 eV/atom and stress exponents which varied from one-to-two with increasing temperature were calculated, no conclusion concerning the controlling mechanism could be reached.

Krishnamachri and Notis[48] have reported an activation energy for "KT" SiC of 1.52 eV/atom employing temperatures and stresses of 1573K-1673K and 34-86 MPa, respectively. The mechanism of creep was concluded to be Coble grain boundary diffusion with Si being the rate controlling species. This was in agreement with their deformation map for polycrystalline SiC having a grain size of 65 µm. Finally, Schnurer et al.[49] have also conducted creep experiments in vacuum and in air on reaction-bonded SiC in the bending mode at 1273-1673K. The stress exponent was found to increase from one to four as the stress was raised from 97 MPa to 190 MPa. This increase in the stress exponent was originally attributed to the increased mobility of dislocations with an increase in applied stress but has since been attributed to free Si at the grain boundaries.

Although several of the previously mentioned researchers believe that creep in reaction-bonded SiC is caused by an Si phase at the grain boundaries, electron microprobe analysis by Marshall[50] and transmission electron microscopy by the present authors have revealed no grain boundary second phase. Furthermore, if the grain boundaries contain a Si phase, it seems very unlikely that these materials would have the good creep resistance which we report below at temperatures above the melting point of Si.

The Norton NC-430 reaction-bonded SiC studied in this research was produced by slip casting a plate from a slurry of colloidal graphite and Achenson-derived SiC powder with the latter having a bimodal grain size distribution (≈100 µm and $\leq$10 µm). The plate was then reaction-bonded in a N_2 atmosphere containing Si vapor at temperatures exceeding 2273K. Experiments to determine the temperatures and stresses required to cause measurable creep were conducted in ≈ one atmosphere pressure Ar between 1573-1923K and 69-220 MPa. All values of stress and temperature dependence were computed using steady-state creep rates.

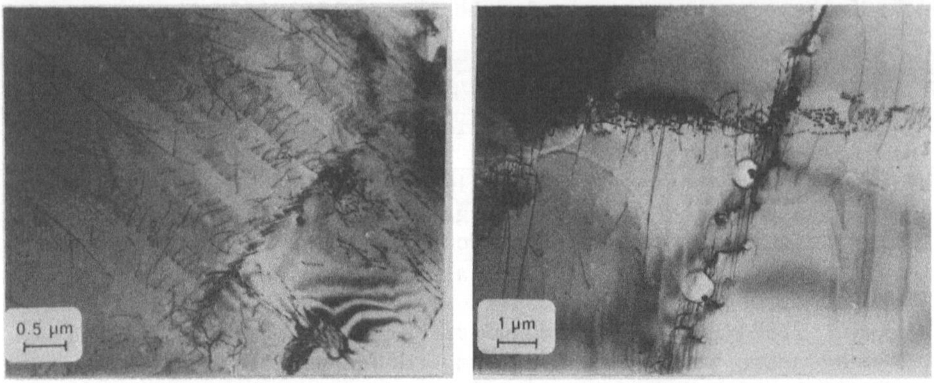

Figure 6. HVTEM micrographs of NC-430 SiC after creep. Slip bands are very evident, a portion of which end in a low angle boundary. (Deformed at 1903K and 220 MPa.) (b) HVTEM micrographs of NC-430 deformed at 1903K and 147 MPa. Two types of low angle boundaries are evident; the horizontal one without cavities is at a high angle to the one with cavities. Loops are also evident in the former boundary.

Transmission electron microscopy revealed that the as-received material contained a fairly high concentration of dislocations. Low angle boundaries were also observed on the interfaces between the Achenson-derived SiC and the SiC that formed during reaction bonding. In addition, all of these boundaries contain cavities, many of which contain small black objects believed to be unreacted C.

The slip traces seen in Fig. 6-A are of great importance in this creep study because they indicate that considerable dislocation glide is occurring. Burgers vector analysis of these slip band dislocations has shown them to be Shockley partials in the hexagonal system, which have the Burgers vector $\bar{a}/3\langle 0110\rangle$. All dislocations lie in the basal plane.

A significant concentration of additional low angle boundaries formed as a result of deformation. Those shown in Fig. 6-B are believed to be produced as a result of dislocation climb and the subsequent formation of very irregular cell walls. The boundaries contain no cavities but contain many dislocation loops.

A comparison of the stress exponent of 5.7 calculated from the experimental log ε versus log σ/G curve of the present research with the theoretical stress exponents of various creep mechanisms reveals that dislocation climb by pipe diffusion, which has a stress exponent of 5, and dislocation glide/climb controlled by climb, with a stress exponent 4.5, are the mechanisms having values closest to the experimental number. In addition, correlation of the value of 7.37 $\pm$ 0.2 eV/atom for the activation energy of creep to that of self-diffusion of Si in SiC[51] reveals that the former value is 0.90 that for lattice diffusion. Dislocation pipe diffusion, however, ranges from 0.3 to 0.7 that of lattice diffusion.

A consideration of all of the above quantitative results together with those derived from the TEM of the deformed samples leads us to conclude that creep in this reaction-bonded SiC is governed by a dislocation glide/climb mechanism that is controlled by climb.

Chemically Vapor Deposited Silicon Carbide

This nominally single phase SiC used in this "first of its kind" research program was produced by chemical vapor deposition via pyrolysis of methyltrichlorsilane ($CH_3SiCl_3 \rightarrow SiC + 3HCl$) at 1673K. The material was deposited on a 5.1×10^{-2} m square graphite substrate (also at 1673K) to a thickness from $(8.9\text{-to-}12.7)10^{-3}$ m and annealed in purified Ar at 1.034×10^{7} MPa and 2373K to remove the residual stresses. The substrate was subsequently ground away.

Experiments to determine the temperatures and stresses required to cause measurable creep were conducted in ≈ one atmosphere Ar between 1573-2023K and 69-220 MPa. As before, all values of stress and temperature dependence were computed using steady-state creep rates.

TEM of the as-deposited material revealed complete faulting; successive annealings at 2300K at the manufacturer and at NCSU to relieve residual stresses followed by deformation at 1973K and 182 MPa allowed a considerable decrease in the faulting, as may be seen in the lattice image of Fig. 7-A. The primary α polytype is 6 H. There is ≈ a 60:40 ratio of β to α polytypes in the experimental samples.

Dislocation slip bands were the predominant dislocation structure observed in the as-crept material at all stresses and temperatures. At the left side of Fig. 7-B, the dislocations within the slip band are shown to be forming an ≈500 nm cavity at a grain boundary. This cavity formation occurs at grain boundaries which have a high angle between the "basal" planes in samples which have undergone considerable deformation. The α lathes in this material restrict the number of slip systems. Thus when the dislocations reach these high angle boundaries, they are unable to move into the

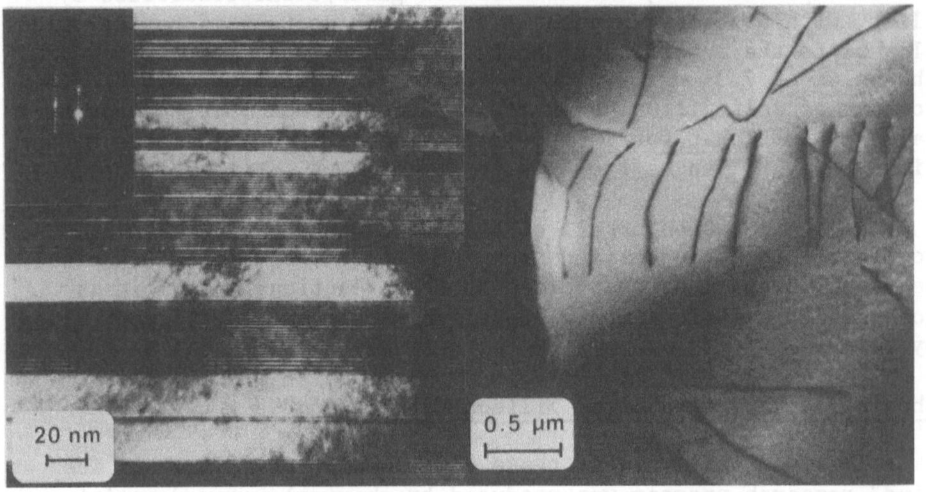

Figure 7. (A) Lattice image of CVD SiC grain after deformation at 1973K, 182 MPa. Dark areas are 6H α-SiC and light areas are β-SiC. (B) TEM micrograph of dislocations in a slip band forming a cavity at the grain boundary.

next grain since no "properly oriented" slip planes are available, and the applied stress is sufficient to overcome the increase in energy caused by the creation of new surface in the form of a cavity. Small angle x-ray scattering revealed that the volume fraction of the cavities was not greater than 10^{-4}. Thus, their existence does not contribute measurably to the deformation.

A Burgers vector analysis of the dislocations in the slip bands again revealed the presence of Shockley pairs of the type a/6 $\langle 2\bar{1}\bar{1} \rangle$ and a/6 $\langle \bar{1}2\bar{1} \rangle$. (The cubic system is used to index the dislocations in this material because of the predominance of the cubic β-SiC.)

The calculated value of the activation energy for steady-state creep in CVD SiC of 1.81 eV/atom is much lower than the activation energy values for self-diffusion or grain boundary diffusion of Si or C in α- or β-SiC[51]. On this basis, it is deemed unlikely that the controlling creep mechanism is diffusion-controlled. The range of values for the stress exponent--2.7 at temperatures below 1923K and 3.7 at 1923K--indicates that the controlling creep mechanism is either dislocation motion or grain boundary sliding. However, the results of the TEM study prove that the creep mechanism definitely involves dislocation glide and also rules out grain boundary sliding.

From the results of all of the aforementioned deformation and TEM research conducted on CVD SiC, it is concluded that below 1923K the controlling creep mechanism is dislocation glide with the glide rate being controlled by a mechanism with a relatively low activation energy. Concerning this mechanism, Weertman[52] has proposed that for materials with high Peierls stresses such as SiC, dislocation glide can be controlled by overcoming the Peierls stress hills. A rough approximation of the activation energy has been calculated from Seeger's expression[53] to be 1.15 eV/atom or 0.64 of the activation energy for creep in CVD SiC. This shows that the measured value is in the proper range to be attributed to double kink formation.

The increase in the stress exponent to 3.7 at 1923K and the presence of additional a/6<114> type dislocations only at this temperature indicate that the mechanism for creep may be changing to one involving climb. It is thus postulated that at these higher temperatures, the controlling mechanism for deformation is again becoming dislocation glide/climb controlled by climb.

Other Forms of Silicon Carbide

Commercial hot pressed SiC is produced from powders of alpha and normally contains approximately two vol% Al_2O_3 as a sintering aid. Farnsworth and Coble[54] found that a dense (>99% of theoretical density) form of this material undergoes viscous deformation in four point bending in argon between 2173K and 2473K at stresses up to ~200 MPa. From the calculated stress exponent value of 1 and the additional data of Francis and Coble,[55] grain boundary diffusion of C was tentatively concluded to be the rate controlling mechanism.

More recently investigators at the General Electric Company and the Carborundum Corp. have employed the additions of B and C as alternative additions to Al_2O_3 to promote densification of hot pressed and sintered β- and α-SiC, respectively. In compressive creep studies on variations of the former material between 1573K and 1773K at stresses between 500 MPa and 1700 MPa, Djemel et al. observed two stages in the $\ln\varepsilon$ vs $\ln\sigma$ curves. Below a critical stress, the values of the stress exponent and activation energy were 1 and ≈300 eV/atom, respectively. As the latter value is considerably less than the activation energies for lattice diffusion of C or Si in SiC,[51] the controlling mechanism in this stress regime was attributed to grain boundary diffusion (i.e., Coble creep). Some grain boundary separation also occurred which was caused by diffusion-controlled cavitation resulting from tensile stresses produced at suitably oriented boundaries. Above the critical stress, Coble creep was not sufficient to maintain the cohesion of the grains and boundary separation, again produced by cavitation, occurred and contributed increasingly to the deformation as a function of stress.

Somewhat similar creep research has been conducted in vacuum at temperatures between 1743-1973K by Grathwohl[56] using sintered SiC with small additions of B and Al (in the form of Al_4C_3 or AlN) and hot pressed SiC containing only the Al-containing additives. It was concluded that the creep behavior of the B-doped materials is governed by a lattice diffusion process while the results from the Al-doped sintered and hot pressed materials suggested to these investigators a boundary diffusion-controlled deformation process.

Summary

At this point, sufficient deformation studies and electron microscopy have been conducted on one reaction-bonded SiC (Norton NC-430) and CVD-SiC to elucidate the mechanisms of deformation in these two materials, albeit in a rather limited temperature range. Additional work with somewhat similar materials containing greater amounts of Si and over a wider temperature range would provide most valuable information for the commercial use of the materials at high temperatures. The kinetics of deformation and creep for various sintered SiC materials containing sintering aids such as B and Al have also now been reported for a range of conditions of stress (and resulting strain rate) and temperature. However, the mechanism(s) governing this deformation has so far been primarily inferred from the kinetic data without sufficient complimentary electron microscopy as additional evidence. Additional research must also be conducted on these last materials in order to determine the effect of the sintering aids on the deformation.

ACKNOWLEDGEMENTS

Portions of the research described in this manuscript were partially supported by the Basic Energy Sciences Division, USDOE, through the SHARE program under contract EY-76-C-05-0033 with Oak Ridge Associated Universities and under contract No. W-7405-eng-26 with Union Carbide Corporation, by the National Science Foundation under contract No. DMR-8022197 and by the Army Research Office under contract No. DAAG29-79-G-006.

REFERENCES

1. D.J. Rowcliffe, p. 49, these Proceedings.
2. L.E. Toth, Transition Metal Carbides and Nitrides, Academic Press, New York, 1971, p. 173.
3. P. Schwartzkopf and R. Kieffer, Refractory Hard Metals, MacMillan, New York, 1953.
4. J.H. Westbrook and E.R. Stover, "Carbides for High Temperature Applications," High Temperature Technology, ed. by I.E. Campbell and E.M. Sherwood, Wiley, New York, 1967.
5. (a)G. Hollox, Mat. Sci. Engr. 3, 121 (1968).
(b)G. Hollox, Nat. Bur. Stand. (U.S.), Sepec. Pub. 303, 201 (1969).
6. W.S. Williams, "Transition-metal Carbides," Progress in Solid State Chemistry 6, 57 (1971).
7. L.E. Toth, Transition Metal Carbides and Nitrides, Academic Press, New York, 1971.
8. G.V. Samsanov, Refractory Carbides, Consultants Bureau, New York, NY, 1974.
9. E.K. Storms, The Refractory Carbides, Academic Press, 1967.
10. W.S. Williams and R.D. Schaal, J. Appl. Phys. 33, 955 (1962).
11. W.S. Williams, J. Appl. Phys. 35, 1329 (1964).
12. K.H.G. Ashbee and R.E. Smallman, Proc. Roy. Soc. A274, 195 (1963).
13. G.E. Hollox and R.E. Smallman, Proc. Brit. Ceram. Soc. 6, 317 (1966).
14. (a)G.E. Hollox and R.E. Smallman, J. Appl. Phys. 37, 818 (1966).
(b)G.E. Hollox and R.E. Smallman, Proc. Brit. Ceram. Soc. 1, 211 (1964).
15. (a)G. Das, J. Less - Comm. Metals 83, L7 (1982).
(b)G. Das, K.S. Mazdiyasni and H.A. Lipsitt, J. Am. Cer. Soc. 65, 104 (1982).
16. D.K. Chatterjee, M.G. Mendiratta and H.A. Lipsitt, J. Mat. Sc. 14, 215 (1979).
17. H. Kurishita, K. Nakajima and H. Yoshinaga, Mat. Sci. Engr. 54, 177 (1982).
18. W.S. Williams in "Prop. Thermo. Phy. Struc. Derives Semi-metalliques," Central Nat. Rech. Scient., Paris, 1967, p. 181.
19. A. Kelly and D.J. Rowcliffe, J. Am. Cer. Soc. 50, 253 (1967).
20. J. Chermant, G. Leclerc and B.L. Mordike, Zeit. Metallknd. 71, 465 (1980).
21. (a)S. Sarian, J. Appl. Phys. 39, 3305 (1968).
(b)S. Sarian, Ibid. 40, 3515 (1969).
22. A. Kelly and D.J. Rowcliffe, Phys. Status Solidi 14, K29 (1966).
23. D.W. Lee and J.S. Haggerty, J. Am. Cer. Soc. 52, 641 (1969).
24. Yu. V. Miloserdin, K.V. Naboichenko, L.I. Laveikin and A.G. Bortsov, Strength. Mater. 4, 302 (1972).

25. S. Sarian and J.M. Criscione, J. Appl. Phys. 38, 1794 (1967).
26. R. Darolia and T.F. Archbold, J. Mat. Sci. 11, 283 (1976).
27. G.E. Hollox and J.D. Venables, Trans. Jap. Instit. Metals 9, 295 (1968).
28. J.L. Martin, B. Jouffrey and P. Costa, Phys. Stat. Sol. 22, 349 (1967).
29. J.L. Martin and B. Jouffrey, J. Phys. 29, 911 (1968).
30. J.L. Martin, P. Lacour-Gayet and P. Costa, J. Phys. E. 4, 615 (1971).
31. J.L. Martin, P. Lacour-Gayet and P. Costa, Proc. Fifth Int. Mat. Symp. "Electron Microscopy and Structure of Materials," ed. by G. Thomas, R.M. Fulrath and R.M. Fisher, Univ. of Calif. Press, Berkeley, 1972, p. 1131.
32. J.L. Martin, J. Micros. 98, 209 (1973).
33. S. Chevacharoenkul, Ph.D. Dissertation, North Carolina State University, Raleigh, NC, 1983.
34. A. Seeger in "Work Hardening," ed. by J.P. Hirth and J. Weertman, Gordon and Breach, New York, 1968, p. 27.
35. P.B. Hirsch and J.S. Lally, Phil. Mag. 12, 595 (1965).
36. N.F. Kocks, T. Hasegawa and R.O. Scattergood, Scripta Met. 14, 449 (1980).
37. D. Kuhlmann-Wilsdorf, Trans. AIME 224, 1047 (1962).
38. (a)B.B. Yu and R.F. Davis, J. Phys. Chem. Solids 40, 997 (1979).
(b)B.B. Yu and R.F. Davis, Ibid., 42, 83 (1981).
39. K. Yvon and E. Parthe, Acta Cristallogr. B26, 149 (1970).
40. J. Billingham and M.H. Lewis, Phil. Mag. 24, 231 (1971).
41. D.J. Rowcliffe and G. Thomas, Mat. Sci. Engr. 18, 231 (1975).
42. C. Allison, M. Hoffman and W.S. Williams, J. Appl. Phys. 53, 6757 (1982).
43. W.S. Williams, Trans. Metall. Soc. AIME 236, 211 (1966).
44. J.D. Venables, Phil. Mag. 16, 873 (1967).
45. J.D. Venables, Phys. Stat. Sol. 15, 413 (1967).
46. J.C.V. Rumsey and A.L. Roberts, Proc. Brit. Cer. Soc. 2, 233 (1967).
47. P. Marshall and R.B. Jones, Powder Metal. 12, 193 (1969).
48. V. Krishnamachari and M.R. Notis, Mat. Sci. Engr. 27, 83 (1977).
49. K. Schnurer, G. Grathwol, and F. Thummler, "Science of Ceramics 10," Berchtesgaden Vol. 2-4 (1979).
50. P. Marshall, "Special Ceramics," British Ceramics Association, 1967, p. 47.
51. J.D. Hong, M.N. Hon and R.F. Davis, Ceramurgia International 5, 155 (1979).
52. J. Weertman, J. Appl. Phys. 28, 1185 (1957).
53. A. Seeger, Phil. Mag. 1, 651 (1956).
54. P. Farnsworth and R.L. Coble, J. Am. Cer. Soc. 49, 264 (1966).
55. T.L. Francis and R.L. Coble, Ibid. 51, 115 (1968).
56. G. Grathwohl, private communication.
57. L.N. Demet'yev, Phys. Met. Metall. 46, 142 (1978).

STRUCTURE OF DISLOCATIONS IN OXIDES

T.E. Mitchell, W.T. Donlon*, K.P.D. Lagerlof and
A.H. Heuer

Department of Metallurgy and Materials Science
Case Western Reserve University
Cleveland, Ohio 44106

ABSTRACT

The atomic structure of dislocations in oxides, especially sapphire and spinel, are discussed. Observations of dislocations using high resolution electron microscopy (structure imaging and weak-beam imaging) show that dislocations introduced by deformation are dissociated in most cases by climb; glide dissociation is rarely observed. The observed stacking faults are such as to fault the cation lattice and leave the anion lattice undisturbed. This observation also applies to dislocation loops formed by defect agglomeration during radiation damage. Very little atomistic modelling of dislocations in oxides have been performed to help understand these effects. However some isotropic and anisotropic elasticity calculations have been performed in the present paper which show that the climb dissociated configuration has a slightly lower energy than the glide configuration and that the latter is metastable with respect to the former. Other factors influencing the structure and behavior of dislocations are discussed.

INTRODUCTION

The last two workshops on the interaction between dislocations and defects in oxides[1,2] were responsible for stimulating a great deal of discussion and assessment of our knowledge and understanding of this subject. There is no doubt that our knowledge in this area has increased enormously during this period. At the same time there are certain critical issues where a fundamental understanding is still lacking. One of these is the atomic structure of dislocations in oxides and its relationship to the stress re-

*Ford Motor Company, Dearborn, Michigan

quired to move dislocations and cause plastic deformation.

As described previously[1,2], the slip systems in oxides are generally well-behaved. The Burgers vector of the primary slip system is usually the shortest lattice vector (exceptions are TiO_2 and Y_2O_3): thus $\vec{b} = \frac{1}{2}\langle 110\rangle$ in fcc anion lattices (MgO, spinel) and $\vec{b} = 1/3\langle 11\bar{2}0\rangle$ in hcp anion lattices (Al_2O_3, BeO). The primary slip plane is usually the closest-packed plane although this is not always well defined. The primary slip systems are therefore $\{110\}\ \langle 1\bar{1}0\rangle$ in fcc anion lattices, the secondary systems being $\{111\}\ \langle 1\bar{1}0\rangle$ (spinel) and $\{001\}\ \langle 1\bar{1}0\rangle$ (MgO), and $(0001)\ \langle 11\bar{2}0\rangle$ in hcp anion lattices such as Al_2O_3 with $\{1\bar{2}10\}\ \langle 10\bar{1}0\rangle$ prismatic plane slip as the secondary system. With regard to dislocation dissociations, the shorter Burgers vector dislocations reveal no discernible dissocations (MgO, TiO_2, UO_2). Dissociations have been observed in dislocations with larger Burgers vectors, especially Al_2O_3 and $MgAl_2O_4$, but these have almost always been climb dissociations. The observed dissociations are the Kronberg reaction in Al_2O_3,

$$\frac{1}{3}\langle 11\bar{2}0\rangle \rightarrow \frac{1}{3}\langle 10\bar{1}0\rangle \text{ and } \frac{1}{3}\langle 10\bar{1}0\rangle \qquad (1)$$

and the Hornstra reaction in $MgAl_2O_4$,

$$\frac{1}{2}\langle 110\rangle \rightarrow \frac{1}{4}\langle 110\rangle + \frac{1}{4}\langle 110\rangle \qquad (2)$$

Further dissociations of these partials have not been seen. In both reactions the partials produce a fault only on the cation sub-lattice, leaving the anion sub-lattice undisturbed. This is also true of the long (8.2Å) $\langle 10\bar{1}0\rangle$ dislocation in sapphire which dissociates into 3 collinear partials[1]

$$\langle 10\bar{1}0\rangle \rightarrow \frac{1}{3}\langle 10\bar{1}0\rangle + \frac{1}{3}\langle 10\bar{1}0\rangle + \frac{1}{3}\langle 10\bar{1}0\rangle \qquad (3)$$

It is noteworthy that, in the smaller unit cell oxides such as MgO, TiO_2 and UO_2, dissociation is impossible without disturbing both sub-lattices.

There appears to have been little progress in the area of atomistic calculations of the structure of dislocations in oxides apart from the work of Puls[3] on MgO, although a good deal has been done on point defect calculations in various oxides, e.g.[4]. field appears to be ripe for such dislocation computations and for equivalent experimental work on structure imaging of the dislocation core using the latest generation of high resolution electron microscopes. This has been applied successfully to determining the core structure of dislocations in semi-conductor materials[5] and some preliminary results on sapphire and spinel will be presented in this paper. We will use these results and some additional observations as the basis for some further thoughts on climb and glide dissociation of dislocations in oxides, especially Al_2O_3 and $MgAl_2O_4$.

SPINEL

Deformation in spinel is a strong function of stoichiometry and temperature. In stoichiometric spinel ($MgAl_2O_4$), temperatures $\sim$ 1800°C are required for plasticity, while in non-stoichiometric crystals ($MgO \cdot nAl_2O_3, n>1$) plastic deformation can be achieved at lower temperatures and stresses. For stoichiometric (n=1) spinel, the primary slip system is {111} $<1\bar{1}0>$ followed closely by the {110} $<1\bar{1}0>$ secondary system which has almost the same critical resolved shear stress. For n>1, {110} $<1\bar{1}0>$ becomes the primary system. Transmission electron microscopy reveals that under most circumstances a network structure forms rapidly in these materials, a consequence of the relative ease of climb and multiple slip at these high temperatures. Such a network is seen in Fig. 1, which also shows that most of the dislocations are dissociated into colinear partials according to Eq. 1.

Detailed analysis of dislocation networks containing mixed edge and screw dislocations shows that the {100} stacking fault plane predominates regardless of stoichiometry, while other planes are commonly observed, even irrational ones. All of these faults are produced partly by climb and partly by glide. Annealing the dislocation networks tends to produce extensions on {112} and {113} fault planes. Isolated 60° dislocations in n>1 crystals formed via {111} slip have been observed to dissociate on $\{11\bar{2}\}$ planes by pure climb.

The nodes of the dislocation network are usually extended by climb into two partial nodes involving three types of ¼ <110> Burgers vector, as described by Donlon et al.[6]. A different type of extended node is shown in Fig. 2 in which all of the partials have the same ¼ $<1\bar{1}0>$ Burgers vector. In this case the slip plane is {110} and n = 1.7. This also shows a rare example of a screw dislocation dissociated by glide into ¼ $<1\bar{1}0>$ partials in the {110} slip plane. In fact the nodes appear to have formed by the climb dissociation of the edge portion of the dislocation which will be connected to the screw portion via a jog in one of the partials; further glide will then lead to a trailing faulted dipole giving the appearance of an extended node. Edge dislocations are also observed sometimes to be dissociated by glide; however this occurs in short segments ($\sim$ 0.2μm) (Fig. 3) and alternates with segments that are dissociated by a combination of glide and climb on the

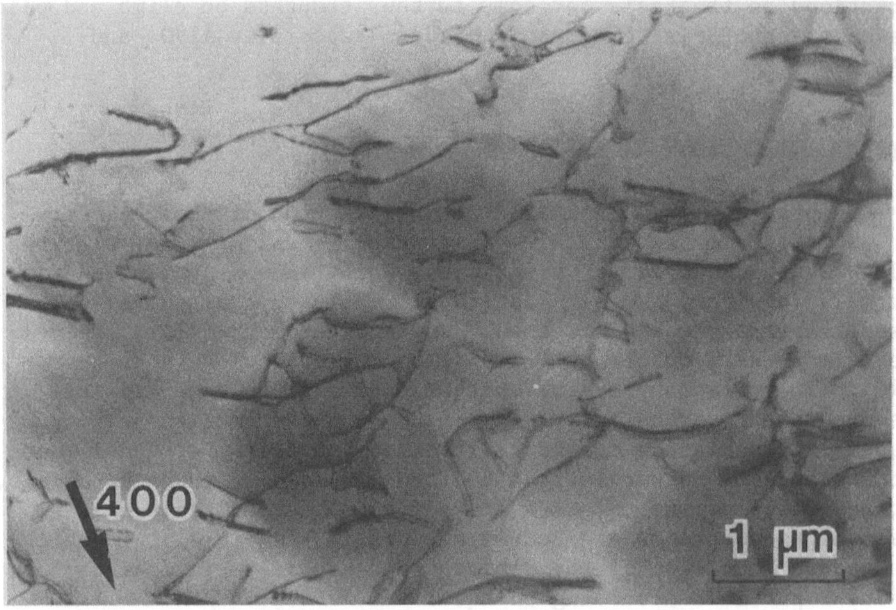

Fig. 1: Bright field micrograph of a network of dissociated dislocations in spinel (n=3.5) deformed at 1400°C.

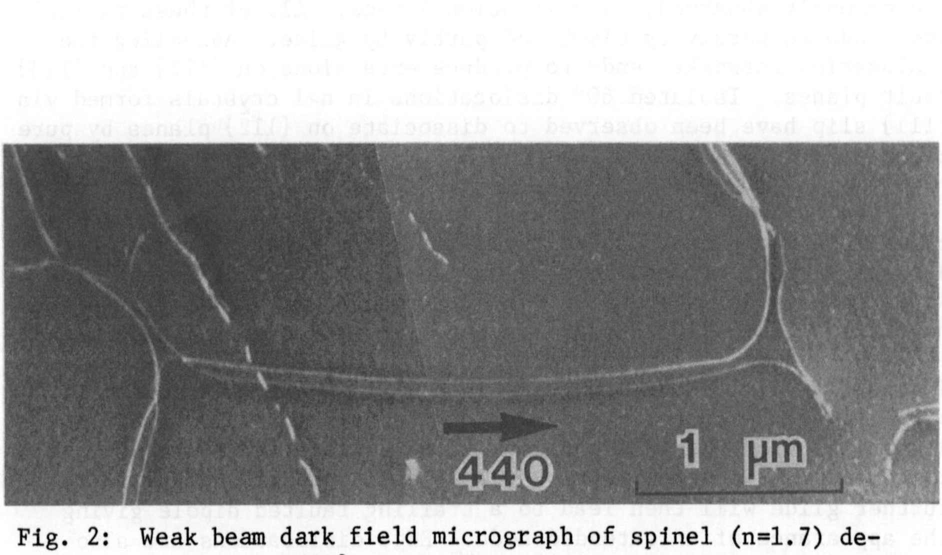

Fig. 2: Weak beam dark field micrograph of spinel (n=1.7) deformed at 1350°C. The long dislocation is an extended screw connecting two nodes.

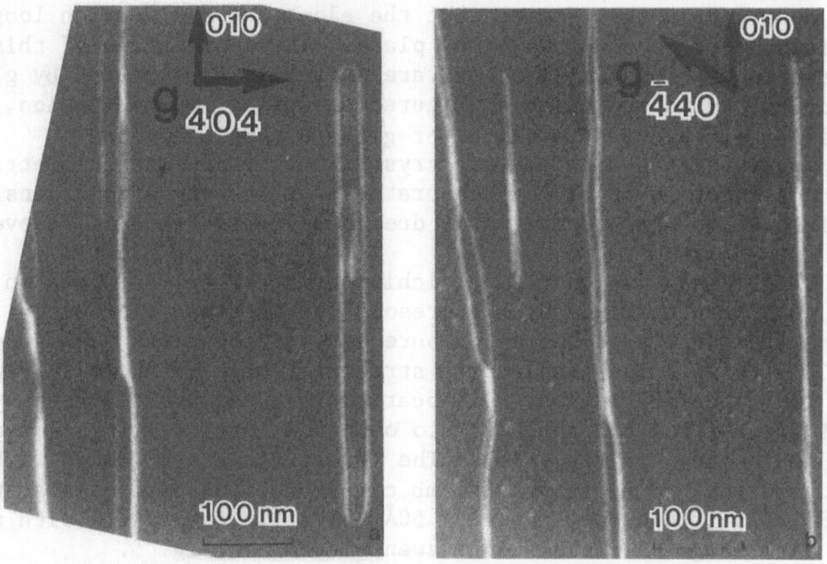

Fig. 3: Weak beam micrograph of spinel (n=1.7) deformed at 1350°C. In (a), segments of edge dislocation are seen dissociated by climb/glide on {100}. In (b) which is tilted by ∿ 45°, the other segments are seen to be dissociated by glide on {110}.

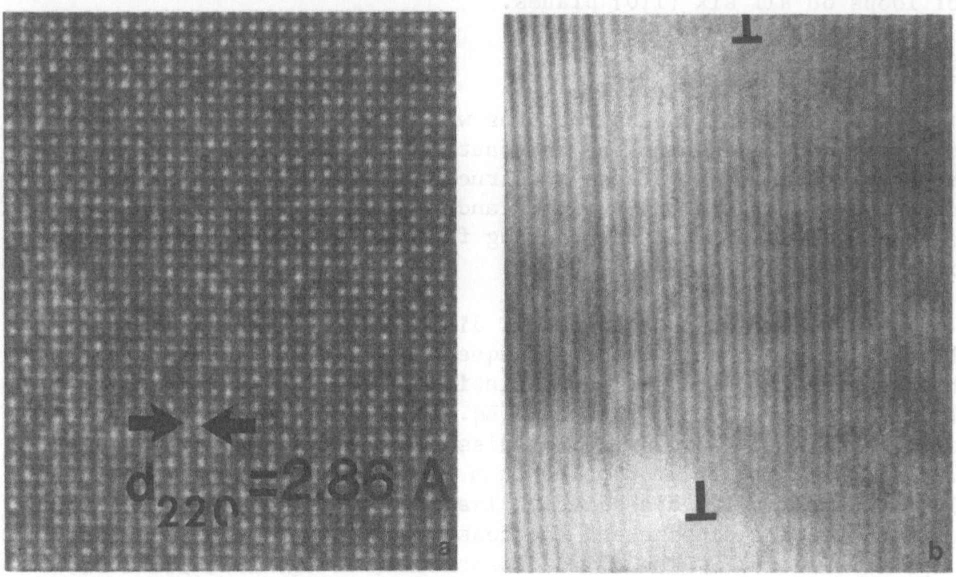

Fig. 4: (a) Structure image of perfect spinel region along <100>. (b) Lattice fringe image tilted slightly off <100> showing end-on view of climb dissociated dislocation.

{100} plane (inclined 45° with the slip plane). In addition it is also observed in this figure that the elongated dislocation loop is dissociated on the same {100} plane. The implication of this observation is that dislocations are initially dissociated by glide and transform to the climb configuration during immobilization, as discussed later. No evidence of glide dissociation has been observed in stoichiometric crystals or in non-stoichiometric crystals deformed at higher temperatures or showing a high density of dislocation nodes, presumably due to a greater rate of recovery.

Climb dissociation in a stoichiometric crystal deformed on {110} has been examined by high resolution electron microscopy (HREM) from structure images of pure edge dislocations seen end-on in <001> foils. An example of a structure image is shown in Fig. 4(a) for a perfect region. Dislocations are more easily seen by tilting away from the zone axis to obtain a lattice fringe image, such as is shown in Fig. 4(b). The two partials are marked with a ⊥ and are seen to be in pure climb configurations with a separation of 18 {220} fringe spacings or ∿ 50Å. This is compatible with the weak beam image of dislocations seen side-on in Fig. 5.

A final example of the formation of faults in spinel is under irradiation conditions where the primary interstitial defects combine together in stoichiometric proportions to form faulted ¼ <110> dislocation loops on {110} planes[7]. The loops themselves are clustered together around a common nucleus to give rosettes of loops on all six {110} planes.

Sapphire

The dislocation substructures which results from the plastic deformation of sapphire are as beautiful and varied as the natural crystals themselves. These substructures are described in more detail elsewhere in this volume[8] and we will concentrate here on extended dislocations and stacking faults which have been observed in sapphire.

Under basal slip conditions, dislocation motion is limited by dipole formation and the subsequent creation of loops by diffusive break-up of the dipoles. Dissociation of the 1/3 <11$\bar{2}$0> dislocation by glide into Kronberg partials (Eq. 1) has not yet been detected with certainty. However, climb dissociation occurs quite readily although more often in dipoles than in isolated dislocations. In fact their climb dissociation leads to annihilation of the inner partials of the dipoles and the consequent formation of faulted dipoles (Fig. 6)[9]. The climb dissociation of the individual dislocations of the unfaulted dipole is shown in Fig. 7. Initial faulting occurs on the {11$\bar{2}$0} plane but the faulted dipoles rotate onto the {10$\bar{1}$0} plane into the edge configuration. All of these processes

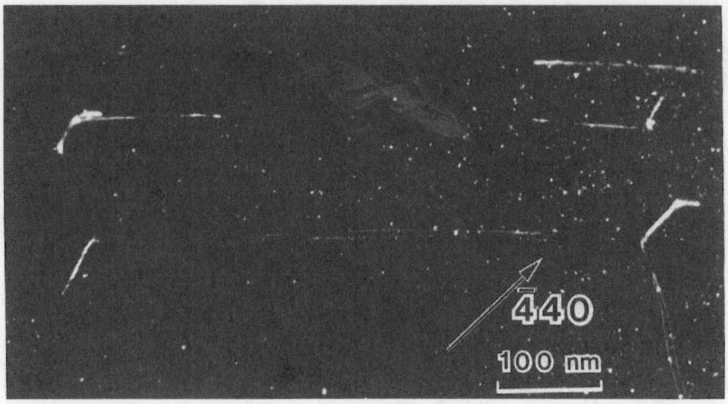

Fig. 5: Weak beam micrograph showing edge-on view of climb dissociated dislocation in stoichiometric spinel.

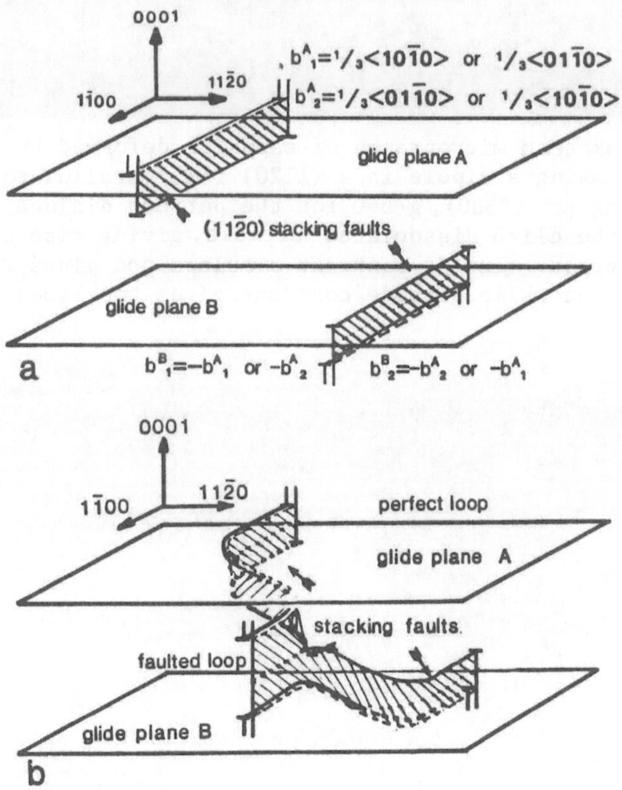

Fig. 6: Diagrams illustrating (a) climb dissociation of the individual edge dislocations in a dipole, and (b) annihilation of a section of the central partials to form a faulted loop at one end of the dipole.

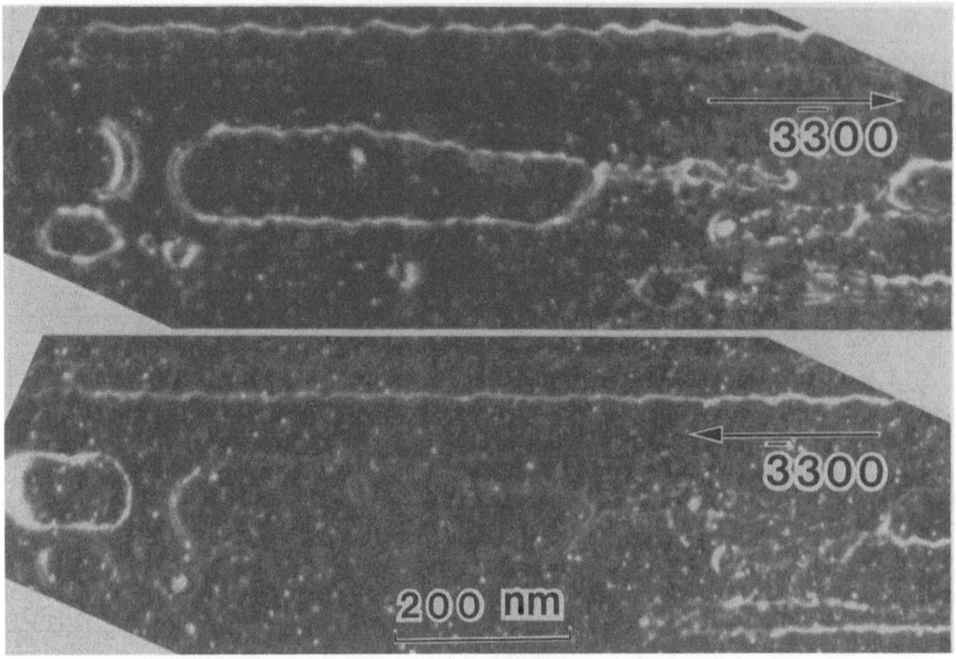

Fig. 7: Weak beam micrographs of sapphire deformed by basal slip at 1400°C showing a dipole in a (11$\bar{2}$0) foil parallel to the fault plane. Using $\vec{g}=\pm(3\bar{3}00)$, $\vec{g}\cdot\vec{b}=0$ for the perfect dislocation but $\vec{g}\cdot\vec{b}=\pm1$ for the climb dissociated partials, giving rise to inside/outside contrast. Inside contrast overlaps and gives rise to only one bright line while outside contrast gives two lines.

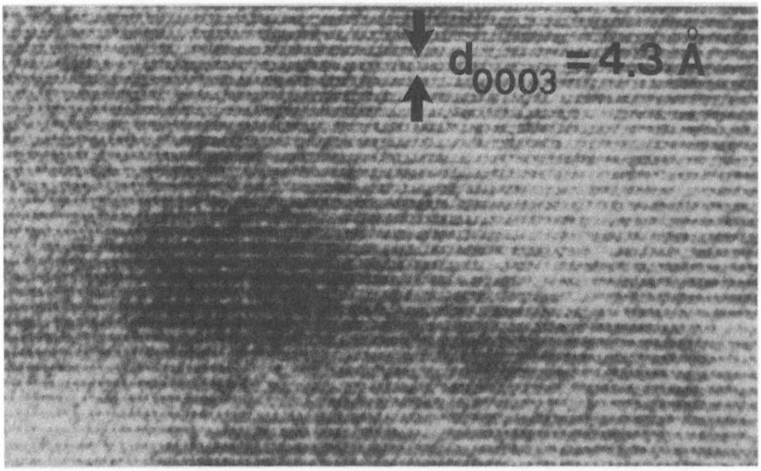

Fig. 8: Structure image of sapphire along <1$\bar{1}$00> showing strain contrast due to a dislocation but the dissociation cannot be discerned.

can occur by self-climb (pipe diffusion) or by fast cation diffusion across the faults.

Prism plane slip on $\{1\bar{2}10\}$ planes gives rise to $\langle 10\bar{1}0\rangle$ dislocations which are able to dissociate by glide into 3 collinear $1/3\ \langle 10\bar{1}0\rangle$ partials according to Eq. 3. Other configurations have been observed and analyzed which show the same dissociation by pure climb or part glide and climb on $\{11\bar{2}0\}$ and $\{10\bar{1}0\}$ planes. Calculated stacking fault energies are all in the range of 0.15 to 0.35 J/m^2, independent of fault plane[12].

Stacking faults are also observed under irradiation conditions in which the interstitials combine together in stoichiometric proportions to form both 1/3 [0001] dislocation loops on (0001) planes and $1/3\ \langle 10\bar{1}0\rangle$ loops on $\{10\bar{1}0\}$ planes[10]. In both cases the loops tend to unfault at a certain size (∿ 50 nm diameter) by the reaction

$$\frac{1}{3}\,[0001] \;+\; \frac{1}{3}\,\langle 10\bar{1}0\rangle \;\rightarrow\; \frac{1}{3}\,\langle 10\bar{1}1\rangle$$

Basal faults are thus able to form under irradiation conditions but not by glide dissociation, as discussed in the next section.

HREM imaging was also attempted on sapphire by structure imaging in the $\langle 1\bar{1}00\rangle$ direction. An example is shown in Fig. 8; the perfect regions and strained regions are seen very clearly. However, we have been unsuccessful so far in interpreting these and other images to see whether glide or climb dissociation is present. Presumably it is necessary to image perfectly straight edge dislocations end-on to obtain understandable structures. We have also searched for dissociation in sapphire deformed at lower temperatures (800°C) under high confining pressures. Under these conditions, extensive twinning on the basal and pyramidal planes occur as well as basal and prismatic slip. In spite of the high stresses involved (∿ 1500 MPa), glide dissociation of basal dislocations is still not apparent (Fig. 9). Very narrow dipoles are observed as well as strings of loops, many of which are faulted, implying that climb dissociations can still occur.

INTERACTION ENERGY BETWEEN PARTIALS

One of the major puzzles of the structure of dislocations in sapphire and spinel is the preponderance of climb over glide dissociation. Let us first examine the interaction energy between partials to see whether elasticity theory can in any way explain this behavior. The interaction energy per unit length is[11]:

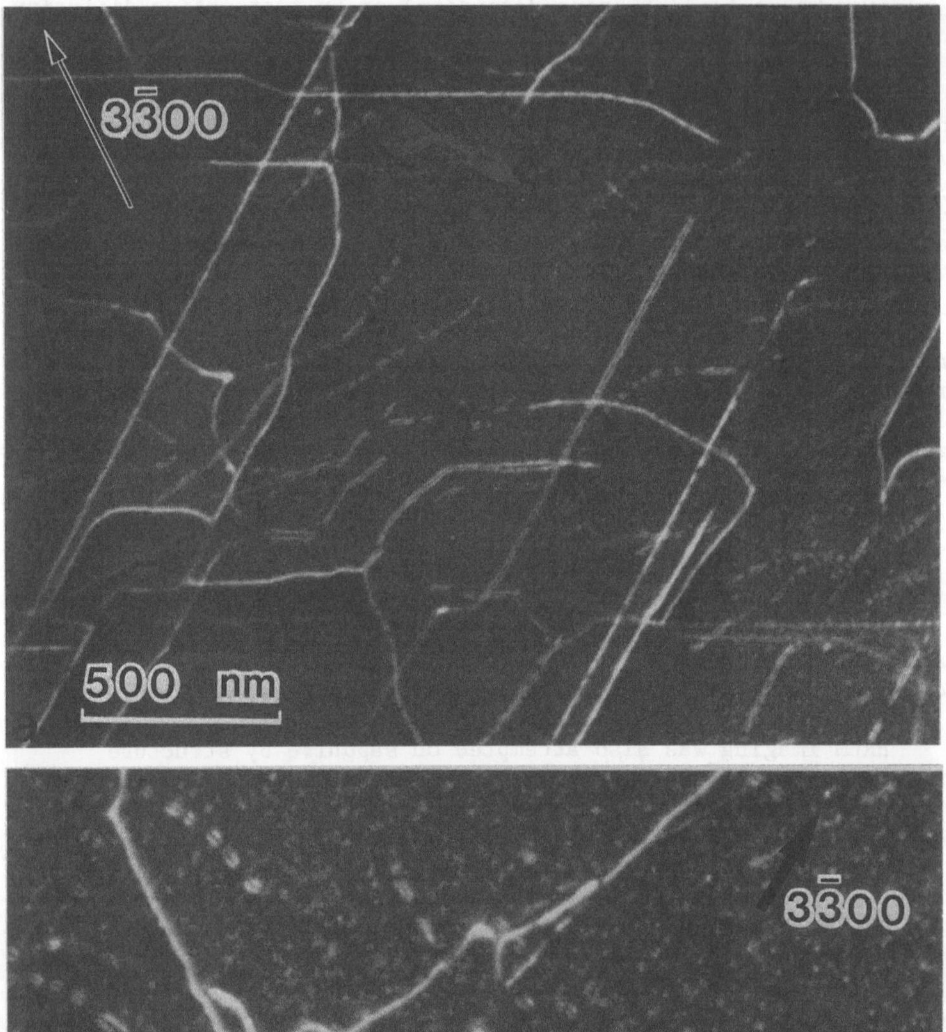

Fig. 9: Weak beam micrographs from basal foil of sapphire deformed at 800°C: (a) edge and 30° dislocations parallel to <10$\bar{1}$0> plus narrow dipoles and strings of mostly faulted loops; (b) high magnification view with no discernible dissociation of dislocations.

$$\frac{W_{12}}{L} = \frac{-\mu(\vec{b}_1\cdot\vec{\xi})\ (\vec{b}_2\cdot\vec{\xi})}{2\Pi} \ln\frac{R}{R_a} - \frac{\mu}{2\Pi(1-\nu)R^2}[(\vec{b}_1 x\vec{\xi})]\cdot(\vec{b}_2 x\vec{\xi})] \ln\frac{R}{R_a}$$

$$-\frac{\mu}{2\Pi(1-\nu)R^2}[(\vec{b}_1 x\vec{\xi})\cdot\vec{R}]\ [(\vec{b}_2 x\vec{\xi})\cdot\vec{R}] \qquad (4)$$

where μ is the shear modulus, ν is Poisson's ratio, $\vec{\xi}$ is the line direction, b_1 and b_2 are the partial Burgers vectors, R is the separation, and R_a is the outer cut-off. For edge dislocations dissociating into partials with an angle $\emptyset$ to the dislocation line (i.e. $\emptyset$ = 90° for spinel and 60° for sapphire) on a plane at an angle θ to the glide plane (θ = 0 for glide dissociation and 90° for climb dissociation), Eq. 4 becomes:

$$\frac{W_{12}}{L} = \frac{\mu b_1 b_2}{2\Pi}[(\cos^2\emptyset - \frac{\sin^2\emptyset}{1-\nu}) \ln\frac{R}{R_a} - \frac{\sin^2\emptyset\sin^2\theta}{1-\nu}] \qquad (5)$$

To this should be added the term due to the stacking fault energy γ

$$\frac{W_{SF}}{L} = \gamma R \qquad (6)$$

The self energy terms are constant and so the equilibrium separation, R_{eq}, can be determined by minimizing the sum of Eq. 5 and 6, giving

$$R_{eq} = \frac{\mu b_1 b_2}{2\Pi\gamma}[\frac{\sin^2\emptyset}{1-\nu} - \cos^2\emptyset] \qquad (7)$$

Note that R_{eq} is independent of θ, i.e. independent of the proportion of glide and climb dissociation except, of course, for the dependence of γ on θ.

The sum of the two variable energy terms is then:

$$\frac{W_{12}}{L} + \frac{W_{SF}}{L} = \gamma R_{eq}(\ln\frac{R_a}{R_{eq}} + 1) - \frac{\mu b_1 b_2\sin^2\emptyset\sin^2\theta}{2\Pi(1-\nu)} \qquad (8)$$

If it is assumed for the moment that γ is independent of θ, Eq. 8 clearly shows that the energy is a minimum for θ = 90°, i.e. for the climb configuration. Furthermore, we can take the glide configuration and show that is is unstable with respect to climb: the climb force on one of the partials is

$$\frac{F_y}{L} = \frac{\mu b_1 b_2 y\ (3x^2 + y^2)}{2\Pi(1-\nu)\ (x^2+y^2)^2} \qquad (9)$$

where (x,y) are the separation coordinates in the glide and climb directions respectively. Equation 9 shows that $F_y = 0$ at $y = 0$ but that $\partial/\partial y\ (F_y) > 0$ at $y = 0$, implying that the glide configuration is a position of unstable equilibrium with respect to climb. Likewise it can be shown that, for climb dissociation, the glide force $F_x = 0$ at $x = 0$ and that $\partial/\partial x\ (F_x) < 0$ at $x = 0$, so that the climb configuration is a position of stable equilibrium.

ANISOTROPIC ELASTICITY

The configurations can also be examined with anisotropic elasticity theory for particular geometries. One relatively easy example is the ½ [1$\bar{1}$0] edge dislocation in spinel lying on the (110) plane in the [001] direction. The stress-fields are given by equations (13-118) in Hirth and Lothe[11] from which the equilibrium glide separation is given when

$$\gamma = \sigma_{xy} b_2 = \frac{M b_1 b_2 c_{44}}{2\Pi\ x_{eq}} \tag{10}$$

where

$$M = \left[\frac{(\bar{c}'_{11} + c_{12})(\bar{c}'_{11} - c_{12})}{c'_{22}\, c_{44}\, (\bar{c}'_{11} + c_{12} + 2\, c_{44})}\right]^{1/2}$$

$$\bar{c}'_{11} = (c_{11} \cdot c'_{22})^{1/2},$$

$$\text{and } c'_{22} = c_{11} + 1/2H = 1/2\, c_{11} + 1/2\, c_{12} + c_{44} \tag{11}$$

Similarly the equilibrium climb separation is given when

$$\gamma = -\sigma_{xx} b_2 = \frac{M b_1 b_2 c_{44}}{2\Pi\ y_{eq}} \tag{12}$$

which is identical to Eq. 10 providing, as before, that γ is the same. This implies that anisotropic elasticity considerations are unlikely to make significant changes to the conclusions from the isotropic elasticity analysis, namely that the climb configuration has the lowest energy and that the glide configuration is unstable with respect to climb.

ANISOTROPY OF THE STACKING FAULT ENERGY

The above discussion has ignored the important issue of the anisotropy of the stacking fault energy. For example, in sapphire there is a strong preference for stacking faults to lie on $\{11\bar{2}0\}$, $\{10\bar{1}0\}$ and (0001) planes: edge dislocation dipoles on the basal planes tend to dissociate by climb on $\{11\bar{2}0\}$ planes to form faulted dipoles which then rotate into $\{10\bar{1}0\}$ planes; prism plane dislocations dissociate into 3 collinear $1/3 \langle 10\bar{1}0 \rangle$ partials by glide on the $\{1\bar{2}10\}$ plane and by climb on the $\{10\bar{1}0\}$ plane; and the interstitial loops formed during irradiation are on either the (0001) or $\{10\bar{1}0\}$ planes. Lagerlof, et al.[12] have analyzed these various configurations and have estimated stacking fault energies in the range 0.15-0.35 J/m^2 with little anisotropy, especially between $\{11\bar{2}0\}$ and $\{10\bar{1}0\}$. Howitt and Mitchell[10] use the unfaulting reaction

$$1/3 \langle 10\bar{1}0 \rangle + 1/3\,[0001] \rightarrow 1/3 \langle 10\bar{1}1 \rangle$$

of the faulted loops formed by irradiation to show that γ on the $\{10\bar{1}0\}$ planes should be about twice that on the (0001) plane. Lagerlof, et al.[12] also point out that there are various types of faults that can form; for example one kind of vacancy fault and two kinds of interstitial faults can form by climb on $\{10\bar{1}0\}$ planes. The same applies to faults on the (0001) plane; this suggests a possible reason for the apparent absence of basal glide dissociation, namely that the shear fault on the basal plane has a higher energy than the interstitial fault on the basal plane or the various faults on the prism planes. However, in fact, the interstitial fault on the basal plane is intrinsic[10] and the same fault can be formed by shear. There may be some mechanical reason for the apparent absence of basal glide dissociation such as a very high Peierls stress on the $1/3 \langle 10\bar{1}0 \rangle$ partial dislocation compared with the $1/3 \langle 11\bar{2}0 \rangle$ perfect dislocation. If this is the case, the situation apparently persists to low temperatures where the stresses are extraordinarily high ($\sim$ 1500 MPa) but still insufficient to produce glide dissociation.

With regard to anisotropy of the stacking fault energy in spinel, the experimental observations are a strong preference for {100} fault planes, frequent occurrence of {110}, {112} and {113} fault planes and the apparent ease with which faults curve and twist from one plane to another. In addition, for a given stoichiometry, the partial dislocation separation appears to be independent of the fault plane. The faults in spinel are thus much less crystallographic than in sapphire and are rather reminiscent of anti-phase boundaries in ordered alloys. As already described, the search for glide dissociation was successful in spinel (at least for n = 1.7 deformed at lower temperatures) and the interaction energy analysis given above allows us to explain the much more common observation of climb dissociation. It is apparent that glide dislocations in spinel <u>are</u> dissociated, unlike sapphire, and that, once glide dislocations stop, they transform into the lower energy climb configuration either by bulk diffusion or by pipe diffusion. The stacking fault energy is approximately isotropic and so there is no barrier to this transformation. However, there is a preference for {100} faults implying that there is a minimum in the fault energy for this plane. The same applies to the other planes, especially {110} which is the dominant plane for 1/4 <110> interstitial loops in irradiated spinel.

A final point about the relative importance of the anisotropy of the fault energy and the interaction energy can be made by rewriting Eq. 7 as:

$$\frac{W_{12}}{L} + \frac{W_{SF}}{L} = \gamma R_{eq} \left(\ln \frac{eR_a}{\gamma R_{eq}} + \ln \gamma - \sin^2 \theta\right) \qquad (13)$$

(This is for spinel; the equation for sapphire is slightly different.) γR_{eq} is constant in Eq. 13 (see Eq. 6) and the driving force for climb is represented by the $\sin^2 \theta$ term; the presence of the $\ln\gamma$ terms shows that climb dissociation will occur if $\ln\gamma_c - \sin^2 90^0 < \ln\gamma_g - \sin^2 0$, i.e. the ratio of the climb to glide fault energy $\gamma_c/\gamma_g < 2.7$. The fact that the {100} 45° fault is observed in many instances rather than the pure climb {100} fault means that $\gamma_{100}/\gamma_{100} \lesssim 0.6$.

GLIDE OF DISSOCIATED DISLOCATIONS

The glide dissociation of dislocations is generally considered to make glide easier at low temperatures by reducing the Peierls stress. This is probably the case for the <10$\bar{1}$0> dislocation in sapphire. Climb dissociation of dislocations, on the other hand, should make glide more difficult. In fact Duclos et al.[13] propose that such dislocations in spinel will be forced to glide by a viscous motion involving cation diffusion across the fault and that this can explain the softening of spinel with increasing n. However, it seems more likely that the climb configuration can transform into the glide configuration simply by one partial moving by glide and the other partial climbing into the same plane (Castaing's "flag" mechanism[14]). Another possibility is that glide motion can be nucleated at the remaining screw segments in the dislocation network which by their nature can only be dissociated by glide, as suggested by Hirsch[15]. The reason for compositional softening of spinel remains obscure but must be related to the interaction of the structural vacancies with the dislocation core.

ACKNOWLEDGEMENT

This research was supported in part by the NSF under Grant No. INT 76-10829 (sapphire) and in part by the DOE under Grant No. EG-77-S-02-42]7,000 (spinel).

REFERENCES

1. T.E. Mitchell, L.W. Hobbs, A.H. Heuer, J. Castaing, J. Cadoz and J. Philibert, "Interaction Between Point Defects and Dislocations in Oxides", Acta Met. 27; 1677 (1979).
2. M.G. Blanchin, J. Castaing, G. Fontaine, A.H. Heuer, L.W. Hobbs and T.E. Mitchell, "The Interactions Between Dislocations and Defects in Oxides", J. de Phys. 42:C3-1 (1981).
3. M.P. Puls, "Atomistic Models of Point Defect/Dislocation Interactions in Oxides", J. de Phys. 42:C3-13 (1981).
4. C.R.A. Catlow, R. James, W.C. Mackrodt and R.F. Stewart, "Defect Energetics in α-Al_2O_3 and Rutile TiO_2", Phys.Rev. B25:1006 (1982).
5. A. Olsen and J.C.H. Spence, "Distinguishing Dissociated Glide and Shuffle Set Dislocations by High Resolution Electron Microscopy", Phil. Mag. A43:945 (1981).
6. W.T. Donlon, T.E. Mitchell and A.H. Heuer, "Climb Dissociation of Network Dislocations in Non-stoichiometric Mg-Al Spinel", Phil. Mag. 40:351 (1979).
7. F.W. Clinard, G.F. Hurley and L.W. Hobbs, "Neutron Irradiation Damage in MgO, Al_2O_3 and $MgAl_2O_4$ Ceramics", J. Nucl. Mater. 108 & 109:655 (1982).
8. K.P.D. Lagerlof, T.E. Mitchell and A.H. Heuer, "The Deformation of and Lattice Diffusion in Sapphire (α-Al_2O_3), this volume.
9. T.E. Mitchell, B.J. Pletka, D.S. Phillips and A.H. Heuer, "Climb Dissociation of Dislocations in Sapphire (α-Al_2O_3)", Phil. Mag. 34:441 (1976).
10. D.G. Howitt and T.E. Mitchell, "Electron Irradiation Damage in α-Al_2O_3", Phil. Mag. A44:229 (1981).
11. J.P. Hirth and J. Lothe, "Theory of Dislocations", 2nd Edition, J. Wiley & Sons, New York (1982).
12. K.P.D. Lagerlof, T.E. Mitchell, A.H. Heuer, J.P. Riviere, J. Cadoz, J. Castaing and D.S. Phillips, "Stacking Fault Energy in Sapphire (α-Al_2O_3)", submitted to Acta Met.
13. R. Duclos, N. Doukhan and B. Escaig, "Study of the Origin of the Composition Influence on the Mechanical Properties of $MgO \cdot nAl_2O_3$ Spinels", Acta Met. 30:1381 (1982).
14. J. Castaing, private communication.
15. P.B. Hirsch, private communication.

THE EFFECTS OF NONSTOICHIOMETRY ON THE DEFORMATION OF OXIDES

J. Castaing, A. Dominguez-Rodriguez* and C. Monty

Laboratoire de Physique des Matériaux - C.N.R.S.

1, Place Aristide Briand, 92195 MEUDON Cedex (France)

ABSTRACT

The role of non-stoichiometry during the high temperature deformation of oxides is examined, in relation to the various creep mechanisms. It is shown that in NaCl-type oxides, Cu_2O and UO_2, the creep-rate is controlled by the slowest diffusing species. The relation between creep and P_{O2} helps to characterize the diffusion mechanism.

INTRODUCTION

Many investigation on the plastic deformation of oxides have been performed during the past ten years. On the occasion of two workshops (Mitchell et al. 1979, Blanchin et al. 1981), the role of point defects has been discussed for both high and low temperature deformation. Isolated or clustered point defects can act either as obstacles to dislocation glide or as elementary steps for diffusion. This diffusion process then allows for dislocation climb and recovery, or for transport of matter and creep (Nabarro or Coble creep).

Point defects are often associated to departure from stoichiometry in many compounds where it is often viewed as non-ideal composition, characterized by x in $A_{1-x}B$. The thermodynamic equilibrium of such a compound is fixed by two parameters. When the components A and B have a low vapour pressure (e.g. metal and graphite for a carbide), x is usually fixed at the stage of preparation and cannot be changed. The situation is quite different for oxides where the

*Permanent adress : Departamento de Estado Solido, Facultad de Fisica, Apartado 1065, Sevilla (Spain).

oxygen pressure P_{O2} is large and the two thermodynamic conditions T and P_{O2} can be easily controlled. In most of the usual experimental conditions, the composition can be changed by a modification of either T, or P_{O2} ; experiments at constant x require variations of both T and P_{O2} and should be avoided (see appendix 1).

At low enough T, x cannot be changed because of low diffusivity. This leads to uncertain and irreproducible conditions. During cooling, the crystal surface equilibrates with the surrounding gas and the point defect concentrations vary. Equilibration, which occurs by diffusion, is finally stopped and a cold crystal is heterogeneous in defect concentrations in a way depending on the cooling rate. We, therefore, do not discuss the low T properties.

We restrict ourselves to review the properties of non-stoichiometric oxides which are in thermodynamic equilibrium, that is at high enough temperatures, so that the point defect diffusivity allows the solid to equilibriate. These defects can be isolated or in associated form (defect complexes) (Table 1). The association of defects is not easy to observe at high temperature ; it has been found in some cases at large x (FeO, UO_2) and suggested for small x on the basis of calculations (NiO) (Catlow 1981). Alternatively, long range ordering or crystallographic shearing occurs, in particular to accomodate large departures from stoichiometry ; we do not consider this case in the following.

Small departures from stoichiometry are limited by impurity levels, which can dominate the point defect behaviour. Although this may lead to "extrinsic" or "simulated" non-stoichiometry, we also leave these aspects out of the present review. We only discuss the case of oxides presenting departure from stoichiometry larger than the impurity content ($\sim 10^{-5}$). Among the various examples, we have taken NaCl-type transition metal oxides, Cu_2O and UO_2 (Table 1). The studies on these compounds have been recently reviewed in particular by T. Bretheau et al. (1979), J. Philibert (1984), A. Dominguez-Rodriquez and J. Castaing (1983) and at the two already mentioned workshops (Mitchell et al. 1979 ; Blanchin et al. 1981). Therefore, we do not intend to cover extensively the results on deformation of non-stoichiometric oxides, but rather discuss the relations between deformation and elementary processes. Among them, diffusion is most often invoked to be the rate controlling one. Although this is supported by some experimental evidence , it does not give a general picture of the situation and we discuss alternative processes which may determine the relation between deformation rate and parameters of the mechanical test.

For that purpose, besides experimental observations, it is necessary to have an appropriate equation which allows to characterize and analyse the mechanical data. We, thus, discuss first the aspect of "creep" models and creep equations.

TABLE 1. Non-stoichiometric oxides MO : majority and creep determined minority point defects ; metal D_M and oxygen D_{ox} diffusion coefficients at 0.85 T_M (Melting temperature) and indicated oxygen chemical activity. Also shown the ratio R of the coefficients of the low over fast diffusing species. The ionization α of majority point defects varies from zero to one or two.

	T_M °C	O_M Min.	O_M Max.	Majority Defect	Min. Def.	D_M cm^2/s	D_{Ox} cm^2/s	R
$Mn_{1\pm x}O$	1842	1.00	1.18	$V_{Mn}^{\alpha'}$ complexes		9.5×10^{9} (4×10^{-12})		
$Fe_{1-x}O$	1374	1.045	1.2	complexes	O_i''	2×10^{-7} (10^{-9})	10^{-11}	5×10^{-5}
$Co_{1-x}O$	1810	1.000	1.012	$V_{Co}^{\alpha'}$	O_i^{*} $V_O^{\cdot\cdot}$(?)	9×10^{-8} (1)	3×10^{-12}	3×10^{-5}
$Ni_{1-x}O$	1957	1.000	1.001	$V_{Ni}^{\alpha'}$	$V_O^{\cdot\cdot}$	4×10^{-9} (1)	7×10^{-14}	2×10^{-5}
$Cu_{2-x}O$	1242	.500	.501	$V_{Cu}^{\alpha'}$	O_i'	6.3×10^{-8} (1)	1.8×10^{-9}	0.03
$UO_{2\pm x}$	2840	1.65	2.25	complexes	V_U	10^{-11} ($x \simeq 0$)	3×10^{-6}	4×10^{-6}

REQUIREMENT FOR CREEP DATA ACQUISITION

Creep equations

During steady state deformation, the fit of the mechanical data with a creep law is one of the tools to identify the rate controlling mechanism. It has to be supported by microstructural observations to detect the absence or occurence of grain boundary sliding, recrystallization, dislocation glide and multiplication, recovery, ... Among the large number of equations (Bretheau et al. 1979; Walser et al. 1982), the most used one has the form :

$$\dot{\varepsilon} = A \left(\frac{b}{d}\right)^h \frac{\mu b}{kT} \left(\frac{\sigma - \sigma_o}{\mu}\right)^n D \qquad (1)$$

where b is the Burgers vector, μ the elastic shear modulus ; σ_o is a threshold stress below which the creep rate $\dot{\varepsilon}$ is null ; D is an effective diffusion coefficient and d is the grain size.

Plastic deformation by transport of matter or by processes involving dislocation climb occurs without decomposition of the oxides ; the flux of the components must be in the same ratio as in the compound ($J_A/J_B = 1$ in AB). To such diffusion, it corresponds a molecular diffusion coefficient which is generally simplified to that of the slowest species. Bretheau et al. (1979), Philibert (1984) discussed this point in relation with plastic deformation. In the bulk, oxygen diffuses the most slowly in all non-stoichiometric oxides discussed in this article, except in UO_2. The ratio R of diffusivities of the two species is shown in table 1.

The complexity increases considerably when one takes into account the various paths for the diffusion ; for bulk (D_B) and pipe (D_p) diffusion, D can be written, with a number of assumptions,

$$D = D_B\ (1 + f\ D_p/D_B) \qquad (2)$$

where f is the fraction of atom sites for pipe diffusion. Ashby et al. (1975) have reviewed this aspect for plastic deformation. The factor f can be evaluated for dislocation creep as

$$f \sim 10\ (\sigma/\mu)^2 \qquad (3)$$

if diffusion occurs along a cylinder of cross-section area b^2 around dislocations in density proportional to σ^2. For pure diffusional creep of polycrystals, the path is along boundaries of thickness δ between grains of size d, giving :

$$f \sim \pi\delta/d \qquad (4)$$

(for spherical grains, $f = 6\delta/d$).

In binary non-stoichiometric oxides $M_{1-x}O$ the bulk diffusion coefficient is of the form :

$$D_B = c_o \exp - \frac{\Delta G_m}{kT} = D_B^o \left(\frac{P_{O2}}{P^*_{O2}}\right)^m \exp - \frac{\Delta H}{kT} \qquad (5)$$

P^*_{O2} = reference pressure

In simple cases, m is characteristic of a single type of point defects. The concentration c_o of these point defects associated to the slow species is much smaller than that of the point defects associated to the fast species. We call them, respectively, minority and majority defects. There is now a reasonable amount of data on bulk diffusion in oxides, which have been reviewed by Monty (1983) for cubic oxides and Matzke (1981). Much less is known about pipe diffusion. It is generally accepted that pipe diffusion is largely different from bulk diffusion. Experiments on nickel self-diffusion in NiO gave an activation energy of 2.56 eV for bulk and 2 eV for pipe-diffusion : the P_{O2} dependence of the two coefficients are similar, suggesting that diffusion takes place by the same vacancy mechanism (Atkinson et al. 1979). On this basis, one may conclude that D_B and D_p show similar variations with P_{O2} and T ; therefore, even if pipe diffusion is active, the relation between $\dot{\varepsilon}$ and P_{O2} may be only slightly affected. The situation might be complicated if D_B and D_p in eq. (2) do not correspond to the diffusion of the same component. This was not found in NiO ; it has been observed that although oxygen diffusion is accelerated along grain-boundaries (Gervais et al. 1980 ; Atkinson et al. 1983), it still diffuses much more slowly than Ni. Roughly, the results are : $D_B^{Ox} \sim 10^{-4} D_B^{Ni}$, $D_B^{Ni} \sim 10^{-4} D_p^{Ni}$ and $D_p^{Ox} \sim D_B^{Ni}$.

When considering the 2 components and the various diffusion paths, the creep equation (1) is already complicated, but it represents a single creep mechanism ; h and n are constant and σ_o, which decreases with increasing temperature, is generally negligible. If various mechanisms are operative with different $\dot{\varepsilon}_i$ given by eqs. of type (1), the overall creep rate $\dot{\varepsilon}$ is a combination of the $\dot{\varepsilon}_i$ (Bretheau et al. 1979 ; Li 1975) and it is impossible to unambiguously identify them on the basis of mechanical data (Mohamed et al. 1981; Ruano et al. 1982). It is therefore much better to perform experiments in conditions where only one mechanism dominates. Microstructural observations are necessary to prove that these conditions have been encountered.

There are two prominent mechanisms involved in the deformation of oxides : (i) pure diffusion creep, with h = 2, n = 1, $\sigma_o = 0$, Ω the molecular volume and d the grain-size (Ashby et al. 1975) :

$$\dot{\varepsilon} = 42 \frac{\Omega}{b^3} \left(\frac{b}{d}\right)^2 \frac{\mu b}{kT} \frac{\sigma}{\mu} D \qquad (6)$$

(ii) dislocation glide and recovery where d can be viewed as the cell-size of the polygonized structure inversely proportional to the stress (Burton et al. 1981) giving n=3, or larger :

$$\dot{\varepsilon} = A' \frac{\mu b}{kT} \left(\frac{\sigma}{\mu}\right)^n D \tag{7}$$

Eq. (7) has been derived in various microscopical models where it is made use of the following eq. :

$$\dot{\varepsilon} = B\, v_c \tag{8}$$

v_c is the velocity of climb allowing for dislocation annihilation. A derivation of this climb velocity has been made by Bretheau et al. (1979) ; avoiding the usual approximations, they found :

$$v_c = c_o \frac{1 - \exp - [(F_a b^2 + W_o)/kT]}{(y/D_c) + \Omega \ln (R/b)/2\pi b^2 D_1} \tag{9}$$

F_a is the climb component of the applied force. W_o is the dislocation-point defect interaction. y is the average distance between jogs which are the point defect sources/sinks and D_c is for the thermally-activated jog mobility through the emission/absorption of point defects. Ω is the volume of the diffusing species and D_1 is the point defect diffusion coefficient for diffusion between two cylinders of radii b and R. From eq. (9), one can deduce through several approximations (Bretheau et al. 1979) ;

$$v_c = C\, D_1 c_o F_a/kT \tag{10}$$

Where C is a constant and $D_1 c_o = D$ is the self-diffusion coefficient of the rate controlling species. Eq. (7) is found using eq. (10) instead of the complicated eq. (9). However, it can be seen that at each stage of approximation, v_c is proportional to c_o. Creep rate is proportional to c_o for both cases of eq. (6) and (7) ; even if creep is affected by pipe diffusion, this is probably valid. Therefore, the following relation is quite general and relatively insensitive to the creep mechanism :

$$\dot{\varepsilon} = E\, c_o = F\, P_{O2}^m \tag{11}$$

High temperature deformation appears as a remarkable tool to study the point defect populations as a function of P_{O2} at constant temperature since E and F in eq. (11) depend on T. Creep gives information on minority point defects which are related to the slowest diffusing species.

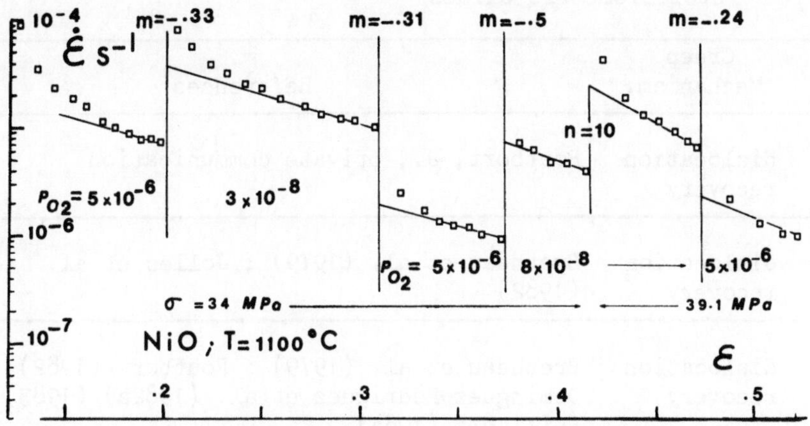

Fig. 1. Constant load compression of NiO single crystal along <001>. The test shows various P_{O_2} changes between CO/CO_2 and Ar/O_2 mixtures. Strain-rate $\dot{\varepsilon}$ versus strain ε.

Experiments

Mechanical tests have generally been performed in compression despite several draw-backs of this mechanical test (Bretheau et al. 1979). In single crystals of the cubic oxides that we review, the deformation occurs by dislocation glide since the C.R.S.S. is always small at high T (Bretheau et al. 1979). Because of the high symmetry, deformation by pure climb has not been observed. Polycrystals of NaCl structure materials cannot deform only by activation of the easy glide system, because it does not provide the five independent slips to accomodate any grain deformation. It is, therefore, possible to study pure diffusional creep (Nabarro or Coble creep) at moderate stress on polycrystals. The preparation of good specimens is not easy. In particular the unavoidable grain size distribution may impede the observation of pure diffusional creep (Raj. et al. 1981).

Eq. (6) and (7) can be used to describe creep behaviour once the steady-state is established. For dislocation creep, this requires the building-up of a substructure during the transient creep (Bretheau et al. 1979). For diffusional creep, no transient should be observed if no other mechanism is operative.

The constants n, m, ΔH of eq. (5) (6), (7) and (11) can be determined by measuring the variations of $\dot{\varepsilon}$, keeping all parameters fixed except one. Different tests may be used for each value of $\dot{\varepsilon}$, or interrupted tests may be performed on a specimen (fig: 1). For

Table 2. Investigations and references on deformation of non-stoichiometric oxides

	Creep Mechanism.	References
MnO	dislocation recovery	Routbort, J., private communication
FeO	dislocation recovery	Bretheau et al. (1979) ; Jolles et al. (1982)
CoO	dislocation recovery	Bretheau et al. (1979) ; Routbort (1982) ; Dominguez-Rodriguez et al. (1982a) (1983) ; Philibert (1984)
NiO	dislocation recovery	Bretheau et al. (1979) ; Cabrera-Cano et al. (1982) ; Dominguez-Rodriguez et al. (1983) ; Philibert (1984)
	diffusion	Bretheau et al. (1979) ; Jimenez-Melendo et al. (1983)
Cu_2O	dislocation recovery	Bretheau et al. (1979) (1981) ; Dominguez-Rodriguez et al. (1983) ; Philibert (1984)
	diffusion	Gervais et al. (1978) ; Dominguez-Rodriguez et al. (1983)
UO_2		Bretheau et al. (1979) ; Mohamed et al. (1982) ; Philibert (1984)

T and P_{O2}, it is necessary to reach thermodynamic equilibrium without excessive structural recovery before resuming test (Cabrera-Cano et al. 1982) ; this is done by annealing for an appropriate duration. A-priori estimations of maximum experimental error give 2 for n, 0.05 for m and 1 eV for ΔH ; these estimations are based on the assumption that creep rates are measured with a 20 % accuracy (Bretheau et al. 1979 ; Dominguez et al. 1983). Creep curves generally show a short transient when the new conditions are applied (fig. 1). If the specimen is submitted to a new P_{O2} and deformed before reaching the new thermodynamic equilibrium, $\dot{\varepsilon}$ shows a gradual variation which has been called <u>transitory creep</u> by Dominguez-Rodriguez et al. (1982b, 1984).

RESULTS

Many recent publications describe the deformation of non-stoichiometric oxides at high temperature, in thermodynamic equilibrium. Table 2 gives an account of these publications together with a few recent ones. We refer the readers to these publications and give only a very brief account of the results.

In addition to earlier work on NiO (Cabrera-Cano et al. 1982) creep has been measured at very low P_{O2} (see results fig. 1) and high temperatures (1450-1500°C) where m has been found equal to - 0.12 (6 measurements, P_{O2} between 10^{-5} atm and 10^{-2} atm) and m = - 0.05 (6 measurements, P_{O2} between 10^{-2} atm and air), in agreement with the earlier results. There has been a number of studies on CoO deformation. We have plotted a few results in fig. 2, where a fairly large scatter can be seen for strain-rates. This probably comes in part from differences in crystal purity. For Cu_2O, the results are remarkably insensitive to test conditions for various crystal orientations and for polycrystals, suggesting a single mechanisms for high temperature creep (Bretheau et al., 1981). Other compounds do not display the same behaviour, rendering the identification of creep mechanisms rather unsafe.

The microstructural observations are generally limited to dislocation polygonization of single crystals. Transitory creep after a P_{O2} change has been studied in CoO (Dominguez-Rodriguez et al. 1982b ; 1984) and in Cu_2O (C. Clauss, unpublished result). In both cases, it seems to be dominated by point defect equilibration kinetics, a result that support the link between $\dot{\varepsilon}$ and c_o established above (eq. 11).

DISCUSSION

<u>Comparison with diffusion</u>

The relation between creep and diffusion is well accepted and it has been discussed in details by Bretheau et al. (1979), Dominguez-Rodriguez et al. (1983), Philibert (1984) for non-stoichiometric oxides. Diffusion data for the slowest species are generally known with a poor precision. There is a large scatter for diffusion data of UO_2 (Matzke, 1981). The cation self diffusion in NaCl-type oxides is known, but for oxygen diffusion, there are disagreements between the various determinations, when they exist (Monty 1983). Recent data for cation and anion diffusion are shown in Table 1. The ratio R is always very small, justifying that the self-diffusion rather than "molecular" diffusion may control the creep rate. The activation energy for creep has always been found substantially larger than for the fast diffusing species, but closer to that of the slowest ones. However, there is never a good agreement between the corresponding activation energies, except for Cu_2O where the two following equa-

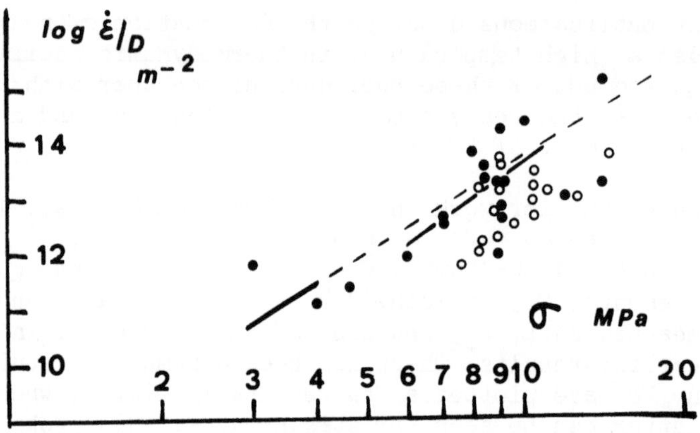

Fig. 2. Comparison of mechanical properties of CoO single crystals deformed in air, by compression along <001>. Strain-rate $\dot{\varepsilon}$ is divided by $D = 8.76 \times 10^{-8} \exp [3\ (eV)/kT] (m^2/s)$ for oxygen self-diffusion (Yamaguchi et al. 1982). Full circles : Dominguez-Rodriguez et al. 1982a. Open circles : Nehring et al. 1977. Lines : Routbort 1982 ; 1400°C and in O_2 at 1200°C.

tions have been determined for oxygen diffusion (Perinet et al. 1982) and creep (Bretheau et al. 1981) :

$$D_{ox}(cm^2/s) = 1.3 \times 10^{-3}\ [P_{O2}(atm)]^{0.32\pm0.05} \exp - 1.5\ 0.3(eV)/kT \quad (12)$$

$$\dot{\varepsilon} = \dot{\varepsilon}_o \frac{\mu b}{kT} \left(\frac{\sigma}{\mu}\right)^{5\pm0.4} (P_{O2})^{0.4\ \pm0.1} \exp - \frac{1.8\pm0.3}{kT} \quad (13)$$

For NiO and CoO, the activation energy is not independant of T, indicating that deformation at high temperature generally involves a mixture of different thermally activated mechanisms.

The P_{O2} dependence may help to clarify the situation in spite of the scarcity of diffusion data. In UO_{2+x}, it has been observed that both $\dot{\varepsilon}$ and D varies as x^2 (Bretheau et al., 1979 ; Matzke, 1981; Philibert, 1984). The relation of $\dot{\varepsilon}$ and D with P_{O2} are also very similar for Cu_2O (see eq. (12) and (13)). The creep-diffusion relation which is supported by these two examples, may fail in the case of NiO, where opposite variations with P_{O2} have been observed for $\dot{\varepsilon}$ and D in the highest temperature range (T = 1400-1500°C) (Monty, 1983 ; Philibert, 1984).

Transitory creep gives, in principle, a way to measure the mobility of minority point defects. It was concluded from experiments on CoO that the kinetics of reequilibration of minority point defects is controlled by that of the majority point defects, local equilibrium between them being very fast (Dominguez-Rodriquez et al. 1982b ; 1984). The kinetics of reequilibration was found independant of strain and was correctly predicted by a bulk diffusion model for crystals of various sizes. This shows that diffusion short circuits are not efficient during the deformation of CoO at $T > 1100°C$ in spite of the dislocations and sub-boundaries present in high density (Dominguez-Rodriguez et al. 1984).

Point defect characterization

Creep appears to be directly related to the concentration of minority defects (eq. (11)). It is then tempting to try to deduce the nature of these point defects from the (P_{O2}) relation. Some knowledge about majority charged defects is necessary (Appendix 2). Results on the nature of minority defects, which have been discussed in detail by Dominguez-Rodriguez et al. (1983) and Philibert (1984), are summarized in Table 1. They are based on experiments performed on pure materials which are believed to show an intrinsic behaviour. In the example of fig. 1, $Ni_{1-x}O$ (low P_{O2}, $x \gtrsim 10^{-5}$) should be intrinsic and if the behaviour of nickel vacancies can be extrapolated at low P_{O2}, we deduce that $V_O^{\bullet}$ rather than $V_O^{\bullet\bullet}$ (Table 1) should be the creep related minority point defect. If extrinsic behaviour is assumed, the negative m value of fig. 1 can be ascribed to two possibilities : $V_O^{\bullet}$ or $V_O^{\bullet\bullet}$ for monovalent impurity and O_i'' for trivalent impurities which both can give $m = -0.5$. This shows the interest of creep results on doped crystals when majority point defects are known, a set of results which has not yet been encountered.

A unified point defect model would be nice for NaCl-type oxides, at least for CoO and NiO which are similar. CoO contains a much larger amount of metal vacancies than NiO. From this, one can deduce that, since $O_i^{\neq}$ and, to a less extent, $V_O^{\bullet\bullet}$ (or $V_O^{\bullet}$) controls the creep of CoO, in the case of NiO the role of $V_O^{\bullet\bullet}$ should be even smaller (see appendix 2) by comparison with $O_i^{\neq}$ whose concentration does not depend on metal vacancy concentration (see appendix 1). This is at variance with experimental results (table 1). The behaviour of minority point defects in NiO and CoO does not seem to fit in the same frame. It is less likely to find a correlation with FeO and MnO where complexes may dominate the majority point defects (Table 1).

The formation of complexes is found in compounds with large nonstoichiometry. It is a consequence of defect interactions which has not been considered for minority point defects. Their expected concentration is very small. The results of table 1 do not take into account the possibility of defect complex formation, although models with majority-minority defect complexes can be conceived.

Comparison with computed defect models

There have been recently a number of computations to determine the point defect properties (Catlow, 1981 ; Vempati et al. 1983, Makrodt, 1984). The computed formation and migration energies for defects are rather sensitive to the details of the models (Table 3). However, they have shown that the formation of majority point defect complexes is important. It is not clear how high in temperature this description is valid, since entropy, which has not been computed, may change the configuration of minimum free energy. The complex formation is not believed to be important for minority point defect.

The values calculated for the activation energy of oxygen-vacancy and oxygen-interstitial mechanism diffusion are collected in Table 3. Their spread is such that one cannot deduce which mechanism is favoured for oxygen diffusion, although CoO creep suggests a much lower energy for interstitial (Table 3). The comparison of creep and diffusion with computation, does not yet shine any additional light on the mechanism of oxygen diffusion in NaCl type non-stoichiometric oxides.

Other deformation mechanisms.

Two other classes of mechanisms can be invoked to explain the deformation properties of non-stoichiometric oxides. The first one is based on the point defect seggregation at dislocations which can be perfect or dissociated. The second one is related to the electronic properties of the crystals.

(i) It can be expected that, in a metal deficient oxide, there is a non-uniform distribution of metal vacancies around dislocations (Hirth and Lothe, 1982). During plastic deformation, dislocations can glide after unpinning from the vacancy clouds ; this requires an extra-stress to initiate deformation (sigmoidal creep or yield point) and cannot control the steady state. An alternative is that the clouds are dragged by the gliding dislocations ; steady state creep would be then related to the vacancy diffusion (chemical diffusion) which is almost invariant with P_{O_2}, contrary to creep results and has an activation energy between 1 and 1.5 eV for NiO and CoO (Monty 1983), much smaller than for creep. Defect cloud dragging is therefore unable to explain the creep properties of Cu_2O, NiO and CoO.

Another possible effect of vacancies would be to increase the width of dislocation dissociation by adsorbtion on the stacking fault, lowering its energy. Such effect has been observed in carbides and spinels (Bretheau et al., 1979 ; Philibert, 1984). Recovery by cross-slip would be more difficult as it requires stacking fault constriction. The cross-slip of dislocations becoming more difficult at large vacancy concentrations, this could increase the

TABLE 3. Comparison of activation energies for the diffusion of slowest species through minority point defects deduced from creep ΔH and diffusion ΔH_D experiments and from computations ΔH_T. By difference between ΔH_T and the computed migration energies $\Delta H_T{}^m$, we have deduced the formation energies $\Delta H_T{}^F$ of minority defects. $\Delta H_T{}^F$ correspond to the energies which appears in the equations of the appendices. All energies in eV. References : a) B.L. Hutchinson, University of London, unpublished result ; b) Yamaguchi et al., 1982 ; c) Reviewed by Monty, 1983 ; d) Results compiled by Vempati et al. 1983 ; e) Makrodt 1984

	ΔH Creep	ΔH_D Diffusion	Defect	ΔH_T	$\Delta H_T{}^m$	$\Delta H_T{}^F$
MnO			V_O	4.82 d	1.82	3
				4.59 d		
				4.8 d	2.7	2.1
				5.1 d	2.8	2.3
			O_i	4.55 d	0.54	4
				5.27 d		
FeO	3.5	2.1[a]	V_O	6.16 d	2.32	3.84
				6.17		
	3	3.9[b]	O_i	5.55 d	0.42	5.13
				5.92 d		
CoO	5.2 ($V_O^{\bullet}$)	4.1[c] ($V_O^{\bullet}$)	V_O	5.35 d	1.99	3.36
				5.89 d		
	∼ 2 (O_i^{*})	3[b]	O_i	5.24 d	0.26	4.98
				6.34 d		
NiO	5.5 to 8.5	2.5[c]	V_O	8.41 e	2.24	6.17
				5.91 d	1.98	3.93
				6.02 d		
		5.6[c]	O_i	5.57 e	1.27	4.3
				5.24 d	0.18	5.06
				6.54 d		

the stage III stress or lower $\dot{\varepsilon}$. In NiO, CoO and Cu_2O, one would always observe a decrease or $\dot{\varepsilon}$ for an increase in P_{O2}, i.e. $m < 0$ which has been observed only for NiO at high temperatures. This, together with the fact that dislocations have never been seen dissociated (Bretheau et al. 1979), even at nodes in sub-boundaries (Gervais et al., 1980 ; Bretheau et al., 1981), allows to discard the role of cross-slip during steady state creep

(ii) Up to this point, the discussion is based upon the elastic interactions of dislocations and point defects. We now turn to electric charge effects. At the temperature of creep tests, the defect structure of crystals is dominated, in a simplified view, by electronic holes and negatively charged vacancies.

In p-type $Ni_{1-x}O$ with $x = [V_{Ni}'] = 10^{-3}$, there is a hole concentration of 5×10^{19} cm^{-3}, giving a Debye length of 1 nm at 1500 K ; in these conditions, screening effects strongly limit the influence of electric charges. If we consider NiO as a divalent ionic compounds jogs have a $\pm e$ charge which can be easily neutralized by electronic carriers or point defects (Hirth and Lothe, 1982). We conclude that dislocations behave as neutral lines in contradistinction with alkalihalides (Whitworth, 1975). Actually the binding has a marked covalent character in non-stoichiometric oxides, and the description used for semi-conductor may be relevant. A P_{O2} variation at constant T is equivalent to a shift of the Fermi level E_F. We take the classical approximation as a rough description of semiconducting properties of oxides. In the example of appendix 1, we can write (Kröger, 1974, ch. 14) : $E_F \propto - kT \ln p \propto - (kT/4) \ln P_{O2}$. A P_{O2} change by a factor of 1000 is usual (fig. 1) and leads to a shift in E_F of 0.2 eV which is negligible compared to the gaps of oxides. Although this shift affects the ionization of majority point defects, it is likely that it does not change substantially the ionization of minority defect or dislocation-associated states which are probably at other levels in the gap. Therefore, this does not seem able to explain the variation of mechanical properties with P_{O2}.

CONCLUSIONS

A relation exists between the high temperature deformation and the point defect properties of non-stoichiometric oxides. This is best shown by the dependence of creep rate on P_{O2}. The influence of stress and T can be more difficult to understand. The comparison of diffusion and plastic deformation is the best way to check the models. In the case of Cu_2O, data are available and show a nice agreement. In other oxides, deformation may have helped to characterize the diffusion of the slowest species.

APPENDIX 1 : Defect chemistry and non-stoichiometry

Let us consider a p-type oxide $M_{1-x}O$ dominated by V'_M metal vacancies, and O^*_i oxygen interstitials which are in much lower concentration. We use the Kröger-Vink notation for defects (Kröger, 1974, ch. 7). The chemical reactions for the formation of defects are :

$$1/2\ O_2 \rightleftarrows O_O + V'_M + h \qquad (\Delta G_1) \qquad (A1)$$

$$1/2\ O_2 \rightleftarrows O^*_i \qquad (\Delta G_2) \qquad (A2)$$

Laws of mass action give :

$$[V'_M]\,p = P_{O2}^{1/2} \exp - (\Delta G_1/kT) \qquad (A3)$$

$$[O^*_i] = P_{O2}^{1/2} \exp - (\Delta G_2/kT) \qquad (A4)$$

The neutrality equation can be written :

$$[V'_M] = p \qquad (A5)$$

The departure from stoichiometry x is :

$$x = [V'_M] + [O^*_i] \simeq [V'_M] \qquad (A6)$$

with $[V'_M] \gg [O^*_i]$

x and p are accessible to many experiments. Constant stoichiometry means that there is a relation between T and P_{O2} which can be calculated by eq. (A6) (A5) and (A3) :

$$[V'_M]^2 = x^2 = P_{O2}^{1/2} \exp - (\Delta G_1/kT) \qquad (A7)$$

With x constant, we can calculate the variation of $[O^*_i]$ using eq. (A4) :

$$[O^*_i] = x^2 \exp - [(\Delta G_2 - \Delta G_1)/kT] \qquad (A8)$$

Contrary to what has often been done, we find it much better to use the parameters T and P_{O2} (oxygen chemical potential) which are directly controlled and involve no defect model and approximations (see eq. (A6)). More complicated cases have been described by Kröger (1974) ch. 14, where it is shown the extent of changes which can take place at constant composition.

APPENDIX 2 : Charged majority and minority defects

The only coupling between majority and minority defects is through charged defects. Let us develop the model of appendix 1, by introducing $V_O^{\bullet\bullet}$ as possible minority defects :

$$O_O + 2h \rightleftarrows 1/2\ O_2 + V_O^{\bullet\bullet} \qquad (\Delta G_3) \qquad (A9)$$

The presence of $V_O^{\bullet\bullet}$ in low concentration does not change the neutrality equation (A5). The law of mass action gives :

$$[V_O^{\bullet\bullet}] = p^2\ P_{O2}^{1/2}\ \exp - \Delta G_3/kT \qquad (A10)$$

The variation of $[V_O^{\bullet\bullet}]$ is related to that of p which can change with P_{O2} as calculated in appendix 1. If other defects such as V_M'' $(V_{Fe}\ Fe_i\ V_{Fe})'$ ... are present, the relation between p and P_{O2} is not simple but it can be calculated, and the relation between minority charged defect concentration and P_{O2} can be deduced. Any alteration of p (doping, alloying, ...) also affects this relation.

ACKNOWLEDGEMENTS

The authors acknowledge the Financial support of Ministère des Relations Extérieures which made a 5 day visit in 1983 possible for A.D.R..

REFERENCES

Ashby, M.F., and Frost, H.J., 1975, The kinetics of inelastic deformation above 0°K, in "Constitutive equations in plasticity" A.S. Argon, ed., The MIT Press, Canbridge.

Atkinson, A., and Taylor, R.I., 1979, The diffusion of Ni in the bulk and along dislocations in NiO single crystals, Phil. Mag. A 39:581

Atkinson, A., Dubois, C., Monty, C., and Taylor, R.I., 1983, Evidence for oxygen grain-boundary self-diffusion in polycrystalline NiO, J. Micros. Spectrosc. Electron. 8:41a

Blanchin, M.G., Castaing, J., Fontaine, G., Heuer, A.H., Hobbs, L.W., and Mitchell, T.E., 1981, The interactions between dislocations and defect in oxides, J. Phys. Colloque C3 42:1

Bretheau, T., Castaing, J., Rabier, J., and Veyssiere, P., 1979, Dislocation motion and high temperature plasticity of binary and ternary oxides, Adv. Phys. 28:829

Bretheau, T., Pellissier, B., and Sieber, B., 1981, Plastic properties of Cu_2O, mechanical tests and transmission electron microscopy - II. high temperature, Acta Met. 29:1617

Burton, B., and Knowles, G., 1981, The relationship between dislocation recovery creep and vacancy diffusion creep, Phil. Mag.A 44:987

Cabrera-Cano, J., Dominguez-Rodriguez, A., Marquez, R., Castaing, J., and Philibert, J., 1982, Point defects and high temperature creep of non-stoichiometric NaCl-type oxide single crystals, I. NiO, Phil. Mag. A 46:397

Catlow, C.R.A., 1981, Defects clustering in non-stoichiometric oxides, in "non-stoichiometric oxides" O.T. Sorensen, ed., Academic Press, New York.

Dominguez-Rodriguez, A., Sanchez, M., Marquez, R., Castaing, J., and Monty, C., 1982a, Point defects and high temperature creep of non-stoichiometric NaCl-type oxide single crystal II. CoO, Phil. Mag. A 46:411

Dominguez-Rodriguez, A., Monty, C., and Philibert, J., 1982b, Creep and diffusion : transitory creep and equilibrium kinetics of population of point defects in $Co_{1-x}O$, Phil. Mag. A 46:869

Dominguez-Rodriguez, A., and Castaing, J., 1983, Point defect and diffusion properties in oxides from high temperature creep in "Proc. Europhys. Conf. Lattice Defects in ionic crystals", Dublin, Sept. 1982, Rad. Effects, to be published

Dominguez-Rodriguez, A., Monty, C., Castaing, J., and Philibert, J., 1984, Creep measurements of chemical diffusion in $Co_{1-x}O$ in "Proc. Int. Conf. Non-stoich. Comp.", Alenya, June 1982, Solid State ionics, to be published

Gervais, H., Pellissier, B., and Castaing, J., 1978, Machine de fluage pour essais en compression à hautes températures de matériaux céramiques, Rev. Int. Htes Temp. et Refract. 15:43

Gervais, A., Tertian, L., Hokim, D., and Deschamps, J., 1980, Observation en microscopie électronique de la migration d'un sous-joint dans l'oxyde de Nickel, J. Microsc. Spectrosc. Electron. 5:309

Hirth, J.P., and Lothe, J., 1982, "Theory of dislocations", 2nd Ed., John Wiley and Sons, New York.

Jimenez-Melendo, M., Cabrera-Cano, J., Dominguez-Rodriguez, A., and Castaing, J., 1983, Pure diffusional creep of NiO polycrystals, J. Phys. Lettres 44:339

Jolles, E., Berthon, J., Pellissier, B., and Philibert, J., 1982, Fluage de la wüstite $Fe_{1-x}O$ à des températures voisine de 1000°C, Rev. Int. Htes Temp. et Refract. 19:393

Kröger, F.A., 1974, "The chemistry of imperfect crystals" 2nd revised edition, North Holland Publ. Co, Amsterdam

Li, J.C.M., 1975, Nature of simultaneous processes in plastic deformation, in "Rate processes in Plastic deformation of materials", Proc. J.E. Dorn Symposium, Cleveland, oct. 1972, Li, J.C.M., and Mukherjee, A.K., ed. American Society for metals.

Mackrodt, W.C., 1984, Defects energetics and their relation to non-stoichiometry in oxides, in "Proc. Int. Conf. Non-Stoich. Comp., Alenya, June 1982", Solid State Ionics, to be published.

Matzke, H.J., 1981, Diffusion in non-stoichiometric oxides, in "Non-stoichiometric oxides" O.T. Sorensen, ed., Academic Press, New York.

Mitchell, T.E., Hobbs, L.W., Heuer, A.H., Castaing, J., Cadoz, J., and Philibert, J., 1979, Interaction between point defects and dislocations in oxides, Acta Met. 27:1677

Mohamed, F.A., and Ginter, T.J., 1981, On the identification of creep processes at low stresses, J. Mat. Sci. 16:2890

Mohamed, F.A. and Soliman, M.S., 1982, On the creep behavior of uranium dioxide, Mat. Sci. Eng. 53:185

Monty, C., 1983, Diffusion in stoichiometric and non-stoichiometric cubic oxides, in "Proc. Europhys. Conf. Lattice Defects in ionic crystals", Dublin, Sept. 1982, Rad. Effects, to be published.

Nehring, V.W., Smyth, J.R., and McGee, T.D., 1977, Compressive creep of CoO single crystals, J. Amer. Ceram. Soc. 60:328

Perinet, F., Barbezat, S., and Philibert, J., 1982, mechanism of oxygen self-diffusion in Cu_2O, in. "Proc. 9th Symp. on Reactivity of Solids, Cracow 1980, Elsevier

Philibert, J., 1984, Creep and diffusion in "Proc. Int. Conf. non-stoich. comp., Alenya, June 1982", Solid state ionics, to be published

Raj. R., and Ghosh, A.K., 1981, Micromechanical modelling of creep using distributed parameters, Acta Met. 29:283

Routbort, J.L., 1982, The stoichiometry dependence of the deformation of $Co_{1-\delta}O$, Acta Met. 30:663

Ruano, O.A., and Sherby, O.D., 1982, Low stress creep of fine-grained materials at intermediate temperatures : diffusional creep or grain boundary sliding ?, Mater. Sci. Eng. 56:167

Vempati, C.S., and Jacobs, P.W.M., 1983, Defect properties of oxides II. Manganese, iron, cobalt and nickel oxide, Cryst. Latt. Def. and Amorph. Mat. 10:19

Walser, B., and Sherby, O.D., 1982, The structure dependence of power law creep, Scripta Met. 16:213

Whitworth, R.W., 1975, Charged dislocations in ionic cyrstals, Adv. Phys. 24:203

Yamaguchi, S., and Someno, M., 1982, The tracer diffusivity of oxygen in Wüstite and Cobaltous oxide, Trans. Jap. Inst. Met. 23:259

HIGH TEMPERATURE DEFORMATION MECHANISMS IN $MgO\cdot n\ Al_2O_3$ Spinels

R. Duclos

Laboratoire de Structure et Propriétés de l'Etat Solide*
Université de Lille 1. 59655 Villeneuve d'Ascq
Cedex, France

ABSTRACT

High Temperature plasticity of $MgO\cdot n\ Al_2O_3$ spinels is strongly dependent on the composition ratio $n=Al_2O_3/MgO$. This effect is due to the dissociation mode of dislocations which are generally split off their slip plane.

From experiments carried out on samples of composition n = 1.1 and 1.8, it's appeared that the evolution of mechanical properties didn't result from an hypothetic slip plane change with n but from different deformation modes; by dislocation climb for n=1.1, after a pre-yield slip state required for dislocation multiplication while for n = 1.8 deformation results from dislocation glide. Several arguments are given in favour of this hypothesis.

The different possibilities to move such sessile dislocations, with the object of understanding the higher mobility of those dislocations when n increases, are then considered. The most likely consists in a viscous glide mechanism of dislocations without partial recombination, equivalent to a climb of the fault ribbon. This needs a local rearrangement of the cationic sublattice in between the partials which is favoured by the presence of constitution vacancies. This model is consistent with a higher slip rate of dislocations with n.

*L.A. C.N.R.S. 234

I. INTRODUCTION

Magnesium aluminate spinel $MgAl_2O_4$ forms with alumina a large range of solid solutions $MgO.n\ Al_2O_3$ the composition ratio n of which varies from 1 to 5 [1]. The structure of those solid solutions is a f.c.c. one and the Burgers vector of perfect dislocation is a/2 $\langle 110 \rangle$. When the composition ratio is different than one the excess of alumina introduces cationic vacancies predominantly located on the octahedral aluminium sites [2]. These vacancies have a strong influence on the high temperature mechanical properties of those spinels (figure 1) as the first deformation experiments carried out by Mc Brayer [3], Radford and Newey [4] and Lewis [5] have shown.

The purpose of this paper is to review the various origins which have been put forward, since firsts experiments, to explain the influence of the composition ratio on the mechanical properties, to discuss them by taking into account ulterior results and to propose a deformation mechanism consistent with all experiments.

2. EXPERIMENTAL PROCEDURES

The different deformation tests reported in the literature have been performed in the temperature range 1300-1900°C except for the equimolar composition where the lowest test temperature was about 1500°C.

Various environments have been used, during tests ; vacuum [3] argon [7] and air [4-6]. It does not seem that this has an influence on the results.

Generally samples have been deformed at constant strain rates, the order of magnitude of which is 10^{-5}-$10^{-4}s^{-1}$ but creep experiments have been also performed [6] ; the strain rate can then reach values as low as $10^{-6}s^{-1}$. The comparison between the two kinds of tests will be made in the discussion.

3. RESULTS AND DISCUSSION

3.1. Review of some models

The first model has been proposed by Radford and Newey [4] : they suggest as an explanation a possible decrease of the self diffusion energy with the increase in deviation from stoichiometry, this resulting in a decrease of the activation energy for climb, this process controlling the high temperature plastic deformation. Since, that hypothesis has been contradicted by the diffusion measurements of oxygen anion, probably the slowest moving species, which do not show any clear dependence of self diffusion energy with the composition ratio [8].

In a second class of model the increase in plasticity was explained by the change of slip systems with the composition ratio : {111} < 1$\bar{1}$0 > for n ≃ 1 and {110} < 1$\bar{1}$0 >for the other compositions.

For Mitchell et al [9] the composition vacancies introduced as soon as n differs from one, induce a lowering of Peierls barriers on the {110} planes but not on the {111} ones. This hypothesis does not explain why {111} glide is also observed for n ≃3. Lewis [5] has explained "the increase in flow stress for nearly equimolar spinels by the decreasing mobility of partial dislocations on {110} planes due to the difficulty in maintaining charge neutrality at the dislocation core as the vacancy concentration decreases, with a possibility for equimolar crystals that the synchro shear mechanism [10] for dislocation motion on {111} was predominant".

The assumption according to which {111} glide was the equimolar spinel glide was based partially on the following prediction of Hornstra : dislocation glide on {111} planes favours the reduction in strain energy by a possible dissociation into quarter partials, but principally on the fact that the compression axes used by these authors systematically favoured the {111} glide. Since it has been established that {110} slip is also active in equimolar spinel [11,12] and is the easiest one [11].

An other model takes into account the dislocation core structure. If the dissociation of dislocations in spinel into two colinear partials according to the relation :

(1) $$a/2<110> \rightarrow a/4<110> + a/4<110>$$

is known since 1966 [13] the first determination of the dissociation plane by N. Doukhan [11] was only made 12 years after. She observed in a nearly equimolar spinel crept along a [100] direction that edge dislocations gliding on {110} planes were dissociated in the {110} plane perpendicular to their slip plane. These authors [11] then proposed as possible explanation a difference in the dissociation modes : sessile in equimolar or nearly equimolar spinels and glissile in Al_2O_3 rich spinels. But ulterior determinations of the dissociation planes in spinels of composition 3.5 [14] and 1.8 [15, 16] showed that dislocations split also off their slip plane in non equimolar spinels.

Moreover T.E.M. observations of dissociated dislocations have showed that the fault ribbon width increases with composition ratio n.

Then the problem has consisted in understanding the possible

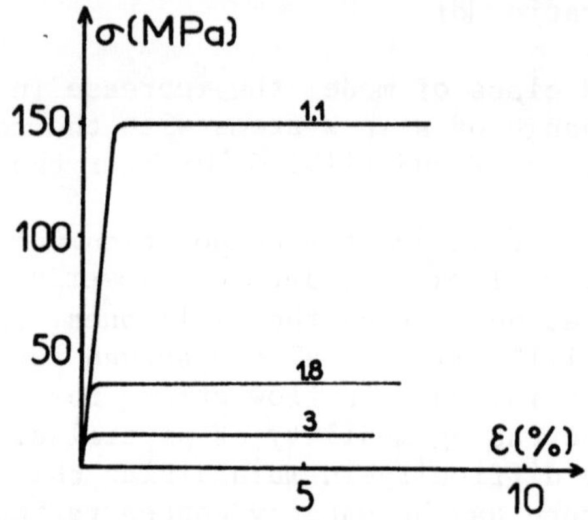

Figure 1 : Stress-strain curves for < 100 > samples.
$\dot{\varepsilon} \sim 1.5\ 10^{-5} s^{-1}$; $T \sim 1630^{o}C$
Curves n = 1.1 and 1.8 from [6] ; n = 3 from [3].

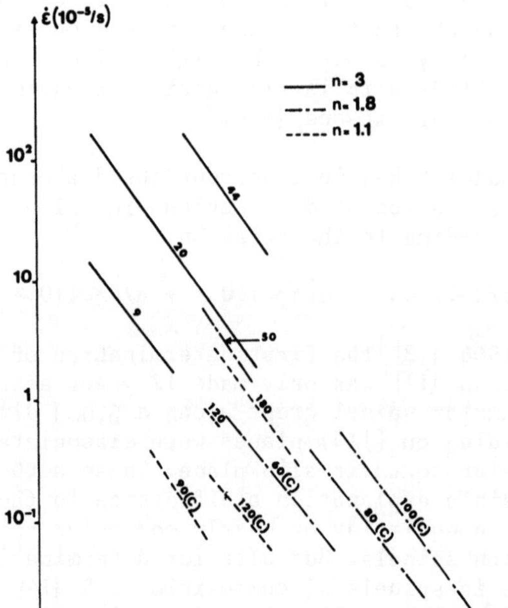

Figure 2 : Plot of the strain rate versus the reciprocal temperature. The number above each curve refers to the applied stress (in MPa). The letter C means that the curve is deduced from creep tests.

influence of the sessile dissociation mode on the mechanical properties of spinels. But before that and for a better understanding it is necessary to examine the influence of composition on the macroscopic deformation laws. That analyse will allow us to define the parameter which accounts for the mechanical effect of composition but particularly the parameters which are not influenced by the composition and thus to produce a more credible model.

3.2. Deformation laws

The deformation laws have been determined in the case of <100> samples, for which {110}<1$\bar{1}$0>slip systems are the active ones [5,11], deformed at constant strain rates or constant stresses.

At high temperature single crystal deformation is generally controlled by dislocation climb and the measured activation energy is only slightly stress dependent. So the temperature and stress dependence of strain rate is analysed here according to the relation, currently observed at high temperature :

$$\dot{\varepsilon} = \dot{\varepsilon}_o \left(\sigma/\mu\right)^s \exp\left(-Q/kT\right) \tag{2}$$

(In the absence of precise data at high temperatures the shear modulus μ has been assumed to be independent on the composition as it is the case at room temperature).

The previous relation (2) has been applied to the steady state of creep curves or to the plateau flow stress of σ-ε curves. For samples of composition $n = 3$ we have used the data published by Mc Brayer [3]. The analyse is not here related in detail, it can be consulted in references [6,15]. Only the principal results are summarized below.

For the three compositions $n = 1.1$, 1.8 and 3 activation energy values equal to 5.4, 5.5 and 5.7 ev and stress exponent values of 4, 3.7 and 3.4 have been respectively obtained.

These values of activation energy are near the self diffusion energy of oxygen anion. Moreover the stress exponent values are consistent with a number of high temperature deformation models involving dislocation climb as recovery process. The results are gathered in figure 2 as a plot of strain rate versus reciprocal temperature.

Several remarks can be made about those results :

Firstly creep tests and constant strain rate tests lead to identical results for $n = 1.1$ and 1.8 except in the parameter $\dot{\varepsilon}_o$ which is about five times higher at constant strain rate than at constant stress. The explanation is obtained by studying the dislocation substructure which shows that the steady state structure at constant strain rate is comparable to some given state of transient creep ; this means that the substructure is an important parameter of the deformation law.

The second remark concerns the values of stress exponent s and activation energy Q in relation (2) which are relatively insensitive to spinel composition and cannot account for differences in mechanical behaviour.

It appears that in fact the effect of composition stems only from the preexponential factor $\dot{\varepsilon}_0$ in relation (2) the values of which are :

$$\dot{\varepsilon}_0 \simeq 10^{19}\,s^{-1} \;;\; n = 1.1$$
$$\dot{\varepsilon}_0 \simeq 10^{21}\,s^{-1} \;;\; n = 1.8$$
$$\dot{\varepsilon}_0 \simeq 10^{23}\,s^{-1} \;;\; n = 3$$

that is $\dot{\varepsilon}_0(n) = 10^{(2n+17)}\ s^{-1}$

corresponding to ratios :

$$\dot{\varepsilon}_0(3)/\dot{\varepsilon}_0(1.1) \simeq 10^4 \quad \text{and} \quad \dot{\varepsilon}_0(1.8)/\dot{\varepsilon}_0(1.1) \simeq 10^2$$

the variation of which cannot be completely ascribed to the corresponding variation of the oxygen self diffusion coefficient since the maximum ratios measured by Reddy [8] are :

$$D_0(3.5)/D_0(1) \simeq 10^2 \quad \text{and} \quad D_0(1.8)/D_0(1) \simeq 7.$$

The higher values of strain rates when n increases must then be explained otherwise and to this end interesting informations about the deformation mode can be deduced from the macroscopic study of deformation substructure.

3.3. Macroscopic deformation substructure

The first information lies in the shape of strained samples. In the case of composition $n = 1.8$ dislocation glide on two orthogonal {110} planes, or four, produces barreled specimens (figure 3) where the ends keep the same area than before deformation. Contrary to this observation, for the same <100> orientation, nearly equimolar samples present after deformation an unusual inverted barrel shape (figure 4c) inconsistent with a deformation produced by dislocation glide.

An other information can be obtained from the study of dislocation substructure by X-ray topography [5,11,17]. For composition $n = 1.8$ a well defined cellular substructure is observed, the subgrain boundaries lying in the {110} planes perpendicular to the slip plane when two orthogonal slip systems are active or in the {110} planes parallel to the compression axis when four systems are at the

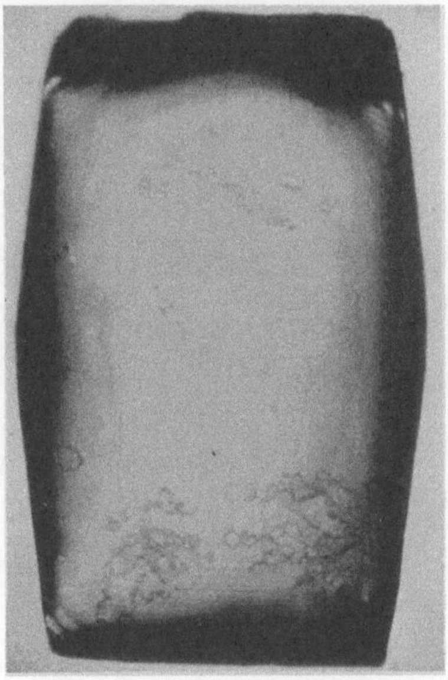

Figure 3 : [001] specimen deformed by dislocation glide on two orthogonal {110} planes.

origin of the deformation (figure 4a and b). On the contrary only some random subgrain boundaries are present in the $n = 1.1$ samples after deformation.

These different observations suggest to us two different deformation modes for $n = 1.1$ and $n = 1.8$. For $n = 1.8$ deformation results from dislocation glide while for $n = 1.1$ observations are consistent with a deformation mechanism where dislocation climb would produce a large part of the total strain.

The sample shape can be then explained by the introduction of supplementary half planes of edge dislocations from the specimen heads [18]. The ends of samples are effectively more enlarged than the center by this process.

On the other hand this point of view agrees very well with the strain rate which can be calculated from the orowan equation :

$$\dot{\varepsilon} = \rho\, v\, b \quad (3)$$

or in the case when dislocation climb controls the deformation

$$\dot{\varepsilon} = \rho \left(L/h\right) v_c\, b \quad (4)$$

where L is the mean distance covered by dislocations, h the climb distance in order to free a new dislocation, v_c the climb velocity, ρ the dislocation density and b the Burgers vector.

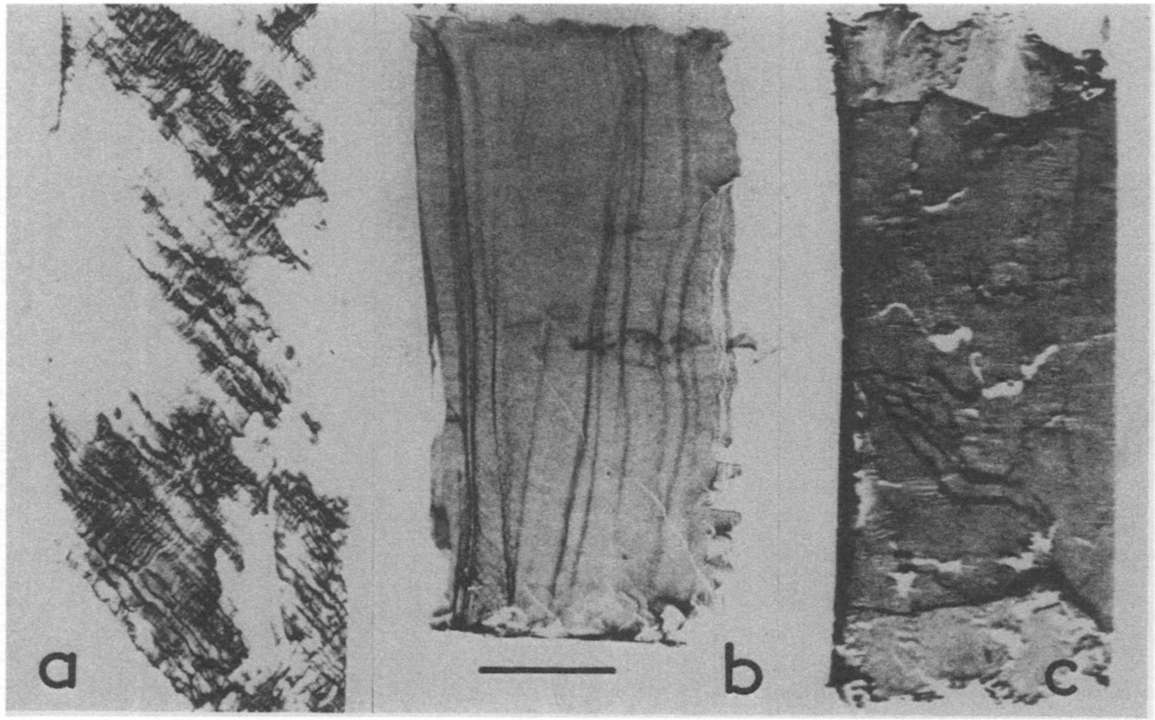

Figure 4 : Dislocation substructure (Berg Barrett topography) of [001] specimens after deformation (100) face ; scale bar = 1 mm.

a) n = 1.8 ; T = 1400°C ; ε = 7 %

b) n = 1.8 ; T = 1650°C ; ε = 10 %

c) n = 1.1 ; T = 1610°C ; ε = 5 % (note the concave shape of the left side)

An usual expression of climb velocity [19] gives here :

$$v_c = \frac{4\Pi\, D_{ox}}{-b\, \ln\left(b\sqrt{\rho}\right)} \frac{\sigma_m\, \Omega}{kT}$$

by assuming that oxygen anion is the controlling species (σ_m is the climb stress on the supplementary half plane and Ω the volume of oxygen anion).

The resulting strain rate is, in the case of a specimen of composition $n = 1.1$ crept at 1920 K under an applied stress of 90 MPa $(\sigma_m = 45\ \mathrm{MPa})$, the measured dislocation density is $\rho \sim 10^{11}\ \mathrm{m}^{-2}$, $D_{ox}(1920\ \mathrm{K}) = 10^{-16}\ \mathrm{m}^2/\mathrm{s}$ [20] and taking $L/h \sim 1$ for a pure climb strain : $\dot{\varepsilon} \sim 3.10^{-7}\ \mathrm{s}^{-1}$ while the experimental one is $\sim 7.10^{-7}\ \mathrm{s}^{-1}$ (figure 2).

This shows that quantitatively a pure climb strain can account for the measured experimental rates.

The strain rate for a sample of composition $n = 1.8$ can be evaluated by the same method ; L is then approximated to the cell size (the slip distance) and h to $\rho^{1/2}$.

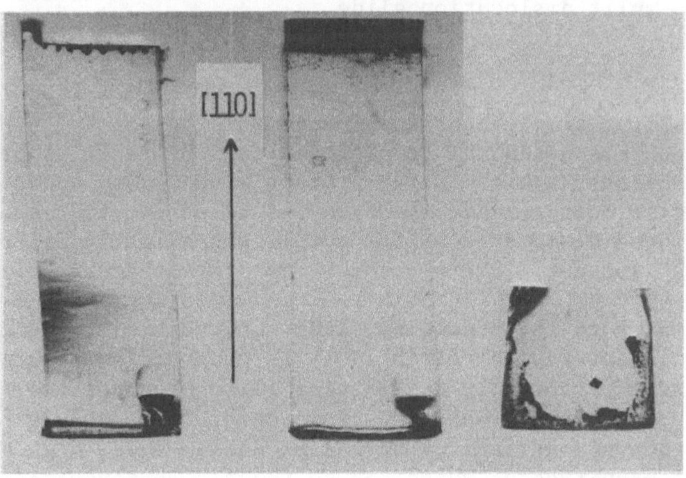

(001) $(1\bar{1}0)$ $(\bar{1}\bar{1}0)$

Figure 5 : Shape of a [110] specimen deformed at T = 1630°C
$\varepsilon = 8\ \%$

We see that the major difference between the two composition ratios can be explained by a greater slip distance for $n = 1.8$.

An other argument in favour of a low glide mobility of dislocations for $n = 1.1$ is given by the shape of samples compressed along a $\langle 110 \rangle$ direction (figure 5). This orientation of the compression axis favours the dislocation glide on the {111} planes. In the case when deformation largely results from dislocation glide, samples compressed along a $\langle 100 \rangle$ direction or a $\langle 110 \rangle$ one must have very different shapes which cannot be deduced from each other. This is not the case as figure 6 shows : the shape of the $\langle 110 \rangle$ sample can be easily deduced from the one of a $\langle 100 \rangle$ specimen.

At last dislocations moving in their climb plane, i.e. by pure climb, have been also observed [15,16] in nearly equimolar spinels.

For the composition $n = 1.1$ the following mechanism can be considered : a pre-yield slip of screw parts of dislocation loops needed for edge multiplication followed by the climb of sessile split dislocations ; the strain then results from screw glide and edge climb.

We see that the increasing plasticity with n can be explained by a larger slip distance of edge dislocations and we have now to understand why the mobility of dislocations is higher in non equimolar spinels than in equimolar one. With this object we examine at first how a sessile split dislocation can move by glide.

3.4. Sessile split dislocation glide

Several possibilities exist :

1. the first simply consists in recombining the two partials before glide. This needs the formation along the fault ribbon of some constrictions, a high energy process without any doubt. Moreover it is difficult to understand why it is easier to recombine largely dissociated dislocations (for $n = 3$) than slightly dissociated dislocations (for $n = 1$). This possibility seems thereby unlikely.

2. Contrary to the preceding situation, partial glide can be considered without recombination as it is proposed in figure 7 : the climb fault is immobile and partial glide produces two faults of energy γ_g in the {110} plane perpendicular to the fault plane. The propagation of partials by such a mechanism needs high stresses the order of magnitude of which can be roughly approximated to :

$$\sigma \sim 2\gamma_g/b \geqslant 2\gamma_c/b \qquad (6)$$

in so far as the glissile fault has a higher energy than the climb one [21].

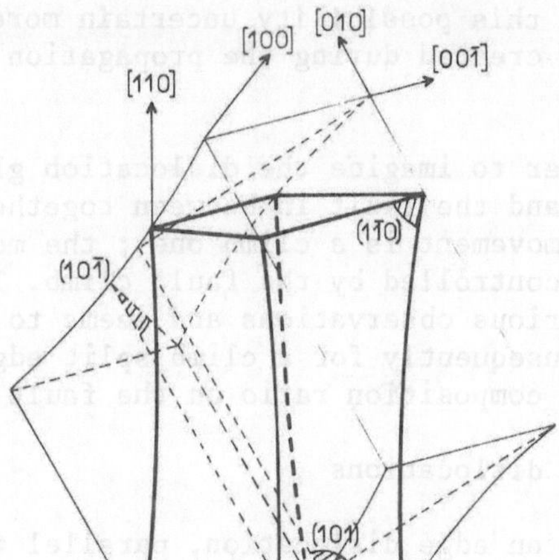

Figure 6 : Geometrical construction of the shape of a [110] specimen from the one of a [100] sample.

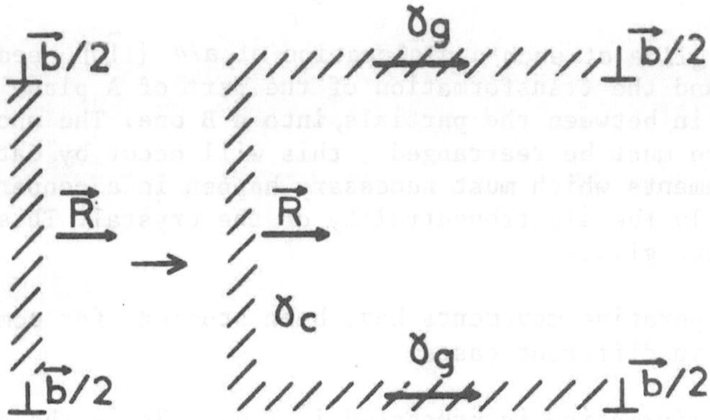

Figure 7 : Partial glide without recombination : two glide faults are created.

In our case, for nearly equimolar crystals, where $\gamma_c \sim 200$ mJ/m^2 [11] stresses higher than 700 MPa, five times the ones currently used, are needed. Identical conclusion is obtained for composition n=1.8. This renders this possibility uncertain more especially as the faults necessary created during the propagation of partials have never been observed.

3. A third manner to imagine the dislocation glide consists in moving the partials and the fault in between together. Note that in this case the fault movement is a climb one ; the mobility of the dislocation is then controlled by the fault climb. This possibility explains well the various observations and seems to be the most likely ; we examine consequently for a climb split edge dislocation the influence of the composition ratio on the fault mobility.

3.5. Viscous slip of dislocations

Let us consider an edge dislocation, parallel to the [001] direction, gliding on the (110) plane ; it has been climb dissociated in the $(1\bar{1}0)$ plane. Figure 8 represents the corresponding projection on the (001) plane of the $(1\bar{1}0)$ planes in the vicinity of the core. The stacking of $(1\bar{1}0)$ planes comprises two kinds of planes ; the γ ones having the $(AlO_2)^-$ composition and the A or B planes of $(Al\,Mg\,O_2)^+$ composition. B planes are deduced from the A ones by the translation a/4 $[11\bar{2}]$. Partial dislocations P_1 and P_2 corresponding to pairs of extra half planes Bγ and Aγ limit the fault ribbon.

While in the perfect crystal the sequence of $(1\bar{1}0)$ planes is ...Aγ Bγ Aγ Bγ ..... it becomes ...Bγ Aγ $\downarrow$ Aγ Bγ ... in between the partials. The climb fault affects only the cationic sublattice since a/4 $[1\bar{1}0]$ is a translation of the oxygen sublattice.

Further glide of such a dislocation of a/4 $[1\bar{1}0]$ needs the glide of partials and the transformation of the part of A plane adjacent to the fault, in between the partials, into a B one. The whole cationic sublattice must be rearranged ; this will occur by cationic diffusional movements which must necessary happen in a cooperative way to keep locally the electroneutrality of the crystal. This is what we call viscous glide.

Such cooperative movements have been studied, for compositions n = 1 and 1.8 in different cases.

The transformation is presented in figure 9a in the case of an ideal equimolar spinel [22], i.e. when aluminium and magnesium cations occupy respectively the octahedral and tetrahedral sites. The ions belonging to neighbouring γ planes have been also figured in full circles. The necessary movements, in the A plane, which transform this one into a B plane are represented by arrows. This shows clearly that in some places as J or K such movements bring the

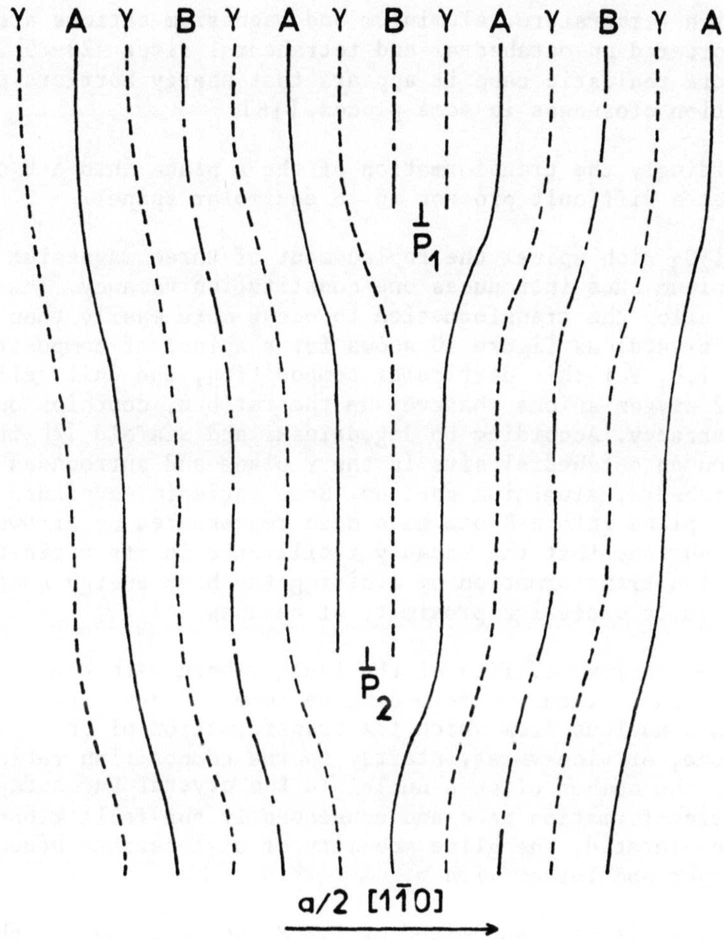

Figure 8 : Stacking of the (1$\bar{1}$0) planes in the vicinity of a climb split edge dislocation. The projection plane is the (001) one.

cations nearer involving thereby high energy barriers.

That situation still persists when the transformation is produced with cations of both A and γ planes (figure 9b). The first jump from I to J produces locally a high deviation from electroneutrality unless the movements of cations are synchronized which is problematic.

At high temperature, aluminium and magnesium cations are generally disordered on octahedral and tetrahedral sites [23-25]. Even in this more realistic case it appears that energy barriers persist due to cation closeness in some places. [16]

Accordingly the transformation of the A plane into a B one seems to be a difficult process in an equimolar spinel.

In Al_2O_3 rich spinel the replacement of three magnesium ions by two aluminimum ones introduces one constitution vacancy. These vacancies will allow the transformation to occur more easily than in an equimolar crystal as figure 10 shows for a spinel of composition ratio $n = 1.8$. For this particular composition, the unit cell, comprising 32 oxygen anions whatever be the ratio n, contains one constitution vacancy. According to Jagodzinski and Saafeld [2] it has been put on an octahedral site in the γ plane and surrounded by three tetrahedral aluminium cations. Some cationic movements changing the A plane into a B one have been represented by arrows 1 to 13. It is obvious that the vacancy facilitates in its vicinity the outset of the transformation by avoiding too high energy configurations due to an excessive proximity of cations.

The increasing mobility of the fault ribbon with the composition ratio can be then understood if we compare each constitution vacancy to a nucleus from which the transformation of th A plane into a B one, or vice-versa, starts. As the composition ratio n increases, the number of such nuclei in the crystal augments and thus the transformation rate and consequently the fault ribbon mobility is accelerated. The glide mobility of dislocations becomes therefore larger and larger with n.

This view is well supported by T.E.M. observations which show that the fault ribbon is frequently curved and twisted for compositions $n = 1.8$ [16] and 3.5 [14] contrary to equimolar composition where it is confined in the climb plane. This behaviour is in agreement with our model. In Al rich spinels where the cationic rearrangement is relatively easy, this one can go faster in some places, due to some local divergences to the macroscopic composition for instance, giving thus to each partial a more independant behaviour than in equimolar crystal.

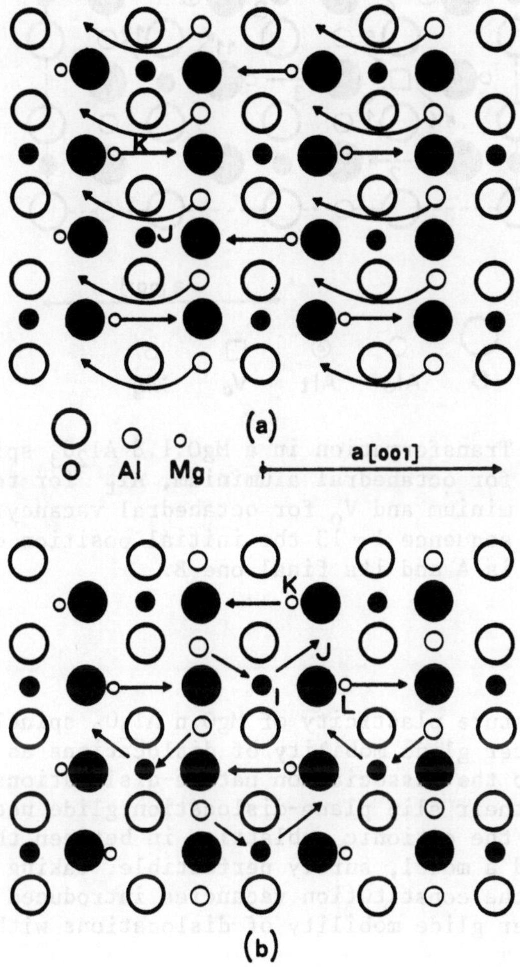

Figure 9 : Cationic movements changing an A plane into a B one in an ordered equimolar spinel.
a) Cations of A plane alone are involved in the transformation.
b) Transformation occurs with cations of A and γ planes.

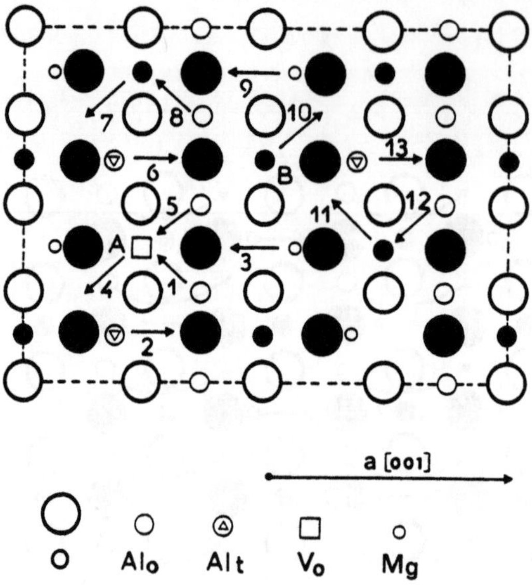

Figure 10 : Transformation in a MgO.1.8 Al_2O_3 spinel. Al_O stands for octahedral aluminium, Al_t for tetrahedral aluminium and V_O for octahedral vacancy. For the jump sequence $1 \rightarrow 13$ the initial position of the vacancy is A and its final one B.

4. CONCLUSION

High temperature plasticity of MgO.n Al_2O_3 spinels is consistent with a greater glide mobility of dislocations as the ratio n increases. Due to the dissociation nature-dislocations are generally dissociated off their slip plane-dislocation glide needs a local rearrangement of the cationic sublattice in between the partials. We have presented a model, surely perfectible, taking into account the presence of the constitution vacancies introduced when $n \neq 1$ to explain the higher glide mobility of dislocations with the composition ratio n.

REFERENCES

[1] D.M. Roy, R. Roy, E.F. Osborn, Am. J. Sci. 251, 837 (1953)
[2] H. Jagodzinski, H. Saafeld, Z. Kristallogr. 125, 197 (1958)
[3] R.D. Mac Brayer, Doctoral dissertation, North Car. state Univ. Raleigh (U.S.A.) (1965)
[4] K.C. Radford, C.W.A. Newey, Proc. Br. Ceram. Soc. 9, 131 (1967)
[5] M.H. Lewis, Phil. Mag. 17, 481 (1968)
[6] R. Duclos, Thèse d'Etat. Lille (France) (1980)
[7] L. Hwang, A.H. Heuer, T.E. Mitchell, Deformation of Ceramic Materials, R.C. Bradt. R.E. Tressler Editors. Plenum Press. N.Y. (1975).
[8] K.P.R. Reddy, A.R. Cooper, J. Am. Ceram. Soc. 64, 368 (1981)
[9] T.E. Mitchell, L. Hwang, A.H. Heuer, J. Mater. Sci. 11, 264 (1976)
[10] J. Hornstra, J. Phys. Chem. Solids, 15, 311 (1960)
[11] R. Duclos, N. Doukhan, B. Escaig, J. Mater. Sci. 13, 1740 (1978)
[12] W.T. Donlon, T.E. Mitchell, A.H. Heuer, Phil. Mag. A.45, 1073 (1982)
[13] M.H. Lewis, Phil. Mag. 14, 1003 (1966)
[14] W.T. Donlon, T.E. Mitchell, A.H. Heuer, Phil. Mag. A.40, 351 (1979)
[15] R. Duclos, N. Doukhan, B. Escaig, Acta Mat. 30, 1381 (1982)
[16] N. Doukhan, Thèse d'Etat. Lille (France) (1980)
[17] R. Duclos, Mem. Sci. Rev. Met. 75, 199 (1978)
[18] J. Grilhé, Workshop "Interaction dislocation-point defects in oxides" Paris-Bellevue (1978) unpublished work
[19] B. Escaig, Rev. Phys. Appl. 14 469 (1979)
[20] K. Ando, Y. Oishi, J. Chem. Phys. 37, 625 (1974)
[21] P. Veyssière, J. Rabier, J. Grilhé, Phys. Stat. Sol. A.24, 605 (1975)
[22] E.J.W. Verwey, E.L. Heilmann, J. Chem. Phys. 15, 174 (1947)
[23] E. Brun, S. Hafner, P. Hartman, F. Laves, die Natur. 12, 277 (1960)
[24] E. Stoll, P. Fisher, W. Halg. G. Maier, J. Physique 25, 447 (1964)
[25] U. Schmocker, H.R. Boesch, F. Waldner, Phys. Lett. 40, 237 (1972)

SLIP SYSTEMS IN MANGANESE ZINC FERRITE CRYSTALS

S. L. Callahan, R. E. Tressler
Pennsylvania State University, University Park, PA

D. W. Johnson, Jr.
Bell Laboratories, Murray Hill, NJ

M. J. Reece
University of Essex, Colchester, England

ABSTRACT

As-received and annealed manganese zinc ferrite crystals were deformed in uniaxial compression along <100>, <110> and <111> compression axes at 1100°C. The compositions of the two types of crystals were calculated to be $Mn^{2+}_{0.536}\,Zn^{2+}_{0.326}\,Fe^{2+}_{0.077}\,Fe^{3+}_{2.040}\,\square_{0.020}O_4$ and $Mn^{2+}_{0.538}\,Zn^{2+}_{0.327}\,Fe^{2+}_{0.099}\,Fe^{3+}_{2.024}\,\square_{0.0119}O_4$ based on the Fe^{2+} content and assuming that the Mn is all in the 2+ state. At 1100°C the easy slip system is $\{111\}\langle 1\bar{1}0\rangle$ as determined by slip trace analysis on the crystals compressed along <110> and <111> axes. This behavior differs from nonstoichiometric magnesium aluminate spinels where $\{110\}\langle 1\bar{1}0\rangle$ slip dominates. From these tests, the lower cation defect concentration, annealed specimens are softer at yield than the as-received, higher cation defect concentration specimens, again differing in behavior from the aluminate spinels which have a much larger range of defect concentrations. TEM examination of deformed specimens revealed a 3-D network of dislocations as a consequence of easy climb under these conditions.

INTRODUCTION

Deformation in spinel structures has attracted much attention since Hornstra[1] predicted that slip should occur on the $\{111\}\langle 1\bar{1}0\rangle$ system, and anticipated that these oxide materials should exhibit ductility at high temperatures. Definitive plastic deformation and microstructure studies on stoichiometric[2] and nonstoichiometric[3] $MgO \cdot nAl_2O_3$ have illustrated the complex considerations needed to rationalize the apparent change in slip plane from {111} to {110} and the resultant softening in nonstoichiometric materials. These include cation vacancy location, stacking fault energies and Peierls forces.

Although {110} planes are closest packed, {111} slip is easier than {110} slip in stoichiometric magnesium aluminate spinel. The reason for this situation is clear since {110} slip would involve both considerable electrostatic repulsion and expansion perpendicular to the slip plane. However, in nonstoichiometric spinel, {110} slip is dominant. This is because the critical resolved shear stress decreases making slip easier on both {110} and {111} systems as nonstoichiometry increases, but the Peierls energy for {110} slip decreases more rapidly. The stacking fault energy for the {110} planes also decreases with increasing nonstoichiometry. The decrease in the stacking fault energy with increasing nonstoichiometry is probably due to the increasing number of cation vacancies and their ability to rearrange to minimize the electrostatic potential. Thus, while the observation of {111} slip in stoichiometric spinel is in accord with dislocation theory, the preference for {110} slip in nonstoichiometric spinel can be explained in terms of the enhanced diffusion due to excess vacancies[2].

For transition metal spinels - especially ferrites - there are very little data on the effects of nonstoichiometry on mechanical properties. It is reasonable to assume that the major concern is oxygen nonstoichiometry, whereby excess oxygen is accommodated largely by the oxidation of Fe^{2+} to Fe^{3+} with the incorporation of cation vacancies. Morineau[4] and Morineau and Paulus[5] have shown that the vacancy accommodation process in manganese zinc ferrites is more complex than the one for $MgO \cdot nAl_2O_3$. They propose that Mn^{3+} and Fe^{3+} are produced thus creating both octahedral and tetrahedral vacancies with $\Delta Fe^{3+}/\Delta Mn^{3+} \cong 2$. This ratio, however, is dependent upon the distribution of Mn between A and B sites which is in turn affected by the Zn concentration.

While there has been no exhaustive study, it is known that the oxygen content of ferrites markedly affects their room temperature mechanical properties. Ishino et al.[6] studied the effects of rapid switching from a higher P_{O_2} during firing to a lower P_{O_2} during cooling. This treatment resulted in an increase in the bending strength at room temperature for some samples. The change in bending strengths was a function of the time and, accordingly, of the furnace temperature at which the oxygen partial pressure was switched. Increases in bending strength were attributed to the formation of an oxidized layer whereas decreases in strength were attributed to hematite formation. It was assumed that the effects of hematite precipitation are analogous to the precipitation of α-Al_2O_3 in nonstoichiometric spinel[7]. Tanaka[8] similarly found that the room temperature strengths of ferrites initially increased with increasing oxygen content and then decreased. Studies by Kosinski et al.[9] have shown that the fracture toughness at room temperature of polycrystalline ferrites decreased with increasing oxygen content even when there was no evidence of hematite precipitation.

A Mn-Zn ferrite single crystal was obtained from Fuji Electrochemical to determine the effect of oxygen nonstoichiometry on the slip systems in this material. The choice of crystal orientations have been guided by the previous deformation studies of $MgO \cdot Al_2O_3$ in which <100>, <110>, and <111> compressive axes were used. The <111> axes result in zero resolved shear stress on the {110} planes and permit the activation of {111} slip planes when they can act as the secondary slip system. One may speculate that the effect of nonstoichiometry in this ferrite will be similar to the case of the aluminate spinel: a transition from {111} to {110} slip as the cation vacancy concentration increases. The choice of oxygen partial pressure was guided by the phase stability studies of Slick[10] in an attempt to maintain the high initial vacancy concentration. In this paper, we summarize the preliminary studies of slip systems and deformation behavior for as-received and annealed crystals.

EXPERIMENTAL

The composition of the ferrite crystal, as determined by wet chemical analysis, was Fe-55.1 mole %, Mn-27.94 mole %, and Zn-16.96 mole %. The Fe content reported is the total (Fe^{2+} and Fe^{3+}) iron content. The Fe^{2+} content, determined by a coulometric titration technique[11] was 1.85 weight %. The vacancy concentration of the ferrite was determined using a site and charge balance approach. The assumptions made were as follows: (1) that the Fe^{3+} concentration was equal to the difference between the total iron concentration and the Fe^{2+} concentration and (2) that all the manganese was present as Mn^{2+}. The charge balance requires that the material remains neutral while incorporating excess oxygen, and the site balance requires that the 4:3 ratio of total cation sites to total oxygen content be maintained. This approach yielded $Mn^{2+}_{0.536}Zn^{2+}_{0.326}Fe^{2+}_{0.077}Fe^{3+}_{2.040}\square_{0.020}O_4$ as the composition of the as received crystal. The specimens were cut to 3mm x 3mm x 10mm from the single crystal, and all faces were oriented within 3° of the intended plane ((110), (111), or (112)) as determined by back reflection Laue. Some of the specimens were annealed at 1100°C in evacuated, sealed fused silica tubes for 30 days. The other specimens were tested as-received, unannealed. Guinier X-ray precession powder diffraction patterns revealed a homogeneous, single phase material before and after annealing. The lattice parameters of the annealed and unannealed specimens were determined to be 8.4887Å and 8.4866Å respectively, using an internal silicon standard. The material remained single phase after deformation.

The compression tests were performed in a vacuum/controlled ambient furnace. The oxygen partial pressure in the nitrogen ambient was maintained by the use of gas flow meters and monitored by a zirconia oxygen sensor. The specimens were isolated from the alumina rams by 1 mil platinum foil and high purity alumina platens and were deformed at strain rates ranging from 4.13×10^{-6} to 2.08×10^{-5} sec^{-1} using an Instron testing machine. During heating, either N_2 or a N_2-O_2 mixture with a lower P_{O_2} than the ultimate test P_{O_2} was used to ensure that the cation vacancy concentration did not change. As the temperature approached the testing temperature, the oxygen partial pressure was increased to the appropriate level. After deformation, the specimens were cooled in the ambient of the particular test by turning off the furnace power; the furnace temperature dropped from 1100°C to 700°C in 5 miniutes and cooled to 100°C in 30 minutes. The deformed specimens were examined by optical and transmission electron microscopy. Optical specimens were polished with γ-alumina and lightly etched in HF-HCl to highlight the active slip planes. The TEM foils were prepared by mechanical and ion beam thinning.

RESULTS

The increase in lattice parameter upon annealing indicates a decrease in the vacancy concentration, and thus, a change in the Fe^{2+} concentration. A linear relationship between the lattice parameter and the Fe^{2+} concentration has been found[12], and using this earlier relationship, the Fe^{2+} concentration in the annealed specimens in the present study was estimated to be approximately 2.39 weight %. Using the site and charge balance method previously described, the composition for the annealed ferrite was determined to be $Mn^{2+}_{0.538}Zn^{2+}_{0.327}Fe^{2+}_{0.099}Fe^{3+}_{2.024}\square_{0.0119}O_4$. The reduction in vacancy concentration (0.020 to 0.0119) apparently occured through oxygen loss and vacancy recombination.

The results of the compression tests at 1100°C for the annealed and as-received specimens are presented in Table 1. The 0.02% offset yield stress for the annealed specimens is lower than that of the corresponding unannealed specimens. As seen in Figure 1, the [001] compression axis is softest, followed by [111] and [110]. The stress-strain curve in Figure 2 shows the yield drop

Table 1. Compression Test Data

Load Axis	Condition	Temperature °C	P_{O_2} ppm	$\dot{\varepsilon}$/sec	0.2% Offset σ_{ys} (MPa)
111	annealed	1100	6800	4.56×10^{-6}	72.83
110	annealed	1100	6500	8.58×10^{-6}	85.56
001	annealed	1100	7000	8.40×10^{-6}	57.70
111	as received	1100	7000	8.56×10^{-6}	83.65
110	as received	1100	7000	8.09×10^{-6}	116.01
001	as received	1100	7000	7.01×10^{-6}	70.68
110	as received	1100	8000	8.05×10^{-6}	126.38
001	as received	1100	8000	8.24×10^{-6}	50.71

behavior of as-received and annealed specimens. The as-received specimens tend to strain soften while annealed specimens exhibit steady-state flow behavior. Differential strain rate tests performed on the annealed specimens gave a value of approximately eight for the stress exponent for all three load axes.

Slip trace analysis by optical microscopy indicates that slip occurs on the $\{111\}\langle 1\bar{1}0\rangle$ system for both the [110] and [111] in annealed and unannealed crystals. Figure 3 shows {111} slip traces on the (110) face making a 70° angle with the [111] compression axis; there is also evidence of interpenetrating slip. Figure 4 shows {111} slip traces on the $(1\bar{1}2)$ face making a 54° angle with the [110] compression axis. Specimens compressed in the [110] direction underwent a shape change (lateral displacement of the ends of the specimen) which is not uncommon[3] in cubic materials. Slip traces for the [001] load axis could not be interpreted. Multiple, interpenetrating slip traces were observed, but the density of the traces was such that determination of the slip system was not possible.

Transmission electron microscopic examination revealed a very low initial dislocation density ($<10^{-4}$ cm^{-2}). In the deformed

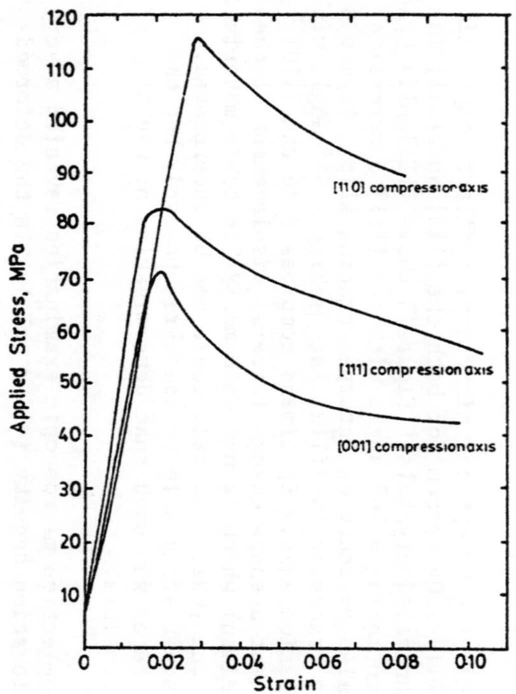

Figure 1. Stress-strain curves for [001], [110] and [111] compression axes of as received specimens tested at 1100°C, 7000 ppm O_2.

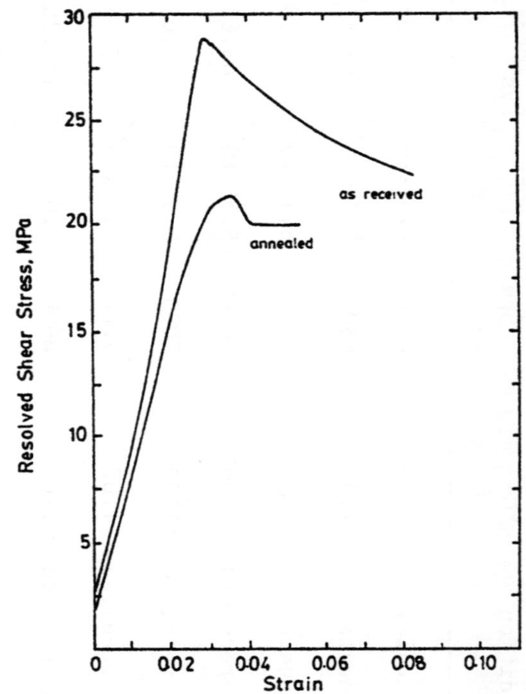

Figure 2. Stress-strain curves for the [110] compression axes for annealed and as received specimens tested at 1100°C, 6500 ppm and 7000 ppm O_2 respectively.

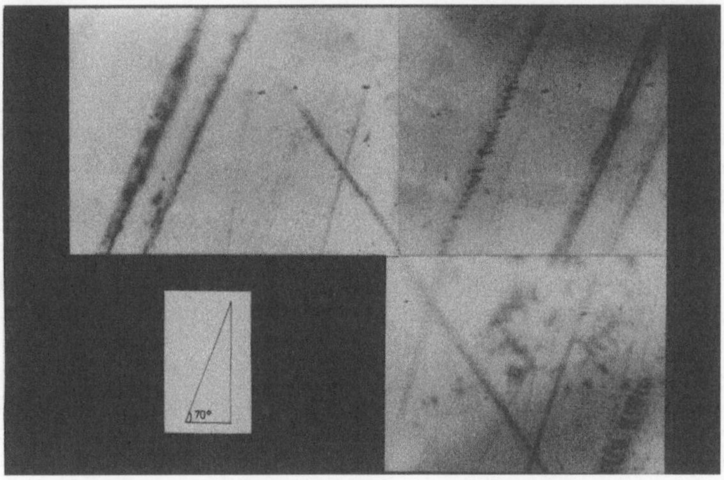

Figure 3. Optical micrograph of {111} slip traces on the ($\bar{1}10$) face of an annealed specimen compressed along the [111] axis of 1100°C, 6800 ppm O_2.

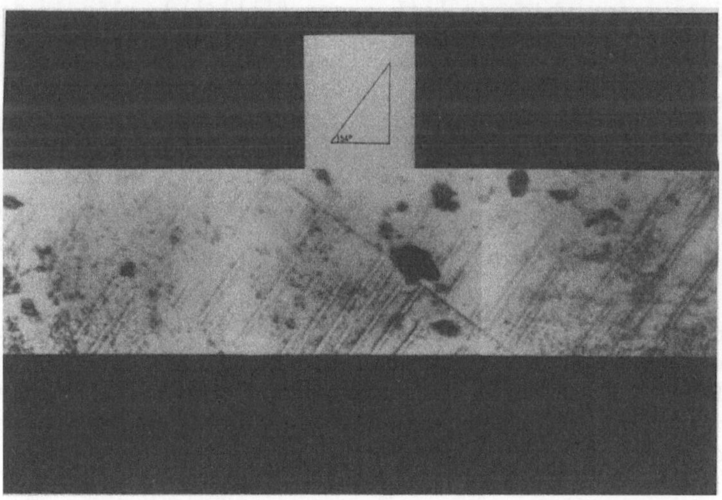

Figure 4. Optical micrograph of {111} slip traces on the ($1\bar{1}2$) face of an annealed specimen compressed along the [110] axis at 1100°C, 6500 ppm O_2.

specimens, a loose 3-d network of dislocations associated with a climb dominated substructure was observed.

DISCUSSION

From the data in Table 2, for the <110> and <111> load axes, one concludes that the {111}<1$\bar{1}$0> is the easy slip system. The slip traces for crystals deformed along the <001> load axis were indistinct and reminiscent of multiple interpenetrating slip which would be consistent with the {110}<1$\bar{1}$0> system being a secondary system. These results contrast markedly with the behavior of magnesium aluminate spinel where {111}<1$\bar{1}$0> slip is marginally easier than {110}<1$\bar{1}$0> slip only for stoichiometric spinel. (Duclos states that {110}<1$\bar{1}$0> slip is always favored[13].) With excess Al_2O_3, {110}<1$\bar{1}$0> is the favored slip system. The composition $MgO \cdot 1.1\ Al_2O_3$, which is the least Al_2O_3 excess that has apparently been studied corresponds to a vacancy concentration nearly equal to the as-received ferrite crystals in the present study. Therefore, although electrostatic repulsion should be of less concern in these semi-conducting spinels, {111} slip planes are favored in nonstoichiometric ferrites rather than the closest packed {110} planes observed for nonstoichiometric magnesium aluminate spinels.

The values of the critical resolved shear stresses for a given stoichiometry differ by a factor of two for the <110> and <111> load axes for which in both cases {111}<1$\bar{1}$0> slip is initiated. We have no explanation for this discrepancy, but we suspect that local stress concentrations in the <111> load axis orientation may be responsible for artifically low values of the 0.2% offset yield stress. Also, if the CRSS for {110}<1$\bar{1}$0> slip is not much higher than that for {111}<1$\bar{1}$0> slip, multiple slip may be initiated in the <110> load axis orientation due to the similarity of the Schmid factors. However, much more extensive microscopic investigation of the dislocation substructure at low strains in these specimens is necessary to explain this apparent discrepancy.

The other surprising aspect of these results is the much higher yield stresses for the as-received crystals with the higher cation defect concentration than the annealed crystals. Based on the behavior of the magnesium aluminate crystals where softening is observed with increasing concentration of defects in the octahedral sites, we expect a moderate softening of the as-received crystals compared to the annealed crystals. However, the defect concentration range in the present study is much smaller than the very large range of defects studied in the magnesium aluminate spinels. However, the post yield behaviors of the two stoichiometries are consistent with a larger defect concentration in the as-received crystals which causes faster

Table 2. Slip System Analysis

Load Axis	Slip System	τ	avg. 0.2% offset σ_{ys} (MPa) annealed	avg. 0.2% offset σ_{ys} (MPa) as received	CRSS (MPa) annealed	CRSS (MPa) as received
001	{110}<1$\bar{1}$0>	0.5σ	57.70	60.70	(28.85)*	(30.35)
	{111}<1$\bar{1}$0>	0.41σ			(23.66)	(24.89)
110	{110}<1$\bar{1}$0>	0.25σ	85.56	116	---	---
	{111}<1$\bar{1}$0>	0.41σ			35.08	47.56
111	{110}<1$\bar{1}$0>	0	72.81	83.67	---	---
	{111}<1$\bar{1}$0>	0.27σ			19.66	22.59

*parenthesis indicate uncertainly in determining the active slip system diffusion and, thus, strain softening due to more rapid recovery than is possible in the annealed crystals with the lower defect concentration. One suspects that the yield behavior is controlled by different processes or that the starting dislocation concentrations are markedly different in the two cases. In the steady state flow region the high stress exponent and the dislocation network structure observed by TEM point to a climb controlled flow behavior. Studies are in progress to resolve these apparent discrepancies and to more thoroughly document the mechanical behavior and dislocation substructures.

REFERENCES

1. J. Hornstra, Dislocations, "Stacking Faults and Twins in the Spinel Structure", J. Phys. Chem. Solids. 15:311 (1960).
2. T. E. Mitchell, L. Hwang, and A. H. Heuer, "Deformation in Spinel", J. Mat. Sci. 11:264 (1976).
3. M. H. Lewis, "The Defect Structure and Mechanical Properties of Spinel Single Crystals", Phil. Mag. 17:481 (1968).
4. R. Morineau, "Structural Defects and Oxidation-Reduction Equilibrium in Mn, Zn Ferrites", Phys. Stat. Sol. (a) 38:559 (1976).
5. R. Morineau and M. Paulus, "Oxygen Partial Pressures of Mn-Zn Ferrites", Phys. Stat. Sol. (a) 20:373 (1973).
6. K. Ishino, M. Makino, and E. Ikehata, "The Effect of Ambient Atmosphere upon the Bending Strength of Mn-Zn Ferrite", in Ferrites: Proceedings of the International Conference, 1970, Y. Hoshino, S. Iida and M. Sugamota, editors, University Park Press, Baltimore (1971).
7. W. T. Donlon, T. E. Mitchell, and A. H. Heuer, "Precipitation in Non-stoichiometric Spinel", J. Mat. Sci. 17:1389 (1982).
8. T. Tanaka, Young's and Shear Moduli, "Hardness and Bending Strength of Polycrystalline Mn-Zn Ferrites", Jap. J. Appl. Phys. 14:1897 (1975).
9. S. G. Kosinski, S. Vaidya, D. W. Johnson, Jr., and R. E. Tressler, "Oxygen Stoichiometry Effects on the Fracture of Mn-Zn Ferrites", Amer. Cer. Soc. Bull. 58:616 (1978).
10. P. I. Slick, "A Thermogravimetric Study of the Equilibrium Relations between a Mn-Zn Ferrite on an O_2-N_2 Atmosphere", in Ferrites: Proceedings of the International Conference, 1970, Y. Hoshino, S. Iida and M. Sugamota, editors, University Park Press, Baltimore (1971).
11. P. K. Gallagher, "A Coulometric Analysis of Iron (II) in Ferrites Using Chlorine", Amer. Cer. Soc. Bull. 57:576 (1978).
12. P. Blann, D. W. Johnson, Jr., and P. K. Gallagher, "Measurement of Oxygen Gradients and Resultant Strains in Ferrites", Amer. Cer. Soc. Bull. 59:553 (1980).
13. R. Duclos, "High Temperature Deformation Mechanisms in $MgO \cdot nAl_2O_3$ Spinels", this volume.

PLASTIC DEFORMATION OF OXIDES WITH GARNET STRUCTURE

J. Rabier and H. Garem

Laboratoire de Métallurgie Physique, L.A. 131 du CNRS

Faculté des Sciences 86022 Poitiers, France

INTRODUCTION

Garnet oxides show very good mechanical properties at high temperatures. This is the case for synthetic garnets such as $Y_3Fe_5O_{12}$ (YIG), $Gd_3Ga_5O_{12}$ (GGG) and $Y_3Al_5O_{12}$ (YAG) as well as natural garnets. Indeed silicates with garnet structure are usually found to be not deformed in the earth mantle when the surrounding minerals have undergone a plastic deformation.

The need for synthetic garnets in the electronic industry has generated extensive studies on the crystal growth of such oxides. GGG, used as a substrate in bubble memory devices, is now available as large single crystals nearly free of defects. Such crystals are useful for application in the high temperature range where good mechanical properties are required. Fundamental work on the plastic properties of this compound is therefore of interest.

This paper deals with the plastic deformation of GGG at high temperatures. Most of the features of the plastic properties of GGG can be extended to other synthetic garnets (ex. : YIG[1,2]). The high temperature plastic deformation of GGG single crystals is investigated in compression between 1300°C and 1500°C. The dislocation substructure is studied by transmission electron microscopy (TEM, accelerating voltage : 200 kV) with the purpose of determining the role of dislocation dissociations on the plastic properties of synthetic garnet oxides.

PLASTIC DEFORMATION TESTS

Samples (2 x 2 x 5 mm^3) with <001> , <011> compression axis were cut form a Czochralski boule of GGG (Union Carbide, San Diego, CA). Creep tests as well as experiments at constant strain rate were performed in air in the temperature range 1300°C-1550°C (0.79 - 0.92 T_M). Even in this high temperature range, high stresses in creep (∿ 200 MPa) on low strains ($\dot{\varepsilon}$= 3.3 $10^{-6}s^{-1}$) are required to induce plastic deformation.

Creep curves exhibit a stage typical of crystals with a low initial density of mobile dislocations. These curves display an acceleration of the strain rate and no decrease in creep rate up to large deformation[3].

When deformed at constant strain rate, samples show high initial work-hardening which leads to a maximum in the stress-strain curves between 0.01 - 0.03 strain. Following this, there is a region of softening until a steady state flow stress is achieved (Fig.1).

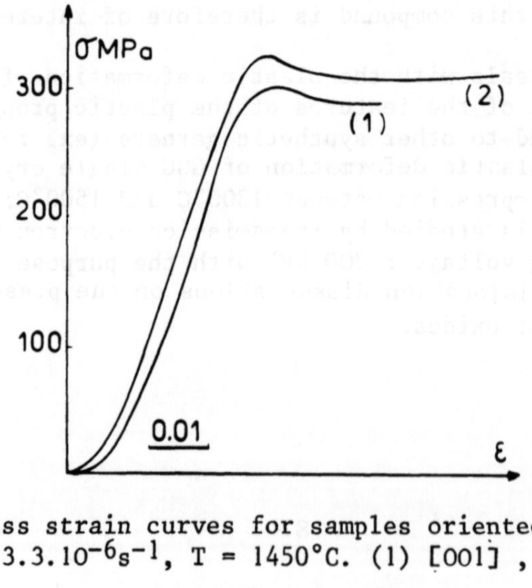

Figure 1 : Stress strain curves for samples oriented along [001] and [$\bar{1}$10]. $\dot{\varepsilon}$ = 3.3.$10^{-6}s^{-1}$, T = 1450°C. (1) [001] , (2) [$\bar{1}$10].

The obtained results for these two types of plastic deformation tests are consistent with the fact that multiplication of dislocations is difficult in GGG and that the dislocations have to overcome a lattice friction stress at high temperatures.

Glide planes as a function of the applied stress orientation

Experimental evidence for the glide planes activated has been obtained by etching and by stress-induced bi-refringence patterns[3]. The glide planes observed at high temperatures are {110}, {112}, {123}, similar to body-centred-cubic (BCC) crystals. The type of activated glide system depends on the orientation of the compression axis. The glide system with the highest Schmid factor is not always activated since {110} planes for a compression along [001] and {112} planes for a compression along [$\bar{1}$10] are observed, whereas the glide planes would be the same since the Schmid factors on {110} and {112} plane respectively are identical for these two orientations (see table 1).

Table 1. Calculated Schmid factors and observed slip planes depending on the orientations of the compression axes.

Compression axis	Potential glide plane {hkl}	Schmid factor on (11) hkl	Glide planes experimentally observed
[001]	{112}	0.47	
	{123}	0.46	
	{110}	0.41	(011) (101) (01$\bar{1}$) ($\bar{1}$01)
[$\bar{1}$10]	{112}	0.47	(1$\bar{1}$2) ($\bar{1}$12)
	{123}	0.46	($\bar{1}$23) (2$\bar{1}$3)
	{110}	0.41	

The observed violation of the Schmid law is likely due to the properties of the dislocations core. The anomalous behaviour can be considered in the light of the slip asymetry observed in BCC metals when a reversal in the direction of the applied stress yields a flow stress asymetry and in some cases, a difference in the activated slip systems[4]. Indeed a compression along <011> produces a shear in the twinning sense (TS, easy direction) in the {112} plane whereas a compression along <001> produces a shear in the anti-twinning sense (ATS, hard direction) (employing the terminology for the BCC metals). This asymetry is related in BCC metals to sessile core structure of screw dislocations. The observation of straight screw segments usually found in BCC metals deformed at low temperatures provides evidence for this sessile splitting. Transmission electron microscope (TEM) observations of such configurations in GGG would confirm the analogy with BCC metals.

Mechanical properties as a function of the applied stress orientation

In order to check the values of the critical resolved shear stress (CRSS) in the different activated glide planes, constant strain rate experiments have been performed on samples with [100] and [$\bar{1}$10] compression axes. The two types of samples show the same kinds of stress-strain curves (Fig. 1). However, samples deformed with [$\bar{1}$10] compression axes are harder.

The resolved shear stresses calculated from the experimentally measured yield stresses follow the inequalities

$$\tau_{(110)} < \tau_{(112)_{TS}} < \tau_{(112)_{ATS}}$$

Glide on {110} planes has the lower CRSS in GGG. These planes appear to be the easy glide planes. For a compression along <011> axis, {112} glide planes have the highest Schmid factor in the "twinning" sense with respect to {110} glide planes. This glide system is then operative since the glide direction is in the soft direction. For a compression along <001> a {112} glide plane can be activated in the antitwinning sense. Although it has the highest Schmid factor, this glide cannot be operative since the shear is in the hard direction and a {110} plane is activated. The analogy with BCC metals helds for the mechanical properties as a function of the orientation of the applied stress.

OBSERVATIONS OF DISLOCATION SUBSTRUCTURES

For the two orientations, slices have been cut parallel to the observed glide plane and perpendicular to the stress axis.

Samples deformed by creep

The deformation substructure of crept sample is characterized by large polygonal glide loops with a/2 <111> Burgers vectors. This segmentation may be ascribed to large lattice friction stresses.

For samples deformed along <001>, the TEM work confirms the operation of the a/2 <111>{110} glide system[3]. Glide loops found in {110} glide planes have straight segments usually aligned in <001> and <111> orientations. These directions are found in most of the configurations and can be attributed to dislocation splitting out of the glide plane.

In the same foil, i.e. parallel to the primary glide system, climb configurations have also been observed. Large climb loops with a/2 <111> Burgers vectors have been found to lie in {110} planes. The Burgers vector is out of the loop plane and the character is of the vacancy type (Fig.2).

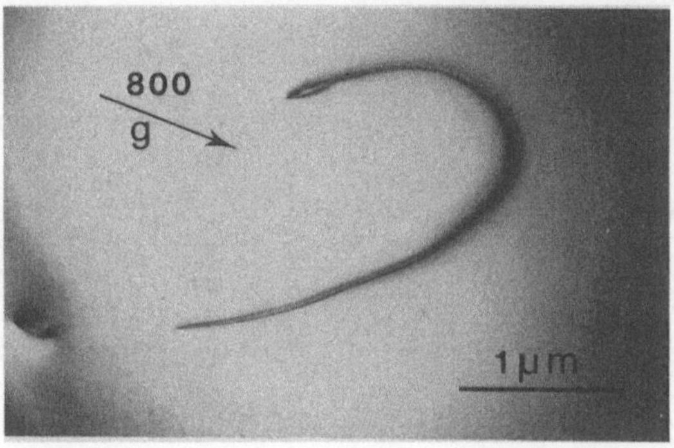

Figure 2 : Climb configuration in the (011) glide plane : vacancy loops a/2 [$1\bar{1}\bar{1}$] (011). Bright field $\vec{g}$ = [800], compression axis [001] .σ = 190 MPa, T = 1550°C,ε = 0.016.

The glide configuration typical of a sample deformed along the [110] axis can be seen in figure 3a. The operation of the ($\bar{1}12$) plane has been confirmed by TEM observations. Straight segments lie along [110], [131] and [311] directions whereas screw segments are seen to be smaller than the other dislocation orientations. Foils cut perpendicular to the compression axis show numerous climb loops with a/2 [$1\bar{1}\bar{1}$] Burgers vector lying in the ($\bar{1}10$) plane.

For these two orientations of the compression axis, the segmentations are consistent with a dissociation out of the glide plane in a {110} plane which does not contain the Burgers vector. This is in agreement with the observation of extensive climb in {110} planes as seen in figures 2 and 3 b. Furthermore a dissociation of a/2 <111> dislocation into two collinear partials has been observed in dislocation subboundaries. This dissociation involves climb of the partials[5].

Samples deformed at constant strain rate

Foils have been cut in samples deformed at constant strain rate and after low strain. It is probable that diffusion is less important at constant strain rate and that the configurations are more relevant to glide properties than those found in crept sam-

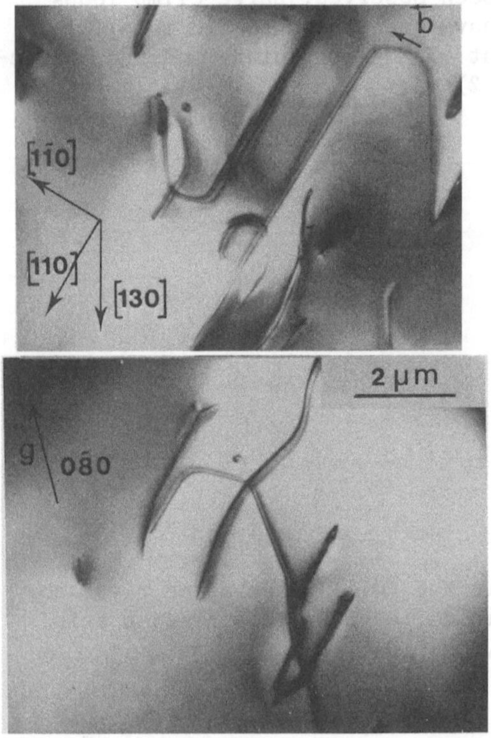

(a) Glide configurations in the ($\bar{1}$12) glide plane. $\vec{b}$ = a/2 [1$\bar{1}$1], Bright field g = [0$\bar{8}$0]. The directions are the projections in the (001) observation plane of [13$\bar{1}$] and [1$\bar{1}$1] directions in ($\bar{1}$12). (σ = 290 MPa, T = 1450°C, ε = 0.03).

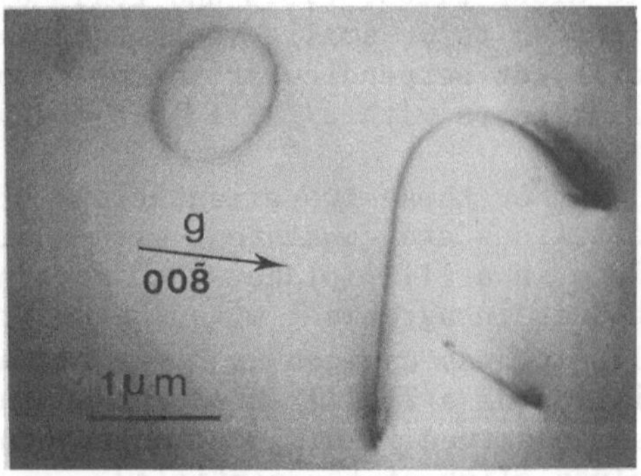

(b) Dislocation configurations in a plane perpendicular to the compression axis vacancy loop a/2 [1$\bar{1}\bar{1}$] ($\bar{1}$10) Bright field g = [008]. (σ = 280 MPa, T = 1550°C, ε = 0.025).

Figure 3 : Dislocation substructures in samples deformed by creep along the [$\bar{1}$10] orientation.

ples. The substructure is mainly the same as in crept samples. However, we can also find long straight screw dislocations in the primary glide plane belonging to Frank-Read sources (Fig.4). These kinds of configurations are analogous to those found in BCC metals at low temperatures.

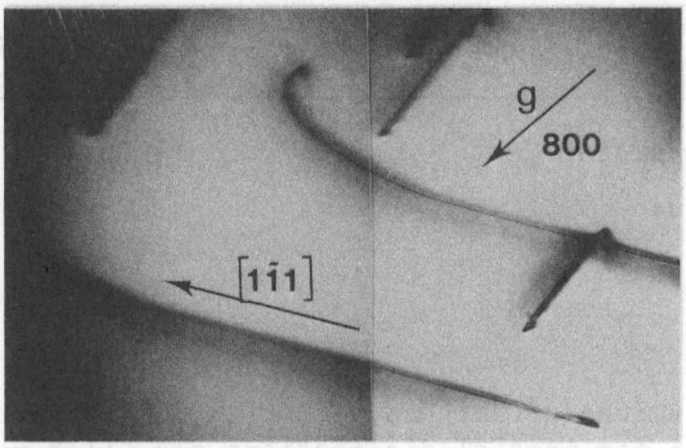

Figure 4 : Frank-Read sources with straight screw dislocation segments : $\vec{b}$ = a/2 [1$\bar{1}$1], Brigh field $\vec{g}$ = [800] , compression axis [001] ,$\dot{\varepsilon}$ = 3.3.$10^{-6}s^{-1}$, T = 1300°C, ε = 0.004.

DISLOCATION SPLITTING AND PLASTICITY

Analogy with BCC metals

The violation of the Schmid law can be explained by the dissociation of screw dislocations as proposed in BCC metals. The basic feature of these models was proposed by Hirsch in 1960[6] : the screw dislocations are weakly dissociated into a non-planar sessile configuration. The Peierls force is then the force required at the absolute temperature to transform this configuration into a glissile planar one by recombinaison of the partials.

The energy required to recombine the dislocation core is of the order of μb^3. Following Kubin[7], one can define for any BCC crystal a homologous temperature. The physical variable is the elastic energy stored in the dislocation($\sim\mu b^3$) versus the thermal energy of the lattice ($\sim$kT). It is then possible to define a transition between different mechanisms as a function of μb^3 in BCC crystals. Studies on various BCC metals have determined that the sessile core structure of screw dislocations controls the plastic properties for values of the ratio $kT/\mu b^3$ lower than 5.10^{-3} [7]. Table 2 gives the values of μb^3 for different BCC metals and for GGG.

Table 2 : Values of μb^3 for various BCC metals and GGG.

	Fe	V	Cr	Nb	Mo	Ta	W	GGG
T_M °C	1575	1710	1890	2500	2620	2995	3370	1720
b^3 eV	7.9	5.3	7.2	5.5	16	10	20	616 76*

T_M : melting temperature

μ : shear modulus

b : a/2 |<111>|

*b : a/4 |<111>|

The large values of Burgers vector in GGG for perfect dislocation as well as for partial dislocations a/4 <111> yield very large values of μb^3, bigger than those found usually for BCC metals. A calculation of $kT/\mu b^3$ for GGG at the melting temperature gives :

$$\frac{kT}{\mu b^3} = 2.7\ 10^{-4} \quad \text{with } b = \frac{a}{2}\ |\langle 111\rangle|$$

$$\frac{kT}{\mu b^3} = 2.10^{-3} \quad \text{with } b = \frac{a}{4}\ |\langle 111\rangle|$$

At the melting temperature, this value is lower than the critical value 5.10^{-3} below which the low temperatures behaviour of BCC metals is observed, i.e. : plastic behaviour controlled by the sessile splitting of the screws. This analysis shows that a dissociation-recombination mechanism for screw dislocations can be operative in GGG up to the melting point.

In agreement with the predictions of such a mechanism are :

- the values of the measured CRSS between 1300 and 1500°C which are of the order of magnitude of the stresses found for BCC metals at low temperature.
- the observations of long straight segments of screw dislocations.
- the observations of the dissociation of a/4 <111> dislocations in experiments performed at high stresses and lower temperature under confining pressure [8].

It should be noted that diffusion mechanisms are obviously competitive with the locking of screw segments in the investigated range of temperature and could hide or modify part of the "BCC features".

Dissociation and climb of dislocations

Polygonal loops have been observed showing dislocation segments confined in directions consistent with a dissociation out of the glide plane on a {110} plane. Dissociations into two collinear partials have been observed only in dislocation networks[5]. Dissociation on a {110} plane involves climb and glide of the partials, since the fault vector is at 55° of the fault plane (a/4 [111] (110)). This kind of dissociation has been discussed for Yttrium Iron Garnet [1,2]. As indicated by the observation of dislocation loops moving on {110} planes which do not contain the Burgers vector (figures 2, 3b), a possible deformation mechanism may be the displacement of the dislocation in the fault plane.

Consequences of dislocation splitting on plasticity

From the above analysis two classes of dissociation can be distinguished depending on the character of the dislocation :

i) Screw dislocations :

(1 S) $\frac{a}{2}[111] \rightarrow \frac{a}{4}[111] + \frac{a}{4}[111]$

This first dissociation occurs necessarily by a glide process.

(2 S) $\frac{a}{4}[111]$ partials are dissociated into a sessile core configuration

ii) Non screw dislocations :

(1 E) $\frac{a}{2}[111] \rightarrow \frac{a}{4}[111] + \frac{a}{4}[111]$ on the (110) plane

This dissociation requires the climb of partial dislocations

(2 E) $\frac{a}{4}[111]$ partials dissociated into a glissile configuration on the $(1\bar{1}0)$ plane.

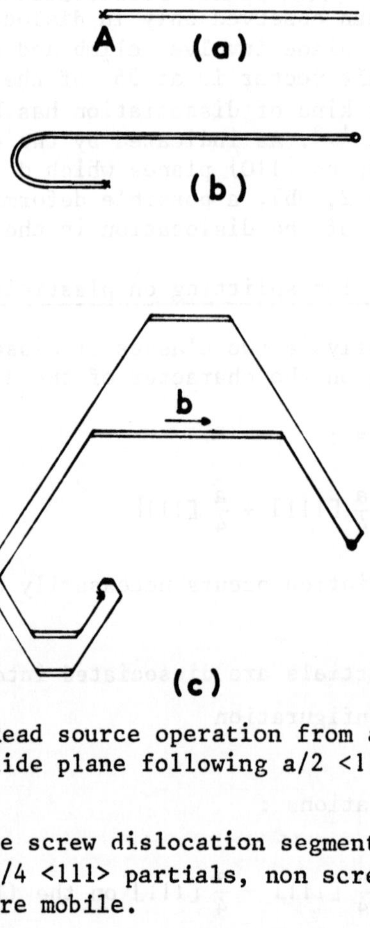

Figure 5 : Frank-Read source operation from a screw segment dissociated in the glide plane following a/2 <111> → a/4 <111> + a/4 <111>.

b) : locking of the screw dislocation segments by the sessile splitting of the a/4 <111> partials, non screw segments glide dissociated are more mobile.

c) : locking of the non screw segments by a climb dissociation a/2 <111>→ a/4 <111> + a/4 <111> in {110} planes out of the glide plane, screw dislocation segments become more mobile.

Reactions (2 S) and (2 E) are the "dissociations" analogous to those for BCC metals for screw and non screw dislocation components.

With respect to dissociation 1, screw dislocation are more mobile, the displacement by glide of this dislocation does not require point defect diffusion. With respect to dissociation 2, non screw segments are more mobile since they are dissociated by a glide mechanism.

Dissociation 1 for the non screw dislocation components controls the loop expansion since the screw segments are the shorter segments of a glide loop (see fig. 3a). However, screw segments can control the nucleation of the dislocation loop.

The expansion of a glide loop is shown schematically in figure 5. The screw segment is supposed dissociated by glide (1S) and will be easier to move by glide than a non screw segment. This segment can expand, dissociation (2S) controlling the propagation of the screw. This scheme is typical of BBC crystals i.e. long screw segments, small edge segments moving faster. When diffusion can be efficient (ex. : modification of the source geometry by obstacles), non screw segments can dissociate by climb by rotation of the stacking fault plane. Screw segments are then more mobile, their length will be reduced and the configuration will be that of figure 3a. In this scheme, the dissociation of screw segments does not control the plastic deformation, rather it controls the choice of the glide plane due to the sessile splitting of the partials. This dissociation takes an active part in the first stage of dislocation multiplication. However, the dissociation of non screw segments controls the recovery processes : displacement of dislocation in their dissociation plane and important contribution of climb to the macroscopic deformation.

CONCLUSION

The high temperature deformation of GGG single crystals has been investigated in compression between 1300°C and 1550°C. Creep experiments as well as constant strain rate tests in air were performed on samples with <100> and <110> compression axis.

A glide asymetry has been observed : according to the orientation of the compression axis (<100> or <110>), {110} or {112} glide planes are activated. This behaviour is analogous to that of BCC metals and can therefore be related to the sessile glide dissociation of a/4 <111> screw dislocations. Straight screw segments have been observed by TEM in agreement with this expected dissociation.

The high temperature deformation substructure is usually built at low strain with segmented glide loops as well as climb loops. The directions of the locked parts of the glide loops are consistent with the climb dissociation of a/2 <111> dislocations into two collinear partials on {110} planes. This is in agreement with the observation of extensive climb in {110} planes at low strain.

This present dissociation model for garnet structure explains the glide asymetry (due to glide dissociation) as well as the recovery process (due to climb dissociation) at high temperature.

Acknowledgments : The authors thank Dr. L.P. Kubin for fruitful discussions and Dr. M.P. Puls for reading the manuscript.

REFERENCES

1. J. Rabier, Dissociation des dislocations dans les oxydes de structure grenat.Application à l'étude de la déformation plastique du grenat de fer et d'yttrium (YIG), Thèse University of Poitiers (1979).
2. J. Rabier, P. Veyssière and H. Garem, Dissociation of dislocation with a/2 <111> Burgers vectors in YIG single crystals deformed at high temperature, Phil. Mag.A 44:1363 (1981).
3. H. Garem, J. Rabier and P. Veyssière, Slip systems in Gadolinium Gallium Garnet single crystals, J. of Mat. Sci. 17:878 (1982).
4. J.W. Christian, Plastic deformation of BCC metals, in "Second international conference on the strength of metals and alloys" Am. Soc. for metals, Metals Park Ohio, 1:31 (1970).
5. H. Garem and J. Rabier, to be published.
6. P.B. Hirsch, Oral communication, in "3th Int. Congress on Crystallography", Cambridge (1960).
7. L.P. Kubin, Deformation plastique et microscopie électronique des métaux et alliages de structure cubique centrée, J. Microsc. Spectrosc. Electron. 1 : 585 (1976).
8. H. Garem and J.Rabier, to be published.
9. F. Louchet, L.P. Kubin and D. Vesely, Phil. Mag. A, 39 : 433 (1979).

DEFORMATION AND DIFFUSION IN SAPPHIRE(α-Al_2O_3)

K.P.D. Lagerlof, T.E. Mitchell and A.H. Heuer

Department of Metallurgy and Materials Science
Case Western Reserve University
Cleveland, Ohio 44106

ABSTRACT

Basal slip in sapphire, α-Al_2O_3, has been investigated in undoped, Mg-doped, and Ti-doped crystals over a wide temperature range. The changes in the stress-strain curves caused by aliovalent dopants are explained by changes in the density of small prismatic loops formed from breakup of edge dipoles. Oxygen lattice diffusion data were obtained for these crystals from loop annealing experiments. A significant doping effect is observed on the diffusion kinetics, which in turn accounts for the change in loop density in the microstructure.

INTRODUCTION

Plastic deformation of sapphire, α-Al_2O_3, has been studied extensively during the last two decades. At elevated temperatures (>1200°C), attention has been given to determining the principal slip systems (1-5), and for studying work hardening and dynamic recovery (6-9). Microhardness studies have primarily been used to examine the plastic behavior at room temperature (10). In recent years, low temperature plastic deformation experiments (11), below the commonly- defined ductile to brittle transition temperature, have utilized hydrostatic confining pressure.

This latter work has shown that basal and rhombohedral twinning are a significant component of plastic behavior at low temperatures. Kronberg (2) has discussed basal twinning in terms of a synchroshear process while Hockey (12) suggested a model involving uniform shear; rhombohedral twinning has been analyzed by Heuer (13) and an atomistic model involving a shuffling mechanism has been proposed by Scott (14).

Lattice diffusion studies of aluminum (15) and oxygen (16-20) have been carried out using tracers. Oxygen diffusivities in particular are extremely small and most diffusivity measurements have been carried out at temperatures above 1500° C.

The relationship between diffusion and plastic deformation in fine-grained polycrystalline ceramics, including Al_2O_3, is well known (21). A similar close coupling between plastic deformation and diffusion exists for high temperature deformation in single crystals. For example, Pletka et al. (7) suggested a work hardening model for sapphire based upon the shrinkage of small loops by bulk diffusion as a rate-controlling process.

This paper will further expand on the relationship between diffusion and plastic deformation in sapphire, particularly the kinetics of dynamic recovery. The discussion of plastic deformation will be restricted to basal slip, since prism plane slip will be treated elsewhere in these Proceedings (22). Our concern with diffusion has involved the kinetics of annihilation of dislocation debris, particularly small prismatic dislocation loops formed from break-up of dislocation dipoles. Loop annealing studies have been carried out between 1200° and 1500° C for both undoped and impurity doped crystals.

PLASTIC DEFORMATION

Plastic deformation of sapphire involves two slip systems of importance: (0001) 1/3<1120> basal slip and $\{11\bar{2}0\}\langle 1\bar{1}00\rangle$ prism plane slip, although pyramidal slip ($\{10\bar{1}2\}\langle\bar{1}011\rangle$) (4, 23) can also be activated in carefully oriented crystals or in polycrystalline Al_2O_3 (4).

The resolved yield stress versus temperature for these several slip systems are plotted in Figure 1. Basal slip is the easy slip system at temperatures above 700°C, while prism plane slip is the easy slip system below this temperature. The yield stress for pyramidal slip is quite high in the temperature range where this slip system was activated.

DEFORMATION SUBSTRUCTURE ARISING FROM BASAL SLIP

Figure 2 shows a transmission electron micrograph of the dislocation substructure of sapphire deformed by basal slip at 1400°C. The sample contains mainly edge-type dislocations, often in the form of dislocation dipoles. Such dipoles are formed by edge-trapping (6), a process in which the strain fields of two edge dislocations of opposite sign gliding on parallel planes interact in such a way that the dislocations trap one another unless the

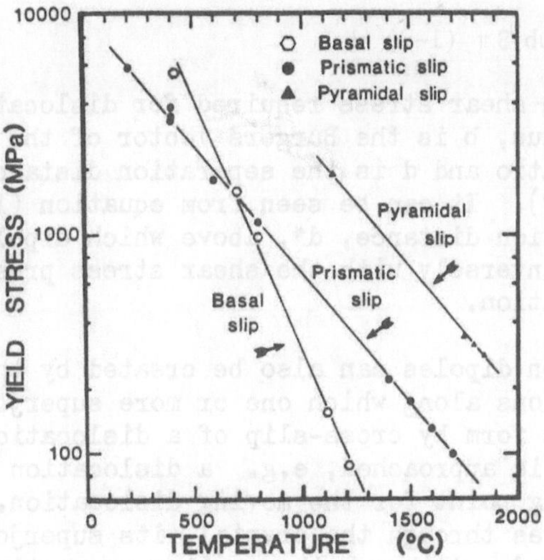

Figure 1: The yield stress for basal slip (1, 2, 5), prism plane slip (3, 5) and pyramidal slip (4, 23) versus temperature.

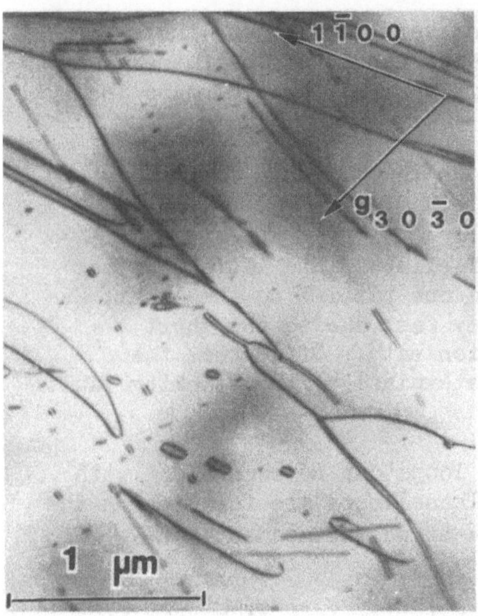

Figure 2: A typical microstructure of sapphire deformed by basal slip at 1400°C to 5% strain.

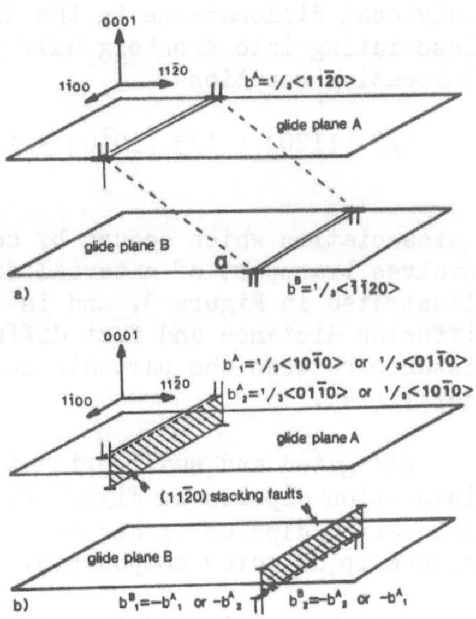

Figure 3: An edge trapped dislocation dipole a) before and b) after dissociation by self-climb.

resolved shear stress, τ, is given by:

$$\tau_o > \tau_o' + \mu b/8\pi(1-\nu)d \qquad (1)$$

where τ_o is the shear stress required for dislocation motion, μ is the shear modulus, b is the Burgers vector of the dislocations, ν is Poisson's ratio and d is the separation distance between the two glide planes (7). It can be seen from equation (1) that the maximum separation distance, d*, above which dipoles break-up by glide, varies inversely with the shear stress present during plastic deformation.

Dislocation dipoles can also be created by the movement of screw dislocations along which one or more superjogs have formed. These superjogs form by cross-slip of a dislocation segment when a local obstacle is approached, e.g. a dislocation pile-up, and can act as a pinning point for the moving dislocation. As the screw dislocation moves through the crystal, its superjog causes a dipole to be left behind. Although the maximum separation distance between the dislocations forming the dipole does not depend on the applied shear stress (rather it depends on the size and strength of the strain field causing the cross-slip), the requirement for dipole stability is the same as for the case of edge trapping defined by equation (1).

The dislocation dipoles are not very mobile, and the individual dislocations in the dipoles can lower their energy by dissociating into Kronberg half partials (2), according to the dislocation reaction

$$1/3\,[11\bar{2}0] \rightarrow 1/3\,[10\bar{1}0] + 1/3\,[01\bar{1}0] \qquad (2)$$

a dissociation which occurs by self-climb. This latter process involves transport of material from one partial to the other, as illustrated in Figure 3, and is very fast due to the short diffusion distance and fast diffusion path. Of course, the distance between the partials is determined by the stacking fault energy (24).

Elongated and small prismatic loops can be seen along with dislocation dipoles in Figure 2. These loops are formed from the dislocation dipoles either by climb of edge components or cross-slip of screw components.

TEM observation of thin foils between sequential annealings showed that the annihilation of elongated loops occurs by two distinct routes:

1) break-up into smaller loops by a dislocation fluctuation process involving pipe diffusion (25, 26), followed by shrinkage of the small loops by lattice diffusion; and

2) transformation into faulted prismatic dipoles, a process also controlled by pipe diffusion (27). These faulted dipoles are further annihilated by the processes described for perfect loops.

Both small perfect loops and faulted loops contribute to the background dislocation debris. Thus, the 1400°C structure shown in Figure 2 indicates that both bulk and pipe diffusion play important roles in the dislocation annihilation process at elevated temperatures.

The dislocation substructure of a sample deformed at 800°C under hydrostatic confining pressure is quite different, as can be seen in Figure 4. Two crystallographically- equivalent rhombohedral twinning systems, as well as the basal twinning system, were activated along with basal slip. The basal dislocations occur in 90° (perfect edge) and 30° orientations; the dislocations thus lie along close-packed $\langle 10\bar{1}0 \rangle$ directions corresponding to the Peierls valleys.

In regions with low dislocation densities, narrow dipoles can be observed ($d<4$ nm). However, very high dislocation densities can be observed close to the rhombohedral twins, e.g. near R in Figure 4, which act as barriers for basal slip.

Rows of very small perfect and faulted loops can also be seen dispersed in the background. These are believed to form either by break-up of the narrow dipoles by stress-induced diffusion, or by jog-dragging. For the case of 30° dislocations, it appears that faulted loops and dipoles are formed by jog-dragging. This can only be possible if the dislocation is dissociated by glide according to reaction (2), since one half partial would be in perfect screw orientation, and thus be able to cross slip and form a jog. Unfortunately, further evidence of glide dissociation has not yet been obtained using TEM.

At 400°C, the critical resolved shear stress (Figure 1) is such that prismatic slip, basal slip and rhombohedral and basal twinning are all activated. As in the 800°C sample, dipoles and small loops are observed (Figure 5). The major difference in the substructures of samples deformed at 400° and 800°C is that the basal dislocations appear to be less "crystallographic" in the former case. This difference may result from interaction with prism plane dislocations in this sample, resulting in node and jog formation. The prism plane dislocations are inclined in a basal foil, indicating mixed character. They are often observed along

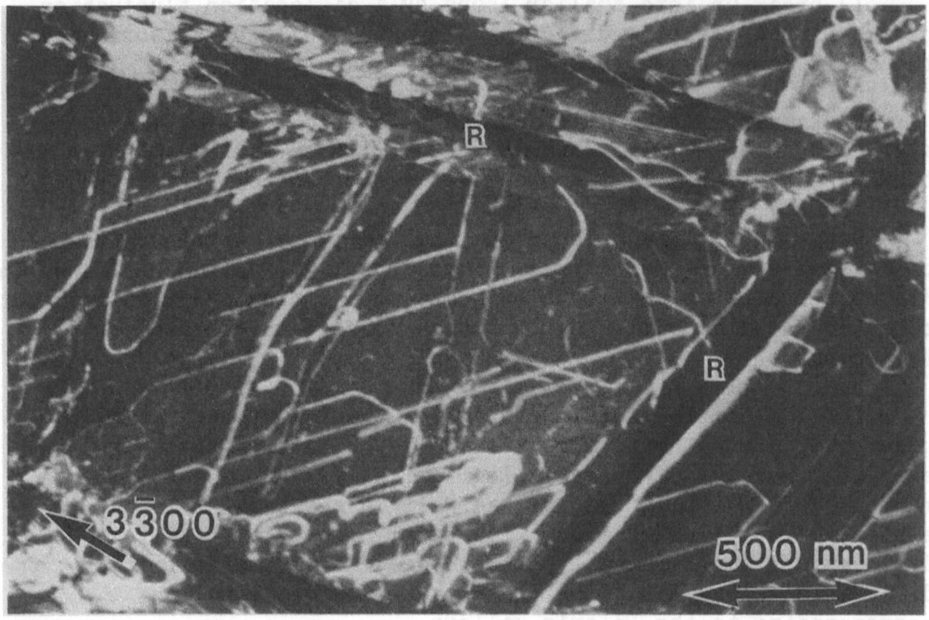

Figure 4: A typical deformation substructure of sapphire deformed by basal slip at 800°C. R indicates a rhombohedral twin.

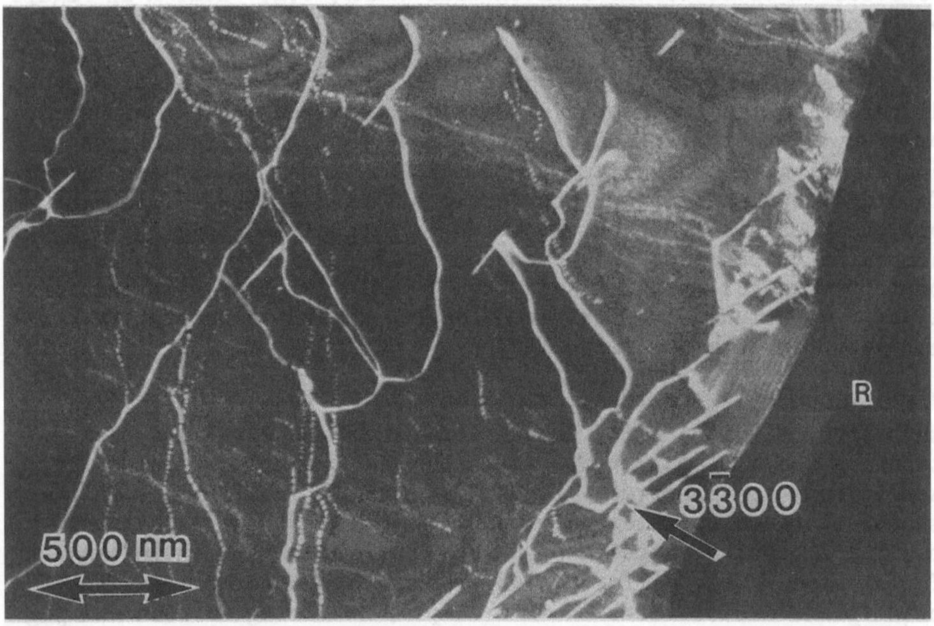

Figure 5: A typical deformation substructure of sapphire deformed by basal slip at 400°C. R indicates a rhombohedral twin.

rhombohedral twin boundaries, which must also act as barriers against their motion. When observed close to a rhombohedral twin, the prism plane dislocations are often dissociated.

It should also be noted that the dislocation substructures for the 800° and 400°C samples are very heterogeneous, and dramatic changes between two different areas of the same foil are often observed (28).

IMPURITY EFFECTS DURING BASAL SLIP

Small amounts of isovalent and aliovalent dopants can have drastic effects on the plastic deformation of sapphire (8). It can be seen from the stress-strain curves for undoped, Mg^{2+}- doped, Ti^{3+}-doped and Ti^{4+}-doped sapphire shown in Figure 6 that Mg^{2+}-doping strengthens the material, but only at small strains, while Ti^{3+}- and Ti^{4+}- doping strengthens sapphire over the entire deformation range.

The strengthening effect for the isovalent solute, Ti^{3+}, can be explained by size and modulus differences between the dopant and the host cation (29). For the case of aliovalent dopants, size and modulus differences are believed to be less important than the anisometric distortion arising from clustering of the charge-compensating point defects associated with aliovalent dopants (30).

The microstructure of the isovalent-doped crystal is the same as for an undoped crystal, but is quite different from both Mg^{2+}-doped and Ti^{4+}-doped crystals. Figure 7 shows the dislocation substructure of a Mg^{2+}-doped crystal deformed to 6% strain. The dislocation debris consists primarily of glide dislocations and unfaulted dipoles with virtually no faulted dipoles and small loops. For comparison, the microstructure of a Ti^{4+}-doped crystal deformed to 3% strain is shown in Figure 8. In this sample the dislocation substructure consists of a few glide dislocations, dipoles, and a high density of small perfect and faulted loops lined up in strings. These strings of loops are believed to form according to the dipole fluctuation models of Groves and Kelly (25) and Phillips et al. (26).

DIFFUSION

The diffusive processes of concern to plastic deformation in sapphire are lattice and pipe diffusion. The traditional way to measure lattice diffusion of oxygen in sapphire is to use stable isotopes as tracer ions. However, since oxygen diffusivities are quite low in the temperature range of interest (1200-1500°C) and since the characteristic diffusion length ought to be at least a

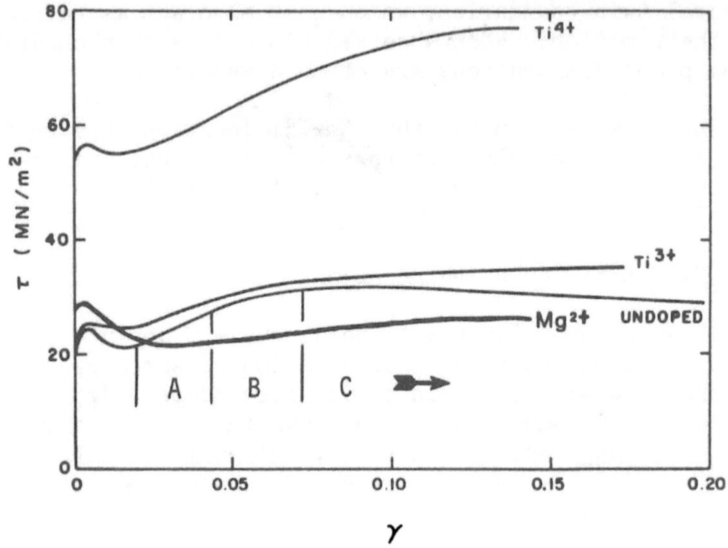

Figure 6: Stress strain curves for undoped, Mg^{2+}-doped, Ti^{2+}-doped and Ti^{4+}-doped sapphire. Regions A, B and C refer to the undoped material.

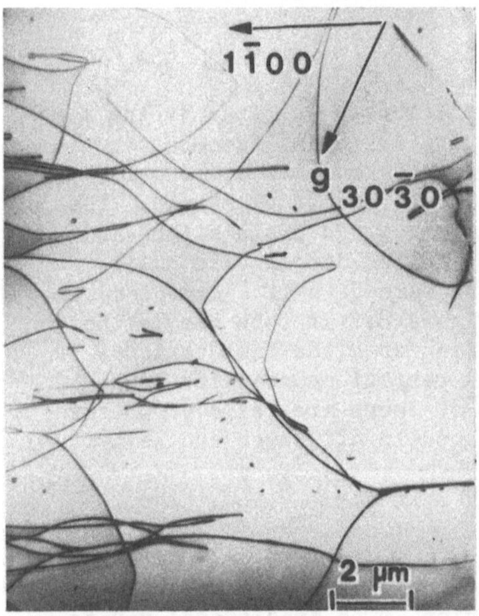

Figure 7: The deformation substructure of Mg^{2+}-doped sapphire deformed by basal slip at 1400°C to 6% strain. Note the low density of loops.

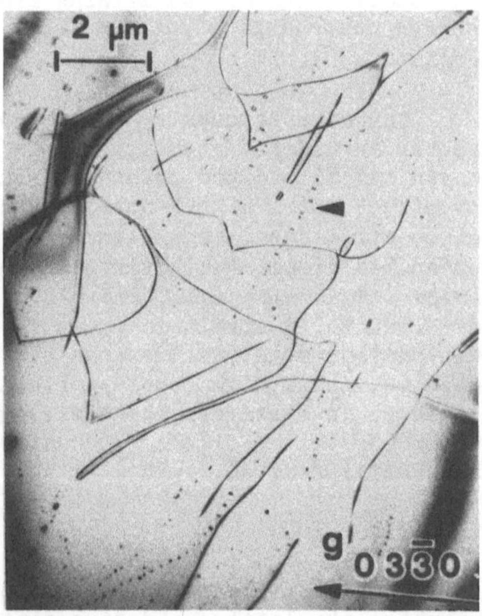

Figure 8: The deformation substructure of Ti^{4+}-doped sapphire deformed by basal slip at 1400°C to 3% strain. Note the prominent strings of loops (arrowed).

few microns to minimize experimental error, extremely long annealing times are required. The linkage between plastic deformation and diffusion in terms of dislocation loop annihilation mentioned above provides an alternative method to obtain diffusion data for this temperature range, and will now be described.

LOOP ANNEALING AND LATTICE DIFFUSION

As shown in Figure 2, the dislocation debris in a crystal deformed at 1400°C contains a high density of small prismatic loops. Pletka et al.(7) were able to relate the density of such loops to the steady state flow stress of sapphire.

After initial yielding occurs, the stress-strain curve can be separated into three different parts, A, B, and C, as shown in Figure 6. Region A is characterized by a constant work hardening rate, $d\tau/d\varepsilon = \theta$ = constant, while region C is characterized by zero work hardening; region B represents the transition between A and C.

In region A, the production rate of small prismatic loops is constant while the annihilation rate (integrated over the volume of crystal undergoing deformation) is small due to the low loop density. During plastic deformation, these small loops act as pinning points for moving dislocations, and give rise to work hardening. At the beginning of region B, as the loop density increases, the annihilation rate of the loops increases and finally in region C, the production rate and the annihilation rate are equal, resulting in a constant loop density and zero work hardening. As the final annihilation of the prismatic loops is controlled by lattice diffusion, the model used to describe the work hardening behavior could be used to estimate the lattice diffusion coefficient D of the rate controlling species, assumed to be oxygen in this case. The relationship between the lattice diffusivity and the plateau - stress at zero work hardening is given by:

$$D_L = (\mu/75\ (1-\nu)\ (\tau_p-\tau_o))^3\ k_B\ T\dot{\varepsilon}/\mu b \qquad (3)$$

where τ_p is the plateau-stress, k_B is Boltzmann's constant, T is the deformation temperature and $\dot{\varepsilon}$ is the strain rate. The plateau stress and the lattice diffusion coefficient are thus directly related at constant temperature and strain rate. Thus, from the stress strain curves shown in Figure 6, we would expect a higher value of D_L for Mg^{2+} -doped sapphire compared to undoped and Ti^{4+} -doped sapphire. This is consistent with the sub-structure shown in Figures 7 and 8, where the absence of small loops for Mg^{2+}-doped sapphire suggests rapid loop annihilation by relatively fast lattice diffusion, while the very high density of small loops for

Ti^{4+}-doped sapphire suggests a reduction in the ratio of pipe to lattice diffusion, presumably arising from a decrease in the lattice diffusion (8).

To estimate the lattice diffusion kinetics with more precision, individual loops within a thin TEM foil can be studied between sequential annealings at various temperatures and times. Silcox and Whelan (31) were the first workers to observe the direct shrinkage of prismatic loops by TEM, and it was later shown that the shrinkage rate was controlled by lattice diffusion (32). Treating the small loops as pure prismatic loops and using the method of Hirth and Lothe (33) to calculate the concentration gradient of point defects around a loop, it can be shown that the change in radius, r, between two observations of the same loop in sapphire before and after annealing is related to the lattice diffusion coefficient, D by:

$$r_1^2 - r_2^2 = \Omega\mu D_L t_a/(1-\nu)\, k_B T \qquad (4)$$

where r_1 and r_2 are the radius before and after annealing, Ω is the volume of an oxygen ion, and t is the annealing time.

Figure 9a shows an annealing sequence of a single loop at 1400°C at different annealing times, while Figure 9b shows the radius squared versus annealing time (the vertical bars represent the estimated error of the actual loop size). It can be seen that the radius squared is proportional to the annealing time, as required by Equation (4); the oxygen lattice diffusion coefficient can readily be calculated from the slope of each curve.

DISCUSSION

The data collected from the loop annealing study of both undoped and impurity-doped sapphire are presented in Figure 10, together with other data obtained from tracer diffusion and zero work hardening experiments. The tracer diffusion data more or less coincide with the loop annealing data, but the results from the zero work hardening model are consistently higher, which can possibly be explained by pipe diffusion contributions from the dynamic interactions between the loops and moving dislocations. The data obtained for undoped sapphire from the loop annealing study are basically an extrapolation of the tracer diffusion data down to lower temperatures and coincide with the recent data obtained by Cawley and Halloran (20) at 1300°C.

The loop annealing study shows a major effect of aliovalent dopants on the oxygen diffusivity; Mg^{2+}-doping increases and Ti^{4+}-doping decreases the oxygen diffusivity compared to undoped sapphire. This can be understood from the possible incorporation reactions of MgO and TiO_2, together with the assumption that oxygen

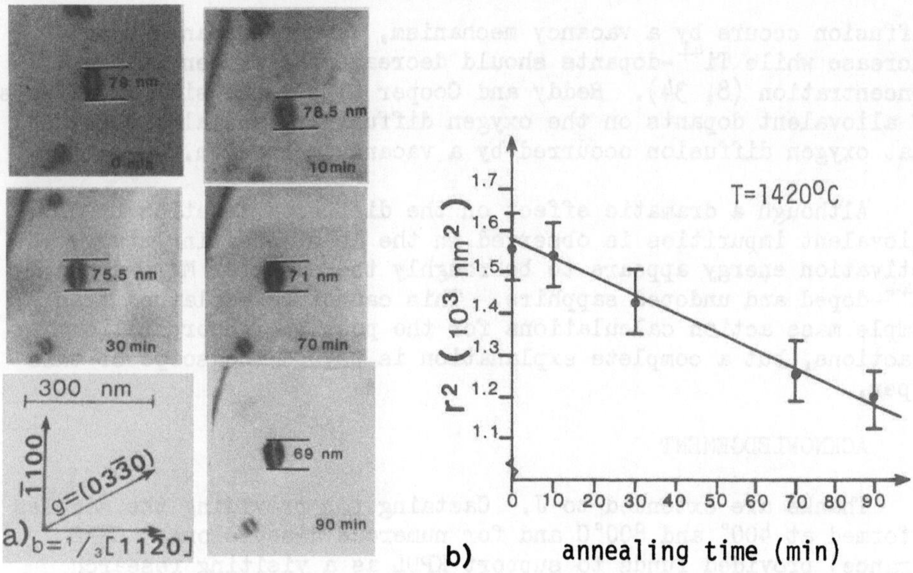

Figure 9: a) Annealing sequence of one loop at 1400°C for various times. b) The radius squared versus annealing time. The vertical bars represent the estimated error.

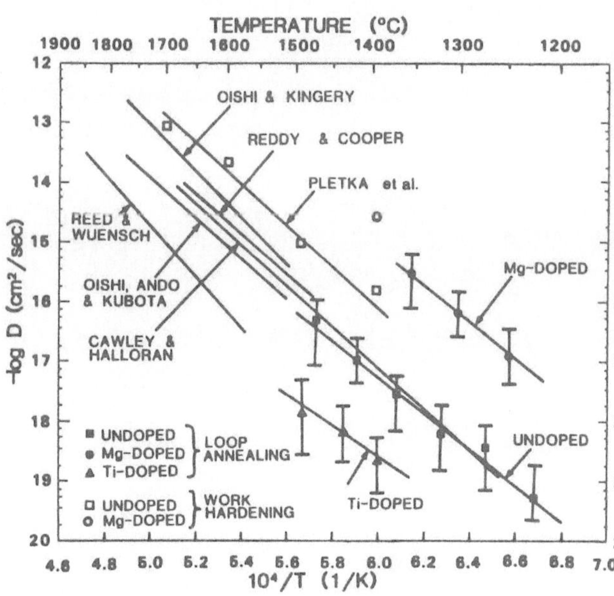

Figure 10: The diffusion coefficient from several investigations plotted versus reciprocal temperature. The vertical bars illustrate the scatter in data from the loop annealing experiments.

diffusion occurs by a vacancy mechanism, as Mg^{2+}-dopants should increase while Ti^{4+}-dopants should decrease the oxygen vacancy concentration (8, 34). Reddy and Cooper (19) found similar effects of aliovalent dopants on the oxygen diffusivity and also concluded that oxygen diffusion occurred by a vacancy mechanism.

Although a dramatic effect on the diffusion kinetics by aliovalent impurities is observed in the loop annealing study, the activation energy appears to be roughly the same for Mg^{2+}-doped, Ti^{4+}-doped and undoped sapphire. This cannot be explained from simple mass action calculations for the possible incorporation reactions, but a complete explanation is beyond the scope of this paper.

ACKNOWLEDGEMENT

Thanks are extended to J. Castaing for providing the samples deformed at 400° and 800°C and for numerous discussions. CNRS (France) provided funds to support KPDL as a visiting research fellow. This research was supported by the National Science Foundation under Grant No. INT76-10829. KPDL is also the recipient of an Alcoa Fellowship.

REFERENCES

1. J.B. Wachtman and L.H. Maxwell, J. Am. Cer. Soc. 40, 377 (1957).

2. M.L. Kronberg, Acta. Met. 5, 507 (1957), ibid. J. Am. Cer. Soc. 45, 274 (1962).

3. R. Scheuplein and P. Gibbs, J. Am. Cer. Soc. 43, 458 (1960).

4. J.D. Snow and A.H. Heuer, J. Am. Cer. Soc. 56, 156 (1973).

5. J. Cadoz, J. Castaing and J. Philibert, Revue Phys. Appl. 16, 135 (1981).

6. B.J. Pletka, T.E. Mitchell and A.H. Heuer, J. Am. Cer. Soc. 57, 388 (1974).

7. B.J. Pletka, A.H. Heuer and T.E. Mitchell, Acta Met. 25 (1977).

8. B.J. Pletka, T.E. Mitchell and A.H. Heuer, Acta. Met. 30, 147 (1982).

9. J. Cadoz, J. Castaing, D.S. Phillips, A.H. Heuer and T.E. Mitchell, Acta Met. 30, 2205 (1982).

10. B.J. Hockey, J. Am. Cer. Soc. 54, 223 (1971).

11. J. Castaing, J. Cadoz and S.H. Kirby, J. Am. Cer. Soc. 64, 504 (1981).

12. B.J. Hockey, Deformation of Ceramic Materials, p. 167, Ed. by R.C. Bradt and R.E. Tressler, Plenum Publishing Corp. (1975).

13. A.H. Heuer, Phil. Mag. 13, 379 (1966).

14. W.D. Scott, Deformation of Ceramic Materials, p. 151, Ed. by R.C. Bradt and R.E. Tressler, Plenum Publishing Corp. (1975).

15. A.E. Paladino and W.D. Kingery, J. Chem. Phys. 37, 957 (1962).

16. Y. Oishi and W.D. Kingery, J. Chem. Phys., 33, 480 (1980).

17. Y. Oishi, K. Ando and Y. Kubota, J. Chem. Phys. 73, 1410 (1980).

18. D.J. Reed and B.J. Wuensch, J. Am. Cer. Soc. 63, 88 (1980).

19. K.P.R. Reddy and A.R. Cooper, J. Am. Cer. Soc. 65, 634 (1982).

20. J.D. Cawley and J.W. Halloran, to be published.

21. R.M. Cannon, W.H. Rodes and A.H. Heuer, J. Am. Cer. Soc. 63, 46 (1980).

22. J. Cadoz, J.P. Riviere and J. Castaing, these Proceedings.

23. R.E. Tressler and D.J. Barber, J. Am. Cer. Soc. 57, 13 (1974).

24. K.P.D. Lagerlof, T.E. Mitchell, A.H. Heuer, J.P. Riviere, J. Cadoz, J. Castaing and D.S. Phillips, Acta. Met., in press.

25. G.W. Groves and A. Kelly, J. Appl. Phys. 33, 456 (1962).

26. D.S. Phillips, B.J. Pletka, A.H. Heuer and T.E. Mitchell, Acta. Met. 30, 491 (1982).

27. T.E. Mitchell, B.J. Pletka, D.S. Phillips and A.H. Heuer, Phil. Mag. 34, 441 (1976).

28. K.P.D. Lagerlof, T.E. Mitchell, A.H. Heuer and J. Castaing, to be published.

29. B.J. Pletka, T.E. Mitchell and A.H. Heuer, Phys. Stat. Sol. 39, 301 (1977).

30. T.E. Mitchell and A.H. Heuer, Mat. Sci. and Eng. 28, 81 (1977).

31. J. Silcox and M.J. Whelan, Phil. Mag. 5, 1 (1960).

32. D.N. Seidman and R.W. Balluffi, Phil. Mag. 13, 649 (1966).

33. J.P. Hirth and J. Lothe, Theory of Dislocations, McGraw-Hill, New York, NY (1968).

34. K.P.D. Lagerlof, T.E. Mitchell and A.H. Heuer, to be published in Radiation Effects (in press).

Suggested Key Words

Sapphire
Basal Slip
Diffusion
Loop Annealing
Self Climb
Edge Trapping
Work Hardening and/or Recovery
Impurity Doped Sapphire
Low Temperature Deformation

T.E.M. OBSERVATIONS OF DISLOCATIONS IN AL_2O_3 AFTER PRISM PLANE SLIP AT LOW TEMPERATURE UNDER HYDROSTATIC PRESSURE

J. Cadoz, J.P. Riviere and J. Castaing

Laboratoire de Physique des Matériaux - C.N.R.S.

1, Place Aristide Briand, 92195 Meudon Cedex (France)

ABSTRACT

Alumina single crystals have been deformed between 200°C and 950°C under hydrostatic pressures up to 1500 MPa. The uniaxial compressive stress was between 6000 MPa and 1000 MPa.

T.E.M. observations have been performed on foils cut out of specimens deformed at 400°C (0.29 T_M). They contain a high density of faulted dipoles and loops, and perfect dislocations with edge and mostly screw character ; they all have Burgers vector in the <10$\bar{1}$0> direction. Glide dissociation in three collinear partials is often observed. This microstructure is very different from that of high temperature prism slip. Some climb related features exist in spite of the low temperature of deformation.

INTRODUCTION

The plastic deformation of alumina single crystals has been performed only at high temperatures in conventional mechanical tests. Below 0.6 T_M (melting temperature = 2320 K), the specimens fail by fracture, and above 0.6 T_M, various slip systems have been activated. (For a review, see Bretheau et al. 1979). Prism plane slip <10$\bar{1}$0> {$\bar{1}$2$\bar{1}$0} has been activated at T > 1150°C by Gooch and Groves (1973), at T > 1400°C by Cadoz et al. (1977) (1981) (1982) and at T > 1550°C by Kotchick and Tressler (1980). The various authors have used, respectively bend, compression and tension tests. In order to deform alumina at T < 1000°C, Castaing et al. (1981a and b ; 1982) have performed compressive tests under hydrostatic pressure which permitted deformation down to 200°C.

Transmission electron microscopy (T.E.M.) has been used to study the dislocations left by prism plane slip at high temperature (Gooch et al.,1973 ; Bilde-Sorensen et al.,1976 ; Cadoz et al. 1977, 1982). We present in this paper observations of the microstructure introduced during tests at about 400°C (0.29 T_M).

EXPERIMENTAL TECHNIQUES

Mechanical test

Constant strain-rate compressions have been performed under hydrostatic pressure p on single crystals with a prism plane $\{1\bar{2}10\}$ at 45° to the uniaxial stress σ. Castaing et al. (1981a) have described in details the mechanical test apparatus and procedure. The use of a solid confining pressure allows to reach high pressures (p ∿ 1500 MPa) in a simple and cheap apparatus. However the corresponding stress tensor is not perfectly hydrostatic ; it may induce deformation, mostly by rhombohedral twinning which requires shear stresses as low as 10 MPa (Scott et al. 1983). To check for this, we

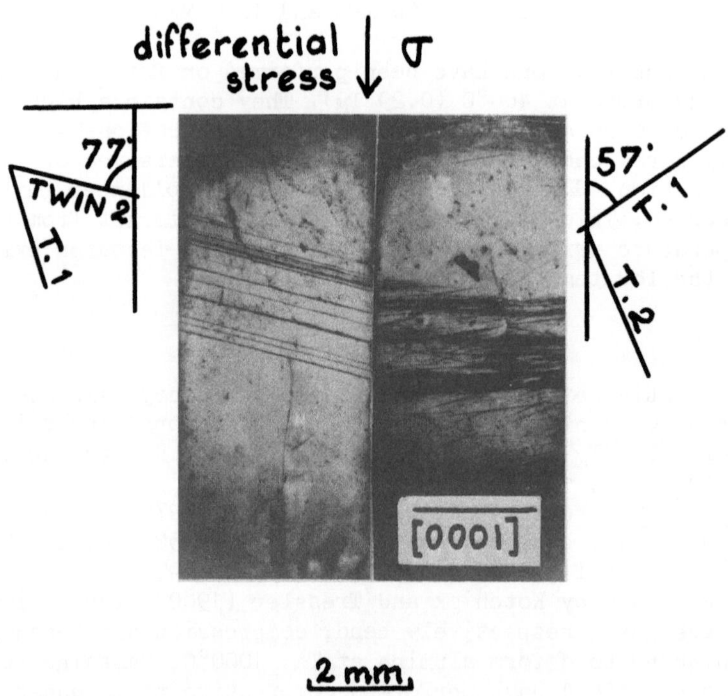

FIG. 1. Optical micrograph of two orthogonal faces of a specimen loaded at 600°C under p = 500 MPa and σ = 1526 MPa. Slip steps and twin traces can be seen.

have performed 3 tests where σ was too low to induce extensive plastic deformation. Figure 1 shows the surface of a crystal which was loaded to 80 % of the yield stress. Steps corresponding to activation of slip from an edge of the specimen (stress concentration) can be seen on the non-basal face of the specimen. Traces visible in both faces, correspond to the activation of three rhombohedral twins, one of them being more visible. Compressive stress σ cannot activate rhombohedral twinning (Scott et al. 1983) which is due essentially to the imperfection of hydrostatic pressurization. According to the various tests these twinning events are generally less frequent than what is shown in figure 1. They did not, apparently, influence the mechanical behaviour of alumina in prism plane slip (Castaing et al., 1981a).

Transmission electron microscopy

Thin foils have been observed by T.E.M. They were cut out of the specimen using a low speed diamond saw, mechanically polished to about 100 μm and ion-milled. Foils must be handled with care because of the high density of cracks. A conventional 100 kV microscope with a tilt-rotation stage was used.

RESULTS AND DISCUSSION

The activation of $<10\bar{1}0>$ $\{1\bar{2}10\}$ slip has been checked by Castaing et al. (1981a) through optical microscopy. The non-basal lateral face of the specimen exhibits steps corresponding to glide bands. Some structure is visible within these steps (fig. 2) which suggests that cross-slip occurs in planes at a small angle to the $\{1\bar{2}10\}$; this cannot correspond to the $\{11\bar{2}3\}$ slip plane observed by Hockey (1975). On the basal face of the specimen, there are no steps since the glide direction is contained in the face. However, the volume between glide bands is heavily cracked (fig. 2), thus revealing prism plane slip. Light is emitted under the electron beam; it also reveals the glide-bands which appear dark on a bright background (λ ∿ 0.7 μm).

Fig. 3 gives a general view of the dislocation structure in a foil cut perpendicular to the Burgers vector $<10\bar{1}0>$. It shows two bands a few microns apart, with cracks between them. These cracks are deviated by the glide bands, which except for a slight-offset, are continuous through the cracks. This is an indication that the cracks were formed after plastic deformation, during unloading of the specimen (Castaing et al., 1981a and b). The dislocation Burgers vector $\vec{b}$ was determined using $\vec{g}.\vec{b}$ analysis ; contrast extinctions were found compatible with $\vec{b}$ along $<10\bar{1}0>$ (fig. 4) in agreement with the macroscopic glide direction. Two types of bands have been observed (Fig. 3). The lower one is made of dislocation debris (faulted dipoles and loop rows aligned with $<20\bar{2}\bar{1}>$; it is

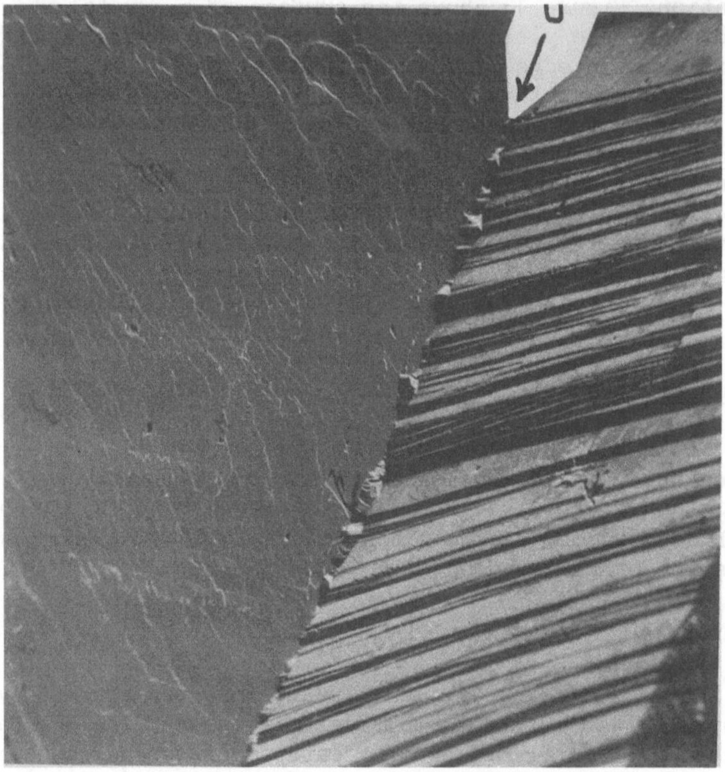

FIG. 2. Scanning electron micrograph of the edge of a deformed specimen showing prism plane slip features.
$T = 400°C$, $p = 1500$ MPa, $\sigma = 3780$ MPa, $\varepsilon = 5.2$ %. The edge is parallel to σ.

bordered by a set of dislocation segments parallel to $<10\bar{1}1>$ (fig. 4). The average distance between these segments is 120 nm, corresponding to an interaction force of about 0.1 N/m, ten times smaller than the applied force ; this configuration probably corresponds to a relaxed dislocation pile-up. This band represents very likely some hardened configuration where very few dislocations are mobile, except on the border.

The dislocation density is much larger in the lower band of fig. 3 than in the upper band. Fig. 5 shows a configuration where the dislocations do not glide in the primary slip plane at 45° to σ. Their Burgers vector has not been determined by $\vec{g}.\vec{b}$ analysis ; based on the contrast variation for the partial dislocations when

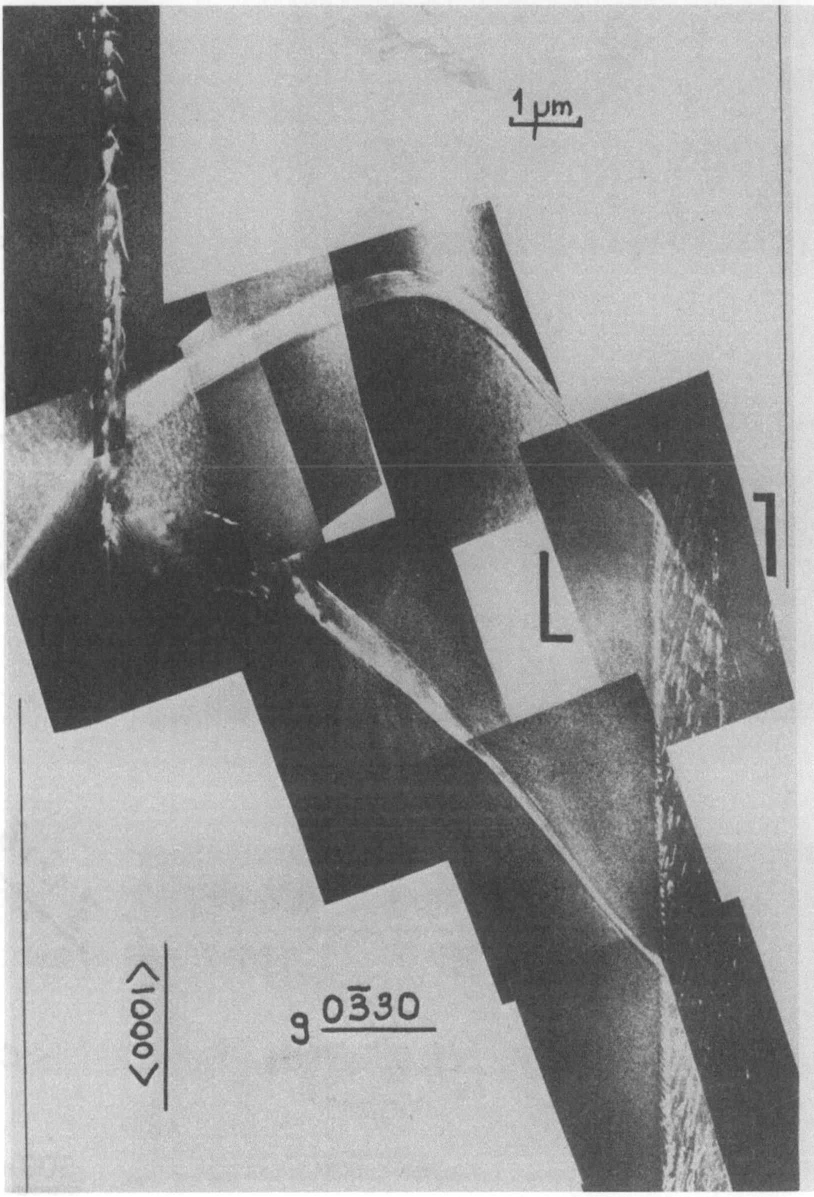

FIG. 3. T.E.M. micrograph showing two glide bands linked by cracks. Foil perpendicular to $\vec{b}$. Conditions of deformation : T = 400°C; p = 1500 MPa, ε = 3.3 %, σ = 3300 MPa.

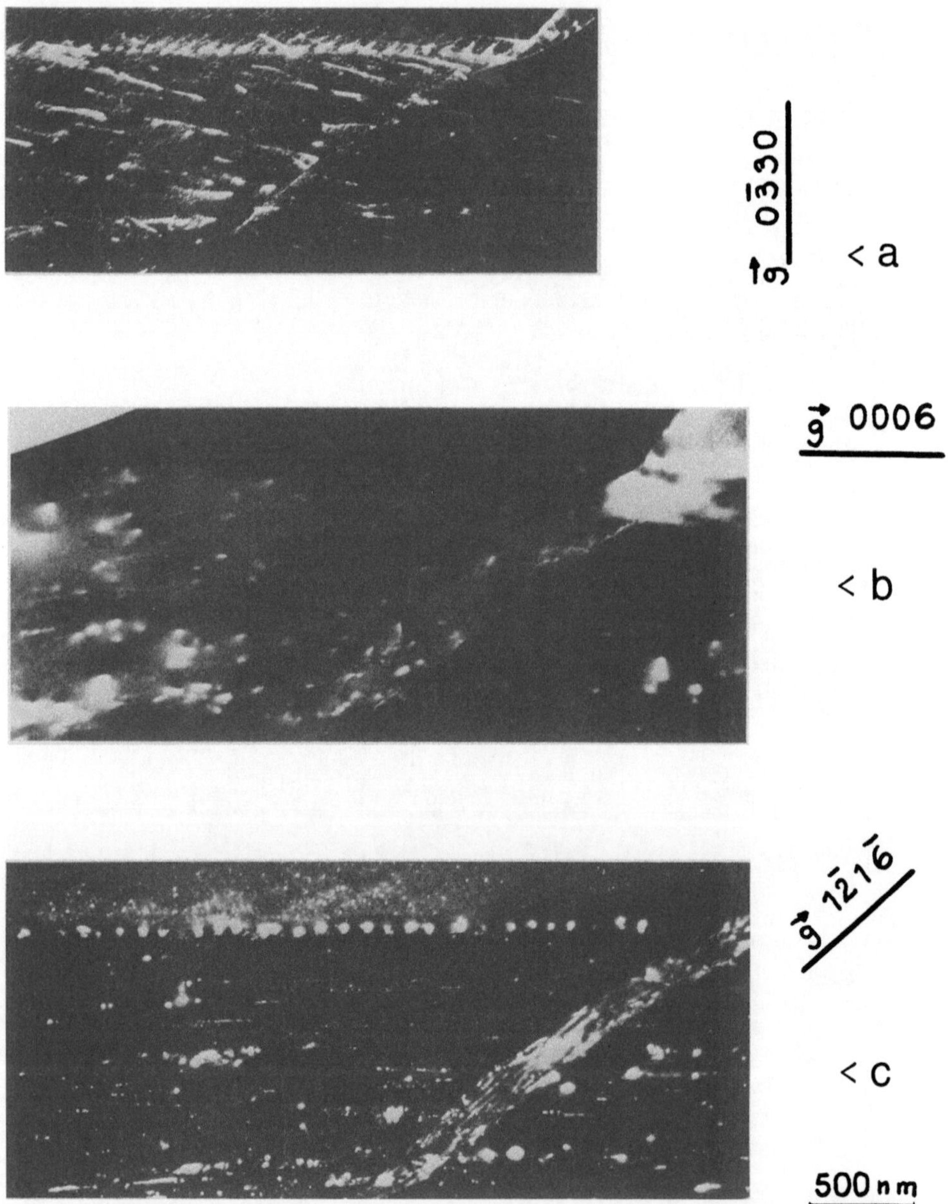

FIG. 4. Detail of the micrograph of fig. 3 for various diffraction conditions : g = 0006 and $1\bar{2}1\bar{6}$ give no contrast for dislocations.

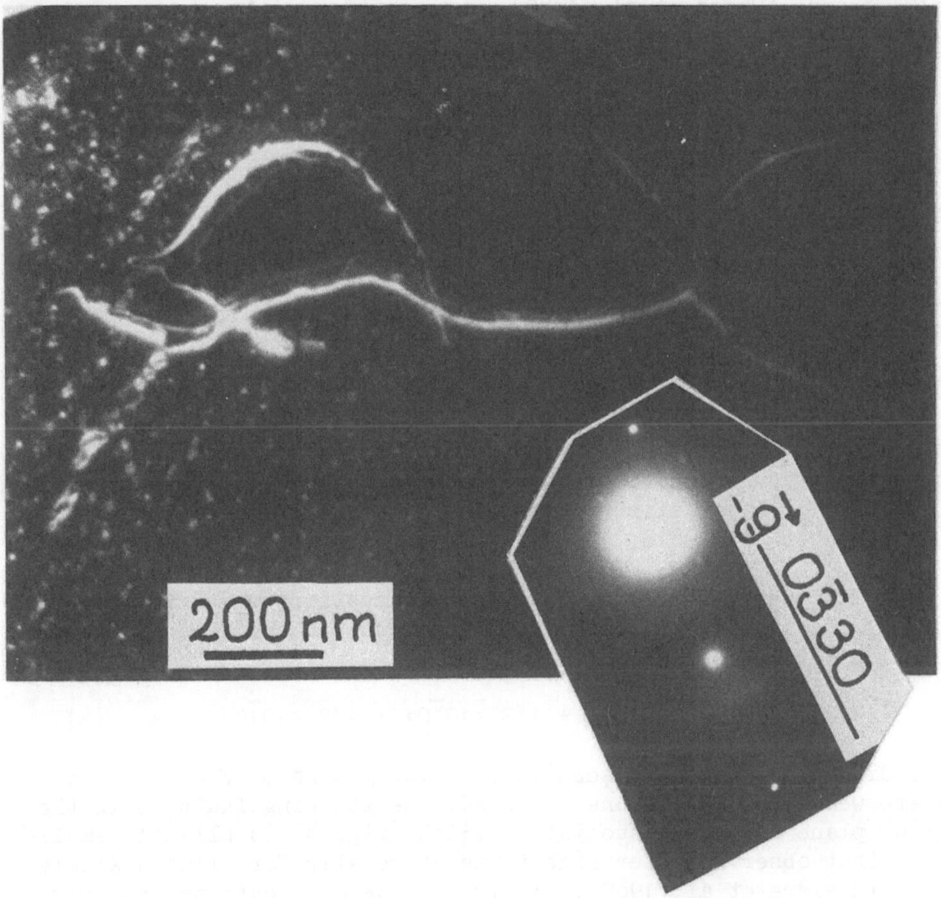

FIG. 5. Detail of the micrograph of fig. 3.

tilting the foil, it has been concluded that none of these dislocations have the primary Burgers vector. They probably have been induced during the pressurization of the material, which can punch out dislocations at elastic discontinuities. Such features should be much less prominent for the experiments performed at low pressures (Castaing et al. 1981a). Intersection of dislocations can be seen, with some dragging effect due to jogs. Dipoles and loops can also be created because of low jog mobility (fig. 5).

The existence of prism plane slip dislocations with a <10$\bar{1}$0> Burgers vector 0.822 nm long is energetically unfavorable by comparison with other dislocations (Cadoz et al. 1982). After high temperature deformation, the microstructure is dominated by 1/3 <1$\bar{2}$10> type dislocations, which have a line energy much lower than the <10$\bar{1}$0> ones (Cadoz et al. 1982). Two reactions allow to generate 1/3 <1$\bar{2}$10> dislocations :

$$[1\bar{1}00] \rightarrow 1/3\ [2\bar{1}\bar{1}0] + 1/3\ [1\bar{2}10] \quad (1)$$

$$[1\bar{1}00] + 1/3\ [\bar{2}110] \rightarrow 1/3\ [1\bar{2}10] \quad (2)$$

We have not observed any 1/3 <1$\bar{2}$10> dislocation after deformation at 400°C. Intersections of <10$\bar{1}$0> with grown-in 1/3 <1$\bar{2}$10> dislocations (reaction (2)) should have the same efficiency at low as well as high temperature, leading to similar microstructures dominated by 1/3 <1$\bar{2}$10> dislocations for both conditions, contrary to observation. This means that reaction (1) is, at high temperature, the dominant mechanism to decrease the energy of <10$\bar{1}$0> dislocations and that, at low temperature, it is not operative because the Peierls stress for the product dislocations is too high. The high temperature behaviour is expected at T > 700°C, when the C.R.S.S. for basal slip becomes smaller than for prism plane slip (Castaing et al. 1981b, 1982).

The energy decrease of <10$\bar{1}$0> dislocations which is not obtained by reactions with perfect dislocations, is achieved by dissociation into three collinear partials:

$$<10\bar{1}0> \rightarrow 1/3\ <10\bar{1}0> + 1/3\ <10\bar{1}0> + 1/3\ <10\bar{1}0> \quad (3)$$

The dissociation can be seen in the micrographs of Fig. 4 and 5 where weak beam conditions are used. The stacking fault is in the glide plane and the dissociation width (fig. 4) is slightly smaller than that observed after high temperature slip for climb dissociation (Rivière et al. 1980). In fig. 5, the dislocations probably are also dissociated by pure glide, but the width of dissociation is variable and can be as large as 100 nm, revealing non-equilibrium condition. This is due to the fact that the glide of partial dislocations is almost possible under the applied stress, and therefore possible at any stress concentration. Indeed, to a stacking fault energy γ of 0.4 J/m^2, it corresponds a critical stress for partial glide of 1500 MPa, equal to the applied shear stress at 400°C ; γ = 0.4 J/m^2 is the upper limit of values determined after high temperature deformation (see Lagerlof et al. 1984).

It is likely that this effect is sufficient at 400°C to explain the faulted dipoles and loops observed in the lower band of fig. 3, their formation being due to dislocation intersection and partial trailing. At T < 400°C, partial glide should be prominent and a large amount of stacking faults should be left in the specimen.

CONCLUSIONS

Preliminary T.E.M. studies of α-Al_2O_3 after prism plane slip at 400°C have revealed that all the observed dislocations have the expected prism plane slip Burgers vector and are dissociated by glide. The applied stress is at the limit where partial dislocation glide can be activated. The dislocation structure is quite different from that introduced at high temperature (T > 1400°C).

Acknowledgement: The mechanical tests have been performed at U.S. Geological Survey (Menlo Park, California) where S.H. Kirby offered laboratory facilities.

REFERENCES

Bretheau, T., Castaing, J., Rabier, J., and Veyssière, P., 1979, Dislocation motion and high temperature plasticity of binary and ternary oxides. Adv. Phys. 28:829

Bilde-Sorensen, J.B., Thölen, A.R., Gooch, D.J., and Groves, G.W., 1976, Structure of the <01$\bar{1}$0> dislocation in sapphire, Phil. Mag., 33:877

Cadoz, J., Hokim, D., Meyer, M., and Rivière, J.P., 1977, Observations des dislocations associées au glissement prismatique dans les monocristaux d'alumine, Rev. Phys. Appl , 12:473

Cadoz, J., Castaing, J., and Philibert, J., 1981, Glissement prismatique de Al_2O_3 déformé par compression, Rev. Phys. Appl., 16:135

Cadoz, J., Castaing, J., Phillips, D.S., Heuer; A.H., and Mitchell, T.E., 1982, Work-hardening and recovery in sapphire α-Al_2O_3 undergoing prism plane deformation, Acta. Met., 30:2205

Castaing, J., Cadoz, J., and Kirby, S.H., 1981a, Prismatic slip of Al_2O_3 single crystals below 1000°C in compression under hydrostatic pressure, J. Amer. Ceram. Soc., 64:504

Castaing, J., Cadoz, J., and Kirby, S.H., 1981b, Deformation of Al_2O_3 single crystals between 25°C and 1800°C : basal and prismatic slip, J. Phys. Colloq. C-3, 42:43

Castaing, J., Cadoz, J., and Philibert, J., 1982, Deformation plastique de l'alumine mono- et polycristalline entre 200°C et 1800°C, Rev. Int. Htes Temp. Refract., 19:377

Catlow, C.R.A., James, R., Makrodt, W.C., and Stewart, R.F., 1982, Defect energetics in α-Al_2O_3 and rutile TiO_2, Phys. Rev. B, 25:1006

Gooch, D.J., and Groves, G.W., 1973, Non-basal slip in sapphire, Phil. Mag., 28:623

Hockey, B.J., 1975, Pyramidal slip on {11$\bar{2}$3} <$\bar{1}$100> and basal twinning in Al_2O_3, in "Deformation of Ceramic Materials" R.C. Bradt and R.E. Tressler, Ed., Plenum Press, New-York.

Kotchick, D.M., and Tressler, R.E., 1980, Deformation behavior sapphire via the prismatic slip system, J. Amer. Ceram. Soc., 63:429

Lagerlof, K.D.P., Pletka, B.J., Mitchell, T.E., and Heuer, A.H., 1983, Deformation and diffusion in sapphire α-Al_2O_3, Rad. Effects, to be published

Lagerlof, K.D.P., Mitchell, T.E., Heuer, A.H., Rivière, J.P., Cadoz, J., Castaing, J., and Phillips, D.S., 1984, Stacking fault energy in sapphire α-Al_2O_3, Acta Met., submitted

Rivière, J.P., Cadoz, J., and Philibert, J., 1980, Climb dissociation of $[1\bar{1}00]$ dislocations in $(11\bar{2}0)$ prismatic slip in sapphire, in Proc. 7th Europ. Cong. Elect. Micro., The Hague, P. Brederoo and G. Boom, ed., 1:410

Scott, W.D., and Orr, K.K., 1983, Rhombohedral twinning in alumina J. Amer. Ceram. Soc., 66:27

PROCESSES CONTROLLING THE DISLOCATION MOBILITY IN MgO SINGLE CRYSTALS CONTAINING ALIOVALENT CATIONIC IMPURITIES

U. Messerschmidt, F. Appel and M. Bartsch

Institute of Solid State Physics
Academy of Sciences of the GDR
Halle/Saale, DDR-4020, GDR

MgO single crystals were deformed inside the high voltage electron microscope and in macroscopic compression tests. Several kinematic and dynamic data were evaluated from the micrographs. It is concluded that the dislocation motion is controlled by localized obstacles, the Peierls stress, long-range dislocation interactions and jog-dragging.

INTRODUCTION

It was the aim of the present investigation to identify the rate controlling mechanisms of the deformation of MgO single crystals at room temperature. A number of papers have been published on this subject[1 to 8] but the interpretation was based only on macroscopic deformation tests[1 to 5] or on measurements of the dislocation velocity by the stress pulse etching technique.[6,7] The present work employs these methods, too. The main emphasis, however, is laid on in-situ straining experiments in the high voltage electron microscope (HVEM). Experimental details and general results are described elsewhere.[9] The dislocation velocities were measured on the same material by Valkovskii and Nadgornyi.[10,11] As the results are concluded from different experimental methods they are relatively reliable though some inconsistencies are still existent. Several methods of evaluating the electron micrographs from the in-situ experiments have been designed, which can be applied also to other materials.

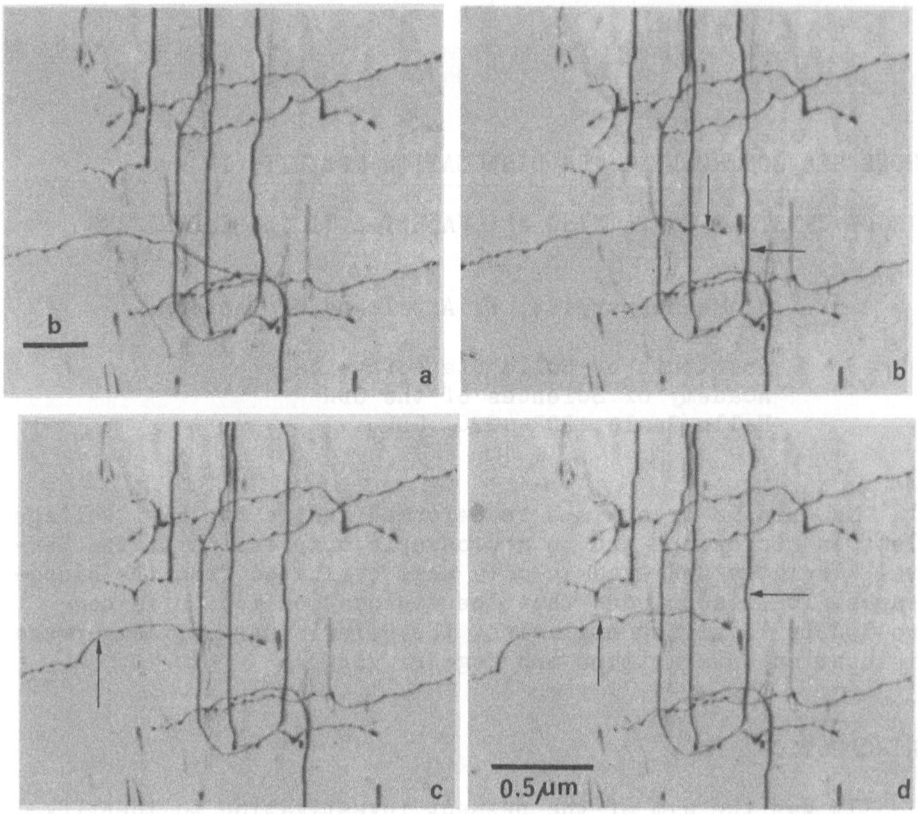

Fig. 1. Four successive micrographs taken at in-situ deformation in the HVEM showing the motion of edge and screw dislocations. Parts of the dislocations which had moved are labelled by arrows.

THE MOTION OF EDGE AND SCREW DISLOCATIONS

The processes controlling the mobility of dislocations may affect their shape under load on the electron microscopical scale. Fig. 1 shows a sequence of images of very slowly moving edge and screw dislocations. The slip planes as well as the ⟨110⟩ Burgers vectors are inclined by 45° with respect to the foil plane. Very frequently edge dislocations exhibit straight segments in cube directions which are connected by superkinks. The dislocations move by shifting these kinks along the dislocations. Several examples of changes in the positions of edge dislocations are imaged in Fig. 1. The mode of motion points out the action of the Peierls mechanism.

Screw dislocations show a curly shape (Fig. 1). Their motion is therefore impeded by localized obstacles. Under load the dislocations bow out between the obstacles. The overcoming of the obstacles is probably an essential contribution to the yield stress. Data on the statistics of the process are described below. Further details on the general dislocation behaviour, e.g. also on the dislocation multiplication, were given in an earlier paper.[9]

The different mechanisms controlling the motion of edge and screw dislocations result in different mobilities. According to the shapes of expanding dislocation loops in the in-situ experiment, edge dislocations move faster than screws by a factor of less than ten. The measurements of dislocation velocities by etch pitting[10,11] point out a much higher ratio. The discrepancy is probably due to a size effect in the thin electron microscopical specimens. X-ray topographic in-situ experiments show also the heads of slip bands of edge type to move much faster than those of screw type.[12]

In terms of the usual models of the interaction between dislocations and the elastic field of inclusions[13] it is difficult to understand that screw dislocations are more strongly impeded than edges. In MgO crystals doped with high concentrations of iron, the impurities precipitate in large octahedral particles of spinel structure. The dislocations cutting the particles have to create antiphase boundaries.[14,15] A similar mechanism is supposed to act also in the present material. Due to the different cutting heights, particles of the same size produce a spectrum of interaction profiles with the dislocations, which is entirely different for edge and screw dislocations.[16] For screw dislocations all particles are active as obstacles while to edges a part is transparent at intermediate stresses. The diameter of the precipitates can be estimated to be in the order of magnitude of 10b.[17]

KINEMATICS OF THE DISLOCATION MOVEMENT

Due to the discrete nature of the localized obstacles, the dislocation motion is discontinuous on a microscopic scale. Different parts of a dislocation line are resting for different time intervals. During a certain observation time the moving segments cover different forward distances. The respective kinematic data may be obtained from video recordings of the dislocation movement. Photographs were taken from the video screen at intervals of about 4 s each. An initial state (Fig. 2a) and a state after some observation time t_0 (Fig. 2b) were selected. A positive copy of the first state and a negative one of the later state were superimposed (Fig. 2c). Those parts of the dislocations which did not move during t_0 show a reduced contrast. The segments having moved appear in a black (initial

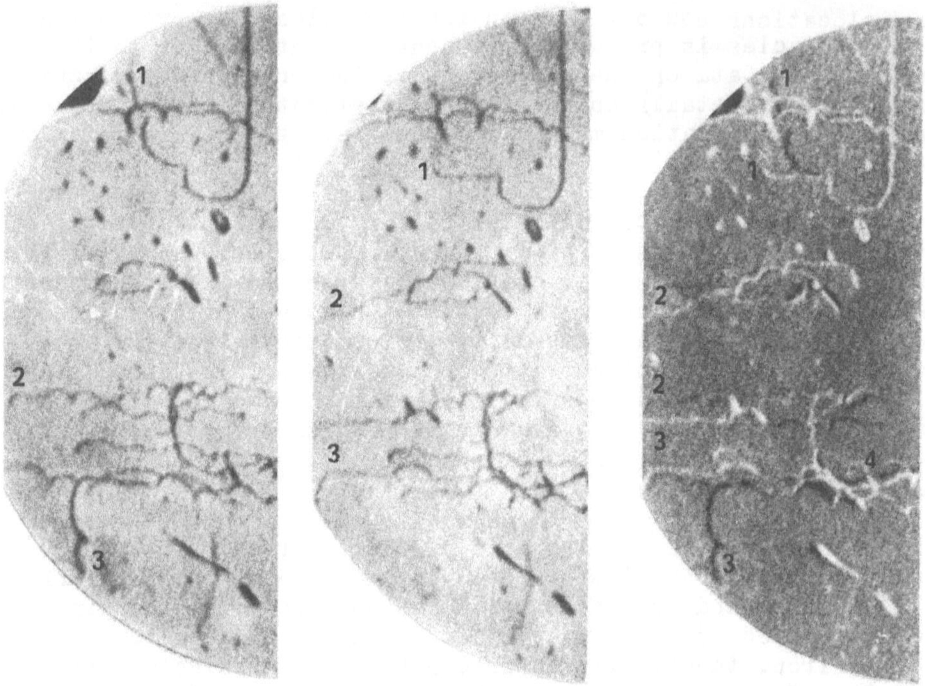

Fig. 2. Kinematic evaluation of video tapes of moving dislocations. Initial position, position after an observation time t_0, and superposition of both images. Dislocations having moved are marked by figures.

state) and a bright line (final state). The superimposed images were projected onto the screen of a coordinate digitizer in high magnification. The initial and final positions were represented by polygonal lines the coordinates of which were fed to a calculator. The details of the method will be published later.[18]

One movie sequence of moving screw dislocations and another one of edge dislocations were analyzed at observation time intervals between about 4 and 70 s. The following data were obtained. For each t_0 the frequency distribution of the local slip paths d of forward motion of the dislocation segments was established. Three examples are given in Fig. 3. For a short t_0 a strongly decreasing distribution is obtained, which becomes more flat with increasing t_0. At a sufficiently long t_0 a Gaussian distribution should be observed but this was not possible since the dislocations move beyond the imaged area. The average slip path $\bar{d}$ is an increasing function of t_0 approaching a linear dependence, which, however, does not meet the origin (Fig. 4). The slip path d is related only to the moving parts of the dislocations. This part is continuously increasing during the elaps-

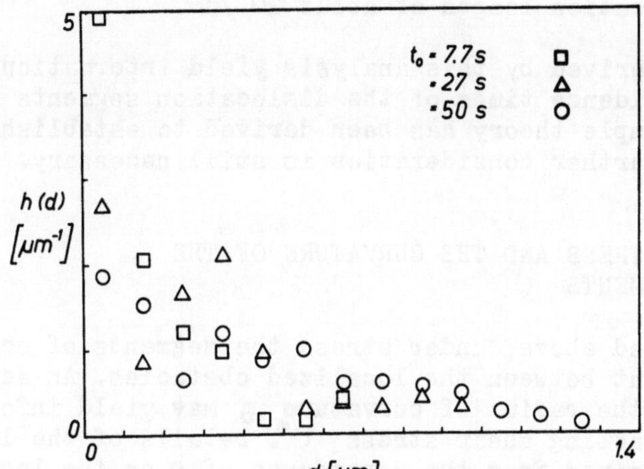

Fig. 3. Histograms of slip paths with different observation times t_o.

ing time. The product of d and the dislocation length having moved is the swept area ΔF. For the total ensemble of dislocations, ΔF should depend linearly on t_o if the dislocation velocity is homogeneous on the scale of observation, even though a spectrum of segments is always resting. Fig. 4 shows that this condition is fulfilled fairly well already for the comparatively small

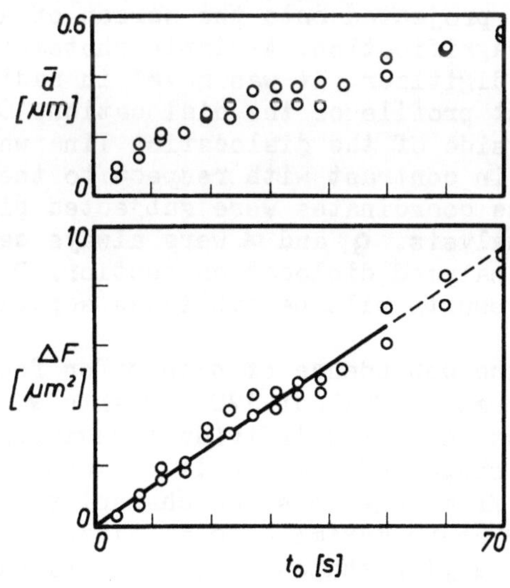

Fig. 4. Dependence of the average slip path $\bar{d}$ and the swept area ΔF on the observation time t_o.

evaluated dislocation length of about 20 /um.

The data derived by this analysis yield information on the spectrum of residence times of the dislocation segments on the obstacles. A simple theory has been derived to establish this spectrum, but further consideration is still necessary.

THE EFFECTIVE STRESS AND THE CURVATURE OF THE DISLOCATION SEGMENTS

As mentioned above, under stress the segments of screw dislocations bow out between the localized obstacles. An accurate measurement of the radius of curvature ϱ may yield information on the locally acting shear stress τ^*. Details of the line tension may be obtained from the dependence of ϱ on the local dislocation orientation α (angle between the tangent of the dislocation line and the Burgers vector b). In the line tension approximation in an isotropic elastic medium, the local radius of curvature is given by

$$\varrho = (1/\tau^*)E_o(1-2\nu+3\nu\cos^2\alpha)\ \ln(1/\beta r_o), \qquad (1)$$

where E_o is the energy factor of an edge dislocation, ν Poisson's ratio, l the distance between the obstacles (segment length), and r_o the inner cut-off radius. The numerical factor β was determined by self-stress calculations[19] to be about 5.

For measuring the dislocation curvature the micrographs of bowed segments were projected onto the screen of a coordinate digitizer in high magnification. A simple photometer was attached to the head of the digitizer. It was moved in radial direction through the contrast profile of the dislocation. Coordinates were measured on either side of the dislocation line which showed a defined difference in contrast with respect to the contrast maximum (Fig. 5). These coordinates were subjected piecewise to a cubic regression analysis. ϱ and α were always calculated for the centre of the analyzed dislocation section. Details of the treatment and the results will be published separately.[20]

Fig. 6 shows the dependence of ϱ on α for four dislocation segments. A total number of about 170 segments was evaluated. General characteristics of the individual segments are
- strongly varying shapes of the single segments,
- an increase of ϱ from edge to screw character, and
- frequently rather sharp maxima of ϱ at different orientations.

The reason for the varying shapes of the individual segments may be the influence of the different configurations of the neighbouring segments (a self-stress problem). After averaging the data of a greater number of segments, the course of ϱ over α

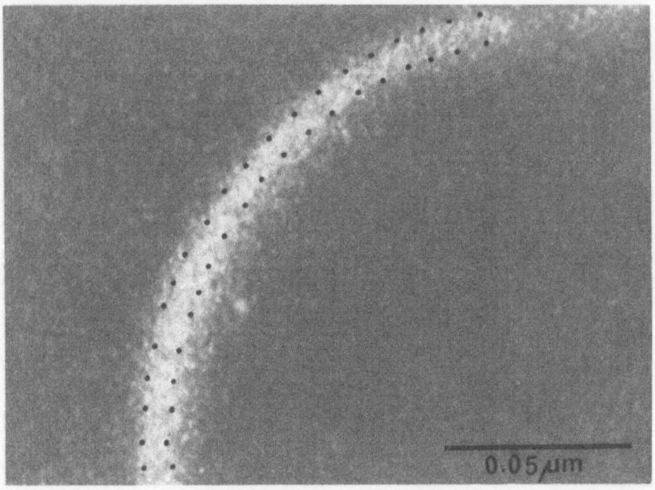

Fig. 5. Method of evaluating the radius of curvature of bowed-out dislocation segments. Negative copy of a micrograph and measuring points.

becomes smooth and very similar to the prediction of the line tension model (eq. (1)). The ratio between the radii of curvature of screw and edge dislocations is somewhat greater than the theoretical value (values between 1.84 and 2.70 instead of 1.86). Nevertheless, the result justifies the use of the line tension shape for routine evaluations.

According to eq. (1) ϱ should logarithmically depend on the

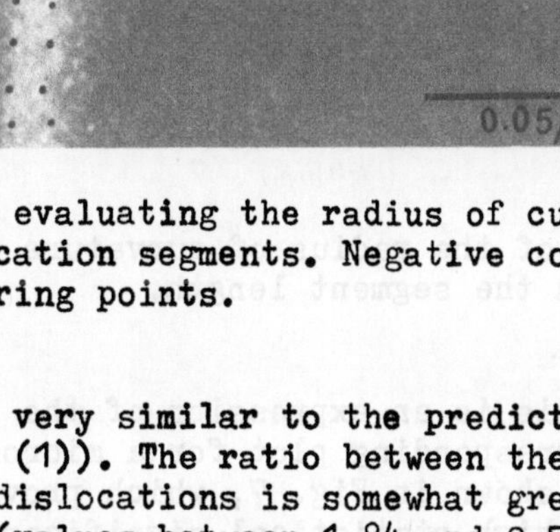

Fig. 6. Dependence of the local radius of curvature ϱ of individual bowed-out dislocation segments on the orientation angle α.

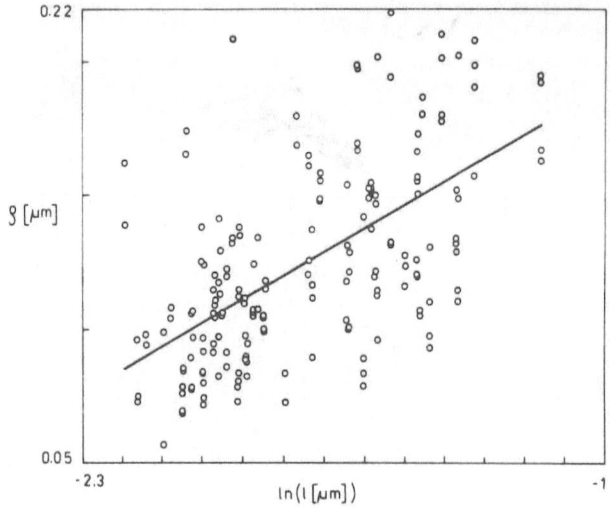

Fig. 7. Dependence of the radius of curvature of dislocation segments on the segment length.

segment length l. This is an expression of the self-stress of the dislocation. The corresponding plot for a micrograph with a high effective stress is shown in Fig. 7, which proves the dependence with a sufficiently high statistical significance. Other micrographs yielded similar plots. Fig. 7 implies a method of determining experimentally βr_0 in eq. (1). While the theoretical value should be about 2b, the experimental value turns out to be about 100 times larger. This is of great importance for the determination of the effective stress from the dislocation curvature after eq. (1).

GEOMETRICAL STATISTICS OF MOVING SCREW DISLOCATIONS

From the geometrical course of a dislocation bowing out between obstacles important dynamical parameters can be derived. The dislocation curvature (or the effective stress τ^*) and the obstacle distance l have already been defined in the preceding section. The dislocation branches touching an obstacle form a cusp. The respective tangent angle ϕ defines the acting force $f = 2T_l F = 2T_l \cos(\phi/2)$. T_l is the line tension and F a dimensionless force. While moving through the non-regular arrangement of obstacles the dislocations take characteristic configurations with specific frequency distributions of the acting forces f and the obstacle distances l. Under simplifying assumptions this problem was treated theoretically by statistical theories and computer simulation. Only one paper is quoted here.[21]

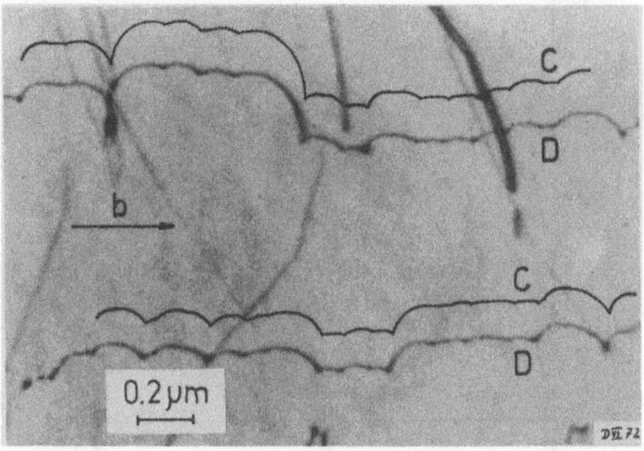

Fig. 8. Method of evaluating geometrical statistical parameters of dislocations moving through an array of localized obstacles. D electron microscopical images of dislocations, C calculator plot representing the measured data. After Ref. 23.

In order to determine the respective distributions from the micrographs, ellipses were matched visually to the individual dislocation segments and the respective coordinates were digitized. The reconstructed sequences of elliptical arcs were plotted and compared with the original micrograph (Fig. 8). Afterwards τ^*, l, and f were calculated. Detailed descriptions of the treatment[22] and of the results[23] are given elsewhere.

Fig. 9 shows a comparison of the distribution of the acting forces for different external stresses in a normalized form with the prediction of computer simulation for a single obstacle type.[24] The general shape of the experimental and theoretical distributions is similar, but the maximum of the experimental distribution is shifted to the left. On principle, such a shift can be explained by the presence of a great number of weak obstacles and a few very strong ones.[25] The numerical values of the averages of τ^*, l, and f are discussed in the following section.

DISCUSSION

As described in the present paper, the dominant mechanisms controlling the dislocation mobility in impure MgO are the action of localized obstacles and most probably the Peierls mechanism. Several problems have, however, been omitted and shall at least be mentioned now. In the in-situ experiment the specimens are subjected to a homogeneous radiation hardening. It does not seem

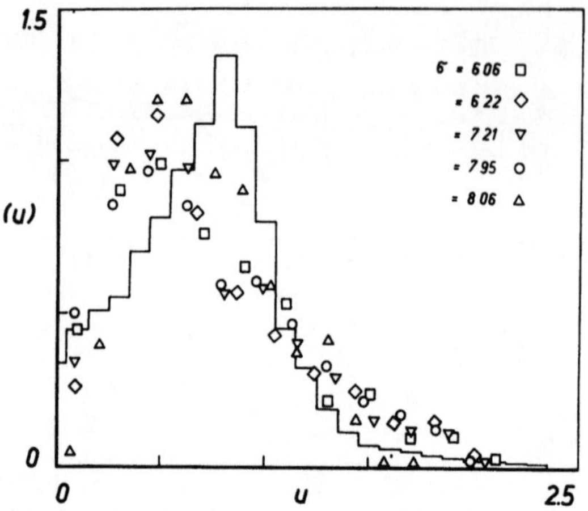

Fig. 9. Frequency distributions of forces acting on obstacles. Different symbols correspond to different applied stresses. The drawn histogram is the result of computer simulation for a single obstacle type. The forces are normalized according to $u = F\bar{u} / \bar{F}$, with $\bar{F}$ average value of F and $\bar{u} = 0.71$ average value of the theoretical histogram. After Ref. 24.

to change the dislocation behaviour substantially but has to be considered in quantitative evaluations. In addition to the mechanisms mentioned above some other processes contribute to the yield stress. The dislocations contain a number of jogs. Most of them are generated by cross slip. They contribute to the yield stress by trailing dipoles in screw dislocations and by a lattice friction in edges. The dislocations stored in the crystal cause a long-range part of the stress which was estimated from the dislocation density to be about 23 N/mm^2.[17]

Table 1 shows a comparison of the average parameters derived from the in-situ experiments with those from the measurement of dislocation velocities by stress pulse etching[10,11] and from macroscopic compression tests.[26] $\tilde{\tau}^*$ and $\bar{f}$ are derived with $\beta r_o = 2b$ in the energy factor. These values are unreasonably high. They will become smaller by a factor of three if βr_o is taken from Fig. 7. Whether this is correct or not needs further theoretical consideration. v is the activation volume and ΔG the Gibbs free energy of activation. The total obstacle strength is given by $\Delta F^* = \Delta G + v\tau^*$. Most probably the true τ^* is somewhat greater than $\tau_o/2$. This would result in $\Delta F^* \approx 2$ eV. The parameters in Table 1 characterize the obstacles to dislocation motion as rather strong ones. Nevertheless, they fit to the model of thermal activation which is also in accord with the kinematic behaviour of the dislocations. The present investigations show

Table 1. Room temperature activation parameters measured by different experimental methods

in-situ deformation	stress pulse etching[10,11]	macroscopic compression tests[26]
$\bar{l} = 460$ b	$v = 240\ b^3$	$v = 300\ b^3$
$\bar{\tau}^* = 84$ N/mm^2	$\tau = 40$ N/mm^2	$\tau_o = 52$ N/mm^2
$\bar{f} = 3.1\times10^{-9}$N	$\Delta G = 0.59$ eV	$\Delta G = 0.66$ eV

that in such a simple material as MgO the processes controlling the dislocation mobility are rather complex and can only be deduced by a combination of several experimental techniques.

REFERENCES

1. G. W. Groves and M. E. Fine, J. Appl. Phys. 35:3587 (1966)
2. R. W. Davidge, J. Mater. Sci. 2:339 (1967)
3. M. Srinivasan and T. G. Stoebe, J. Appl. Phys. 41: 3726 (1970)
4. M. N. Sinha, D. L. Lloyd, and K. Tangri, Philos. Mag. 28:1341 (1973)
5. C. N. Ahlquist, J. Appl. Phys. 46:14 (1975)
6. R. N. Singh and R. L. Coble, J. Appl. Phys. 45:981 (1974)
7. R. N. Singh and R. L. Coble, J. Appl. Phys. 45:990 (1974)
8. F. Sato and K. Sumino, J. Mater. Sci. 15:1625 (1980)
9. F. Appel, H. Bethge, and U. Messerschmidt, Phys. Status Solidi (a) 42:61 (1977)
10. S. N. Valkovskii and E. M. Nadgornyi, Soviet Phys. Solid State 17:1733 (1975)
11. S. N. Valkovskii and E. M. Nadgornyi, Soviet Phys. Solid State 17:2036 (1975)
12. U. Messerschmidt, Y. Nishino, T. Imura, and H. Saka, Phys. Status Solidi (a) 76:277 (1983)
13. T. E. Mitchel and A. H. Heuer, Mater. Sci. Eng. 28: 81 (1977)
14. B. Reppich, Acta Metall. 23:1055 (1975)
15. H. Knoch and B. Reppich, Acta Metall. 23:1061 (1975)
16. U. Messerschmidt, submitted to Mater. Sci. Eng.
17. F. Appel, M. Bartsch, U. Messerschmidt, E. M. Nadgornyi, and S. N. Valkovskii, to be submitted to Phys. Status Solidi

18. U. Messerschmidt and F. Appel, to be publ.
19. R. O. Scattergood and D. J. Bacon, Philos. Mag. 31:179 (1975)
20. U. Messerschmidt, H. Schmid, and F. Appel, to be publ.
21. J. W. Morris and D. H. Klahn, J. Appl. Phys. 45:2027 (1974)
22. U. Messerschmidt and F. Appel, Kristall und Techn. 14:1331 (1979)
23. U. Messerschmidt and F. Appel, in: "Strength of Metals and Alloys", P. Haasen, V. Gerold, and G. Kostorz, eds., Pergamon Press, Oxford 1979, p. 1305
24. F. Appel, U. Messerschmidt, E. M. Nadgornyi, and S. I. Zaitsev, Mater. Sci. Eng. 52:69 (1982)
25. A. I. Landau, private communication
26. M. Bartsch, to be publ.

DEFORMATION TWINNING IN CERAMICS

William D. Scott
Ceramic Engineering
University of Washington
Seattle, Washington 98195

ABSTRACT

Deformation twinning is a process which usually occurs when glide deformation is inhibited. In some ceramic materials and minerals, deformation twinning is the lowest temperature deformation process observed. Details of twin formation and growth in alumina are presented and a surface defect model for a twin boundary is suggested.

INTRODUCTION

The permanent deformation of crystals in response to mechanical stress may be through diffusional processes, dislocation glide or deformation twinning. All of these processes may be interrelated and occurring at the same time. However, as a general trend, deformation twinning is often the only macroscopic deformation process observed in complex ceramic crystals at low homologous temperatures and relatively high strain rates. Deformation twinning is the homogeneous shear of a portion of a crystal lattice such that the product structure is unchanged. The twinned region is oriented differently but in a specific relation to the parent lattice. The shear properties of deformation twinning have led to development of models based on dislocation glide. Other properties of the process such as the high speed of twin formation and the ability of the process to occur at very low temperatures has led to modeling as a diffusionless martensitic transformation. This paper will review some observations on the occurrence and growth of deformation twins in non-metallic crystals. We will emphasize those aspects of the process which must be considered in any unified model for deformation twining in compounds with relatively complex structures.

Majahan and Williams (1) reviewed deformation twinning in metals and alloys with copious references which relate to twinning in both metals and non-metals. Cahn (2,3) has published extensive reviews which include observations on calcite and other non-metals. Russian work was reviewed by Klassen-Neklyudova (4) and the book by Hall (5) is a standard reference.

TWINNING CRYSTALLOGRAPHY

A schematic illustration of a twin-matrix relation and a unit advance of the twin boundary is shown in Figure 1. The twin and matrix remain in contact at a mutual boundary, there is no volume change, and the process is equivalent to a homogeneous shear of the original lattice in plane strain. Figure 1b is a schematic diagram showing a step in the twin boundary when the unit advance is incomplete. This step is called a twinning dislocation, and its role in twin growth will be discussed below.

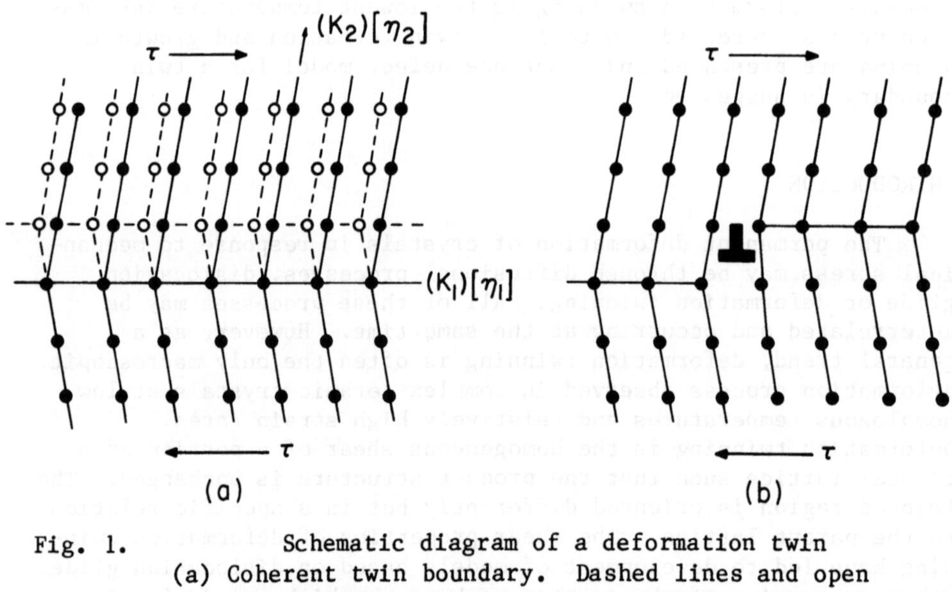

Fig. 1. Schematic diagram of a deformation twin
(a) Coherent twin boundary. Dashed lines and open circles indicate lattice points before the movement of the twin boundary.
(b) A twinning dislocation at a step in a twin boundary. The diagram does not show lattice relaxations away from the discontinuity.

Figure 2 shows the crystallographic elements which are used to define the twin orientation and shear. The reader is referred to Christian (6) or Cahn (2) for a complete description of the element selection. In addition to the homogeneous shear, it is required that the twin lattice is reconstructed in a fixed relationship to the matrix. Two different orientation relationships are described here, and the reader is again referred to Christian (6) and Cahn (2) for details. The orientation relation of Type I twins may be produced either by a reflection across the K_1 plane or by rotation 180° around the normal to K_1. The orientation relation of Type II twins is through 180° rotation around η_1 or by reflection in the plane perpendicular to η_1. Christian (6) points out that within each type, the two operations are equivalent if the structure is centrosymmetric. He also states that only the primitive lattices of the twin and matrix are related by simple shear. If the structure consists of more than one atom per lattice point, then some atoms must move in different directions, and these movements out of the shear direction are called shuffles.

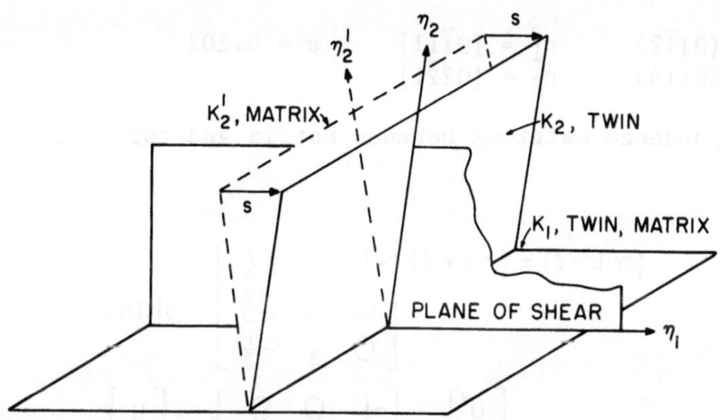

Fig. 2. The twinning elements. The twinning plane, K_1, the reciprocal twinning plane, K_2, also called the second undistorted plane. The twinning direction η_1 and the reciprocal twinning direction η_2. At unit height, s is the deformation shear.

TRANSFORMATIONS OF PLANES AND DIRECTIONS BY TWINNING

Taking a Type I twin as an example, the positions of crystallographic elements can be described by a reflection across the twin composition plane, or by a 180° rotation about the twin plane normal. However, it is clear that a physical element in the lattice like a plane consisting of a sheet of individual atoms, a dislocation line, a Burgers vector, or a scratch in the surface are not displaced in this same way. It is of interest to know how the twin transformation alters physical elements in the matrix.

For a plane or a vector lying in the matrix lattice, the twinning process can be considered in two steps. First, the homogeneous shear displaces the plane or vector in space relative to its matrix position. Second, there is a reorientation of crystal axis around the new, fixed plane or vector position so that they are described by a new set of indices relative to the twin structure.

The problem was first considered by O. Mügge in 1889 and reviewed more recently by Bell (7) and Pabst (8). Equations are given for transformations in terms of K_1 and η_2 or K_2 and η_1. The author has rearranged these equations into matrix form for twinning in alumina, and the details will be published elsewhere.

For alumina, the twin elements reported by Heuer (9) (converted to the structural unit cell, c/a 2.730) are:

$K_1 = (01\bar{1}2)$ $\quad \eta_1 = [0\bar{1}11]$ $\quad s = 0.202$

$K_2 = (0\bar{1}14)$ $\quad \eta_2 = [02\bar{2}1]$

The correspondence matrices between matrix and twin are:

$$(h'k'\cdot\ell) = (hk\cdot\ell)\begin{bmatrix} -1 & \frac{2}{3} & 1\frac{1}{3} \\ 0 & \frac{1}{3} & 2\frac{2}{3} \\ 0 & \frac{1}{3} & -\frac{1}{3} \end{bmatrix} \quad \text{planes}$$

$$\text{directions} \quad \begin{bmatrix} u' \\ v' \\ \cdot \\ w' \end{bmatrix} = \begin{bmatrix} -1 & 0 & 0 \\ \frac{2}{3} & \frac{1}{3} & 1\frac{1}{3} \\ \frac{1}{3} & \frac{2}{3} & -\frac{1}{3} \end{bmatrix} \begin{bmatrix} u \\ v \\ \cdot \\ w \end{bmatrix}$$

The primed indices refer to the twin crystal lattice and unprimed indices to the matrix in the 4-axis 3-index hexagonal notation.

The (0001) plane of the matrix becomes $(01\bar{1}\bar{1})$ in the twin. Three glide dislocation Burgers Vectors in the matrix basal plane transform as follows during twinning:

$[\bar{2}\bar{1}10] \rightarrow [2\bar{1}\bar{1}0]$

$[1\bar{2}10] \rightarrow [\bar{1}01\bar{1}]$

$[11\bar{2}0] \rightarrow [\bar{1}011]$

Two of the three dislocations lose their mobility through the transformation because the Burgers vectors are transformed to a secondary slip system. The exception is a screw dislocation with Burgers vector [$\bar{2}$110] which only changes sign and which can cross glide onto the new basal plane of the twin. Yoo and Wei (10) have discussed this same cross glide in zinc. Two consequences of the transformation produced by twinning are that formerly mobile dislocations are rendered sessile and that mobile dislocations cannot cross a twin boundary unless very specific relations between the Burgers vector, the dislocation line orientation, and the twin matrix lattice orientations are satisfied.

TWIN NUCLEATION AND GROWTH

Figure 3 shows the stress displacement curve obtained for twinning in an alumina single crystal compressed along the c-axis at 700°C. This pattern is typical for the formation of large, individual twins in many materials. The usual interpretation is that the stress necessary for the initiation of the twin (a peak on the curve) is greater than the stress necessary to broaden the twin. (In a hard testing machine, the falling stress accompanies large rapid specimen deformation.) The problem of twin nucleation has been considered by many authors (see Reference 1) and is still not resolved in detail. In alumina, the appearance of twins is strongly influenced by the surface condition of the specimen. For example, specimens with damaged, as-ground surfaces consistently initiate twins at only 12.5 MPa RSS at 700°C (11), but one specimen which had been chemically polished in sodium tetraborate

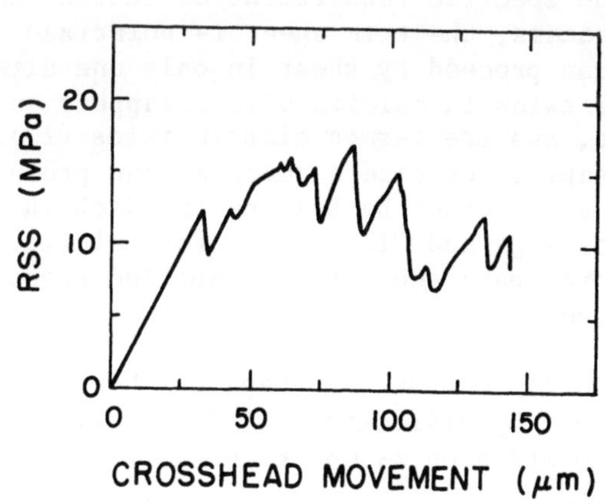

Fig. 3. Twinning in an alumina single crystal loaded in compression along the c-axis at 700°C, and an axial strain rate of $10^{-5}s^{-1}$. See Ref. 11 for experimental details.

withstood a RSS of over 160 MPa before twinning, a factor of 13 greater stress (12). The heterogeneous nature of twin nucleation is probably one reason for the wide variation in the "twinning stress" reported for nominally identical materials (13).

It is well established that the rapid load drop associated with a twin event is the spreading of a wedge shaped twin and/or the broadening of a parallel sided twin. This process is termed twin growth to distinguish it from the nucleation process. The homogeneous shear of the twinning process can be produced by the passage of partial dislocations or twinning dislocations sequentially on successive lattice planes parallel to the twin composition plane. Any model for twin growth by this process must include a mechanism for the generation of the twinning dislocations on sequential planes. The pole-mechanisms (14-16) and the twin-slip interactions (10,17,18) require the movement of glide dislocations to initiate or continue twin growth. All of the above models have been developed for metals where it is reasonable to assume that substantial deformation by dislocation glide is occurring at the same time as the twinning. However, there are many features of the twinning process which argue against the participation of glide dislocations.

Figures 4 and 5 show the effect of temperature on the twinning stress in alumina and calcite. Twinning is the deformation process requiring the lowest stress at low temperature and, over a wide temperature range, does not exhibit the strong temperature dependence normally found with dislocation glide. In metals, twinning is favored in conditions where slip is difficult, and in some materials twins become the dominant deformation mode only under impact loading at crygogenic temperatures.

Because of the specific requirement of lattic reorientation in the twinning process, the twin shear is polarized, i.e., the growth of a twin can proceed by shear in only one direction. Small wedge shaped twins in calcite will disappear on removal of the applied stress, and are termed elastic twins (20). Parallel-sided twins are stable, but stress reversal can produce de-twinning or twin shrinkage as reported by Bolling and Richman (21) for Fe-Be alloys and by Kaga and Gilman (22) for calcite. Thus a twin boundary can be moved back and forth by applied stress much like a low angle tilt boundary.

This was shown to be true for graphite where movement of the twin boundary was not by dislocation glide parallel to K_1, but by movement of a dislocation on K_2 which is the basal plane (23).

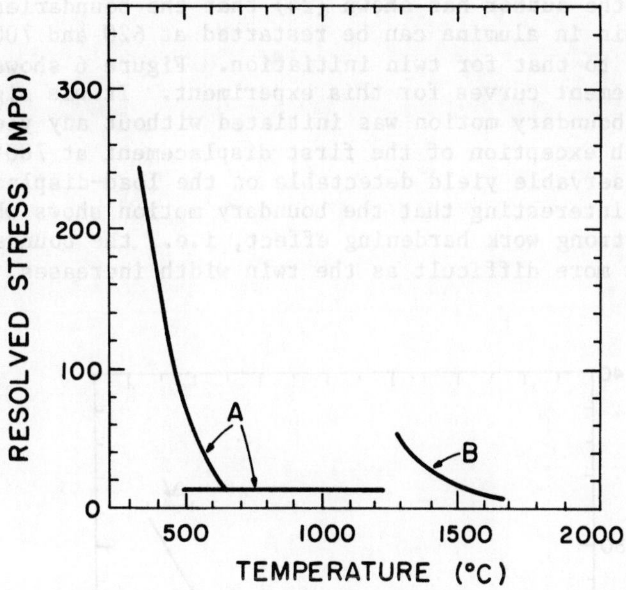

Fig. 4. Resolved shear stress for rhombohedral twinning (A) and basal slip (B) in alumina single crystals at approximately the same strain rate after Scott and Orr (11).

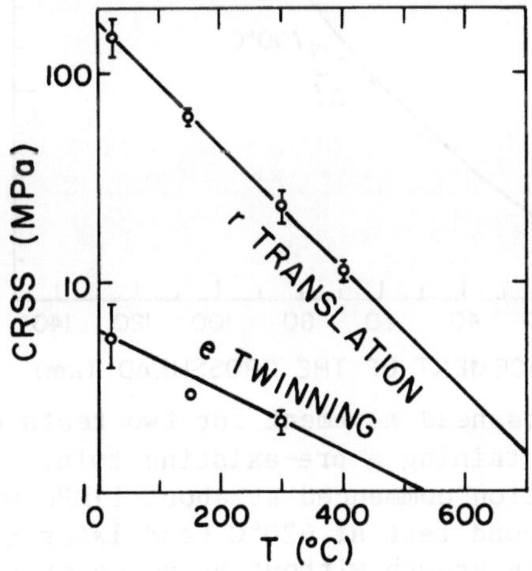

Fig. 5. Twinning and Translation in Calcite

Critical resolved shear stress at yield point versus temperature. Bars indicate limit of values for each temperature. Circles are average values. Ordinate is logarithmic. After Turner et al. (19).

Recently the author has shown (24) that the boundaries of a preexisting twin in alumina can be restarted at 620 and 700°C at a stress similar to that for twin initiation. Figure 6 shows typical stress-displacement curves for this experiment. It was significant that the boundary motion was initiated without any yield drop, and, with exception of the first displacement at 700°C, without any observable yield detectable on the load-displacement curve. It is interesting that the boundary motion shows what is apparently a strong work hardening effect, i.e., the boundary motion becomes more difficult as the twin width increases.

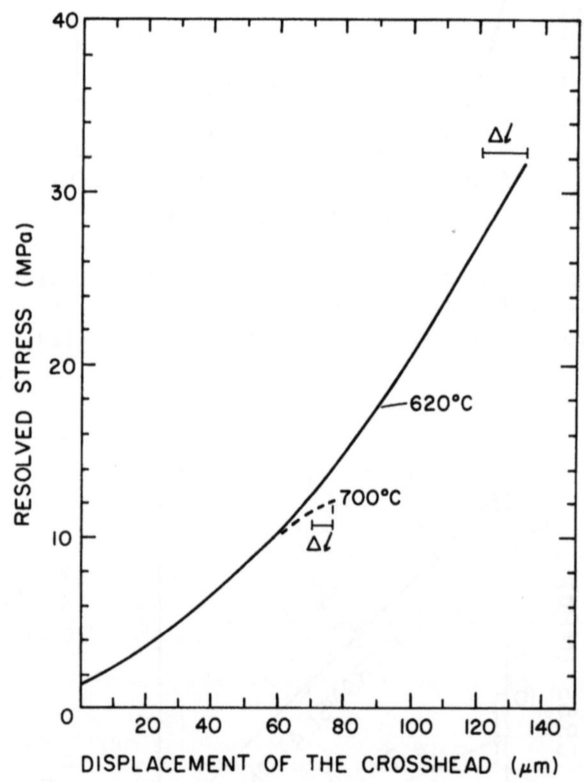

Fig. 6. Stress - crosshead movement for two tests on the same specimen containing a pre-existing twin. At 700°C the boundary motion commenced at about 11MPa without a yield drop. A second test at 620°C (and later at 700°C) produced twin growth without evidence of yielding. The distance $\Delta \ell$ is the change in length expected from the twin growth as measured after the test. The curvature is caused primarily by soft grafoil pads inserted to reduce end effects in the compression test.

This effect is shown more clearly in Figure 7 where the displacment of the two boundaries of a preexisting twin were measured after increments of deformation. Furthermore, the total movement of an individual boundary can be quite large. Gilman moved boundaries in calcite as much as 4.3 mm, and 200 μm wide twins in alumina are common (11). The rapid work hardening indicates that the boundary motion is not by the generation of new dislocations on successive slip planes and their movement through undeformed material. Furthermore, rapid twin formation over widths typically 100 to 200 μm, would, in the case of alumina, require extensive perfect pole dislocations. As seen in Figure 7, the two twin boundaries seem to move independently, and this would require a double-acting source.

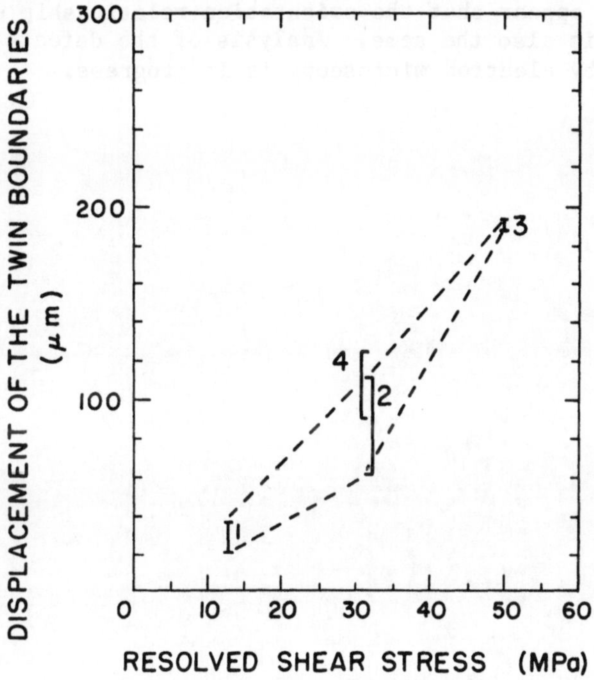

Fig. 7. Movement of individual twin boundaries as a function of applied stress in repeated load tests. (No. 1 and 2 correspond to Fig. 6). The specimen contained a 10 μm wide, parallel sided twin before the first test, and movement was measured from this origin. Test #2 was at 620°C and all others were at 700°C. The specimen was annealed 14 hr at 1400°C, between tests 3 and 4 and the measurement origin reset to zero.

ETCH PITS AND TWIN FORMATION

Dislocation etchants have revealed a complex dislocation structure associated with twin formation in calcite (3,4,22,25). Similar etch pits have been observed by the author in rhombohedral twins in alumina. An example of these features are shown in Figure 8, and the twin matrix orientation in Figure 9. The lines of dislocation pits in the twin coincide with the [0001] direction in the twin. Sequential polish and etch experiments show that they lie on the (01$\bar{1}$0) plane of the twin. They often, but not always appear from a preexisting network dislocation as revealed by etching prior to twinning. In some instances, the dislocation pits appear as pairs indicating that they may be loops. The lines terminate at a defect in the twin boundary. The dislocation etch pits seen here in alumina are similar to those seen in calcite (4,22,25). From the orientation information given in the calcite work, it would appear that the orientation relationship of the dislocation pits is also the same. Analysis of the defects in alumina twins by electron microscopy is in progress.

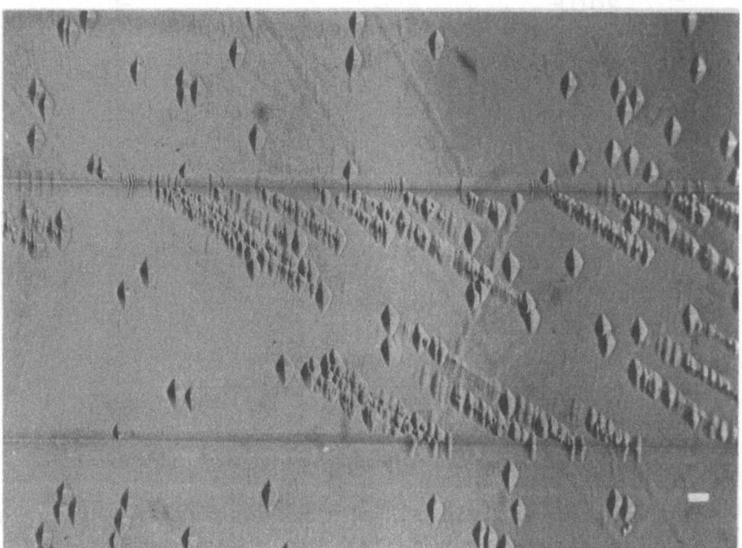

Fig. 8. Dislocation etch pits on {$\bar{2}$110} face of twinned alumina crystal. Etch was 20 to 60 sec, sodium tetraborate, 820°C. The two twin boundaries moved approximately equal distances from an initial thin twin formed at 700°C and MPa (12). Note how each line terminates in a boundary defect and that large sections of twin boundary are free of etch pits. The bar = 10 μm.

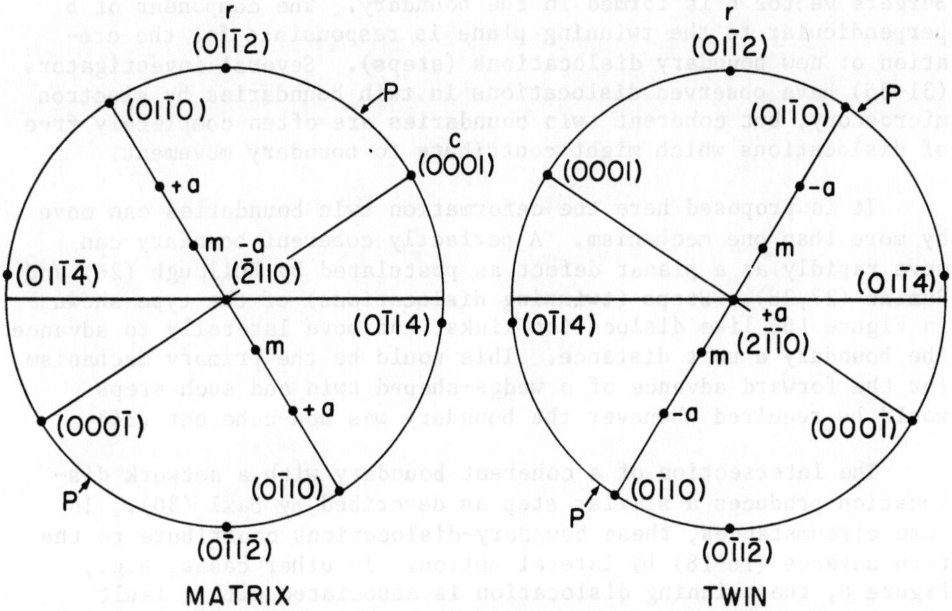

Fig. 9. Orientation of matrix and twin in Fig. 8. Arrows marked P indicate location of compressive load.

AN ALTERNATE TWIN BOUNDARY

Several authors have considered a twin boundary as an extended two dimensional defect. Bullough (26) calls it a "simple glissile surface dislocation," and Sumino (27,28) a "(twinning) surface dislocation." Surface dislocations were also described by Marcinkowski (29) in a general treatment of grain boundaries. The basic feature of these models is that the surface dislocation, moving normal to itself, transforms the lattice into a new orientation. Sumino (28) points out that steps in the boundary are the defects usually called twinning dislocations. The advance of the boundary may be uniform, analogous to the uniform advance of a dislocation, or by the formation and lateral movement of steps, analogous to the nucleation and lateral movement of kinks on a dislocation line (11). The rate of boundary advance may be limited if the steps are present and their lateral movement is impeded. The most rapid movement postulated by Sumino is when the boundary is free from steps.

Saxl (30), Sleeswyk (17) and Yoo (10) have considered the interaction of a moving dislocation with a twin boundary. If one considers the boundary to be moving and the dislocation stationary, the same effects are produced. Following Saxl, when a dislocation is incorporated into a twin, a step equal to the

Burgers vector b is formed in the boundary. The component of b perpendicular to the twinning plane is responsible for the creation of new boundary dislocations (steps). Several investigators (31-35) have observed dislocations in twin boundaries by electron microscopy, but coherent twin boundaries are often completely free of dislocations which might contribute to boundary movement.

It is proposed here the deformation twin boundaries can move by more than one mechanism. A perfectly coherent boundary can move rapidly as a planar defect as postulated by Bullough (26) and Sumino (27,28). Steps (twinning dislocations) of the type shown in Figure 16, like dislocation kinks, can move laterally to advance the boundary a unit distance. This would be the primary mechanism for the forward advance of a wedge-shaped twin and such steps would be required whenever the boundary was non-coherent (34).

The intersection of a coherent boundary with a network dislocation produces a similar step as described by Saxl (30). In some circumstances, these boundary-dislocations contribute to the twin advance (16-18) by lateral motion. In other cases, e.g., Figure 8, the twinning dislocation is associated with a fault plane, and is not free to move laterally. If the planar boundary advance is continued, then the sessile defects constitute an impediment to the advance which increases as boundary motion continues. There are indications from the present work that such interactions are enhanced during slower boundary motion, and that rapidly moving boundaries leave behind fewer dislocation trails and, when they stop, etch pits show that fewer defects are present in the boundary.

DEFORMATION TWINS IN NON-METALS

Twinning in naturally occurring minerals, both growth twins and deformation twins, has been studied for over 100 years, and extensive reviews are available (3,4). Quantitative measurements of deformation twinning by the application of loads in the laboratory are less common. The most widely studied material is calcite (4,22,25,33,35,36) because of its geological importance and ease of twin formation at room temperature. Deformation twins have been produced in many minerals at high temperature and under confining pressure (37) and the author (11) has studied deformation twins produced in alumina under simple compressive axial loading. In this latter case, twins have been produced at RSS as low as 12 MPa at 600°C and at RSS of 240 MPa at 350°C. A very interesting study by Coe and by Coe and Kirby on coherent phase transitions in $MgSiO_3$ may be applicable to deformation twinning (38,39). They have shown that the equilibrium of the first order transition between a high and low temperature form is displaced by shear stress. A similar "equilibrium" can be postulated for a twin-matrix transition, i.e., a shear assisted phase transition. Since ΔV and ΔS are zero, there is no temperature dependence for the process, and the only energy terms are for the initial creation of the twin boundaries and the creation of internal defects by boundary movement.

Recent observations on the transition of tetragonal to monoclinic particles of ZrO_2 contained in a constraining matrix have demonstrated the close connection between martensitic phase transitions and deformation twinning (39-40). Twins produced during the phase transition assist the accommodation of the particle in the constraining matrix. The frequency of the twins and the regularity of the twinned structure suggest nucleation and growth mechanisms which do not depend on relatively rare or unusual dislocation events.

SUMMARY AND CONCLUSIONS

Individual rhombohedral twin boundaries in alumina were moved in repeated tests at 620 to 720°C at low resolved shear stresses. Slowly moving boundaries generated progressively increasing residual dislocation damage in the twin as revealed by etch pit observations. The boundaries appeared to work harden so that the distance of boundary motion was proportional to the applied stress.

The low temperature and stress at which large scale deformation occurred suggests that glide dislocations were not participating significantly in the deformation process. A model proposed previously by others in which a twin boundary is considered to be a two dimensional surface dislocation may be useful to describe twin boundary motion in alumina and other ceramic materials.

REFERENCES

1. S. Mahajan and D. F. Williams, "Deformation Twinning in Metals and Alloys, Intl. Met. Rev. 18:43 (1973).
2. R. W. Cahn, "Soviet Work on Mechanical Twinning, Nuovo Cimento (Supl.) 10:350 (1953).
3. R. W. Cahn, "Twinned Crystals," Adv. Phys. 3:363 (1954).
4. M. V. Klassen-Neklyudova, "Mechanical Twinning of Crystals," Consultants Bureau, New York (1964).
5. E. O. Hall, "Twinning and Diffusionless Transformations in Metals," Butterworth, London (1954).
6. J. W. Christian, "The Theory of Transformations in Metals and Alloys," Pergamon, New York (1975).
7. J. F. Bell, "Morphology of Mechanical Twinning in Crystals," Am. Mineral. 26:247 (1941).

8. A. Pabst, "Transformation of Indices in Twin Gliding," Bull. Geol. Soc. Am. 66:897 (1955).
9. A. H. Heuer, "Deformation Twinning in Corundum," Phil. Mag. 13:379 (1966).
10. M. H. Yoo and C. T. Wei, "Growth of Deformation Twins in Zinc Crystals," Phil. Mag. 14:573 (1966).
11. W. D. Scott and K. K. Orr, "Rhombohedral Twinning in Alumina," J. Am. Ceram. Soc. 66:27 (1983).
12. F. B. Karp, "Rhombohedral Twinning in Aluminum Oxide Single Crystals," M.S. Thesis, University of Washington (1981).
13. R. L. Bell and R. W. Cahn, "The Nucleation Problem in Deformation Twinning," Acta Met. 1:752 (1953).
14. A. H. Cottrell and B. A. Bilby, "A Mechanism for the Growth of Deformation Twins in Crystals," Phil. Mag. 42:573 (1951).
15. N. Thompson and D. J. Millard, "Twin Formation in Cadmium," ibid. 43:422 (1952).
16. A. W. Sleeswyk, "1/2 <111> Screw Dislocations and the Nucleation of {112}<111> Twins in the B.C.C. Lattice," ibid. 8:1467 (1963).
17. A. W. Sleeswyk and C. A. Verbaak, "Incorporation of Slip Dislocations in Mechanical Twins-I," Acta Met. 9:917 (1961).
18. D. I. Tomsett and M. Bevis, "The Incorporation of Basal Slip Dislocations in $\{10\bar{1}2\}$ Twins in Zinc Crystals," Phil. Mag. 19:129 (1969).
19. F. J. Turner, D. T. Griggs and H. Heard, "Experimental Deformation of Calcite Crystals," Bull. Geol. Soc. Am. 65:883 (1954).
20. I. V. Obreimov and V. I. Startsev, "Work of Formation of an Elastic Twin in Calcite," Soviet Phys JETP (Trans), 35:743 (1959).
21. G. F. Bolling and R. H. Richman, "Continual Mechanical Twinning," Acta Met. 13:709 (1965).
22. H. Kaga and J. J. Gilman, "Twinning and Detwinning in Calcite," J. Appl. Phys. 40:3196 (1969).
23. E. J. Freise and A. Kelly, "Twinning in Graphite," Proc. Roy. Soc. A264:269 (1961).
24. W. D. Scott, J. Philibert, "Croissance des Macles dans l'Alumine," Rev. Intl. Hautes Temp. 19:373 (1982).
25. R. E. Keith and J. J. Gilman, "Dislocation Etch Pits and Plastic Deformation in Calcite," Acta Met. 8:1 (1960).
26. R. Bullough, "Deformation Twinning in the Diamond Structure," Proc. Roy. Soc. A241:568 (1957).
27. K. Sumino, "Surface Dislocations and the Growth of Deformation Twins," Acta Met. 14:1607 (1967).
28. K. Sumino, "Experimental Determination of Intrinsic Resistive Stress for Twinning Surface Dislocations," Phys. Stat. Sol. 33:327 (1969).

29. M. J. Marcinkowski, "Unified Theory of Grain Boundaries," Phys. Stat. Sol. A73:409 (1982).
30. I. Saxl, "The Incorporation of Slip Dislocations into Twins," Czech. J. Phys. B18:39 (1968), B19:1605 (1969).
31. B. J. Hockey, "Pyramidal Slip on $\{11\bar{2}3\}<\bar{1}100>$ and Basal Twinning in Al_2O_3," Deformation of Ceramic Materials, R. C. Bradt and R. E. Tressler, Ed. Plenum (1975).
32. L. A. Bursill and R. L. Withers, "Twinning Dislocations in Hematite Iron Ore," Phil. Mag. A 40:213 (1979).
33. D. J. Barber and H. R. Wenk, "Deformation Twinning in Calcite, Dolomite, and Other Rhombohedral Carbonates," Phys. Chem Minerals 5:165 (1979).
34. B. Yangui et al. "High Resolution Study of Incoherent Twin Boundaries and of Isolated Wedge Microtwins in Rare-Earth Monoclinic Sesquioxides," Phil. Mag. A 45:443 (1982).
35. P. Braillon and J. Serughetti, "Mechanical Twinning in Calcite," Phys. Stat. Sol. (A) 33:405 (1976).
36. V. Z. Bengus, S. N. Komnik, and I. T. Ponomarenko, "Studies of Discontinuous Flow Mechanism Under Expansion of Single Twin Layer in Calcite," Cryst Lattice Defects 6:101 (1975)
37. J. J. Kollé, "Plastic Deformation of Two Single Crystal Clinopyroxenes," Ph.D. Thesis, University of Washington (1980).
38. R. S. Coe, "The Thermodynamic Effect of Shear Stress on the Ortho-Clino Inversion in Enstatite and Other Coherent Phase Transitions Characterized by Finite Simple Shear," Contr. Mineral. and Petrol. 26:247 (1970).
39. R. W. Coe and S. H. Kirby, "The Orthoenstatite to Clinoenstatite Transformation by Shearing and Reversion by Annealing," ibid. 52:29 (1975).
40. J. E. Bailey, "The Monoclinic-Tetragonal Transformation and Associated Twinning in Thin Films of Zirconia," Proc. Roy. Soc. A 279:359 (1964).
41. W. M. Kriven, "Martensite Theory and Twinning in Composite Zirconia Ceramics," Adv. Ceramics. 3 (1980).
42. A. G. Evans, N. Burlingame, M. Drory and W. M. Kriven, "Martensitic Transformations in Zirconia - Particle Size Effects and Toughening," Acta Met. 29:447 (1981).

HIGH-TEMPERATURE CREEP OF SILICATE OLIVINES

D. L. Kohlstedt and D. L. Ricoult

Department of Materials Science and Engineering
Cornell University
Ithaca, NY 14853

INTRODUCTION

The high-temperature, high-pressure rheological properties of magnesium-iron silicate olivines--$(Mg,Fe)_2SiO_4$--have been investigated in a number of rock mechanics laboratories over the past ten years. Constitutive equations describing the plastic flow of this mineral/ceramic and of rocks in which it is the primary phase are necessary for modeling large-scale solid-state convective flow in the earth's upper mantle as well as more localized deformations resulting from bending of a subducting lithospheric plate, volcanic loading of the lithosphere at a sea-mount, or unloading during melting of a glacier (post-glacial rebound). As in many engineering applications, flow laws determined from laboratory experiments on rocks and minerals must be extrapolated several orders-of-magnitude in strain rate, in this case from laboratory strain rates which are generally greater than $10^{-6} sec^{-1}$ to geologic strain rates which are typically less than $10^{-12} sec^{-1}$. Thus, a physical understanding of the effects of pressure, temperature, stress, and chemical environment on creep rate and mechanisms of deformation is important.

Before data were available for the self-diffusion of oxygen and silicon in forsterite, Mg_2SiO_4, the high-temperature creep data for natural (Mg, Fe)-silicate olivine crystals appeared to be in good accord with models for high-temperature creep rate-limited by dislocation climb, and extrapolation to mantle conditions seemed complicated only by the need to extend creep experiments to polycrystalline, polyphase aggregates (i.e., rocks) and to determine the effects of water and carbon dioxide on flow strength. The activation energy, $Q \cong 525 kJ/mole$, lay between values for oxygen self-diffusion in MgO and Al_2O_3 while the stress exponent, $n \cong 3$, was comfortably in the range observed in metals undergoing climb-controlled deformation (Kohlstedt and Goetze, 1974; Kohlstedt et al., 1976a). In addition, the deformation-induced microstructures indicated that diffusion and dislocation climb must be important (Goetze and Kohlstedt, 1973). Subgrain boundaries formed readily during creep experiments and rapidly under static annealing conditions (Ricoult, 1979). However, diffusion experiments on Mg_2SiO_4 yielded activation energies of only 290-380 kJ/mole for both silicon and oxygen (Jaoul et al., 1981; Reddy et al., 1980; Jaoul et al., 1980) while creep experiments on Mg_2SiO_4 gave even higher values, 500-750 kJ/mole, than had been determined for $(Mg,Fe)_2SiO_4$.

To provide new information which might help constrain the possible mechanisms of creep in olivine, recent experiments have examined the dependence of creep rate in (Mg,Fe)-silicate olivines on the partial pressure of oxygen, p_{O_2}, and the activities of the constituent oxides, a_{ox}. In addition, deformation experiments on single crystals of Fe_2SiO_4 and Co_2SiO_4 have been carried out under controlled oxygen partial pressure, oxide activity, and temperature conditions in an attempt to develop a systematic description of deformation behavior and deformation mechanisms in a range of silicate olivines. In the present paper, the high-temperature creep behavior of the Mg, (Mg,Fe), Fe, and Co silicate olivines is reviewed and compared. The results are examined specifically in terms of the point defect chemistry of these materials with particular emphasis on modifications proposed to constitutive laws for high-temperature steady-state creep to account for the nucleation and migration of jogs and kinks as one possible resolution of the discrepancy between activation energies for self-diffusion and those for creep. (Hereafter in this paper the term olivine will be used for brevity in place of silicate olivine; however, it should be noted that other ternary compounds, for example magnesium germanate (Mg_2GeO_4), also crystallize with the olivine structure.)

STRUCTURE

Olivine crystallizes with orthorhombic symmetry and has lattice parameters of $a \sim 0.5$nm, $b \sim 1.0$nm, and $c \sim 0.6$nm. The olivine structure is built-up of SiO_4 tetrahedra which are independent, that is the tetrahedra are connected only by metal cations and are not linked together by corner sharing of oxygen ions. Thus, olivine is sometimes referred to as an island silicate. The oxygen sublattice is approximately hexagonally close-packed with the basal plane parallel to (100). Metal ions (e.g., Me = Mg, Mn, Ni, Co, and/or Fe) occupy one-half of the octahedral interstitial sites and silicon ions one-eighth of the tetrahedral sites. This structure is illustrated in an (100) projection in Fig. 1. Complete solid solution exists between, for example, Mg_2SiO_4 and Fe_2SiO_4, as well as between other olivine pairs.

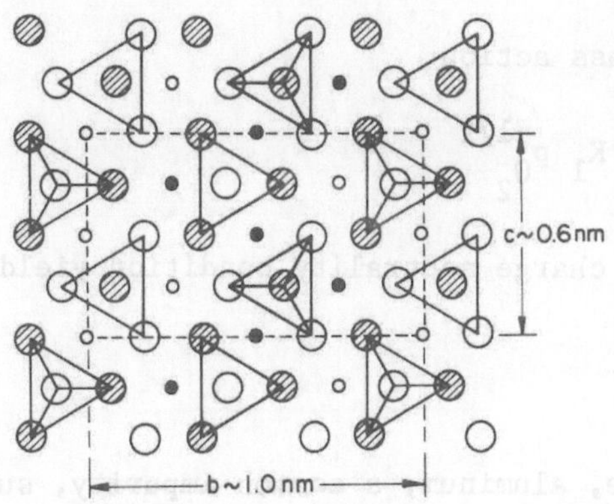

Fig. 1. Projection of olivine structure onto (100). Large circles are oxygen ions; small circles are metal cations. Silicon ions in the SiO_4 tetrahedra are not shown. The SiO_4 tetrahedra point alternatively into and out of the page.

POINT DEFECT CHEMISTRY

For discussion of point-defect thermodynamics, the silicates can be divided into two major classes, those with transition-metal cations and those without. At one end of the spectrum then is Mg_2SiO_4; at the other end are Fe_2SiO_4, Co_2SiO_4, Ni_2SiO_4, and Mn_2SiO_4 with cations with variable oxidation states. In the present paper, the Kröger-Vink notation is used to denote the site and charge of a given point defect.

The main point defects that have been considered for Mg_2SiO_4 are the ionized vacancies and interstitials of the constituent species, that is, V''_{Mg}, $Mg_i^{\cdot\cdot}$, V''''_{Si}, $Si_i^{\cdot\cdot\cdot\cdot}$, $V_O^{\cdot\cdot}$, and O''_i (Schmalzried, 1978). The main experimental evidence bearing on point defects in Mg_2SiO_4 is found in the work of Pluschkell and Engell (1968) and is discussed in some detail in Smyth and Stocker (1975). A brief summary is presented here. (i) Because forsterite dissolves some excess SiO_2 but no excess MgO, the majority defects for stoichiometric (Mg/Si = 2) materials are $V''_{Mg} + Mg_i^{\cdot\cdot}$ and for non-stoichiometric (Mg/Si > 2), $V''_{Mg} + Si_i^{\cdot\cdot\cdot\cdot}$. (ii) Ionic conductivity increases with increasing SiO_2 content indicating that $[V''_{Mg}]$ is increasing and that diffusion of magnesium occurs by a vacancy mechanism. (iii) At low oxygen partial pressures, the conductivity is proportional to $P_{O_2}^{-1/6}$ suggesting an oxygen deficient regime governed by the charge neutrality condition $[e'] = 2[V_O^{\cdot\cdot}]$ in which the reaction

$$O_O^x = 1/2\ O_2(g) \rightleftharpoons V_O^{\cdot\cdot} + 2e' \qquad (1)$$

with the law of mass action

$$[V_O^{\cdot\cdot}]\,[e']^2 = K_1\,p_{O_2}^{-1/2} \tag{2}$$

combined with the charge neutrality condition yields

$$[e'] \propto p_{O_2}^{-1/6}.$$

In forsterite, aluminum, a common impurity, substitutes for silicon on tetrahedral sites. In crystals containing aluminum, electrical conductivity studies indicate two charge neutrality regimes: $[Al_{Si}'] = 2[Mg_i^{\cdot\cdot}]$ at lower temperatures and $[Al_{Si}'] = [h^{\cdot}]$ at higher temperatures (Morin et al., 1977; 1979; Hobbs, 1983).

When transition-metal cations replace magnesium, the point defect chemistry is controlled by the aliovalent ions. The primary published experimental observations which place constraints on the possible majority defects are contained in Greskovich and Schmalzried (1970) for Co_2SiO_4, Schwier et al. (1973) for $(Co,Mg)_2SiO_4$ with Co/Mg > 1/3, and Sockel (1974) for Fe_2SiO_4. Both electrical conductivity and thermogravimetric weight change are proportional to $p_{O_2}^{1/6}$. These olivines are all cation deficient with cation vacancy concentrations approximately 1% of those in the corresponding simple oxides. In Fe_2SiO_4 as an example, charge neutrality is set by

$$[Fe_{Fe}^{\cdot}] = 2[V_{Fe}''] \tag{3}$$

The influence of changes in the oxygen partial pressure on fayalite, Fe_2SiO_4, can be expressed through the relation

$$1/2\ O_2(g) + 3Fe_{Fe}^{x} = 2Fe_{Fe}^{\cdot} + V_{Fe}'' + FeO \tag{4}$$

The law of mass action then gives the equilibrium condition

$$[V_{Fe}'']\,[Fe_{Fe}^{\cdot}]^2 = K_2 a_{FeO}\,p_{O_2}^{1/2} \tag{5}$$

which combines with Eq. 3 to yield

$$[Fe_{Fe}^{\cdot}] = 2[V_{Fe}''] \propto a_{FeO}^{-1/3}\,p_{O_2}^{1/6}\ . \tag{6}$$

Thus in experiments in which the activity of FeO is held constant by buffering with excess FeO or SiO_2,

$$[Fe^{\cdot}_{Fe}] = 2[V''_{Fe}] \propto p_{O_2}^{1/6} \quad . \tag{7}$$

This result accounts for the dependence of the weight loss on oxygen partial pressure, which involves addition or removal of metal cation vacancies, and of the electrical conductivity, which occurs via hopping of holes. This point defect model is also supported by Mg-Fe interdiffusion studies which give an interdiffusion coefficient proportional to $p_{O_2}^{1/6}$, suggestive of a simple vacancy mechanism for Fe/Mg > 1/10 (Buening and Buseck, 1973).

To this point, point defect associations have been ignored. Results of thermogravimetric measurements on (Mg,Fe) olivines in progress in Schmalzried's laboratory challenge this assumption (Nakamura and Schmalzried, 1983a). In unbuffered samples (no excess FeO or SiO_2), the authors observe weight change proportional to $p_{O_2}^{1/5}$ for Fe/Mg > 0.25. They interpret this result in terms of three majority defects $Fe^{\cdot}_{Me}$, V''_{Me}, and Fe'_{Si} with the defect association $\{Fe'_{Si}Fe^{\cdot}_{Me}\}$. At Fe/Mg = 0.25, they find a $p_{O_2}^{2/11}$ dependence which is consistent with unassociated defects.

BUFFERED VERSUS UNBUFFERED CONDITIONS

To completely define the thermodynamic state of a ternary oxide such as Fe_2SiO_4, four independent variables need to be controlled. For experiments carried out at one atmosphere, three of the four variables are usually oxygen partial pressure, total pressure, and temperature. One of two approaches is then used to specify the thermodynamic state: (i) the ternary sample is buffered with one of its constituent oxides, or (ii) the ratio of the two cation species is (assumed) fixed (i.e., the sample is run unbuffered). The former arrangement corresponds to an open-system experiment, while the latter to a closed-system experiment. In the first case, the buffered case, crystals are annealed in either FeO or SiO_2 powders, and the experiment (deformation, diffusion, electrical conductivity, etc.) is carried out with small amounts of the same powder either in direct or vapor contact with the sample. If an Fe_2SiO_4 sample is in equilibrium with FeO, for example, then the activity of FeO is unity, $a_{FeO} = 1$, and the activity of SiO_2 is given by $a_{SiO_2} = \exp(\Delta G^{\circ}_{Fe_2SiO_4}/RT)$. $\Delta G^{\circ}_{Fe_2SiO_4}$ is the Gibbs energy of formation of fayalite from the oxides according to the reaction $2FeO + SiO_2 = Fe_2SiO_4$ (Schmalzried, 1965; 1981). In this case, the thermodynamic state is well defined because four thermodynamic variables are fixed. In the second case, the unbuffered case, the cation species are assumed to be sufficiently nonvolatile that the cation ratio, Fe:Si, is constant during an experiment. If the Fe:Si ratio is fixed, then the activities of both constituent oxides are functions of temperature and in general neither activity will be unity (Smyth, 1976). (The unbuffered or open system arrangement is often called the fixed-cation ratio case.)

The variation of the concentrations of the various possible point defects with oxygen partial pressure and enstatite, $MeSiO_3$, activity have been worked out and tabulated for Mg_2SiO_4 and $(Mg,Fe)_2SiO_4$ olivines for both the buffered (Stocker and Smyth, 1978) and the unbuffered (Stocker, 1978) cases. For Mg-olivine and the (Mg,Fe) olivines, the constituent oxides are $MgO/MgSiO_3$ and $(Mg,Fe)O/(Mg,Fe)SiO_3$, respectively. $FeSiO_3$ is not stable at one atmosphere so that the constituent oxides are FeO/SiO_2.

Three specific aspects concerning buffered or unbuffered experiments are commented on here. First, with experiments which are intended to be unbuffered, solid-state reactions with parts of the apparatus--for example, the pistons in a deformation run or the alumina furnace tube via vapor transport in an electrical conductivity experiment or a creep run (Durham et al., 1983)--can change the test conditions from unbuffered to buffered. One outstanding example of this situation involves the use of Pt in contact with Fe-bearing silicates. Because Fe is soluble in Pt, Fe diffuses from the sample into the Pt liberating O_2 to the gas. As a result, a layer of SiO_2 or $(Mg,Fe)SiO_3$ can form to buffer the sample (cf. Grove, 1981).

Second, in buffered experiments the question arises as to the time required to equilibrate the sample with an externally imposed buffer. In the case of changes of oxygen partial pressure, the creep rate for (Mg,Fe)-olivine reaches a new steady state level in less than 15 minutes at 1325°C (Kohlstedt and Hornack, 1981). The new equilibrium is reached quickly because it requires only the diffusion of cation vacancies, V''_{Me} (and holes, $h^{\cdot}$, to maintain charge neutrality), and not a flux of cations or anions. In contrast, internal oxidation as is involved in decoration of dislocations in olivine (Kohlstedt and Vander Sande, 1975) requires removal of iron atoms from the bulk to the surface or to grain boundaries. This point is supported by the fact that a careful examination of the surface of a single crystal of (Mg,Fe)-olivine which has been oxidized to decorate dislocations reveals a very thin layer of iron oxide. The relaxation time governing the changes in Me:Si ratio upon change in buffer from,say MgO to $MgSiO_3$, has not yet been determined experimentally, although experiments are in progress in our laboratory. If a change in the ratio of the cations on the two different sublattices requires only vacancy diffusion, then it should be possible to establish a new thermodynamic state quite quickly upon a change in the buffer.

Third, if precipitates of one of the constituent oxide phases exist within a crystal, then the crystal will be self-buffered or internally buffered. In this case, it will be difficult to equilibrate the sample with another oxide buffer as cation diffusion will be required over distances comparable to the sample size. We have found, as one example, small precipitates in crystals of forsterite grown by zone melting by Dr. John Haggerty at MIT. Electron micrographs show in Fig. 2 the local density and size of the precipitates. Analytical scanning transmission electron microscopy observations by our group indicate that regions containing precipitates

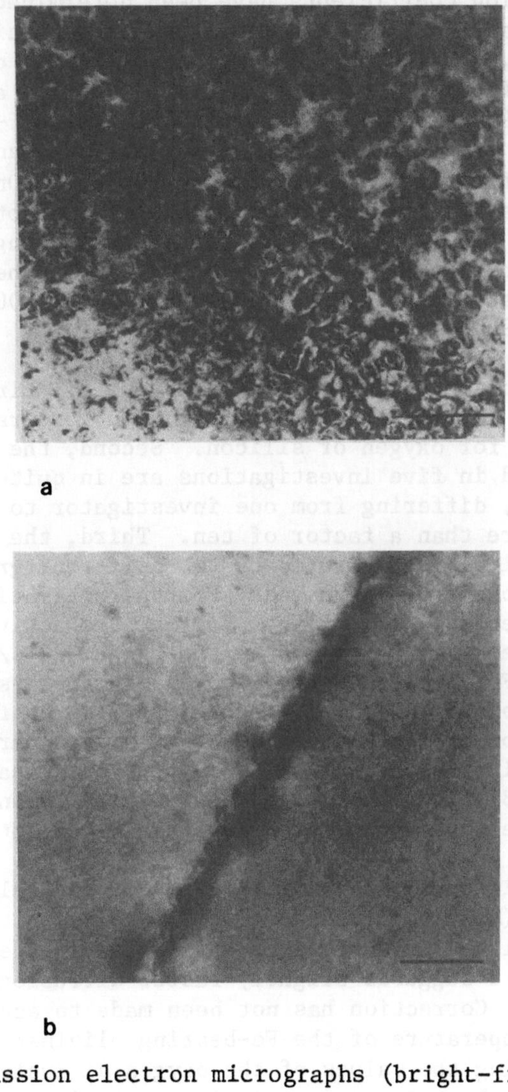

Fig. 2. Transmission electron micrographs (bright-field) showing precipitates believed to be MgO which nucleated homogeneously (a) and heterogeneously (b) on cooling a Mg_2SiO_4 crystal from the melting temperature. Scale bars represent 0.2 μm.

are magnesia-rich suggesting that the precipitates are MgO. Based on the self-diffusion data for Mg (Hallwig et al., 1980), annealing times of approximately six days would be required to equilibrate millimeter-sized samples of these forsterite crystals with enstatite at the melting temperature or two years at 1600°C!

DIFFUSION

Self-diffusion coefficients have been determined for all three ions in Mg_2SiO_4 and Co_2SiO_4. Interdiffusion coefficients have been reported for (Fe,Mg), (Co,Mg), (Ni,Mg), and (Mn,Mg) olivines as well as the tracer diffusivity of Ca in Mg_2SiO_4 (Buening and Buseck, 1973; Morioka, 1980; 1981). The results of the self-diffusion measurements are summarized in Fig. 3 for Mg_2SiO_4, and references for specific diffusivities are listed in Table I. One point not made in Fig. 3 is that diffusion in Mg_2SiO_4 is anisotropic. Diffusion of Mg along [100] and [010] is faster than along [001] (Hallwig et al., 1980); the latter data are plotted here. The oxygen diffusivity is somewhat higher along [010] than along [100] and [001] (Reddy et al., 1980).

Several points should be noted from Fig. 3. First, the diffusion coefficient for magnesium is roughly four orders-of-magnitude larger than that for oxygen or silicon. Second, the oxygen diffusivities reported in five investigations are in quite good agreement with one another, differing from one investigator to the next by just slightly more than a factor of ten. Third, the diffusion coefficient for silicon as measured by two different groups differ by more than a factor of ten thousand. Fourth, activation energies for all three species are similar. For Mg, the activation energy is about 400 kJ/mole. For O, values in the range 320-420 kJ/mole have been reported. For Si, the activation energy appears to be between 195 and 375 kJ/mole. In addition, the diffusion coefficients of oxygen and silicon are both independent of oxygen partial pressure when excess MgO is present during the diffusion anneals (Jaoul et al., 1983; 1981); and the oxygen diffusivity was not observed to differ from buffered to unbuffered crystals (Jaoul et al., 1983).

One data point has been published for oxygen self-diffusion in (Mg,Fe)-olivine (Muehlenbachs and Kushiro, 1974). It is in reasonable agreement with values of oxygen diffusion in the pure Mg-olivine, though it suggests slightly faster diffusion in the Fe-bearing olivine. Correction has not been made to account for the lower melting temperature of the Fe-bearing olivine; such a correction would bring the values of the oxygen diffusivities in the two olivines into better agreement. A systematic study of self-diffusion in (Mg,Fe)-olivines is in progress in Jaoul's laboratory and on interdiffusion in these olivines in Schmalzried's laboratory (Nakamura and Schmalzried, 1983b).

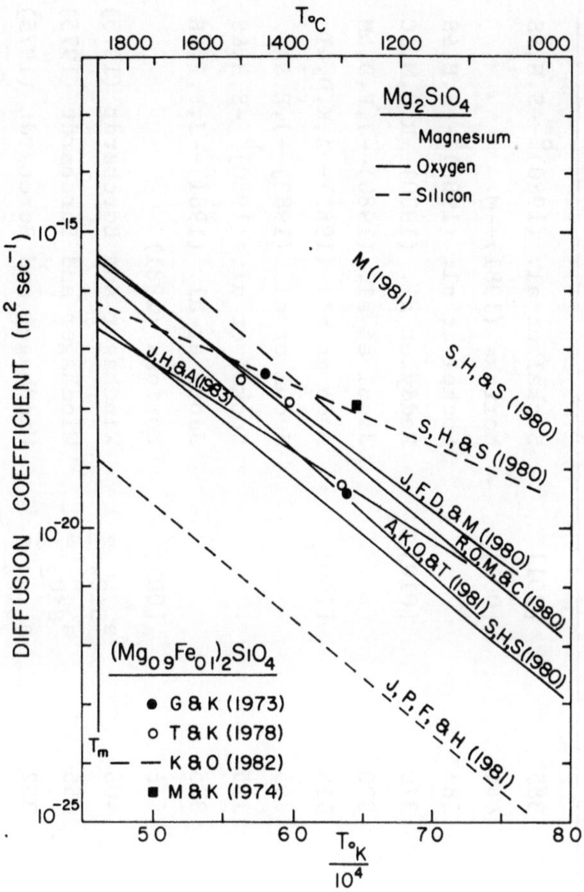

Fig. 3. Self-diffusion coefficient versus inverse temperature for magnesium, oxygen, and silicon in Mg_2SiO_4. The diffusivity of the slowest species in $(Mg,Fe)_2SiO_4$ has been estimated from the collapse rate of sessile dislocations and the recovery kinetics for dislocation annihilation. References are given in Table I.

Self-diffusion coefficients for Co, Si, and O in Co_2SiO_4 are also presented in Table I. As in the Mg-olivine, cobalt diffuses much more rapidly than oxygen or silicon. Dependencies on oxygen partial pressure were not reported. However, the silicon diffusivity increases with increasing SiO_2 activity consistent with an interstitial mechanism of diffusion.

The self-diffusion coefficients for Co_2SiO_4 are compared with those for Mg_2SiO_4 in a semi-log plot of diffusion coefficient versus

Table I. Self-Diffusion Coefficients for Silicate Olivines

Olivine	Species	$D_o (m^2/sec)$	Q(kJ/mole)	Comments	Reference[a]
Mg_2SiO_4	Mg	1.0×10^{-3} [c]	385[c]	[001]	Sockel et al. (1980)[b]--S,H,&S
Mg_2SiO_4	Mg	1.5×10^{-1}	444	[001]	Morioka (1981)--M
Mg_2SiO_4	O	1.0×10^{-7} [c]	385[c]		Sockel et al. (1980)[b]--S,H,&S
Mg_2SiO_4	O	3.5×10^{-7}	370	[010]	Reddy et al. (1980)--R,O,M,&C
Mg_2SiO_4	O	1.6×10^{-8} [c]	320		Jaoul et al. (1980)--J,F,D,&M
Mg_2SiO_4	O	2.9×10^{-6}	416	[100]	Ando et al. (1981)--A,K,D,&T
Mg_2SiO_4	O	2.3×10^{-10}	295		Jaoul et al. (1983)--J,H,&A
Mg_2SiO_4	Si	2.5×10^{-12}	195		Sockel et al. (1980)[b]--S,H,&S
Mg_2SiO_4	Si	1.5×10^{-10}	375		Jaoul et al. (1981)--J,P,F,&H
Mg_2SiO_4	Ca	7.2×10^{-3}	416	[001]	Morioka (1981)
Co_2SiO_4	Co	6.5×10^{-1}	406	$a_{SiO_2} = 1$	Windhager and Borchardt (1975)
Co_2SiO_4	O	8.5×10^{-3}	456	$a_{SiO_2} = 1$	Windhager and Borchardt (1975)
Co_2SiO_4	Si	6.0×10^{-6}	352	$a_{SiO_2} = 1$	Windhager and Borchardt (1975)

[a]Letter designations refer to Fig. 2.
[b]See also Hallwig et al. (1980).
[c]Calculated by present authors.

T_m/T in Fig. 4. At a given value of T_m/T, the magnesium diffusivity is about a factor of ten larger than that for cobalt. This result is somewhat surprising because cobalt diffuses more rapidly than magnesium in Mg_2SiO_4 (Morioka, 1981). Possibly it suggests that the concentration of V''_{Mg} in Mg_2SiO_4 is larger than the concentration of V''_{Co} in Co_2SiO_4. The oxygen diffusivities are in quite good agreement in their region of overlap. The silicon diffusivity in Co_2SiO_4 lies between the two values published for Mg_2SiO_4, and its magnitude is similar to that for oxygen diffusion in Co_2SiO_4.

DISLOCATION RECOVERY

To supplement the direct determinations of self-diffusion coefficients, a diffusivity was calculated from the rate of collapse of sessile dislocation loops in $(Mg_{0.9}Fe_{0.1})_2SiO_4$ at two temperatures (Goetze and Kohlstedt, 1973). This diffusivity should correspond to that of the slowest moving species. In addition, we have estimated a diffusion coefficient from Toriumi and Karato's (1978) observations on the decrease in density of dislocation loops, combined with the

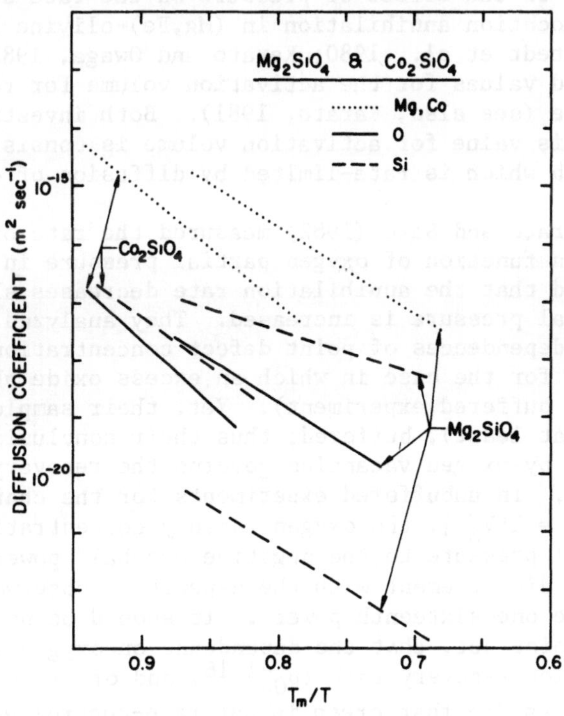

Fig. 4. Self-diffusion coefficients versus T_m/T for Co_2SiO_4 and Mg_2SiO_4. References are given in Table I.

analysis of Kirchner (1973) and an estimated value of 1μm for the starting maximum radius of a loop (Karato, private communication), 1983). The two sets of data for diffusivity are in reasonably good agreement with each other and fall in the band of values reported for oxygen self-diffusion in Mg_2SiO_4. They are plotted in Fig. 3 as G & K (1973) and T & K (1982).

Static annealing experiments on (Mg,Fe)-olivines have also been used to study the reduction of free dislocation density by formation of tilt boundaries (Goetze and Kohlstedt, 1973) and dislocation-dislocation annihilation (Toriumi and Karato, 1978; Kohlstedt et al., 1980, Karato and Sato, 1982). Karato and Sato (1982) calculated the diffusivity from an analysis of their dislocation-dislocation annihilation kinetics which is labeled T & O (1982) in Fig. 3. This diffusivity is slightly larger than those for self-diffusion of oxygen in Mg-olivine. Goetze and Kohlstedt (1973) demonstrated that recovery by formation of low-angle tilt boundaries proceeds at a rate four orders of magnitude faster than can be accounted for by self-diffusion (based on their loop collapse results), and they concluded that climb of dislocations into subgrain boundaries occurs by diffusion along dislocation cores.

Two studies of the effect of pressure on the rate of dislocation recovery by dislocation annihilation in (Mg,Fe)-olivine have been published (Kohlstedt et al., 1980; Karato and Owaga, 1982). These experiments yield values for the activation volume for recovery of about 14 cm^3/mole (see also, Karato, 1981). Both investigations conclude that this value for activation volume is consistent with dislocation climb which is rate-limited by diffusion of oxygen.

Finally, Karato and Sato (1982) measured the rate of dislocation annihilation as a function of oxygen partial pressure in (Mg,Fe)-olivine and found that the annihilation rate decreases slightly as the oxygen partial pressure is increased. They analyzed this result in terms of the dependences of point defect concentrations on oxygen partial pressure for the case in which an excess oxide phase is present (i.e., a buffered experiment). Yet, their samples were not (intentionally, at least), buffered; thus their conclusion that oxygen diffusion by oxygen vacancies governs the recovery rate needs to be reanalyzed. In unbuffered experiments for the charge neutrality condition $[Fe^{\bullet}_{Me}] = 2[V''_{Me}]$, the oxygen vacancy concentration varies as oxygen partial pressure to the negative one-half power (Stocker, 1978), in marked disagreement with the dependence observed (approximately a negative one-sixteenth power). It should be noted, however, as the authors point out, that the dependence on oxygen partial pressure of dislocation recovery rate ($p_{O_2}^{-1/16}$) and of creep ($p_{O_2}^{+1/6}$) are not the same suggesting that creep is not rate-controlled by a recovery process which is dominated by dislocation annihilation.

SLIP SYSTEMS

Dislocations with Burgers vector b of either [100] or [001] completely dominate in plastically deformed crystals. The high self-energy of dislocations with b = [010] ($|b| \sim 1.0$nm) makes them highly unfavorable defects, although they have been observed at facets in low-angle tilt boundaries as have dislocations with b = [101] (Goetze and Kohlstedt, 1973; Ricoult and Kohlstedt, 1983a). With an emphasis on the hexagonal close-packed character of the anion sublattice, both glissile and sessile splittings of b = [100] dislocations have been proposed which avoid breaking silicon-oxygen bonds (Poirier, 1973; Poirier and Vergobbi, 1978; Gueguen, 1979). However, weak-beam transmission electron microscopy (TEM) analyses have been unable to confirm the proposed dissociations and place an upper limit on the separation between partial dislocations near 3nm (Vander Sande and Kohlstedt, 1976; Gueguen and Darot, 1982a).

Based on optical microscopy observations, the slip systems identified in polycrystalline olivine samples are (010)[100], (001)[100], (0kl)[100], plus (100)[010], (001)[010], plus (100)[001], (010)[001], {110}[001], low index planes being strongly preferred. Dislocations with b = [001] are dominant at low temperatures (high stresses) and those with b = [100] at high temperatures (low stresses) (Raleigh, 1968; Raleigh and Kirby, 1970; Carter and AvéLallemant, 1970). Primary and secondary slip systems have been determined from analyses of changes in the shape of single crystals compressed along the $[110]_c$, $[101]_c$, and $[011]_c$ directions (Durham and Goetze, 1977). The $[uvw]_c$ direction refers the compression axis to cubic axes superimposed on the orthorhombic crystallographic axes; for example, $[110]_c$ specifies a compression axis 45° to [100] and 45° to [010].) For the $[110]_c$ orientation, the observed shape changes can be accounted for by operation of the (010)[100] slip system alone. For the $[011]_c$ orientation, slip on only the (010)[001] system is required. However, for the $[101]_c$ orientation, glide on the primary slip system (001)[100] and the secondary slip system (100)[001] cannot produce the observed shape change; climb is necessary to account for approximately 20% of the strain.

DISLOCATION STRUCTURES

The dislocation structures in experimentally and naturally deformed olivine grains have been studied both by transmission electron microscopy (TEM) and optical microscopy. While the TEM investigations have multiplied rapidly since the advent of commercial ion-thinners, optical analyses have blossomed with the discovery that dislocations in (Mg,Fe)-olivines can be decorated by oxidation at temperatures low enough, ∿900℃, to preclude any recovery (Kohlstedt et al., 1976b). This technique was extended to Mg-olivine by diffusing iron into the crystal before the oxidation treatment (Jaoul et al., 1979). The two techniques have proven to be complementary, the TEM providing detailed information on Burgers vectors and dislocation interactions and the optical observations revealing three dimensional "finger print" structures across entire deformation samples. Dislocation structures in deformed single crystals of olivine have been carefully examined for $(Mg,Fe)_2SiO_4$ (Durham et al., 1977) and for Mg_2SiO_4 (Darot and Gueguen, 1981; Gueguen and Darot, 1982b). These results are emphasized because the experiments were carried out at much lower stresses than most work on polycrystalline samples and thus bear most directly on deformation and possible mechanisms of deformation in the earth's mantle. The samples studied were deformed at $0.7 < T/T_m < 0.85$ and $10^{-4} < \sigma/\mu < 10^{-3}$, where μ is the shear modulus.

Three characteristic features are commented on here. First, the microstructure which is produced during deformation depends on the direction along which the stress is applied. For example, in samples compressed along $[110]_c$, long, somewhat curved edge dislocations with $\underline{b}$ = [100] dominate, Fig. 5a, while in samples compressed along $[011]_c$ very straight screw dislocations with $\underline{b}$ = [001] dominate, Fig. 5b. Second, dislocation glide loops are composed of straight segments which indicate a marked degree of crystallographic control on dislocation line directions, Fig. 5c, that is, the Peierls force appears to be significant for the olivine structure. A value of 8.8 GPa has been determined for the Peierls stress from hardness measurements on (Mg,Fe)-olivine (Evans and Goetze, 1979). Third, low-angle tilt and twist boundaries develop readily at strains of less than 10%, Fig. 5d. The twist boundaries formed from screw dislocations with $\underline{b}$ = [100] and those with $\underline{b}$ = [001]. The tilt boundaries are primarily (100) boundaries.

CREEP AND POINT DEFECTS

Under the assumption that the creep rate, $\dot{\varepsilon}$, at high temperatures is determined by the rate of climb of dislocation, steady state creep data are usually analyzed with an equation of the form (Weertman, 1968)

$$\dot{\varepsilon} = A_1 \frac{\mu b}{RT} \left(\frac{\sigma}{\mu}\right)^n D \qquad (8)$$

where A_1 is a materials parameter, μ the shear modulus, σ the applied or differential stress, n the stress exponent, D a diffusion coefficient and RT has the usual meaning. In multicomponent materials such as the olivines, diffusion of the slowest-moving species controls the rate of dislocation climb. To explicitly indicate that the diffusion coefficient for a given species is proportional to the concentration of the appropriate point defects, which will be function of the partial pressure of oxygen and the activity of a

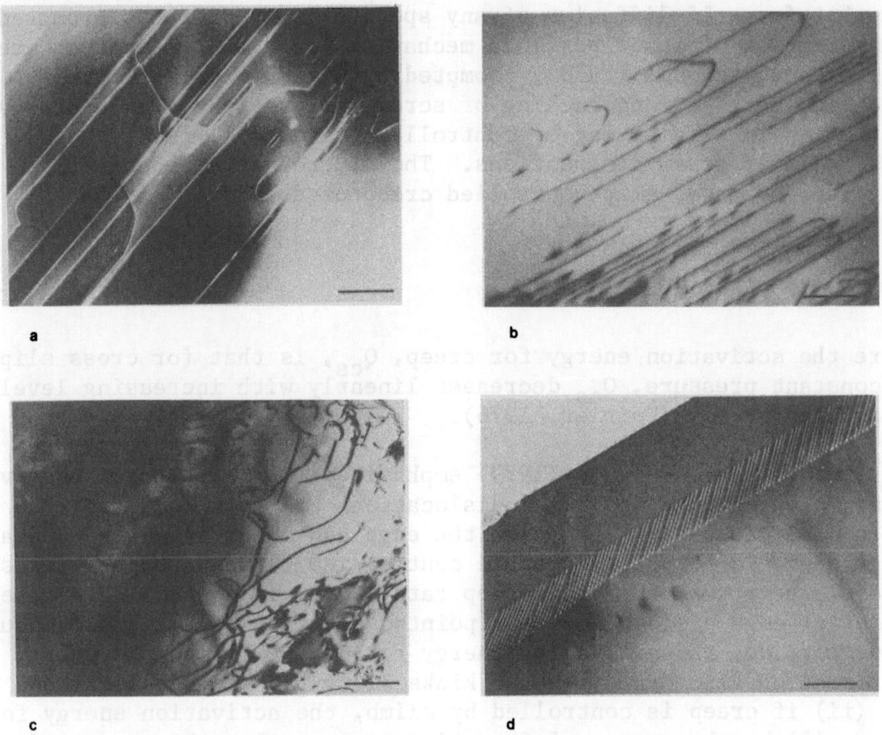

Fig. 5. Transmission electron micrographs of dislocation structures in deformed olivine. (a) Weak-beam image of b = [100] edge dislocations dipoles and loops in synthetic forsterite crystal deformed at 1600°C and 30MPa. The sample, which is viewed in the (010) plane, was compressed along $[110]_c$. (b) Bright-field image of straight screw dislocations with b = [001] in deformed polycrystalline (Mg,Fe)-olivine. (c) Bright-field micrograph of glide loops in deformed polycrystalline (Mg,Fe)-olivine. (d) Weak-beam micrograph of tilt boundary composed of b = [100] dislocations in naturally deformed olivine. Scale bars represent 1μm.

constituent oxide, Eq. 8 can be rewritten as

$$\dot{\varepsilon} = A_2 \frac{\mu b}{RT} \left(\frac{\sigma}{\mu}\right)^n p_{O_2}^m \, a_{ox}^q \, \exp\left(\frac{-Q_d}{RT}\right) \qquad (9)$$

where Q_d is the activation energy for diffusion. The values of m and q will be indicative of the point defects involved.

The lack of correlation between the activation energy for creep and that for self-diffusion of any species in Mg_2SiO_4 has prompted consideration of other possible mechanisms of high-temperature creep. Poirier and Vergobbi (1978), prompted by dislocation structures which indicate "strong locking of screw dislocations," have proposed that creep in olivine may be controlled by thermally activated cross-slip of screw dislocations. They introduce a constitutive equation for cross-slip controlled creep of the form

$$\dot{\varepsilon} = A_3 \left(\frac{\sigma}{\mu}\right)^2 \exp\left(\frac{-Q_{cs}}{RT}\right) \tag{10}$$

where the activation energy for creep, Q_{cs}, is that for cross slip; at constant pressure, Q_{cs} decreases linearly with increasing level of applied stress (Poirier, 1976).

In contrast, Gueguen (1979) emphasized microstructural observations in which edge and mixed dislocations are more numerous than screw dislocations and in which the edge and mixed segments show a large degree of crystallographic control indicating a high Peierls stress. He argued that the creep rate, then, is controlled by the motion of edge dislocations and pointed out that (i) if creep occurs by pure glide, the activation energy for creep will be the energy of formation (and migration) of kinks along the edge dislocations and (ii) if creep is controlled by climb, the activation energy for creep will be the energy of formation of jogs plus the energy of migration of jogs (i.e., the activation energy for diffusion of the slowest species). Hobbs (1981, 1983) extended this argument by noting that jogs and kinks may be charged in non-metals. As a result an additional energy term is required to take into account the energy to ionize a kink or a jog (see also Hirsch, 1979; 1981). Equation 9 must now take on the general form

$$\dot{\varepsilon} = A_4 \frac{\mu b}{RT} \left(\frac{\sigma}{\mu}\right)^n p_{O_2}^m a_{ox}^q \exp\left(\frac{-Q_c}{RT}\right) \tag{11}$$

The activation energy for creep, Q_c, is now composed of (i) the energies for formation, ionization, and migration of kinks, if creep is controlled by the drift of kinks along dislocations or (ii) the energies for the formation, ionization, and migration of jogs, if creep is controlled by climb of jogs. In neither case, then, is the activation energy for creep equal to that for self-diffusion of the slowest species. In addition, in the former case, the oxygen partial pressure and the oxide activity terms enter because the kinks are ionized; for negative kinks,

$$K^x = K' + \dot{h} \tag{12}$$

so that by the law of mass action for a constant concentration of neutral kinks,

$$[K'] \propto [h^{\cdot}]^{-1} \tag{13}$$

Likewise for positively charged kinks,

$$[K^{\cdot}] \propto [h^{\cdot}]^{+1} \tag{14}$$

Because the concentration of holes is a function of oxygen partial pressure and oxide activity, so also are the concentrations of positively and negatively charged kinks. In the latter case, when creep is controlled by climb of jogs, the oxygen partial pressure and oxide activity terms enter for two reasons. First, as indicated in Eq. 9, the self-diffusion coefficient depends on the oxygen partial pressure and the oxide activity. Second, the concentration of ionized jogs will depend also upon these quantities through expressions similar to Eqs. 13 and 14,

$$[J'] \propto [h^{\cdot}]^{-1} \tag{15}$$

$$[J^{\cdot}] \propto [h^{\cdot}]^{+1} \ . \tag{16}$$

It is instructive to consider two examples for (Mg,Fe)-olivine with the majority point defects $[Fe^{\cdot}_{Me}] = 2[V''_{Me}]$. (1) If creep is controlled by drift of positively charged kinks, then from Eq. 14

$$\dot{\varepsilon} \propto [K^{\cdot}] \propto [h^{\cdot}] \propto p_{O_2}^{1/6} \ a_{en}^{1/3} \ . \tag{17}$$

(2) If creep is controlled by climb of positively charged jogs whose mobility is limited by the diffusion of oxygen by a vacancy mechanism, then

$$\dot{\varepsilon} \propto [J'][V_o^{\cdot\cdot}] \propto p_{O_2}^{o} \ a_{en}^{1} \tag{18}$$

DEFORMATION

High-temperature compressive creep experiments have been carried out by several investigators on single crystals of Mg_2SiO_4, $(Mg,Fe)_2SiO_4$, Fe_2SiO_4, and Co_2SiO_4. Values for the stress exponent and activation energy are summarized in Table II with references. The Mg, Fe, and Co-olivine crystals were synthetic; the (Mg,Fe)-olivine crystals were natural (San Carlos, Arizona).

Table II. Parameters in Power-Law Creep Equation[a]

Olivine	Orientation	n	Q(kJ/mole)	m	Comments	Reference
Mg_2SiO_4	$[110]_c$	2.6	460	-	$\sigma > 15$ MPa	Darot and Gueguen (1981)
Mg_2SiO_4	$[110]_c$	-	-	0		Jaoul et al. (1980)
Mg_2SiO_4	$[110]_c$	-	~730	0	$\sigma = 30$ MPa	Ricoult and Kohlstedt (1983b)
Mg_2SiO_4	$[101]_c$	2.6	586	-		Durham and Goetze (1977b)
Mg_2SiO_4	$[101]_c$	3.5	575	-	$\sigma > 15$ MPa	Darot and Gueguen (1981)
Mg_2SiO_4	$[101]_c$	-	585	0	$\sigma = 40$ MPa	Ricoult and Kohlstedt (1983b)
Mg_2SiO_4	$[011]_c$	2.7	600	-	60 MPa	Darot and Gueguen (1981)
Mg_2SiO_4	$[011]_c$	-	730	0	$\sigma = 70$–90 MPa	Ricoult and Kohlstedt (1983b)
Mg_2SiO_4	$[111]_c$	2.9	670	-		Durham et al. (1979)
$(Mg,Fe)_2SiO_4$	-	~3	525	-	P_{O_2} not constant	Kohlstedt and Goetze (1974)
$(Mg,Fe)_2SiO_4$	$[110]_c$	3.6	525	-	P_{O_2} not constant	Durham and Goetze (1977a)
$(Mg,Fe)_2SiO_4$	$[110]_c$	-	445	-	$P_{O_2}^{1/6}$ corrected	Darot (1980)
$(Mg,Fe)_2SiO_4$	$[110]_c$	-	-	1/6	unbuffered	Jaoul et al. (1980)
$(Mg,Fe)_2SiO_4$	$[110]_c$	-	-	1/6	unbuffered	Ricoult and Kohlstedt (1983b)
$(Mg,Fe)_2SiO_4$	$[110]_c$	-	-	0	buffered	
$(Mg,Fe)_2SiO_4$	$[101]_c$	3.7	525	-	P_{O_2} not constant	Durham and Goetze (1977a)
$(Mg,Fe)_2SiO_4$	$[101]_c$	-	455	-	$P_{O_2}^{1/6}$ corrected	Darot (1980)
$(Mg,Fe)_2SiO_4$	$[101]_c$	-	448	1/6	unbuffered	Kohlstedt and Hornack (1981)
$(Mg,Fe)_2SiO_4$	$[101]_c$	2.5	670	-	P_{O_2} not constant	Karato et al., 1982[a]
$(Mg,Fe)_2SiO_4$	$[101]_c$	-	540	0	buffered	Ricoult and Kohlstedt (1983b)
$(Mg,Fe)_2SiO_4$	$[101]_c$	-	-	1/6	unbuffered	Ricoult and Kohlstedt (1983b)
$(Mg,Fe)_2SiO_4$	$[011]_c$	3.5	525	-	P_{O_2} not constant	Durham and Goetze (1977a)
Fe_2SiO_4	$[110]_c$	5-8	~800	0	buffered	Ricoult and Kohlstedt (1983b)
Co_2SiO_4	$[110]_c$	4-7	~600	1/6	buffered	Ricoult and Kohlstedt (1983b)

[a]Samples in the Karato et al. (1982) experiments had length to width ratios of 3:2 compared to values of ~2:1 used by other investigators.

The stress exponent for $(Mg_{0.9}Fe_{0.1})_2SiO_4$ is consistently near 3.5, independent of sample orientation. For Mg_2SiO_4, the stress exponent lies in the range 2.6 to 3.5 and depends upon the orientation of the sample relative to the load axis. The stress exponent for both Fe_2SiO_4 and Co_2SiO_4 is somewhat larger, 4 to 7.

The activation energy for creep in $(Mg,Fe)_2SiO_4$ falls quite consistently in the range 440-525 kJ/mole. Values for the activation energy for creep of Mg_2SiO_4 are more varied with values lying between 450 and 750 kJ/mole. Darot and Gueguen (1980) observed a variation of activation energy with sample orientation, the $[110]_c$ direction with the lowest value and the $[011]_c$ with the highest. They also report a decrease in activation energy with increasing differential stress. Although the data are more limited, activation energies for creep in Fe_2SiO_4 and Co_2SiO_4 appear to be roughly 600 to 800 kJ/mole.

The creep results for the four olivine compositions are compared in Fig. 6 in a log-log plot of strain rate versus stress for single crystals compressed parallel to $[110]_c$ at $T/T_m = 0.9$. The temperature was chosen to minimize extrapolation of the Fe and Co-olivine data. The curves for $(Mg,Fe)_2SiO_4$, Fe_2SiO_4, and Co_2SiO_4 are in reasonable agreement with one another; however, at a given stress level the curve for Mg_2SiO_4 lies about two orders of magnitude higher

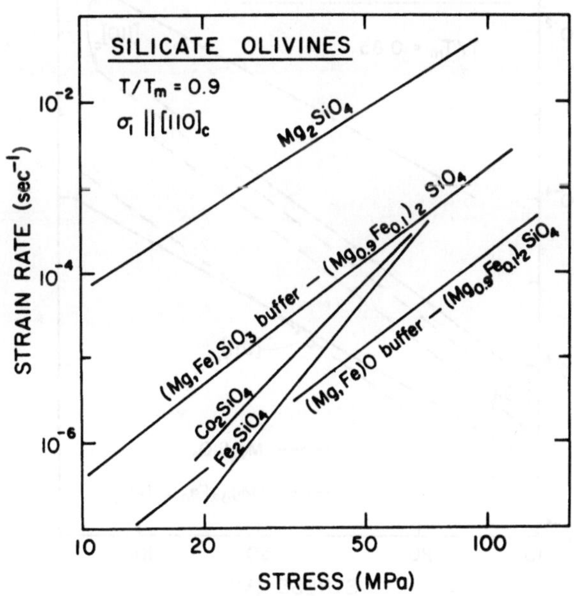

Fig. 6. Strain rate versus stress for olivines of four different compositions. The compression axis was $[110]_c$. Curves are plotted for $T/T_m = 0.9$. References are given in Table II.

in strain rate (i.e., the Mg_2SiO_4 crystals are significantly weaker than the olivines with transition-metal cations).

This comparison is extended to other orientations in Fig. 7 for Mg_2SiO_4 and $(Mg,Fe)_2SiO_4$. Data are not available for the Fe and Co-olivines for the $[101]_c$ and $[011]_c$ orientations. The temperature is now 0.85 T_m to minimize extrapolation of the experimental results. The marked difference in strength between Mg_2SiO_4 and $(Mg,Fe)_2SiO_4$ crystals compressed along $[110]_c$ is reemphasized. However, in the $[101]_c$ orientation the Mg-olivine is only slightly weaker than the (Mg,Fe)-olivine, while in the $[011]_c$ orientation the two olivines show virtually identical strengths. The results of Durham and Goetze (1977b) also show that, at a given homologous temperature, Mg_2SiO_4 and $(Mg,Fe)_2SiO_4$ compressed along $[111]_c$ have similar strength.

The effect of oxygen partial pressure on creep rate has also been explored. For (Mg,Fe)-olivine crystals compressed along $[110]_c$ and $[101]_c$, the creep rate varies as $p_{O_2}^{1/6}$ in unbuffered (closed system) experiments (Jaoul et al., 1980; Hornack and Kohlstedt, 1981).

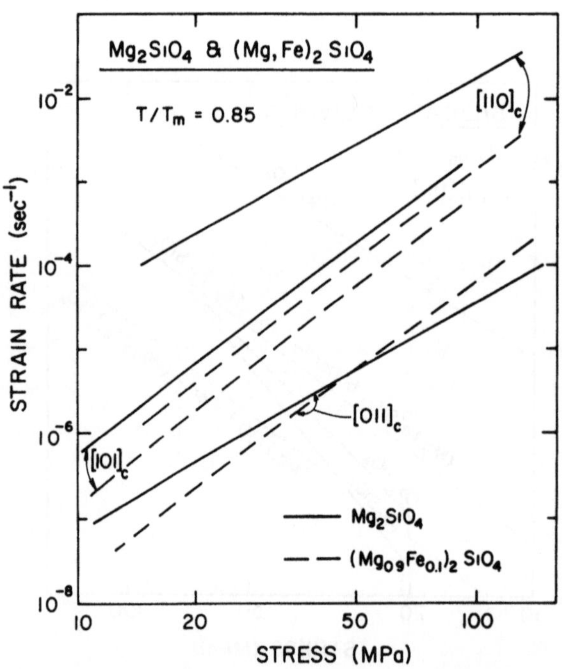

Fig. 7. Strain rate versus stress for Mg_2SiO_4 and $(Mg,Fe)_2SiO_4$ crystals at T/T_m = 0.85. Creep strength is compared for three orientations. References are given in Table II.

and it varies as $p_{O_2}^{0}$ in buffered (open system) runs (Ricoult and Kohlstedt, 1983b). For Mg-olivine, the creep rate is independent of oxygen partial pressure in both buffered and unbuffered experiments (Jaoul et al., 1980; Ricoult and Kohlstedt, 1983). For Fe-olivine, the creep rate is also independent of oxygen partial pressure when samples are buffered. However, for Co-olivine, the creep rate varies as $p_{O_2}^{1/6}$ if the crystals are buffered with one of the oxide constituents (Ricoult and Kohlstedt, 1983c).

Recent experiments have also demonstrated a pronounced effect of using Si-rich and (Mg,Fe)-rich buffers on the creep behavior of $(Mg,Fe)_2SiO_4$ samples. Crystals buffered with enstatite--$(Mg,Fe)SiO_3$--creep ten times faster than those buffered with magnesiowustite--(Mg,Fe)O--Ricoult and Kohlstedt, 1983). Stated in terms of component activities, the strain rate increases as the activity of enstatite, a_{en}, is increased: roughly, $\dot{\varepsilon} \propto a_{en}^{1.2}$. The difference in creep strength between crystals buffered with enstatite and those buffered with magnesiowustite is clearly shown in Fig. 6.

DISCUSSION

The activation energy for creep in olivine is consistently larger than that for self-diffusion of the slowest (or any) species. As a result, the high-temperature creep behavior of olivine cannot be explained in terms of a simple model for diffusion-controlled dislocation creep such as that described by Eqs. 8 and 9. Further constraints are placed on the possible rate-controlling mechanisms of deformation by the observed dependences of creep rate on oxygen partial pressure and oxide activity. For the (Mg,Fe)-olivine samples deformed under buffered conditions, the experimental creep data yield

$$\dot{\varepsilon} \propto p_{O_2}^{0}\ a_{en}^{1.2} .$$

This result suggests, as indicated by Eq. 18, that the high-temperature creep of olivine is governed by climb of positively charged jogs with oxygen diffusing as the slowest species. The same conclusion can be reached for the unbuffered crystals for which

$$\dot{\varepsilon} \propto p_{O_2}^{1/6} ,$$

although it requires a change in the charge neutrality condition from $[Fe_{Me}^{\bullet}] = 2[V_{Mg}'']$ to $[Mg_i^{\bullet\bullet}] = [V_{Mg}'']$ (Ricoult and Kohlstedt, 1983b). A jog-oxygen diffusion model is also consistent with the results for Mg-olivine, provided the jogs are neutral. However, a strain rate proportional to oxygen partial pressure to the one-sixth power for buffered samples of Co-olivine indicates a glide mechanism controlled by the drift of positively charged kinks (Ricoult and Kohlstedt, 1983c).

Two observations stand out in examining the effects of oxygen partial pressure and oxide activity on synthetic olivine crystals. First, a change in buffer from MeO to $MeSiO_3$ (or SiO_2) has no effect on the creep strength, and second, both buffered and unbuffered crystals show the same dependence of strain rate on oxygen partial pressure. These results could be indicative of the operative creep mechanism. Yet, for natural olivine crystals a change in buffer produces a significant (factor of ten) change in strain rate, and the oxygen partial pressure dependence changes from one-sixth to zero in going from unbuffered to buffered crystals. Possibly the synthetic crystals are internally (self-) buffered by precipitates that formed on cooling from the melting temperature during growth of the crystals; high-resolution TEM examinations, however, have not revealed precipitates in the deformed or starting samples (Ricoult, private communication, 1983).

The conclusion that oxygen diffusion controls the climb rate of jogs would appear to contradict the diffusion data in Fig. 3 which indicates that silicon is the slowest diffusion ion (Jaoul et al., 1981). Three points indicate that the inconsistency may lie with the silicon diffusion data for Mg_2SiO_4. First, the dislocation climb velocity, v_c, based on the silicon diffusion results would have to be exceedingly small. An upper limit (assuming a near-equilibrium concentration of vacancies along the dislocation core) can be calculated from the expression (Schöck, 1980).

$$v_c = \frac{2\pi D\sigma\Omega}{bkT \, \ell n(R/b)}$$

where R is half the distance between dislocations emitting and absorbing vacancies and Ω is an atomic volume. At 1600°C for a stress of 100 MPa, $v_c \sim 10^{-6}$µm/sec. In a three hour experiment with a dislocation density of 2×10^{12} m^{-2}, the Orowan equation predicts that the strain produced by climb will be $\sim 10^{-5}$, a value substantially less than that measured in shape-change experiments (Durham and Goetze, 1977a). If oxygen diffusion is slowest, strains of magnitude comparable to those determined by Durham will be produced by climb. Second, the rate of collapse of sessile dislocation loops yields a diffusivity in good agreement with that for oxygen. Third, the silicon diffusivity in Co_2SiO_4 is larger than the oxygen diffusivity. Based on these three observations, we believe that a diffusion-controlled dislocation climb mechanism of creep with oxygen as the slowest species is consistent with the self-diffusion results.

A number of creep experiments have been conducted in solid-medium apparatus on polycrystalline samples composed dominantly of (Mg,Fe)-olivine (Carter and AveLallemant, 1970; Raleigh and Kirby, 1970; Zeuch and Green, 1979). Large frictional forces inherent to the solid confining medium, as well as the large temperature and stress gradients somewhat limited the reliability of the creep data. Recently, these technical problems have been reduced by using a gas-medium high-pressure apparatus to deform natural (Mg,Fe)-olivine polycrystals (Chopra and Paterson, 1981; 1983). These results are compared with single crystal data in Fig. 8. Whereas grain boundary sliding may be enhanced by the presence of water in the so-called "wet" experiments, intra-crystalline plastic deformation by dislocation motion is dominant when the samples are dry (Chopra and Paterson, 1983). Indeed, the flow law for the dry polycrystals is in excellent agreement with that for single crystals (Durham et al.,1983).

Creep experiments were also performed on synthetic polycrystalline Mg_2SiO_4 at 1 atm confining pressure (Relandeau, 1981). At high

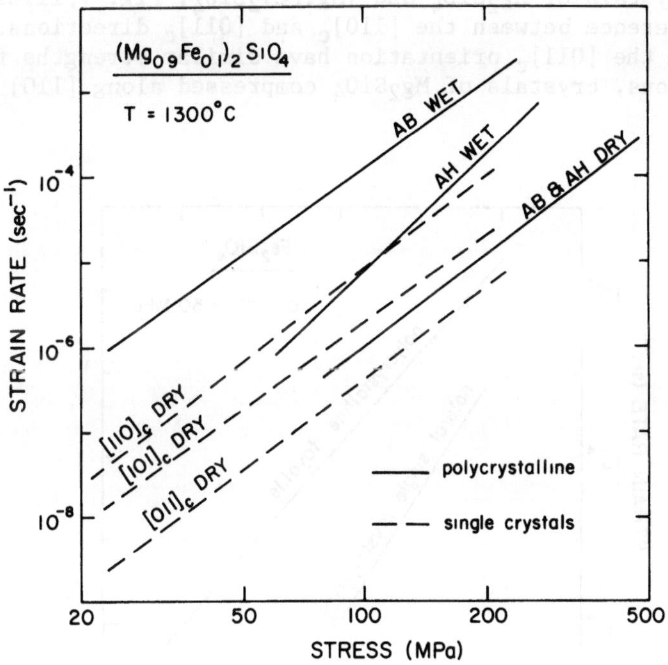

Fig. 8. Strain rate versus stress for single crystal and polycrystalline samples of $(Mg,Fe)_2SiO_4$. The two polycrystalline samples tested were from Anita Bay, New Zealand (AB) and Åheim, Norway (AH) (Chopra and Paterson, 1981; 1983). References for the single crystal results are given in Table II. The results of Zeuch and Green (1979) are in good agreement with the AB and AH dry results.

temperature, the creep rate is diffusion controlled. In the low temperature range, where dislocation mechanisms are operating, the flow law matches very well the flow law determined for single crystals of the same material.

Creep data have also been published for polycrystalline Fe_2SiO_4; these experiments were carried out at one atmosphere (Ilschner et al., 1977). In contrast to the results for $(Mg,Fe)_2SiO_4$, the unconfined polycrystalline Fe_2SiO_4 deformed one to two orders of magnitude faster than the single crystals at a given stress and temperature (Fig. 9). For a material with only three independent slip systems such as olivine, this result is expected. We suggest that it is quite likely that the polycrystalline samples are weaker than the single crystals because of microcracking which occurs in the polycrystalline samples due to stresses set up by thermal and elastic anisotropies of the individual grains, as well as intragranular stress resulting from the limited number of slip systems.

A comparison of the creep data for the three $\{110\}_c$ orientations for single crystals of Mg_2SiO_4 and $(Mg,Fe)_2SiO_4$, Fig. 7, illustrates a marked difference between the $[110]_c$ and $[01\bar{1}]_c$ directions. While crystals with the $[011]_c$ orientation have similar strengths for the two compositions, crystals of Mg_2SiO_4 compressed along $[110]_c$ are

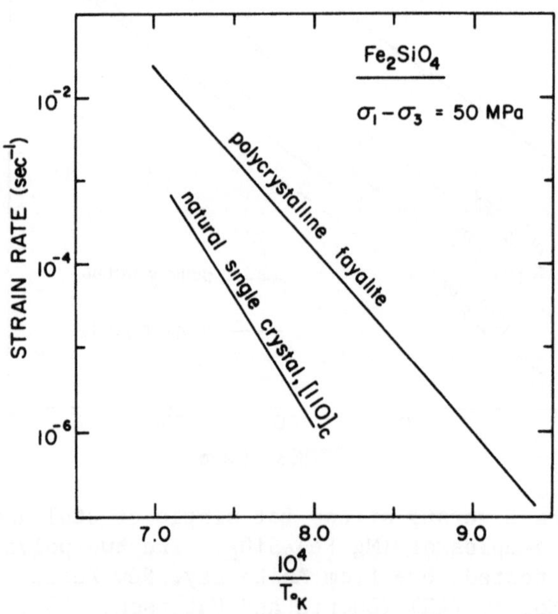

Fig. 9. Strain rate versus inverse temperature for single crystal and polycrystalline samples of Fe_2SiO_4.

significantly weaker than those of $(Mg,Fe)_2SiO_4$ compressed along $[110]_c$. Deformation in these two orientations is controlled by the motion of dislocations of quite different nature, screw dislocations with b = [001] versus edge and mixed dislocations with b = [100]. One explanation of the large observed difference for the two olivines is that the density of charged defects changes with iron content. If the [001] screw dislocations are neutral, their motion does not involve charged defects. In contrast, the edge dislocations are quite likely charged so that the addition of iron to Mg_2SiO_4 results in a change in the concentration of charged kinks and jogs. If the concentration of defects controlling the motion of edge dislocations is decreased (eg., oxygen vacancies and jogs) then the climb rate will decrease. One might anticipate from this argument that the creep rate of crystals compressed along $[011]_c$ will be independent of oxygen partial pressure. Experiments to test this point are in progress.

Additional experimental and theoretical work on point defect thermodynamics and creep in the olivine family is needed. Deformation experiments on crystals doped with acceptor or donors might clarify the role of charged kinks and jogs. Defect associations as well as multiple ionization states for point defects need to be considered. Yet, new information provided by systematic control and variation of the oxygen partial pressure and oxide activity have provided new constraints on the possible models of high-temperature creep in olivines.

Acknowledgments: This research was supported by a grant, #M45-8441, from the National Science Foundation through the Materials Research Center at Cornell University and by a fellowship from the Guggenheim Foundation. Two electron micrographs were kindly provided by Dr. John Fitz Gerald.

REFERENCES

Ando, K., Kurokawa, H., Oishi, Y., and Takei, H., 1981, Self-diffusion coefficient of oxygen in single-crystal forsterite, Commun. Am. Ceram. Soc., C-30.

Buening, D. K. and Buseck, P. R., 1973, Fe-Mg Lattice diffusion in olivine, J. Geophys. Res., 78: 6852-6862.

Carter, N. L. and AvéLallemant, H. G., 1970, High-temperature flow of dunite and peridotite, Geol. Soc. Am. Bull., 81: 2181-2202.

Chopra, P. N. and Paterson, M. S., 1981, The experimental deformation of dunite, Tectonophys., 78: 453-473.

Chopra, P. N. and Paterson, M. S., 1983, The role of water in the deformation of dunite, preprint.

Darot, M., 1980, Déformation expérimentale de l'olivine et de la forstérite, thèse d'Etat, Univ. de Nantes, Nantes, France.

Darot, M., and Gueguen, Y., 1981, High-temperature creep of forsterite single crystals, J. Geophys. Res., 86: 6219-6234.
Durham, W. B., and Goetze, C., 1977a, Plastic flow of oriented single crystals of olivine, 1. mechanical data, J. Geophys. Res., 82: 5737-5753.
Durham, W. B., and Goetze, C., 1977b, A comparison of creep properties of pure forsterite and iron-bearing olivine, Tectonophys., 40: 15-18.
Durham, W. B., Froidevaux, C., and Jaoul, O., 1979, Transient and steady-state creep of pure forsterite at low stress, Phys. Earth Planet. Inter., 19: 26 -274.
Durham, W. B., Goetze, C., and Blake, B., 1977, Plastic flow of oriented single crystals of olivine, 2. observations and interpretations of the dislocation structures, J. Geophys. Res., 82: 5755-5770.
Durham, W. B., Ricoult, D., and Kohlstedt, D. L., 1983, Interaction of slip systems in olivine, in: "Point Defects in Minerals," R. N. Schock, ed., accepted for publication.
Evans, B., and Goetze, C., 1979, The temperature variation of hardness of olivine and its implications for polycrystalline yield stress, J. Geophys. Res., 84: 5505-5524.
Goetze, C., and Kohlstedt, D. L., 1973, Laboratory study of dislocation climb and diffusion in olivine, J. Geophys. Res., 78: 5961-5971.
Greskovich, C., and Schmalzried, H., 1970, Nonstoichiometry and electronic defects in Co_2SiO_4 and $CoAl_2O_4$-$MgAl_2O_4$ crystalline solutions, J. Phys. Chem. Solids, 31: 639-646.
Grove, T. L., 1981., Use of FePt alloys to eliminate the iron loss problem in 1 atmosphere gas mixing experiments: theoretical and practical considerations, Contrib. Mineral. Petrol., 79: 298-304.
Gueguen, Y., 1979, High-temperature creep of olivine: evidence for control by edge dislocations, Geophys. Res. Lett., 6: 357-360.
Gueguen, Y., and Darot, M., 1982a, Les dislocations dans la forstérite déformée à haute température, Phil. Mag., A45: 419-442.
Gueguen, Y., and Darot, M., 1982b, Upper mantle plasticity from laboratory experiments, Phys. Earth Planet. Inter., 29: 51-57.
Hallwig, D., Schachtner, R., and Sockel, H. G., 1980, Diffusion of magnesium, silicon and oxygen and formation of the compound in the solid state, in: "Science of Ceramics," H. Hausner, ed., Deutsche Keramische Gesellschaft, Berlin, Vol. 10: pp. 385-393.
Hirsch, P. B., 1979, A mechanism for the effect of doping on dislocation mobility, J. de Physique, Colloque C6, 40: 117-121.
Hirsch, P. B., 1981, Plastic deformation and electronic mechanisms in semiconductors and insulators, J. de Physique, Colloque C3, 42: 149-160.
Hobbs, B. E., 1981, The influence of metamorphic environment upon the deformation of minerals, Tectonophys., 78: 335-383.

Hobbs, B. E., 1983, Constraints on the mechanism of deformation of olivine imposed by defect chemistry, Tectonophys., 92: 35-69.

Ilschner, B., Sockel, H. G., and Streb, G., 1977, Der Ernfluss der Nichtstöchiometrie auf die Plastizität von Oxiden und Silikaten unter besonderer Berücksichtigung der Erebnisse am Fe_2SiO_4, Zeitsch. Physik. Chemie Neue Folge, 108: 235-245.

Jaoul, O., Gueguen, Y., Michaut, M., and Ricoult, D., 1979, A technique for decorating dislocations in forsterite, Phys. Chem. Miner., 5: 15-19.

Jaoul, O., Froidevaux, C., Durham, W. B., and Michaut, M., 1980, Oxygen self-diffusion in forsterite: implications for the high-temperature creep mechanism, Earth Planet. Sci., 47: 391-397.

Jaoul, O., Poumellec, M., Froidevaux, C., and Havette, A., 1981, Silicon diffusion in forsterite: a new constraint for understanding mantle deformation, in: "Anelasticity in the Earth," F. D. Stacey, M. S. Paterson, and A. Nicolas, eds., Am. Geophys. Union and Geol. Soc. Am., Washington and Boulder, pp. 95-100.

Jaoul, O., Houlier, B., and Abel, F., 1983, Study of ^{18}O diffusion in magnesium orthosilicate by nuclear microanalysis, J. Geophys. Res., 88: 613-624.

Karato, S., 1981, Comment on "The effect of pressure on the rate of dislocation recovery in olivine" by D. L. Kohlstedt, H. P. K. Nichols, and Paul Hornack, J. Geophys. Res., 86: 9319.

Karato, S., and Ogawa, M., 1982, High-pressure recovery of olivine: implications for creep mechanisms and creep activation volume, Phys. Earth Planet. Inter., 28: 102-117.

Karato, S., and Sato, H., 1982, Effect of oxygen partial pressure on the dislocation recovery in olivine: a new constraint on creep mechanisms, Phys. Earth Planet. Inter., 28: 312-319.

Karato, S., Toriumi, M., and Fujii, T., 1982, Dynamic recrystallization and high-temperature rheology of olivine, in: "Advances in Earth and Planetary Sciences: High-Pressure Research in Geophysics, S. Akimoto and M. H. Manghnani, eds., Vol. 12: pp. 171-189.

Kirchner, H, O. K., 1973, Size distribution of dislocation loops, Acta Met., 21: 85-91.

Kohlstedt, D. L., and Goetze, C., 1974, Low-stress high-temperature creep in olivine single crystals, J. Geophys. Res., 79: 2045-2051.

Kohlstedt, D. L., and Vander Sande, J. B., 1975, An electron microscopy study of naturally occurring oxidation produced precipitates in iron-bearing olivines, Contrib. Mineral. Petrol., 53: 13-24.

Kohlstedt, D. L., and Hornack, P., 1981, Effect of oxygen partial pressure on the creep of olivine, in: "Anelasticity in the Earth," F. D. Stacey, M. S. Paterson, and A. Nicolas, eds., Am. Geophys. Union and Geol. Soc. Am., Washington and Boulder, pp. 95-100.

Kohlstedt, D. L., Goetze, C., and Durham, W. B., 1976a, Experimental deformation of olivine single crystals with application to flow in the mantle, in: Petrophysics: "The Physics and Chemistry of Rocks and Minerals," R. G. J. Strens, ed., John Wiley and Sons, London, pp. 35-49.
Kohlstedt, D. L., Goetze, C., Durham, W. B., and Vander Sande, J. B., 1976b, New technique for decorating dislocations in olivine, Science, 191: 1045-1046.
Kohlstedt, D. L., Nichols, H. P. K., and Hornack, P., 1980, The effect of pressure on the rate of dislocation recovery in olivine, J. Geophys. Res., 85: 3122-3130.
Morin, F. J., Oliver, J. R., and Housley, R. M., 1977, Electrical properties of forsterite, Mg_2SiO_4, Phys. Rev. B, 16: 4434-4445.
Morin, F. J., Oliver, J. R., and Housley, R. M., 1979, Electrical Properties of forsterite, Mg_2SiO_4. II, Phys. Rev. B, 19: 2886-2894.
Morioka, M., 1980, Cation diffusion in olivine--I. cobalt and magnesium, Geochim. Cosmochim. Acta, 44: 759-762.
Morioka, M., 1981, Cation diffusion in olivine--II. Ni-Mg, Mn-Mg, Mg and Ca, Geochim. Cosmochim. Acta, 45: 1573-1580.
Muehlenbachs, K., and Kushira, I., 1974, Oxygen isotope exchange and equilibrium of silicates with CO_2 and O_2, "Carnegie Inst. Washington, Yearbk 73," pp. 232-236.
Nakamura, A., and Schmalzried, H., 1983a, On the nonstoichiometry and point defects in olivine, preprint.
Nakamura, A., and Schmalzried, 1983b, On the Fe^{2+}-Mg^{2+} interdiffusion in olivine, preprint.
Pluschkell, W. and Engell, H. J., 1968, Ionen and Elektronenleitung im Magnesiumorthosilikat, Ber. Dtsch. Keram. Ges., 45: 388-394.
Poirier, J. P., 1975, On the slip systems of olivine, J. Geophys. Res., 80: 4059-4061.
Poirier, J. P., 1976, On the symmetrical role of cross-slip of screw dislocations and climb of edge dislocations as recovery processes controlling high-temperature creep, Rev. Phys. Appl., 11: 731-738.
Poirier, J. P., and Vergobbi, B., 1978, Splitting of dislocations in olivine, cross-slip controlled creep and mantle rheology, Phys. Earth. Planet. Int., 16: 370-378.
Raleigh, C. B., 1968, Mechanisms of plastic flow in olivine, J. Geophys. Res., 73: 5391-5406.
Raleigh, C. B., and Kirby, S. H., 1970, Creep in the upper mantle, Min. Soc. Am. Spec. Paper, 3: 113-121.
Reddy, K. P. R., Oh, S. M., Major, L. D., and Cooper, A. R., 1980, Oxygen diffusion in forsterite, J. Geophys. Res., 85: 322-326.
Relandeau, C., 1981, Etude de la déformation plastique de la forstérite polycristalline, Geophys. Res. Lett., 8:733-736.
Ricoult, D. L., and Kohlstedt, D. L., 1983a, width of low-angle grain boundaries in olivine, Phys. Chem. Minerals, 9: 133-138.
Ricoult, D. L., 1979, Experimental annealing of a natural dunite, Bull. Minéral., 102: 86-91.

Ricoult, D. L., and Kohlstedt, D. L., 1983b, Experimental evidence for the effect of chemical environment upon the creep rate of olivine, in: "Point Defects in Minerals," R. N. Schock, ed., accepted for publication.

Ricoult, D. L., and Kohlstedt, D. L., 1983c, Creep of Co_2SiO_4 and Fe_2SiO_4 single crystals in a controlled thermodynamic environment, preprint.

Schmalzried, H., 1965, Point defects in ternary ionic crystals, in: "Progress in Solid State Chemistry," Vol. 8II, H. Reiss, ed., Pergamon Press, Oxford, pp. 265-308.

Schmalzried, H., 1978, Reactivity and point defects of double oxides with emphasis on simple silicates, Phys. Chem. Minerals, 2: 279-294.

Schmalzried, H., 1981, "Solid State Reactions," Academic Press, New York, p. 48.

Schöck, G., 1980, Thermodynamics and thermal activation of dislocations, in: "Dislocations in Solids," Vol. 3, F. R. N. Nabarro, ed., North-Holland Publishing Comp., New York, p. 118.

Schwier, V. G., Dieckmann, R., and Schmalzried, H., 1973, Punktfehlstellen in Oxidmischphasen. I. Fellstellenthermodynamik der Mischphasen $(Co_xMg_{1-x})O$ und $(Co_xMg_{1-x})_2SiO_4$, Ber. Bunsenges Phys. Chem., 77: 402-408.

Smyth, D. M., 1976, Thermodynamic characterization of ternary compounds. I. The case of negligible defect association, J. Solid State Chem., 16: 73-81.

Smyth, D. M., and Stocker, R. L., 1975, Point defects and nonstoichiometry in forsterite, Phys. Earth Planet. Inter., 10: 183-192.

Sockel, H. G., 1974, Defect structure and electrical conductivity of crystalline ferrous silicate, in: "Defects and Transport in Oxides," M. S. Smeltzer and R. J. Jaffee, eds., Plenum Press, New York, pp. 341-354.

Sockel, H. G., Hallwig, D., and Schochtner, R., 1980, Investigations of slow exchange processes at metal and oxide surfaces and interfaces using secondary ion mass spectrometry, Mater. Sci. Eng., 42: 59-64.

Stocker, R. L., 1978, Influence of oxygen partial pressure on defect concentrations in olivine with fixed cationic ratio, Phys. Earth Planet. Inter., 17: 118-129.

Stocker, R. L., and Smyth, D. R., 1978, Effect of enstatite activity and oxygen partial pressure on the point defect chemistry of olivine, Phys. Earth Planet. Inter., 16: 145-156.

Toriumi, M. and Karato, S., 1978, Experimental studies of the recovery process of deformed olivines and the mechanical state of the upper mantle, Tectonophys., 49: 79-95.

Vander Sande, J. B., and Kohlstedt, D. L., 1976, Observation of dissociated dislocations in deformed olivine, Phil. Mag., 34: 653-658.

Weertman, J., 1968, Dislocation climb theory of steady state creep, Trans. Amer. Soc. Metals, 61: 681-694.

Windhager, H. J., and Borchardt, G., 1975, Tracerdiffusion und Fehlordnung in dem Orthosilikat Co_2SiO_4, Ber. Bunsenges. Phys. Chem., 79: 1115-1119.
Zeuch, D. H., and Green, H. W. II, 1979, Experimental deformation of an "anhydrous" synthetic dunite, Bull. Mineral., 102: 185-187.

INFLUENCE OF pO_2 AND pH_2O ON THE

HIGH TEMPERATURE PLASTICITY OF OLIVINE

B. Poumellec* and O. Jaoul

Laboratoire de Géophysique et Géodynamique interne
Bâtiment 510, Université Paris Sud, Centre d'Orsay
91405 Orsay, France

INTRODUCTION

Deformation processes in the Earth's upper mantle control large-scale tectonic events. Their understanding requires deep knowledge of the deformation mechanism of olivine, which is the major component of the upper mantle. This paper deals with the dependence of the creep rate of olivine single crystals (San Carlos, Arizona) on pO_2 and pH_2O, the oxygen and water partial pressures in the experimental atmosphere. Up to now, the effect of the stress and temperature upon the strain rate and the dislocation microstructures of this material, have been largely documented. (e.g. Kohlstedt and Goetze, 1974, and Durham and Goetze, 1977). But the influence of the atmosphere surrounding the single crystal was less understood. This study brings new data after the recent investigation of Kohlstedt and Hornack (1981), who already showed an increasing creep rate when pO_2 is increased ($\dot{\varepsilon} \propto pO_2^m$ with $1/10 < m < 1/5$) in a rather narrow range of oxygen partial pressure. Another attempt to determine m was performed by Jaoul et al. (1980) who found m = 1/5 - 1/7, but for only one temperature, and with an atmosphere containing 2% water. The present work which corresponds to part of the thesis of B. Poumellec (1981) is an attempt to investigate the $\dot{\varepsilon}$ vs pO_2 dependence over a large range of T and pO_2, where pO_2 is varied as an independent parameter. Moreover the influence of different gas mixtures (H_2/H_2O and CO/CO_2), giving the same values of pO_2, has been tested: one corresponds to "wet" conditions, the other to "dry" conditions.

---

* Now at: Laboratoire des Composés non stoechiométriques, Bât. 415, Université Paris Sud, Centre d'Orsay, 91405, Orsay, France

These results bring one more brick to built our understanding of the creep mechanism of olivine and give more constraints to help to make a choice between different possible mechanisms controlling creep (e.g. dislocation glide or climb). We also try to propose a schematic model of point defects to interpret the difference in behavior between samples submitted to dry or to wet conditions.

EXPERIMENTAL

A dead load apparatus has been used to deform the specimens under stresses ranging from 10 MPa to 75 MPa. The temperature (1250 to 1550 °C) was provided by a rhodium furnace and controlled within ±1°C. The pistons (Ø=14mm), made of c//σ oriented corundum, were protected by platinum foils lying between them and the sample. Two thin rods of corundum, 30 cm long, coupled with displacement monitors, gave the "in situ" true length of the sample within an accuracy of .1 μm. The recorded deformation rates ranged within 3×10^{-8}-$10^{-5}s^{-1}$. Any change of σ or T was followed by a transient extending over 0.3 to 1% strain. The data reported here do correspond to steady state creep runs, i.e. with a constant deformation rate ($\dot{\varepsilon}$). Many pO_2 increments, decrements and cycles were performed on a same sample (each time, for strain increments of few 10^{-3}) so that the total strain of the specimen may have been relatively high (up to 18% for one specimen). The creep tests have been performed by stress increments. Thereby one was paying attention never to reduce the stress (unless one exception), in order to avoid a reduction of the dislocation densities by recovery processes.

The origin of the specimens is San Carlos (Arizona). Most of the samples, 2x2x4 mm^3, were oriented with σ_1// $[110]_c$, some with σ_1// $[101]_c$ to compare the effects of pO_2 on different slip systems. The subscript c means: in reference with a pseudo-cubic lattice superimposed to the orthorhombic Pbnm system (a<c<b).

We only focused on the pO_2 exponent. However, the whole set of data obtained with different applied stresses and temperatures is in good agreement with the stress exponents and activation energies published by Durham and Goetze (1977), Durham (1975), and Kohlstedt et al. (1976). The oxygen partial pressure was obtained by the use of a gas mixer (patent ANVAR-CNRS) already described by Jaoul et al. (1983). It consists in mixing two different gas mixtures prepared in advance and made of high quality and dried Ar-10%H_2 and Ar-2%H_2O vapor (obtained by bubbling argon in water at a temperature less than the ambient temperature) in various proportions. By the same process, CO and CO_2 could also be mixed in a wide range of proportion, but in the absence of argon. The gas was slowly flowing through the alumina tubing around the sample at a rate of some liters per hour. A zirconia gauge monitored the pO_2, which was in good agreement with its calculated value at the temperature of the sample. If some leak occured, so that the calculated and monitored

pO_2 values did not agree and departed from each other by a factor of more than 1.1, the data point was systematically rejected.

RESULTS

Transients in pO_2

Each small change in pO_2 at constant T and σ occurring after a modification of the gas composition was followed by a transient in the creep rate which lasted typically 20 to 60 mn, always less than 100 minutes for pO_2 changes of less than one order of magnitude. However the new steady state took two to three hours to be achieved after a pO_2 change of six orders of magnitude. The same long transient periods were observed after a change from a CO/CO_2 mixture to H_2/H_2O or reversely. Fig. 1 shows a rather long transient, but also that the time recorded for the creep transient includes a long period for the renewal of the working atmosphere: it is due to a

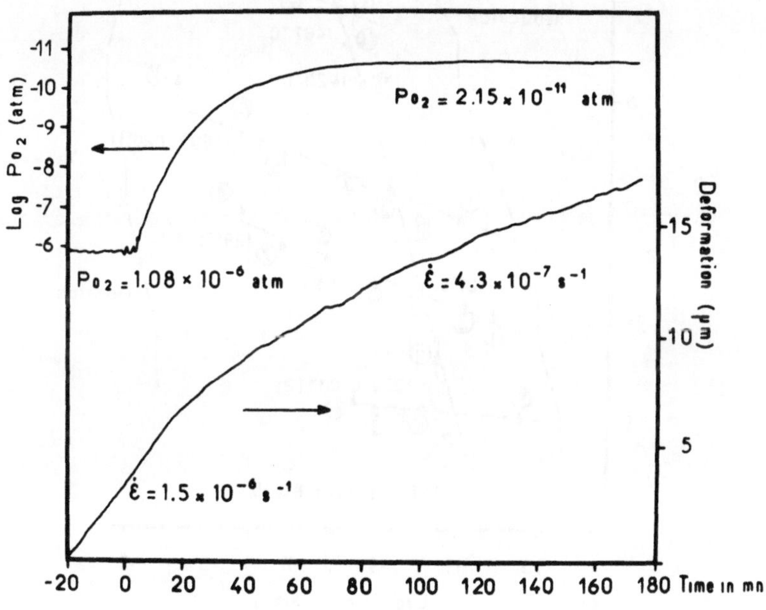

Fig. 1. A rather long transient in the creep rate after an important pO_2 change in a H_2/CO_2 mixture; 43.1 Mpa, 1287°C and $[101]_c$ orientation. pO_2 has been monitored by a zirconia gauge. 70 minutes were needed to completely renew the atmosphere in the working tubing (flow: 3 liters/hr). The transient creep rate lasted 90 mn, including the above 70 mn. (Sample B11).

low gas flow such that the temperature remains unaffected. In fact, the transients occur over small amounts of strain (say, always much less than $\Delta\varepsilon = 10^{-3}$) before a new steady state is achieved, which will take place over much larger strains.

Steady state deformation

The creep data in steady state conditions have been obtained under different stresses and temperatures applied to each specimen. This makes the representation of the data somewhat difficult.

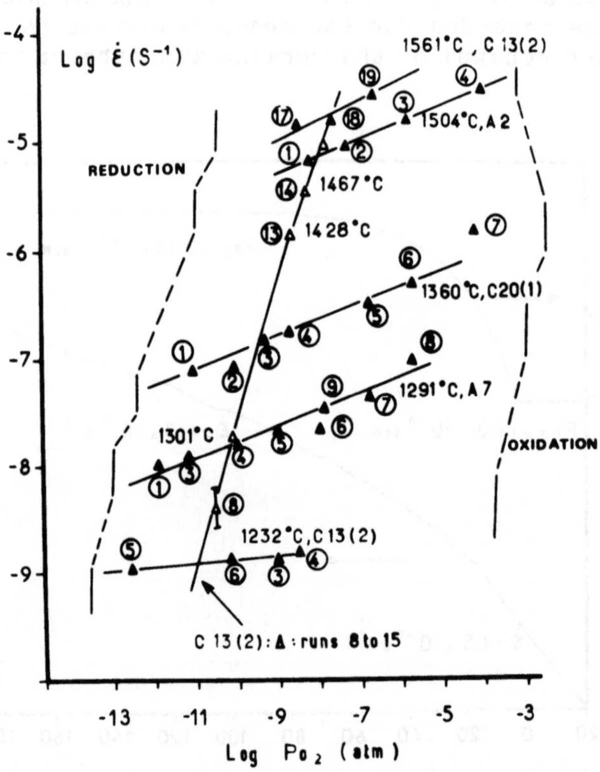

Fig. 2. Log$\dot{\varepsilon}$ vs Log pO_2. 4 specimens; CO/CO_2 atmosphere; $\sigma = 14.9$ MPa; $\sigma // [110]_c$. The circled numbers correspond to the sequence of the runs (see appendix II). The dashed lines indicate the boundaries of the stability field of the olivine Fo_{92} (Nitsan, 1975). Black triangles: the CO/CO_2 ratio has been varied at constant T; Open triangles on the steep line: the temperature has been varied and the CO/CO_2 ratio kept fixed.

For clarity we report these data, all reduced to a same stress (14.9 MPa), which only corresponds to vertical translations on a plot of $\dot{\varepsilon}$ versus pO_2. For that purpose, we used stress exponents $n = [\partial Log\varepsilon / \partial Log\sigma]_{T,pO2}$ determined from some samples, including the samples C13(1) and C14 for low stresses, data from Poumellec's thesis (1981) and other unpublished data. Our values of n are in relatively good agreement with the data of Durham (1975) and Durham and Goetze (1977). However we found a certain trend for n to decrease for increasing stresses: n varies from 4.9 at 6 Mpa to 3.4 at 75 Mpa with an average corresponding to the value given by Durham (1975) (n=3.6), as deduced from data at 1600°C and in a stress range (20 to 100 Mpa) which does not overlap enough with ours. It must be pointed out that our low stress data often correspond to high temperatures, and high stresses to lower temperatures. This was done intentionally in order to keep a good sensitivity in the strain rate determinations. Therefore it is not possible to draw any conclusion about this evolution of n at low stress.

The original data can be found in Appendix II where the n values further used for the plots in Fig 2 and Fig 3 are reported.

Use of CO/CO_2. On Figure 2, the data are reported for 4 specimens deformed by using two different experimental procedures. Varying the gas composition (the CO/CO_2 ratio: black triangles) at constant temperature for a given stress yields the partial derivative $m = [\partial Log\, \dot{\varepsilon} / \partial Log \partial pO_2]_{T,\sigma}$, which we measured at 5 temperatures (Table 1). The steep curve obtained by using a fixed gas composition (open triangles) intersects the others but corresponds to a different slope which combines both the activation energy $E=-R[\partial Log\, \varepsilon / \partial (1/T)]_{\sigma,pO2}$ and the dependence of $\dot{\varepsilon}$ with $pO_2(T)$ for a given CO/CO_2 ratio. The function $pO_2(T)$ involves the enthalpy of equilibrium for the reaction $CO + 1/2O_2 \rightleftarrows CO_2$. The apparent activation energy for creep is therefore:

$$E_{ap} = -R[\partial Log\dot{\varepsilon} / \partial (1/T)]_{\sigma,CO2/CO} = E - mR[\partial Log\ pO_2 / \partial (1/T)]_{CO2/CO}$$

m can also be deduced from this equation. Kohlstedt and Hornack (1981) give the details of this calculation. For temperatures between 1290 and 1560°C one finds $0.14<m<0.22$ (1/7.1 to 1/4.6), relatively independent of the temperature. This corresponds to an average : m=0.18 (=1/5.6). However it seems that this m value tends to decrease for the lowest temperatures down to 1250°C.

The use of a constant CO/CO_2 ratio with varying temperatures (the steepest line on Fig 2) gives a good agreement with the other data at constant temperatures. One deduces E_{ap} to vary from 220 kcal/mol at 1250°C to 110 kcal/mol at 1550°C, as shown in Table 1. This variation cannot be explained by the only dependence of the free enthalpy of the CO/CO_2 reaction with the temperature. Our

Table 1. Apparent activation energy* and pO_2 exponent versus temperature.

Temperature (°C)	1232	1291	1360	1400	1504	1561
E_{ap}(Kcal/mol)	220±60	180±10	150±10	112±7		
m (±.005)	.038	.141	.143		.159	.22

*Determined from the steep line on Figure 2 with a constant CO/CO_2 composition.

Table 2. Apparent activation energy* and pO_2 exponent versus temperature.

Temperature (°C)	1250	1360	1472	1522
E_{ap}(Kcal/mol)	176±13	170±10	140±5	125±10
m		.154 ±.04	.161 ±.005	.169 ±.005

*Determined from the steep line on Figure 3 with a constant H_2/H_2O composition.

Table 3. Ratios $\dot{\varepsilon}(H_2O/H_2)$ to $\dot{\varepsilon}(CO_2/CO)$ for different samples, stresses and temperatures.

Sample	T(°C)	σ(MPa)	pO_2(atm)	$\frac{\dot{\varepsilon}(H_2O/H_2)}{\dot{\varepsilon}(CO_2/CO)}$
C13(2)	1183	74.6	10^{-11}	1.6
C13(2)	1232	74.6	10^{-11}	1.8
B11	1287	43.1	10^{-12}-10^{-8}	1-1.6
A7	1291	30	10^{-12}	1
C20(1)	1360	14.9	10^{-11}-10^{-8}	2.27-1.39
C19(2)	1492	10.4	10^{-9}-10^{-5}	1.93
A2	1504	12.7	10^{-10}-10^{-6}	2
B8	1504	12.7	10^{-9}	2
C13(2)	1561	9.8	10^{-8}	1.93

values for E_{ap} differ a little from the results of Kohlstedt and Hornack (1981) who indicate 145 ± 15 Kcal/mol between 1220 and 1420°C in a H_2/CO_2 gas mixture which was rich in CO_2 (70%) and for a much higher applied stress (100 MPa).

Use of H_2/H_2O. The results are reported in Figure 3. Three specimens were studied using this gas mixture. The pO_2 range investigated is narrower than in the previous case, mostly because the gas mixture used here was less efficient to buffer the pO_2. This is due to the method of preparing the mixture itself: in those experiments, pH_2O was kept fixed and low, whatever the pO_2, so that high pO_2 values did correspond to very small amounts of hydrogen, therefore somewhat difficult to control. However in this low-pO_2 domain , the sensitivity of $\dot{\varepsilon}$ vs pO_2 is very close to that for the CO/CO_2 atmosphere. It yields $0.15<m<0.17$ (i.e. 1/6.7 to 1/6.0). The values of m and the apparent activation energies E_{ap} deduced with this type of gas have the same values than in the CO/CO_2 case , as well as a similar temperature dependence: they are given in Table 2.

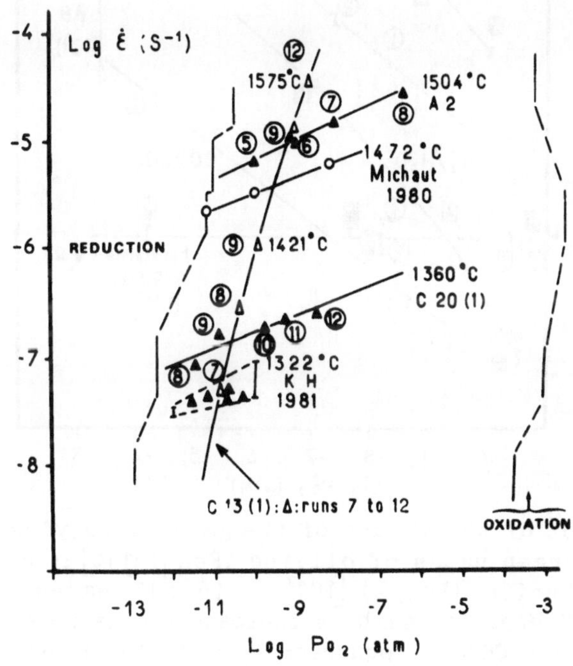

Fig. 3. Log $\dot{\varepsilon}$ vs Log pO_2. 3 specimens; H_2/H_2O mixtures; σ=14.9 MPa; σ// $[110]_c$. The circled labels correspond to the sequence of the runs (see appendix II). The dashed lines indicate the stability field boundaries of Fo_{92}. The data points of Kohlstedt and Hornack (1981) are taken from their Figure 6 (1322°C) where $0.1<m<0.25$ (in H_2/CO_2).

Comparison between CO/CO_2, H_2/H_2O, H_2/CO_2. This is shown on Figure 4. The use of two different gas mixtures (CO/CO_2 or H_2/H_2O) give almost the same dependence of the strain rate $\dot{\varepsilon}$ with the oxygen partial pressure. This holds for different stress orientations $[101]_c$ or $[110]_c$ which activate different slip systems. Although the alignment of the data points for the $[101]_c$ direction

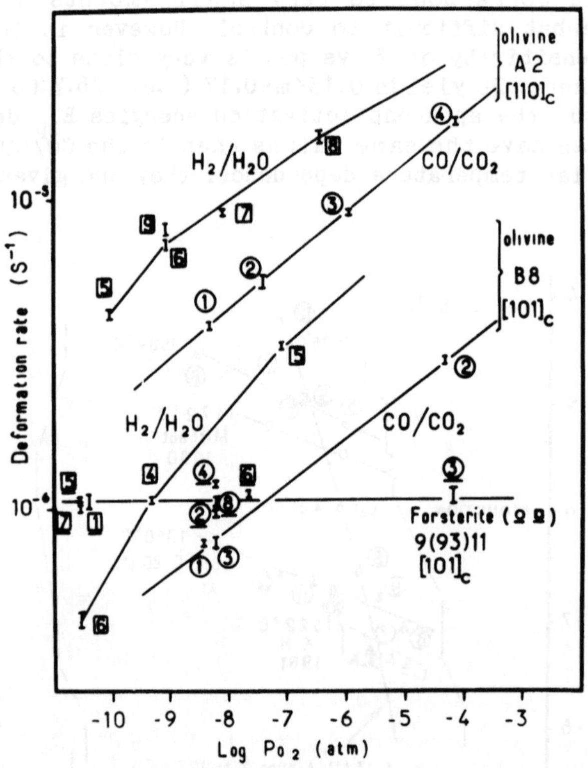

Fig. 4. Effect of the nature of the gases supplying the pO_2 upon the creep rates of olivine (Fo_{92}) $[110]_c$; olivine $[\bar{1}01]_c$; forsterite (Fo_{100}) $[101]_c$. In all cases, T = 1504°C, σ=12.7 Mpa. The numbers indicate the sequence of the runs; circles: CO/CO_2; squares: H_2/H_2O. The deformation rate of the forsterite is independent of pO_2 and insensitive to the nature of the surrounding atmosphere. Although olivine has almost the same m values whatever the stress orientation or the nature of the gases, it deformes about 2 to 3 times faster in the H_2/H_2O medium than in CO/CO_2, for a given set of T, σ , and pO_2.

is a little poorer than for the others, a mean value of m = 0.17 ± 0.03 (1/5 to 1/7) seems to be appropriate to fit the data, for any orientation or gas composition.

The striking difference between the two atmospheres is the increase of $\dot{\varepsilon}$ for a given pO_2 when CO/CO_2 is replaced by H_2/H_2O. Table 3 gives the enhancement factor for 7 samples successively deformed in both atmospheres (H_2/H_2O and CO/CO_2) at a same pO_2. The general trend is that the wet atmosphere (H_2/H_2O) induces creep rates about twice faster than in dry conditions (CO/CO_2). For most of the tests, the wet conditions are such that the atmosphere contains 2% H_2O vapor in 1 atm. total pressure.

At this point of the presentation of the results, an important difference must be pointed out between the H_2/CO_2 mixtures (e.g. used by Durham and Goetze, 1977, and Kohlstedt and Hornack, 1981) and H_2/H_2O or CO/CO_2. In the last two cases, when the proportions of the two constituents are fixed in advance, all partial pressures are practically T-independent, except, obviously pO_2. The opposite occurs when H_2/CO_2 is used. In that situation, the CO_2/H_2 ratio which was fixed in advance at room temperature significantly evolves when this gas passes through the working tubing at T. It produces CO and H_2O in large amounts so that the oxygen partial pressure is achieved by the equilibrium of 4 constituents H_2, H_2O, CO, CO_2. For instance, the ratios pH_2/pCO_2 are different at T than initially at room temperature. Furthermore it might produce tremendous amounts of water. In the example shown in Figure 5, the partial pressures of the constituents of the gas at a given temperature are shown as a function of the initial composition $(CO_2/H_2)^o$ at room temperature. The production of water by this mixture might be high for initial ratios $(CO_2/H_2)^o$ between 0.1 and 10 and can amount up to 7% water in the atmosphere at the working temperature.
Moreover, the amount of produced water is a function of the temperature. In fact, the H_2/CO_2 mixture closely behaves as a H_2/H_2O system when $(pCO_2/pH_2)^o < 0.1$, and as a CO/CO_2 system when $(pCO_2/pH_2)^o > 10$. This makes the apparent activation energies E_{ap} difficult to compare between authors using H_2/CO_2, H_2/H_2O, or CO/CO_2.

III INTERPRETATION

It is based on the fact that the creep rate is sensitive to the point defect chemistry. For example, the behaviour of forsterite is very different from that of iron bearing olivine. Forsterite is insensitive to pO_2 (Jaoul et al., 1981, 1983) in a similar range of pO_2 and T. Moreover it is insensitive to all the partial pressures as shown in Figure 4 where the strain rate does not depend on the nature of the gas which fixes pO_2.

On the other hand, natural olivine has a behavior sensitive to

pO_2 ($\dot{\varepsilon} \propto pO_2^m$). The m exponent is the same, within the precision of the data, for different slip systems activated by different stress orientations, $[101]_c$ or $[110]_c$. (Note: the anisotropy of the creep rate in the present or in similar experiments on forsterite (Michaut, 1981) are in agreement respectively with Durham and Goetze (1977) and Gueguen and Darot (1982)). But, for both stress orientations, the presence of H_2 or H_2O around the specimen enhances the strain rates by the same amounts relatively to the dry conditions.

Therefore not only is natural olivine sensitive to pO_2, obviously because of the presence of iron substituting magnesium, but it is also sensitive to other partial pressures. The aim of this

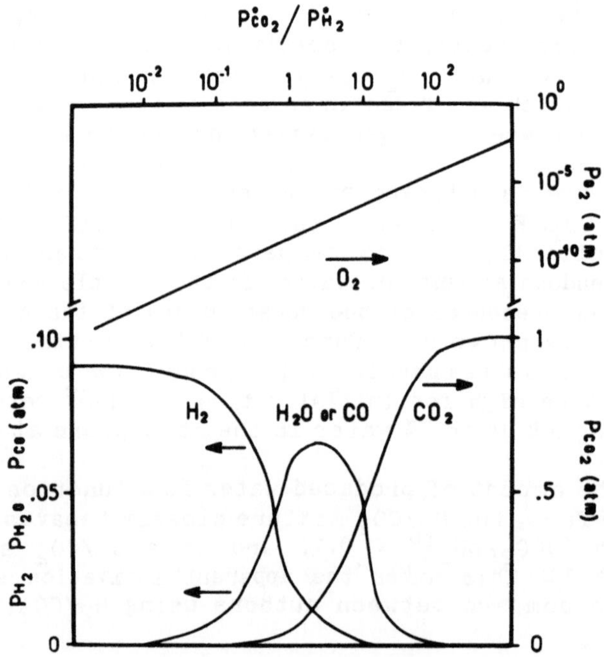

Fig. 5. Computed partial pressures (atm.; 1 atm ~ 10^5 Pa) at 1486°C in a mixture made of argon-10% H_2 and CO_2 with a given initial proportion at room temperature (pCO_2^o/pH_2^o). Log pO_2 is proportional to Log pCO_2^o/pH_2^o; the other partial pressures are strongly dependent on this ratio. The production of water is maximum when $pCO_2^o \sim pH_2^o$ (in this particular case, the small shift of the peak from $pCO_2^o/pH_2^o=1$ towards the right is due to the existence of an argon partial pressure. The mass action constants utilized are: $K(H_2O/H_2) = 3.66 \times 10^{-5}$ atm$^{1/2}$ and $K(CO_2/CO) = 1.44 \times 10^{-4}$ atm$^{1/2}$ (Handbook of Chemistry and Physics, 1979).

section is to discuss our mechanical data in terms of point defects whose nature and concentrations are determined by the thermodynamic conditions (T, partial pressures...). It focuses on the steady state regime for deformation, but also addresses the question of the existence of a very rapid transient regime.

Transients in pO_2

The kinetics for reaching a new equilibrium in the crystal after a change in the external gaseous composition involves the migration of point defects from the surface to the interior, following a surface exchange between the gas and the solid. This exchange has been shown to be rapid in comparison to the former for oxygen incorporation into various silicates like spinels and forsterite (Reddy et al. 1980, Reddy and Cooper, 1981). The time to reach the steady state in our deformation experiments is of the order of a few tens of minutes for specimens of 2 millimeters in size. This corresponds to a chemical diffusion coefficient D about 10^{-5} $cm^2 s^{-1}$ at 1400°C. The question is to explain which kind of point defect can migrate so fast to achieve of a new steady state creep so rapidly. Following Karato and Sato (1982) who assume that the chemical diffusion is predominantly controlled by only one specie of defects, one good candidate is the magnesium vacancy V''_{Mg}. This implies that other silicon or oxygen defects are considered as much less mobile than the Mg-defects. This seems to be a reasonable assumption. But it means that the magnesium-sublattice reaches a new equilibrium state largely before the silicon and oxygen sublattices. It remains to explain why a change in pO_2 at the surface can rapidly affect the interior without any easy migration of the oxygen itself. An experiment where the kinetics of oxidation of olivine was studied out of the stability field suggests an answer. It is described in appendix I, where it is shown that the chemisorption of oxygen onto the surface is associated with the creation of negatively charged magnesium vacancies and positive holes. Within the stability field of the olivine, this surface reaction can be written:

$$1/2\ O_2(g) + n\ Me^x_{Mg} \rightleftarrows Me_nO + n\ V''_{Mg} + 2n\ h^{\bullet}$$

with n = 1, 1, 3/4 respectively if the metal Me on a Mg-site precipitates into MgO, FeO or Fe_3O_4. This can explain why a thin precipitate of magnetite sometimes occurs in the form of a skin covering the sample. This has been often observed for specimens annealed within the stability field of olivine in the high pO_2, low T part of it.

The superficial excess of these defects (V''_{Mg}, $h^{\bullet}$) corresponds to a disequilibrium situation so that they both will migrate towards the interior. These charged defects known to be mobile (electrical conductivity) are now available to oxidize Fe^{2+} into Fe^{3+} everywhere in the bulk material.

$$Fe^{x}_{Mg} + h^{\bullet} \rightleftharpoons Fe^{\bullet}_{Mg}$$

The newly formed $Fe^{\bullet}_{Mg}$ is charge-compensated with the magnesium vacancies: $2[V''_{Mg}] \sim [Fe^{\bullet}_{Mg}]$. This last approximation assumes that the above reaction is strongly displaced to the right so that the hole concentration is small compared to the oxidized iron: $[h^{\bullet}] << [Fe^{\bullet}_{Mg}]$ (Morin et al. 1979, Parkin 1972). Optical absorption spectroscopy (unpublished data) and a Mössbauer study (Shinno 1981) indicate a $Fe^{\bullet}_{Mg}/(Fe+Mg)$ ratio of the order of 1/2% (or less) for highly oxidized olivines, but still within their stability field. This estimate gives an order of magnitude for the thickness of the oxide layer Me_nO which forms at the surface in high pO_2 conditions : 1 μm for a specimen of olivine Fo_{92} (2mm in thickness).

Steady state

Up to now, interpretations of the steady state plastic behavior of olivine were based on both electron microscopy and discussions on the activation energies for creep compared to that for diffusion. The pO_2 sensitivity of the deformation is another constraint to be accounted for. Two main models can be opposed. The first assumes a climb controlled mechanism where dislocations of the main slip system (010) [100] pile up against tilt walls. The second implies that the deformation is glide controlled like in covalent or semi-covalent materials with a strong lattice friction, where the dislocation velocity could be controlled by the concentration and mobility of kinks.

Recovery (diffusion) controlled creep. In this first model, the annihilation time of dislocations piling up against the subboundaries (see Blum, 1971) is a function of the mean distance between dislocations in the walls and of the length of the pile-ups. It also implies the selfdiffusion of the slowest diffusing specie, silicon (Jaoul et al., 1981), which diffuses much more slowly, at least in forsterite, than the other atomic species, like oxygen (Jaoul et al., 1980, 1983) or the other metallic cations (Buening and Buseck, 1973). This pile-up of dislocations and their annihilation in the tilt walls has been studied by Michaut (1981). He systematically observed, by decoration and TEM, such dislocations arrays in olivine and forsterite deformed at low stress (3 MPa) so that the dislocation and tilt wall densities are low enough to allow their easy observation. This work will be published, in collaboration with L. Kubin. Michaut (1981) and Gueguen and Darot (1982) showed this annihilation mechanism to control the high temperature deformation of synthetic forsterite. They write the strain rate to be proportional to D_{sd}, the silicon selfdiffusion coefficient, and to a jog concentration C^{o}_{j} (assumed to be pO_2 independent): $\dot{\varepsilon} \propto D_{sd} C^{o}_{j}$ so that the activation energy E for creep is the sum of the diffusion energy W_{sd} and the jog formation energy

F_j. The jog migration energy is supposed to be small and is neglected. Taking $W_{sd} = 4 \pm 0.4$ eV for the silicon diffusion in forsterite (Poumellec, 1979, Jaoul et al., 1981) and $E = 4.8 \pm 0.5$eV for the (010)[100] slip system of forsterite (Gueguen and Darot, 1982), this yields an estimate of $F_j \sim 0.8$ eV. (1eV ~ 23 Kcal/mol)

As the dislocation microstructures are very similar in forsterite and in iron bearing olivine (Jaoul et al., 1979, Durham et al., 1977), the same kind of discussion may be done for olivine. However, as pointed out by Hobbs (1981, 1983), the jog population may be made of neutral, positive or negative jogs, with concentrations:

$$C_j^+ = C_j^\circ \exp\,(E_j^D - E_F + W_E)/kT$$

$$\text{or} \quad C_j^- = C_j^\circ \exp\,(E_F - E_j^A - W_E)/kT \qquad .....(1)$$

where E_j^D, E_j^A ($E_j^D < E_j^A$) represent energy levels for the jogs (donors or acceptors respectively) within the band gap, and E_F the Fermi level. W_E represents the electrostatic energy of the dislocation line and will be assumed to be small and thus, further neglected. Depending on the value of E_F, one of these two concentrations is dominant. This yields the general expression:

$$\dot{\varepsilon} \propto D_{sd}(C_j^\circ + C_j^+ + C_j^-) \qquad .....(2)$$

The total jog concentration is related to the Fermi level which in turn is related to the hole or electron concentrations:

$$[h^\bullet] = N_V \exp(-E_F/kT)$$

$$[e'] = N_C \exp(-E_g + E_F)/kT \qquad .....(3)$$

N_V and N_C are respectively the electronic densities of states of the top of the valence and the bottom of the conduction bands. If the energy origin is taken at the top of the valence band, the bottom of the conduction band corresponds to the gap energy E_g ($E_g > 0$).

On the other hand, Morin et al. (1979) interpreted their conductivity data on forsterite as following : at low T (T<1000°C), the conductivity is ionic and pO_2-independent. At higher T the dominant carriers are holes and the conductivity is pO_2-dependent. Other data on the conductivity of low-Fe olivines are available (Erve, 1976). Their behavior is similar over a very wide domain of pO_2 and T : Erve's data (1000-1450°C) also suggest a conductivity by holes at high pO_2, by Mg-vacancies in an intermediate pO_2 range and by electrons at low pO_2. For olivines containing more iron, the ionic conductivity becomes overtaken by the holes (at high pO_2) or electrons (at low pO_2) when the iron content amounts 0.2% or more at 1000-1450°C (data of Parkin, 1972, cited in Shankland, 1975).

Fe-rich olivines (Fo_{90}) studied by Duba (1976) exhibit the same behavior in conductivity, which can be interpreted in the same way.

This shows that the holes or the electrons are numerous (and mobile) defects ; if it is assumed that their mobility μ_h or μ_e is unaffected by pO_2 and T, however their concentrations are T- and pO_2-dependent.

$$\text{At high } pO_2 \quad \sigma_{el}=[h^\cdot]\, q\, \mu_h \propto pO_2^{c} \exp(-E_D/kT)$$

$$\text{and low } pO_2 \quad \sigma_{el}=[e']\, q\, \mu_e \propto pO_2^{-c'} \exp(-E'_D/kT) \qquad .....(4)$$

where E_D and E'_D are the measured activation energies for the conductivity in both regimes. Here q is the elementary charge, c and c' are experimental. Parkin's (1972) and Duba's (1976) results can be coherently fitted with $c \approx c' \approx 1/6$ in both low and high pO_2 ranges. E_D and E'_D are of the order of 3eV. Therefore c and c' will be undistinguished in further discussions, as well as E_D and E'_D. Combining equations (3) and (4) yields an expression for E_F:

$$\exp(-E_F/kT) \propto pO_2^{c} \exp(-E_D/kT) \qquad \text{at high } pO_2$$

$$\exp(-E_F/kT) \propto pO_2^{c} \exp(E_D-E_g)/kT \qquad \text{at low } pO_2 \quad .....(5)$$

A comparison of (1) and (5) shows that the jog-concentrations are pO_2-dependent as follows:

$$C_j^+ \propto pO_2^{c}, \quad C_j^o \propto pO_2^{o}, \quad C_j^- \propto pO_2^{-c}. \qquad .....(6)$$

The creep rate of iron bearing olivine is also proportional to D_{sd}, the selfdiffusion coefficient of the slowest moving atomic specie. In olivine the oxygen self diffusion is very close to that observed in forsterite, as shown by our own recent measurements (still unpublished), but however the silicon diffusion data are still lacking. As in forsterite, it might be possible that silicon is also the slowest diffusing specie in olivine. Anyway D_{sd} could be pO_2-dependent:

$$D_{sd} \propto pO_2^{s} \exp(-W_{sd}/kT).$$

According to the above model with the assumption that only one kind of jog dominates (C_j^+ or C_j^o or C_j^-) the strain rate is therefore:

$$\dot\varepsilon \propto pO_2^{s+c} \exp(-W_{sd}+E_j^D-F_j-E_D)/kT \quad \text{for } C_j^+, \text{ when } E_F<E_j^D$$

$$\dot\varepsilon \propto pO_2^{s} \exp(-W_{sd}-F_j)/kT \quad \text{for } C_j^o, \text{ when } E_j^D<E_F<E_j^A \qquad .....(7)$$

$$\dot\varepsilon \propto pO_2^{s-c} \exp(-W_{sd}-E_j^A-F_j+E_D)/kT \quad \text{for } C_j^-, \text{ when } E_j^A<E_F$$

This can be compared to the experimental data:

$$\dot{\varepsilon} \propto pO_2^m \exp(-E/kT) \quad .....(8)$$

with $0.14<m<0.22$ and $E \simeq 4.6$ eV in agreement with Kohlstedt and Hornack's data (1981). The conductivity data (Parkin, 1972; Duba, 1976) indicate $0.18<c<0.23$, so that the dependence of the diffusion of the slowest specie with pO_2 has to be : $-0.09<s<0.04$ (a value close to zero) if the deformation is dominated by C_j^+. If C_j^o controls, then one expects to have $s=m$. In the case favouring C_j^-, it follows that $s=m+c$ and therefore one has $0.32<s<0.43$.

As described above, the conductivity shows a minimum versus pO_2 (for example at 1400°C for $pO_2 = 10^{-7}$ atm)(Duba, 1976). In these conditions, the Fermi level is close to the middle of the gap E_g, and it remains close to it, within ± 0.2eV, when pO_2 is varied from its maximum to its minimum value ($E_F^{max}-E_F^{min}$= $c.RT.Log(pO_2^{min}/pO_2^{max})$, (see equation 5). If one makes the hypothesis of a dominant population of neutral jogs, as it is in forsterite ($m=0$, $s=0$ and $c \sim 1/6$ for its electronic conductivity), thus, because $E_j^D<E_F<E_j^A$, it follows that $E_j^A-E_j^D>0.4$eV. In these conditions, the activation energy for the diffusion of the slowest specie would be W_{sd}=3.8eV (if F_j=0.8eV, as in forsterite) and its pO_2-dependence, the same as for creep. On the other hand, if C_j^+ or C_j^- were dominating, the corresponding Fermi levels would lie below or above their respective levels E_j^D or E_j^A. Up to now, there is no information about the positions of E_j^D and E_j^A within the band gap, thus we cannot choose between the three possibilities of neutral or charged jogs. Data on diffusion of the slowest atomic specie, its pO_2-dependence and activation energy are needed to make a choice, within the assumption of a jog controlled dislocation climb.

Glide controlled creep. Karato and Sato's experiments (1982), where the kinetics of static recovery has been studied as a function of pO_2, lead to the conclusion that "the creep in olivine single crystals is not rate-controlled by recovery", at least in their experimental conditions (1400°C, pO_2 ranging from 10^{-4} to 10^{-11} atm). This kinetics is pO_2-independent or is even slightly slowed by increasing pO_2. According to Karato and Sato, creep in olivine single crystals could be controlled by dislocation glide and/or multiplication. This possibility should not be ruled out in this kind of material with strong Peierls forces: the crystallography strongly influences the shape of the dislocations as demonstrated by decoration techniques (e.g. Durham et al., 1977, for olivine, and Jaoul et al., 1979, or Gueguen and Darot, 1982, for forsterite): dislocations have a trend to be straight and elongated along some principal directions. However the recent TEM observations in forsterite by Gueguen and Darot (1982) do not clearly prove (neither deny) the above statement. Hence, in this section, we will consider the possibility for creep to be glide-controlled. As Hirsch (1981) does for covalent or semicovalent materials, it will be assumed that glide itself is controlled by the nucleation rather than the migration of kinks. The model developed below is very similar to the formalism detailed in the previous section (recovery control). Again, charged and/or neutral kinks can exist, according to possible electronic transfers from the band structure.

Equations (1) to (6) remain formally similar if the subscript j is replaced by k (for kinks). However equation (2) must be modified as follows :

$$\dot{\varepsilon} \propto (\mu_k^- C_k^- + \mu_k^o C_k^o + \mu_k^+ C_k^+) \quad .....(9)$$

where μ_k are the mobilities of the neutral or charged kinks with concentrations C_k. For simplification the mobilities are assumed to be pO_2 independent and not very sensitive to the temperature (low activation energy).

The modified equation (6) is:

$$C_k^+ \propto pO_2^c \,, \qquad C_k^o \propto pO_2^o \,, \qquad C_k^- \propto pO_2^{-c}$$

A comparison of this set of equations and equation (9) to the experiments (equation 8) shows that the only possibility is to have positive kinks K^+ dominating the others . The strain rate is therefore written:

$$\dot{\varepsilon} \propto pO_2^c \exp(E_k^D - F_k - W_k - E_D)/kT$$

where F_k and W_k are the energies for the formation and migration of the kinks, and where the exponent c of the conductivity is the same as for the creep rate (m=c). This implies that the kink energy level E_k^D is always higher than the Fermi level E_F (see the modified equation 1), whatever the oxygen partial pressure is. The same kind of numerical estimate, as in the previous section, yield $E_k^D > E_g/2+0.2eV$. The formation and migration energies F_k and W_k are such that: $F_k+W_k = E+E_k^D-E_D > E+E_g/2-E_D+0.2eV$, with $E=4.6eV$, $E_D=3eV$, $E_g \sim 6eV$ for the energies respectively for creep, conductivity and the band gap. Then, $F_k + W_k > 4.8eV$. Assuming $W_k \sim 1eV$ for the kink migration, one finds $F_k > 3.8eV$. This value for the kink formation is very high. However it represents an order of magnitude, deduced from a comparison between conductivity data and creep, if the latter is controlled by the nucleation and migration of positive kinks, which is the only possible kind of kink in this glide-controlled model.

<u>The particular effect of the water on creep</u>. In the case where the mobility of the dislocations is controlled by positively charged jogs or kinks, the creep rate $\dot{\varepsilon}$ is explicitely correlated to E_F, pO_2, the electronic conductivity, and is proportional to the hole concentration. However this interdependence is different if

the jogs were neutral or negative. Unfortunately, in the previous section, we have been unable to make a definite choice, mainly because of the lack of data on the pO_2-dependence of the diffusion coefficient of the slowest atomic specie.

However, there is a strong similarity between the pO_2-dependences of $\dot{\varepsilon}$ and the concentration of the holes. Experimentally, $\dot{\varepsilon} \propto pO_2^m$ and $[h^\bullet] \propto pO_2^c$ over the whole range of pO_2, and $m \approx c$. (equations 3, 5 and 8). Then, independently of the above discussion, $\dot{\varepsilon}$ will be regarded as proportional to $[h^\bullet]$. Thus $[h^\bullet]$ will be used as the relevant parameter to describe the influence of the nature of the surrounding atmosphere (wet or dry) upon creep rate.

The first step to be looked at is the interaction between the gas and the surface. The production of point defects in the vicinity of the surface is dominated by the nature and the partial pressures of the gas. This production and the oxygen incorporation has already been described in a previous section dealing with transient phenomena : oxygen incorporates to olivine, producing mobile doubly charged magnesium vacancies and holes, associated with a precipitation of MgO lying onto the surface :

$$1/2\ O_2(g) + Mg_{Mg}^x \rightleftarrows MgO(surf) + V_{Mg}'' + 2h^\bullet \quad .....(10)$$

For a given value of imposed pO_2, this chemical equation reaches an equilibrium : this is the main reaction occuring in a CO/CO_2 medium. In the case where pO_2 is produced by an H_2/H_2O atmosphere, one more equation could be written :

$$H_2O(gas) + V_{Mg}'' + Mg_{Mg}^x \rightleftarrows 2V_{Mg}'(H) + MgO(surf) \quad .....(11)$$

This new equilibrium takes into account the formation of a new defect : a proton trapped in an empty octrahedral site, a magnesium vacancy. This defect is of the same kind as described by Morin et al. (1977) in magnesium oxide and is proposed to be likely in olivine (N.M.R. or I.R. could be a test to check its existence). This complex defect could form at the surface and be mobile enough to migrate towards the interior. It also could form in the bulk after separate migrations of $H^\bullet$ and V_{Mg}''. Whatever the exact nature of this defect, it yields the following result : in the presence of water vapor, equation (11) is displaced towards the right, so that an amount of V_{Mg}'' is converted into $V_{Mg}'(H)$. This change induces a displacement of equation (10) to partly compensate the lack of V_{Mg}''. The net result is an increase of $2[V_{Mg}''] + [V_{Mg}'(H)]$. Still assuming that the magnesium vacancies and Fe^{3+} are the majority point defects, the new neutrality condition is :

$$[Fe_{Mg}^\bullet] = 2[V_{Mg}''] + [V_{Mg}'(H)]$$

The ionization equilibrium of iron, with the constant K_{Fe},

$$[Fe_{Mg}^\bullet] = K_{Fe}[Fe_{Mg}^x]\cdot[h^\bullet]$$

shows $[Fe^{\bullet}_{Mg}]$ to be proportional to $[h^{\bullet}]$ because $[Fe^{\bullet}_{Mg}] \ll [Fe^{x}_{Mg}]$. Therefore $[h^{\bullet}]$ is an increasing function of pH_2O (the water amount in the gas) at fixed pO_2, and consequently, so is the creep rate. The following consequences can be deduced from the above description:

1) In a CO/CO_2 atmosphere where $V'_{Mg}(H)$ is not produced, one expects:

$$[h^{\bullet}] \propto pO_2^{1/6}$$

2) On the contrary, when $[V'_{Mg}(H)] \gg [V''_{Mg}]$ (an extreme situation), one predicts:

$$[h^{\bullet}] \propto pO_2^{1/8} . pH_2O^{1/4}$$

3) In our experiments, the real situation for the H_2/H_2O gaseous atmosphere (low pressure) is intermediate between the first and second cases.
It must be outlined that our experiments in H_2/H_2O had increasing pH_2O values when pO_2 was increased. This could explain the continuous variation of the pO_2 exponent (with a trend to decrease) when increasing pO_2 and pH_2O.

CONCLUSION

This set of experiments definitely confirms that the creep rate of iron bearing olivine varies as $pO_2^{1/7}$ to $pO_2^{1/5}$ (when increasing T) in CO/CO_2, or $pO_2^{1/7}$ to $pO_2^{1/6}$ in H_2/H_2O, and therefore is not depending very much on the nature of the gas providing the oxygen partial pressure. It agrees with Kohlstedt and Hornack's results (1981) obtained in H_2/CO_2 ("wet" conditions, as demonstrated above). In our experiments we focused on the comparison between the creep rates for different values of pO_2, or with a change in the nature of the gas supplying the desired pO_2. Care was taken to perform the runs with a same stress and temperature on a same specimen. This allows to extract with confidence the true sensitivity of $\dot{\varepsilon}$ with pO_2 and the nature of the gas. In particular we undoubtely find a difference in the creep rates when the tests are performed on a same sample using CO/CO_2 (dry) or H_2/H_2O (wet), and with all the other parameters kept constant (pO_2, T). This is in contradiction with the conclusion drawn by Justice et al. (1982), but not with their experimental results because the scatter in their data points (one unit on a decimal log-scale) can hide the weak difference in $\dot{\varepsilon}$ (maximum half a unit on a log-scale) for the two gaseous atmospheres. This scatter could be attributed to different initial microstructures of each specimen.

With our experimental conditions, the change in $\dot{\varepsilon}$ when one passes from CO/CO_2 to H_2/H_2O is small, but definitely detectable. An increase of the strain rate by a factor of two to three, as we observed, corresponds to a similar increase in the concentrations of $[h^{\cdot}]$ or $[Fe^{\cdot}_{Mg}]$. This gives an estimate of the concentration of the complex defect $V'_{Mg}(H)$ formed in the wet experiments, whose abundance is of the same order as the magnesium vacancies in the dry experiments. Unfortunately, the amount of hydrogen incorporated during the deformation runs in H_2/H_2O remains unknown. Perhaps the I. R. determinations made by Justice et al. (1982) at room temperature on samples similar to ours (San Carlos olivine also deformed under wet conditions) give an order of magnitude: they indicate 0 to 55 ppm H/Si atoms when passing from dry to wet specimens.

Another interesting feature is the short duration of the transient creep after each change in pO_2. Practically, it gave us the possibility to make numerous tests in steady state conditions on a same sample. To explain the transients, we have proposed a process involving chemical diffusion. Any oxidation phenomenon ($Fe^{2+} \rightarrow Fe^{3+}$ on octahedral sites, decoration of dislocations, formation of precipitates) occuring in olivine can be explained by a rapid migration of magnesium vacancies and electronic defects from or towards the surfaces in contact with the oxygen.

Once the concentration at equilibrium of these two major point defects is achieved in the bulk, the creep becomes stationary. The latter is therefore related to the populations of the magnesium vacancies and electronic defects. In order to interpret the influence of the point defects upon the creep rate, we used two different models based on a control by climb (jog formation and nucleation) or by glide (kinks). The lack of data on the selfdiffusion of silicon, supposed to be the slowest moving specie, does not allow to make any choice between the models. However, whatever the model utilized, we have suggested that the stationary creep rate is proportional to the concentration of the electronic defects $[h^{\cdot}]$. Thus the discussion of the dependence of $\dot{\varepsilon}$ with pO_2 simply reduces to that of $[h^{\cdot}]$ with pO_2, independently of the mechanism controlling the mobility of the dislocations. This yields an $\dot{\varepsilon}$ vs pO_2 law which is in good agreement with the experimental data, in both wet and dry atmospheres.

APPENDIX I: DESCRIPTION OF AN OXIDATION EXPERIMENT

To help the understanding of the oxidation process in olivine, we did the following experiments (Poumellec, 1981): samples of olivine Fo_{92} have been annealed in air at 1536°C for different durations (1 mn to 8 hrs), so that the kinetics of the oxidation could be followed. Then the samples were quenched in air after the heat treatment ($100°C.s^{-1}$). The details of this experiment will be published elsewhere. However the description of the samples after the quench is useful to support the model of point defects developed in this article. The above conditions (T and pO_2) are such that olivine partly transforms in magnetite and a silica rich phase (Nitsan, 1975). The remaining olivine is impoverished in Fe.

In fact, all samples which were annealed for durations of more than 1 hour show three different zones lying parallel to the surface. The following description corresponds to a sample which was annealed for 4 and 1/2 hours. Close to the surface, a 2μm-thick hematite layer formed. Behind it, one finds a Fe-impoverished region (50 μm) with very few opaque precipitates (magnetite). The rest of the bulk is made of numerous and large precipitates of magnetite (typical size 5 μm). The formation of these magnetite precipitates is in agreement with Nitsan's diagram (1975). (It is recalled that the same kind of oxidation works when dislocations are decorated (1/2 hr, 800°C), but in this case, hematite forms along the dislocations lines, also in agreement with Nitsan). The 50 μm-thick layer, where Fe-precipitates are lacking, is correlated with the 2 μm layer of pure hematite. However, hematite should not have formed at this temperature and pO_2, under equilibrium conditions.

The chief point is that the oxidation reaction corresponds to a disequilibrium situation: the trend of the bulk to reequilibrate under the new (say higher) pO_2 conditions let it playing the role of a sink for the Mg-vacancies and the holes generated at the surface. This forces the magnetite precipitates lying on the surface to incorporate oxygen from the gas, which corresponds to an oxidation reaction into hematite .

$$O_2(g) + Fe^x(ol) + Fe_3O_4(mag) \rightarrow 2Fe_2O_3(hem) + (V''_{Mg}+2h^{\cdot})(ol)$$

This also creates a force attracting the iron ions (Fe^x_{Mg}) from their octahedral site in the olivine lattice in the vicinity of the surface. (The cross means: uncharged, relatively to the olivine lattice). This chemisorption reaction produces Mg-vacancies (V''_{Mg}) which carry two negative charges (Morin et al., 1979), electrically compensated by electronic holes $h^{\cdot}$. Those two mobile defects migrate towards the bulk and oxidize the Fe^{2+} into Fe^{3+}, so that magnetite and a silica-rich phase precipitate at places, e.g. along the dislocations, by a combination of (Fe^{2+}, Fe^{3+}) and magnesium vacancies, without any migration of silicon neither oxygen from the surface:

$$Fe_2^{III}SiO_4 + V_{Mg}Fe^{II}SiO_4 \rightarrow Fe_3O_4 + SiO_2$$

The same kind of mechanism can also explain how dislocations can be decorated at 800°C in air. In this latter case, the oxide which precipitates is hematite stable at this temperature:

$$V_{Mg}V_{Mg}SiO_4 + 2Fe_2^{III}SiO_4 \rightarrow 2Fe_2O_3 + 3SiO_2$$

This agrees with the observations of Kohlstedt and Van der Sande (1975) who analyzed the precipitates along decorated dislocations.

APPENDIX II: EXPERIMENTAL DATA

The following Table indicates, for each sample, the sequence of the runs (N°), the strain (Δl in μm), the duration in minutes or hours (h), the strain rate, stress, temperature, pO_2, the nature of the gas, and the stress exponent n.

N°	100 ×	Δl μm	Δt mn.	$\dot{\varepsilon}$ s^{-1}	T °C	σ MPa	pO_2 atm.	Gas mixture	n
				Sample orientation : $[110]_c$					
C13(2)									
1	1.16	15	13h	$6.0\ 10^{-8}$	1183	74.6	$1.1\ 10^{-11}$	H_2/H_2O	3.4
2	1.23	3	226	$3.7\ 10^{-8}$			$1.4\ 10^{-11}$	CO/CO_2	
3	1.47	10	158	$3.3\ 10^{-7}$	1232		$7.1\ 10^{-10}$		
4	1.72	11	119	$3.9\ 10^{-7}$			$2.8\ 10^{-9}$		
5	3.04	55	15.7h	$2.8\ 10^{-7}$			$2.6\ 10^{-13}$		
6	3.21	7	103	$3.6\ 10^{-7}$			$6.3\ 10^{-11}$		
7	9.21	86	14.4h	$6.1\ 10^{-7}$			$8\ 10^{-12}$	H_2/H_2O	
8	10.6	5	16h	$2.8\ 10^{-8}$	1261	25.4	$2.6\ 10^{-11}$	CO/CO_2	3.6
9	10.63	2	125	$1.54\ 10^{-7}$	1301		$7.9\ 10^{-11}$		
10	10.73	4	80	$2.84\ 10^{-7}$	1330		$1.6\ 10^{-10}$		
11	10.92	7	72	$7.15\ 10^{-7}$	1349		$2.6\ 10^{-10}$		
12	11.26	13	51	$2.53\ 10^{-6}$	1380		$6.8\ 10^{-10}$		
13	12.27	39	40	$9.22\ 10^{-6}$	1428		$1.9\ 10^{-9}$		
14	14.99	101	36	$2.45\ 10^{-5}$	1467		$4.6\ 10^{-9}$		
15	18.48	126	23	$6.07\ 10^{-5}$	1517		$1.3\ 10^{-8}$		
16	19.04	20	18h	$1.48\ 10^{-6}$	1561	8.4	$6.8\ 10^{-9}$	H_2/H_2O	4.4
17	19.49	16	194	$7.12\ 10^{-7}$			$2.7\ 10^{-9}$	CO/CO_2	
18	19.69	7	94	$7.66\ 10^{-7}$			$1.9\ 10^{-8}$		
19	19.92	8	82	$1.78\ 10^{-6}$			$1.7\ 10^{-7}$		
A7									
1	8.66	100	80h	$1.58\ 10^{-7}$	1291	30	$1.4\ 10^{-12}$	CO/CO_2	
2	9.31	15	22h	$1.76\ 10^{-7}$			$2.1\ 10^{-12}$	H_2/H_2O	
3	9.39	2	122	$1.98\ 10^{-7}$			$6.4\ 10^{-12}$	CO/CO_2	4
4	9.48	2	83	$2.56\ 10^{-7}$			$8.8\ 10^{-11}$		
5	9.65	4	132	$3.44\ 10^{-7}$			$8.3\ 10^{-10}$		
6	9.78	3	80	$3.66\ 10^{-7}$			$9.2\ 10^{-9}$		
7	10.00	5	101	$6.82\ 10^{-7}$			$1.5\ 10^{-7}$		
8	10.40	9	101	$1.45\ 10^{-6}$			$1.4\ 10^{-6}$		
9	12.97	35	14h	$4.35\ 10^{-7}$			$1.2\ 10^{-8}$		
C20(1)									
1	1.02	22	14h	$1.07\ 10^{-7}$	1360	14.9	$7.1\ 10^{-12}$	CO/CO_2	4
2	1.06	2	85	$1.01\ 10^{-7}$			$4.3\ 10^{-11}$		

(Continued)

APPENDIX II Cont.

N°	100 ×	Δl μm	Δt mn.	$\dot{\varepsilon}$ s^{-1}	T °C	σ MPa	pO_2 atm.	Gas mixture	n
3	1.15	4	98	1.87 10^{-7}			4.2 10^{-10}		
4	1.22	3	69	2.21 10^{-7}			1.6 10^{-9}		
5	1.33	5	129	4.00 10^{-7}			1.3 10^{-7}		
6	1.49	7	75	6.05 10^{-7}			1.3 10^{-6}		
7	1.97	22	93	1.89 10^{-6}			4.4 10^{-5}		
8	2.46	22	15h	1.25 10^{-7}			3.1 10^{-12}	H_2/H_2O	
9	2.53	3	82	2.39 10^{-7}			1.3 10^{-11}		
10	2.60	3	111	2.69 10^{-7}			1.6 10^{-10}		
11	2.64	2	88	3.14 10^{-7}			4.8 10^{-10}		
12	2.69	2	70	3.75 10^{-7}			2.9 10^{-9}		
<u>A2</u>									
1	6.44	45	60	3.93 10^{-6}	1504	12.7	5.2 10^{-9}	CO/CO_2	4
2	7.41	33	44	5.38 10^{-6}			3.8 10^{-8}		
3	9.27	62	60	8.92 10^{-6}			1.2 10^{-6}		
4	12.25	97	57	1.73 10^{-5}			7.0 10^{-5}		
5	15.19	93	100	4.15 10^{-6}			8.8 10^{-11}	H_2/H_2O	4.5
6	16.26	33	68	7.01 10^{-6}			9.2 10^{-10}		
7	17.23	30	45	9.0 10^{-6}			8.3 10^{-9}		
8	18.82	48	50	1.55 10^{-5}			3.4 10^{-7}		
9	23.07	48	78	7.93 10^{-6}			8.9 10^{-10}		
<u>MICHAUT(1980)</u>					1472	10.2		H_2/H_2O	4.5
<u>KOHLSTEDT</u> & <u>HORNACK</u> <u>(1980)</u>					1325	100		H_2/CO_2	3.4
<u>C13(1)</u>									
1	.52	6	93	2.35 10^{-7}	1472	10.9	3.2 10^{-10}	H_2/H_2O	4.9
2	.71	8	55	6.41 10^{-7}		13.2			
3	1.24	22	50	1.97 10^{-6}		16.5			
4	2.42	49	43	5.32 10^{-6}		20.1			
5	5.69	133	48	1.49 10^{-5}		24.9			
6	6.75	42	15h	1.96 10^{-8}	1277		4.0 10^{-12}		3.2
7	6.83	3	71	2.73 10^{-7}	1325		1.3 10^{-11}		
8	7.21	15	42	1.72 10^{-6}	1380		4.5 10^{-11}		
9	8.23	40	38	5.97 10^{-6}	1421		1.1 10^{-10}		
10	10.99	106	26	2.11 10^{-5}	1475		3.2 10^{-10}		
11	19.45	307	38	6.85 10^{-5}	1522		8.1 10^{-10}		
12	26.87	249	15	1.78 10^{-4}	1572		2.0 10^{-9}		
<u>C14</u>									
1	.82	2	115	6.07 10^{-8}	1480	6.27	3.8 10^{-8}	H_2/H_2O	4.9
2	1.03	9	120	2.47 10^{-7}		8.56			
3	1.25	9	105	3.95 10^{-7}		9.57			
<u>C19(2)</u>									
1	1.54	23	90	1.76 10^{-6}	1492	10.4	1.4 10^{-4}	CO/CO_2	
2	2.49	8	16.5h	4.01 10^{-7}			2.7 10^{-9}		
3	2.96	16	160	5.62 10^{-7}			3.1 10^{-8}		
4	3.32	12	86	9.82 10^{-7}			1.1 10^{-6}		
5	4.01	23	74	2.45 10^{-6}			6.4 10^{-4}		
6	5.64	45	14h	4.31 10^{-7}			5.7 10^{-11}	H_2/H_2O	

N°	100 ×	Δl μm	Δt mn.	$\dot{\varepsilon}$ s^{-1}	T °C	σ MPa	pO_2 atm.	Gas mixture	n
				Sample orientation : $[101]_c$					
B8									
1	3.90	19	85	7.8 10^{-7}	1504	12.7	4.1 10^{-9}	CO/CO_2	
2	4.67	25	82	3.04 10^{-6}			4.6 10^{-5}		
3	5.50	27	110	7.69 10^{-7}			6.5 10^{-9}		
4	5.87	12	80	1.04 10^{-6}			4.8 10^{-10}	H_2/H_2O	
5	6.83	31	80	3.57 10^{-6}			3.5 10^{-8}		
6	7.40	18	65	6.37 10^{-7}			3.1 10^{-11}		
B11									
1	2.3	89	9h	1.14 10^{-6}	1287	43.1	2.4 10^{-9}	H_2/H_2O	
2	4.92	33	16h	1.83 10^{-7}			5.9 10^{-13}		
3	8.91	117	228	1.88 10^{-6}			9.8 10^{-6}	CO/CO_2	
4	9.81	32	13h	2.02 10^{-7}			1.2 10^{-12}		
5	12.09	40	185	2.39 10^{-6}			6.4 10^{-5}		
6	13.08	34	135	2.86 10^{-6}			2.7 10^{-4}	H_2/CO_2	
7	13.60	18	135	1.35 10^{-6}			10^{-7}		
8	14.66	36	16h	2.18 10^{-7}			10^{-12}		
9	15.02	12	140	9.29 10^{-7}			10^{-8}		
10	15.52	17	125	1.51 10^{-6}			1.1 10^{-6}		
11	15.94	14	178	4.32 10^{-7}			2.2 10^{-11}		
12	16.36	14	15h	1.48 10^{-7}			6.6 10^{-7}		
FORSTERITE	9(93)11								
0	2.28	55	16h	1.72 10^{-7}	1504	7.7	3.5 10^{-11}	H_2/H_2O	
1	2.95	18	109	1.11 10^{-6}		12.7			
2	3.39	12	77	9.85 10^{-7}			5.9 10^{-9}	CO/CO_2	
3	4.02	17	96	1.12 10^{-6}			5.7 10^{-5}		
4	4.40	10	50	1.2 10^{-6}			5.6 10^{-9}		
5	4.75	9	50	1.06 10^{-6}			2.4 10^{-11}	H_2/H_2O	
6	5.15	11	56	1.15 10^{-6}			2.1 10^{-8}		
7	5.53	10	56	1.06 10^{-6}			2.6 10^{-11}		
3	5.83	8	49	1.05 10^{-6}			5.6 10^{-9}	CO/CO_2	
j	9.66	99	14h	7.18 10^{-7}			>4.0 10^{-5}	H_2O only	

REFERENCES

Blum, W., 1971, Role of dislocation annihilation during steady-state deformation, Phys. Stat. Solid.(B), 45:561.

Buening, D. K., and Buseck, P. R., 1973, Fe-Mg lattice diffusion in olivine, J. Geophys.Res., 79(29):6852.

Duba, A., 1976, Review of the relevance of laboratory electrical conductivity data of the earth, Proc. of the third int. workshop on electromagnetic induction, Sopron, Hungary.

Durham, W. B., 1975, Plastic flow of single crystal olivine, PhD. thesis, M. I. T., Mass.

Durham, W. B., and Goetze C., 1977, Plastic flow of oriented single crystals of olivine, I: Mechanical data, J. Geophys. Res., 82:5737.

Durham, W. B., Goetze, C., and Blake, B., 1977, Plastic flow of oriented single crystals of olivine, II: Observation and interpretation of the dislocation structure, J. Geophys. Res., 82:5755.

Erve, H., 1976, Ionic conductivity and diffusion in Mg_2SiO_4 and crystalline SiO_2, ERAMS-I/TR 29, (may 1976), thesis, University of Erlangen, F. R. G.

Gueguen, Y., and Darot, M., 1982, Les dislocations dans la forsterite déformée à haute température, Phil. Mag., A. 45(3,4):419.

Handbook of chemistry and physics, 1979, p. D-45, C. R. C. Press, Boca Raton, Florida.

Hirsch, P. B., 1981, Plastic deformation and electronic mechanisms in semiconductors and insulators, J. de Physique, C-3, 42:149.

Hobbs, B. E., 1981, The influence of metamorphic environment upon the deformation of minerals, Tectonophysics, 78:335.

Hobbs, B. E., 1983, Point defect chemistry of minerals under a hydrothermal environment, J. Geophys. Res., 88:in press.

Hornack, P., and Kohlstedt, D. L., 1979, The effect of oxygen partial pressure on the creep of olivine, E. O. S. (Trans. Am. Geophys. Union), 60:369.

Jaoul, O., Froidevaux, C., Durham, W .B., and Michaut, M., 1980, Oxygen self diffusion in forsterite : implications for the high temperature creep mechanism, Earth Planet. Sci. Lett., 47:391.

Jaoul, O., Michaut, M., Gueguen, Y., and Ricoult, D., 1979, Decorated dislocations in forsterite, Phys. Chem. Minerals, 5:15.

Jaoul,O., Poumellec, M., Froidevaux, C., and Havette, A., 1981, Silicon diffusion in forsterite : a new constraint for understanding mantle deformation, in: " Anelasticity in the Earth," Geodyn. Ser., vol. 4, p. 95, F. D. Stacey et al., ed., AGU, Washington, D. C.

Jaoul, O., Houlier, B. and Abel, F., 1983, Study of ^{18}O diffusion in magnesium orthosilicate by nuclear microanalysis, J. Geophys. Res., 88:613.

Justice, M. G. Jr., Graham, E. K., Tressler, R. E., and Tsong, I. S. T., 1982, The effect of water on high-temperature deformation in olivine, Geophys. Res. Let., 9(9):1005.

Karato, S., and Sato, H., 1982, Effect of oxygen partial pressure on the dislocation recovery in olivine: a new constraint on creep mechanisms, published, Phys. Earth Planet. Inter.

Kohlstedt, D. L., and Goetze, C., 1974, Low stress high temperature creep in olivine single crystals, J. Geophys. Res., 79:2045.

Kohlstedt, D. L., Goetze, C., and Durham, W. B., 1976, Experimental deformation of single crystal of olivine with application to flow in the mantle, in: "The Physics and Chemistry of Minerals and Rocks," p. 35, R. G. J. Strens, ed., John Wiley, London.

Kolhstedt, D. L., and Hornack, P., 1981, Effect of oxygen partial pressure on the creep of olivine, in: "Anelasticity in the Earth," Geodyn. Ser., vol. 4, p. 101, F. D. Stacey et al., ed., Washington.

Kohlstedt, D. L., and Van der Sande, J. B., 1975, An electron microscopy study of naturally occurring oxidation produced precipitates in iron-bearing olivines, Contrib. Mineral. Petrol., 53:13.

Michaut, M., 1981, Etude du fluage de la forsterite et d'olivines monocristallines, Thèse, Université Paris-Sud, 91405 Orsay, France.

Morin, F. J., Oliver, J. R., and Housley, R. M., 1977, Electrical properties of forsterite Mg_2SiO_4, I, Phys. Rev. B., 16(10):4434.

Morin, F. J., Oliver, J. R., and Housley, R. M., 1979, Electrical properties of forsterite Mg_2SiO_4, II, Phys. Rev. B., 19(6):2886.

Nitsan, U., 1975, Stability field of olivine with respect to oxidation and reduction, J. Geophys. Res., 79(5):706.

Parkin, T., 1972, The electrical conductivity of synthetic forsterite and periclase, Thesis, School of Physics, The University of Newcastle, Newcastle upon Tyne.

Poumellec, B., 1981, Fugacité d'oxygène et déformation plastique de l'olivine à haute température, Thèse, Université Paris Sud, 91405 Orsay , France.

Poumellec, M., 1979, Autodiffusion du silicium 30 dans la forstérite Mg_2SiO_4, Thèse, Université Paris Sud, 91405 Orsay, France.

Reddy, K. P. R., Oh, S. M., Major, L. D. Jr., and Cooper, A. R., 1980, Oxygen self-diffusion in forsterite, J. Geophys. Res., 85:322.

Reddy, K. P. R., and Cooper, A. R., 1981, Oxygen diffusion in magnesium aluminate spinel, J. Am. Ceram. Soc., 64(6):368.

Shankland, T. J., 1975, Electrical conduction in rocks and minerals: parameters for interpretation, Phys. Earth Planet. Inter., 10:209.

Shinno, I., 1981, A Mössbauer study of ferric iron in olivine, Phys. Chem. Minerals, 7:91.

A TEM MICROSTRUCTURAL STUDY OF EXPERIMENTALLY DEFORMED ANITA BAY DUNITE

N. Doukhan[1], J. C. Doukhan[1], J.D. Fitz Gerald[2], P.N. Chopra[3] and M.S. Paterson[2]

[1]University of Lille I, 59655 Villeneuve d'Ascq Cedex, France
[2]R.S.E.S., Australian National University, Canberra A.C.T. 2601, Australia
[3]Bard Hall, Cornell University, Ithaca, NY 14853, U.S.A.

ABSTRACT

Six specimens from an earlier study of deformation behaviour under wet and dry conditions at 1000°C-1400°C and 300 MPa confining pressure were examined in transmission electron microscopy (TEM) and their microstructures related to those seen in optical microscopy. As the flow stress decreases and the temperature increases an increasing role of climb is indicated by a gradual change in structure from high densities of long straight apparently lattice-friction-controlled dislocations to lower densities of dislocations organized partly into subgrain boundaries and leaving curved free dislocations within the subgrains. Although comparison at given temperature and stress is not possible with this suite of specimens, it can be indirectly inferred that there is probably also an increased role of climb in wet specimens compared with dried and that there may therefore be some intragranular hydrolytic weakening as well as grain boundary weakening in the wet specimens. Little information on the processes in the grain boundaries has yet been gained from TEM.

INTRODUCTION

The flow behaviour and optical microstructures of two dunites, deformed at various strain rates and temperatures under "wet" and "dry" conditions in a gas-medium apparatus at 300 MPa confining pressure, has been described elsewhere (Chopra and Paterson, 1981,

1983; Fitz Gerald et al., 1983). The marked weakening and grain size dependence in "wet" specimens was initially attributed to grain boundary effects related to the release of water from accessory hydrous minerals, a process which leads to optically visible melt above 1100°C and a specimen bulk water content of a few hundred ppm by weight. Elimination of the weakening effect requires pre-drying of the specimens to a bulk water content of a few tens ppm by weight or less.

Here we describe additional microstructural studies on six selected specimens of one of the dunites, Anita Bay dunite. We concentrate particularly on transmission electron microscopy (TEM) on dislocation structures in the olivine grains, but we comment later on the relationship of inter- and intra-crystalline processes in the deformation. Earlier studies on the dislocation structures in deformed olivine have been made by Green (1976), Durham et al. (1977) and Gueguen (1979).

Anita Bay dunite, although somewhat heterogeneous, has a mean grain size of about 100 μm and consists dominantly of olivine with minor accessory phases and trace amounts of hydrous layer silicate minerals (Fitz Gerald and Chopra, 1982). Details concerning the deformed specimens studied are given in Table 1 and their stress-strain curves in Figure 1. The specimens were sectioned parallel to the compression axis, and ion-beam thinned for TEM. TEM observations were made at 200 kV (JEOL 200B) and at 100 kV (Philips 300).

MICROSTRUCTURE AS-RECEIVED

The matrix olivine grains, as distinct from porphyroclast grains (Boland et al. 1971; Boland, 1977), show few signs of internal optical features associated with prior natural deformation. In TEM, the dislocation density (Fig. 2) is observed to vary widely, from 0.5 to 30 x $10^{12}m^{-2}$, both within grains and from grain to grain (note: all dislocation densities reported here are approximate only). The dislocations are rarely organized, except for some rough (100) walls with spacings as small as 2 μm and free dislocation densities of 5 x $10^{12}m^{-2}$ between, and their shapes vary from straight (mainly parallel to [001]) to sharply curved and tangled, with some small loops in places. Most have $\underline{a}$ [100] or $\underline{c}$ [001] Burgers vectors, and glide loops indicate $(0\bar{1}0)$[100] as one active glide system (Burgers vectors were identified using the criteria of minimum contrast for $g \cdot \underline{b} = 0$ and $g \cdot \underline{b} \times u = 0$).

Many of the grain boundaries have thin (<0.5 μm) continuous coatings (Fig. 2), and rare infilled narrow cracks have also been seen within the olivine grains. Such thin coatings would be invisible in the optical microscope. It is possible that the grain boundary films could be as thin as 10 nm, as reported by White and White (1981), and that they are hydrated aluminosilicates, as seen optically, although we have not had opportunity to verify the compositions. However, grain boundaries that do not appear to be contaminated by hydrated material are also present.

MICROSTRUCTURE AFTER PRE-DRYING

One specimen was examined after the 1200°C pre-drying treatment (Chopra and Paterson, 1983) but before deformation. At the optical scale, planar arrays of tiny bubble-like features are visible optically within matrix grains but the main effects were parting and local brown discolouration at grain boundaries. Layer silicate grains were seen to be decomposed into aggregates of extremely fine grain size.

TEM reveals numerous well-organized dislocation walls parallel to (100), with fewer parallel to (010) and (001). These walls are composed of moderate numbers of dislocations spaced at ~ 50 nm and define cells containing low densities of dislocations, generally less than $10^{12}m^{-2}$ but rising locally to $5 \times 10^{12}m^{-2}$. Many of the free dislocations are connected to the walls. Small faceted "bubbles" (< 300 nm) linked by complex dislocation structures and similarly-sized interconnected "tubules" lie in quasi-planar arrays throughout the grains. Small "bubbles" also occur in segments of dislocation walls that curve away from low index orientations (Fig. 3). Diffuse dark-field imaging (Clarke, 1979) suggests that all "bubbles" and "tubules" contain some amorphous material. Possibly they represent the healing of cracks formed in the initial heating to the drying temperature (cf. Wanamaker et al., 1982).

In spite of difficulties in TEM because of parting and preferential ion etching at the grain boundaries, it is clear that many of these are lined with up to 0.5 μm of fine-grained polycrystalline material (Fig. 4), accompanied commonly by glass and evidently derived from original hydrous grains. $MgAl_2O_4$ spinel and pyroxene are tentatively identified in this material which, perhaps with some glass, often lies "sandwiched" between almost straight-edged grains of olivine or can fill cracks of up to 0.2 μm width in the grains. The glass is always less than a few tenths of a micrometre in thickness and is occasionally seen isolated from any obvious source region. In otherwise clean, coherent grain boundaries, thin (~ 50 nm) films of "foam-textured" material (cf. White and White, 1981) are occasionally seen and are stable under electron irradiation.

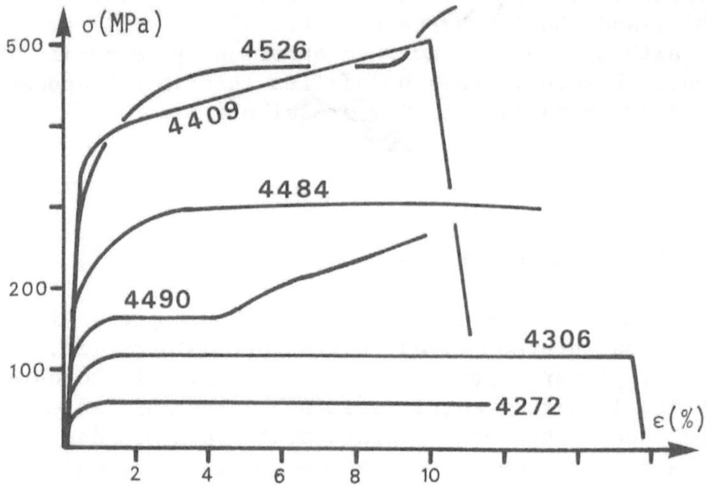

Fig. 1. Stress-strain curves for the six experiments selected for TEM study.

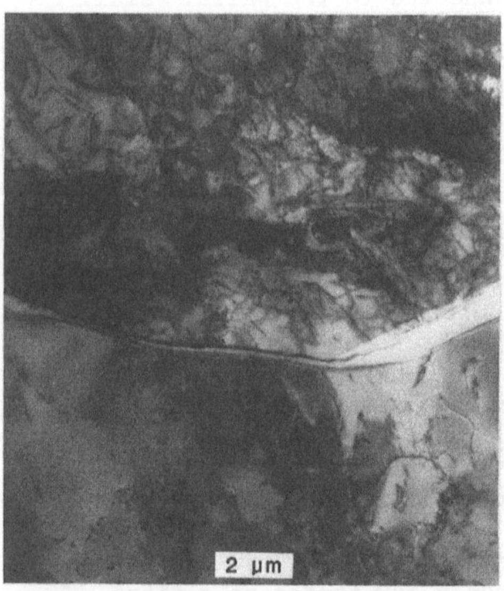

Fig. 2. Variation in dislocation density, and thin layer of hydrated material in a grain boundary region. As-received Anita Bay Dunite.

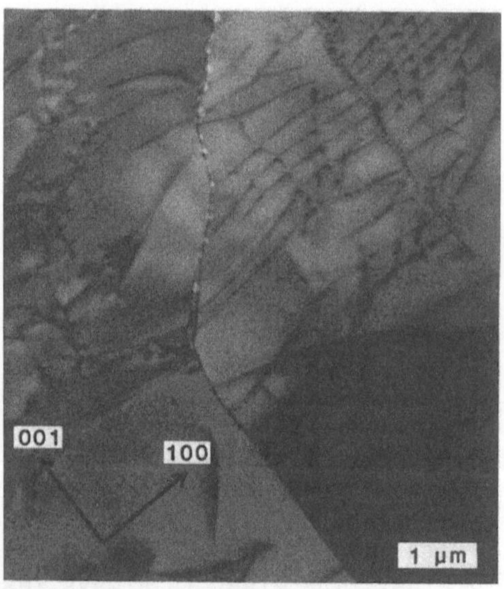

Fig. 3. Low densities of free dislocations, some twist SGB's on (010), and a near-vertical SGB marked by tiny bubbles where it does not lie parallel to (100). Predried Anita Bay Dunite.

MICROSTRUCTURES OF DEFORMED SPECIMENS

Related papers (Chopra and Paterson, 1981, 1983; Fitz Gerald and Chopra, 1982; Fitz Gerald et al., 1983) describe the optical features such as multiple deformation-band structures predominantly parallel to (100), the details of which vary with deformation conditions but not, in general, between wet and dry specimens. Heterogeneously distributed glass from decomposition of hydrous phases was seen in wet specimens deformed above about 1100°C, appearing as thin films in some grain boundaries, linked to larger pockets (~1 μm) at grain triple-junctions. We now describe TEM observations.

Intragranular microstructures

The six specimens (Table 1, Fig. 1) are grouped in the following descriptions so as to bring out the roles of various factors, including flow stress, temperature and water conditions which are listed in brackets after the specimen numbers. In view of the amounts of strain involved and the changes in microstructure seen optically, the dislocation structures now to be described can generally be taken to have totally overprinted those originally present and are believed not to have been unduly affected by the unloading history.

Specimen 4526 (482 MPa, 1200°C, dry): There is a fairly homogeneous distribution of free straight screw dislocations with $\underline{a}$ or $\underline{c}$ Burgers vectors (Fig. 5), both types having densities around $10^{13}m^{-2}$. Larger kinks and other non-screw components lie in orientations consistent with (010)[100] and (010)[001] glide systems. There is no evidence for climb other than a very few subgrain boundaries (SGB's).

Specimen 4409 (501 MPa, 1000°C, wet, relaxed to ~100 MPa): On the one hand, some grains exhibit well organized cell structures such as shown in Fig. 6, typical of all wet specimens, with very low densities of free dislocations between the SGB's. If we assume that the SGB's are formed primarily by gliding dislocations which continue to lie in glide planes even if rearranged somewhat by climb, an analysis based on Frank's formula (e.g. Amelinckx, 1979) can be carried out to determine which of the known common glide planes, viz.(001), (010) and (0kl) for $\underline{a}$ glide and (100), (010) and (110) for $\underline{c}$ glide, have been active. Application to this specimen indicates that the (100) pure tilt SGB's result from (010)[100] glide and the (010) pure twist SGB's from (010)[001] and (010)[100] glide. The grains have orientations such that the Schmid factors for these slip systems are rather low (~0.1), and the cells defined by the SGB intersections appear in thin section as elongated rectangles about 1 μm wide. On the other hand, there are many grains containing almost no SGB's but having high densities of free straight screw dislocations, a morphology suggestive of lattice

Table 1. Deformation Conditions

Specimen number	Run conditions	Temperature °C	Strain rate s^{-1}	Final strain %	Final flow stress MPa	Remarks
4272	wet	1195	1.1×10^{-6}	11.5	63	Total duration ~30 hours.
4306	wet	1300	1.2×10^{-4}	15.4	118	Followed by stress relaxation test.
4409	wet	1000	1.1×10^{-5}	11.2	501	Followed by stress relaxation test.
4484	wet	1300	1.1×10^{-3}	14.2	291	Total duration ~5 minutes.
4490	wet then drier	1300	1.1×10^{-4}	9.9	147 then to 263	After 4% strain jacket leak led to progressive drying.
4526	pre-dried	1200	0.5×10^{-4}	10.3	549	Mostly at constant flow stress of 480 MPa; late increase attributed to minor buckling.

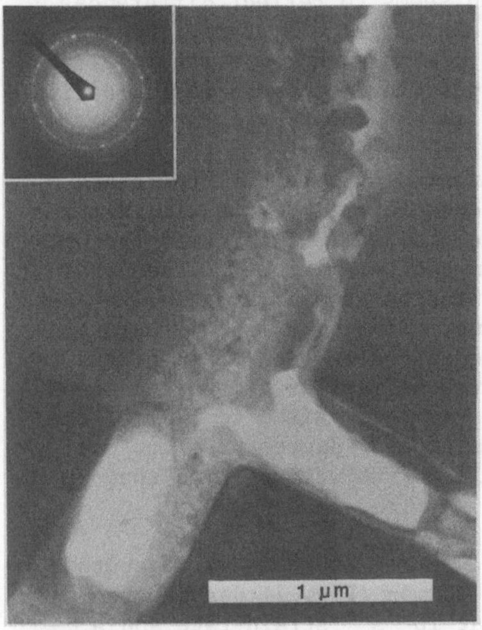

Fig. 4. Grain boundary region (see text). Predried Anita Bay Dunite.

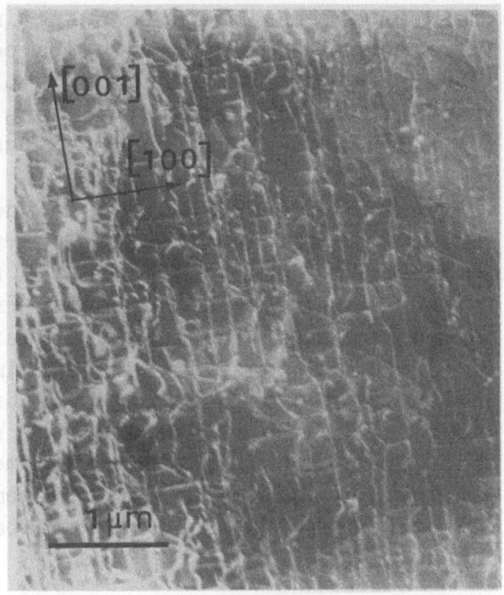

Fig. 5. Dislocation microstructure of high and homogeneous density of straight a and c screw dislocations. Sample 4526 deformed at 1200°C under dry conditions.

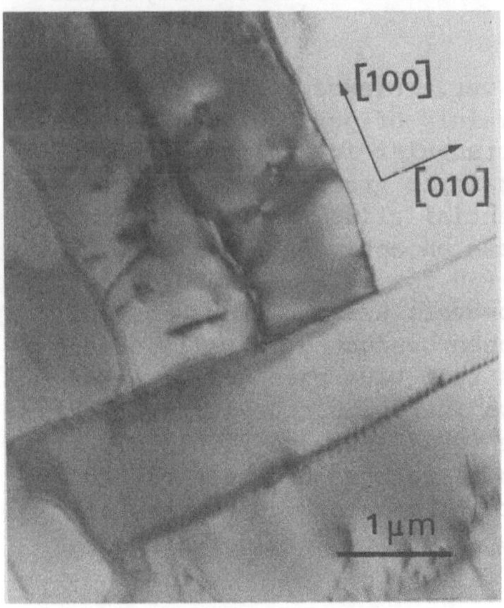

Fig. 6. Well-organized cell structure with low density of free dislocations. Sample 4409 deformed wet at 1000°C and relaxed.

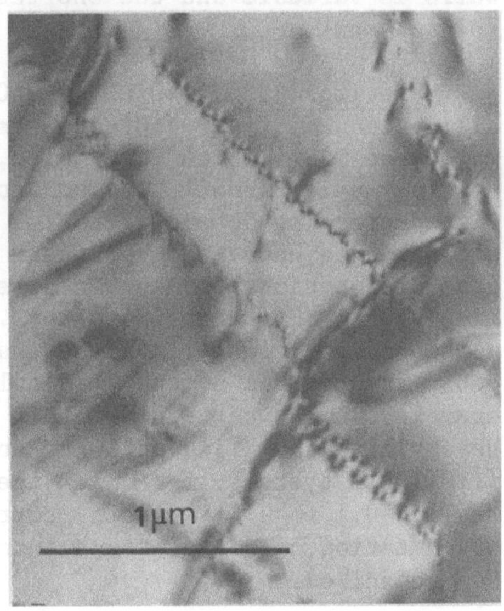

Fig. 7. Rough and irregular subgrain structure. Sample 4484 deformed wet at 1300°C at a fast strain rate.

friction control, with no evidence of recovery. The marked grain-grain heterogeneity in this specimen seems to reflect differences in resolved shear stress on the potential glide systems. Recovery appears to be more effective in the grains with low stressed glide systems but this recovery process could have occurred mostly during the relaxation stage, under low stress and low strain rate.

4526 vs. 4409. Both being dominated by straight free a and c screw dislocations with little evidence for climb, the similarity in microstructure in 4526 and in the grains of high Schmid factor in 4409 supports the conclusion from optical microstructures that the mechanism of intragranular deformation is similar under the wet and dry conditions in these specimens. The comparison is unfortunately complicated here by the subsequent relaxation of specimen 4409. If, while maintaining the stresses similar in the two cases, the temperature of the wet specimen had been raised to 1200°C, the same as for the dry specimen, then a greater degree of climb and recovery activity would be expectable in the wet specimen than in the dry.

Specimen 4484 (291 MPa, 1300°C, wet): Subgrain walls containing relatively few dislocations are common (Fig. 7) despite the short duration of this run (3 min. at $10^{-3}s^{-1}$ strain rate and a further 2 min. after stress removal before cooling at about 2°C s^{-1}). The rough and irregular nature of the walls may indicate an early stage of development. The density of free dislocations between walls is variable and the shortest cell dimension is usually small (<1 to 4 μm).

Specimen 4306 (120 MPa, 1300 C, wet, relaxed to ~7 MPa): The dislocation microstructure consists mainly of parallel and regularly spaced SGB's which form cells, rectangular in section. The SGB spacing varies from 1 to 5 μm from grain to grain and increases with decreasing Schmid factor on potential glide systems, suggesting a dependence upon resolved shear stress on active glide planes. The free dislocations are low in density (~0.5 x $10^{-12}m^{-2}$) and do not lie in well defined glide planes. However, using the assumptions mentioned previously, analyses have been conducted on well formed SGB's in order to deduce glide planes that have been active. For example, in the particular grain shown in Fig. 8, the structure can be explained by glide activity on the (001)[100] and (100)[001] or (110)[001] systems, for which the Schmid factors had been rather small (~0.1 to 0.2). Again, correlation between high degrees of organization of dislocations into SGB's and low Schmid factors can be recognized.

Specimen 4272 (63 MPa, 1200°C, wet): Here the SGB's are very well-formed, with relatively large spacing (~10 μm) and an extremely low density of free dislocations, indicating that recovery processes have been very effective in the long duration of this very low strain rate run (30 hrs at $10^{-6}s^{-1}$).

4484 vs. 4306 and 4272. The effectiveness of climb processes increases and the degree of lattice friction control on glide clearly decreases as the flow stress decreases in these runs, while the additional opportunity for re-organization of the dislocation structure during relaxation down to a much lower stress (~7 MPa) in 4306 has evidently not resulted in a drastic overprinting, during the additional 0.9% strain, of the previously established structure. From these experiments alone the effects of the decrease in stress and increase in duration cannot be distinguished. However, comparison of 4306 and 4526, of similar duration but a factor of four different in flow stress, indicates that the flow stress is an important factor determining the dislocation structure, although this comparison involves different temperatures and so does not enable direct conclusions to be drawn concerning the relative roles of stress, temperature and water content.

Specimen 4490 (147/263 MPa, 1300°C, wet/dry): The rapid strain hardening that occurred beyond 5% strain is attributed to progressive drying out of the specimen owing to a jacket leak (Chopra and Paterson, 1983). The TEM microstructure is heterogeneous. In some regions (Fig. 9a) there are rough SGB's regularly spaced at about 1 μm and a moderate density of free dislocations ($\sim 5 \times 10^{12} m^{-2}$). In other regions (Fig. 9b), the free dislocations occur in considerably higher density and exhibit strong preferred line orientations, while SGB's are still present; free dislocations appear to have interacted with these SGB's and the increase in dislocation density and decrease of SGB's may have been due in part to SGB deknitting.

4490 vs. 4484, 4526 etc. The effectiveness of climb in the wet specimens appears to increase and the degree of lattice friction control to decrease as the flow stress decreases at given temperature (4484, 4306). The relaxation down to a much lower stress (~7 MPa) in 4306 may have contributed to the effect but it is questionable whether the additional 0.9% strain would produce a drastic overprinting of the previously established structure. Further decrease in flow stress in 4272 compared to 4306 appears to have more or less compensated for the decrease in temperature in respect of the favouring of recovery processes.

Grain Boundaries

TEM observations on grain boundaries (GB's) in the deformed dunite are of particular interest for two reasons. Firstly, intergranular glass is known to occur in wet specimens at the optical scale (Fitz Gerald et al., 1983) and it would therefore be interesting to know its submicroscopic distribution. Secondly, it has been inferred that the relative weakness of wet specimens

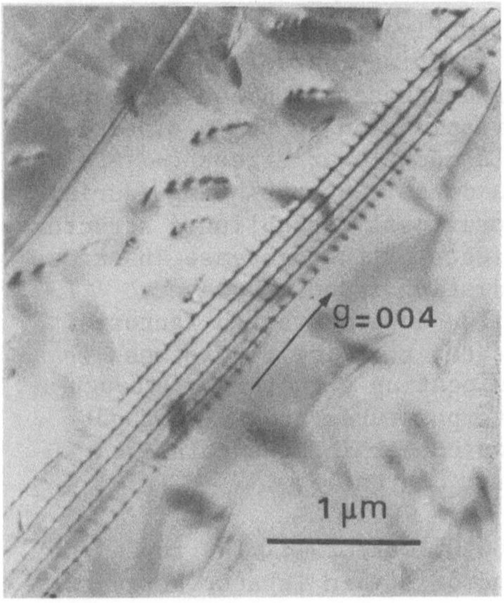

Fig. 8. Well-formed SGB constituted by c screws (in the plane of the foil) and a dislocations with ~65° character. Sample 4306 deformed wet at 1300°C and relaxed.

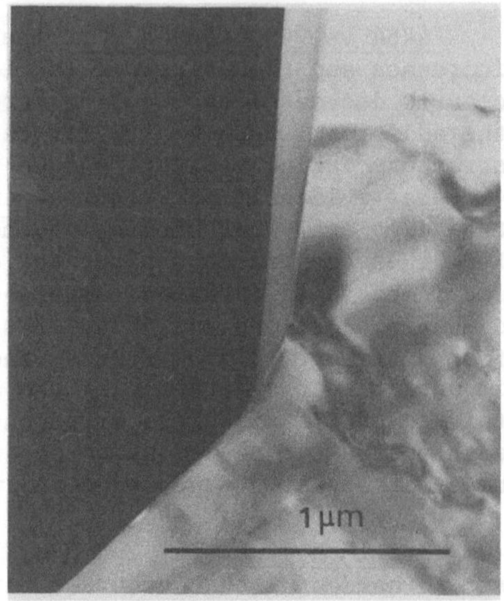

Fig. 10. Thin film of glassy phase in a grain boundary region. Sample 4526 deformed at 1200°C under dry conditions.

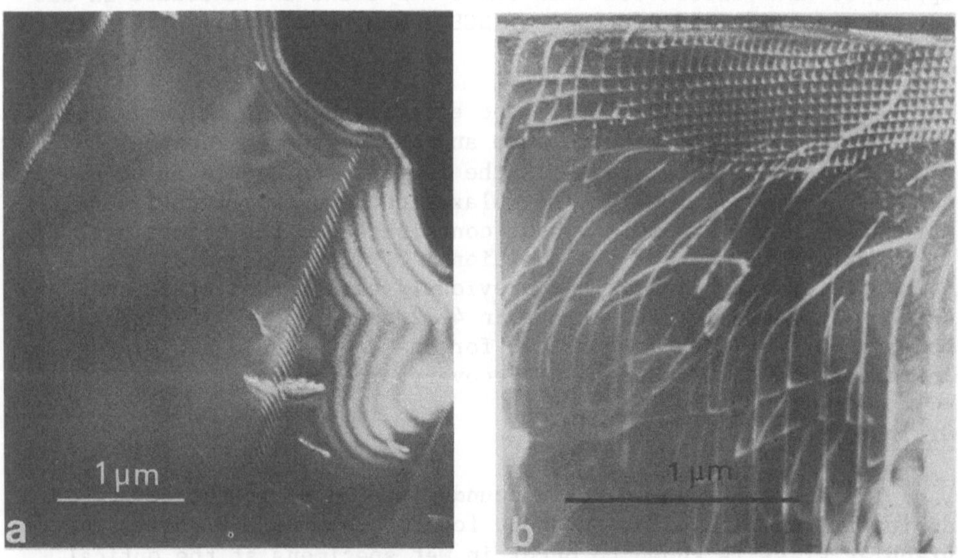

Fig. 9. (a) SGB's approximately 1 micron apart (b) SGB's and high density of free dislocations, many of which interact with SGB. Both regions from sample 4490, deformed at 1300°C, which involved late-stage progressive drying due to a jacket leak.

involves accommodation processes at GB's, facilitated by the presence of a water-related species (Chopra and Paterson, 1983), and it would be desirable to have observations bearing on these processes. Unfortunately, no features apparently throwing light on the latter have been found in TEM and only slight progress can be reported in respect of the former. A serious obstacle in TEM studies on GB's is the preferential etching of the boundary regions that occurs when preparing foils, and this is compounded in Anita Bay Dunite by the sampling problem arising from the relatively large spacing of GB's and the heterogeneity of glass distribution. However, the following observations have been made.

Dry specimen 4526. Some small crystals are seen at GB's and are presumably either inherited from the pre-drying phase or arise in further crystallization from earlier formed melt. Some glass also persists (Fig. 10).

Wet specimens. In general, the presence in GB's of glass and some microcrystalline material has been recorded and serves to extend the optical observations (Fitz Gerald et al., 1983) to a finer scale, but at this stage no detailed analysis of the extent of GB wetting or penetration by melt can be given. In specimen 4409 (1000°C) no evidence was found of melt or dehydration products from layer silicates such as had been seen optically but, in view of the heterogeneity at the optical scale, this absence may only reflect the sampling problem in TEM.

DISCUSSION

In view of the marked heterogeneity at the grain scale in the dunite specimens (Fitz Gerald et al., 1983), caution is needed in generalizing from the present TEM observations because the grains sampled were relatively few and therefore possibly not fully representative. In particular, the TEM observations appear not to reveal the coarser features such as the deformation band boundaries seen optically which presumably represent a level of structural organization on a scale larger than is sampled by TEM in a single foil. However, some interesting deductions can be made from the observations relating to dislocation glide and climb.

Glide based on both a and c Burgers vectors has occurred in all specimens, as evidenced by common occurrences of both throughout the free dislocation population and in the subgrain boundaries. Glide plane determinations are less certain, either because the free dislocations often do not lie in one plane at the lower stresses or because of the assumptions necessary to analyse those in subgrain boundaries, but all the six common glide systems of olivine are deduced to have been active. However, these only contribute three independent glide systems (Paterson, 1969) and some contribution from at least one additional independent deformation mechanism is probably necessary for the polycrystalline deformation (Hutchinson, 1977, shows that the von Mises requirement of five can be reduced to four when heterogeneous deformation is allowed). Grain sliding, climb, kinking, microcracking, etc are possible additional deformation mechanisms. It is difficult to predict which mechanism is controlling the plastic deformation (i.e. the flow stress at constant strain rate) in wet and dry samples. Furthermore, it is not necessarily the same across the whole temperature range tested. However, TEM observations indicate that climb is active in the wet specimens and may have contributed somewhat to overall strain (cf. significant amounts of strain attributed to climb in the deformation of dry single crystals of olivine by Durham and Goetze, 1977).

An outstanding question is whether there is an intragranular weakening in the wet specimens relative to the dry. This question cannot be settled with the present observations but the comparisons given above, especially 4526 vs. 4409, suggest that in the wet specimens there could be increased climb, and therefore presumably some intragranular weakening, compared with dry specimens at the same stress and temperature. However an intragranular weakening in which grains of 100 μm diameter are strongly affected within a few minutes would require a diffusion coefficient of 10^{-11} to 10^{-12} m^2s^{-1} at 1300°C. Such a value is higher than usual for diffusion coefficients in minerals (Freer 1981), but it is possible that diffusion is somewhat enhanced by pipe diffusion in dislocation cores.

Lastly, it should be mentioned that the flow stress (63 MPa) for specimen 4272 is below the 100 MPa level at which Chopra and Paterson (1981) deduce, on the basis of relaxation observations, that there is a transition in rheological regime to a lower stress exponent n, suggestive of an increase in grain boundary sliding. In the absence of clear evidence for such sliding in this study we cannot throw further light on it but there does not seem to be any obvious change in the general trend in intragranular microstructures with decreasing stress at these levels.

ACKNOWLEDGEMENT

This work has been mainly supported by the Research School of Earth Sciences, Australian National University. Discussions with many colleagues, especially Dr. Shun Karato, have been stimulating and valuable.

REFERENCES

Amelinckx, S., 1979, Dislocations in particular structures, in: "Dislocations in Solids", Vol 2, F.R.N. Nabarro, ed., North Holland.

Boland, J.N., 1977, Deformation mechanisms in alpine type ultramafic rocks from New Zealand, Tectonophysics, 39:215.

Boland, J.N., McLaren, A.C. and Hobbs, B.E., 1971, Dislocations associated with optical features in naturally-deformed olivine, Contrib. Mineral. Petrol., 30:53.

Chopra, P.N., and Paterson, M.S., 1981, The experimental deformation of dunite, Tectonophysics, 78:453.

Chopra, P.N., and Paterson, M.S., 1983, The role of water in the deformation of dunite. Submitted to J. Geophys. Res.

Clarke, D.R., 1979, On the detection of thin intergranular films by electron microscopy, Ultramicroscopy, 4:33.

Durham, W.B., and Goetze, C., 1977, Plastic flow of oriented single crystals of olivine. 1. Mechanical Data, J. Geophys. Res., 82:5737.

Durham, W.B., Goetze, C., and Blake, B., 1977, Plastic flow of oriented single crystals of olivine. 2. Observations and interpretations of the dislocation structures, J. Geophys. Res., 82:5755.

Fitz Gerald, J.D., and Chopra, P.N., 1982, Deformation mechanisms in dunite - the results of high temperature testing, p.735-740, in: Proc. Sixth Intl. Conf. Strength Metals and Alloys, R.C. Gifkins, ed., Pergamon.

Fitz Gerald, J.D., Chopra, P.N., and Paterson, M.S., 1983, Microstructures of experimentally deformed dunite. To be submitted to Tectonophysics.

Freer, R., 1981, Diffusion in silicate minerals and glasses: a data digest and guide to the literature, Contrib. Mineral. Petrol., 76:440.

Green, H.W., 1976, Plasticity of olivine in peridotites, in: "Electron Microscopy in Mineralogy", H.R. Wenk, ed., Springer Verlag, Berlin.

Gueguen, Y., 1979, Dislocations in naturally deformed terrestrial olivine: classification, interpretation, applications, Bull. Minéral., 102:178.

Hutchinson, J.W., 1977, Creep and plasticity of hexagonal polycrystals as related to single crystal slip, Met. Trans. A., AIME, 8A:1465.

Paterson, M.S., 1969, The ductility of rocks, in: "Physics of Strength and Plasticity", A.S. Argon, ed., MIT Press.

Wanamaker, B.J., Bergman, S.C., and Evans, B., 1982, Crack healing in silicates: observations on natural lherzolite nodules (abstr.), Trans. A.G.U. EOS, 63:437.

White, J.C., and White, S.H., 1981, On the structure of grain boundaries in tectonites. Tectonophysics, 78:612.

DISTRIBUTION OF STRAIN IN SOME GEOLOGICAL MATERIALS EXPERIMENTALLY DEFORMED AT HIGH PRESSURE AND TEMPERATURE

J.D. Fitz Gerald[1] and P.N. Chopra[2]

[1]R.S.E.S., Australian National University, Canberra A.C.T. 2601, Australia
[2]Bard Hall, Cornell University, Ithaca, NY 14853, U.S.A.

ABSTRACT

Ion-etching of fiducial markers has allowed, for the first time, direct observation and measurement of strain distribution in some materials of geological interest deformed in compression at high temperatures (above 700°C). Deformation of two polycrystalline materials, one olivine rich, the other consisting of calcite, has been briefly investigated on the basis of this new practical method.

INTRODUCTION

Characterization of the distribution of strain developed in experimentally deformed polycrystalline materials of geological interest is difficult. This is primarily due to the fact that such materials have very low tensile strength and can only be deformed to significant strains in compressive tests under confining pressures of 10^8 to 10^9 Pa. As a result of this experimental limitation, there are consequently no free sample surfaces on which fiducial markers (Atwood and Hazzledine, 1976) can be imprinted and recovered in a useful state. A small number of experiments using the split cylinder technique have however been performed; these have involved deformation of composite specimens each consisting of two matched pieces separated by a thin inert metal foil. Previously, marker patterns have been recovered successfully from 'internal' surfaces of split cylinders only in low temperature experiments (400°C, Spiers,1979). Unfortunately, at higher temperatures passive markers (scratches) do not survive experiments and stable, inert, thin-film markers (such as copper grid meshes) potentially have sufficient strength to affect significantly the local deformation at the internal surfaces.

We report here on a new experimental procedure through which passive fiducial markers can routinely be recovered from experiments at temperatures up to 1300°C.

PROCEDURE

The method of surface etching used here involves atom sputtering induced by bombardment with high-energy inert ions, a technique widely used in the preparation of thin foils of refractory materials for transmission electron microscopy. While this method does produce some preferential grain boundary (GB) etching, it has the advantage that the relict GB regions which extend into the sample from the bases of the etch zones are chemically pristine.

The ion-etching technique is quite simple. A metallic grid mesh (as widely used for specimen support in electron microscopy) of desired size is firmly attached to the polished surface of one of the split cylinder pieces and the region is then exposed to an ion beam at an incidence angle of 90 degrees. The grid mesh acts as a mask which shields parts of the specimen surface until the mesh itself is sputtered away. In practice we have found that a sharp pattern can be etched into samples to a depth of approximately 5 μm using Ni grid material and 5 kV Ar ions. GB's have usually been etched up to 5 μm deeper into the sample. Standard grid meshes are sufficiently even and regular for good fiducial markers to be etched at the 100 μm scale.

RESULTS

Reliable recovery of markers has been achieved on a routine basis for two materials of geological origin which have been experimentally deformed in split cylinder experiments in a gas-medium apparatus operated at a confining pressure of 300 MPa.

Anita Bay Dunite

This is a rock with mean grain size of 100 μm which is rich in the silicate mineral, olivine. Split-cylinder experiments were conducted with long Fe sample jackets (Paterson et al., 1982) enclosing the composite specimens and with thin Fe foils inserted between sample pieces. Markers derived from hexagonal meshes of 85 μm spacing were recovered from experiments at temperatures of 1200°C and 1300°C. Following deformation, a heterogeneity of strain could be clearly recognized at the grain scale through local grid distortion (fig. 1), but the definition of markers was considered insufficient for reliable determination of strain parameters.

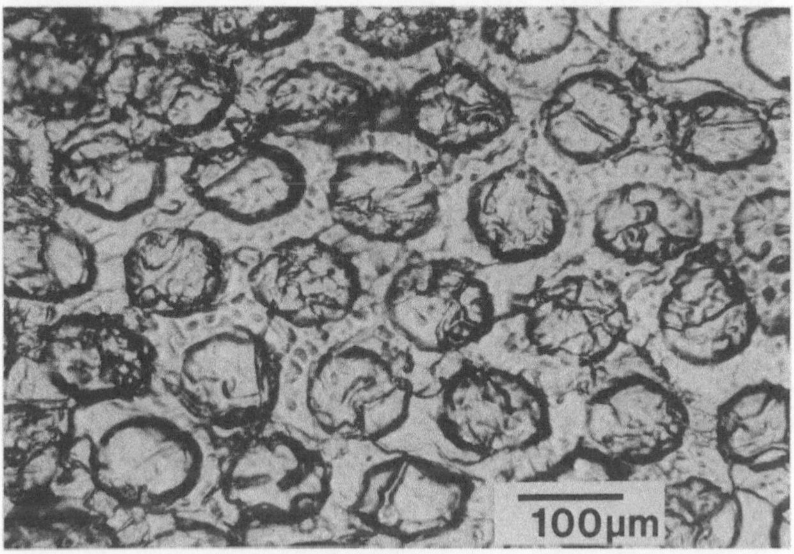

Figure 1. Optical micrograph of deformed hexagonal grid from a sample of Anita Bay dunite (see text).

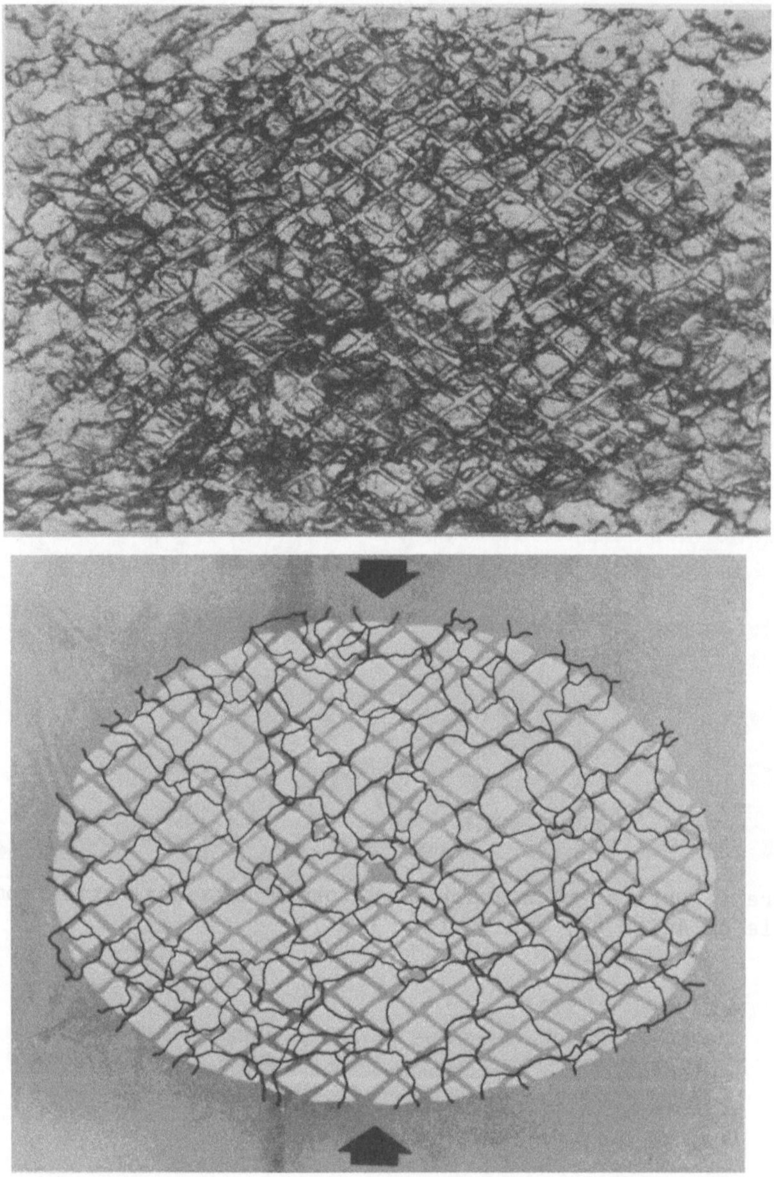

Figure 2. Reflected light micrograph with sketch of deformed marker region on Carrara marble surface. The outline of the marker region was originally circular in shape but is now elliptical (major axis dimensions 2.95 mm by 2.08 mm). Compression axis aligned N-S.

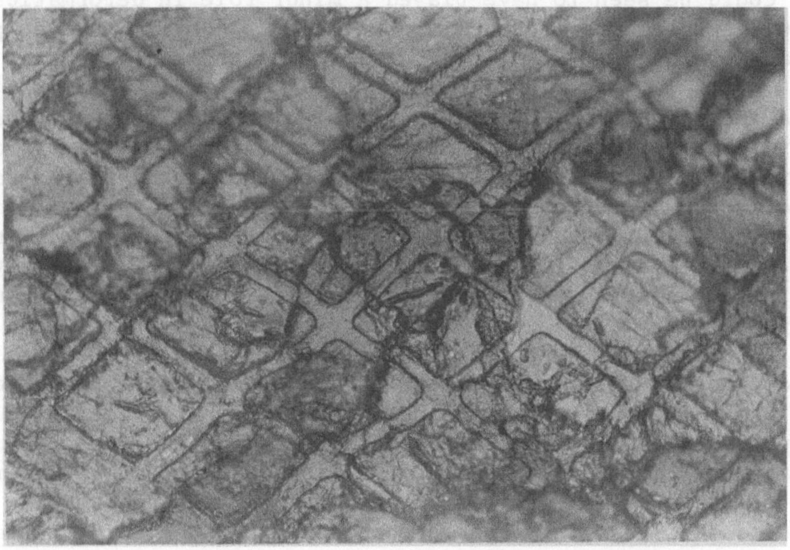

Figure 3. Detail of figure 2. The ion-etched grid pattern was originally square with a repeat distance of 170 μm and grid 'bar' thickness of 25 μm. The pattern is clearly deformed as seen by the (variable) angular distortion and the sharp offsets at some GB. Compression axis aligned N-S.

Carrara Marble

This material is comprised of grains of $CaCO_3$ (mineral name calcite) of mean grain size 200 μm. Split cylinder experiments were conducted with Cu jacket material and Cu or Au separating foils (dependent upon the experimental temperature). Reliable recovery was achieved for all experiments in the temperature range 700°C to 1000°C.

One sample, deformed at 800°C, $\dot{\varepsilon} \sim 2 \times 10^{-5}$ s^{-1} to a total ε of 0.21 was selected for detailed study. This sample was considered particularly suitable for measurement as twinning and recrystallization could be seen to have played a minor role in deformation, and careful measurements of density were made to confirm that no porosity had developed in the polycrystal during the experiment.

A marker pattern, initially circular in outline, was conspicuously deformed into an oval shape (fig. 2) during the experiment. Measurement of the deformed dimensions of the marker region, and of the gross sample shape, confirmed that an axially-symmetric shortening of 20.8% had been produced. Close inspection of individual marker lines (fig. 3), originally of spacing 170 μm, revealed: (i) distortion of the grid pattern from its initially orthogonal configuration and (ii) distinct, sharp marker offsets at some GB. Since the marker pattern was initially carefully imprinted so that the compression axis would bisect the angle between the two sets of grid lines, the observed grid angles in the final specimen could be interpreted directly in terms of local strain. The measurement of displacements at GB's was performed using a reflecting light microscope to observe: (i) offset of markers in the plane of the surface and (ii) grain offset perpendicular to the surface measured using the calibrated fine focus control.

Several conclusions can be drawn from detailed examination of the deformed grid in this specimen:
(a) Induced strain is relatively homogeneous within single grains and there are few slip bands and no core-mantle effect (Gifkins, 1976). However, distinct differences are observed in the internal strains preserved in neighbouring grains. These differences are presumably related to variation in crystallographic orientation relative to the applied stress direction.
(b) Grid offsets are sharp and the positions of individual grain boundaries relative to the grid reference frame are apparently unchanged by the experimental deformation and temperature history. In other words, GB migration during deformation is undetectable in the marked surface region.
(c) GB offsets rarely appear to result from strain incompatibility between adjacent grains. Instead, the distribution of offset in terms of the angle between the compression axis and the surface trace of the GB's (fig. 4) indicates a strong shear stress control.

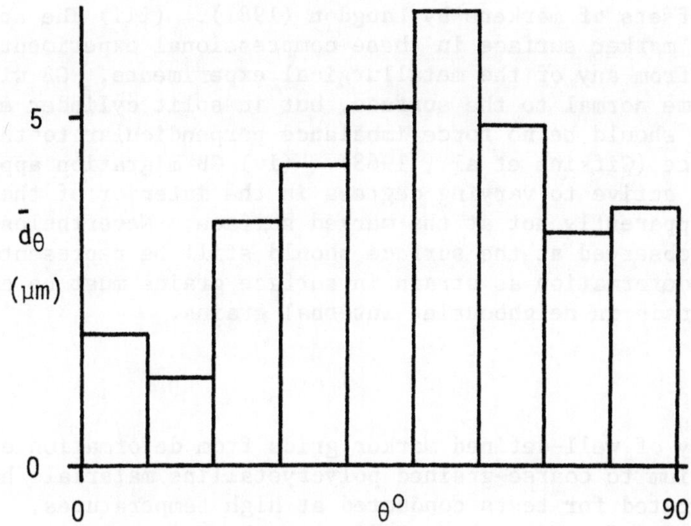

Figure 4. Histogram showing distribution of marker displacement in the plane of the Carrara split cylinder surface. θ = angle between compression axis and trace of a GB. $\bar{d}_\theta$ = average displacement along all GB's falling within specified θ interval. Data presented here from 391 observations. The peak towards the centre of the θ-range indicates shear stress control.

(d) Quantification of the amount of strain due to GB sliding (GBS) is somewhat of a problem. The specimen grain size was too large to permit adequate statistical sampling of grain displacements using single longitudinal or transverse traverses (Langdon, 1972). Consequently, measurements of u, v and w (displacement components - see Langdon 1972 for a review) could only be carried out in a 'random' fashion at each of the 391 sites where GB's intersected marker lines. Calculation using average values for these components and the relationships derived from numerous metallurgical studies (see again Langdon, 1972) yield values for the fraction of total strain due to GBS, ξ, ranging from 0.10 to 0.21. Paradoxically, the value of ξ resulting from the average deformed grid angle is 0.39 (assuming that cavitation is absent as indicated by the density measurements, and that the total strain is therefore the sum of the internal grain strain and the strain due to GBS). Several factors may contribute to, but do not completely explain, this discrepancy: (i) The observed grid angles do not follow a statistically normal distribution and the average angle measured here may not be a good indicator of average grain strain. (ii) In compressional experiments, grain sliding will result in some parts of the surface markers adjacent to GB becoming hidden as the total surface area is decreased on longitudinal sections. Observed displacements will therefore underestimate the true offsets. This effect is the reverse of the "new surface" effect considered for tensional offsets of markers by Langdon (1981). (iii) The state of the plane marker surface in these compressional experiments is unlike that from any of the metallurgical experiments. GB will tend to become normal to the surface, but in split cylinder experiments there should be no force imbalance perpendicular to the marker surface (Gifkins et al., 1968). (iv) GB migration appears to have been active to varying degrees in the interior of the sample but apparently not at the marked surface. Nevertheless, deformation observed at the surface should still be representative of internal deformation as strain in surface grains must be compatible with strain in neighbouring internal grains.

SUMMARY

Recovery of well-defined marker grids from deformation experiments of medium to coarse-grained polycrystalline materials has been demonstrated for tests conducted at high temperatures. Measurement of deformed markers indicates that some caution is nevertheless required in interpretations concerning the distribution of strain throughout the aggregates.

ACKNOWLEDGEMENT

We gratefully acknowledge the assistance given by J.P. Shore and P.E. Willis.

REFERENCES

Atwood, D.G. and Hazzledine, P.M., 1976, A fiducial grid for high-resolution metallography, Metallography, 9:483.
Gifkins, R.C., 1976, Grain boundary sliding and its accommodation during creep and superplasticity, Met. Trans. A, 7A:1225.
Gifkins, R.C., Gittins, A., Bell, R.L. and Langdon, T.G., 1968, The dependence of grain-boundary sliding on shear stress, J. Mater. Sci., 3:306.
Langdon, T.G., 1972, The effect of surface configuration on grain boundary sliding, Met. Trans., 3:797.
Langdon, T.G., 1981, The significance of grain boundary sliding in creep and superplasticity, Met. Forum., 4:14.
Paterson, M.S., Chopra, P.N., and Horwood, G.R., 1982, The jacketing of specimens in high-temperature, high-pressure rock-deformation experiments, High Temp. High Press., 14:315.
Spiers, C.J., 1979, Fabric development in calcite polycrystals, Bull. Mineral., 102:282.

PLASTIC DEFORMATION OF Al_2SiO_5 POLYMORPHS,

INFLUENCE OF STRAIN ON THEIR POLYMORPHIC TRANSFORMATIONS

J.C. Doukhan[1], P.S. Koch[2] and J.M. Christie[2]

[1]University of Lille I, 59655 Villeneuve d'Ascq Cedex, France
[2]Dept. of Earth & Space Sci., UCLA Los Angeles Ca 90024 USA

ABSTRACT

The crystal structures of the three Al_2SiO_5 polymorphs (sillimanite, kyanite, and andalusite) are built up with parallel chains of edge sharing AlO_6 octahedra. The easy glide systems keep intact the Si-O bonds as well as these chains. The observed differences in the mechanical properties of these polymorphs are related to their different chain bondings. Furthermore the core structures of the mobile dislocations (dissociations) strongly depend on the experimental temperature and confining pressure. For T, P values near the sillimanite-kyanite boundary, plastic deformation can induce a martensitic type transformation, a dislocations in sillimanite become widely dissociated and the associated fault is the preferential nucleation site of the new phase (kyanite).

INTRODUCTION

The phase diagram of the three Al_2SiO_5 polymorphs has been extensively studied but remains poorly defined (see the critical reviews by Zen[1], Holdaway[2] and Day and Kumin[3]). The polymorphic transformations are sluggish because of the very low differences between their Gibbs free energies. Lattice defects like impurities, dislocations ... might contribute to the observed discrepancies among studies of aluminium silicate stability. These defects (especially dissociated dislocations) could also be the sites for preferential nucleation of the other polymorphs.

From a crystallographic point of view the three crystal structures can be seen as a frame work of chains of edge sharing AlO_6 octahedra. These chains are parallel to the **c** axes and are bound to one

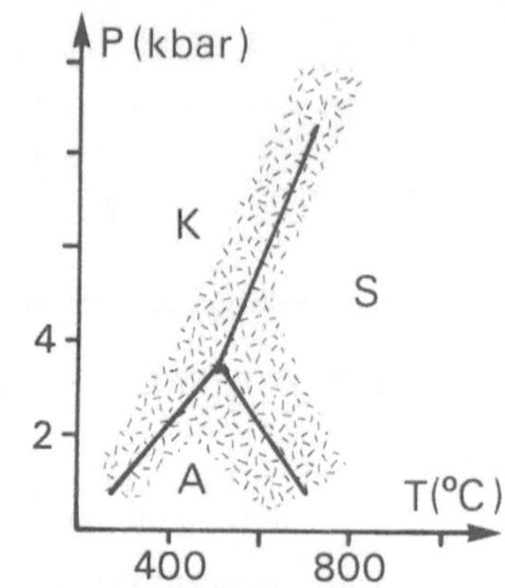

Fig.1. Equilibrium phase diagram A, K and S are the stability fields of andalusite, kyanite and sillimanite respectively. Solid lines: phase boundaries after Holdaway[2]; the other experimental results spread out in the whole stippled area.

another by various polyhedra : SiO_4, AlO_4, AlO_5 and AlO_6 (see in Vaughan and Weidner[4] a critical study of the relations between these structures and their elastic properties). Idealised views of these structures are shown in fig.2. From purely geometrical considerations it was suggested some time ago[8] that a phase change from one orthorhombic structure (S or A) to the triclinic one (K) could be achieved by simple shear in a plane parallel to the chains. Such

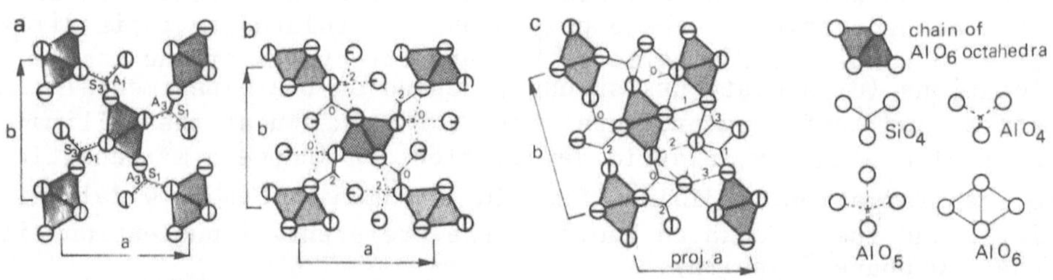

Fig.2. Idealised view of the three structures projected on a plane perpendicular to the chain axes **c**. The open circles are the oxygen atoms. Their level above the projection plane is indicated by a bar : ⊖ level 0 ; ⦶ level c/4 ; ⊖ level c/2 ; ⦶ level 3c/4 (in kyanite this level does not coïncide with the atom abscissa along the c axis), the levels of cations are indicated in c/4 units.

(a) Sillimanite is orthorhombic[5] (Pbnm) ; a=7.48 Å, b=7.68 Å, c=5.78 Å. The cross linking polyhedra between adjacent chains are SiO_4 and AlO_4 tetrahedra.

(b) Andalusite is orthorhombic[6] (Pnnm) : a=7.80 Å, b=7.90 Å, c=5.56 Å. The cross linking polyhedra are SiO_4 tetrahedra and AlO_5 bipyramids.

(c) Kyanite is triclinic[7] ($P\bar{1}$) ; a=7.12 Å, b=7.85 Å, c=5.57 Å, α=90°, β=101.1°, γ=106°. The cross linking polyhedra are SiO_4 tetrahedra and AlO_6 octahedra.

shear transformations may be common in minerals. Coe and Kirby[9,10] have shown that this mechanism operates in the ortho to clino enstatite inversion owing to partial dislocations. This renders attractive the study of the plastic properties and of the induced lattice defects of the Al_2SiO_5 polymorphs.

EXPERIMENTAL RESULTS

Kyanite

Experimental deformation of kyanite in various TP conditions[11,12,13,14] have clearly shown that this mineral has only one easy glide system (100) [001] with a low elastic limit. For compression orientation //**c** i.e. with a zero Schmid factor for the above glide system, kyanite is much stronger and deformation proceeds by mechanical twinning and kinking[12]. Kyanite thus appears strongly anisotropic. It can be considered as a layered structure, the limits of the layers being the easy glide planes for the dislocations. In these planes they do not cut the strong Si-O bonds or the chains. Mechanical twins are generated with periodic shear by partial dislocations 1/2[001] and 1/2[011] which also glide in the (100) planes. These partials are probably generated by two dissociation modes of the [001] perfect dislocations[12,14]:

$$[001] \rightarrow 1/2\,[001] + 1/2\,[001] \quad \text{and} \quad [001] \rightarrow 1/2\,[011] + 1/2\,[0\bar{1}1]$$

Both dissociations introduce very small perturbations in the structure. They suggest the following criteria for low fault energies in these alumino silicates : (i) the anionic sublattice (which is an

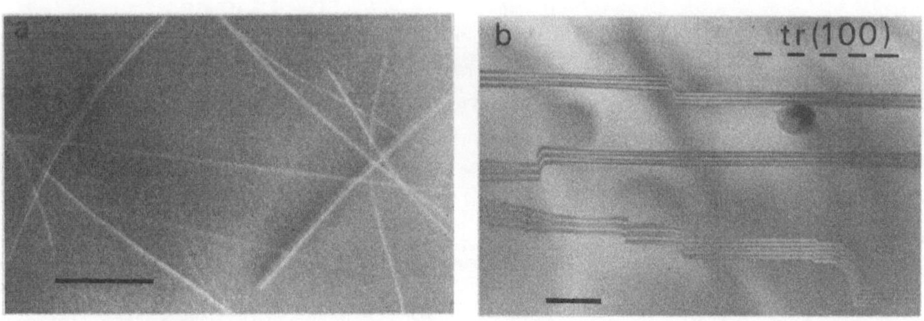

Fig.3. Influence of (T,P) conditions on the dissociation width in kyanite. (a) sample experimentally deformed of 5% at 20°C, P=1 bar. Dissociation into two colinear **c**/2 partials, dissociation width ≃ 300 Å. In samples deformed at higher T and P=1 bar this width increases up to ≃ 1μm. (b) natural deformation in estimated (T,P) conditions near to the triple point. The widely dissociated dislocations cross-slip in planes in zone with the [011] direction.

almost perfect FCC in kyanite) is not affected ; (ii) the coordinations of the various cations are not modified and (iii) the distances between nearest cations are not shortened. Finally it is to be noted that the dissociation widths strongly vary with the TP deformation conditions (fig.3).

Andalusite

Very few deformation experiments have been done on andalusite. TEM investigations on indented samples reveal the following glide systems[15] :{110}[001] which is the easy glide system, (010) [100] and (100) [010]. In uniaxial compression tests,only {110}[001] could been activated[16]. At constant strain rate this glide system exhibits a very peculiar behaviour throughout the tested temperature range (500 to 900°C) : flow stress increases with increasing temperature (fig.4). In TEM one observes very straight c screw dislocations, the density of which seems to increase with the deformation temperature. In fact, at the highest temperature, most of the observed screws are the arms of very long and narrow screw dipoles. This suggests that **c** screws are slightly dissociated out of their {110} glide plane. The screw orientation would thus correspond to that of lowest dislocation mobility and the long screw dipoles would be left in their wake by the mobile edges. Furthermore the flow stress increase with T suggests that the dissociation width increases with T leading to a lower mobility of these screws and to a larger density of screw dipoles. It has not been possible to observe this dissociation directly in TEM, but it has been suggested[15] that **c** screws would be dissociated into two colinear partials c/2 in (010). In the faulted plane the AlO_5 bipyramids would become distorted AlO_4 tetrahedra and this coordina-

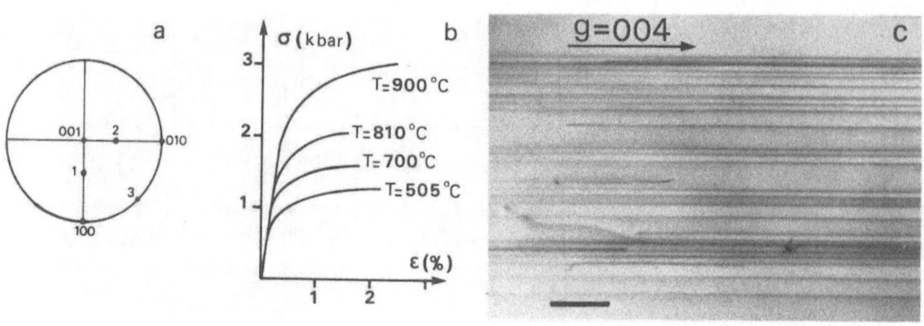

Fig.4. Uniaxial compression of andalusite (at P=1 bar).(a) chosen compression orientations ; 1 and 2 activate {110}[001] only while 3 leads to brittle fracture. (b) stress-strain curves at $\dot{\varepsilon} = 2.10^{-5}s^{-1}$ for orientations 1 and 2. Note the flow stress increase with increasing T. (c) typical dislocation substructure. Sample strained 2% at 900°C, (010) foil. All the dislocations are **c** screws.

tion change would lead to a large fault energy and to a correspondingly low dissociation width.

Sillimanite

c dislocations are the predominant defects in naturally deformed sillimanite. They are widely dissociated into two colinear partials **c**/2 in (100) or (010). Both faults verify the stated criteria for low fault energy[17]. Apparently **c** glide is the easy glide system in sillimanite. Experimental deformation has been performed only along the [110] compression orientation[18]. It corresponds to a zero Schmid factor for **c** glide and to a maximum one for (100) [010] and (010) [100] (**a** and **b** glides in planes parallel to the chains). Experiments have been conducted in a Griggs-Blacic solid medium deformation apparatus in T, P conditions lying in both S and K fields. The stress strain curves are reported on fig.5. It can be seen on it that sillimanite is quite strong at moderate temperature (say below 700°C). The flow stress does not seem to be significantly temperature dependent (run A). At experimental conditions it is comparable to that of dry quartz[19] (which originates in the necessity for the mobile dislocations to break the strong Si-O bonds).

TEM investigations on deformed samples show contrasting dislocation structures.

(i) For samples deformed at high temperature (900°C, samples B and D) recovery processes are very active and the dislocation structures mostly consist of small cells separated by subgrain boundaries. Within the cells the dislocation density is generally low. In some misoriented regions **c** glide in (100) and (010) has been activated. These dislocations are not dissociated. Of course one also observes (010) [100] glide which has a Schmid factor $\phi = 0.5$. **a** dislocations exhibit a preferred edge character and are not dissociated. **b** dislocations have not been observed although $\phi = 0.5$ for (100) [010].

(ii) In samples A and C the predominant feature is the occurrence of **a** dislocations widely dissociated into colinear partials **a**/2 in (010). Both partials are straight and parallel to **c** (edge orientation, fig.6). The dissociation width reaches 0.2 μm in sample A and 20 μm in sample C. Again on time one notes that dissociation widths drastically change with the T, P deformation conditions. However the (010) 1/2 [100] fault considered here does not verify the criteria previously proposed. The fault energy which apparently is high in the S field drastically decreases near the S-K boundary. This suggests that the fault structure could be very similar to an elementary layer of kyanite. Dissociation of **a** dislocations would thus be the premonitory stage of a strain induced transformation, the new polymorph being nucleated on the lattice defects. Fig.7 is a tentative representation of this transformation. One sees that this transformation is not simple and Si-O bonds must be broken. This could explain why the flow stresses of A and C samples which are governed by the low mobility of partial **a**/2 dislocations are comparable to the flow stress of dry quartz.

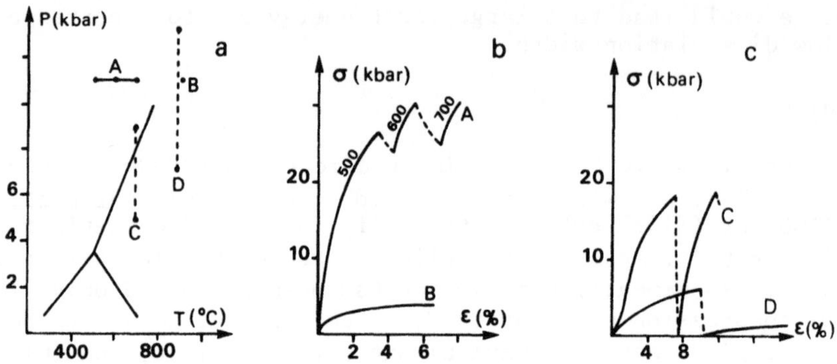

Fig.5. Experimental deformation of sillimanite ; compression axis //[110]. (a) T, P conditions for the four studied samples. (b) stress-strain curves for samples A and B. Run A is a succession of three runs at the same P=11 kbar but at T=500, 600 and 700°C respectively. Between two runs a stress relaxation (broken line) allows recovery of the dislocation structure. The flow stress does not decrease as T increases. In contrast sample B deformed at similar $\dot{\varepsilon}$ and P but at T=900°C shows a flow stress which is an order of magnitude lower. (c) stress-strain curves for samples C and D. The tests consist of two successive runs at constant T but at different P such that the TP conditions move from the S field (1st run) to the K one (2nd run). In sample C the deformation is followed by a stress relaxation at the final T, P conditions which results in a very slight deformation (0.5 % at an approximately constant strain rate $\dot{\varepsilon} \simeq 8\ 10^{-8} s^{-1}$ and a constant stress $\sigma \cong 19$ kbar).

The complicated deformation of sample C has been made with a view to test the possibility of a strain induced transformation. After a first deformation in S field, P was increased, thereby bringing the sample into the K field. After a subsequent shortening of ≃4% this sample underwent a final 15^H stress relaxation before quenching under stress. In TEM one observes very widely dissociated **a** dislocations (up to 20 μm). Furthermore some thin lamellae of kyanite are nucleated on the faults (fig.6d). Diffraction patterns clearly show the epitaxial relation $(100)_K // (2\bar{1}0)_S$ as suggested by the model fig.7. These observations are to be compared with the ones on sample D which had a similar history but at higher T. It was first deformed in the S field, at conditions far from the S-K boundary (i.e. at conditions where the **a** dislocations are not dissociated and the resultant flow stress σ was low ; at this stage of its deformation history, sample D should exhibit a dislocation structure similar to the one of sample B). P was then increased at constant T bringing conditions

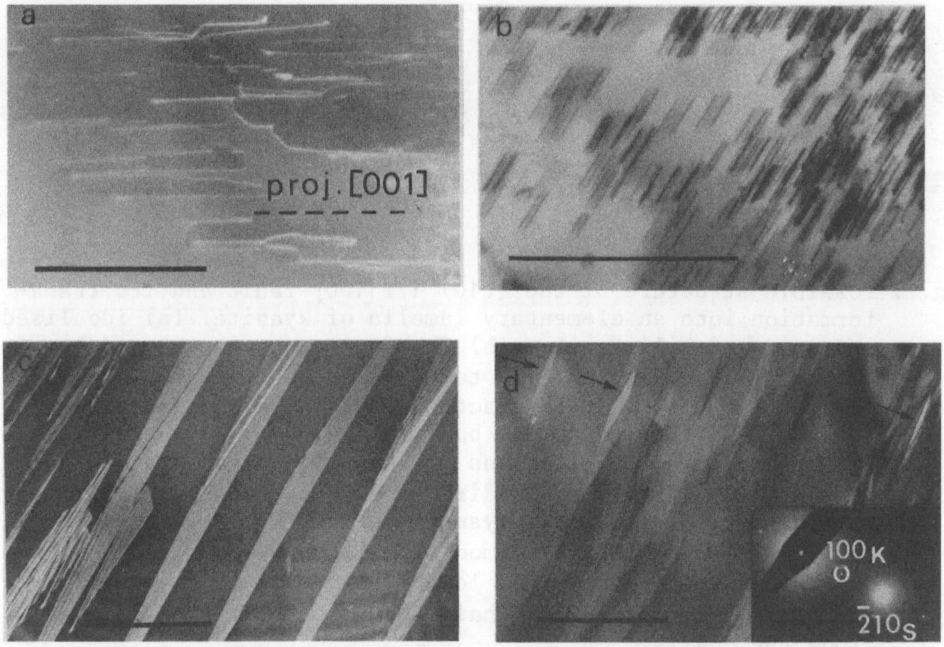

Fig.6. Defect structures in deformed sillimanite. (a) sample B (T=900°C) the **a** dislocations have a marked edge character but they are not dissociated. (b) sample A (T=700°C), the **a** dislocations are dissociated. The **a**/2 partials are straight and // [001] (edge character). (c) sample C (T=700°C, slightly redeformed in K field). The **a** dislocations are widely dissociated (up to 20 μm). (d) sample C ; thin lamellae of kyanite grown on some defects. They are imaged with their $(100)_K$ diffraction spot

near the phase boundary (in principle in K field but the phase diagram is not well known at these T and P values). Upon subsequent deformation of ≃10% the flow stress σ remained low. In TEM one observes a typical climb controlled deformation configuration. The a dislocations are not dissociated. Comparison of samples C and D thus shows that (i) **a** dislocations become widely dissociated only in the case of glide deformation at T, P values near the S-K boundary and at moderate temperature (at T=900°C and $\dot{\varepsilon} \cong 10^{-6} s^{-1}$ deformation appears to be governed by climb) ; (ii) the occurence of widely dissociated **a** dislocations considerably enhances the S→K transformation (or at least its nucleation stage). These features are the ones generally associated with martensitic transformations[20]. However, in the present case the transformation remains very slow and no transient mechanical weakness is observed in contrast with most cases of transformation plasticity[21,22]. We believe that both phenomena are related to a dislocation core effect. Indeed as the **a** dislocations become dissociated, they suffer a much higher Peierls friction

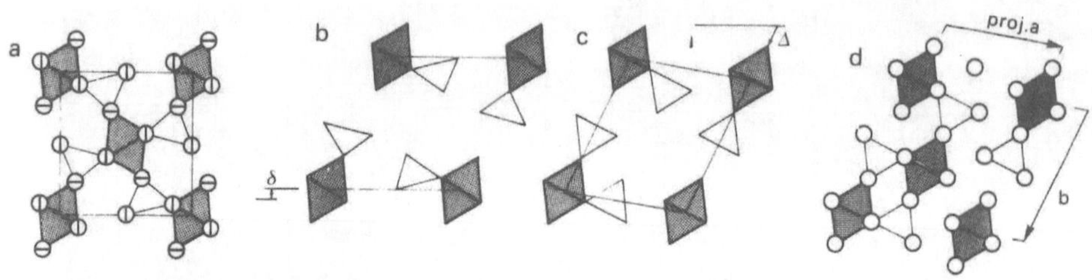

Fig.7. Possible structure of the (010) 1/2[100] fault and its transformation into an elementary lamella of kyanite. (a) idealised structure of sillimanite. (b) The fault runs in the middle of the cell. It breaks no SiO_4 tetrahedron but the central chain is destroyed. A small contraction $\delta/b \simeq 4\%$ has been added to fit the K lattice parameter b_K=7.85 Å. (c) the previous situation is modified by rotations of the SiO_4 tetrahedra and by a shear Δ along the $[2\bar{1}0]_S$ direction to fit the other lattice parameters. (d) idealised kyanite structure. Comparison with (c) shows that two further modifications have to be done (i) one Si has to move from the lower left chain to the lower right and (ii) both right chains have to be shifted of ≃ c/5 along the **c** axis.

because they have to cut the strong Si-O bonds. A drastic decrease of their mobility results. At a given stress level, the strain rate (which governs the transformation rate) also decreases. Furthermore there is no obvious mechanism for generating periodic shearing by partial dislocations in the (010) plane. The growth of the new phase is thus very slow when compared to martensitic transformation in steels.

MICROSCOPIC MODEL OF THE STRAIN INDUCED SILLIMANITE-KYANITE TRANSFORMATION

In metals, periodic shearing by partial dislocations is assumed to be achieved by cross-slip or by rotation of a mobile partial around a fixed screw dislocation perpendicular to the glide plane[23]. Such processes do not operate in the present case and another mechanism is to be considered : ex - nihilo nucleation of a small loop of partial dislocation just above an already faulted plane. Let r be its radius, the volume of the new phase increases of $v = \pi r^2 b$. The nucleation free enthalpy ΔG is equal to the self energy of the dislocation loop W_D minus the work done by the external stress W_S and minus the free enthalpy change $v\,\Delta g(T,P) = v\,(g_S - g_K)$ where g_S and g_K are the free enthalpies per unit volume of sillimanite and kyanite respectively

$$\Delta G = W_D - W_S - \pi r^2 b\, \Delta g(T,P) \qquad (1)$$

The self energy of a small dislocation loop is[24]

$$W_D = \frac{\mu a^2}{16} \frac{2-\nu}{1-\nu} r \ln\left(\frac{4r}{\rho} - 2\right) \tag{2}$$

where $a/2$ is the Burgers vector, ρ the core radius (one can take $\rho=a$). μ and ν are the shear modulus and the Poisson ratio of the material considered as an isotropic elastic medium. The work done by the applied stress is

$$W_S = \frac{\sigma}{2} \frac{a}{2} \pi r^2 \tag{3}$$

where σ is the compressive stress (and $\sigma/2$ the resolved shear stress in the (010) loop plane). Taking $x=r/a$, ΔG is written.

Fig.8. (a) Nucleation free enthalpy ΔG of a small dislocation loop of radius r as a function of the parameter B (see text) (b) maximum free enthalpy ΔG^* and critical radius r^* versus B. r^* is almost constant while ΔG^* varies rapidly with B.

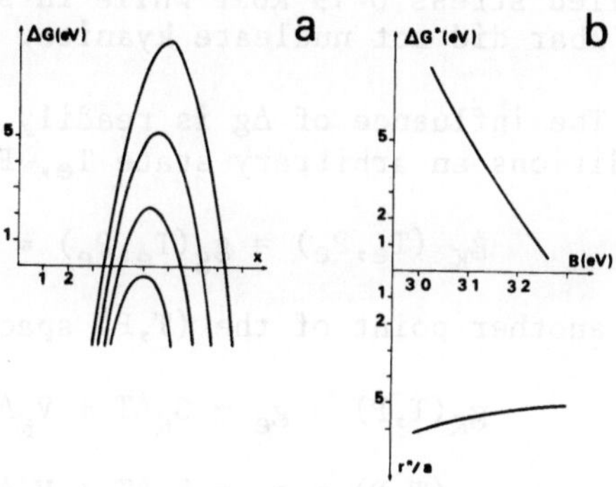

$$\Delta G(T,P,x) = Ax\ln(4x-2) - B(T,P)x^2 \tag{4}$$

The factor A is easily estimated. For (010) 100 glide one can take as a first approximation $\mu=1/s_{66}=40$ GPa*, $\nu=1/3$ and $a=7.5$ Å which leads to $A=16.5$ eV and

$$\Delta G(\text{eV}) = 16.5\ x \ln(4x-2) - bx^2 \tag{4'}$$

* Sillimanite is markedly anisotropic and, depending upon the crystallographic direction, its shear modulus at room temperature varies from 50 to 77 GPa[4]. The temperature variations of the elastic coefficients are not known. In most refractory materials μ decreases by approximately 20% between room temperature and melting temperature. This justifies a 10% decrease of μ at 700°C. We have arbitrarily added a further 10% decrease (leading to $1/s_{66}$ (700°C, 9 kbar) = 40 GPa) because the considered T, P conditions are near a phase transition boundary.

This equation is represented on fig.8 for various B values. ΔG reaches a maximum value ΔG^* for a critical loop radius r^*. Both ΔG^* and r^* depend on B. Their variation is also shown on fig.8. For $\Delta G^*>0$, the probability of nucleating a loop varies as $\exp-(\Delta G^*/kT)$ i.e. the strain induced phase transformation can only occur if ΔG^* is small, say $\Delta G^*<5$ eV. This corresponds to $B\simeq3.1$ eV and $r^*=5a$ i.e.

$$B(eV) = 3.1 = \pi\, a^2\, (\frac{\sigma a}{4} + b\, \Delta g) \qquad (5)$$

On the phase boundary or in its immediate vicinity, Δg is negligible and a compressive stress of $\sigma=4B/\pi a^3=17$ kbar is required to nucleate the small loop. This value is in qualitative agreement with our experimental results. In sample C, kyanite lamellae grew under an applied stress $\sigma\simeq19$ kbar while in sample D a compressive stress < 6 kbar did not nucleate kyanite.

The influence of Δg is readily estimated too, using as standard conditions an arbitrary state T_e, P_e on the phase boundary, one has

$$g_K\,(T_e,P_e) = g_S(T_e,P_e) = g_e \qquad (5)$$

For another point of the (T,P) space i.e. $T=T_e+\Delta T$, $P=P_e+\Delta P$, one gets

$$g_K(T,P) = g_e - S_K\Delta T + V_K\Delta P$$

$$g_S(T,P) = g_e - S_S\Delta T + V_S\Delta P$$

$$\Delta g = g_S - g_K = -(S_S - S_K)\Delta T + (V_S - V_K)\Delta P \qquad (6)$$

As a rough approximation one can neglect the variations with T and P of the S's and V's and use their values in normal conditions which are[25]

$S_S=96.11 \pm 0.42$ J mole^{-1}K^{-1} ; $V_S=4.99 \pm 0.004$ J mole^{-1}bar^{-1}

$S_K=83.76 \pm 0.34$ J mole^{-1}K^{-1} ; $V_K=4.41 \pm 0.007$ J mole^{-1}bar^{-1}

Noting that there are four Al_2SiO_5 molecules per unit cell, these values lead to

$$\Delta g(eV\,\text{Å}^{-3}) \simeq 1.5\ 10^{-6}\ \Delta T + 0.7\ 10^{-4}\ \Delta P\ (kbar) \qquad (7)$$

So Δg is negligible when compared to σ in eq.(4). Thus for the investigated range of stress-strain curves, the kyanite growth process proposed here can only operate with compressive stresses of the order of 17 kbar.

CONCLUSION

The remarkable feature in the plasticity of Al_2SiO_5 polymorphs is the influence of the T, P deformation conditions on the dissociation width of the dislocations. Such a change in their core structure can strongly affect the plastic properties of these minerals. In andalusite the flow stress increases with increasing temperature. In sillimanite the a dislocations become widely dissociated near the S-K boundary. A larger flow stress results because the partials suffer a much larger Peierls friction than the perfect dislocations. Under high compressive stresses, the very large widening of these dislocations can even induce a pseudo martensitic (strain induced) transformation into kyanite.

ACKNOWLEDGMENT

The authors thank Dr E.GREW, UCLA for providing the sillimanite single crystals. They also thank Prof. W. BENOIT, Lausanne for helpful discussions on the microscopic model of strain induced transformation.

REFERENCES

1. E. An Zen, The stability relations of the polymorphs of Al silicates: a survey and some comments; Amer.J.Sci., 267:297 (1969).
2. M.J. Holdaway, Stability of andalusite and the aluminium silicate phase diagram, Amer.J.Sci., 271:97 (1971).
3. H.W. Day & H.J. Kumin, Thermodynamic analysis of the aluminium silicate triple point, Amer.J.Sci., 280:265 (1980).
4. M.T. Vaughan & D.J. Weidner, The relationship of elasticity and crystal structure in andalusite and silimanite, Phys.Chem.Minerals 3:133 (1978).
5. C.W. Burnham, Refinement of the crystal structure of sillimanite, Z. Krist. 118:127 (1963).
6. C.W. Burnham & M.J. Buerger, Refinement of the crystal structure of andalusite, Z.Krist. 115:269 (1961).
7. C.W. Burnham, Refinement of the crystal structure of kyanite, Z. Krist., 118:337 (1963).
8. E.S. Larsen & P.W. Bridgman, Shearing experiments on some selected minerals and mineral combinations, Amer.J.Sci., 36:81 (1938).
9. R.S. Coe & S.H. Kirby, The orthoenstatite to clinoenstatite transformation by shearing and reversion by annealing ; mechanism and potential applications, Contrib.Mineral.Petrol., 52:29 (1975).
10. S.H. Kirby, The role of crystal defects in the shear induced transformation of orthoenstatite to clinanoenstatite, in "Electron microscopy in mineralogy", H.R.Wenk ed. Springer Berlin (1976).
11. C.B. Raleigh, Glide mechanisms in experimentally deformed minerals, Science, 150:739 (1965).

12. J.N. Boland, B.E. Hobbs & A.C. Mc Laren, The defect structure in natural and experimentally deformed kyanite, Phys.Stat.Sol.(a) 39:631 (1977).
13. D. Menard, J.C. Doukhan & J. Paquet, Uniaxial compression of kyanite, Bull.Mineral., 102:159 (1979).
14. A. Lefebvre & D. Menard, Stacking faults and twins in kyanite, Acta Cryst.A, 37:80 (1981).
15. A. Lefebvre, Transmission electron microscopy of andalusite single crystals indented at room temperature, Bull.Mineral., 105:347 (1982).
16. J.C. Doukhan & J. Paquet, Plastic deformation of andalusite single crystal Al_2SiO_5, Bull.Mineral., 105:170 (1982).
17. D. Menard & J.C. Doukhan, Defauts de reseau dans la sillimanite Al O_3-SiO_2, J.Physique, 39:L19 (1978).
18. J.C. Doukhan & J.M. Christie, Plastic deformation of sillimanite Al_2SiO_5 single crystals under confining pressure and TEM investigation of the induced defect structure, Bull.Mineral. 105:583 (1982).
19. L. Heard & N.L. Carter, Experimentally induced "natural" intragranular flow in quartz and quartzite, Amer.J.Sci. 266:1 (1963).
20. C.N.R. Rao & K.J. Rao, Phase transformations in solids, Mc Graw Hill New-York(1978).
21. G.W. Greenwood & R.J. Johnson, The deformation of metals under small stresses during phase transformations, Proc.R.Soc. London, A 283:403 (1965).
22. J.P. Poirier, On transformation plasticity, J.Geophys.Res., 87:6791 (1982).
23. J. Christian, Twinning and martensitic transformations, J.Physique, C7, 35:65 (1974).
24. J.P. Hirth & J. Lothe, Theory of crystal dislocations, Mc Graw Hill, New-York (1968).
25. R.A. Robie, B.S. Hemingway & J.R. Fisher, Thermodynamic properties of minerals and related substances at 298.15°K and 1 bar and at higher temperature, U.S.Geol.Surv.Bull. 452 (1978).

CREEP OF DOPED POLYCRYSTALLINE MAGNESIUM AND ALUMINUM OXIDES

Ronald S. Gordon and Yasuro Ikuma[1]

Department of Materials Science and Engineering
University of Utah
Salt Lake City, Utah

INTRODUCTION

It is well known that under conditions of small stress, fine grain size and enhanced lattice and/or grain boundary diffusion polycrystalline ceramics will deform at elevated temperatures by one or more diffusional creep mechanisms.[1-9] In this paper creep deformation maps will be utilized to illustrate the role of soluble (aliovalent) dopants, grain size, and oxygen fugacity on the relative contributions of (1) lattice diffusion, (2) grain boundary diffusion, (3) interfacial defect reactions at grain boundaries, and (4) power law or dislocation deformation mechanisms on the creep deformation of polycrystalline magnesium and aluminum oxides.

POLYCRYSTALLINE MgO

Three compositions have been studied [1-3]: (1) pure, (2) iron-doped (0.05-5.3 cation % Fe), and (3) hot pressed material using a LiF sintering aid.[3] When LiF processed material is annealed at 1000°C, nearly all of the lithium is lost except for a small residual amount (~75 ppm). Significant amounts (10-40 atomic percent) of fluorine remain at grain boundaries in annealed material. Some residual fluorine (0.002-0.07 anion%) may even remain in the lattice.[3]

---

[1]Present address: Ikutoku Institute of Technology, Atsugi, Japan.

At low stresses and small grain sizes, polycrystalline MgO deforms by diffusional creep mechanisms. Pure (>99.98%) and iron-doped (~0.53%) MgO [2,3] deform by the Nabarro-Herring creep equation, i.e.

$$\dot{\varepsilon}(h^{-1})=44\times10^{14}\Omega_V\sigma D^{\ell}_{Mg}/\pi kT(GS)^2 \quad (1)$$

in which Ω_V is the molecular volume of MgO (1.86×10^{-29} m^3), k is Boltzmann's constant (1.38×10^{-23} Nm/^oK molecule), σ is the stress (MPa), (GS) is the grain size (μm), and D^{ℓ}_{Mg} is the magnesium lattice diffusion coefficient (cm^2/h). For iron doped MgO [2], the cation diffusion coefficient is a function of oxygen fugacity (P_{O_2} in Pa) and iron in solid solution ($[Fe_T]$ in cation%), according to the following equation:

$$D^{\ell}_{Mg}-(D^{\ell}_{Mg})_o=1.38\times10^7\exp[-490{,}000 Jmol^{-1}/RT]P_{O_2}^{1/7}[Fe_T]^{0.85} \quad (2)$$

in which R is the gas constant (8.31J/^oK mol) and $(D^{\ell}_{Mg})_o$ is the cation lattice diffusion coefficient in pure MgO inferred from diffusional creep. The value of $(D^{\ell}_{Mg})_o$ at 1350°C was determined experimentally from diffusional creep studies.[3]

In MgO either processed with the LiF additive and or doped with iron in small concentrations (~0.05 cation%), diffusional creep was a combination of both Nabarro-Herring and Coble creep controlled by magnesium lattice and grain boundary diffusion [1,3], respectively, according to the following relation:

$$\dot{\varepsilon}(h^{-1})=44\times10^{14}\Omega_V\sigma[D^{\ell}_{Mg}+\pi\delta_{Mg}D^{b}_{Mg}/(GS)]/\pi kT(GS)^2 \quad (3)$$

D^{b}_{Mg} is the magnesium grain boundary diffusion coefficient (cm^2/h) and δ_{Mg} is the effective thickness (μm) for grain boundary diffusion.

For MgO [2] doped at higher iron concentrations ($\geq$2.65% Fe), diffusional creep in oxidizing atmospheres is a combination of Nabarro-Herring creep controlled by magnesium lattice diffusion and Coble creep controlled by oxygen grain boundary diffusion, i.e.

$$\dot{\varepsilon}(h^{-1})=44\times10^{14}\Omega_V\sigma[D^{\ell}_{Mg}/(1+GSD^{\ell}_{Mg}/\pi\delta_O D^{b}_{O})]/\pi kT(GS)^2 \quad (4)$$

D^{b}_{O} is the oxygen grain boundary diffusion coefficient (cm^2/h) and δ_O is the effective thickness (μm) for grain boundary diffusion.

In Table 1 representative values of D^{ℓ}_{Mg}, $\delta_{Mg}D^{b}_{Mg}$ and $\delta_O D^{b}_{O}$ are given for various compositions in MgO at 1350°C.

At higher stresses and relatively large grain sizes [12], viscous creep deformation gives way to power law or non-viscous deformation, i.e.:

Table 1. Lattice and Grain Boundary Diffusivities in Polycrystalline MgO at 1350°C

Composition (cation%)	D^{ℓ}_{Mg} (cm^2/h)	$\delta_{Mg}D^{b}_{Mg}$ (cm^3/h)	$\delta_O D^{b}_O$ (cm^3/h)	Reference
Undoped	1.3×10^{-9}	$>10^{-14}$	$>10^{-15}$	3
LiF Processed	4.7×10^{-9}	7.2×10^{-12}	–	3
0.05% Fe	2.8×10^{-9}	7.2×10^{-12}	–	1
0.53% Fe	8.3×10^{-9}	–	–	2
2.65% Fe	3.2×10^{-8}	–	5.8×10^{-11}	2
5.30% Fe	5.4×10^{-8}	–	1.2×10^{-10}	2
Magnesium Tracer	4.1×10^{-9}	–	–	10
Oxygen Tracer	1.4×10^{-13}	–	–	11

Pure MgO

$$\dot{\varepsilon}(h^{-1}) = 1300\sigma^{3.0}\exp[-335{,}000 Jmol^{-1}/RT] \tag{5}$$

Fe-doped MgO

$$\dot{\varepsilon}(h^{-1}) = 31.2\sigma^{3.0}\exp[-297{,}000 Jmol^{-1}/RT] \tag{6}$$

In contrast to the diffusional creep regimes, power law creep in pure and iron-doped polycrystalline MgO is characterized by a higher stress exponent and creep rates which are independent of grain size, oxygen fugacity and iron dopant concentration.

In Fig. 1 creep deformation maps at 1350°C are given for pure (Fig. 1A) and iron-doped (Fig. 1B) MgO along with MgO fabricated with the LiF sintering aid (Fig. 1C). These maps were constructed from Eqs. (1-6) and the data in Table 1. Several interesting features are readily apparent: (1) Diffusional creep regimes are enhanced at low stresses and small grain sizes. (2) Residual fluorine enhances both magnesium lattice and grain boundary diffusion; hence the transition stress to power law creep at a given grain size is increased by factors of two to four depending on the grain size. (3) Trivalent iron in solid solution, which enhances magnesium lattice diffusion via the creation of cation lattice vacancies, significantly increases the transition stress to power law creep.

To gain additional quantitative insight on the factors which promote the dominance of diffusion creep controlled by cation lattice diffusion, the transition stress (σ_T) between Nabarro-Herring and power law creep in iron-doped MgO was derived by substituting Eq. (2) into Eq. (1) (with $D^{\ell}_{Mg} \gg (D^{\ell}_{Mg})_O$), equating to Eq. (6) and solving for σ_T, i.e.

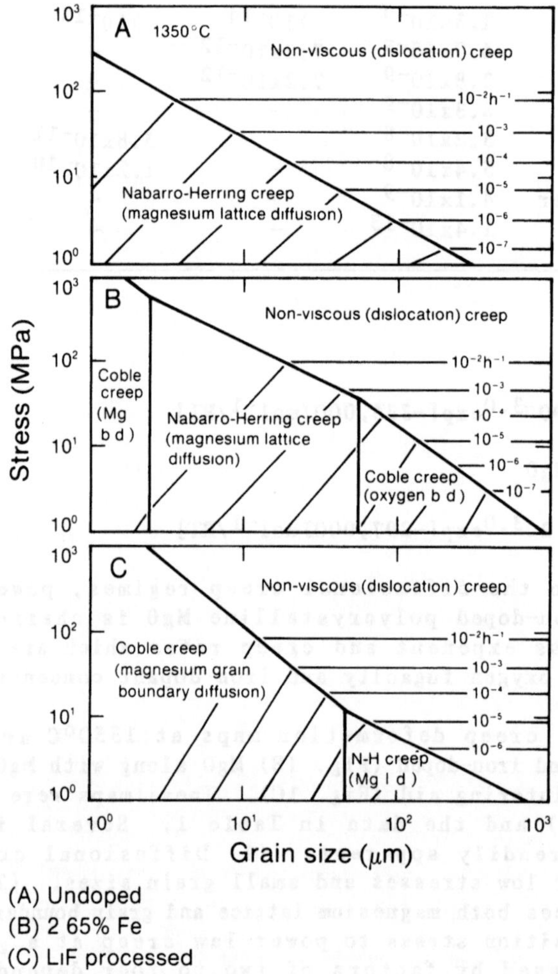

Figure 1: Creep Deformation Maps for Polycrystalline Magnesium Oxide

$$\sigma_T(\text{MPa})=2.49\text{x}10\text{x}^{10}\left(\frac{\Omega_V}{kT(GS)^2}\right)^{1/2}\exp\left(\frac{-96,500}{RT}\right)P_{O_2}^{\ 1/14}(\text{FeT})^{0.43} \quad (7)$$

The effects of iron doping, oxygen fugacity, grain size, and temperature can be determined readily from an examination of Eq. (7). (1) A decrease in the iron dopant concentration by a factor of five will decrease σ_T by a factor of approximately two, (2) A decrease in oxygen fugacity by a factor of 10^7 will decrease σ_T by a factor of three, (3) A decrease in grain size by a factor of ten will increase σ_T by the same amount (a larger increase in σ_T occurs when diffusional creep is controlled by grain boundary diffusion, i.e. $\sigma_T \alpha (GS)^{-3/2}$), and (4) An increase in the temperature from 1350 to 1450°C will increase σ_T by a factor of 1.6.

The power law creep regime with an exponential stress relationship and an absence of any grain size dependence is probably an intragrannular process involving some type of dislocation climb.[12] A process such as this should be rate limited by anion and not cation lattice diffusion. Oxygen diffusion in single crystals of MgO has been reported [11] to be independent of iron doping with an activation energy of 536 kJ mol^{-1}. Power law creep rates were also independent of the iron dopant concentration.[12] Assuming that the power law creep regime is due to Nabarro climb within subgrains formed by dislocations, then

$$\dot{\varepsilon}(h^{-1}) \approx 10^{18}\pi\Omega_V\sigma^3 D_O^{\ell}/10b^2G^2kT \quad (8)$$

In Eq. (8) b is the Burgers vector (~$3\text{x}10^{-10}$m) and G is the shear modulus (~$7\text{x}10^9$ N/m^2). Setting Eq. (8) equal to Eq. (6) led to an estimate of D_O^{ℓ} (oxygen lattice diffusivity) of $2\text{x}10^{-10}$ cm^2/h at 1350°C, approximately three orders of magnitude higher than oxygen tracer diffusion in single crystals with very small dislocation densities (Table 1). Enhanced oxygen lattice diffusion is frequently encountered at these low temperatures (~1350°C) due to the presence of dislocation networks or subgrain-boundaries.[11]

Due to coupling of cation lattice and anion grain boundary diffusion [13,14], as the grain size and/or the cation lattice diffusivity are increased relative to the oxygen grain boundary diffusivity (i.e. increased temperature, oxygen fugacity, and/or iron dopant concentration), as predicted by Eq. (4) diffusional creep becomes dominated by anion boundary transport. Diffusional creep controlled by magnesium grain boundary diffusion is dominant at small iron dopant concentrations, small grain sizes, low temperatures and under conditions where cation grain boundary transport is enhanced (e.g. iron or fluorine doped versus undoped MgO) as predicted by Eq. (3).

POLYCRYSTALLINE Al_2O_3

Several compositions will be examined: (1) pure [5] and MgO saturated [15,16], (2) iron dopant [4,5,8] (0.2-2 cation%), (3) titanium dopant [6,8] (0.05-0.2 cation%), (4) mixed Fe-Ti dopant [8] and (5) mixed Mn-Ti dopant.[7] At all compositions with the exception of the single titanium dopant, diffusional creep in polycrystalline alumina could be described as Nabarro-Herring and/or Coble creep rate limited by aluminum lattice and grain boundary diffusion, respectively, according to a relation similar to Eq. (3) multiplied by a factor, $1/\alpha$. α is a stoichiometric factor equal to 2. Ω_V in this case is the molecular volume of Al_2O_3 (4.25×10^{-29} m^3), D^{ℓ}_{Al}, the aluminum lattice diffusion coefficient, and $\delta_{Al}D^{b}_{Al}$, the aluminum grain boundary diffusion parameter, are the appropriate diffusion terms.

In Table 2, representative values of D^{ℓ}_{Al} and $\delta_{Al}D^{b}_{Al}$ are given at various compositions at 1350°C.

Using activation energies of 628 kJ mol^{-1} and 418 kJ mol^{-1} for aluminum lattice and grain boundary diffusion [5,8], respectively, several creep deformation maps (Fig. 2) were constructed using modified versions of Eq. (3) and the data in Table 2.

Table 2. Lattice and Grain Boundary Diffusivities in Polycrystalline Al_2O_3 at 1350°C

Composition (cation%)	D^{ℓ}_{Al} (cm^2/h)	$\delta_{Al}D^{b}_{Al}$ (cm^3/h)	Reference
Pure	2.5×10^{-10}	–	5
MgO-Saturated	1.4×10^{-10}	9.0×10^{-14}	15,16
0.2% Fe	4.7×10^{-10}	–	8
1% Fe	7.6×10^{-10}	3.5×10^{-13}	8
2% Fe	1.4×10^{-9}	5.0×10^{-12}	8
2% Fe (P_{O_2} ~10^{-1} Pa)	–	2.4×10^{-11}	5
0.05% Ti*	4.3×10^{-9}	–	8
0.05% Mn	5.0×10^{-10}	–	7
0.1% Fe+0.1% Ti	4.0×10^{-9}	–	8
1% Fe+0.05% Ti	4.3×10^{-9}	–	8
Mn+Ti (0.35%)	8.6×10^{-8}	9.1×10^{-11}	7
Aluminum Tracer	5.0×10^{-11}		17
Oxygen Tracer	1.5×10^{-16}		18

*Calculated from creep of 100 μm grain size specimens where interfacial effects are less significant.[6]

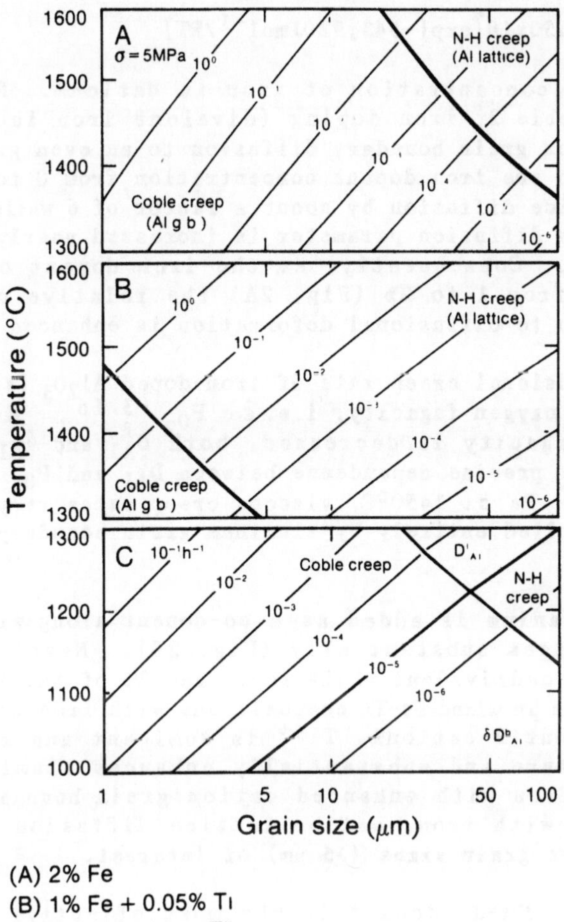

(A) 2% Fe
(B) 1% Fe + 0.05% Ti
(C) ~0.35% (Mn + Ti)

Figure 2: Creep Deformation Maps of Polycrystalline Aluminum Oxide Doped with Transition Metal Oxides

Iron in solid solution enhances D^{ℓ}_{Al} in oxidizing atmospheres [8] (P_{O_2} ~10^5 Pa) according to the relation

$$D^{\ell}_{Al}-(D^{\ell}_{Al})_o=9.29\times10^{10}\exp[-628{,}000\mathrm{Jmol}^{-1}/RT][Fe_T]^{0.90} \quad (9)$$

with the aluminum diffusivity in pure alumina $(D^{\ell}_{Al})_o$ given by

$$(D^{\ell}_{Al})_o=7.50\times10^{7}\exp[-543{,}920\mathrm{Jmol}^{-1}/RT] \quad (10)$$

$(Fe_T]$ is the concentration of iron in cation%. Referring to the data in Table 2, iron doping (divalent iron in particular) enhances cation grain boundary diffusion to an even greater extent. An increase in the iron dopant concentration from 0 to 2% enhances aluminum lattice diffusion by about a factor of 6 while the aluminum grain boundary diffusion parameter is increased nearly two orders of magnitude. Consequently, as the iron dopant concentration is increased from 1 to 2% (Fig. 2A) the relative contribution of Coble creep to diffusional deformation is enhanced considerably.

The diffusional creep rate of iron-doped Al_2O_3 [4] is sensitive to changes in oxygen fugacity, i.e. $\dot{\varepsilon}\ \alpha\ P_{O_2}^{-3/16}$. However, as the oxygen fugacity is decreased, both D^{ℓ}_{Al} and $\delta_{Al}D^{b}_{Al}$ increase [8], thus, the precise dependence between D^{ℓ}_{Al} and P_{O_2} is uncertain. When P_{O_2} ~10^{-1} Pa at 1450°C, viscous creep rates vary with $(GS)^{-3}$ and are controlled entirely by aluminum grain boundary diffusion. [5,8]

When titanium is added as a co-dopant along with iron, the situation changes substantially (Fig. 2B). Nearly 80% of the titanium is quadrivalent while less than 2% of the iron is divalent.[8] Hence in mixed Fe-Ti compositions with iron concentrations less than about 2 cation%, Ti^{4+} is dominant and controls the defect structure and substantially enhances aluminum lattice diffusion. Even with enhanced cation grain boundary diffusion due to doping with iron, cation lattice diffusion is dominant for most of the grain sizes ($\gtrsim$5 μm) of interest.

The mixed Mn-Ti dopant is the most effective in enhancing both aluminum lattice and grain boundary diffusion.[7] The enhancement is so effective that Coble creep is dominant over most of the useful temperature range (1000-1300°C) for grain sizes below about 20 μm (Fig. 2C). Cation lattice diffusion becomes important at larger grain sizes (>50 μm) and at temperatures over about 1200°C.

When titanium is present as a single dopant (0.05-0.2 cation%) [6] creep rates, although significantly enhanced in the limit of large (~100 μm) grain size, are directly proportionalto the stress, vary inversely with the grain size, and are independent of changes in oxygen fugacity. These characteristics are consistent

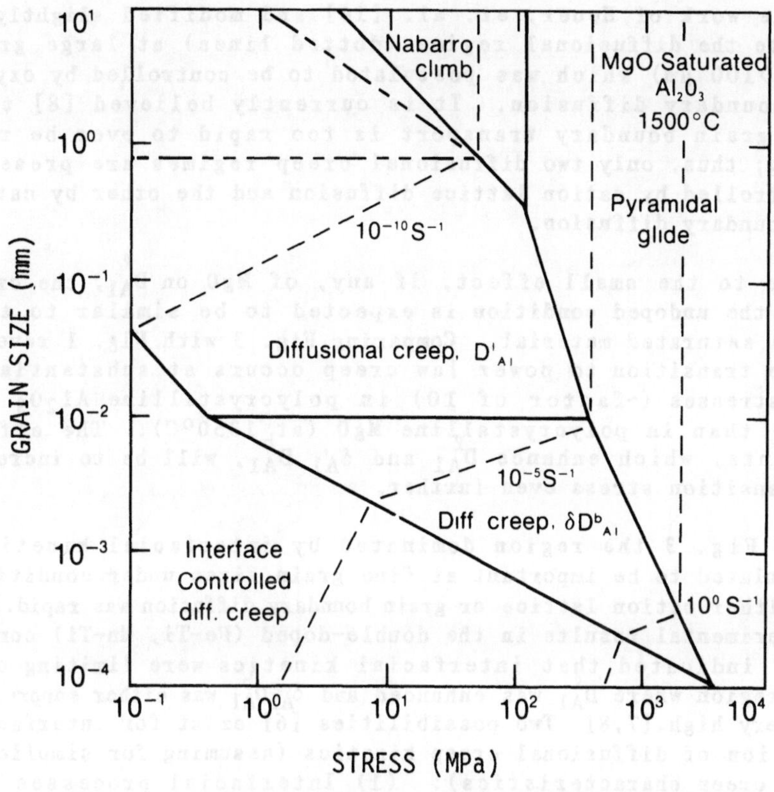

Figure 3: Creep Deformation Map of Polycrystalline Aluminum Oxide Saturated with MgO

with kinetics controlled by the creation and/or annililation of point defects at grain boundaries. Apparently cation lattice diffusion is enhanced to the degree that interfacial processes at grain boundaries become limiting kinetically. In the mixed Fe-Ti compositions even though cation lattice diffusion is also significantly enhanced (Table 2), cation grain boundary diffusion is also enhanced and interfacial processes are not a factor in the overall transport kinetics.[8] In this case, creep rates are rate limited by cation lattice diffusion. This result suggests that interfacial kinetics become limiting under conditions when cation lattice diffusion is enhanced and cation grain boundary diffusion is suppressed (or is not very fast).[6] Since D^{ℓ}_{Al} is about forty times faster in Ti-doped alumina compared to undoped material at 1475°C, interfacial controlled creep would not become important in pure Al_2O_3 until the grain size dropped below about 2 µm (a result in reasonable agreement with the work of Heuer, et.al [16]).

In Figure 3 a creep deformation map at 1500°C is shown for polycrystalline Al_2O_3 saturated with MgO. This map was taken from the work of Heuer, et. al. [16] and modified slightly to eliminate the diffusional regime (dotted lines) at large grain sizes (>100 μm) which was postulated to be controlled by oxygen grain boundary diffusion. It is currently believed [8] that oxygen grain boundary transport is too rapid to ever be rate limiting; thus, only two diffusional creep regimes are present, one controlled by cation lattice diffusion and the other by cation grain boundary diffusion.

Due to the small effect, if any, of MgO on D^{ℓ}_{Al}, the creep map for the undoped condition is expected to be similar to that for MgO saturated material. Comparing Fig. 3 with Fig. 1 reveals that the transition to power law creep occurs at substantially higher stresses (~factor of 10) in polycrystalline Al_2O_3 (at 1500°C) than in polycrystalline MgO (at 1350°C). The effect of dopants, which enhance D^{ℓ}_{Al} and $\delta_{Al}\ D^{b}_{Al}$, will be to increase this transition stress even further.

In Fig. 3 the region dominated by interfacial kinetics[2] was postulated to be important at fine grain sizes under conditions where either cation lattice or grain boundary diffusion was rapid.[16] The experimental results in the double-doped (Fe-Ti, Mn-Ti) compositions indicated that interfacial kinetics were limiting only in the region where D^{ℓ}_{Al} was enhanced and $\delta_{Al}D^{b}_{Al}$ was either suppressed or not very high.[7,8] Two possibilities [6] exist for interfacial limitation of diffusional creep kinetics (assuming for simplicity viscous creep characteristics): (1) Interfacial processes act in series only with cation lattice diffusion and

$$\dot{\varepsilon}(h^{-1})=\frac{44\times10^{14}\Omega V\sigma}{\alpha kT(GS)^2}\left[\frac{D^{\ell}_{Al}K_{Al}(GS)/\pi 44}{D^{\ell}_{Al}/\pi+K_{Al}(GS)/44}\right] \qquad (11)$$

in which K_{Al} is the interfacial rate constant for the creation and/or annihilation of defects at grain boundaries ($K_{Al}\simeq 1.8\times10^{-4}$ cm h^{-1} at 1475°C in Ti-doped alumina, 0.05-0.2% Ti), (2) Interfacial processes act in series with both cation lattice and grain boundary diffusion and

$$\dot{\varepsilon}(h^{-1})=\frac{44\times10^{14}\Omega V\sigma}{\alpha kT(GS)^2}\left[\frac{1}{44/K_{Al}(GS)+1/(D^{\ell}_{Al}/\pi+\delta_{Al}D^{b}_{Al}/(GS))}\right] \qquad (12)$$

Heuer, et. al. [16] postulated that the slightly non-viscous stress exponents ($\dot{\varepsilon}\alpha\sigma^N$ with $1 < N < 2$) were the result of partial limitation of the creep kinetics by interfacial reactions at grain boundaries. Slightly non-viscous stress exponents have always been observed in the work of the present authors [4,5,8]

---

[2]In this particular map [18] a σ^2 dependence was used instead of a linear $\dot{\varepsilon}$-σ relation.

with one notable exception. When the creep kinetics were controlled essentially by cation grain boundary diffusion,[5,7] pure viscous creep behavior was observed. This observation, along with the absence of any interfacial effects in double doped compositions [8] where D^{ℓ}_{Al} and $\delta_{Al}D^{b}_{Al}$ were very high, led to the postulate that interfacial reactions act in series only with lattice diffusion[6] according to Eq. (11). However, if dopants (e.g. Fe, Mn) can enhance interfacial kinetics to a greater degree than either cation lattice or grain boundary diffusion, then the assumption invoked in the construction of the creep map in Figure 3 would be valid and Eq. (12) would be governing.

SUMMARY

Creep deformation maps have been used to illustrate the variables which control the relative contributions of diffusional and intragrannular power law creep in pure and doped polycrystalline MgO and Al_2O_3. The transition stress between viscous and power law creep is sensitive to material type, grain size, oxygen fugacity and dopant concentration. Kinetic limitation of diffusional creep rates by defect creation and/or annihilation at grain boundaries is important when (1) lattice diffusion is significantly enhanced relative to grain boundary diffusion and/or (2) the grain size is very small.

ACKNOWLEDGMENT

This work was supported by the Department of Energy under Contract No. EY-76-S-02-1591.

REFERENCES

1. G. R. Terwilliger, H. K. Bowen and R. S. Gordon, "Creep of Polycrystalline MgO-Fe_2O_3 Solid Solutions at High Temperatures", J. Am. Ceram. Soc. 53 (5) 241-251 (1970).
2. R. T. Tremper, R. A. Giddings, J. D. Hodge, and R. S. Gordon, "Creep of Polycrystalline MgO-FeO-Fe_2O_3 Solid Solutions", J. Am. Ceram. Soc., 57 (10) 421-428 (1974).
3. J. D. Hodge and R. S. Gordon, "Grain Growth and Creep in Polycrystalline Magnesium Oxide Fabricated with and without a LiF Additive", Ceramurgia International, 4 (1) 17-20 (1978).
4. G. W. Hollenberg and R. S. Gordon, "Effect of Oxygen Partial Pressure on the Creep of Polycrystalline Al_2O_3 Doped with Cr, Fe or Ti", J. Am. Ceram. Soc., 56 (3) 140-147 (1973).
5. P. A. Lessing and R. S. Gordon, "Creep of Polycrystalline Alumina, Pure and Doped with Transition Metal Impurities", J. Mater. Sci., 12 (11) 2291-2302 (1977).

6. Y. Ikuma and R. S. Gordon, "Role of Interfacial Defect Creation - Annihilation Processes at Grain Boundaries on the Diffusional Creep of Polycrystalline Alumina" in Surfaces and Interfaces in Ceramic and Ceramic Metal Systems, J. Pask and A. Evans. Ed., pp. 283-294, Plenum Press N.Y. (1981).
7. Y. Ikuma and R. S. Gordon, "Enhancement of the Diffusional Creep of Polycrystalline Al_2O_3 by Simultaneous Doping with Manganese and Titanium", J. Mater. Sci., 17 2961-2967 (1982).
8. Y. Ikuma and R. S. Gordon, "Effect of Doping Simultaneously with Iron and Titanium on Creep of Polycrystalline Al_2O_3", J. Am. Ceram. Soc., 66 (2) 139-147 (1983).
9. J. D. Hodge, P. A. Lessing and R. S. Gordon, "Creep Mapping in a Polycrystalline Ceramic: Application to Magnesium Oxide and Magnesiowustite" J. Mater. Sci 12 1598-1604 (1977).
10. B. J. Wuensch, W. C. Steele and T. Vasilos, "Cation Self Diffusion in Single Crystal MgO", J. Chem. Phys., 58 (12) 5258-66 (1973).
11. (a) Y. Oishi, Ken Ando, H. Kurokawa and Y. Hiro, "Oxygen Self-Diffusion in MgO Single Crystals", J. Am. Ceram. Soc., (in press).
(b) Ken Ando, Y. Kurokawa and Y. Oishi, "Oxygen Self-Diffusion in Fe-Doped MgO Single Crystals", J. Chem. Phys. (in press).
12. Paul A. Lessing and Ronald S. Gordon, "Impurity and Grain Size Effects on the Creep of Polycrystalline Magnesia and Alumina" in Deformation of Ceramic Materials, Edited by R. C. Bradt and R. E. Tressler, pp. 271-296, Plenum Press, N.Y. (1975).
13. R. S. Gordon, "Mass Transport in the Diffusional Creep of Ionic Solids", J. Am. Ceram. Soc., 56 (3) 147-152 (1973).
14. R. S. Gordon, "Ambipolar Diffusion and Its Application to Diffusion Creep" in Mass Transport in Ceramics, A. R. Cooper and A. H. Heuer, Eds., pp. 445-464, Plenum Press, N.Y. (1975).
15. R. M. Cannon and R. L. Coble, "Review of Diffusional Creep of Al_2O_3", pp. 61-100 in Deformation of Ceramic Materials, Edited by R. C. Bradt and R. W. Tressler, Plenum, N.Y. 1975.
16. (a) R. M. Cannon, W. H. Rhodes, and A. H. Heuer, "Plastic Deformation of Fine Grained Alumina (Al_2O_3): I, Interface-Controlled Diffusional Creep" J. Am. Ceram. Soc. 63 (1-2) 46-53 (1980).
(b) A. H. Heuer, N. J. Tighe and R. M. Cannon, "Plastic Deformation of Fine Grained Alumina (Al_2O_3): II, Basal Slip and Nonaccomodated Grain-Boundary Sliding" J. Am. Ceram. Soc. 63 (1-2) 53-58 (1980).
17. A. E. Paladino and W. D. Kingery, "Aluminum Ion Diffusion in Aluminum Oxide", J. Chem. Phys., 37 (5) 957-62 (1962).
18. (a) Y. Oishi and W. D. Kingery, "Self-Diffusion of Oxygen in Single Crystal and Polycrystalline Aluminum Oxide", J. Chem. Phys., 33 (2) 480-86 (1960).
(b) K. P. R. Reddy and A. R. Cooper, "Oxygen Diffusion in Sapphire", J. Am. Ceram. Soc., 65 (12) 634-38 (1982).

SUPERPLASTIC FLOW IN CERAMICS ENHANCED BY A LIQUID PHASE

R. Raj, R.L. Tsai[†], J.-G. Wang[*] and C.K. Chyung[††]

Department of Materials Science and Engineering
Bard Hall
Cornell University
Ithaca, New York 14853

ABSTRACT

Grain boundaries are paths for rapid solid-state diffusion in most polycrystalline materials. If a liquid phase is segregated to the grain interfaces then the diffusivity is further enhanced, in some cases by several orders of magnitude, provided that the crystal is soluble in the liquid. In stressed polycrystals the enhanced rate of matter transport can lead to unusually fast strain-rates, which may render ceramics amenable to superplastic forming. We show here that in fine grained glass ceramic materials a near ideal superplastic behavior, charaterized by strain rate sensitivity of one, is obtained. The ductility of these ceramics is found to be limited by the nucleation and growth of cavities at grain boundaries, or by the propagation of cracks from the surface in tensile testing. Nevertheless tensile elongations greater than one hundred percent are easily obtained. The results from these experiments are described. In addition a theoretical basis for predicting superplastic behavior in fine grained ceramics is developed.

1. INTRODUCTION

Liquid phase sintering of ceramic materials is frequently practiced for two reasons. First, solid state sintering of

†Now with Hewlett Packard Computer Division, 3404 East Harmony Road, Fort Collins, CO 80525. *Now with Motorola Incorporated, 3501 Ed Blustin Boulevard, Austin, TX 78721, ††With Corning Glass Works, Corning, NY 14830.

ceramics often requires very high temperatures where the species may be volatile and where grain growth may be excessive. Second, in most ceramics an intergranular liquid phase is easily obtained by the addition of alkaline and alumino silicates. Silica frequently forms a eutectic liquid phase with most ceramics with a melt temperature much lower than the melt temperature of the pure crystal phase. Thus, in the presence of the liquid phase sintering becomes possible at relatively low temperatures.

The mechanism of liquid phase sintering, first proposed by Kingery[1], pointed out the importance of surface tension of the liquid, which was assumed to form at meniscus at the junction of two spheres, in providing a driving force for sintering. Recognition of the fact that the liquid phase may also enhance the self diffusivity along grain interfaces was recognized first in the creep literature[2] rather than in the sintering literature. It is now clear that the most significant aspect of liquid phase sintering is the enhanced rate of matter transport along the grain interfaces promoted by the liquid phase.

The first models[3] for diffusion creep, enhanced by the liquid phase, simply substituted the coefficient for boundary diffusion, δD_b, in the Coble creep equation[4]

$$\dot{\varepsilon}_e = 44 \frac{\sigma_e \Omega}{kT} \frac{\delta D_b}{L^3} \tag{1}$$

by the viscosity, η, of the liquid phase through the Stokes-Einstein equation

$$\eta = \frac{kT}{6\pi(\delta D_b)_{\text{effective}}} \tag{2a}$$

In equation (1), $\dot{\varepsilon}_e$ and σ_e are deviatoric strain-rates and stress, (defined later) Ω is the atomic volume and L is the "equiaxed" grain size. Such a substitution, however, is simplistic because it does not recognize the importance of solubility of the crystal in the liquid, and it does not clarify the atomistic mechanism by which a liquid promotes crystal dissolution and re-precipitation at two grain junctions. The second question is particularly important in view of the observation that the liquid phase always segregates to the triple grain junctions in polycrystals (although one or two mono-layers of the fluid phase may penetrate the two grain junctions), yet matter must be removed from and added to the two grain junctions to effect deformation. We must also explore the connection between the dihedral angle and the penetration of the liquid to the two grain junctions. In summary the viscosity of the liquid phase, the kinetics of dissolution and precipitation of the crystal from the liquid phase, the solubility of the crystal in the liquid, and the dihedral angle are the factors which are important in the phenomenon of deformation enhanced by a liquid phase. All of these will be reviewed and discussed in this paper.

Once deformation by enhanced matter transport becomes a viable mechanism of deformation it will give a strain-rate sensitivity index, defined as follows:

$$m = \left(\frac{\partial \ell n \sigma_e}{\partial \ell n \dot{\varepsilon}_e}\right)_{\varepsilon_e}, \qquad (2)_b$$

of m = 1. This is equivalent to pseudo Newtonian viscous flow meaning that the polycrystal should deform as if it were a fluid or a glass. The difference is that deformation in the polycrystal is inhomogeneous. Grain boundary sliding leads to stress concentration at triple grain junctions and promotes creep-cavitation. Also, cracks may propagate from the surface along the thin liquid film in the grain boundary or the glass may decompose at the surface at high temperature and develop into cracks, ultimately leading to fracture. Further experimental and theoretical research is needed in understanding the relative importance of these failure mechanisms. In the present paper only the creep-cavitation failure mechanism is discussed since that has become reasonably well understood in recent years.

In the following sections models for deformation and fracture in ceramic polycrystals which contain some liquid phase are described. Next, results from experiments with two β-spodumene lithia-alumino-silicate glass ceramics will be presented. Rather than attempting to give a wide literature review we will discuss mostly our own work in this field. Perhaps this is a one sided view of the subject but it will allow for a unified presentation of the concepts and the phenomenon.

2. MICROSTRUCTURE

An idealized microstructure of a polycrystal with a liquid phase segregated at the triple grain junctions is shown in Fig. 1. The geometry of this microstructure is completely defined by three parameters: L, the grain size, θ, the half-dihedral angle, and either u, the radius of the interstitial pocket at the triple junction, or r, the radius of the interfaces which form that pocket. It can be shown[5] that u and r are related through the equation:

$$u = r \left[\mathrm{Sec}\, \frac{\pi}{6}\, \mathrm{Cos}\theta - 1 \right] \tag{3}$$

while the volume fraction, v_g, of the liquid phase in the polycrystal is given by:

$$v_g = \frac{8r^2}{\pi L^2} \left\{ 2\cos^2\theta \mathrm{Sin}\, \frac{\pi}{3} - \frac{\pi}{2} + 3\left(\theta - \mathrm{Sin}\, \frac{2\theta}{2}\right) \right\} \tag{4}$$

Note that θ is a thermodynamic material parameter which depends on the energy of the crystal-crystal interface, γ_b, and the crystal-liquid interface, γ_ℓ, in the following manner:

$$\mathrm{Cos}\theta = \frac{\gamma_b}{2\gamma_\ell} \tag{5}$$

Equation (4) is valid only when the volume fraction of the glass is small enough that the perturbation from four grain junctions is negligible. This should be the case if $v_g < 0.1$.

The change in the dihedral angle poses some interesting situations. Examples are shown in Fig. 2. The case $\theta=0$ is interesting because then the glass wets the crystal completely. The case $\theta<30°$ is distinctly different from the case $\theta>30°$ because the radius of curvature switches from convex to concave. If $\theta<30°$ then the liquid is interconnected throughout the polycrystal but if $\theta>30°$ it would be isolated into small pockets interspersed along the three grain junctions. The interconnectivity of the liquid phase gives some unique properties to the polycrystal. It gives the fluid a characteristic pore pressure. Thus if the polycrystal is in contact with a resevoir of fluid at a pressure different than the pore pressure then fluid will flow in or out of the polycrystal until the pressures become equal. The deformation response of a polycrystalline aggregate which contains an interconnected fluid depends on the stress applied to the grains in the polycrystal, σ_{ij}, and the pore pressure of the fluid, p_o. A general stress σ_{ij} may be separated into a deviatoric component, σ_e and mean stress, p, which are related to the principal stress components of σ_{ij}, called here as σ_1, σ_2 and σ_3, by the equations:

$$\sigma_e = \frac{\left\{ \frac{1}{2}\left[\left(\sigma_1 - \sigma_2\right)^2 + \left(\sigma_2 - \sigma_3\right)^2 + \left(\sigma_3 - \sigma_1\right)^2 \right] \right\}^{1/2}}{(1 - x)} \tag{6}$$

and

$$p = -\frac{(\sigma_1+\sigma_2+\sigma_3)}{3(1-x)} \tag{7}$$

The quantity x in (6) and (7) is equal to the area fraction where intimate particle-particle contact does not occur, for example the area fraction occupied by a fluid phase will be equal to x. Thus $1/(1-x)$ becomes an average stress concentration factor for loading the solid grains of the aggregate.

The effective pressure acting on the polycrystal is defined by the equation:

$$p_e = p - p_o \tag{8}$$

Some consequences of equation (8) are interesting: (a) If the effective pressure on the aggregate goes to zero then the polycrystalline aggregate will have no cohesive strength and it will disintegrate. (b) A polycrystal, having the microstructure shown in Fig. 1 is held together merely by the surface tension pressure in the fluid. The net driving force (expressed as a pressure, $\hat{p}$) for densification of the polycrystal has been shown to be[6]:

$$\hat{p} = p_e - \frac{\gamma_e}{r} \tag{9}$$

where γ_ℓ is the specific energy of the crystal-liquid interface and r is defined in Fig. 1. At equilibrium $\hat{p}=0$, otherwise the polycrystal would dilate (if $p<0$) or densify (if $p>0$). Substituting from (8) and (9) and requiring equilibrium, i.e. setting $\hat{p}=0$, and assuming that no external stress is applied ($p=0$), we have that $p_o = -\gamma_\ell/r$, meaning that the aggregate is held together merely by the negative pressure induced by surface tension arising from curved surfaces at the triple junctions. Note that the magnitude of p_o increases as v_g decreases, since $v_g \propto r^2$ from equation (4). Another interesting consequence of equation (9) is that the polycrystal will disintegrate if it is in contact with an infinite reservoir of fluid at zero pressure; setting $p_o=0$ and $p=0$, we get that $\hat{p} = -\gamma_\ell/r$ which means that the polycrystal will expand and eventually disintergrate.

3. EQUATIONS FOR TIME-DEPENDENT DEFORMATION

Mechanisms of deformation in polycrystals at high temperatures fall into two categories, one involves the movement of crystal dislocations, and the other the transport of matter by diffusion from interfaces stressed in compression to those in tension. The mechanisms are additive since one does not preclude the other. In ceramics with an intergranular liquid phase, however, the diffusive mechanism is often dominant. We give here the equations derived in earlier papers[6,7]:

$$\left[\dot{\varepsilon}_e\right]_{ic} = \frac{\sigma_e \Omega}{kT} \frac{k_1' \bar{c}}{L} \tag{10}$$

and

$$\left[\dot{\varepsilon}_e\right]_{dc} = 23 \frac{\sigma_e \Omega}{\eta L} \bar{c}\alpha \tag{11}$$

The symbols in equations (10) and (11) are defined in the next paragraph. The subscripts ic and dc represent interface control and diffusional control in the rate of matter transport. As shown schematically in Fig. 3, deformation requires the transport of matter from interfaces such as AB which have a low tensile stress applied to them to interfaces at high tensile stress - such as BC. The transport of matter occurs by the dissolution of the crystal in the fluid, diffusion through the fluid and re-precipitation on the crystal at a different site. Factors which can influence the net rate of matter transport by this process are the atomic roughness of the interface at sites where atoms are removed or deposited, the solubility of the crystal in the fluid and the viscosity of the fluid. If the solubility is low and the viscosity is high then diffusion through the fluid is likely to be rate limiting. If the interface has a low density of ledge sites then the interface reaction may be slower and, therefore, rate limiting. In the instance that diffusional and interface resistance are equally significant then the following equation for the strain rate is obtained:

$$\dot{\varepsilon}_e = \frac{[\dot{\varepsilon}_e]_{ic}\,[\dot{\varepsilon}_e]_{dc}}{[\dot{\varepsilon}_e]_{ic} + [\dot{\varepsilon}_e]_{dc}} \tag{12}$$

The right hand side in (12) reduces to $[\dot{\varepsilon}_e]_{ic}$ if $[\dot{\varepsilon}_e]_{dc} \gg [\dot{\varepsilon}_e]_{ic}$ and vice-versa. It is, therefore, the general form for the flow equation.

The terms $\dot{\varepsilon}_e$ and σ_e, which are scalar definitions of strain-rate and flow-stress, are given in terms of the principal stress and strain-rate components by equation (6):

$$\dot{\varepsilon}_e = \frac{2}{3}\sqrt{\frac{1}{2}\left[(\dot{\varepsilon}_1-\dot{\varepsilon}_2)^2 + (\dot{\varepsilon}_2-\dot{\varepsilon}_3)^2 + (\dot{\varepsilon}_3-\dot{\varepsilon}_1)^2\right]} \tag{13}$$

The other parameters in (10) and (11) are: η = viscosity of the fluid phase, $\bar{c}$ = molar solubility of the crystal in the liquid, $k_1'\bar{c}$ = the lineal growth velocity of a solid liquid interface for a driving force of 1 kT, Ω = atomic volume, L = grain-size, and $\alpha = xh/\Omega^{1/3}$, where x is the area fraction of the interface occupied by the fluid (the statement in reference 7 that it is the area fraction occupied by the islands is incorrect), h is the height of the ledges at the interface. The concept of ledges and islands at the two grain junctions is necessary for two reasons. First, the dissolution and growth of the crystal from an undersaturated or supersaturated liquid occurs by the consummation or the growth of ledges. The need for the islands has been discussed at length in reference 7; the idea is that islands, where the two crystals meet in intimate contact, can support a gradient in normal traction, while the surrounding fluid allows an interconnected path for rapid transport of matter. The viability of liquid phase enhanced diffusional creep is contingent upon the two grain junctions being able to support a traction gradient and also being able to provide a path for enhanced diffusive transport through the fluid phase: the conceptual island structure meets both these requirements.

4. FURTHER DISCUSSION OF THE ISLAND STRUCTURE

Although the island structure of the two grain junctions is so far an experimentally unverified concept, let us assume that is is viable and consider its consequences on deformation behavior. Two important questions arise: (i) Is the island structure likely to exist in thermodynamic equilibrium, is it induced by applied stress or is it simple frozen in by the structural constraints at the interface? (ii) Are the properties of the island structure related to the dihedral angle θ, shown in Fig. 1?

First we consider the possible impact of the points raised above on deformation. If the islands do not exist in thermodynamic equilibrium then steady state deformation will not be possible. Deformation will come to a halt when all the ledge sites at the two grain junctions have been consumed by dissolution and precipitation reaction; the only exception would be if a crystal screw dislocation emanates at the boundary in which case a spiral motion of the interface can continue in steady state. Even if the interface does not contain ledges, the driving force for crystal growth and dissolution, induced by the applied stress, can induce a roughening transition[8]. In this case a threshold stress for deformation would be observed. The relationship between the dihedral angle, θ, and the structure of the interface is intuitive (although a limited amount of data cited below appears to bear it out). If θ=0 then the fluid may penetrate the two grain junction without a cost in free energy[5]. It would be unnatural to expect that the probability of finding fluid in the two grain junction suddenly drops to zero as θ becomes greater than zero. In fact the configurational entropy of the island structure may allow the fluid to exist in equilibrium even when θ>0. Later in the paper

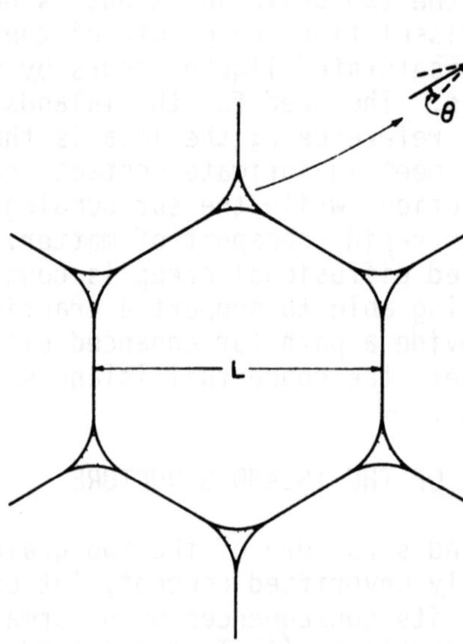

Figure 1. An idealized description of a polycrystal with a liquid phase.

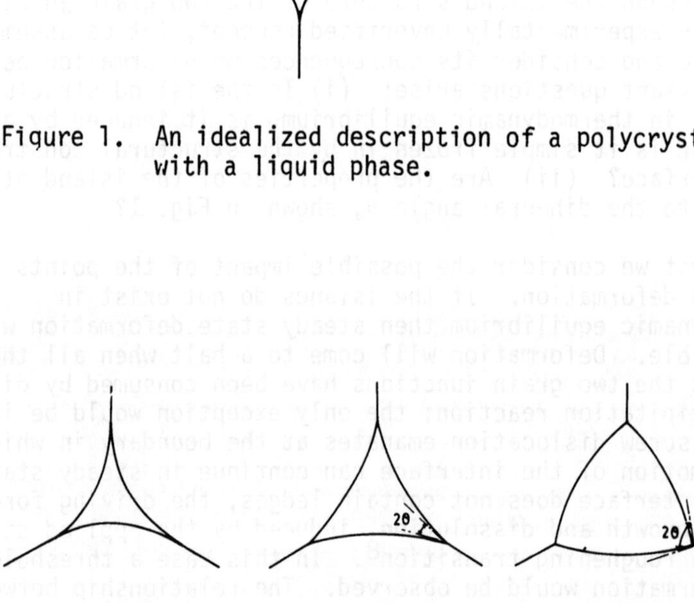

Figure 2. Change in the shape of the three grain junction with a change in the dihedral angle. Note the change in curvature of the interface where θ crosses 30°.

we will attempt to derive a relationship between x and the dihedral angle, expecting that x decreases as θ increases.

Creep data on various glass-ceramic systems provide at least certain trends which reflect some light on the issues raised above. First, so far the data on several systems shows that a threshold stress is not present. Second, steady state deformation leading to several hundred percent strain is possible by the solution/precipitation mechanism. Third, the enhancement in the rate of deformation with and without a liquid phase appears to be dependent on the dihedral angle. These results, which are reviewed in greater detail below, would suggest that the islands exist in thermodynamic equilibrium and that the probability of finding island like structures decreases as the dihedral angle increases.

4.1 Existence of a Threshold Stress

Creep experiments were carried out on three glass ceramic materials at Corning Glass Works. Their composition was:

Table I: Composition of the Glass Ceramics Reported in Figures 4 and 5 and Figure 12

	LAS1	LAS2 (weight percentages)	MAS	9608
SiO_2	69.0	65.7	47.53	69.93
Al_2O_3	20.0	23.5	28.69	17.9
Li_2O	3.5	4.0	---	2.7
MgO	2.0	---	13.45	2.6
Na_2O	---	0.3	---	0.4
ZnO	---	2.6	---	1.0
TiO_2	4.5	---	7.17	4.7
ZrO_2	---	3.5	3.00	---
As_2O_5	1.0	0.4	0.45	0.6

LAS1, LAS2 and 9608 are β-spodumene Li_2O - Al_2O_3 - SiO_2 glass ceramics. The LAS1 and 9608 are nucleated with TiO_2 while ZrO_2 is used as a nucleation catalyst in LAS2. For the same crystallization heat treatment LAS2 has a finer grain size. Both materials contain a glass phase at the triple junction. The

Figure 3. A schematic A schematic for the accrual of deformation by matter transport through the fluid phase (from reference 27).

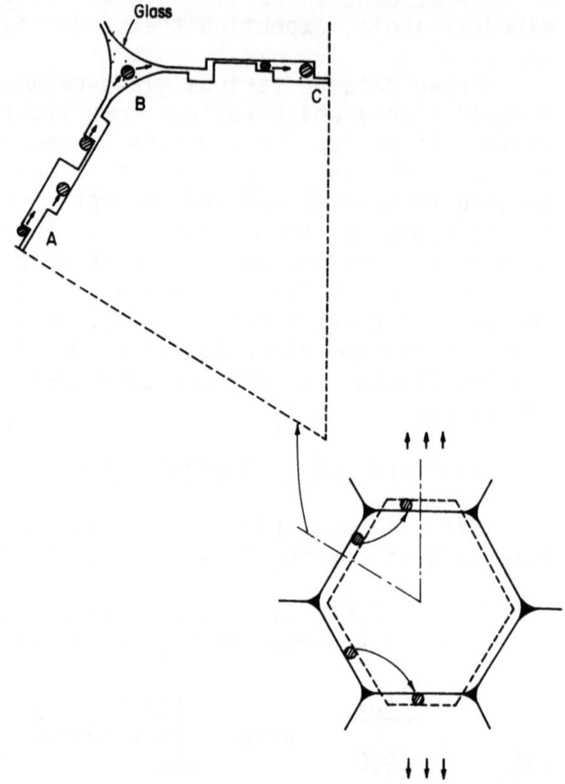

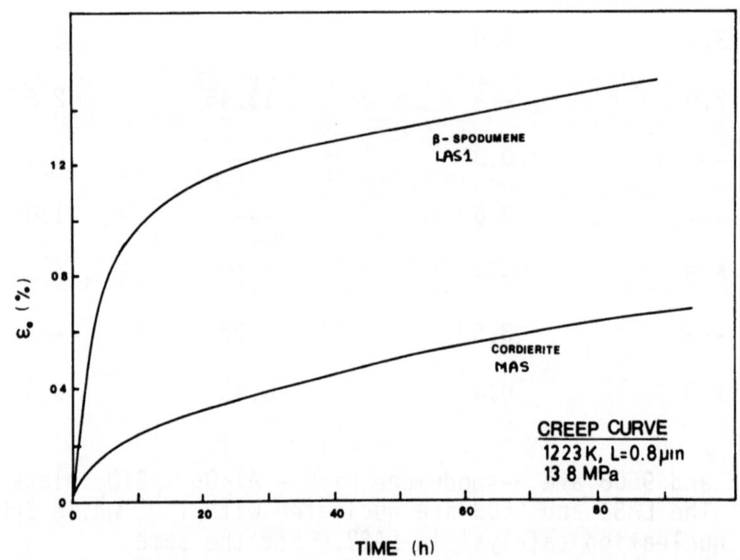

Fiaure 4. Creep curves for glass ceramics LAS1 and MAS.

volume fraction of the glass is about 3% in 9608 and LAS1 and much less in LAS2; the dihedral angle, θ, as defined in Fig. 1 varies between 15 to 30°. The MAS material is a cordierite $MgO - Al_2O_3 - SiO_2$ system. The dihedral angle in MAS is greater, θ being in the range 45° to 60°. The difference in the dihedral angles is significant, because as pointed out in Fig. 2, it means that the fluid phase is interconnected in 9608, LAS1 and LAS2 but is isolated into pockets in MAS.

The remarkable difference in the creep curves for the spodumene and the cordierite systems is in the transient behavior as seen in Fig. 4. The β-spodumene glass ceramic exhibits much greater primary creep strain than does the cordierite glass ceramic. We ascribe this to the interconnectivity of the fluid phase in β-spodumene glass ceramics. Since the creep experiments were done in three point bending, the specimen experienced a mean stress which could have lead to volumetric time-dependent strain. This point is discussed in greater detail in reference 6.

The steady state (deviatoric) creep rate in materials LAS1, LAS2, and MAS, all having a comparable grain size, is plotted as a function of the (deviatoric) applied stress in Fig. 5, at 1223K. Two features of this plot are notable. One that the strain-rate is proportional to stress which is in agreement with equations (10) and (11), and second that the line fitted to the data extrapolates to zero stress at zero strain-rate, that is, the threshold stress is zero. Experiments with 9608 at higher temperature and higher strain-rate[9] also yield a zero threshold stress*. A similar result was obtained in uniaxial compression creep experiments with MgO fluxed hot-pressed silicon-nitride[11,12].

4.2 Influence of Dihedral Angle on the Rate of Creep

Systematic studies are needed to clearly establish the correlation between the dihedral angle and creep-rate. Currently available data merely give a trend, which is not conclusive. Partly, the problem is that crystal-crystal and crystal-liquid interfacial energies can be quite anisotropic leading to a range of dihedral angle in a polycrystal. Another problem is that specimens must be meticulously prepared. Even the slightest pick up of silica and alkaline elements from the surroundings or the furnace can lead to liquid phase effects. Conceptually, at least, even one or two monolayers of the fluid phase may be enough to alter the creep behavior. Yet another point of caution is that

*The usual practice is to plot $\ln(\dot{\varepsilon}_e)$ vs. $\ln(\sigma_e)$ rather than using a linear-linear scale as done in Fig. 5. If a slope of one is obtained on the ln-ln scale then the threshold stress has to be zero because an equation of the form $\dot{\varepsilon}_e = k(\sigma_e - \sigma_o)$ will yield a slope of only if $\sigma_o = 0$.

the validity of the diffusional creep mechanism in the range of experimental variables must be firmly established through stress and grain-size dependence studies, before drawing conclusions from the data.

Although the magnitude of the effect may be questioned, there is strong evidence that an intergranular fluid phase increases the rate of diffusional creep. Two sets of results are cited in Figures 6 and 7 to illustrate the point. In Fig. 6 the creep behavior of reaction bonded silicon, which does not contain a liquid phase, is compared to that of fluxed and hot-pressed silicon nitride[12]. Clearly the fluxed material creeps much faster than the reaction bonded silicon-nitride. Data on deviatoric creep in a polycrystalline aggregate of KCl tested dry, saturated with paraffin oil, and saturated with water are compared in Fig. 7[13]. The dry and paraffin soaked polycrystals behave identically, because the oil does not dissolve KCl. The water saturated polycrystal, however, creeps several orders of magnitude faster. The difference is more remarkable at the lower stresses because the dry material deforms with a stress-exponent of 7.3 (presumably by dislocation mechanisms) while the wet specimen deforms with a stress exponent of about 1 as required by the diffusional transport mechanisms.

In the case of silicon-nitride, the dihedral angle is nearly zero, and the liquid phase enhancement is very pronounced. Similar experiments with polycrystals of olivine, with and without the addition basalt, have shown a much smaller effect on the creep rate. The results of Cooper and Kohlstedt[14] are reproduced in Fig. 8 and we find that the enhancement factor is only about 2. Note that the grain-size dependence of the creep-rate supports the diffusion-controlled solution precipitation mechanism (equation 11). The explanation for the difference between the two sets of results may lie in the dihedral angle, θ, which, on the average, was greater than zero in olivine (about 15° to 30°) but about zero in silicon-nitride. However, the possibility that the olivine polycrystals may have contained some silica as an impurity should also be kept in mind.

4.3 Thermodynamic Stability of the Island Structure

High temperature deformation experiments published so far allow us to make some fairly convincing and some tentative conclusions. It appears probable that there is no threshold stress for the solution/precipitation mechanism, and that large deformations are possible by this mechanism(9). It appears likely that a larger dihedral angle depresses the rate of deformation. If the island structure is indeed real, these observations would imply that it exists in thermodynamic equilibrium and that the relative area fraction of regions of good fit where the crystals are in

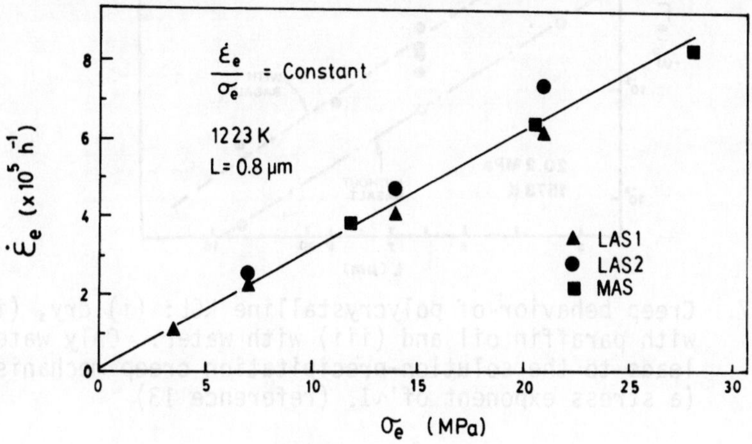

Figure 5. A stress-strain rate plot for three glass ceramics showing linear relationship and a zero threshold stress.

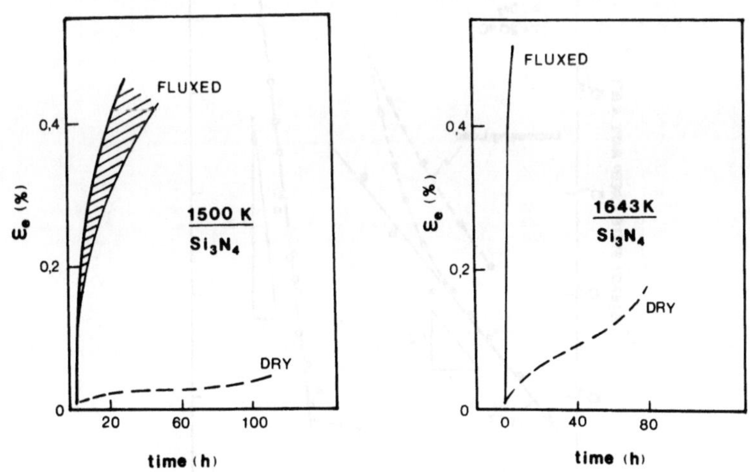

Figure 6. A comparison of creep in Si_3N_4 with and without a liquid phase (reference 12).

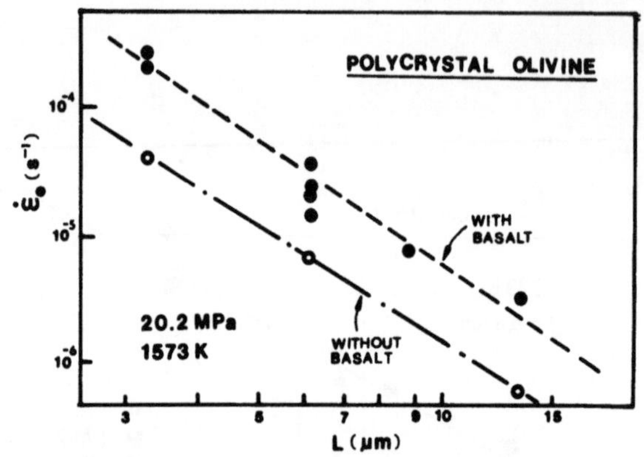

Figure 7. Creep behavior of polycrystalline KCL: (i) dry, (ii) with paraffin oil and (iii) with water. Only water leads to the solution-precipitation creep mechanism (a stress exponent of ∼1. (reference 13)

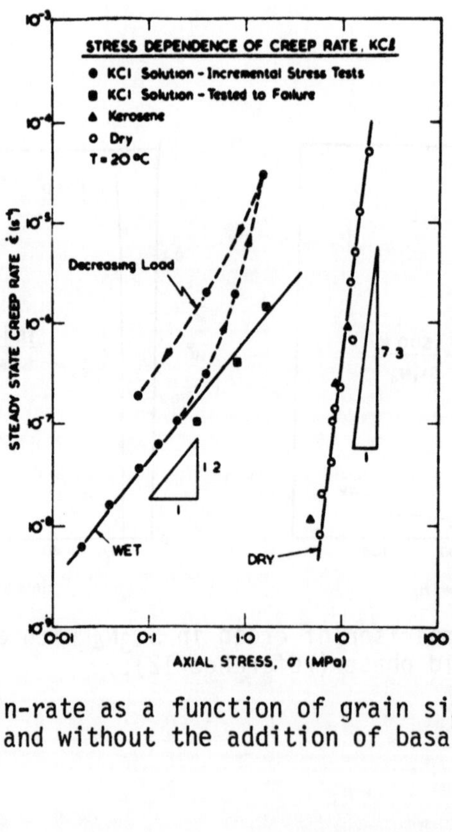

Figure 8. Strain-rate as a function of grain size in olivine, with and without the addition of basalt (from reference 14).

intimate contact (equal to 1-x) and regions of poor fit where the fluid is present (x) is a function of the dihedral angle.

We approach the question of thermodynamic stability of the island structure in terms of two, two dimensional phases, one of them being the crystal phase, where the adjoining crystals come into intimate contact with each other, and the other the fluid phase. We apply the regular solution theory[15,16] to calculate the change in Gibbs Free Energy, ΔG, from the state where an interface continues not fluid phase, to the state where an area fraction x is occupied by the fluid phase.

$$\Delta G = (2\gamma_\ell - \gamma_b)x + W\,x(1-x) + NkT\,[x\,\ell n x + (1-x)\,\ell n(1-x)] \quad (14)$$

In (14), ΔG has units of energy per unit area of the interface, and N is the number of lattice sites per unit area. The first term in the above equation is the cost in energy for inserting the fluid phase into the interface (it goes to zero if the dihedral angle goes to zero, see equation 2), the second term is the heat of mixing and the third term is the entropy of mixing of the two phases. The term W, is defined as:

$$W = Z\,(u_{XY} - \frac{u_{XX} + u_{YY}}{2}) \quad (15)$$

where Z is the coordination number for atoms in the two dimensional structure, and U is the enthalpy of the bond between solid-liquid (XY), liquid-liquid (XX) and solid-solid (YY). If the solid and the fluid phases tend to segregate then $W>0$, but if they prefer to be next to the opposite phase then $W<0$. If the dihedral angle is greater than zero then $W>0$. The third term in (14) is the entropy of mixing.

A composite plot of all three terms in (14) is given in Fig. 9. Note that the entropy of mixing can lower ΔG below zero leading to a thermodynamically stable value for x. Since we are dealing here with an open system, i.e., an infinite resevoir of the fluid phase (present at the three grain junctions) is available to the interface, the stable condition will be given by the point of the lowest free energy in Fig. 9. Note that as the temperature increases, x^*, the equilibrium value of x, also increases. An interesting transition in the curves is when the double minimum in the curves merges into a single minimum; this means that the solid and the liquid phase will be scattered at random rather than being segregated. The temperature at which this occurs can be calculated from equation (14) (We call it the roughening transition temperature, T_r):

$$T_r = \frac{W}{2Nk} \quad (16)$$

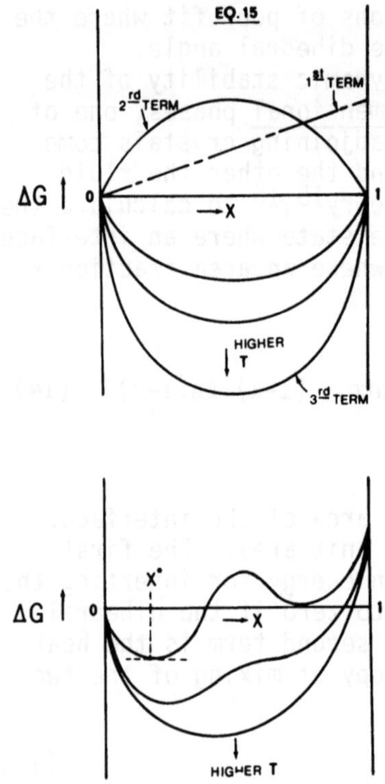

Figure 9.
Top: Curves for the three terms in the Free Energy equation (14); only the entropy term is negative and temperature dependent.
Bottom: The composite curve and the influence of temperature. The composite curve and the influence of temperature. The temperature at which the two minima merge is defined as the roughening transition.

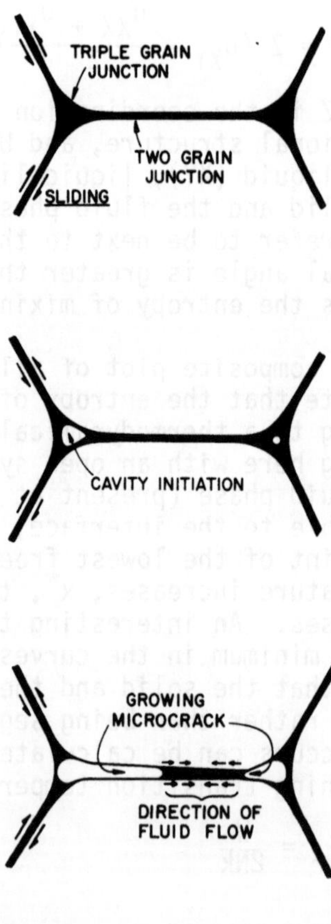

Figure 10.
A model for the nucleation and growth of cavities in the fluid phase. The negative pressure induced by sliding at the triple junction is believed to be the cause of cavity nucleation.

The above analysis has an interesting implication in solution/ precipitation creep. It suggests that interface control should be replaced by diffusion control as the temperature is increased, and that the transition would probably occur at the roughening transition temperature given by equation (17). Useful application of the analysis hinges on the estimate of the parameter W. An attempt to estimate W, even approximately, would be premature at this point given the hypothetical nature of the model. It should be related to the heat of solution of the crystal atoms in the fluid phase. W will be positive if the heat of solution is positive. Even though speculative, the thoughts presented here give some guidance in designing experiments to see if the island structure exists. First a system with a low dihedral angle should be chosen. Second, keeping in mind that the structure may be thermodynamically stable only at high temperature, the specimens should be quenched from the equilibrium temperature.

5. CAVITATION

The model for cavitation first published by Tsai and Raj[17,18] for cavitation and shown schematically in Fig. 10, appears now to be supported by observations in deformation experiments with ceramics containing a liquid phase[19]. The model theorizes that grain boundary sliding induces a hydrostatic tensile stress in the glass pocket at the triple junction. The negative pressure nucleates a cavity which propagates along the two grain junction to form a microcrack. A number of such microcracks may form before the polycrystal suffers total fracture. The growth of cavities and microcracks leads to dilation[20] which can be measured by the change in the density of the specimen[21].

Polycrystals which deform by the solution precipitation mechanism should obey a linear relationship between the stress and the strain-rate. This amounts to Newtonian viscous flow which would lead to infinite ductilities if the growth of a neck were the only mechanism of failure[22]. In fact deformation of polycrystals is inhomogeneous and cavitation at triple junctions may intervene and limit the ductility.

The idea that the suppression of cavitation in glass ceramics should lead to very large ductilities has been discussed in the literature[17,23]. A likely mechanism of cavity suppression is that the stress is relaxed by a dissolution process at the triple grain junction which relaxes the negative pressure induced there by grain boundary sliding. Two mechanisms for relaxation can be envisaged. First, if the fluid is interconnected throughout the polycrystal it can flow from regions of positive pressure to regions of negative pressure to relax the stress. The other mechanism is that the effective volume of the triple junctions increases by the dissolution of the crystal into the liquid phase which neutralizes the negative pressure. Both mechanisms are time dependent, the faster one is likely to occur first. The fluid flow mechanism would depend on the conductance of the channels at the triple junctions through which the fluid must flow. The greater the volume fraction of the channels, the greater the cross-section of the pipes and the faster the stress-relaxation would occur. The observations of Lange et al[21] who found that increasing the fluid volume fraction increased the incidence of cavitation, run contrary to the expectation form this relaxation model, so for the time being we shall not consider it further.

The second relaxation mechanism , which invokes crystal dissolution to relax the local dilation, was analyzed by Tsai[24] and is reproduced here. Consider a change in the size of the glass pocket from u to u+Δu by the dissolution of the crystal into the glass phase. Let $\bar{V}_c$ be the partial molar volume of the crystal in the liquid phase and V_c the molar volume of the crystal phase. Then the net increase in total volume relative to the original volume will be given by:

$$\frac{\Delta v}{v} = 2\,\frac{\Delta u}{u}\,\frac{\bar{V}_c - \bar{V}_c}{V_c} \tag{17}$$

The decrease in the negative pressure, P, as a result of the increase in volume will be given by:

$$\frac{\Delta p}{B_g} = \frac{-\Delta v}{v} \tag{18}$$

where B_g is the bulk modulus of the liquid phase. Substituting from (17) in (18) and differentiating with respect to time we obtain that:

$$\frac{dP}{dt} = -\frac{1}{u}\frac{du}{dt}\cdot B_g\,\frac{(\bar{V}_c - \bar{V})}{V_c} \tag{19}$$

Now

$$\frac{du}{dt} = k_1'\bar{c}\;\frac{P(\bar{V}_c - \bar{V})}{RkT} \tag{20}$$

Since $k_1'\bar{c}$ is the lineal velocity of the dissolution of the interface for a driving force of 1 kT. Here the driving force for dissolution = $P(\bar{V}_c - V_c)$. Combining equations (19) and (20) leads to the result:

$$\frac{dP}{dt} = -\frac{2B_g}{u}\,\frac{(\bar{V}_c - V_c)^2}{V_c}\,\frac{k_1'\bar{c}}{RT}\quad P. \tag{21}$$

which integrates to the equation:

$$P = P_o \exp(-\frac{t}{\tau_R}) \tag{22}$$

where

$$\tau_R = \frac{u}{2B_g} \frac{V_c}{(\bar{V}_c - V_c)^2} \frac{RT}{k_1'\bar{c}} \tag{23}$$

We may substitute for u in terms of the volume fraction of the glass phase, v_g, using equation (4). If we assume that $\theta=0$ then u and v_g are related as follows:

$$u = \left(\frac{v_g}{17.13}\right)^{1/2} L \tag{24}$$

where L is the grain size. Thus we find that the relaxation time, τ_R, becomes shorter as the volume fraction of the glass phase decreases, which is in accordance with the experiments of Lange et al[21].

The ductility of the glass-ceramic will depend on the strain-rate because the stress-relaxation is time dependent. If the applied strain-rate is faster than the stress can relax then fracture would occur. Below a critical strain-rate however, the relaxation would be effective and cavitation would be suppressed. That critical strain rate, $\dot{\varepsilon}^*$, will be approximately given by:

$$\dot{\varepsilon}^* = \frac{\varepsilon_f}{\tau_R} \tag{25}$$

where ε_f is the fracture strain, typically about 0.1.

In summary, the threshold strain-rate for cavitation will be given by:

$$\dot{\varepsilon}^* = \frac{0.2B_g}{RT} \frac{(\bar{V}_c - V_c)^2}{V_c} \frac{k_1'\bar{c}}{u} \tag{26}$$

It is interesting to apply equation (26) to MgO fluxed, hot-pressed Si_3N_4 where data for $k_1\bar{c}$ is available[25]. Substituting $u \sim 20$nm, $B_g = 3.7\times10^4$MPa, $\bar{V}_c - V_c = 0.01\ V_c$ (assumed), $V_c = 0.0731$ nm^3, and $k_1\bar{c} = 1.16\times10^{-12}$ms^{-1}(kT)$^{-1}$ at T = 1673K[25], we find that $\dot{\varepsilon} \sim 1.4\times10^{-7}$s^{-1}. We would expect, therefore, that at strain-rates slower than this Si_3N_4 would flow without cavitation. Note that the activation energy associated with $k_1\bar{c}$ is very high, about 690 kJ mole^{-1} [26]. Therefore the critical strain-rate will increase rapidly with temperature. At higher temperatures, though, and other failure mechanism such as environmental degradation and volatalization of the glass from the surface may become dominant.

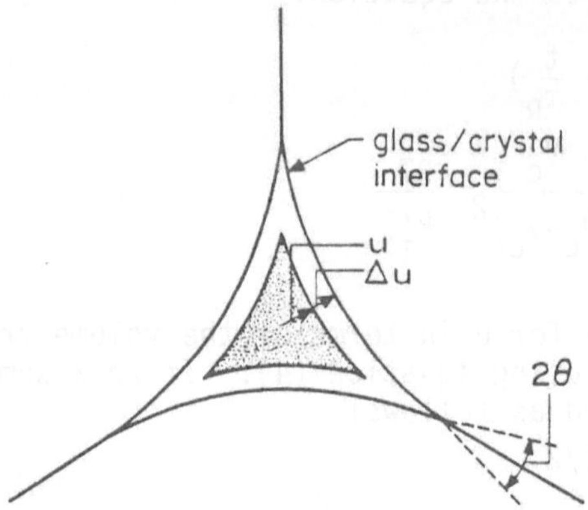

Figure 11. Relaxation of negative pressure by dissolution of the crystal at the triple grain junction.

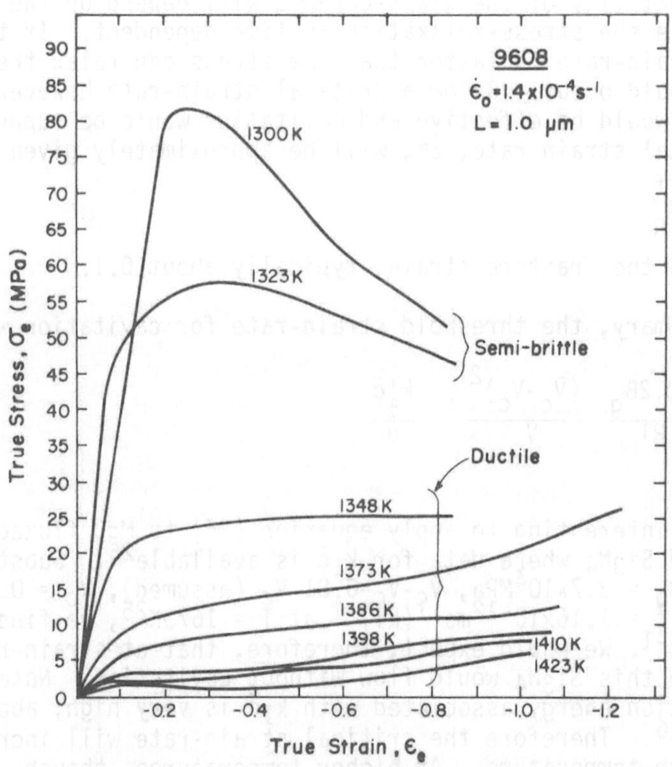

Figure 12. Flow curves as a function of temperature for a β-spodumene glass ceramic (reference 9).

6. EXPERIMENTS WITH TWO β-SPODUMENE GLASS CERAMICS

The results summarized here are being published in detail elsewhere[9,10]. The purpose of this summary is to show i) that under appropriate strain-rate, grain-size and temperature, ceramics which contain a liquid phase can be deformed superplastically, even in uniaxial tension, ii) that the flow stress is intimately related to the viscosity of the liquid phase, as described by equation (11), even though the volume fraction of the glass phase is only a few percent, and iii) that cavities form during deformation but they can be suppressed by the superimposition of hydrostatic pressure.

Two types of β-spodumene glass ceramics with Corning Glass code numbers 9608 and 0329 were tested. The composition of 9608 is given in Table 1. That of 0329 differs in that it contains 2 wt.% B_2O_3. The B_2O_3 lowers the viscosity of the glass phase at the grain boundaries. Thus 9608 contains a glass of high viscosity and 0329 a glass of low viscosity.

The influence of temperature on ductility for 9608 is shown in Fig. 12. We note that at a strain-rate of $1.4 \times 10^{-4} s^{-1}$ the material becomes ductile at temperatures above 1348K. Note the extermely low flow stress of the material. A flow stress of only about 3MPa at 1410 K at such fast strain-rates is quite remarkable. The flow behavior of the material was measured as a function of stress, temperature and the grain-size, and the following relationship was found (9608):

$$\dot{\varepsilon}_e = 2 \times 10^{-5} \frac{\sigma_e \text{ (Pa)}}{L^3 \text{ (m}^3\text{)}} \exp\left(-\frac{635 \pm 35 \text{ kJ mole}^{-1}}{RT}\right) \text{ s}^{-1} \quad (27)$$

for a temperature range of 1373 to 1423K. Parametrically the equation is in near exact agreement with equation (11) suggesting diffusion-controlled solution-precipitation creep as the deformation mechanism. Interestingly, the same material, at lower temperature exhibited interface-controlled deformation[7]. This result is in agreement with the discussion pertaining to the roughening transition (see equation 15), where a switch from interface to diffusion control at higher temperatures was discussed.

Since solubility and viscosity data for the composition of the glass segregated to the triple junctions is not available, the magnitudes of measured and predicted strain rates cannot be compared. The relative flow stress of 9608 and 0329 materials, however, shows the importance of the viscosity of the liquid phase. The relative strain-rate (at a fixed stress) for several different materials is compared in Fig. 13. The 0329 ceramic deforms nearly two orders of magnitude faster than 9608. Other, model lithia-alumina-silicate glass ceramics deform slower than 9608 because the glass phase is relatively free from alkaline impurities. In one case where CaO is deliberately added, the strain rate increases by more than one order of magnitude (data of Bold and Groves and Northover and Groves).

The tensile ductility of the 9608 specimens varied between 45% and 135%. The 0329 exhibited even larger ductilities, presumably because of a lower flow stress. Density measurements revealed a cavitation volume strain of 0.2 to 2% but numerous cracks were also visible on the surface. In order to distinguish between the cavitation and the cracking mode of failure tensile experiments were done under a superimposed hydrostatic pressure in order to suppress cavitation. The results are given in Fig. 14 where the cavitation strain is plotted as a function of the superimposed pressure normalized with respect to the flow stress. The quantities in brackets are the tensile ductilities. The remarkable result in Fig. 14 is that whereas superimposed pressure was successful in abating cavitation, the fracture strain remained unchanged. The conclusion to be drawn is that the dominant failure mechanism was the propagation of flaws from the surface, not the cavitation. (In 0329, cavitation was suppressed and the ductility was infinitely enhanced by the superimposed pressure, suggesting a role for the flow stress in the initiation and propagation of cracks from the surface).

7. SUMMARY

In this review paper, largely of our own work, we have discussed:

i. The morphology and distribution of the liquid phase in grain grain-interfaces of polycrystalline ceramics.

ii. The conditions under which the liquid phase can enhance the diffusional mechanism of deformation by the solution-reprecipitation process. The constitutive equations describing the flow behavior were presented.

iii. The viability of the solution-precipitation mechanism depends on the assumption of an island structure for the two grain junctions consisting of regions of good fit where the crystals are assumed to be in intimate contact and regions which are occupied by the fluid phase.

iv. Theoretical and experimental arguments were given to suggest that the island structure may exist in thermodynamic equilibrium at high temperature. The thermodynamics was

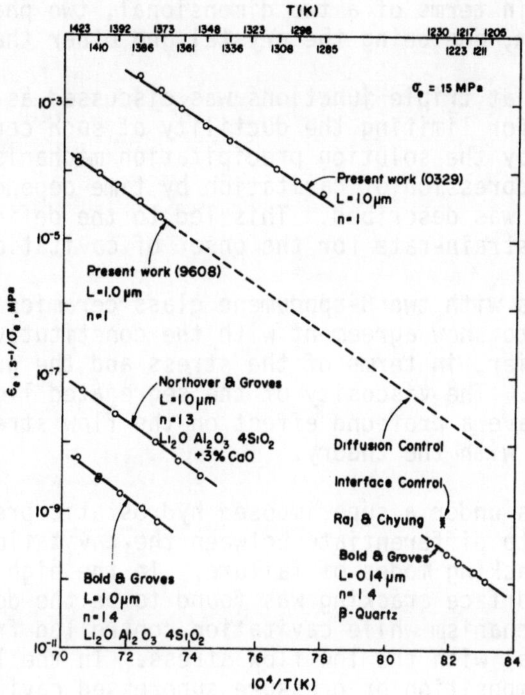

13. Influence of the composition of the glass phase on the flow resistance of polycrystalline β-spodumene (from reference 10).

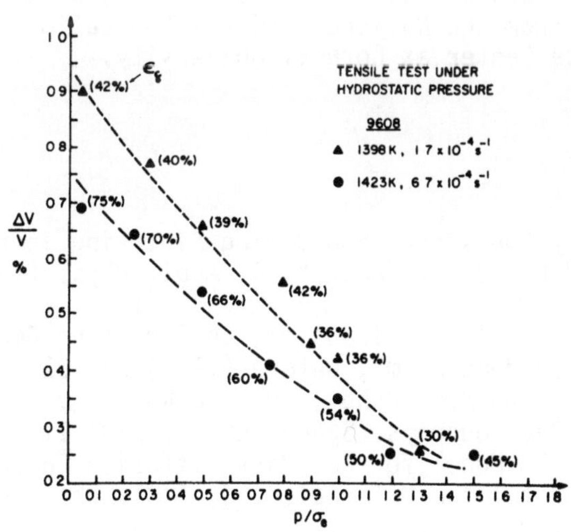

14. Influence of hydrostatic pressure on cavitation (ΔV/V) and fracture strain (in brackets) in 9608 glass ceramic (from reference 10).

discussed in terms of a two dimensional, two phase equilibrium, one being the crystal and other the fluid phase.

v. Cavitation at triple junctions was discussed as a possible mechanism for limiting the ductility of such ceramics when deforming by the solution precipitation mechanism. Criterion for the suppression of cavitation by time-dependent stress relaxation was described. This led to the definition of threshold strain-rate for the onset of cavitation.

vi. Experiments with two β-spodumene glass ceramics were described to show agreement with the constitutive equation given earlier, in terms of the stress and the grain-size dependence. The viscosity of the segregated liquid phase was found to have a profound effect on the flow stress, also in accordance with the theory.

vii. Experiments under a superimposed hydrostatic pressure were described to differentiate between the cavitation and the surface cracking modes of failure. In the high flow stress material surface cracking was found to be the dominant failure mechanism while cavitation controlled fracture in the material with the low flow stress. In the latter case, the superimposition of pressure suppressed cavitation and led to infinite ductility.

ACKNOWLEDGEMENTS

This research was supported by the Department of Energy under Grant De-AC02-77ER04386. J.-G. Wang and R. Raj also received partial support from the National Science Foundation through the Materials Science Center at Cornell University.

REFERENCES

1. W.D. Kingery "Densification During Sintering in the Presence of a Liquid Phase. I. Theory", J. Appl. Phys. (1959) 30 301-306.
2. D.W. Durney, "Solution-Transfer, an Important Geological Deformation Mechanism", Nature (1972) 235 315.
3. R.A. Stocker and M.F. Ashby, "On the Rheology of the Upper Mantle", Rev. Geophys. Space Phys. (1973) 11 391-426.
4. R.L. Coble, "A Model for Boundary Diffusion Controlled Creep in Polycrystalline Materials", J. Appl. Phys. (1964) 34 1679-1684.
5. R. Raj, "Morphology and Stability of the Glass Phase in Glass Ceramic Systems", J. Amer. Ceram. Soc. (1981) 64 [5] 245-248.

6. R. Raj, "Creep in Polycrystalline Aggregates by Matter Transport Through a Liquid Phase", J. Geophys. Res. (1982) 87 [B6] 4731-4739.
7. R. Raj and C.K. Chyung, "Solution Precipitation Creep in Glass Ceramics", Acta Metall. (1981) 29 159-166.
8. J.D. Weeks and G.H. Gilmer, "Dynamics of Crystal Growth: Advances in Chemical Physics" (Editors Prigogine and Rice) (1979) 40 157-118, Wiley, NY.
9. J.-G. Wang and R. Raj, "The Mechanism of Superplastic Flow in a Fine-Grained Ceramic Containing Some Liquid Phase", submitted for publication to Jour. Amer. Ceram. Soc.
10. J.-G. Wang and R. Raj, "Influence of Hydrostatic Pressure and Humidity on Superplastic Ductility of Two β-Spodumene Glass Ceramics", submitted for publication to Jour. Amer. Ceram. Soc.
11. F.F. Lange, B.I. Davis and D.R. Clarke, "Compressive Creep of Si_3N_4-MgO Alloys. I. Effect of Composition", J. Mater. Sci., (1980) 15 [3] 601-610.
12. N.J. Osborne, "Creep Testing of High Temperature Engineering Ceramics" - Proc. British Ceram. Soc. (1975), vol. 25.
13. G.M. Pharr and M.F. Ashby, "On Creep Enhanced by a Liquid Phase", Acta. Metall. (1983) 31 129-138.
14. R.F. Cooper and D.L. Kohlstedt, "Solution Precipitation Enchanced Diffusional Creep of Partially Molten Olivine-Basalt Aggregates During Hot-Pressing", Tectono Phys., in press.
15. K. Denbigh, The Principles of Chemical Equilibrium, Cambridge University Press, 1971, p. 270.
16. R.A. Swalin, Thermodynamics of Solids, John Wiley and Sons, 1972, p. 141.
17. R.L. Tsai and R. Raj, "The Role of Grain Boundary Sliding in Intergranular Fracture of Hot-Pressed Silicon Nitride", J. Amer. Ceram. Soc. (1980) 63 [9-10] 513-517.
18. R.L. Tsai and R. Raj, "Overview No. 18, Creep Fracture in Ceramics Containing Small Amounts of a Liquid Phase", Acta Metall. (1982) 30 1043.
19. J.E. Marion, A.G. Evans, M.D. Drory and D.R. Clarke, "Overview No. 28, High Temperature Failure Initiation in Liquid Phase Sintered Materials", Acta Metall. (1983) 31 [10] 1445-1457.
20. R. Raj, "The Separation of Creep Strain and Cavitation Strain During Deformation", J. Amer. Ceram. Soc., (1982) 65 C-46.
21. F.F. Lange, B. I. Davis and D.R. Clarke, "Compressive Creep of Si_3N_4-MgO Alloys: I. II, III", J. Mater. Sci. (1980) 15 [3] 601-618.
22. E.W. Hart, "Theory of the Tensile Test", Acta Metall., (1967) 15 351.
23. A.G. Evans, J.R. Rice and H.P. Hirth, "Suppression of Cavity Formation in Ceramics, Prospects for Superplasticity", J. Amer. Ceram. Soc. (1980) 63 368.

24. R.L. Tsai, "Studies on High Temperature Mechanical Behavior and Phase Change Kinetics of Magnesium-Silicon-Oxynitride Glass-Ceramics", Ph.D. Thesis, Cornell Univeristy, Ithaca, N.Y., 1981.
25. R.L. Tsai and R. Raj, "Kinetics of Dissolution of β-Si_3N_4 in an Mg-Si-O-N Glass", J. Amer. Ceram. Soc., (1982) 65 [5] 270-274.
26. R. Raj and P.E.D. Morgan, "Activation Energies for Densification Creep and Grain Boundary Sliding", J. Amer. Ceram. Soc. (1981) 64 [10] C-143-145.
27. R.L. Tsai and R. Raj, "A Theoretical Estimate of Solution-Precipitation Creep in MgO-Fluxed Si_3N_4", J. Amer. Ceram. Soc. (1982) 65 [6] C88-90.

CREEP PHENOMENA IN REACTION BONDED AND SINTERED SILICON NITRIDES

T. Soma, M. Matsui and I. Oda

Research and Development Laboratory
NGK INSULATORS, LTD., Nagoya, Japan

ABSTRACT

The creep and fatigue behavior of reaction-bonded silicon nitride were studied in air up to 1500 °C. The creep deformation strongly depends on oxidation during creep. The microstructure of specimen deformed above 1100 °C contained many cavities, suggesting that the creep mechanism is grain-boundary sliding due to the glass phase formed by the oxidation. Below 1100 °C, the process of internal oxidation primarily determined the creep mechanism. The fatigue failure proved to be caused by the slow crack growth mechanism in all the temperature range. These results were compared with those of sintered silicon nitride.

INTRODUCTION

Silicon nitride is a promising material for high-temperature structural ceramics. To assure the reliability of the material for long-period use, the fatigue mechanism must be fully understood. With respect to sintered Si_3N_4 (SSN), the fatigue failure mechanism has been reported to be due to both slow crack growth and creep deformation.[1] In contrast, reaction-bonded Si_3N_4 contains considerable open pores and this tends to lead to internal oxidation at relatively low temperature. This internal oxidation has been reported to affect the creep behavior of RBSN.[2]

In this study, the creep and fatigue failure in RBSN were studied under various flexural stresses and temperatures. The microstructural change during creep and fatigue failure mechanism for RBSN were discussed in comparison with SSN.

EXPERIMENTAL PROCEDURE

The specimen of RBSN was prepared by the following procedure. Two kinds of silicon powder with different particle sizes were mixed in water with Fe_2O_3 as nitriding aid. The mixture was formed into a plate by slip-casting and was isostatically pressed at 2500 Kg/cm^2. The formed Si plate was nitrided for 24 hours at 1400 °C. The open porosity in prepared specimen was 15 %. The bulk density was 2.7 g/cc. The nitrided plate was machined into bars of 2 x 4 x 40 mm for a creep specimen. The surface was finished with a 800 grit diamond wheel.

Creep and static fatigue were measured by a 4-point bending device of 30 mm outer span and 10 mm inner span. The creep test was carried out with stresses ranging from 100 to 300 MPa and at temperatures from 900 to 1500 °C in air.

Microstructural studies were made by a scanning electron microscope (SEM) and a transmission electron microscope (TEM) attached with an EDX device. To prepare foils for TEM, crept specimens were cut with a diamond wheel to 1 mm thick. The slices were mechanically ground down to 50 µm and subsequently by manual grinding down to 30 µm. These thin slices were ion-milled to electron transparency. Before the observation, the foils were coated with evaporated carbon to avoid charging by electrons.

Oxidation test was carried out in air up to 100 hours at temperatures between 900 and 1400 °C. After oxidation, the weight change of each specimen was measured to an accuracy of 0.1 mg.

RESULTS

The strain-time curves at 150 MPa and at temperatures from 900 to 1400 °C are shown in Fig. 1. The transient creep region shows a rapid increase of strain. It is noteworthy that the creep rate at 1000 °C in this region is higher than that at 1100 °C. The creep rates above 1300 °C were determined from the tangential lines at 0.05 % strain in the curves, and the activation energy for creep was estimated as 620 kJ/mol.

The Arrhenius plots of the fatigue failure time, t_f, for the creep test are shown in Fig. 2. Above 1100 °C, the t_f decreases with increasing temperature. Below 1100 °C, the t_f tends to be shorter than those at 1100 °C. The fatigue failure time above 1100 °C can be fitted to the empirical equation,

$$t_f = A \exp\left(\frac{a - b\cdot\sigma}{RT}\right)$$

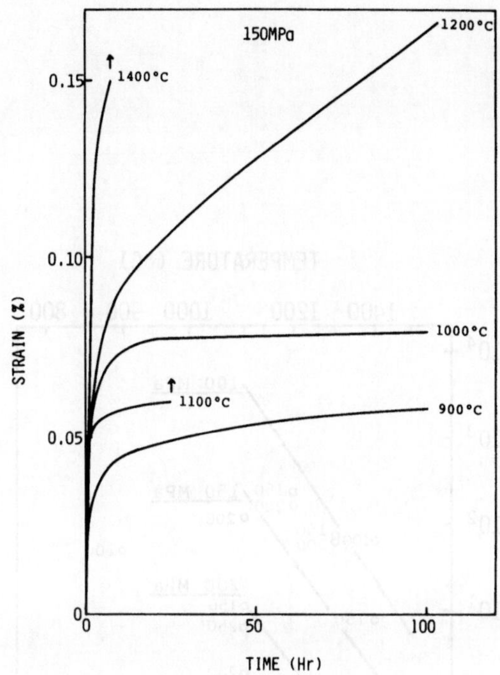

Fig. 1 Strain-time curves for RBSN tested at 150 MPa and various temperature.

where $A = 4.2 \times 10^{-10}$ (hr), a = 411 (kJ/mol), $b = 6.9 \times 10^{-4}$ (m^3/mol) and σ is applied stress. The activation energies for fatigue failure calculated from this equation ranges between 200 and 340 KJ/mol.

In Fig. 3, Arrhenius plots of the strain at failure, ε_f, are compared for RBSN and SSN. The ε_f for SSN tends to increase with increasing temperature above 1100 °C, but RBSN exhibits no such distinct tendency. This suggests that the fatigue failure of SSN occurs at a certain strain which is temperature-dependent, while that of RBSN is independent of strain. It is noticeable that the temperature at which the oxidation become most severe is coincident with the temperature at which anomalous change has occured in transient creep.

Fig. 4 shows the oxidation behavior of RBSN. The weight gain rate shows a peak at 1000 °C. Similar tendency has been generally observed in oxidation of RBSN where oxidation products tend to seal open pores and inhibit internal oxidation through open pores.[3]

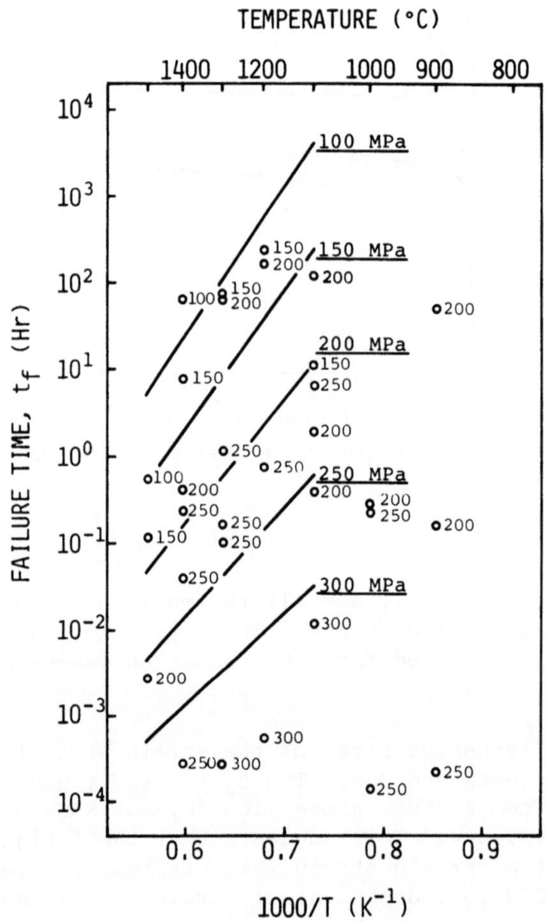

Fig. 2 Arrhenius plot of fatigue failure time for RBSN under static 4-point bending stress.

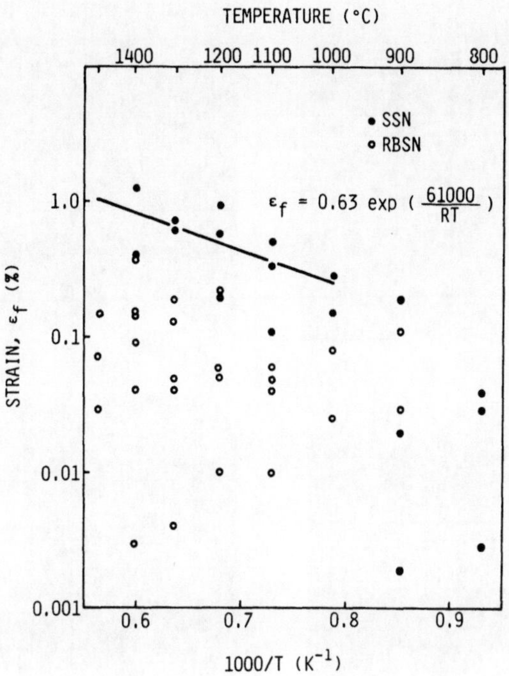

Fig. 3 Arrhenius plots of strain at failure for SSN and RBSN

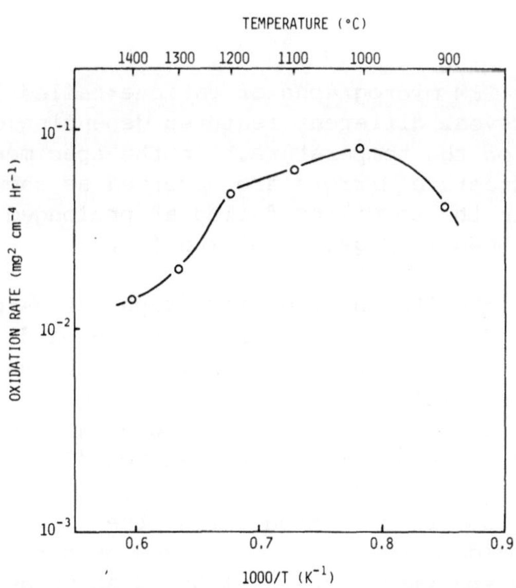

Fig. 4 Effect of temperature on oxidation rate for RBSN

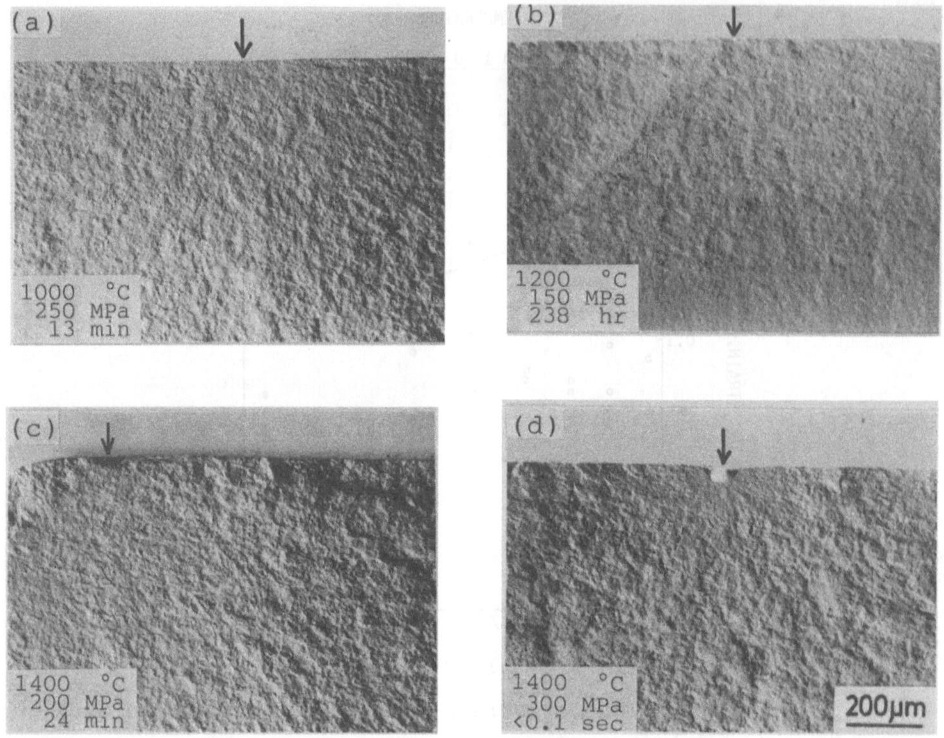

Fig. 5 Scanning electron micrographs showing fractured surfaces for RBSN at various creep conditions.

Fig. 5 shows SEM micrographs of fatigue-failed fracture surfaces, which reveal different features depending on the failure time rather than on the temperature. In the specimens failed at short times, fracture mirrors are observed as shown in Figs. 5 (a) and (d). In the specimens failed at prolonged times, this is not clear as shown in Figs. 5 (b) and (c).

Fig. 6 shows the SEM and TEM micrographs of as-prepared and crept specimens. The observed surfaces are perpendicular to the tensile surfaces. The as-prepared specimen contains large and fine pores, and is composed of large and fine grains. The large grains are mainly β-Si_3N_4 and the fine grains α-Si_3N_4. No glass phase is observed. The specimen crept at 900 °C shows annihilation of fine pores and rounding of initial pores. These changes are interpreted to be due to oxidation during the creep test from the fact that a oxidation layer is observed on surfaces of rounded large pores. The specimen crept at 1200 °C shows development

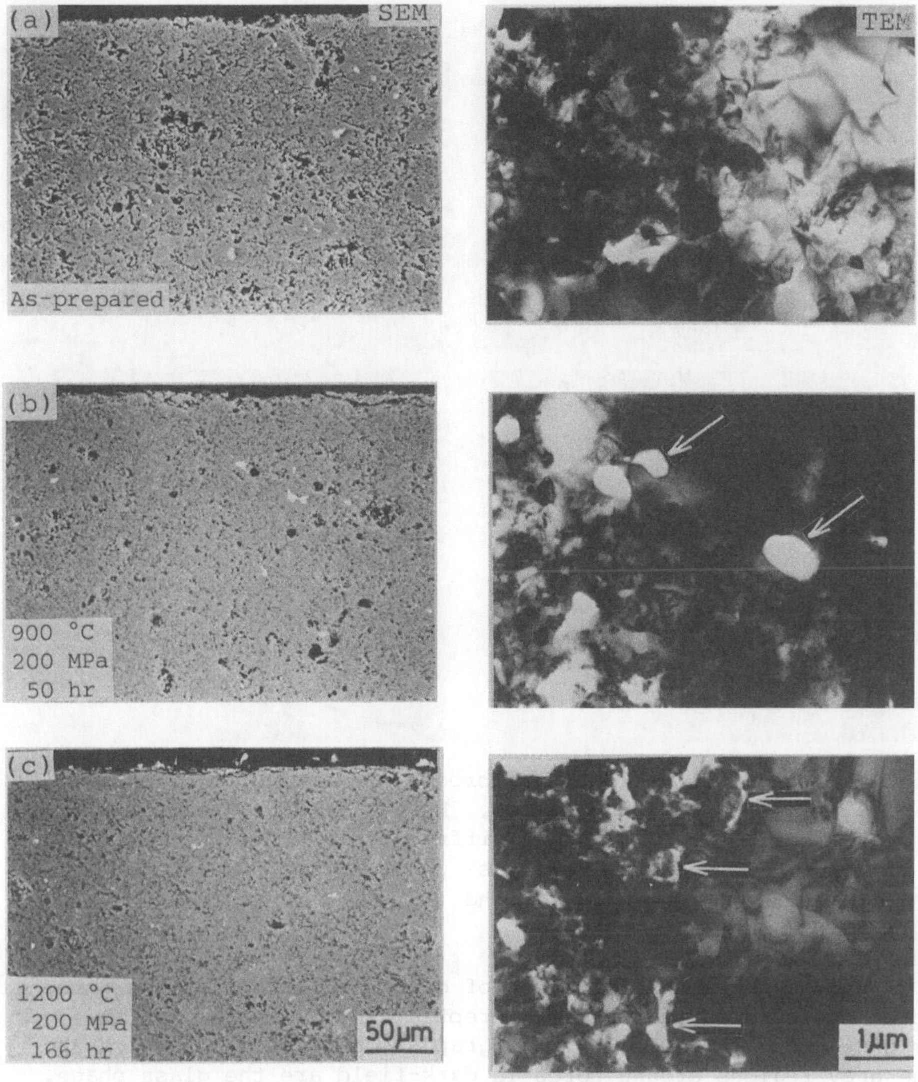

Fig. 6 SEM and TEM photographs of as-prepared and crept RBSN. (a) As-prepared; (b) Crept at 900 °C and 200 MPa; (c) Crept at 1200 °C and 200 MPa.

of cavities. The cavities exist in the fine grain region and have slender shapes. The development of cavities indicates that the grain-boundary sliding occurred during creep. Microstructural changes due to creep deformation at 1300 °C had been studied using mercury porosimeter by Thümmler et al..[4] Their result of increase in the number of small pore by creep deformation is coincident with the formation of cavity observed in present work.

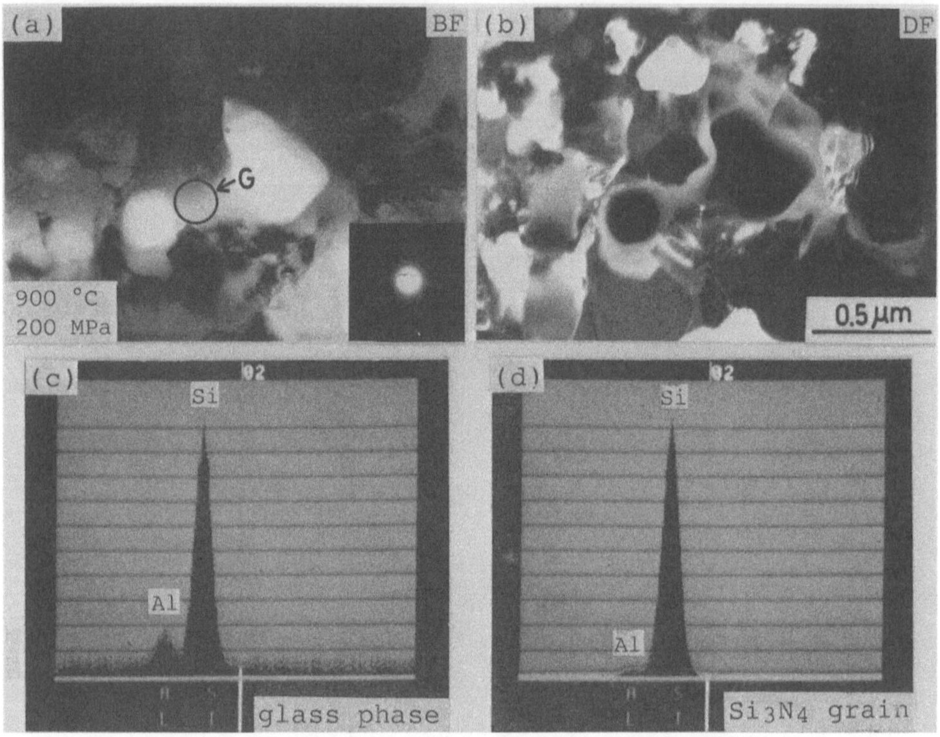

Fig. 7 Transmission electron micrographs showing formation of glass phase around pores. Bright field (a) and dark field (b). The insent diffraction pattern is from region G. EDX spectra from glass phase in region G and Si_3N_4 grain are shown in (c) and (d).

Fig. 7 shows the formation of oxidation layers around internal pores observed in the specimen crept at 900 °C where both the bright- and dark-field TEM micrographs identify the glass phase. The gray regions around pores in dark-field are the glass phase. The inset diffraction pattern of this layer confirms that this layer is amorphous. The EDX spectra from this layer and the Si_3N_4 grain are shown in Figs. 7 (c) and (d), respectively. The Al content of this layer is higher than that of the Si_3N_4 grain. This suggests that Al, which is the main impurity of the starting Si power, was enriched in the oxidation layer.

Fig. 8 shows the grain boundary separation and the cavity observed in the specimen crept at 1200 °C. The grain boundary separation and the cavity formation are interpreted to result from the viscous flow of the grain boundary glass phase, which was presumably formed by oxidation of Si_3N_4 during the creep test.

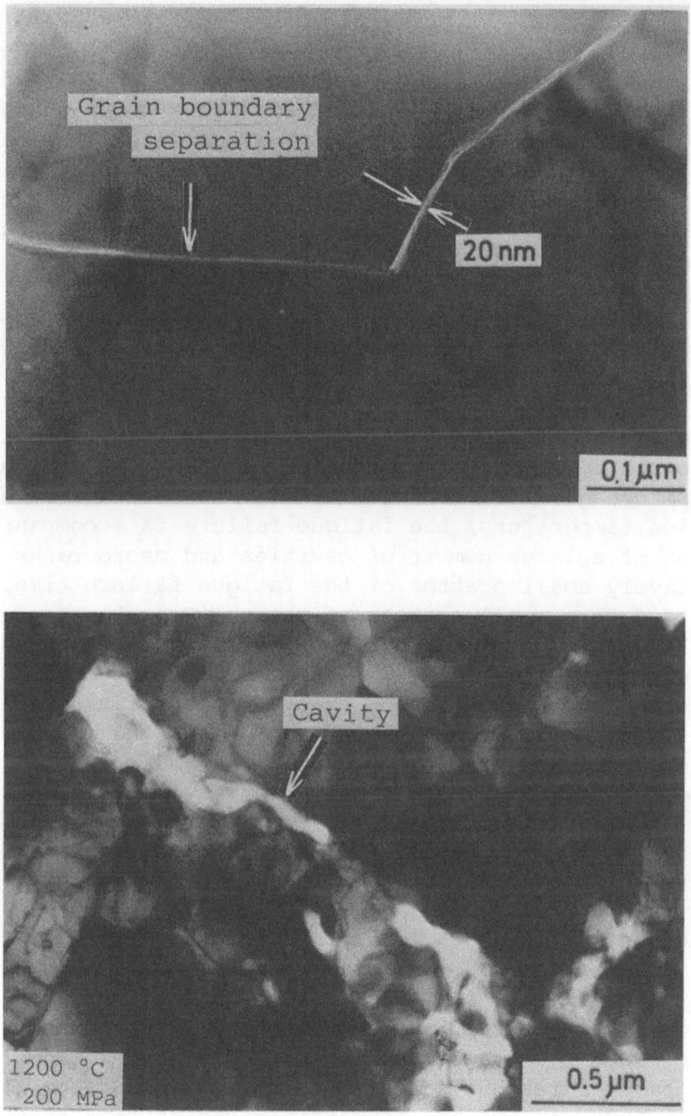

Fig. 8 Transmission electron micrographs showing the grain boundary separation and cavity.

DISCUSSION

The temperature dependence of the creep rate shows a marked change between 1000 and 1100 °C. This anomalous change is attributed to the oxidation behavior as explained by the relationship between the oxidation rate and the temperature (Fig. 4), and also by the microstructural change (Fig. 6). Above 1100 °C, the creep is primarily caused by sliding and movement of fine grains, surfaces of which are covered with a glass phase. Below 1100 °C, the process of internal oxidation determines the creep behavior. Remarkable transient creep was observed in all the temperature range tested in this study. The transient creep has been generally observed in silicon nitride ceramics. This phenomenon can be satisfactorily explained by the viscoelastic model due to the glass phase at the grain boundary.[5] In addition to this viscoelastic effect, internal oxidation kinetics is also responsible for the transient creep especially below 1100 °C in the present work.

The fatigue failure behavior in RBSN differs from that in SSN reported by the present authors. In SSN, the fatigue failure mechanism changed at a transition temperature. Above the transition temperature, the fatigue failure is accompanied by formation of a large number of cavities and macrocracks and results in relatively small scatter of the fatigue failure time, which should lead to a small scatter of the ε_f (Fig. 3). These results shows the existence of the deformation limit for the instantaneous fracture. Below the transition temperature, the mirror surface was observed and a significant scatter of the fatigue failure time resulted. These results indicates that the fatigue failure is caused by a slow crack growth (SCG) of a single crack. In RBSN, no macrocrack formation was observed in crept specimens and the fracture mirror was observed. Consequently, the fracture mechanism for RBSN above 1100 °C is essentially the same as the mechanism for the SSN below the transition temperature. Fig. 9 shows schematically the fatigue failure maps for RBSN and SSN. The fracture map for RBSN is divided into Regions I, I' and II. The fatigue map for SSN is divided into Regions I and II. In Region I, the fatigue failure occurs by SCG of a single crack. In Region II, the creep is accompanied by a large number of cavities and macro cracks, and the fatigue failure occurs when the deformation reaches a certain limit. Region II will also exist for RBSN at a higher temperature region than tested in this study. Region I' shows the regime where the fatigue failure occurs by internal oxidation. In SSN, the apparent activation energy for delayed failure above 1000 °C (Region II) was about 1000 kJ/mol and the temperature dependence of the fatigue failure time was stronger than that of the RBSN, for which the apparent activation energy for delayed failure above 1100 °C (Region I) was about

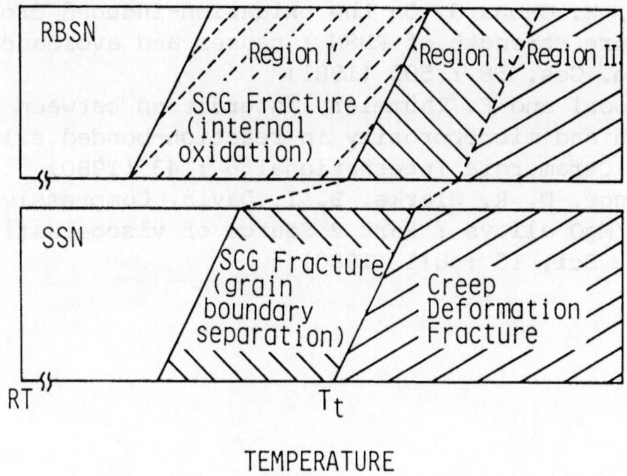

Fig. 9 Fatigue failure maps for SSN and RBSN

300 kJ/mol. This difference is attributed to the difference in fatigue failure mechanism and in property of the grain boundary phase in SSN and RBSN.

CONCLUSION

1. Creep mechanism in RBSN strongly depends on oxidation kinetics. Below 1100 °C, internal oxidation primarily determined creep behaviors. Above 1100 °C, grain boundary sliding primarily determined creep behaviors.

2. The delayed failure in RBSN is mainly caused by slow crack growth mechanism, rather than by creep deformation mechanism which has been observed in SSN.

3. Above 1100 °C, small cavities were formed by migration of fine grains, and separation of large grains occured. Below 1100 °C, small pores disappeared by being filled with silicate glass and large pores were rounded by oxidation of the internal surface.

REFERENCES

1. I. Oda, M. Matsui and T. Soma, High temperature fatigue failure in pressureless sintered silicon nitride, in : " Progress in Nitrogen Ceramics, " F. L. Riley, ed., Martinus Nijhoff Publishers, The Hague (1983)
2. G. Grathwohl and F. Thümmler, Creep of reaction-bonded silicon nitride, J. Mater. Sci. 13 : 1177 (1978)

3. G. Wirth, W. Gebhard, On the oxidation-induced drop in room-temperature strength of RBSN : causes and avoidance, Ber. Dt. Keram. Ges. 58 : 505 (1981)
4. G. Grathwohl and F. Thümmler, Interaction between creep, oxidation and microporosity in reaction-bonded silicon nitride, Ceramurgia International 6 : 43 (1980)
5. F. F. Lange, D. R. Clarke, B. I. Davis, Compressive creep of Si_3N_4/MgO alloys : Part 2 Source of viscoelastic effect, J. Mater. Sci, 15 : 611 (1980)

HIGH TEMPERATURE CREEP OF DENSE

FINE GRAINED SILICON CARBIDES

C. Carry and A. Mocellin

Swiss Federal Institute of Technology
Lausanne (Switzerland)

ABSTRACT

Various grades of commercial hot consolidated Al-doped α-SiC have been creep tested in compression between 1800 and 2060° C up to ~110 MPa in vacuum. Initial grain sizes were found to lie between 1.5 and 2.8 μm and the Al content was variable from sample to sample and probably also within samples. Such a chemical variability is believed to be responsible for the poor reproducibility which was noted in the strain rates of a priori identical test pieces. In a particular group of samples containing a nominal 1.5 % Al, strong evidence for superplastic-like behaviour was found. Strains up to - 0.43 were reached with negligible cavitation but details of the controlling mechanism(s) at the atomic level remain unclear.

I. INTRODUCTION

Interest for silicon carbide has been revived considerably in recent years, as a result mainly of the quest for ceramics showing promise for advanced structural applications. On the whole, however, the currently available literature on the high temperature mechanical behaviour of silicon carbide, either single- or polycrystalline, is still rather incomplete if not somewhat controversial. In broad terms, the published information dealing with mechanical properties of silicon carbide at moderate to high temperatures is concerned with strength and crack growth, dislocation plasticity, and creep. Only the last two will be the object of some consideration in this paper. Dislocation characteristics and slip systems have been investigated, in particular in hardness and fracture experiments[1]. But the deformation mechanisms remain mostly unknown except for a few results. For example, yielding and work-hardening was observed at temperatures as low as ~1200° C in some 3C polycrystals obtained by CVD[2]. In 6H polycrystals, however, dislocations were claimed not to be important for bending deformation at ~2100° C under 200 MPa stress[3]. But more recent observations with 6H single crystals[4] are somewhat at variance with such a statement.

The creep deformation of both hot-pressed[5] and sintered[6] silicon carbides has been investigated in recent years in temperature ranges correspondig to those envisioned for future applications and not exceeding 1600 - 1700° C. Bending has been the preferred mode of testing thus not allowing overall strains beyond a few percent to be reached, unless complications due to creep crack growth would arise. Strain rates recorded in such studies were rather small if not practically unobservable on some material grades. In first approximation they seemed to vary more or less linearly with the applied stress. Also, the corresponding apparent activation energies were found to assume rather dispersed values, so that despite an attempt at organizing some observations into a synthetic deformation mechanism map[7], it must be concluded that so far no coherent picture of the creep behaviour of silicon carbide has emerged yet.

The present work was undertaken with the objective of generating data which could complement those already available. More specifically our initial purpose was to extend beyond 1800° C the testing temperature and to investigate a set of modern fine grained samples in compression. In doing so, it was hoped that deformation could be pursued to more substantial strains and that comparisons could be made with earlier work available for such temperature levels[3-8]. Perhaps advantage also could be taken in interpreting results of recently published diffusivities in SiC[9] and of current research on grain boundary structures and compositions[10].

II. MATERIALS AND PROCEDURES

Three different grades of commercially available silicon carbide polycrystals were obtained from the same source (*). One of them had been manufactured by hot isostatic pressing (to be referred to as:"HIP") whereas the other two were made by uniaxial hot-pressing of Al-containing powders. According to the supplier the aluminium levels were 0.3 % (Ref. "A") and 12.5 % (Ref. "AA") by weight respectively. Consolidation temperatures also were reported to have been in excess of 2000° C. The crystallographic phase distributions have not been evaluated quantitatively but are presumed to be the same as those reported by other workers[10] in samples of

(*) Elektroschmelzwerk - Kempten (West Germany)

the same qualities and origin: 50 % of 4H and 50 % of 6H polytypes. Initial densities were (3.205 ± 0.005) x 10^3 kg/m^3 in all cases.

Test samples for the creep experiments were prepared as parallelepipeds with linear dimensions a x a x 3a where 8.7 < a < 13 mm. They were obtained by diamond cutting and machining from the as-received slugs (50 mm diameter by 100 mm long for the "HIP" grade, and two 75 mm by 25 mm high for each "A" and "AA" grades). It was anticipated that potential structural and/or compositional variations (typically between the cores and outer surfaces) within a given disc might be reflected subsequently on the deformation behaviours. Care thus was taken to record the relative positions and orientations of the individual test samples with respect to the original discs (Figure 1).

Smaller pieces also (labelled "T" in what follows) were cut, usually not too far from the outer disc or rod surfaces. They were to serve as reference samples for the ones being deformed, in that one such small piece was always placed in the furnace next to the bar being tested. It would thus have received exactly the same heat treatment but in a stress-free condition.

Compressive creep tests and related strain and strain rate measurements were conducted in a tungsten resistance vacuum furnace as described previously[11]. Temperatures and stresses were adjusted in the 1800 - 2060° C and 40 - 115 MPa ranges respectively. The former were controlled within ± 10° C through the electric power delivered to the heating element and measured with a standard optical pyrometer. The latter was controlled by maintaining the applied load within ± 20 kg of its nominal value. It was transmitted to the sample by means of high strength graphite pistons, large grain size sintered SiC discs (*) or graphite discs and thin sheets of graph-

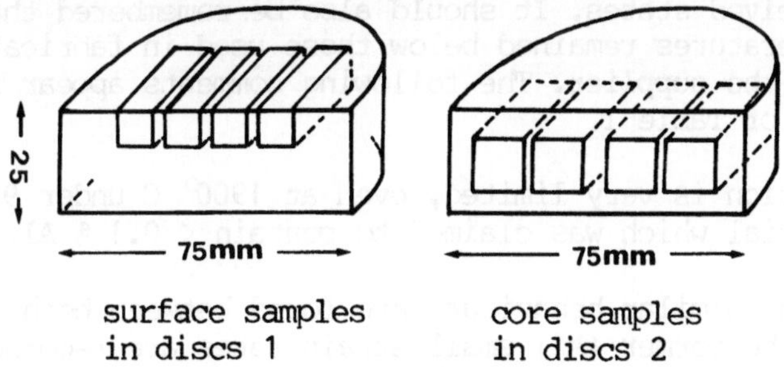

Fig. 1. Test sample positions in the as-received A and AA material

(*) from the Carborundum Company

ite "paper". Strain rate sensitivities "m" on the applied stress were evaluated by the standard instantaneous stress jump ($\pm$ 15 % relative) procedure. Apparent activation energies were obtained in a similar fashion from $\pm$ 50° C fast temperature changes and apparent activation volumes were computed from:

$$V^* = \frac{kT}{\sigma_2 - \sigma_1} \ln \frac{\dot{\varepsilon}_2}{\dot{\varepsilon}_1}$$

Both transmission and scanning electron microscopies were used to observe and evaluate microstructures. Thin slices for the TEM were prepared by diamond grinding followed by ion milling and observed under 200 kV. SEM observations on the other hand were made on polished planes containing the creep stress direction and, when applicable, also the hot-pressing direction. The following etching procedure was used to reveal grain boundaries: 12 minutes at 1200° C in flowing O_2, then 3 - 5 minutes at 20° C in 20 % HF solution. Grain sizes were evaluated from SEM micrographs by measuring the average section area $\bar{s}$ of at least 600 grains per sample. For consistency, the numerical values for the linear grain sizes quoted below will equal 1.38 $\sqrt{s}$. Finally, bulk densities were measured by the Archimedes method of immersion in water.

III. RESULTS AND DISCUSSION

1. Influence of material type

A summary of the creep results obtained with the HIP, A and AA materials will be found in Table I and it should be borne in mind that initial average grain sizes for these three grades were found to be to equal to 1.5, 2.8 and 1.5 µm respectively. No significant grain size variations could be detected between regions in the interior or close to the outer surface in any of our disc or rod in their as-received states. It should also be remembered that our testing temperatures remained below those used in fabrication, according to the supplier. The following comments appear in order on the basis of Table I.

i) Deformation is very limited, even at 1900° C under 96 MPa in the HIP material which was claimed to contain < 0.1 % Al.

ii) Rather dissimilar behaviours are found between both the grade A discs. In the former (A1) small strain rates are recorded with a stress sensitivity exponent slightly above 0.5 and virtually no density change even after ~20 % strain. In the latter (A2) significantly higher strain rates (X4-5) linearly dependent upon stress are accompanied by some cavitation (~4 % after ~30 % strain). In neither case however were the grain size and equiaxiality found to have been modified during testing. It thus seems that in both cases

Table I. Summary of conditions and results of creep deformation (f refers to final strained state and (D_f) is the stress free annealed grain size).

Material	a (mm)	T (°C)	σ (MPa)	t (min)	$\lvert\varepsilon_f\rvert$	$\lvert\dot{\varepsilon}\rvert$ (10^{-6}/s)	ρ_f (gr/cm^3)	D_f (μm)	m	V^* (10^{-28}m^3)	Q (kcal/M)
HIP	12	1900	96	90	< 0.005	< 1	3.205	-			
	12	1900, 2060	45-55	500	0.039	1-2.8	3.194	-	1	6	
A 1	9	1900	85-102	298	0.152	6-11	3.205	-	-	3	
	9	1950	82-111	151	0.127	11-16	3.212	-	0.6		
	9	2000	87-114	125	0.202	21-35	3.208	-	0.6	3	
A 2	13	1900	88-105	146	0.288	29-40	3.079	2.8	1.1	3	
	12	1900	45-55	420	0.260	7-14	3.164	-	1	6	150
	12	1950	45-55	395	0.428	10-29	3.128	-	1	6	
AA 1	8,6	1900, 1965	48-74	340	0.084	2-9	3.211	-			160
	8,6	1950	45-55	545	0.114	2-5	3.194	-	0.6	9	
AA 2	13	1815, 1865	81-94	115	0.190	16-42	3.196	-			156
	11	1850	87-103	147	0.269	24-43	3.183	-	0.9	3	
	13	1900	96	64	0.266	60-82	3.202	1.6 (1.9)			167
	13	1900	88-102	95	0.430	54-100	3.190	1.9 (1.9)	1.1	3	
	11	1900	44-52	200	0.274	16-40	3.193	2.4 (2.7)	1.0	6	

deformation occurred by a significant contribution of a viscous mechanism, probably grain rearrangment via boundary sliding[12]. Stress concentrations at grain junctions however might have been more or less completely and/or differently accommodated in either cases. In the A1 sample, participation of dislocation effects or interface reaction kinetics[13] in accounting for the overall deformation might also be envisaged. Besides unknown variances perhaps processing related, between the a priori identical A1 and A2 discs, it seems that differences in the observed behaviours could be ascribed to non identical overall sizes of the test sample and/or their relative positions in the as-received discs (Figure 1). Obviously such features would be relevant if it is postulated that the structure and/or grain boundary composition either were or became different during creep testing. Figure 2 showing that some exudation from inside the sample takes place even at the moderate thermal etching temperatures brings support to this assumption. And results of EDAX analysis showing aluminium to be present in the exuded drops further suggests that this sintering aid should be given serious consideration in this context.

iii) In the AA material grade a similar, if not more pronounced lack of reproducibility is found between samples taken from two different as-received discs. With the information now available there appears no reason to believe that the arguments just put forth in the case of grade A should not account for variability in AA as well. Indeed, it is interesting to note that although of identical initial grain sizes both the HIP and AA materials exhibited substantially different strain rates for the investigated temperatures and stresses.

2. Observation of a superplastic behaviour

Bearing in mind the existence of difficulties associated with the dispersion of behaviours from sample to sample, it seems nonetheless possible to analyze in a little more detail the results obtained with the AA2 hot-pressed disc (Table I). As illustrated in Figure 3, rather respectable overall strains, e.g. > 40 % under 96 MPa at 1900° C, could be reached with limited (~ 0.4 to 0.5 %) cavitation. The linear stress-strain rate dependence also should be noted, together with the occurrence of normal grain growth during deformation (Figure 2). But contrary to what has recently been reported for fine grained Al_2O_3[11] the application of stress in the present particular case was not seen to affect the grain growth rate constant. Average grain sizes were found equal within experimental uncertainties in related sets of deformed and stress-free annealed materials. Apparent activation energies for deformation between 1850 and 1900° C amounted to ~160 kcal/mol under 96 MPa, whereas activation volumes went from ~15 to ~30 times the molar SiC volume when the applied stress was changed from 48 to 96 MPa. All observations thus appear qualitatively consistent with the assump

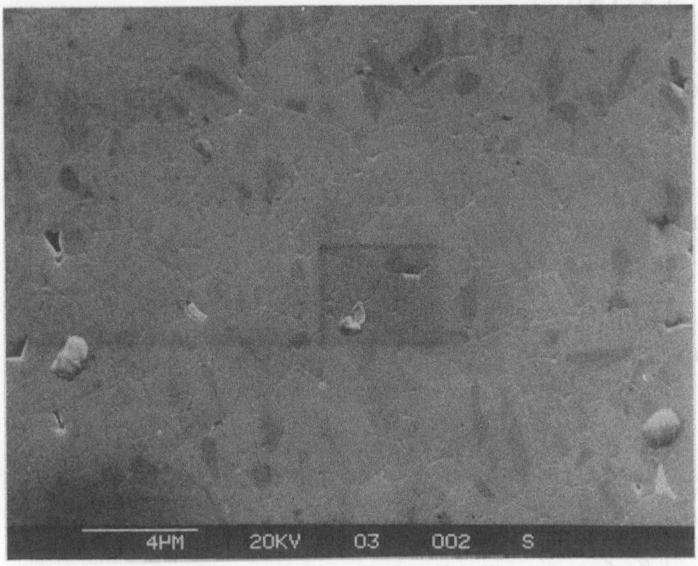

a) before creep test

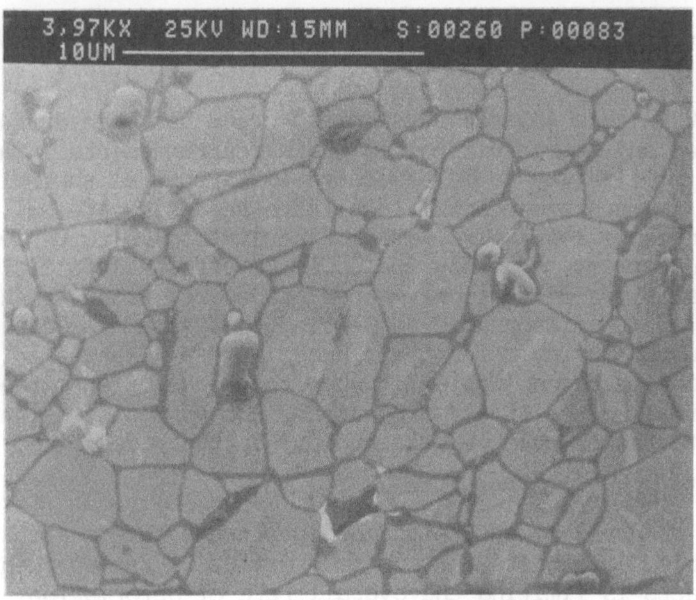

b) ε = 0.266 in 200 min. under 48 MPa

Fig. 2. S.E.M. micrographs of AA2 samples showing grain growth during deformation and evidence of exudation during thermal etching.

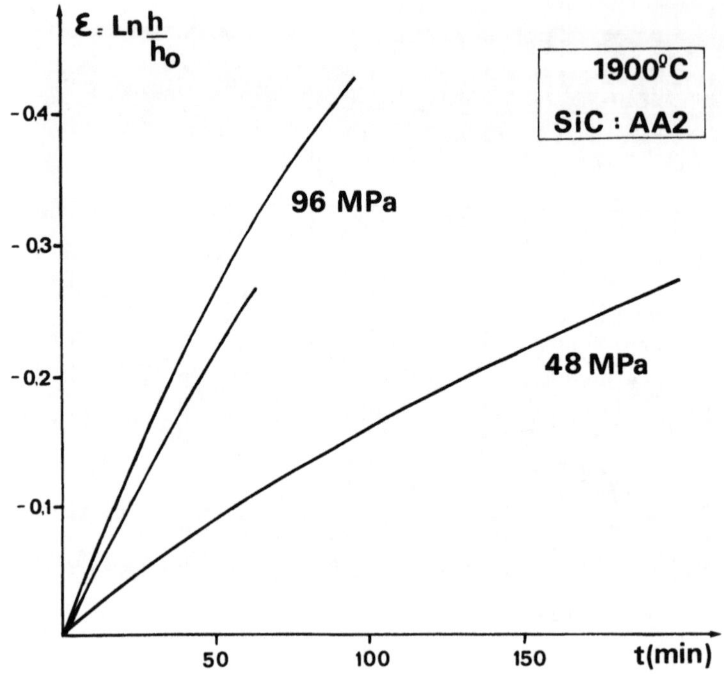

Fig. 3. Recorded strain versus time curves in AA2 samples

tion that in this AA2 set of samples, creep has been diffusion-controlled and probably involved grain rearrangements in a geometrically similar way as is advocated in structural superplasticity[12]. In such a mechanism local stresses which are built up at grain junctions must be relieved if cavitation is to be avoided. In the present case, accommodation could be effected via dislocation processes or via diffusional matter transport. Figure 4 shows examples of TEM micrographs suggesting that some dislocation activity occurred during deformation under 96 MPa. Changes in the X-ray diffraction patterns (Table II) show that basal planes tend to align normal to the stress directions. But no modifications with respect to the as-received state occur in the stress-free annealed material.

The observed linear relationship between stress and strain rate (Table I) suggests that in fact diffusion rather than dislocation effects is limiting the overall rate of deformation. If we now assume, as in a similar study on Al_2O_3[11] that quasi steady state prevailed during deformation, then the classical equation for grain growth:

$$D^n - D_o^n = K_n t \qquad [1]$$

may be substituted in the strain rate expression:

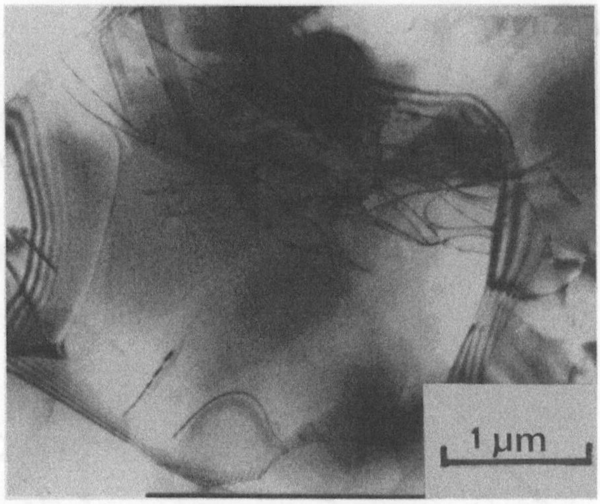

Fig. 4. Dislocations in a AA2 sample deformed (-0.43) under 96 MPa at 1900° C

Table II: Relative X-ray diffraction peak height ($h = I_{hkl} / \Sigma I_{hkl}$) of AA2 material: I as-pressed, T stress free annealed, and strained to - 0.266 and 0.43 recorded normal to stress.

4H h k l	6H h k l	I	T	1900° C 96 MPa ε=-0.27	ε=-0.43
1 0 0		29	28	15	1
	1 0 1	51	43	47	32
1 0 1		111	127	61	53
0 0 4	{1 0 1; 0 0 6}	258	249	368	449
1 0 2	1 0 3	161	157	122	102
	1 0 4	21	18	22	18
1 0 3		19	20	15	17
	1 0 5	9	8	11	7
1 0 4		9	10	9	10
	1 0 7	11	9	13	12
1 0 5		30	34	30	37
1 1 0	{1 1 0; 1 0 8}	124	120	101	76
1 0 6	1 0 9	74	75	94	104
	1 1 6	90	98	89	71

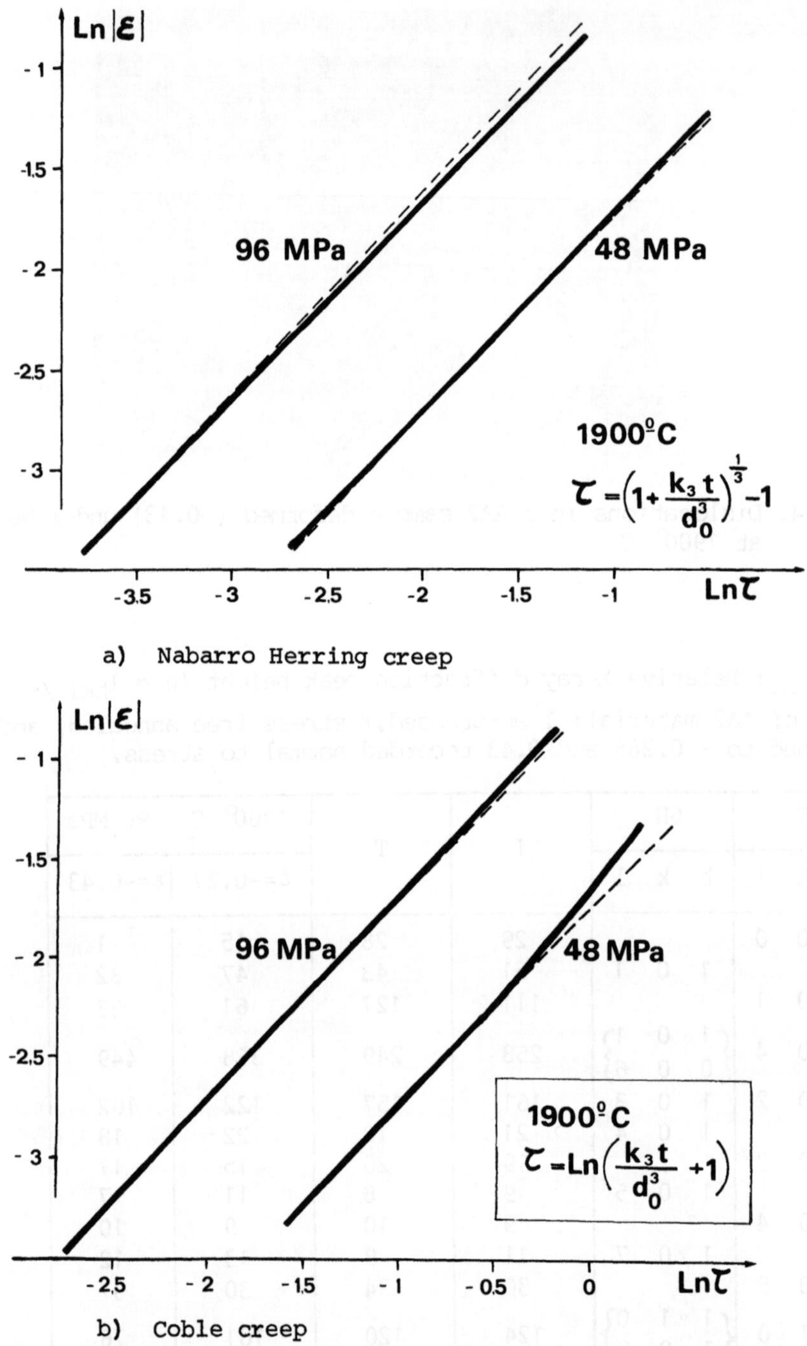

Fig. 5. Ln-Ln plots of observed strain versus reduced time incorporating grain growth kinetics.

$$\dot{\varepsilon} = A_p \cdot \frac{\sigma}{D^p} \qquad [2]$$

Integration with respect to time then yields theoretical expressions for the progress of the overall strain with time, which can be compared with experiment (Figure 3). In particular, for n = 3 and p = 2 (Nabarro-Herring creep)

$$\mathrm{Ln}\ \varepsilon = ct + \mathrm{Ln}\left[\left(1 + \frac{K_3}{D_o^3} t\right)^{1/3} - 1\right] \qquad [3]$$

for n = 3 and p = 3 (Coble Creep)

$$\mathrm{Ln}\ \varepsilon = ct + \mathrm{Ln}\mathrm{Ln}\left(1 + \frac{K_3}{D_o^3} t\right) \qquad [4]$$

making use of the measured initial and final grain sizes (Table I) the grain growth rate constant K_3 can be evaluated and the reduced times τ can be computed therefrom where:

$$\tau = \left(1 + \frac{K_3}{D_o^3} t\right)^{1/3} - 1 \quad \text{or} \quad \tau = \mathrm{Ln}\left(1 + \frac{K_3}{D_o^3} t\right)$$

according to [3] or [4] respectively. Figure 5 shows such experimental $\mathrm{Ln}\varepsilon$ versus $\mathrm{Ln}\tau$ plots for the most extensively strained samples at 1900° C. It is seen that the straight line with slope 1 is best achieved for p = 2 and p = 3 under 48 and 96 MPa stresses respectively. Attempts at adjusting grain growth exponents other than n = 3 have also been made yielding poorer fit. It would appear from Figure 5 that Coble creep was predominant under 96 MPa whereas Nabarro-Herring creep was controlling under 48 MPa stress.

Identifying $A_2 = \dfrac{14 D_v \Omega}{kT}$ and $A_3 = \dfrac{14\pi\delta D_b \Omega}{kT}$ in Equation [2]

Table III: Calculated diffusivities in α-SiC at 1900° C from different assumed deformation mechanisms.

stress (MPa) / diffusivities	48	96
D_{vol} (m^2/s)	1.8×10^{-16}	2.5×10^{-16}
$\delta D_{bound.}$ (m^3/s)	1.2×10^{-22}	1.6×10^{-22}

above and making use of the integral expression for strain and Figure 5 allow numerical estimates to be made for effective diffusivities in α-SiC at 1900° C. These are given in Table III and it may be noted that the present lattice diffusivity values are in good agreement with those recently obtained [10] in direct diffusion experiments.

IV. CONCLUSION

The present work is believed to show that:

i) the high temperature creep of fine grained α-SiC is markedly influenced by the amount and distribution of the aluminium sintering aid. Such amount and distribution may vary during testing at rates fast enough to affect the control.

ii) there exists a range of practical conditions under which deformation is essentially superplastic, i.e. occurring via viscous grain rearrangements. Local stress incompatibilities can be accommodated by diffusion and/or through dislocation generation and displacement. The latter process may be easy enough not to limit the overall strain rate.

iii) due to the limited experimental evidence collected here but also to the unavailability of quantitative information on grain boundary diffusion and dislocation slip in α-SiC, it is not possible to precisely rationalize the atomistics of the deformation. It is likely however that in most cases, more than one elementary mechanisms have contributed to the overall creep.

Finally, the present results perhaps might have some practical implications in that they point to the feasibility of deformation processing or hot forging of silicon carbide. It appears conceivable for example to develop textured materials by systematic recourse to intragranular plastic deformation modes. It is also possible to set up shaping operations, as was shown by the present authors who prepared SiC shells from hot formed discs at 1900° C; in a manner similar to that already reported for fine grained Al_2O_3 [11a].

REFERENCES

1.a) S. Amelinckx, G. Strumane and W.W. Webb, J. Appl. Phys. 31 (1960) 1359-1370

b) R. Stevens, J. Mater. Science 5 (1970) 474-477

c) B.J. Hockey, B.R. Lawn, J. Mater. Science 10 (1975) 1275-1284

d) J. Lankford, D.L. Davidson, J. Mater. Science 13 (1979) 1669-1675

e) Koichi Niihara, J. Less. Com. Metals 65 (1979) 155-166

f) G.R. Sawyer, P.M. Sargent and T.F. Page, J. Mater. Science 15 (1980) 1001-1003

2. T.D. Gulden, J. Amer. Ceram. Soc. 52 (1969) 585-590
3. P.L. Farnsworth, R.L. Coble, J. Amer. Ceram. Soc. 49 (1966) 264-268
4.a) I.N. Frantsevich, V.A. Kravets, K.V. Nazarenko, V.Z. Smushkevich, Sov. Powder Metall. Met. Ceram. 12 (1973) 654-662
b) I.N. Frantsevich, V.A. Kravets, K.V. Nazarenko, Sov. Powder Metall. Met. Ceram. 14 (1975) 679-682
5.a) G. Grathwohl, T.H. Reets and F. Thümmler, Science of Ceramics 11 (1981) 425-431
b) L. Hermansson, R.C. Bradt, Science of Ceramics 11 (1981) 447-452
6.a) P. Marshall, R.B. Jones, Powder Metall. 12 (1969) 193-208
b) S. Prochazka, P.C. Smith, Fin. Report SRD-74-040 Contract N62269-73-C-0356 NADC (1973)
c) C.A. Johnson and S. Prochazka, Fin. Report SRD-77-145 Contract N62269-76-C-0243 NADC (1977) ADAO 47504
d) A. Djemel, J. Cadoz, J. Philibert. Creep and Fracture of Engineering, Materials and Structures, edited by B. Whilshire and D.R.J. Owen, Pineridge Press, Swansea, U.K. (1981) 381-394
7. V. Krishnamachari, M.R. Notis, Mater. Science and Engineering 27 (1977) 83-88
8.a) T.L. Francis, R.L. Coble, J.Amer.Ceram. Soc. 51 (1968) 115-116
b) R.N. Ghoshtagore, R.L.Coble, Phys. Review 143 (1966) 623-626
9.a) J.D. Hong, M.H. Hon, R.F. Davis, Ceramurgia International 5 (1979) 155-160
b) J.D. Hong, R.F. Davis, D.E. Newbury, J. Mater. Science 16 (1981) 2485-2494
10.a) Yo Tajima, W.D.Kingery, Comm. Amer.Ceram.Soc. 65 (1982) C-27
b) Yo Tajima, W.D. Kingery, J. Mater. Science 17 (1982) 2289-2297
c) Yo Tajima, Kazunori Kisina, W.D. Kingery, J. Chem. Phys. 77 (1982) 2592-2598
d) G. Nouet, S. Magege, R. Moussa, F. Osterstock, J. Vicens, J.L. Chermant, Rev. Int. Hautes Temper. Refract. 19 (1982) 335-343
11.a) C. Carry, A. Mocellin, Proceedings of British Ceram. Soc. 33 (1983) 101-115
b) J.D. Fridez, C. Carry, A. Mocellin, MgO/Al_2O_3 Conf. MIT (1983) to be published in Advances of Ceramics, Vol. 12
12.a) R.N. Stevens, Surface Science 31 (1972) 543-565
b) M.F. Ashby, R.A. Verall, Acta Met. 21 (1973) 149-163
c) R.C. Gifkins, J. Mater. Science 13 (1978) 1926-1936
d) O.A. Ruano, O.D. Sherby, Mater. Science Engineering 56 (1982) 167-175
13.a) R.M. Cannon, W.M. Rhodes, A.M. Heuer, J. Amer. Ceram. Soc. 63 (1980) 46-52
b) J.M. Schneibel, P.M. Hazzledine, J. Mater. 18 (1983) 562-570
c) M.F. Ashby, Scripta Met. 3 (1969) 837-842

PLASTIC DEFORMATION OF Mg-DOPED FINE-GRAINED ALUMINUM OXIDE

R. S. Jupp and B. J. Pletka

Department of Metallurgical Engineering
Michigan Technological University
Houghton, MI 49931

INTRODUCTION

The plastic deformation of fine-grained (1-15μm) aluminum oxide has been studied in detail by Heuer et al.[1] and Cannon et al.[2] The dominant flow mechanism was found to be diffusional creep which occurs as a result of stress-directed diffusional transport between grain boundaries.[3-5] For this mechanism, the strain rate, $\dot{\varepsilon}$, due to an applied stress, σ, can be expressed as[6]

$$\dot{\varepsilon} = \frac{14\,\Omega}{kTd^2}\left[D_L + \frac{\pi w D_B}{d}\right]\sigma \tag{1}$$

where D_L and D_B are the lattice and grain-boundary diffusivities, w is the "effective" grain-boundary diffusion width, d is the grain size, Ω is the atomic volume, and kT has its usual meaning. The diffusional term within the brackets of this equation is modified for alumina since both ions must diffuse.[7,8] However, the apparent rate controlling nature of aluminum diffusion during creep of fine-grained alumina leads to an expression identical to equation (1) except that the right hand side of the equation is divided by a factor of two.

An additional observation of Cannon et al.[2] was that diffusional creep became interface-controlled at low stresses. The strain-rate sensitivity, m, defined as

$$m = \frac{\ln\ \sigma_2/\sigma_1}{\ln\ \dot{\varepsilon}_2/\dot{\varepsilon}_1} \tag{2}$$

where σ_1 and σ_2 are the flow stresses at strain rates $\dot{\varepsilon}_1$ and $\dot{\varepsilon}_2$, was found to decrease in value from ~1 to 0.5 with decreasing flow stress for a 1.2 μm grain size alumina. From equation (1), a strain-rate sensitivity of 1 is predicted, and the change in m with decreasing flow stress was attributed to interface-controlled diffusional creep. This mechanism is important when the grain boundaries do not act as perfect sources/sinks for vacancies, and vacancy creation/annihilation rather than vacancy diffusion becomes rate controlling. A model based on the creation and annihilation of vacancies by the glide and climb of grain-boundary dislocations (GBD's) and which takes into account the stress dependence of the GBD density leads to an expression for the strain rate of the form[9,10]

$$\dot{\varepsilon} \approx \frac{M\sigma^2}{d} \tag{3}$$

where M is the GBD mobility. The strain-rate sensitivity value of 0.5 derived from equation (3) was in agreement with the experimental data, and thus supported the contention that diffusional creep in fine-grained alumina becomes interface-controlled at low stresses.

Because diffusional creep in alumina involves the mass transport of both cations and anions, the lattice and grain-boundary diffusivities can be affected by the addition of aliovalent impurities. Cannon and Coble[8] demonstrated that under pure diffusional creep conditions, doping Al_2O_3 with MgO in excess of the solubility limit led to an increase in lattice cation diffusion but did not affect the grain-boundary diffusivity. However, Gordon and coworkers have shown that aluminum grain-boundary diffusion is enhanced by adding the divalent cations, Mn^{2+} and Fe^{2+}.[11,12] Although Mn and Fe are multivalent while Mg is not, there does not seem to be any reason why there should be a difference in the effect of adding these divalent cations on aluminum grain-boundary diffusion. In an attempt to clarify this discrepancy, the present paper presents an investigation in which deformation experiments were performed on a series of fine-grained (<3μm) aluminas containing various levels of magnesium doping. Transmission electron microscopy (TEM) observations in conjunction with mechanical testing were used to study the overall deformation behavior.

EXPERIMENTAL PROCEDURE

Four fine-grained hot-pressed aluminas (all >99% theoretical density) containing different levels of magnesium were studied in the present work (Table I). Samples A, B, and C were produced from Al or Al-Mg alloy powder which had been hydrothermally oxidized in an autoclave. The only significant impurity in these three aluminas is silicon which was unintentionally introduced during the autoclaving process. Alumina D, which is very similar to the fine-grained alumina studied by Heuer et al.[1] and Cannon et al.[2], was fabricated using Linde A powder doped with 0.25 wt% MgO and its chemistry is typical for such a material. The grain sizes, reported

Table I

Microstructural and Chemical Information

Alumina	Grain Size (μm)	Chemistry* (wt ppm) Mg	Si	Ca	Others
A	1.7	1	200	1	1
B	2.8	600	200	1	1
C	2.2	1000	300	1	1
D	1.2	1500	50	10	200

*Mass spectroscopy analysis

as 1.5 times the average lineal intercept, were determined from TEM micrographs.

Constant crosshead-speed compression tests were performed in air on specimens saw cut and ground from the hot-pressed billets to approximate dimensions of 2mm x 2mm x 4.8mm. Deformation experiments were conducted using strain rates of 2×10^{-6} s^{-1} to 2×10^{-5} s^{-1} at 1623 and 1673K; the temperature was controlled to ± 1K during the tests. The testing routine outlined schematically in Figure 1 was utilized. During the first part of the deformation run, a "jump test" was performed by changing the crosshead speed, $\dot{L}$. For the samples tested at 1623K, $\dot{L}_2$ was a factor of ten less than $\dot{L}_1$, while at 1673K it was a factor of five lower. The data obtained can be used to calculate the strain-rate sensitivity, m, using equation (2). At the completion of the jump test, the crosshead was stopped and the sample was allowed to stress relax. The stress relaxation occurs when the elastic strain of the testing machine and the sample is converted into plastic strain in the sample, and it provides another method of determining the strain-rate sensitivity.[13] If m is constant, a plot of $\ln \sigma$ versus $\ln(-d\sigma/dt)$ where t is the time, yields a straight line whose slope is 1/m. Thus, two independent means of determining the strain-rate sensitivity were utilized.

TEM observations were performed on several of the samples to study the effects of deformation on the microstructure. In all cases, foils were cut perpendicular to the stress axis and near the center of the sample; they were examined on a Philips 301 microscope operating at 100 kV.

RESULTS AND DISCUSSION

The general deformation behavior of all four aluminas is shown in the true stress-true strain curves of Figure 2 for 1623K and a

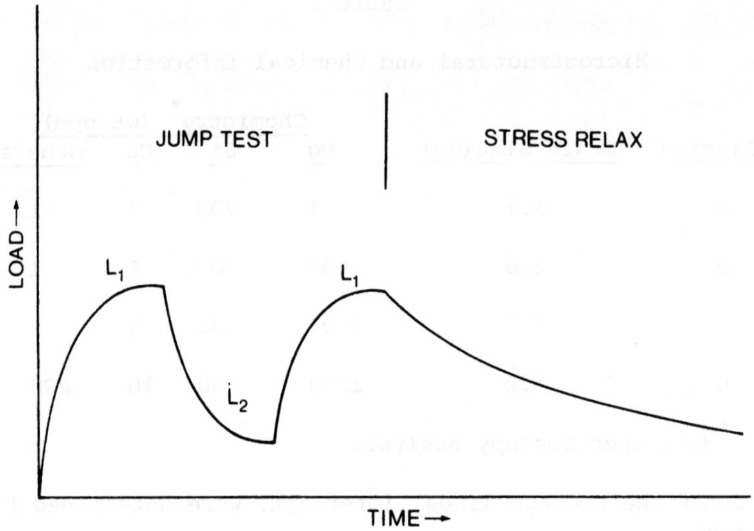

Figure 1. Testing routine used for deformation tests. $\dot{L}$ is the crosshead speed.

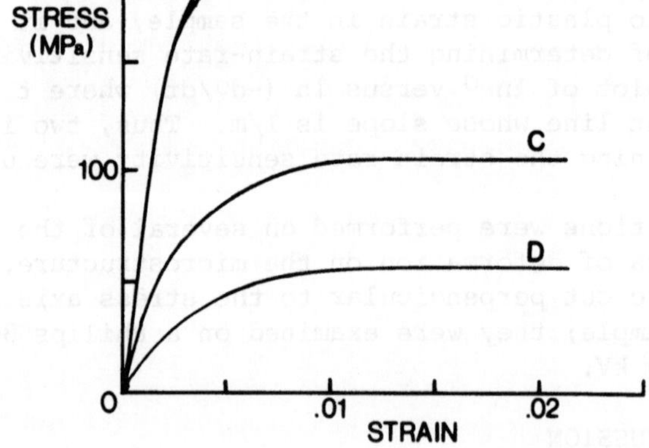

Figure 2. True stress-true strain curves for the four aluminas at a temperature of 1623K and a strain rate of $2x10^{-5}s^{-1}$.

strain rate of $2x10^{-5}s^{-1}$. After yielding, the stress increased at a decreasing rate until a constant flow stress was attained after ~1% strain for all aluminas. (Since constant flow stresses are obtained at approximately constant strain rates, these data are equivalent to constant stress/strain rate results from creep tests.) A direct comparison of the flow stress data in Figure 2 is not possible because of the difference in grain sizes between the various aluminas. However, these data can be used to assess the effect of Mg on the Al grain-boundary and lattice diffusivities via equation (1) once it has been established that diffusional creep is the dominant deformation mechanism.

The results of the deformation tests are summarized in Table II. The flow stresses listed are for the faster of the two crosshead speeds used in the jump test; this crosshead speed corresponds to a strain rate of $2x10^{-5}s^{-1}$ at both temperatures. The strain-rate sensitivity values from stress-relaxation tests were determined by a least squares fit of ln σ versus ln-dσ/dt.

The high strain-rate sensitivities ($0.5<m<1.0$) in Table II indicate that, consistent with previous work, diffusional creep and interface-controlled diffusional creep are the dominant deformation mechanisms. A problem in the interpretation of the data is that lower m values result from the stress-relaxation tests compared to the jump tests. This same result has been observed in metallic systems[13] and in a previous study on alumina.[14] The change in stress in both tests was comparable although any changes in microstructure would evolve more rapidly in the jump tests. In addition, the relative strain induced by the two techniques is different. The jump test normally introduces several percent strain

Table II

Results of Deformation Testing

Sample	σ(MPa)	Strain-rate sensitivity (m): Jump Test	Strain-rate sensitivity (m): Stress Relaxation
1623K:			
A	254.4	.78	.50
B	246.4	.83	.61
C	106.3	.94	.71
D	54.4	.69	.49
1673K:			
A	117.6	.78	.54
B	142.2	.90	.71
C	32.4	.94	.60
D	22.7	.65	.54

into the sample while stress relaxation requires <1%. As a result, stress relaxation yields an approximate isostructural strain-rate sensitivity while for the jump test this is not necessarily the case. Finally, additional spurious relaxation of the testing machine may have occurred. What relative contributions these effects have, if any, on the strain-rate sensitivities is unknown at present. Since the m values for aluminas A, B, and C (from jump tests) are close to 1 at high flow stresses, in agreement with the data of Cannon et al.[2], it will be assumed in the present work that diffusional creep is the dominant deformation mechanism for these materials. The deformation of alumina D at the imposed strain-rates in the present work appears to be controlled by interface-controlled diffusional creep due to the low strain-rate sensitivity values from both jump tests and stress relaxation and the low flow stresses.

TEM observations performed on deformed specimens indicated that cavitation had occurred during deformation. These cavities form at both triple points and along the grain boundaries as shown in Figures 3 and 4, respectively. The cavities appear to be primarily the result of grain boundary sliding although occasionally deformation within a grain is the apparent cause (Figure 5). This cavitation behavior, in conjunction with a general lack of dislocation activity, is again consistent with previous work[2,15] and supports our contention that diffusional creep and interface-controlled diffusional creep are the primary deformation mechanisms.

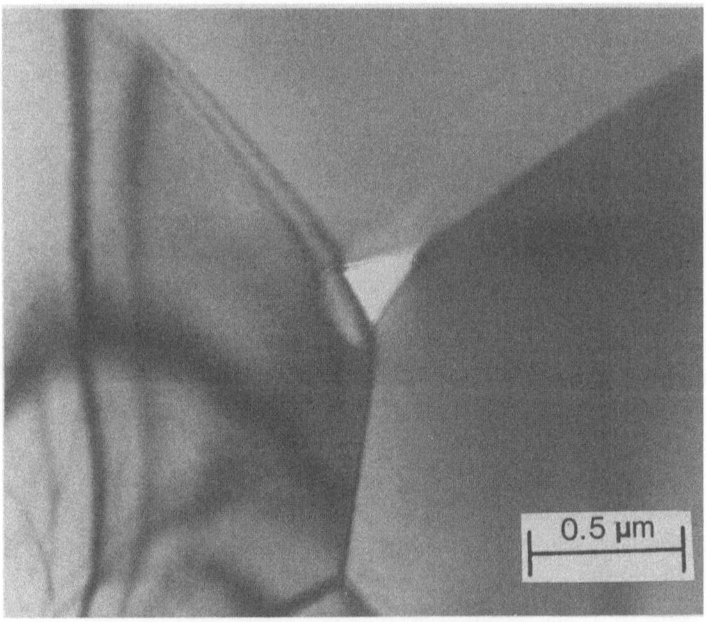

Figure 3. Typical triple point cavity present in a specimen deformed to 5% strain at 1623K.

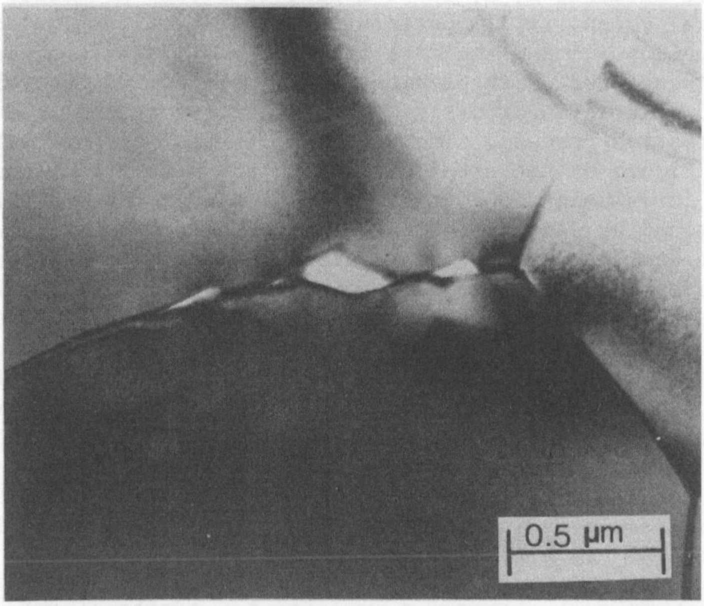

Figure 4. In addition to triple point cavities, deformation induced cavitation also took place along grain boundaries as in this specimen deformed to 5% strain at 1623K.

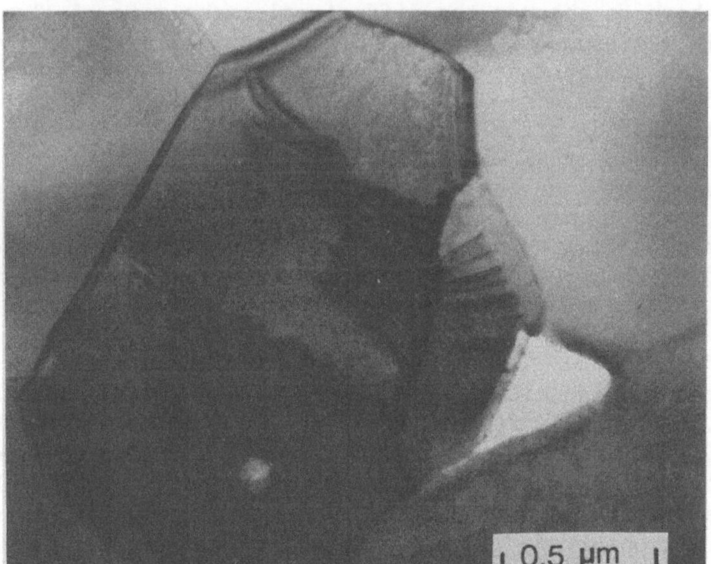

Figure 5. Cavitation in this region appears to be the result of localized dislocation activity.

Since aluminas A, B, and C appear to deform primarily by diffusional creep, it is possible to calculate diffusion coefficients using equation (1) and the data from Table II. This was accomplished, consistent with previous work, by assuming that either Al lattice diffusion or grain-boundary diffusion totally controlled the deformation process. The resulting lattice diffusivities are shown in Figure 6; the straight line is the best fit diffusion equation for alumina doped with Mg in excess of the solubility limit as determined by Cannon and Coble.[8] (Calculated values for alumina D are not included due to the apparent predominant interface-controlled nature of the deformation.) The calculated diffusivity values are consistently higher compared to the plotted line with the doped aluminas B and C yielding larger values than the undoped alumina A. The calculated boundary diffusivities plotted in Figure 7 show a similar trend for undoped versus doped alumina. More importantly, with the exception of alumina C at 1673 K, the diffusivities for the doped aluminas are in excellent agreement with the best fit boundary diffusion equation determined by Cannon and Coble.[8] The latter result is in agreement with Cannon et al.[2] and is anticipated for such fine-grained materials.

The absolute magnitude of the calculated diffusivities in Figures 6 and 7 must be viewed with caution due to the presence of silicon in aluminas A, B, and C. The influence of silicon in solution on the lattice and grain-boundary diffusivities of alumina has not been studied. The presence of the tetravalent ion, Ti^{4+}, has been shown to enhance the diffusional creep rate of alumina.[16] However, it was suggested that interface-controlled diffusional creep was the dominant mechanism and furthermore, that interface-controlled kinetics could be promoted by a number of factors which include slow cation grain-boundary diffusion.[11,17] In the present study, the effect of Si on grain-boundary diffusivities appears to be relatively small as the diffusivities calculated for the doped aluminas, B and C, are in good agreement with previous work. Also, the presence of silicon cannot explain the differences in the grain-boundary diffusivities of the doped and undoped samples because the amount of silicon is approximately constant.

Based on Figure 7, a comparison of aluminas B and C versus A suggests that doping alumina with Mg enhances Al grain-boundary diffusion. While this result is not in agreement with previous work which indicated that Mg increased only lattice diffusion (see Cannon and Coble[8] for a review of this topic), it is consistent with work on alumina containing Mn^{2+} and Fe^{2+} where enhanced grain-boundary diffusion was observed.[11,12] A possible reason for the previous observed discrepancy may be the inherently low solubility of Mg in alumina coupled with the divalent/trivalent states that Mn and Fe exist in. In order to evaluate the effects of divalent Fe or Mn on deformation of alumina, the experiments must be conducted in a reducing atmosphere. However, all the Mn and Fe is not converted

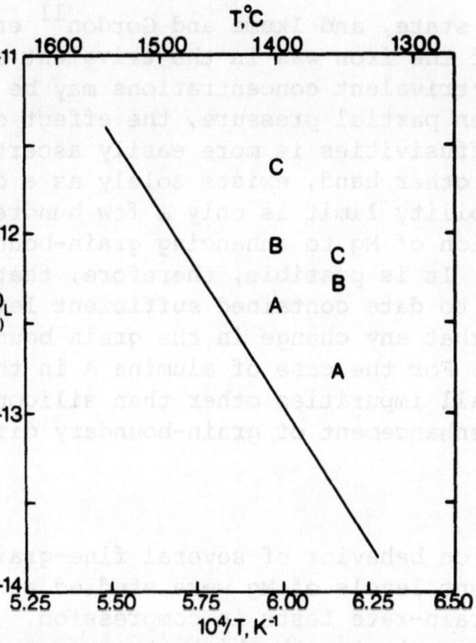

Figure 6. Plot of apparent lattice diffusivities for aluminas A, B, and C. Solid line is the lattice diffusivity equation from Cannon and Coble.[8]

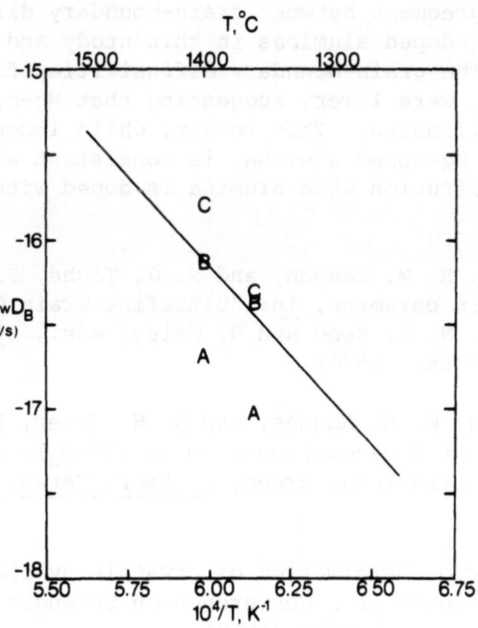

Figure 7. Plot of apparent grain-boundary diffusivities for aluminas A, B, and C. Solid line is the boundary diffusivity equation from Cannon and Coble.[8]

into the divalent state, and Ikuma and Gordon[11] estimated that in their work, 98% of the iron was in the trivalent state. Since a range of divalent/trivalent concentrations may be attained by changing the oxygen partial pressure, the effect of Fe or Mn on the grain-boundary diffusivities is more easily ascertained experimentally. Mg on the other hand, exists solely as a divalent ion and the observed solubility limit is only a few hundred ppm. Thus, the maximum contribution of Mg to enhancing grain-boundary diffusion is quickly attained. It is possible, therefore, that the pure materials studied to date contained sufficient levels of Mg and/or other impurities that any change in the grain-boundary diffusivity was undetectable. For the case of alumina A in the present study, the low level of all impurities other than silicon may have permitted detection of the enhancement of grain-boundary diffusion by Mg.

SUMMARY

The deformation behavior of several fine-grained aluminas containing different levels of Mg were studied at 1623 and 1673K using constant strain-rate tests in compression. At the imposed strain rates in these tests, the high strain-rate sensitivity values and flow stresses for three of the aluminas indicated diffusional creep was the dominate deformation mechanism; as-deformed microstructures were consistent with this mechanism.

Excellent agreement between grain-boundary diffusivities calculated for Mg-doped aluminas in this study and previous work were obtained. The grain-boundary diffusivities for the undoped alumina, however, were lower, suggesting that Mg-doping increases Al grain-boundary diffusion. This result, while inconsistent with previous work on Mg-doped alumina, is consistent with increases in grain-boundary diffusion when alumina is doped with Mn^{2+} and Fe^{2+}.

REFERENCES

1. A. H. Heuer, R. M. Cannon, and N. J. Tighe, Plastic deformation in fine grain ceramics, in: "Ultrafine-Grain Ceramics," J. J. Burke, N. L. Reed and R. Weiss, eds., Syracuse Univ. Press, Syracuse, (1970).

2. R. M. Cannon, W. H. Rhodes, and A. H. Heuer, Plastic deformation of fine-grained alumina (Al_2O_3): I, Interface-controlled diffusional creep, J. Amer. Ceram. Soc., 63:46 (1980).

3. F.R.N. Nabarro, Deformation of crystals by the motion of single ions, in: "Report of a Conference on Strength of Solids," Physical Society, London (1948).

4. C. Herring, Diffusional creep viscosity of a polycrystalline solid, J. Appl. Phys., 21:437 (1950).

5. R. L. Coble, Model for boundary diffusion controlled creep in polycrystalline materials, J. Appl. Phys., 34:1679 (1963).

6. R. Raj and M. F. Ashby, Grain boundary sliding and diffusional creep, Metall. Trans., 2:1113 (1971).

7. R. S. Gordon, Mass transport in the diffusional creep of ionic solids, J. Am. Ceram. Soc., 56:147 (1973).

8. R. M. Cannon and R. L. Coble, Review of diffusional creep of Al_2O_3, in: "Deformation of Ceramic Materials," R. C. Bradt and R. E. Tressler, eds., Plenum Press, New York (1975).

9. B. Burton, Interface reaction-controlled diffusional creep: A consideration of grain boundary dislocation climb sources, Mater. Sci. Eng., 10:9 (1972).

10. M. F. Ashby and R. A. Verrall, Diffusional-accomodated flow and superplasticity, Acta. Metall., 21:149 (1973).

11. Y. Ikuma and R. S. Gordon, Effect of doping simultaneously with iron and titanium on the diffusional creep of polycrystalline Al_2O_3, J. Am. Ceram. Soc., 66:139 (1983).

12. Y. Ikuma and R. S. Gordon, Enhancement of the diffusional creep of polycrystalline Al_2O_3 by simultaneous doping with manganese and titanium, J. Mater. Sci., 17:2961 (1982).

13. J. Hedworth and M. G. Stowell, The measurement of strain-rate sensitivity in superplastic alloys, J. of Mat. Sci., 6:1061 (1971).

14. J. D. Snow, "High Temperature Plastic Deformation of Polycrystalline Aluminum Oxide," M. S. Thesis, C.W.R.U., Cleveland (1971).

15. A. H. Heuer, N. J. Tighe, and R. M. Cannon, Plastic deformation of fine-grained alumina (Al_2O_3): II, Basal slip and nonaccommodated grain-boundary sliding, J. Am. Ceram. Soc., 63:53 (1980).

16. G. W. Hollenberg and R. S. Gordon, Effect of oxygen partial pressure on the creep of polycrystalline Al_2O_3 doped with Cr, Fe, or Ti, J. Am. Ceram. Soc., 56:140 (1973).

17. Y. Ikuma and R. S. Gordon, Role of interfacial defect-creation-annihilation processes at grain boundaries on the diffusional creep of polycrystalline alumina, in: "Surfaces and Interfaces in Ceramic and Ceramic-Metal Systems," J. Pask and A. G. Evans, eds., Plenum Press, New York (1981).

DEFORMATION STRUCTURES IN POLYCRYSTALLINE CaO

W. J. Brindley and B. J. Pletka

Department of Metallurgical Engineering
Michigan Technological University
Houghton, Michigan 49931

ABSTRACT

Polycrystalline CaO was deformed in compression using constant strain-rate tests at temperatures from 1323 to 1523K. The as-deformed surface morphology was examined and the dislocation structures analyzed by transmission electron microscopy. Subgrain dislocation structures reminiscent of the cell structures observed in creep of MgO are present in CaO and support a proposed recovery controlled dislocation creep model for CaO. In addition, a void growth contribution to the total strain at higher temperatures was suggested to be the result of viscous sliding by an intergranular glassy phase.

INTRODUCTION

The steady-state creep rate ($\dot{\varepsilon}$) at high temperatures in ceramic and metallic materials is related in many models to the temperature (T), diffusion coefficient (D), grain size (d), and applied stress (σ) as

$$\dot{\varepsilon} \propto \frac{D}{T} \left(\frac{1}{d}\right)^p (\sigma)^n \qquad (1)$$

where p and n are the inverse grain size and stress exponents, respectively. Frequently, the dominant deformation mechanism in a material is deduced by comparing experimentally measured values of p, n, and D with the predicted values for specific mechanisms. For example, a stress exponent of 1 is taken to indicate a process which depends on the presence of grain boundaries (boundary mechanisms) and is Newtonian viscous in nature. The most common boundary mechanism involves a stress-directed diffusional mechanism either through the lattice (Nabarro-Herring creep) or along the grain boundaries (Coble creep). Higher stress exponents of 3 to 5 are indicative of mechanisms independent of grain boundaries (lattice mechanisms) which involve some form of a dislocation glide and/or climb model. (See references 1, 2 or 3 for detailed discussions of the various creep mechanisms.)

Although a transition in the dominant creep mechanism for a polycrystalline material may occur as the applied stress is changed, a change in stress exponent from 5 to 1 as the stress is lowered in single crystals is not anticipated since the latter value is usually associated with a boundary mechanism. However, such a transition has been observed in the creep behavior of single crystals of aluminum[4] and CaO[5] as well as in polycrystals of these materials. For aluminum, a dislocation subgrain structure develops during creep and the stress exponent transition is attributed to a change in the stress dependence of the network density. Specifically, the dislocation density within the subgrains was found to be independent of stress at low stresses. A similar mechanism was suggested by Dixon-Stubbs and Wilshire to account for the transition in stress exponent for CaO although no experimental evidence of the subgrain structure was presented.[5]

The present paper reports on the as-deformed surface morphology and dislocation substructures which result when polycrystalline CaO is deformed in constant strain-rate tests. Although a constant flow stress was not attained in the tests (which would be equivalent to obtaining a steady state creep rate in a creep test), the approximate stress/temperature range was achieved where $n \simeq 5$ creep was observed in the Dixon-Stubbs and Wilshire study. Thus, an evaluation of the proposed model was possible.

EXPERIMENTAL PROCEDURE

Polycrystalline CaO pellets were prepared from 99.95% CaO powder (Alfa Division, Ventron Corp.) by isostatic pressing at 240 MPa followed by sintering 24 hours at 1853K in air. The sintered pellets had a density of ~95% theoretical as determined by point counting and a grain size of 40 μm. Examination of an as-sintered specimen by transmission electron microscopy (TEM) showed the initial dislocation density was essentially zero.

Compression specimens were saw cut and ground on a 15 μm bonded diamond lap to approximate dimensions of 1.8mm x 1.8mm x 4mm. One face on each sample was polished through 1/4 μm diamond to facilitate observations of the as-deformed surface using scanning electron microscopy (SEM). Compression testing was carried out in air at a nearly constant strain rate of 2×10^{-4} sec^{-1} at temperatures of 1323, 1423, and 1523K. The temperature was controlled to ± 1K during the deformation tests. Following compression testing to ~5% true strain, the samples were unloaded and cooled at approximately 100K/min to below 800K. Cooling below 800K proceeded more slowly.

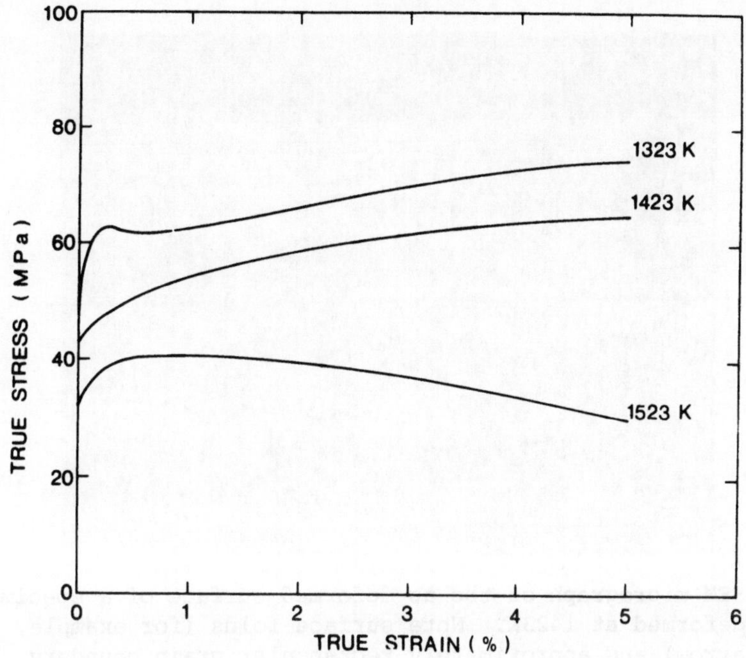

Fig. 1. True stress vs. true strain for constant strain-rate tests at 1323, 1423 and 1523K.

Foils for TEM were produced by mechanical polishing to approximately 0.08mm and ion thinning. TEM observations were made on a JEOL 100CX operating at 120 kV.

RESULTS

True stress-true strain curves for CaO deformed in compression at 1323, 1423, and 1523K are shown in Figure 1. A small yield drop occurred at 1323K but was absent at 1423 and 1523K. An initial work-hardening rate of ~μ/160 was found at 1323 and 1423K, followed by a decrease in the work-hardening rate with increasing strain. At 1523K, in contrast, yielding is followed by an approximate constant flow stress of 40 MPa until work-softening begins at strains of 1.25%.

The change in stress-strain behavior between 1423 and 1523K is accompanied by a change in the as-deformed surface morphology. Surface folds (which are caused by intense local slip) and small grain offsets out of the sample surfaces were observed in specimens deformed at 1423K (Figure 2) and at 1323K. Occasional grain boundary cavities oriented approximately parallel to the stress axis were also observed in specimens deformed at 1423K (Figure 2). Specimens deformed at 1523K, however, exhibit extensive grain offsets out of

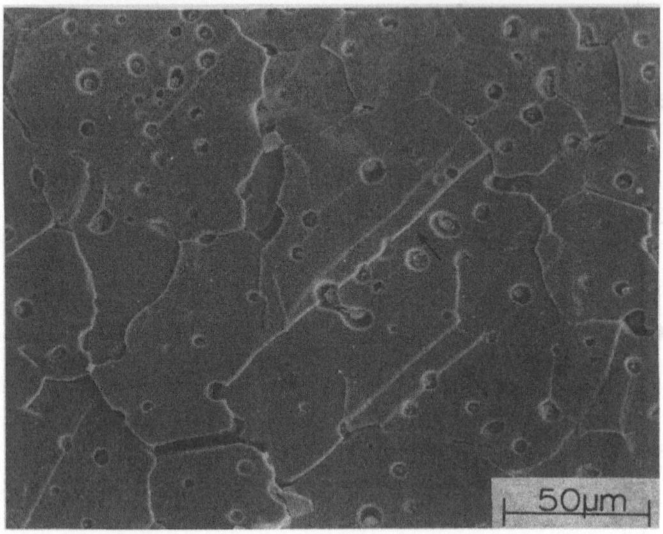

Fig. 2. SEM micrograph of the as-deformed surface of a specimen deformed at 1423K. Note surface folds (for example, see arrow) and approximately rectangular grain boundary cavities (bottom left center). Compression axis is horizontal.

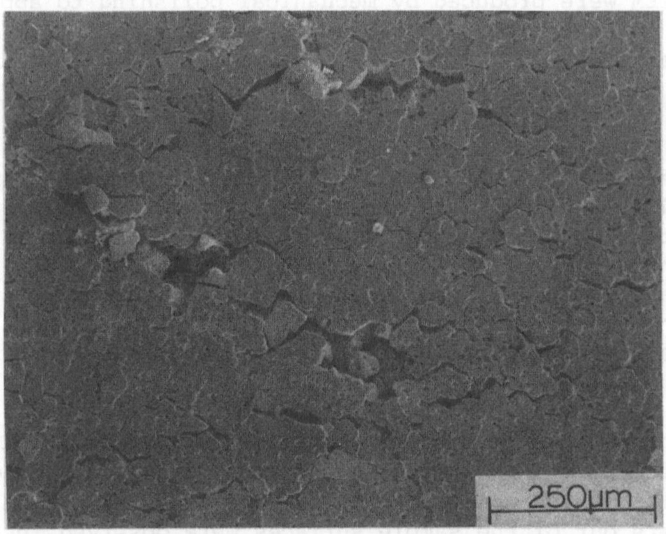

Fig. 3. As-deformed surface of a specimen deformed at 1523K. Large grain offsets and extensive cavity formation are present. Compression axis is horizontal.

the sample surface and profuse grain boundary cavitation, as shown in Figure 3. Only faint traces of surface folds were evident at 1523K.

The drastic change with temperature of the stress-strain behavior and deformed surface morphology did not extend to the observed dislocation substructures. At all three test temperatures, the dislocation substructures are characterized by a cell structure with a random distribution of dislocations in the cell interior, as shown for 1323 and 1523K deformation temperatures in Figures 4 and 5, respectively. All foils examined in the present study were cut from specimens deformed to approximately 5% true strain. Since the constant strain-rate tests at 1323 and 1423K had reached a low work-hardening rate at this strain (Fig. 1) the structures should represent those that would correspondingly form in creep tests during the later stages of transient creep; because work-softening occurred at 1523K, it is more difficult to ascertain whether the structure is equivalent to those formed at lower temperatures during transient creep.

Several different types of cell boundaries were observed. The most prevalent boundaries were composed of parallel arrays of dislocations (Figure 6) or a square cross-grid of dislocations (Figure 7). Diffraction contrast analysis showed that the former were composed of a single <110> Burgers vector while the latter consisted of two families of parallel dislocations with different <110> Burgers vectors. These "simple" cell boundaries have also been observed to form during transient creep in MgO and the details of their formation have been discussed by Reppich and Hüther.[6] In addition, more complex hexagonal networks were present at all three temperatures (Figure 8). The formation of the hexagonal cell boundaries has been explained by a mechanism in which glide dislocations from the cell interior "knit" together with the dislocations in the simple cell boundaries.[6] The hexagonal networks are also observed in the creep of MgO and are characteristic of steady-state deformation.[6]

DISCUSSION

The stress-strain curves for polycrystalline CaO at 1323 and 1423K have a work-hardening rate which decreases linearly with increasing stress after attaining a maximum value. Eventually, a steady state stress is achieved when the work-hardening rate equals zero; this was estimated from a plot of work-hardening rate versus stress to be at 79 MPa and 68 MPa at 1323 and 1423K, respectively. The equivalent creep test stress/strain-rate data for 1423K are found to be in good agreement with the experiments of Dixon-Stubbs and Wilshire[5] at 1473K in the $n \simeq 5$ regime, i.e., an extrapolated strain-rate of 2×10^{-4} s^{-1} at a stress of ~77MPa. This result supports our contention that the dislocation structures observed at true strains of 5% at 1323 and 1423K represent those formed in the

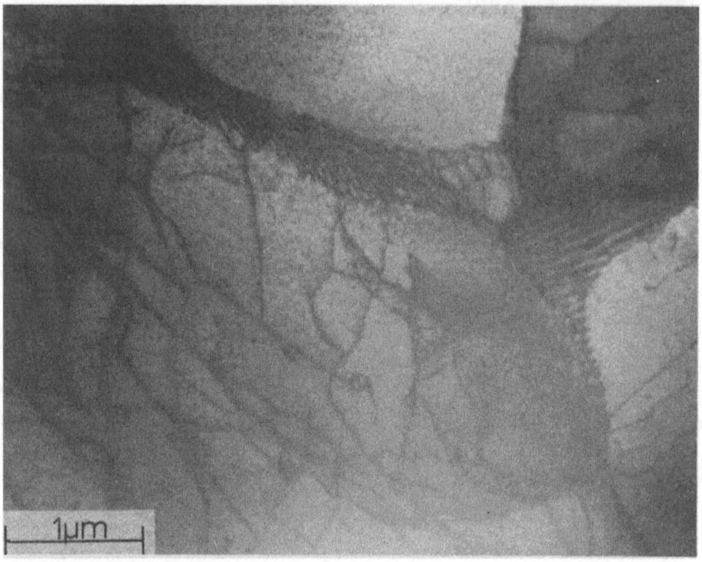

Fig. 4. Dislocation substructure in specimen deformed ~5% true strain at 1323K. Note cell boundaries and random dislocations within cells. Zone axis diffraction condition.

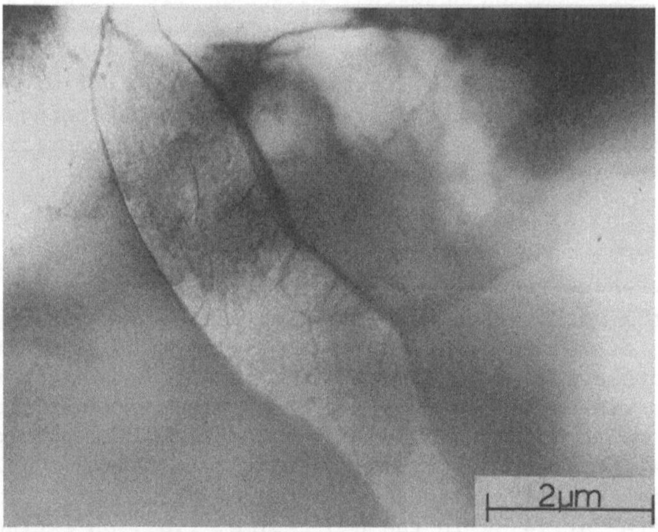

Fig. 5. Dislocation substructure in specimen deformed ~5% true strain at 1523K. Compare to Figure 4. See also Fig. 7.

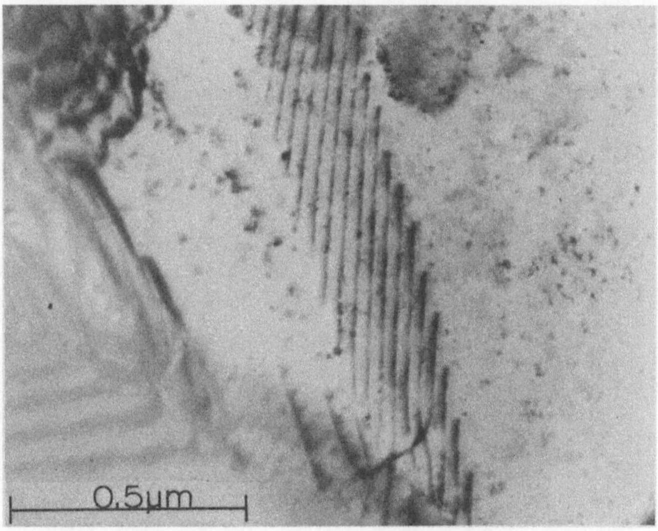

Fig. 6. Parallel dislocation array in specimen deformed at 1523K. Zone axis diffraction condition. Note: Dark spots in Figs. 6-8 are due to electron beam damage.

later stages of transient creep. Routbort has also argued that if a stress-strain curve exhibits a positive work-hardening rate which decreases linearly with increasing stress, a normal creep curve should be observed for that material in creep tests, i.e., a decreasing creep strain rate with increasing strain until steady state is achieved.[7] The predicted creep behavior from constant strain-rate tests have been confirmed in MgO and NiO[7] (which both possess the rocksalt structure, as does CaO) and is also confirmed when the present work is compared to the creep experiments of Dixon-Stubbs and Wilshire.

The work-hardening observed for CaO can be attributed in part to the formation of the cell boundaries. Reppich and Hüther pointed out that the cell boundaries in MgO can act as a "forest" for glide dislocations.[6] However, work hardening can also result from the interaction of the free glide dislocations within the cell interior. The relative contributions of these mechanisms to the work-hardening rate is unknown at present.

From our observations, it is concluded that the observed cell boundary structure is consistent with the structure necessary for the recovery-controlled dislocation mechanism proposed by Dixon-Stubbs and Wilshire for polycrystalline CaO in the $n \simeq 5$ regime. A critical test to ascertain whether the model proposed by Barrett, et al.[4] for aluminum is also appropriate to CaO at low stresses, is

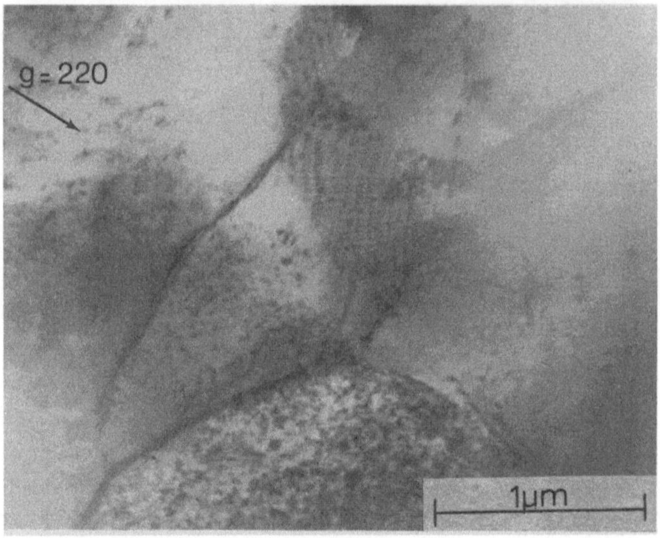

Fig. 7. Square dislocation array in specimen deformed at 1523K.

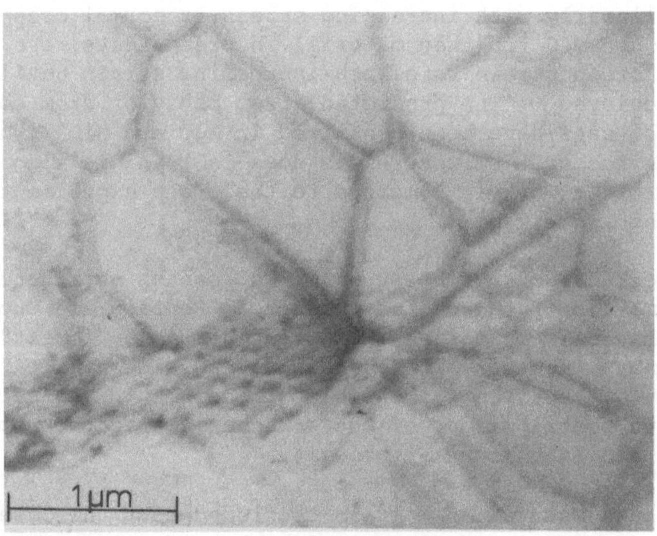

Fig. 8. Hexagonal dislocation array in specimen deformed at 1423K. Zone axis diffraction condition.

to determine whether the dislocation density within the subgrains becomes independent of stress in the n≃1 regime. Barrett et al. were able to measure the dislocation density by etch pitting because the subgrain diameter was large, ~0.01 to 1 cm. The subgrain diameters are smaller in CaO (see Figures 4 and 5) and such measurements would have to be performed using TEM. The deformation experiments at the low stress levels and the subsequent analysis are planned for the future.

The dislocation substructure at 1523K is similar to that formed at 1323 and 1423K, but the stress-strain behavior and surface morphology are quite different. Although extensive grain boundary cavitation occurred at 1523K, cavitation was also evident at 1423K. Since pores already exist on the grain boundaries, it is suggested that the change in behavior is due to more rapid void growth at 1523K than at 1323 or 1423K. This more rapid void growth leads to the link up of grain boundary voids and consequently, unstable void growth as the strain increases. Unstable void growth would lead to a decreasing stress with increasing strain as observed in the stress-strain curve at 1523K.

In view of the dislocation substructure developed even at 1523K, it would be tempting to suggest a dislocation creep-void growth mechanism.[8] This is unreasonable, however, since dislocation creep dominates at the lower temperatures and extensive cavity growth is not developed. Also, it is not obvious how such a rapid change in deformation behavior could be effected over such a narrow temperature range by this mechanism. Although void growth by a diffusional mechanism,[8] another possibility, is temperature dependent, the large grain offsets, surface folds at triple points and general cavity shape are consistent with partially accommodated grain boundary sliding (GBS). Therefore, unaccommodated GBS is suggested as the void growth mechanism.[10] Two GBS models could account for the results: GBS due to a dislocation mechanism which proceeds too rapidly to be accommodated by a general grain strain or a viscous GBS mechanism activated by the softening of an intergranular glassy phase.[10] Again, the deformation behavior at the lower temperatures suggests that the former mechanism is not feasible. The second, however, is plausible for the following two reasons: 1) glassy phase softening in Si_3N_4[9,10] and MgO-$CaMgSiO_4$[11] was observed to occur at approximately 1300 - 1600K, leading to extensive grain boundary sliding, and 2) a transition in deformation behavior (as the temperature was raised from below to above the glass softening temperature) in Si_3N_4 and MgO-$CaMgSiO_4$ also took place over a small range in temperature.[10,11] Although this transition may complicate the deformation behavior of polycrystalline CaO, the additional strain from void growth should not be dominant.[10]

CONCLUSIONS

The dislocation substructures in polycrystalline CaO deformed at 1323, 1423 and 1523K are characterized by cell boundaries and free dislocations within the cells. The cell boundaries were composed of a parallel dislocation array, square cross grids or hexagonal networks. The first two boundary types are of a simple character and have been observed to form during transient creep of MgO. The hexagonal array forms as a result of the interactions of glide dislocations with the simple cell boundaries and is characteristic of steady state creep. These structures are consistent with a proposed model for recovery controlled dislocation creep at 1473K in single and polycrystalline CaO in the high stress regime where $n \simeq 5$.

The surface morphology exhibited small grain offsets out of the surface and intense slip bands at 1323 and 1423K, not inconsistent with deformation controlled by a dislocation glide/climb model. However at 1523K, large grain offsets and extensive grain boundary cavitation were observed. It was suggested that a viscous sliding mechanism activated by the softening of an intergranular glassy phase led to more rapid void growth at 1523K.

ACKNOWLEDGMENTS

This work was supported in part by the National Science Foundation under grant number DMR 78-05741.

REFERENCES

1. W. R. Cannon and T. G. Langdon, Creep of ceramics, part 1 mechanical characteristics, J. Mat. Sci., 18:1 (1983).
2. T. G. Langdon, Grain boundary deformation processes, in "Deformation of Ceramic Materials," R. C. Bradt and R. E. Tressler, eds., Plenum Press, N.Y. (1975).
3. A. G. Evans and T. G. Langdon, Structural Ceramics, Prog. Mat. Sci., 21:171 (1976).
4. C. R. Barrett, E. C. Muehleisen, and W. D. Nix, High temperature-low stress creep of Al and Al+0.5% Fe, Mat. Sci. Eng., 10:33 (1972).
5. P. J. Dixon-Stubbs and B. Wilshire, Deformation processes during creep of single and polycrystalline CaO, Phil. Mag. A, 45:519 (1982).
6. B. Reppich and W. Hüther, Formation and structure of dislocation networks developed during high-temperature deformation of MgO, Phil. Mag., 30:1009 (1974).
7. J. L. Routbort, High temperature deformation of cubic oxides, Submitted to J. dePhysique.
8. A.C.F. Cocks and M. F. Ashby, On creep fractures by void growth, Prog. Mat. Sci., 27:189 (1982).

9. A. G. Evans, Deformation and failure caused by grain boundary sliding and brittle cracking, Acta Met., 28:1155 (1980).
10. O. L. Krivanek, T. M. Shaw, and G. Thomas, The microstructure and distribution of impurities in hot-pressed and sintered silicon nitride, J. Am. Ceram. Soc., 62:585 (1979).
11. W. E. Snowden and J. A. Pask, Creep behavior of a model refractory system $MgO-CaMgSiO_4$, J. Am. Ceram. Soc., 61:231 (1978).

PLASTIC DEFORMATION AND HARDENING OF POLYCRYSTALLINE HALIDES

Werner Skrotzki* and Peter Haasen

Institut für Metallphysik, Universität Göttingen
Göttingen, FRG

ABSTRACT

A review is given of recent advances in the understanding of the mechanical properties of polycrystalline halides. Emphasis is placed on the hardness/temperature relation and the brittle-to-ductile transition of these materials. It is shown that these properties can be clearly correlated with results on the plasticity of single crystals. The hardness of polycrystalline halides is determined by the activation of secondary {100}<110> slip, necessary to fulfil the von Mises criterion. On the other hand the ductility is correlated with the change of the slip mode on {110} planes from planar to wavy. Waviness is generated by cross slip of dissociated screw dislocations. Application of cross slip theory allows a quantitative estimate of the brittle-to-ductile transition temperature of polycrystalline aggregates.

INTRODUCTION

A great deal of interest has arisen recently in understanding the mechanical behaviour of polycrystalline halides. On the one hand these materials represent model substances for ceramics, which are becoming increasingly important from a technological point of view. On the other hand there also exists a great interest in understanding the thermomechanical behaviour of salt domes, which are considered

*present address: Department of Materials Science and Engineering, Cornell University, Ithaca, NY 14853

as potential repositories for radioactive nuclear waste. While in the first case basic research has been done, in the second case for a long time it has been tried to solve the thermomechanical disposal problems only from the engineering point of view. However, it has become evident, that long-term safety can be predicted only if necessary extrapolations are based on physically founded models.

Sprackling[1] has given a brief survey of the plasticity of ionic polycrystals, emphasizing the significance of the secondary slip systems for a general deformation of these materials. The actual role of these systems has however not been demonstrated clearly. A number of extended deformation studies have been made in our laboratory on polycrystalline halides in order to clarify this question[2,3]. The results of these studies will be discussed on the basis of the plastic anisotropy[4] and cross slip[5] investigated on single crystals. Emphasis is on the hardness/temperature relation and the brittle-to-ductile transition of these materials. Experimental details can be obtained by reference to the appropriate papers[2,3].

ROLE OF THE PLASTIC ANISOTROPY

NaCl-type ionic single crystals deform preferentially by slip on {110} and {100} planes along the <110> direction. To activate these slip systems different critical resolved shear stresses (CRSS) are necessary. This plastic anisotropy depends on temperature, the level of impurities and the degree of ionicity of the material. In halides in general, {110}<110> is found to be the primary and {100}<110> the secondary slip system. For pure alkali halides this is demonstrated in Fig. 1. Below a critical temperature T_o, the CRSS increases steeply due to the operation of the Peierls mechanism[6]. The higher Peierls potential for cubic slip leads to a strong plastic anisotropy below T_o{100}. The effect of this characteristic feature on polycrystalline plasticity will be discussed below.

A homogeneous plastic deformation of a polycrystal with a random distribution of grain orientations requires the activation of five independent slip systems[7]. There exist in halides six crystallographically distint slip systems of each type {110}<110> and {100}<110>, but only two and three, respectively, are independent of each other. Therefore the activation of both types of slip systems is necessary to deform a polycrystalline ionic solid. This is clearly seen in Fig. 2. The flow stress at 0.1% strain starts to increase at the temperature at which the harder cubic slip becomes more difficult. The fact that $\sigma_{0.1}(T)$ does not increase as strongly as τ_o{100}(T) presumably means that the deformation is mainly achieved by slip on {110} planes, especially at low temperatures. This effect is due to the double fibre texture existing in the polycrystals used[8]. For NaCl the fibre axis of the texture is mainly <100>, so the predominance of {110} slip is to be expected. AgCl on the other

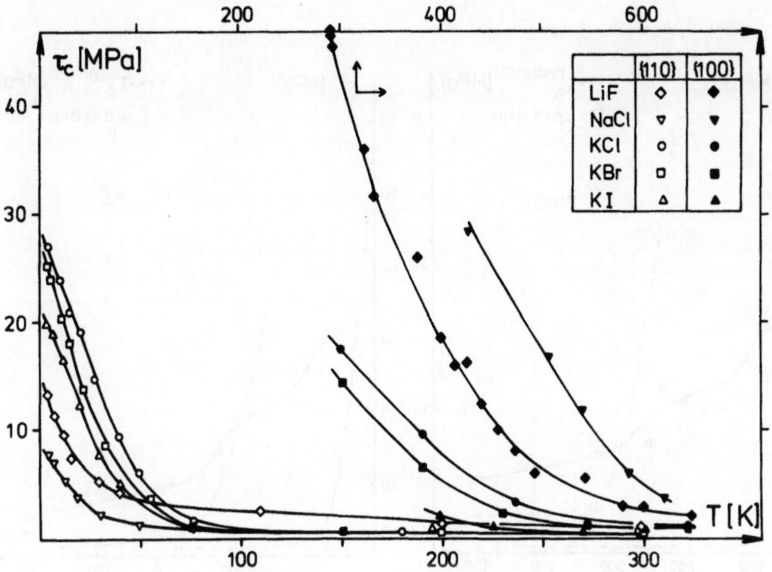

Fig. 1. Temperature dependence of the critical resolved shear stresses for slip on {110} and {100} planes. ($\dot{\varepsilon} \simeq 5 \cdot 10^{-5}$ sec^{-1})

hand has a <100><111> fibre texture. In agreement with this, the critical flow stress rises steeply at T_o{100}.

Brittle behaviour of polycrystals is to be expected, when the stress to activate the required independent slip systems is higher than the fracture stress. In polycrystalline halides this should occur below T_o{100}. Tension tests, however, reveal that the brittle-to-ductile transition temperature T_{bd} of these materials lies at much higher temperatures (Figs. 3 and 4). Slip line observations show that T_{bd} is correlated with T_w, the temperature at which the {110} slip mode changes from planar to wavy (Table 1). This effect is demonstrated in Fig. 5. At low temperatures the slip lines of screw dislocations on {110} planes are straight, while above a certain temperature, slip becomes wavy. Slip bands produced by screw dislocations of the cubic systems are wavy at all temperatures investigated (Fig. 6). For a long time it has been concluded that the change of the slip mode occurs at the temperature at which secondary slip becomes easy[9,10]. Table 1, however, clearly indicates that T_o{100} is smaller than T_w.

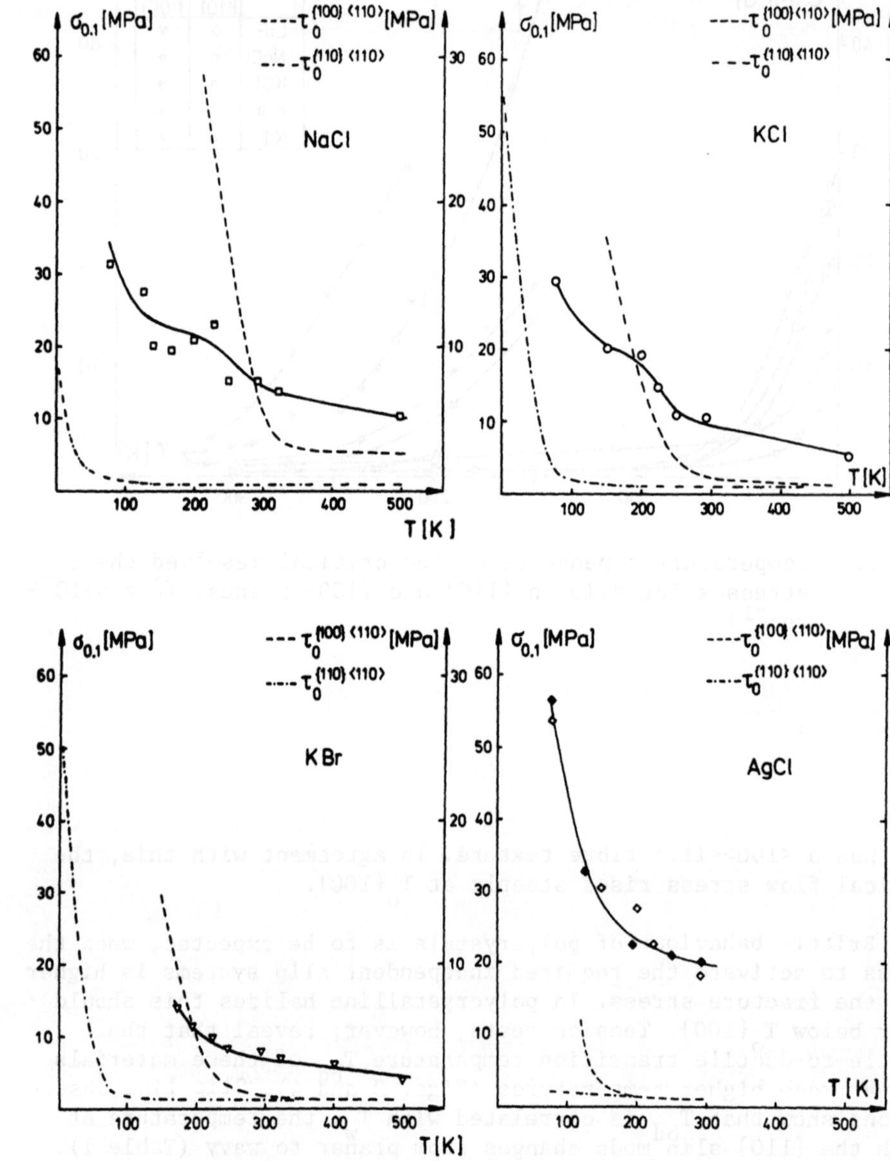

Fig. 2. Temperature dependence of the flow stress of polycrystalline ionic solids at 0.1% strain. Open and filled symbols refer to compression and tension, respectively. The dashed lines represent the temperature dependence of the CRSS for slip on {110} and {100} planes. ($\dot{\varepsilon} = 5\cdot10^{-5}$ sec^{-1})

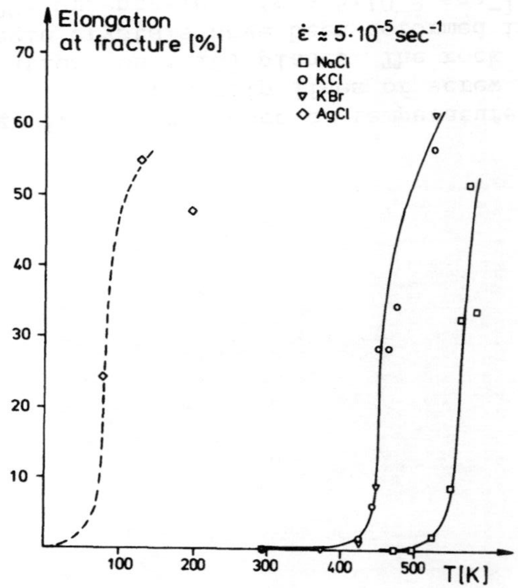

Fig. 3. Influence of temperature on the elongation at fracture of polycrystalline 'pure' ionic solids deformed in tension.

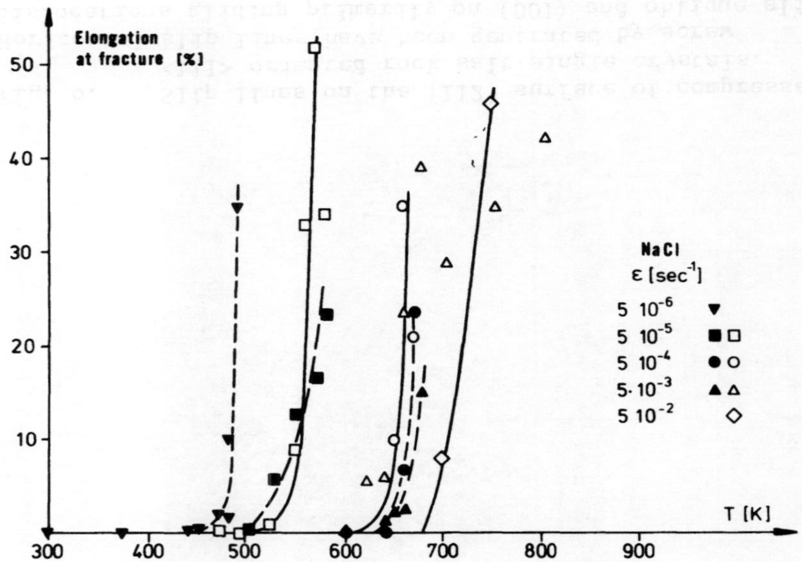

Fig. 4. The same as Fig. 3 for salt at different strain rates. Open and filled symbols refer to 'pure' and natural salt, respectively.

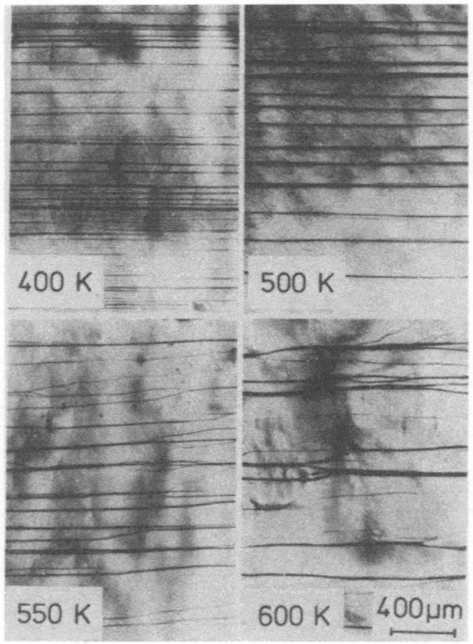

Fig. 5. Influence of temperature on the slip lines of screw dislocations on {110} planes. The rock salt single crystals have been deformed in <100> orientation. ($\dot{\varepsilon} \simeq 5 \cdot 10^{-5}$ sec^{-1}, $\varepsilon \simeq 1\%$)

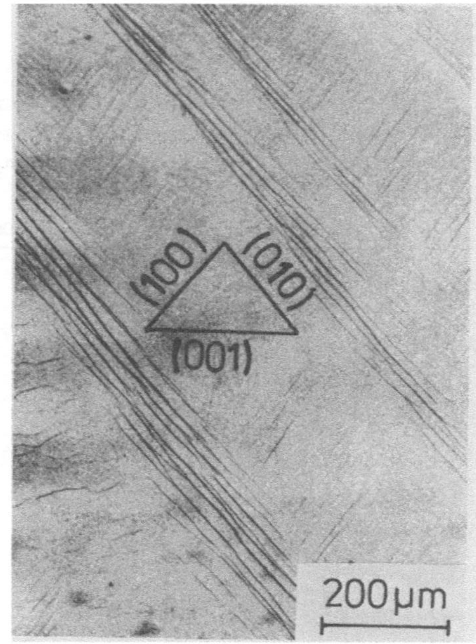

Fig. 6. Slip lines on the {112} surface of compressed <111> oriented rock salt single crystals. Horizontal slip lines have been generated by screw dislocations gliding primarily on (001) and oblique slip lines by dislocations with nearly edge character, gliding on (100) and (010) planes, respectively. ($T = 500$K, $\dot{\varepsilon} \simeq 5 \cdot 10^{-5}$ sec^{-1}, $\varepsilon \simeq 3\%$)

Table 1. Comparison of the brittle-to-ductile transition temperature (T_{bd}), the temperature at which slip of screw dislocations on {110} planes becomes wavy at the beginning of deformation (T_w) and the temperature below which slip on {100}becomes difficult (T_o{100}).

	LiF	NaCl	KCl	KBr	KI	AgCl
T_{bd} [K]		550	450	450		80
T_w [K]	700[30] 630[31]	550 570[32]	470	420	370	100
T_o{100} [K]	520[33]	320	250	230	215	165

ROLE OF CROSS SLIP

The term cross slip refers to the ability of screw dislocations to change their slip plane to any sharing the same slip direction. For screw dislocations on {110} the preferred cross slip plane is {100}[2]. The opposite is observed for {100} slip[11]. Slip line observations suggest that T_{bd} is correlated with the change of the slip mode on {110} from planar to wavy. Wavy slip is generated by cross slip of screw dislocations. If this process is possible at the beginning of deformation, then dislocation pile-ups need not be formed at grain boundaries. As a consequence stress concentrations leading to intergranular fracture[12] and crack nucleation according to the models of Stroh[13] and Cottrell[14] cannot build up. In the following section it will be shown that the latter model best explains the fracture behaviour of salt at temperatures below T_{bd}.

The fact that the plastic anisotropy increases drastically at T_o{100}, far below T_{bd}, shows clearly that just dislocation motion itself on the cross slip plane does not determine wavy slip. In the following it will be argued that the difficult step in the cross slip process is the constriction of the dissociated screw dislocation. Dissociation in ionic crystals has not been observed directly so far but there are a lot of indications supporting such an assumption:

i) Calculations of the stacking fault energy of ionic crystals[15-17] predict dislocations to be less dissociated on {100} than on {110} planes, in agreement with the different slip behaviour of screw dislocations inferred on these planes at low temperatures.

ii) The observed cross slip sequence from {110} to {111} and further to {100} found for NaCl correlates with an increase of the stacking fault energy[11].

iii) The onset of stage III of the stress strain curve of crystals deforming by {110}<110> slip has been shown to be determined by cross slip[11]. The dependence of τ_{III} on temperature, strain rate and hydrostatic pressure is in agreement with cross slip theory[18]. According to Hesse[19] and Skrotzki and Liu[5] in ionic crystals $\ln \tau_{III}/\mu$ ($\mu = (c_{11}-c_{12})/2$ shear modulus of the {110}plane) decreases linearly with increasing temperature as follows from the cross slip theory of Haasen[20] and Seeger et al.[21]. An extrapolation of τ_{III}/μ to high temperatures yields $\tau_{III}/\mu = \tau_o/\mu$ (τ_o = CRSS) at T_w, the temperature at which slip of screw dislocations on {110} planes becomes wavy at the beginning of deformation. According to table 1 this temperature agrees with T_{bd}. Thus there is a possibility to predict T_{bd} with the aid of the cross slip theory.

According to Seeger et al.[21] the strain rate dependence of T_{bd} is given by

$$T_{bd} = \frac{\ln [(\tau_{IIIo}/\mu_o)/\tau_o/\mu)]\ U}{\ln (\dot{\varepsilon}_o/\dot{\varepsilon})\ k} \qquad (1)$$

Fig. 7 shows $T_{bd}(\dot{\varepsilon})$ for pure salt estimated with $\tau_{IIIo}/\mu_o = 4.5\cdot10^{-3}$, $\tau_o/\mu = 0.22\cdot10^{-4}$, $\dot{\varepsilon}_o = (1.2\cdot10^2 - 10^5)\ \sec^{-1}$ and $U = (0.16 \pm 0.03)$ eV[5]. The hatched area gives the scatter resulting from the uncertainty in estimating $\dot{\varepsilon}_o$ and U. The agreement with the measured T_{bd} is good for $\dot{\varepsilon} < 10^{-3}\ \sec^{-1}$ justifying an extrapolation to low temperatures. The temperature range between 300 and 500K is important because such temperatures are expected around repositories of radioactive waste. Our analysis shows that in the case of tensile stresses higher than the critical flow stress, crack formation will occur if the strain rate exceeds a well defined limit, e.g. $10^{-(11 \pm 1)}\ \sec^{-1}$ at room temperature. This limit shifts to higher strain rates with increasing hydrostatic pressure p[18,22]. According to Skrotzki and Liu[5] the activation energy for cross slip is connected with the stacking fault energy γ by

$$U \approx \frac{\mu_s\, b^3}{157\ \gamma/\mu b}\,(\ \alpha - \gamma/\mu b\ )^{1/2} \qquad (2)$$

for $\gamma/\mu b \geq 0.015$, where

$$\gamma = \gamma_o + \delta b p/2 \qquad (3)$$

($\mu_s = (c_{44} \cdot (c_{11} - c_{12})/2)^{1/2}$ effective modulus for the interaction of screw dislocations; b magnitude of the Burgers vector; γ_o(NaCl) = 0.288 J/m^2 [17]; δ(NaCl) = 0.327[18], expansion of the lattice perpendicular to the stacking fault). Combining eqs. (2) and (3) with (1) gives $T_{bd}(p)$. According to this dependence, the pressures of about 0.3 kbar existing at depths considered for the disposal of nuclear waste can be neglected. Impurities lead to an increase of τ_o [4] and τ_{III} [22]. As there is agreement between T_{bd} of natural and 'pure' salt eq. (1) implies that the ratio $(\tau_{IIIo}/\mu_o)/(\tau_o/\mu)$ remains constant with doping. At high temperatures, the measured transition temperatures deviate from those expected from eq. (1) because grain boundary sliding and dynamic recrystallization also contribute to polycrystalline plasticity[10]. Up to the present investigations, the thick marked area in Fig. 7 has to be considered as brittle field.

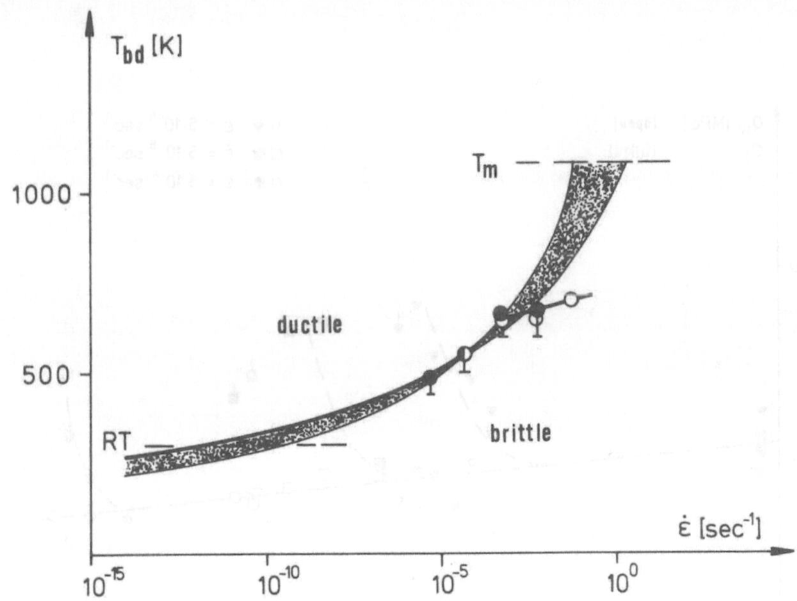

Fig. 7. Strain rate dependence of the brittle-to-ductile transition temperature. Open and filled symbols refer to 'pure' and natural salt, respectively. The sharpness of the transition is marked. The hatched area gives the uncertainty in the estimation of T_{bd} with eq. (1). (RT = room temperature, T_m = melting temperature)

FRACTURE MECHANISMS

The fracture behaviour of ionic polycrystals changes drastically - from intragranular to intergranular - at T_{bd}. This is shown up in the temperature dependence of the fracture stress (Fig. 8) and is supported by observations of the original and fracture surfaces of the specimens (Figs. 9 and 10). Below T_{bd} the fracture stress is about the flow stress $\sigma_{0.1}$ (i.e. fracture occurs at the beginning of deformation), above T_{bd} the fracture stress is larger (Fig. 8). Now the specimens tend to constrict. Below T_{bd} the strains at fracture are too small to produce slip steps large enough to be resolved by an optical microscope. Above T_{bd} large strains are achieved by wavy slip (Fig. 9). Now the elongation is limited by intergranular separations, mainly at grain boundaries perpendicular to the tensile axis. With increasing ductility, the density of 'river' patterns observed on the fracture surfaces rises markedly. When grain separation occurs the fracture surface is no longer planar. In the following, the possible intragranular fracture mechanisms operating below T_{bd} will be discussed in more detail.

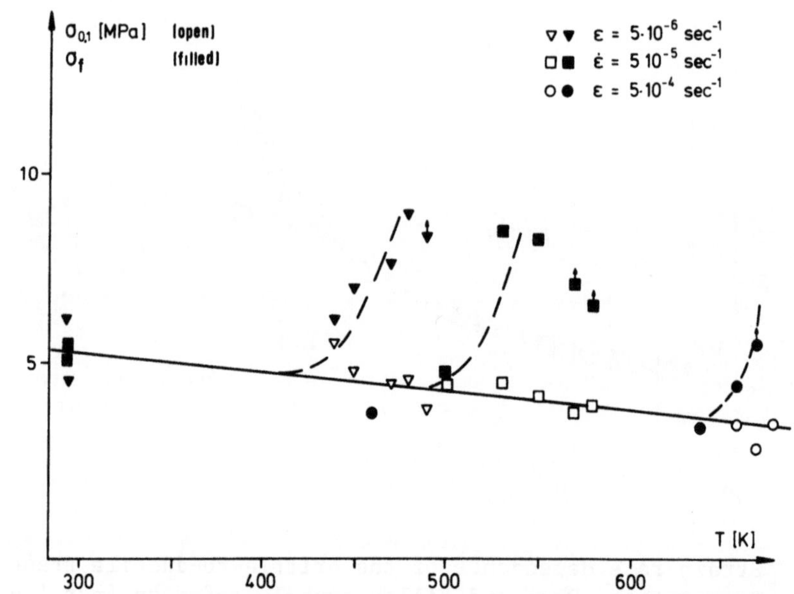

Fig. 8. Temperature dependence of the flow stress at 0.1% strain $\sigma_{0.1}$ and fracture stress σ_f of natural salt. Symbols with arrows mark specimens which showed constriction.

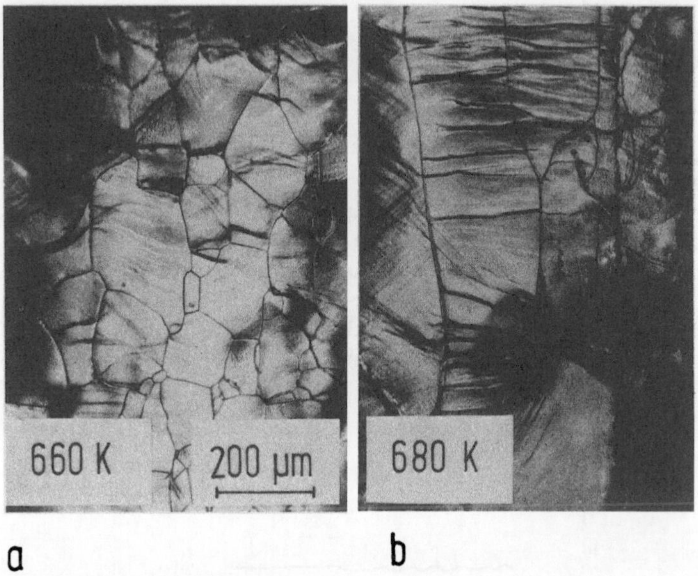

Fig. 9. Slip line structure on the specimen surface of tensile tested natural salt for $T > T_{bd}$. ($\dot{\varepsilon} \approx 5 \cdot 10^{-4}$ sec^{-1})

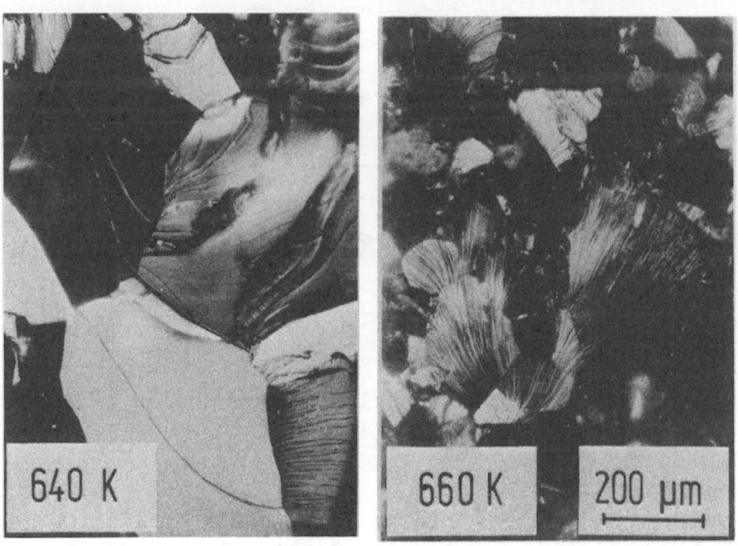

Fig. 10. Structure of the fracture surface of natural salt for $T \approx T_{bd}$. ($\dot{\varepsilon} \approx 5 \cdot 10^{-4}$ sec^{-1})

According to Stroh[13] (Fig. 11b), crack nucleation at a dislocation pile-up of n dislocations leads to fracture if a critical stress

$$\sigma_f = K d^{-1/2} + \sigma_o \qquad (4)$$

is reached, where $K = \left(\frac{3 \pi \gamma^* \mu}{2 \alpha(1-\nu)}\right)^{1/2}$, d = grain size, $\sigma_o = 2 \tau_o$ friction stress, γ^* = specific surface energy, ν = Poisson's ratio, $\alpha = L/d$, L = slip band length. According to this model, crack nucleation occurs by a combination of the two leading dislocations in the pile-up. Combination of further dislocations, i.e. crack growth, requires a lower stress. Stroh cracks are oriented perpendicular to the plane of the pile-up, thus in this case on {110}. Such features have been observed by Johnston [23] in compressed MgO.

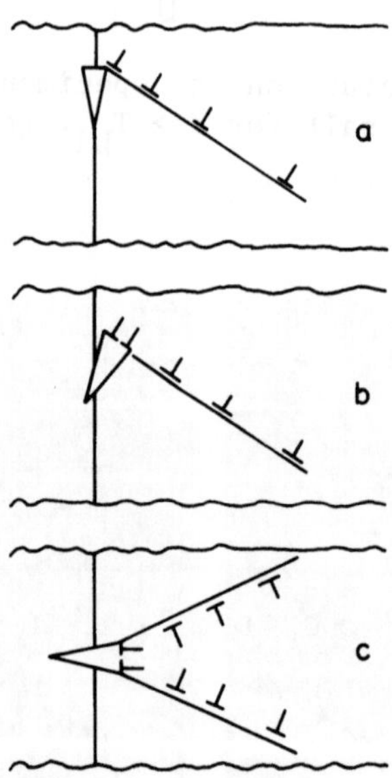

Fig. 11. Mechanisms of crack nucleation according to Zener[12] (a), Stroh[13] (b) and Cottrell[14] (c).

According to Cottrell[14] (Fig. 11c), crack nucleation can occur along the intersection of interfering slip planes, if the slip bands intersect at a grain boundary. A strong obstacle has to exist to prevent the sessile dislocation in front of the pile-up formed by dislocation reaction from dissociation. According to this model the dilated area of the reaction dislocation acts as crack nucleus. However, contrary to Stroh it is not the crack nucleation which determines fracture but the growth to a critical size. Fracture occurs when the total energy of the crack does not increase any more with increasing length. The condition for this situation is

$$\sigma\, n\, b = 2\, \gamma^* \tag{5}$$

With $nb = \beta nb' = \alpha\beta\,\pi d(1-\nu)(\tau-\tau_o)/\mu$ (Eshelby et al.[24]) and $\tau = \sigma/2$ eq. (5) gives

$$\sigma_f = \frac{\sigma_o}{2} + \left[\left(\frac{\sigma_o}{2}\right)^2 + \frac{4\,\mu\,\gamma^*}{\alpha\beta\,\pi(1-\nu)d}\right]^{1/2} \tag{6}$$

(β = constant; b, b' Burgers vectors of the reaction dislocation and the individual dislocations of the pile-up, respectively). In ionic crystals with NaCl-structure two different types of dislocation reactions of the {110}<110> slip systems are possible:

$$\text{I:}\quad \frac{a}{2}[\bar{1}01] + \frac{a}{2}[101] = a\,[001]$$

$$(101)\qquad (10\bar{1})\qquad (100)$$

The intersecting planes are perpendicular to each other and there is no reduction in line energy. The crack nucleus lies on a {100} plane. (a = lattice constant)

$$\text{II:}\quad \frac{a}{2}[10\bar{1}] + \frac{a}{2}[0\bar{1}1] = \frac{a}{2}[1\bar{1}0]$$

$$(101)\qquad (011)\qquad (112)$$

The intersecting planes are oblique (60°) to each other and the elastic energy of the reaction dislocation is reduced by 38%[25]. The crack nucleus lies on a {110} plane.

Fig. 12 shows a comparison of the trend of σ_f with $d^{-1/2}$ obtained from experiment and from eqs. (4) and (6), using the following parameters: σ_o = 1 MPa, $\gamma^*\{100\}$ = 0.150 J/m^2, $\gamma^*\{110\}$ = 0.375 J/m^2 [26], $\mu = 1.8\cdot 10^4$ MPa, ν = 0.2, $\beta(\text{I}) = \sqrt{2}$, $\beta(\text{II})$ = 1. For α a value of 4/3 has been used[27], taking into account that not all grain boundaries have to act as obstacles, especially in textured material.

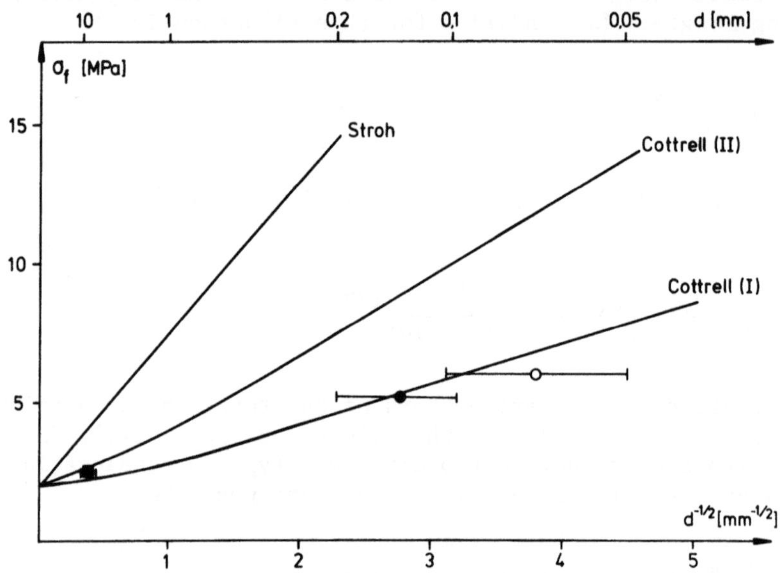

Fig. 12. Dependence of the fracture stress on grain size. Comparison of the experimental results with estimations according to Stroh[13] and Cottrell[14]. (■ Gessler[29])

Fig. 12 shows that the fracture behaviour of NaCl below T_{bd} is best described by Cottrell's model for orthogonal slip. Cottrell cracks on {100} have been found by Washburn et al.[28] in MgO deformed in tension. Fracture induced by interaction of oblique slip systems occurs at higher stresses. Stroh's model yields fracture stresses which are much too high. It might become more favourable if the crack nucleation in front of a pile-up occurs along the grain boundary according to a model of Zener[12] (Fig. 11a). However, using also eq. (4) a 25 times smaller surface energy has to be assumed to obtain agreement with experiment. This seems unreasonable. Moreover no signs have been observed on the fracture surfaces that fracture has been initiated intergranularly.

ACKNOWLEDGEMENT

The authors thank the Gesellschaft für Strahlen- und Umweltforschung mbH München for support.

REFERENCES

1. M. T. Sprackling, "The Plastic Deformation of Simple Ionic Crystals", Academic Press, London (1976)
2. W. Skrotzki, G. Frommeyer and P. Haasen, "Plasticity of Polycrystalline Ionic Solids", phys. stat. sol. (a) 66: 219 (1981)
3. W. Skrotzki, "An Estimate of the Brittle to Ductile Transition in Salt", Proc. of the 'First Conf. on the Mechanical Behaviour of Salt', Penn State Univ. , (1981)
4. W. Skrotzki and P. Haasen, "Hardening Mechanisms of Ionic Crystals on {110} and {100} Slip Planes", J. Physique 42: C3-119 (1981)
5. W. Skrotzki and Z. G. Liu, "Analysis of the Cross Slip Process in Alkali Halides", phys. stat. sol. (a) 73: K225 (1982)
6. W. Skrotzki and T. Suzuki, " Peierls Stresses in Ionic Crystals with the NaCl-Structure", Proc. of the '4. Europhysical Topical Conf. on Lattice Defects in Ionic Crystals', Dublin, (1982)
7. R. von Mises, "Mechanik der plastischen Formänderung von Kristallen", Z. angew. Mech. 8: 161 (1928)
8. W. Skrotzki and P. Welch, "Development of Texture and Microstructure in Extruded Ionic Polycrystalline Aggregates", Tectonophys. 98: (1983), in press
9. G. W. Groves and A. Kelly, "Independent Slip Systems in Crystals", Phil. Mag. 8: 887 (1963)
10. R. J. Stokes, "Mechanical Properties of Polycrystalline Sodium Chloride", Proc. Brit. Ceram. Soc. 6: 189 (1966)
11. K.-H. Matucha, "Elektronenmikroskopie von Gleitlinien auf verformtem NaCl", phys. stat. sol. 26: 291 (1968)
12. C. Zener, in: "Fracturing of Metals", ASM Cleveland (1948)
13. A. N. Stroh, "A Theory of Fracture of Metals", Adv. Phys. 6: 418 (1957)
14. A. H. Cottrell, "Theory of Brittle Fracture in Steel and Similar Metals", Trans. MS AIME 212: 192 (1958)
15. G. Fontaine, "Dissociation des Dislocations sur les Plans (110) dans les Cristaux Ioniques du Type NaCl", J. Phys. Chem. Solids 29: 209 (1968)
16. G. Fontaine and P. Haasen, "Hydrostatic Pressure and Plastic Deformation of the Alkali Halides", phys. stat. sol. 31: K67 (1969)
17. P. W. Tasker and T.J. Bullough, "An Atomistic Calculation of Extended Planar Defects in Ionic Crystals: Application to Stacking Faults in Alkali Halides", Phil. Mag. 43: 313 (1981)
18. P. Haasen, "Dissociation of Dislocations and Plasticity of Ionic Crystals", J. Physique 35: C7-167 (1974)
19. J. Hesse, "Die plastische Verformung von Natriumchlorid", phys. stat. sol. 9: 209 (1965)
20. P. Haasen, "Plastic Deformation of Nickel Single Crystals at Low Temperatures", Phil. Mag. 3: 384 (1958)

21. A. Seeger, R. Berner and H. Wolf, "Die experimentelle Bestimmung von Stapelfehlerenergien kubisch flächenzentrierter Metalle", Z. Phys. 155: 247 (1959)
22. E. Aladag, L. A. Davies and R. B. Gordon, "Cross Slip and the Plastic Deformation of NaCl Single and Polycrystals at High Pressure", Phil. Mag. 21: 469 (1970)
23. W.G, Johnston, "An Observation of Crack Formation in MgO", Phil. Mag. 5: 407 (1960)
24. J. D. Eshelby, F. C. Frank and F. R. N, Nabarro, "The Equilibrium of Linear Arrays of Dislocations", Phil. Mag. 42: 351 (1951)
25. Y. T. Chou and R. V. Whitmore, "Single and Double Pile-up of Dislocations in MgO Crystals", J. Appl. Phys. 32: 1920 (1961)
26. K. Meyer, "Physikalisch-chemische Kristallographie", VEB Verlag, Leipzig, 189 (1977)
27. A. Kochendörfer, "Die Bruchvorgänge in Experiment und Theorie, III", Z. Metallkde 62: 173 (1971)
28. J. Washburn, A. E. Gorum and R. E. Pasker, "Cause of Cleavage Fracture in Ductile Materials", Trans. MS AIME 214: 230 (1959)
29. K. Gessler, in: "Zulässige thermische Belastung von Gesteinen und sich ergebende Konsequenzen für die Methoden der Endlagerung radioaktiver Abfälle, Teil III: Endlagerung im Salz", Synthese Studie erstellt im Auftrag der Kommission der Europäischen Gemeinschaften, 119 (1980)
30. Z. G. Liu, "Gleitung auf {100} Ebenen in LiF", Diploma work, Göttingen (1981)
31. H. L. Fotedar and T. G. Stoebe, "Effect of Temperature, Strain Rate and Impurity Content on the Work-hardening Characteristics of LiF Single Crystals", Phil. Mag. 23: 859 (1971)
32. B. H. Kear, C. E. Silverstone and P. L. Pratt, "Latent Hardening, Work-hardening and Plastic Deformation of NaCl Single Crystals", Proc. Brit. Ceram. Soc. 6: 269 (1966)
33. Z. G. Liu and W. Skrotzki, "Slip on {100} Planes in LiF", phys. stat. sol. (a) 70: 433 (1982)

DEFORMATION PROCESSING OF CESIUM IODIDE CRYSTALS

W.W. Durand, B.G. Koepke* and W.W. Gerberich**

Honeywell Systems and Research Center, Minneapolis, MN
*Honeywell Ceramics Center, New Hope, MN
**University of Minnesota, Minneapolis, MN

I. INTRODUCTION

Cesium iodide (CsI) is a very useful optical material combining a low refractive index with broad spectral transparency. Unfortunately, the mechanical weakness of CsI crystals severely limits practical applicability of the material. Nearly all uses require mechanical strengthening. The strengthening treatment cannot, of course, degrade the desirable single crystal optical properties. The primary goal of this work was, therefore, demonstration of high optical quality, strengthened cesium iodide material.

Various strengthening treatments have been developed and demonstrated for other alkali halide crystals. The most important treatment involves deformation processing. Single crystal potassium chloride (KCl) and KCl alloys have been processed successfully at elevated temperatures (150° to 300°C) to produce strengthened, fine grained polygonized material. Increases in yield strength as large as an order of magnitude have been obtained without optical degradation[1]. Subgrain size, dependent upon process conditions, was as small as 4 μm in the strongest material. Strength decreased as process temperature (and subgrain size) increased. Unconstrained and constrained uniaxial forging as well as rolling were used successfully to strengthen KCl and other alkali halides having the rocksalt structure. Isostatic pressure and deformable constraining rings were frequently used to inhibit edge cracking during deformation, which often involved engineering strains in excess of 60 percent.

Our approach toward strengthening cesium iodide involved application of these same techniques. Initially, there were doubts concerning the applicability of these techniques because of differences in slip characteristics between KCl and CsI. The potassium salts, and other alkali halide crystals having the rock salt structure, slip on {110} <1$\bar{1}$0> systems at room temperature[2]. There are two independent {110} <1$\bar{1}$0> slip systems. According to the Von Mises criterion, five independent slip systems are required to maintain strain continuity (i.e., prevent cracking) during large, general deformations[3,4]. Activation of the three independent {110} <001> (above 200°C) fulfills the strain continuity requirement and allows deformation processing of these materials. CsI, on the other hand, has the body centered cubic cesium chloride structure and slips only on the primary {110} <001> systems at all practical processing temperatures[4,5].

Secondary slip may occur in CsI on the {110} <1$\bar{1}$0> systems. Sizova and Urusovskaya[6] have obtained evidence of secondary slip in the vicinity of hardness indentations using etch pitting techniques. Their calculations showed that stresses needed to activate the secondary system are 40 percent greater than for the primary system. Evidence of secondary slip was obtained only at very low temperatures under crystal orientations that inhibited primary slip. It is doubtful if secondary slip occurs under processing conditions used in our work. As a result, an inadequate number of slip systems operate in CsI and crack formation during deformation processing is expected. The presence of internal cracks or voids causes optical scattering, severely degrading performance. Demonstration of high optical quality, strengthened CsI was, therefore, not straightforward.

Although not straightforward, development of strengthened material is clearly needed. Pure, single crystal CsI is extremely weak. Dobryak, et al.[7] deformed pure CsI crystals along the <110> direction and measured yield stresses near 1.7 MPa (250 psi) at room temperature. Berezhkova and Regel[8] tested crystals having <110> orientations and showed that yield strengths decreased continuously as temperature increased over the range from room temperature to 500°C. Johnson and Pask[9] studied the mechanical behavior of CsBr crystals (also with the CsCl structure) and showed crystals deformed along the <111> direction had slightly lower yield stresses than those deformed along the <110> direction. This was attributed to the higher resolved shear stress on the operating slip systems in this orientation. CsI crystals cannot undergo general deformation along the <100> direction because no resolved shear stress exists on any of the primary {110} <001> slip systems. Under these conditions, crystals deform inhomogeneously by kinking[5,9,10]. At large strains, kinked crystals develop microscopic fractures[11].

Since five independent slip systems do not operate in CsI under most conditions, cracks may nucleate if local stress concentrations within the material cannot be relieved by plastic flow. Stress concentrations typically occur at the intersection of two slip bands.

The intersecting bands act as mutual dislocation glide barriers and cause dislocation pile-ups yielding local stress concentrations. Nucleation of microscopic cracks in this manner has been treated in detail by Stroh and others[12-14], who found that substantial stresses could be generated. Formation of a Stroh crack at the intersection of two {110} slip planes in a CsI crystal is shown schematically in Figure 1. The crystal is under load along the <110> direction. Cracks are expected to lie along the slip planes as shown. The number of cracks is expected to increase as deformation proceeds due to the increase in slip band density. As will be shown later, this occurs in CsI under certain conditions.

There is a major difference between our work and other previously reported CsI deformation studies. In previous studies, crystals were deformed to engineering strains up to about 20 percent[8,9]. In the present study, however, crystals were typically deformed to compressive engineering strains greater than 50 percent in order to maximize substructural refinement, strength and finished billet size. Fractures induced in CsI by these large deformations have not been addressed in the previous studies. Fracture, and its prevention, is a central issue in the present work.

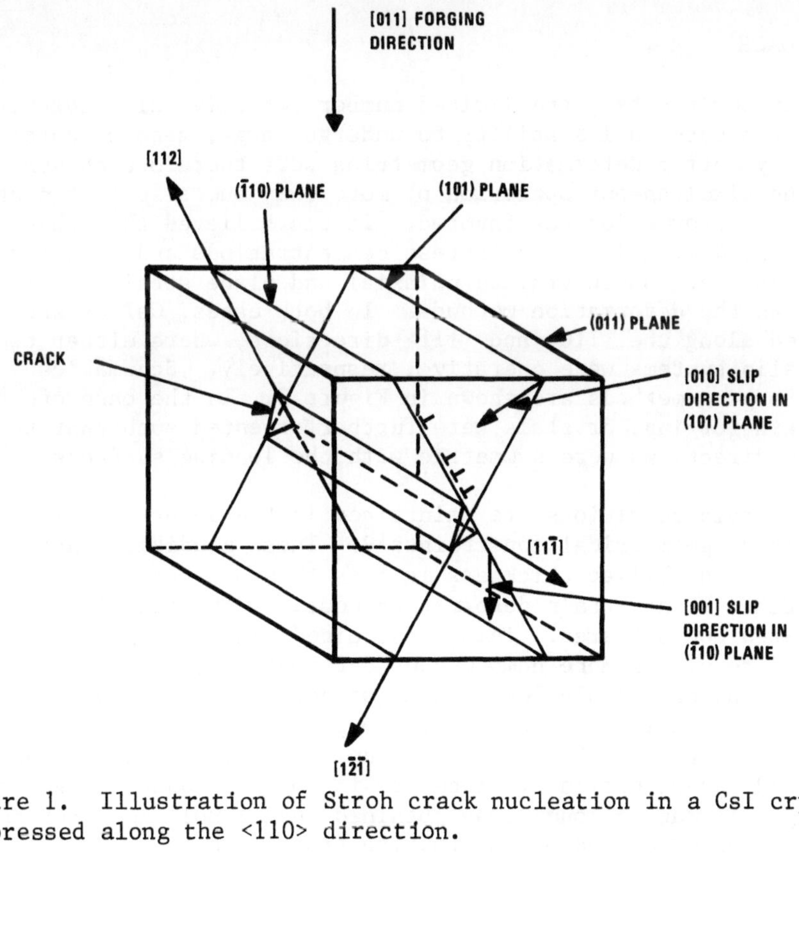

Figure 1. Illustration of Stroh crack nucleation in a CsI crystal compressed along the <110> direction.

The following discussion presents the details of our efforts to produce strengthened CsI by deformation processing of single crystals. The approach and experimental procedures are presented first, followed by a discussion of results. Also included are results of a thermally activated deformation analysis performed for CsI. This analysis provided considerable insight into the deformation mechanisms operating in the material.

As a final note, we wish to point out that our efforts were successful. Under certain conditions, substantially strengthened CsI was obtained with excellent optical performance. These results were obtained from pure, single crystal starting material. Additional strengthening may be possible by alloying. Several investigators have shown that even minor additions of mono or divalent cationic dopants to materials with the rocksalt structure produce appreciable strengthening[15-17]. The effects of alloying on CsI are, however, not well known. Dobryak, et al.[7] have shown considerable strengthening in CsI crystals alloyed with a number of elements but the data must be considered incomplete. We therefore elected to initially study deformation processing of only pure CsI material.

II. APPROACH

As previously noted, the limited number of active slip systems in CsI cast doubts on its ability to undergo large, general deformations. Symmetric deformation geometries were therefore chosen, so that the simultaneous operation of more than one slip system at the onset of deformation was favored. It was believed that this condition would minimize local stress concentrations and associated crack nucleation. Unconstrained uniaxial and plane strain forging were used as the deformation methods. In both cases, CsI crystals were loaded along the <110> and <111> directions, where either two or three slip systems were operative, respectively. Schematics of these forging methods are shown in Figure 2. In the case of plane strain forging, crystals were further oriented such that the <100> slip directions were symmetric with the loading surfaces.

Plane strain conditions are maintained in the deforming billet as a result of geometrical constraints[18]. These conditions are obtained when the billet thickness is less than one tenth of the billet width. Plane strain deformation confines the plastic zone to the platen contact area. This constrained plastic zone of limited size should result in more homogeneous deformation (less slip band separation) and presumably less chance of persistent slip band induced crack nucleation. Plane strain forgings were performed using long, narrow platens configured in the uniaxial forging apparatus. This allowed investigation of plane strain forging capabilities while simulating deformation conditions obtained during rolling. Hot rolling is a preferred method of deformation processing.

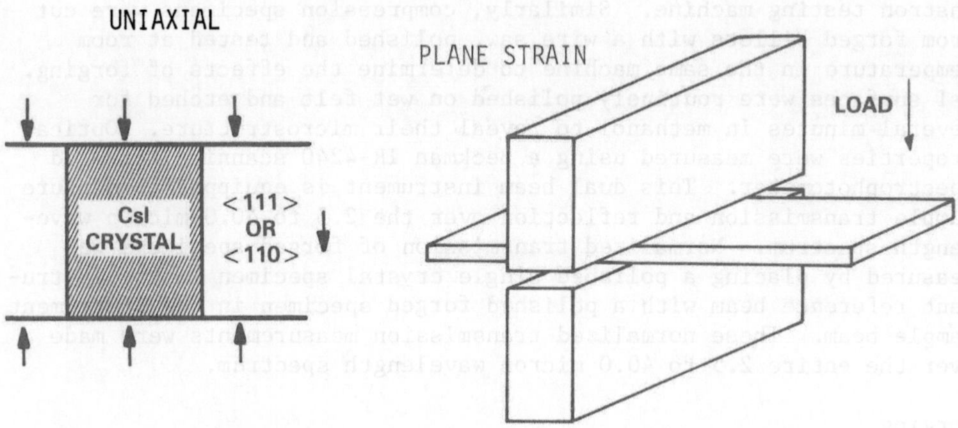

Figure 2. Diagrams showing both uniaxial and plane strain CsI forging approaches.

Deformation temperatures were minimized in all cases in order to produce fine grained microstructures and billets with maximum strength. Earlier alkali halide deformation studies showed that such fine grained microstructures are produced by polygonization and not by recrystallization[1]. This was considered advantageous because low angle grain boundaries are presumably more easily penetrated by slip bands during deformation than high angle grain boundaries. Ease of slip band penetration would, of course, minimize nucleation and growth of Stroh cracks.

III. EXPERIMENTAL PROCEDURE

Materials

Cylindrical, single crystal cesium iodide billets were obtained from the Harshaw Chemical Co., Solon, Ohio. Most crystals had a unity aspect ratio (equal diameter and height) with a 1.3 cm (0.5 inch) diameter. The crystals had a nominal purity of 99.999 percent. The axial crystal orientations were specified as <110> and <111>. Laue x-ray measurements confirmed that these orientations were obtained within several degrees. Plane strain forging specimens were cut from the cylinders with the {111} orientation parallel to the platen surfaces. All specimens were water polished on moist felt to remove machining damage prior to forging. Mechanical properties of the as-received crystals were measured in compression using an

Instron testing machine. Similarly, compression specimens were cut from forged billets with a wire saw, polished and tested at room temperature in the same machine to determine the effects of forging. CsI surfaces were routinely polished on wet felt and etched for several minutes in methanol to reveal their microstructure. Optical properties were measured using a Beckman IR-4240 scanning infrared spectrophotometer. This dual beam instrument is equipped to measure sample transmission and reflection over the 2.5 to 40.0 micron wavelength spectrum. Normalized transmission of forged specimens was measured by placing a polished single crystal specimen in the instrument reference beam with a polished forged specimen in the instrument sample beam. These normalized transmission measurements were made over the entire 2.5 to 40.0 micron wavelength spectrum.

Forging

A schematic diagram and photograph of the stainless steel forging apparatus used for this work are shown in Figure 3. The upper and lower platens were heated by resistive cartridge heaters and the outer cylinder by a resistive band heater. All heaters were separately controlled and temperatures inside the chamber were controlled to within $1^{o}C$ during any forging experiment. The chamber was filled with silicone oil to enhance temperature uniformity and to keep forge surfaces lubricated. During process development, thin teflon sheets were placed between the platens and billet to provide further lubrication. The effects of teflon addition were pronounced and will be discussed later.

Most forgings were carried out at constant crosshead speeds corresponding to initial strain rates, ranging from 10^{-3} min^{-1} to 10^{-2} min^{-1}. Forge temperatures ranged from $25^{o}C$ to $250^{o}C$, but were most commonly maintained between $150^{o}C$ and $250^{o}C$. Strain rate sensitivity tests were also performed during forging and loading increments corresponding to changes in crosshead speed were recorded for experimental analysis of thermally activated deformation in CsI crystals.

IV. RESULTS AND DISCUSSION

Uniaxial Forging

Forging Stress. Stress-strain characteristics for three <111> oriented CsI crystals forged at different temperatures are shown in Figure 4. These typical characteristics indicate that little apparent work hardening occurs before compressive strains greater than 20 percent are reached. Beyond 20 percent strain, forging stresses increase. We speculate that the increase in stress results from an increase in friction at the platen/billet interfaces, rather than

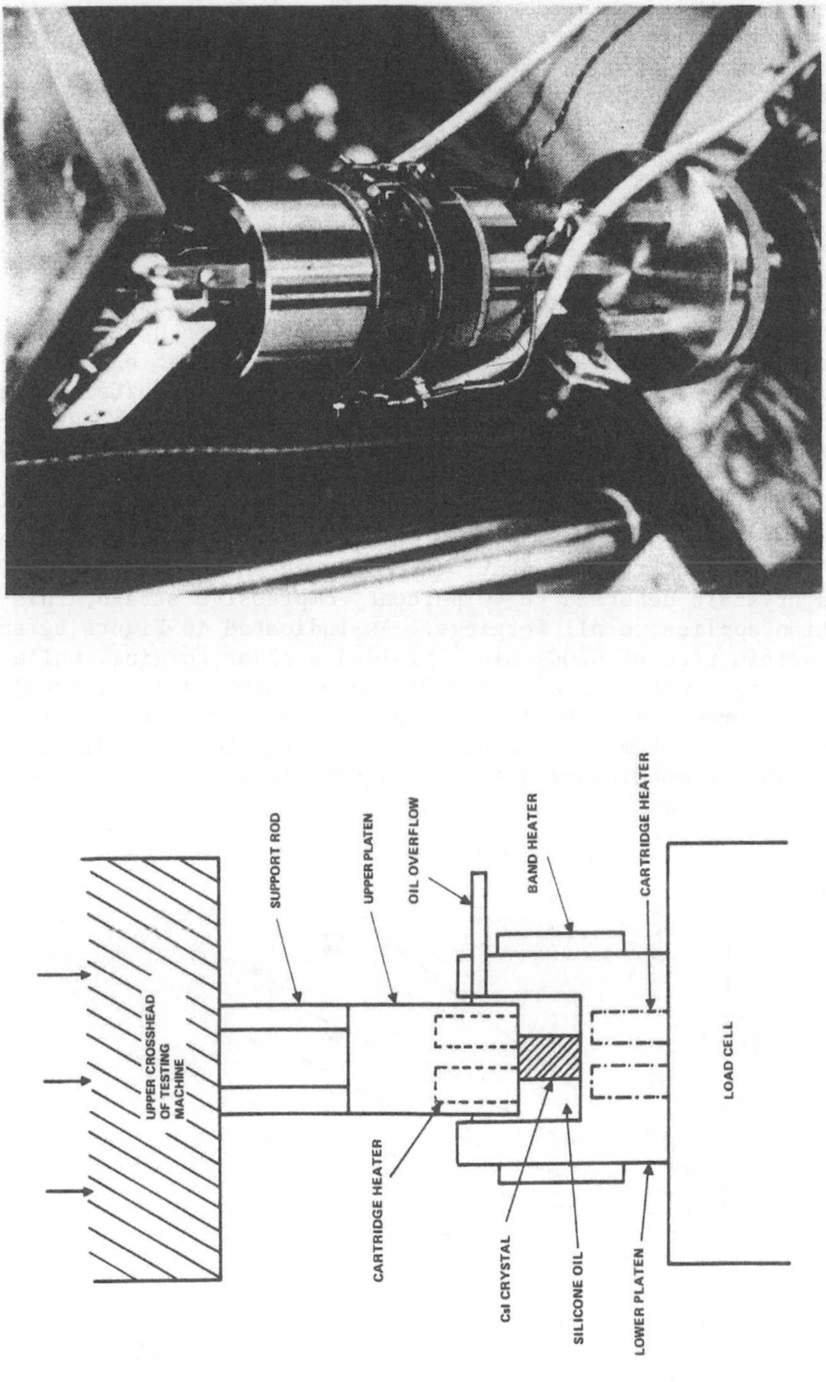

Figure 3. Schematic drawing and photograph of the CsI forging apparatus.

an increase in the work hardening rate. The observed low work harding rates at low total strains agrees with previously reported results[5]. Forging stresses decrease with increasing temperature and increase with increasing strain rate as expected. Similar results were obtained for CsI crystals forged along the <110> direction. In this case, however, forging stresses are slightly higher due to lower resolved shear stresses on the active slip systems[9].

General Appearance. Early in the study, CsI crystals were forged without using teflon sheet lubrication. All forgings appeared extremely cloudy as shown in Figure 5a. The subsequent use of teflon sheet lubrication yielded clear billets under the same forging conditions as shown in Figure 5b. The teflon sheets are believed to provide a deformable layer at the platen/CsI billet interface where slip bands intersect during forging. Such surface deformation relieves local stresses preventing crack nucleation.

Clear forgings were not obtained under all conditions even with teflon sheet lubrication. Figure 6 shows that increased strain rates lead to greater internal fracturing. Although shown only for <110> oriented crystals deformed to 40 percent compressive strain, this observation applies to all forgings. As indicated in Figure 6, an initial strain rate of 0.004 min^{-1} yielded a clear forging, while cloudy forgings were obtained at 0.01 and 0.2 min^{-1} initial strain rates. The forged <110> billets appear oval as a result of only two active slip systems. Crystals forged along the <111> direction under equivalent conditions are more symmetrical.

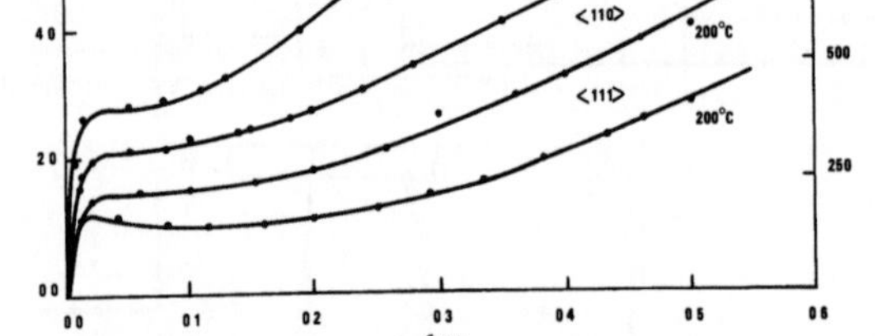

Figure 4. Stress-strain behavior during forging of CsI crystals along the <111> direction. A characteristic obtained by forging along the <110> direction is included for comparison. All forgings were made at the same constant crosshead speed ($\dot{\varepsilon}_0$=0.004 min^{-1}).

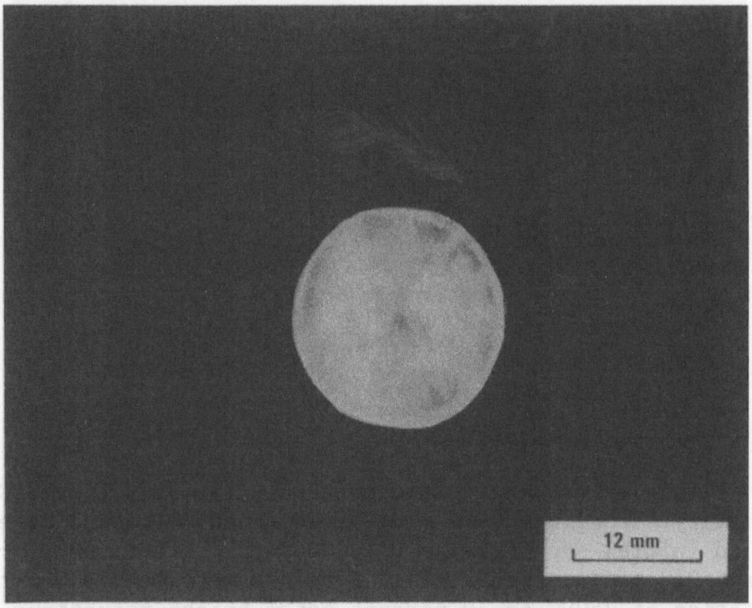

Figure 5a. Photograph of a CsI billet forged without teflon lubrication along the <111> direction. Constant crosshead speed according to an initial strain rate of 0.004 min^{-1} was applied at 250°C.

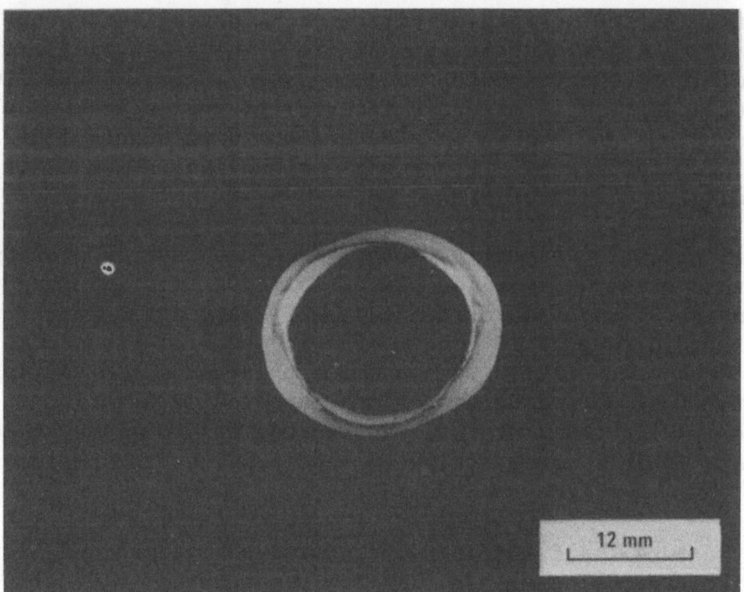

Figure 5b. Photograph of a CsI billet forged with teflon lubrication along the <111> direction. Constant crosshead speed according to an initial strain rate of 0.004 min^{-1} was applied at 250°C.

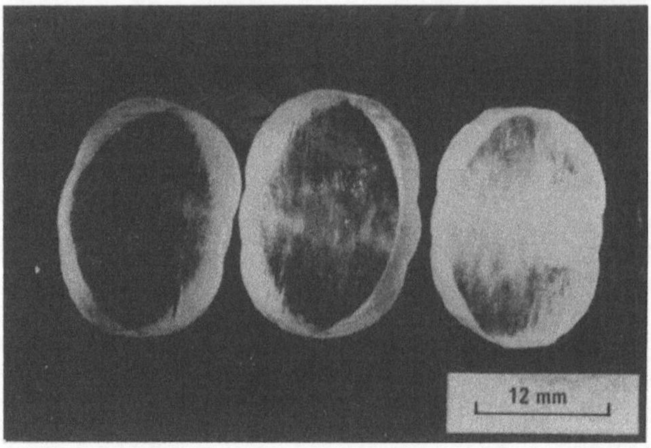

Figure 6. Effect of varying strain rate on uniaxially forged CsI crystal quality. Forgings were made at constant crosshead speed according to initial strain rates of 0.004 min^{-1} (left), 0.01 min^{-1} (center) and 0.2 min^{-1} (right). All forgings were made along the <110> direction at 250°C to 40 percent total height reductions.

Crack morphologies were examined in several forged billets. Forgings with limited internal cracks were polished to form flat surfaces parallel to the forging direction. Internal structure was examined using a light microscope. Figure 7 shows a typical structure obtained from a crystal forged along <110> direction at 200°C to 40 percent compressive strain. Internal fractures form at slip band intersections and extend in sheets along persistent <110> slip bands. This characteristic is taken as strong indication of fracture according to the Stroh mechanism.

A very interesting effect of strain rate was observed during the course of this study. A forging was performed at 25°C in which a <111> oriented CsI crystal was reduced by 70 percent in height at an initial strain rate of 10 sec^{-1}. The forged billet was completely transparent and crack-free. We speculate that adiabatic heating in the vicinity of slip band intersections may provide high local temperatures. High temperatures, as observed for forgings at low strain rates, inhibit crack nucleation and provide transparent billets.

Microstructure. All forgings introduced a fine grained microstructure into the CsI crystals. At low tempeatures, microstructures have a banded appearance corresponding to relatively high strain concentrations within the slip bands. An example is shown in Figure 8a. This structure, without well defined grain delineation, is typical of low temperature crystal forgings. Higher forging temperatures

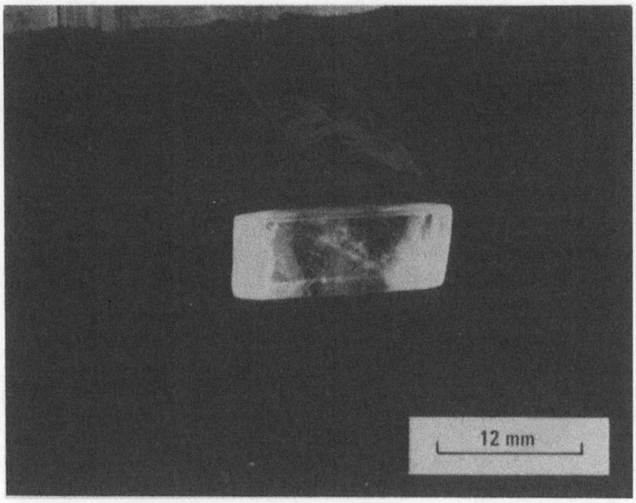

Figure 7. Side view of CsI billet forged along the <110> direction showing Stroh crack extension along slip bands. Billet was forged at a constant crosshead speed according to an initial strain rate of 0.01 min^{-1} at 250°C to 40 percent total height reduction.

yield a more equiaxed microstructure as shown in Figure 8b. Regardless of forging temperature, all billets forged during the study exhibited fine grained microstructures. This was of critical importance in strengthening pure CsI crystals. In contrast to previous work[11], microstructures produced during the study appear stable and do not undergo grain growth or secondary recrystallization.

Strength and Optical Properties. After forging, samples were cut from the billets using a wire saw. These samples were then polished and tested in compression at room temperature to determine their yield strengths. Measured yield strengths are plotted as a function of forging temperature in Figure 9. Yield strengths decrease linearly with forging temperature. The highest strength material obtained in the study was forged at room temperature. Strength increases as large as an order of magnitude above those for single crystals were obtained as shown. Measured strengths for crystals forged to 50 percent and 70 percent strains were approximately the same, indicating that work hardening is relatively independent of strain at large strains. This supports the contention that the flow stress increases shown in Figure 4 for strains beyond about 20 percent result from frictional effects as suggested earlier.

Optical properties were measured for the forged billets according to the previously described procedure. As mentioned, all samples were polished prior to measurement in order to eliminate surface scattering effects and thereby allow accurate characterization of

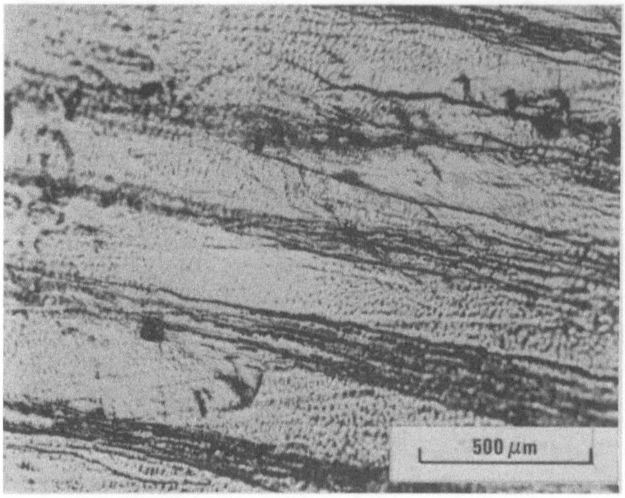

Figure 8a. Photomicrograph showing typical banded microstructures observed for low temperature (100°C) forgings. Billet was forged along the <111> direction to a 70 percent total reduction in height at a constant crosshead speed according to an initial strain rate of 0.004 min^{-1}.

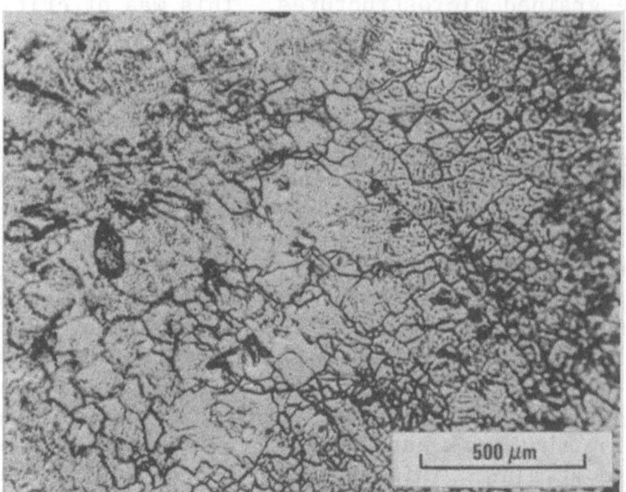

Figure 8b. Photomicrograph showing equiaxed microstructure observed for high temperature (200°C) forgings. Billet was forged along the <111> direction to a 50 percent total reduction in height at a constant crosshead speed according to an initial strain rate of 0.004 min^{-1}.

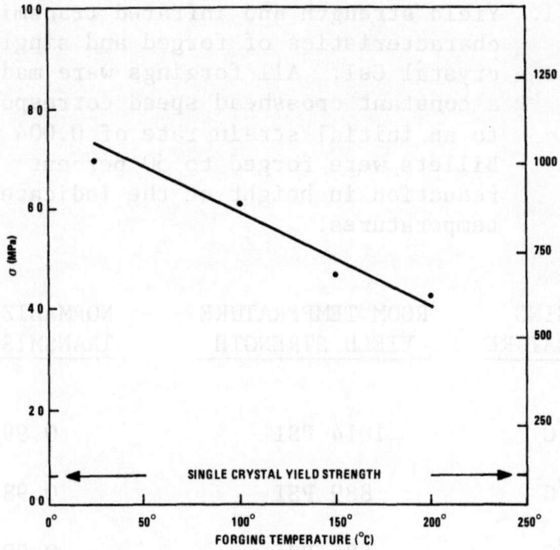

Figure 9. Room temperature yield strengths of forged CsI crystals as a function of forging temperature.

internal optical properties. Polished single crystals were used to perform normalized spectral transmission measurements and to establish a maximum transmission reference. Forged billets selected for optical measurement were equivalent to those characterized for yield strengths. These billets were all processed under optimized conditions and were visibly transparent. Measured infrared spectral transmission characteristics of single crystal and forged CsI samples were virtually identical. Within the accuracy of the analog spectrophotometer output, all measurements matched the maximum single crystal transmission reference value. These results are indicated in Table 1. Normalized transmission values represent an average value over the entire 2.5 to 40.0 micron wavelength spectral range and include a worst case measurement inaccuracy of 1 percent. Single crystal optical properties have clearly been preserved in the forged material.

Plane Strain Forging

As previously mentioned, plane strain or channel forging was investigated because it limits the size of the plastic deformation zone. This was regarded as a favorable condition in minimizing stress concentrations and associated crack nucleation. Single crystals used in the plane strain experiments were 1.3 cm (0.5 inch) diameter discs cut and polished to approximately 1.3mm (0.05 inch) thicknesses. Channel forging platens were also 1.3mm (0.05 inch) wide. According to Backoffen[18], these geometries will yield plane

Table 1. Yield strength and infrared transmission characteristics of forged and single crystal CsI. All forgings were made at a constant crosshead speed corresponding to an initial strain rate of 0.004 min^{-1}. Billets were forged to 50 percent reduction in height at the indicated temperatures.

FORGING TEMPERATURE	ROOM TEMPERATURE YIELD STRENGTH	NORMALIZED IR TRANSMISSION
30°C	1014 PSI	0.99
100°C	889 PSI	0.98
150°C	685 PSI	0.99
200°C	633 PSI	0.99
Single crystal	100 PSI	1.00

strain conditions. After forging, the crystals were reduced only along the narrow strip contacted by the platens. As such they are not useful components. As mentioned, however, channel forging adequately similates the conditions of rolling, which is a practical, production compatible processing method.

Photographs of crystals forged along the <100>, <110> and <111> directions under plane strain conditions are shown in Figure 10. Forged channels are indicated by dotted lines. The crystal orientations are also shown. All three forgings were transparent and showed no evidence of cracking. Most interesting was the clear channel obtained in the <100> oriented crystal. Uniaxial forgings of <100> oriented crystals always yielded cloudy material as a result of kinking and fracture. As previously discussed, homogeneous deformation does not occur in crystals with this orientation because of zero resolved shear stresses on active slip systems. Apparently, the complex stress states present during plane strain forging combined with the confined plastic zones yield crack-free material. Microstructures were examined for all samples forged under plane strain conditions. They appear very similar to those observed for uniaxially forged billets. Based on this limited study, we conclude that plane strain forging, particularly rolling, is a viable deformation processing method to produce strengthened, crack-free CsI.

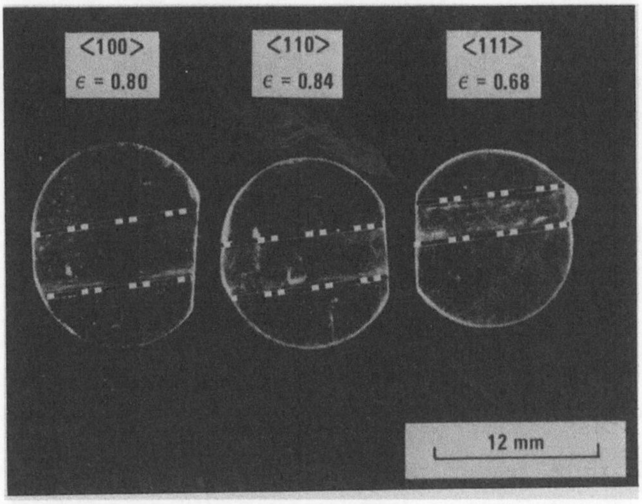

Figure 10. Photographs of channel forged CsI crystals loaded along the indicated directions. Forgings were performed with a constant crosshead speed according to an initial strain rate of 0.004 min^{-1} at 100°C. All forged channels are transparent. Observed blemishes are surface features.

Thermal Activation Analysis

A thermal activation analysis was conducted using data obtained during the various crystal forgings. The purpose of the analysis was to gain some insight into the principal rate-controlling mechanism(s) operating in CsI during deformation. General analysis methodologies are described in detail elsewhere[19,20,21] and will therefore not be treated here. Material strain-rate sensitivity data was obtained during forging at different temperatures. Measurements were made by incrementally varying the crosshead displacement rate of the testing machine. Maximum crosshead speed was always kept sufficiently low to obtain clear, crack-free forgings. Well defined "steady-state" loads were always observed after an incremental change in strain-rate.

The experimental data were analyzed using relations derived from a state equation for the thermal component of the flow stress. These relations provide a strain-rate sensitivity value (m*) and thermal activation volume (V*) according to

$$m^* = \left(\frac{\partial \ln \dot{\varepsilon}_p}{\partial \ln \sigma^*} \right)_{\varepsilon_p \to 0}$$

and

$$V^* = \frac{m^* kT}{\sigma^*}$$

where

σ^* = thermal flow stress component

ε_p = plastic strain

$\dot{\varepsilon}_p$ = plastic strain-rate

k = Boltzmann's constant

T = absolute temperature.

An activation enthalpy (ΔH) may be calculated using the relation

$$\Delta H_{\sigma^*} = -kT^2 \left(\frac{\partial \ln \dot{\varepsilon}_p}{\partial \sigma^*}\right)_T \left(\frac{\partial \sigma^*}{\partial T}\right)_{\dot{\varepsilon}_p}$$

Based on these relations, thermal activation parameters were calculated for CsI. These parameters are summarized in Table 2. Activation enthalpies were calculated near 0.56 eV (12.88 kcal/mole) and remained relatively constant over the temperatures studied. All calculations are consistent with analysis results for other alkali halide materials[22].

Comparison of these results to theoretical models allows detailed barrier profile analysis. This, in turn, defines the nature of the rate-controlling mechanism. Unfortunately, microscopic models are not usually well defined. Consequently, mechanism determinations usually proceed less formally, involving inferences from other related experiments. These inferences can be drawn from Sprackling[4] for CsI. Multiple stage hardening in various ionic solids is discussed. Our calculated thermal activation parameters fit well with a dislocation jog intersection model for work hardening. Jogs in CsI are sessile. Furthermore, calculated activation volumes and enthalpies correspond well with a thermally activated cross-slip mechanism, which also explains the observed general homogeneous deformation of CsI without fracture.

This brief summary of our thermal activation analysis has presented parameters calculated for forgings at different temperatures. These values compare well with both sessile jog intersection mechanisms for work hardening and a thermally activated cross-slip deformation mechanism. Expected macroscopic manifestations of these mechanisms are, in fact, observed experimentally.

Table 2. Summary of Activation Parameters for CsI Crystals

Temperature (°K)	m*	V*
298°	14.1	$313b^3$
373°	12.8	$889b^3$
423°	11.2	$980b^3$
473°	10.0	$1088b^3$
523°	8.8	$2117b^3$

V. REFERENCES

1. B.G. Koepke, R.H. Anderson and R.J. Stokes, in Deformation of Ceramic Materials, R.C. Bradt and R.E. Tressler, eds., Plenum, New York (1975) p. 497.
2. J.J. Gilman, Acta Met. 7, 608 (1959)
3. G.W. Groves and A. Kelly, Phil. Mag., 8, 877 (1963)
4. M.T. Sprackling, The Plastic Deformation of Simple Ionic Crystals, Academic Press, New York (1976)
5. V.R. Regel and G.V. Berezhkova, Soviet Phys. Crystal., 4, 718 (1959)
6. N.L. Sizova and A.A. Urusovskaya, Kristall and Technik 12, 823 (1977)
7. V.M. Dobryak, Ya.A. Zakharin, A.N. Panova, I.A. Rachkov and A.A. Urusovskaya, Soviet Phys. Crystal, 23, 239 (1978)
8. G.V. Berezhkova and V.R. Regel, Soviet Phys. Crystal., 6, 115 (1961)
9. L.D. Johnson and J.A. Pask, Journ. Amer. Ceramic Soc., 47, 437 (1964)
10. M.V. Klassen-Neklyudova, M.A. Chernysheva and G.E. Tomilovskii, Soviet Phys. Crystal., 5, 617 (1960)
11. A.A. Urusovskaya and R. Thyagarajan, Phys. Stat. Sol., 10, 349 (1965)
12. A.N. Stroh, Proc. Roy. Soc., A233, 404 (1954)
13. A.N. Stroh, Adv. Phys., 6, 418 (1957)
14. E. Smith and J.T. Barnby, Metal Sci. Journ., 1, 56 (1967)
15. B.G. Koepke, R.H. Anderson, E. Bernal G and R.J. Stokes, Proc. Fourth Conf. on High Power Infrared Laser Window Materials, ARPA, Arlington, VA (1975), p. 621
16. W.B. Harrison, G.D. Hendrickson and J.E. Starling, Proc. Third Conf. on High Power Infrared Laser Window Materials, C.A. Pitka, A. Armington and H. Posen, eds., Air Force Cambridge Research Laboratories, AFCRL-TR-74-0085 (1974).

17. P.F. Becher, R.W. Rice, P.H. Klein and S.W. Freiman, in Deformation of Ceramic Materials, R.C. Bradt and R.E. Tressler eds., Plenum, New York (1975), p. 517
18. W.A. Backofen, Deformation Processing, Addison-Wesly, Reading, MA., (1972)
19. U.F. Kocks, A.S. Argon and M.F. Ashby, Thermodynamics and Kinetics of Slip, in Progress in Materials Science, Vol. 19, B. Chalmers, J.W. Christian and T.B. Massalski, e s., Pergamon Press, New York, (1975)
20. A.G. Evans and R.D. Rawlings, Phys. Stat. Sol., 34, 9 (1969)
21. J.C.M. Li, Canadian Journ. Phys., 45, 493 (1967)
22. A.G. Evans, in Deformation of Ceramic Materials, R.C. Bradt and R.E. Tressler, eds., Plenum, New York, (1975) p. 127

HIGH TEMPERATURE CREEP OF PURE TUNGSTEN CARBIDE AND WC-Co ALLOYS WITH LOW COBALT VOLUMIC RATIOS

Sabine Lay and Frédéric Osterstock

Equipe Matériaux-Microstructure du L.A. 251
ISMRA - Université
14032 Caen Cedex, France

INTRODUCTION

Until recently, the high temperature deformation and creep behavior of tungsten carbide-cobalt alloys have been studied only in a temperature range which corresponds to the technical use of these materials (1)(2)(3)(4). Very high stresses must be applied and the deformation is often limited. Different cobalt volumic ratios, carbide grain sizes and additions are further parameters which make comparisons difficult. W.S. Williams et al. (5)(6)(7) made pioneering work in raising the testing temperature up to 1400°C.

As the carbide and the cobalt are expected to have very different properties at high temperatures, a systemic study of the influence of the cobalt content on the creep behavior in the temperature range 1050°C-1350°C was started in our laboratory (8)(9). Only alloys with cobalt volumic contents higher than 22 % were investigated. The creep behavior of these materials exhibits a sigmoïdal log $\dot{\varepsilon}$ - log σ plot and, in the range where the strain-rate dependance with stress was the lowest, the stress exponent proved to be solely dependant on the cobalt volumic ratio. The apparent activation energy were 130-135 kcal/mole and stress independant.

The aim of this work is to extend the examined cobalt content range toward lower volumic ratios, in order to verify the conclusions which were drawn and to compare the behavior of materials with and without cobalt, both having been sintered in the same conditions.

MATERIALS AND EXPERIMENTAL PROCEDURES

Some of the materials have been provided by the Eurotungsten Company. They were materials with a mean diameter of carbide crystals, $\bar{D}_{WC}$, of 0.7 μm with cobalt volumic ratios of 37 % (A2025), 22 % (A2015), 5 % (A2003) or 2.2 μm (cobalt volumic ratio 16 %, noted A110) and a pure carbide, with a mean diameter of carbide crystals of 6-8 μm, hot pressed at 2000°C. The two other batches have been kindly provided by Metallwerk Plansee (Reutte, Austria). They have the same mean diameter of the carbide crystals (0.8 μm) and are believed to have the same carbide structure since they have been sintered from the same powders and in the same conditions. The difference is that the one has been tested in the as-sintered state and the cobalt has been extracted from the second using a HCl solution. The extraction has been performed at Metallwerk Plansee. TEM examination on thinned foils showed that practically all of the cobalt had gone.

The creep experiments were run under vacuum (∿ 10^{-6} torr) in a VMDI-Sesame-type furnace. Heating elements were two semi-cylindrical shells in tantalum. The shields are in tantalum, molybdenum and steel respectively as one moves off from the center of the furnace. Temperature is measured with a tungsten-rhenium 6-26 % thermocouple fixed directly on the lower pushrod near the specimen. A second thermocouple, fixed on the first shield, is connected with a regulator which allows the temperature to the maintained constant within ± 2°C during the tests.

The pushrods are in tungsten. For cobalt volumic ratio higher than 10 %, we used compression on parallelepipedic specimens approximatively 8 mm high with a section between 3 x 3 mm^2 and 4 x 4 mm^2, the furnace being fixed between the screws of an universal Tinius-Olsen, type Locap, testing machine. The load was maintained constant using a TD A3 programmator. For the lower cobalt volumic ratios and the pure carbides three point bending was used, constant loading being obtained via a static lever system. Whereas in the compression test stress and strain rate could be continuously corrected as the specimen deformed, in the three point bending we used the simple expression for the strain-rate, $\mathring{\varepsilon}$:

$$\mathring{\varepsilon} = \frac{6\, e\, \mathring{h}}{L^2}$$

with e : specimen thickness (∿ 2 mm).
L : bending span (14 mm).
$\mathring{h}$: deflection-rate.

This leads to some error in the true strain-rate, however previous measurements on TiC (10) with the same set-up showed that the main features of the creep behavior (stress exponent and apparent activation energy) to be approximatively the same. As a further precaution, the deflection for each specimen was limited to a maximum value of 1 mm.

The displacement of the load was recorded as a function of time on a Linseis 0490L flat-bed recorder. It is measured outside of the furnace, on the upper pushrod using two symetrically located Penny & Gilles inductive transducers.

Observation of the deformed specimens was done in a Jeol 20TS scanning electron microscope.

RESULTS AND DISCUSSION

In the previous work (8)(9) the strain rate dependance on stress was shown to have the sigmoïdal shape which is schematically represented in figure 1. In stage II, where the strain-rate sensitivity to stress is lowest the stress exponent was shown to depend solely on the cobalt volumic ratio, tending to a value of 1 as the cobalt content tends to zero. Also in stage II the creep rate is very sensitive to the microstructure and we have found an $\dot{\varepsilon} \propto D_{WC}^{-2}$ dependance.

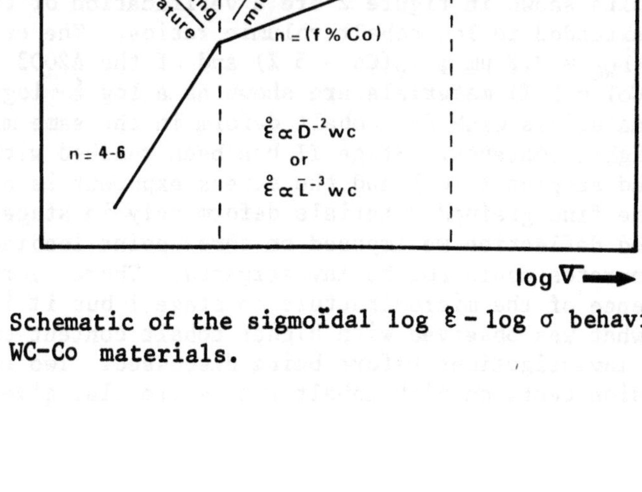

Fig. 1. Schematic of the sigmoïdal log $\dot{\varepsilon}$ - log σ behavior for WC-Co materials.

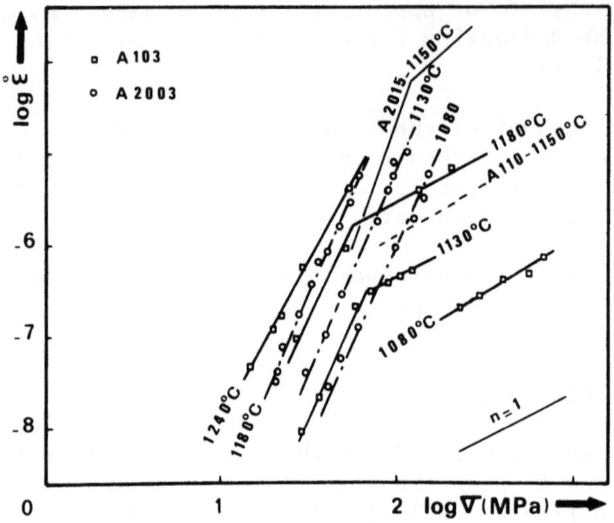

Fig. 2. Creep rate as a function of applied stress and temperature for A103 (V_v(Co) = 5 % ; $\bar{D}_{WC}$ = 2.2 µm) and A2003 (V_v(Co) = 5 % ; $\bar{D}_{WC}$ = 0.7 µm) materials. Data from compression testing are reported as a reference for the A110 (V_v(Co) = 16 % ; $\bar{D}_{WC}$ = 2.2 µm) and A2015 (V_v(Co) = 22 % ; $\bar{D}_{WC}$ = 0.7 µm) materials.

On the other hand in stage I the stress exponent is high ($n \sim$ 4-6) but not dependent on the cobalt content. As a consequence the stage I-stage II boundary is shifted toward high stresses and strain-rates as the microstructure becomes finer whereas it is shifted toward higher strain-rates and lower stresses as the temperature is increased and the microstructure kept constant.

The results shown in figure 2 are a verification of the above observation extended to low cobalt volumic ratios. The creep results of the A103 ($\bar{D}_{WC}$ = 2.2 µm ; V_V(Co = 5 %) and of the A2003 ($\bar{D}_{WC}$ = 0.7 µm ; V_V(Co) = 5 %) materials are shown as a log $\mathring{\varepsilon}$ - log σ plot. Clearly the materials with few cobalt deform in the same manner as those with higher contents. Stage II has been reached with the coarse grained samples (A103) and the stress exponent is $n \sim 1$ as expected. The fine grained materials deform only in stage I. Due to the limited deflection we imposed on three point loading the higher strain-rates could not be investigated. There seems to be a slight influence of the microstructure in stage I but it is in the opposite of what was observed with higher cobalt content and needs thus further investigations before being discussed. Two references from compression tests on high cobalt ratios are also given. It shows that the creep rates from bending tests are over-estimated in comparison with the compressive ones.

A first conclusion which may be drawn from these results is that, since the creep rate may be determined solely from parameters of the carbide phase (microstructure, stress exponent and apparent activation energy) and since the cobalt is believed to have little mechanical resistance at this temperatures, the binder may have no role during the steady-state creep which would be completely controlled by the carbide skeleton.

The results of figure 3 do not confirm this claim. We have compared here the two other fine grained materials (HO3T) both liquid phase sintered and the cobalt extracted for the half of the samples. The two fine grained with cobalt materials (A2003 and HO3T) are compared at 1180°C, only a slight difference exists. The behavior of the material from which the cobalt has been extracted (HO3T-sc) is completely different from the materials containing cobalt. Comparison at 1180°C shows that the cobalt extracted material has, for the same applied stress, a strain rate at least one order of magnitude lower than the one still containing cobalt. Extracting the cobalt has two further effects : the stress exponent in stage I is lower and the stage II exists with a stress exponent of $n \sim 1$.

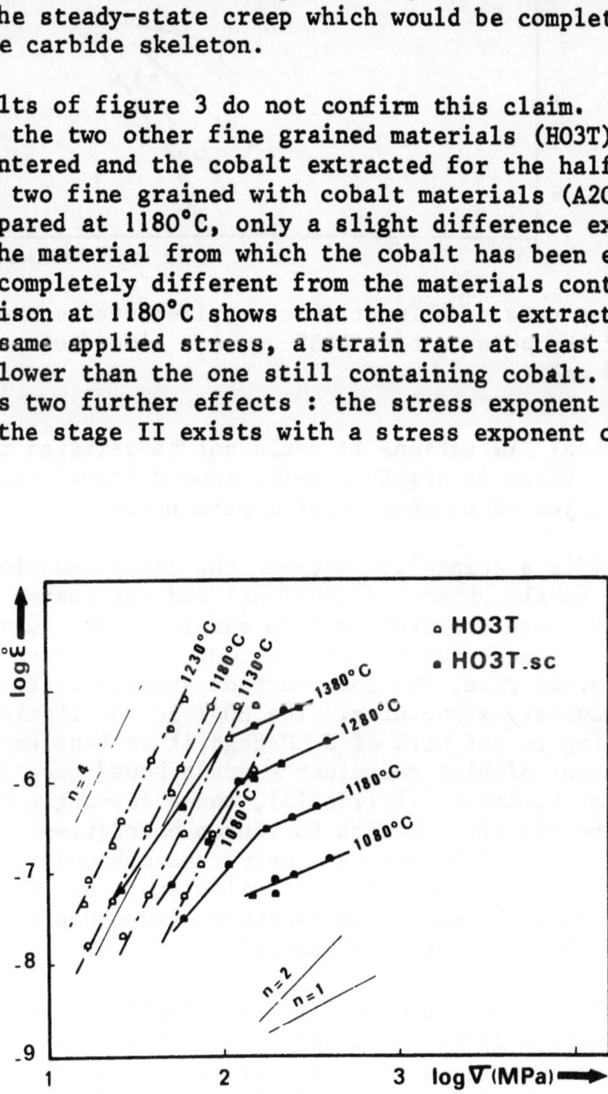

Fig. 3. Creep rate as a function of applied stress and temperature for HO3T (V_V(Co) = 5 % ; $\bar{D}_{WC}$ = 0.8 μm) and HO3T-sc, pure carbides (porosity = 5 % ; D_{WC} = 0.8 μm). Reference at 1180°C is given for A2003 material.

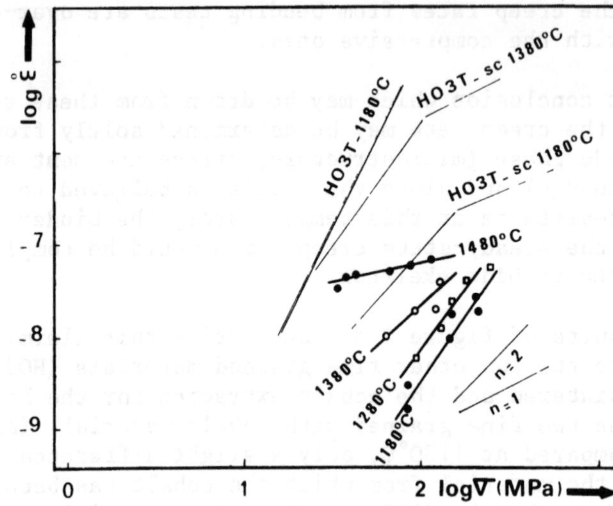

Fig. 4. Creep rate as a function of applied stress and temperature for the pure carbide HO3T-sc, and A100 (hot-pressed ; $\bar{D}_{WC}$ = 6-8 μm).

Due to practical limitations it could not be verified if the appearence of this stage is simply a shift toward lower strain-rate values or is subjected to more complex parameters.

Figure 4 is a comparison between the creep behaviors of the fine-grained cobalt extracted (HO3T-sc) and the coarse grained hot-pressed (A100) pure tungsten carbide samples. The curve noted HO3T-1180°C is a reference for the cobalt containing structure. In addition to the grain size, the pure carbide samples differ by the carbide grain boundary structures. The HO3T-sc was liquid phase sintered. According to the work of S. Hagège, it is thus expected to have a certain amount of high coincidence grain boundaries (Σ = 2 in the CSL-concept of Bollman) (11)(12)(13), these low-angle boundaries may behave differently with respect to random boundaries. On the other hand, the hot-pressed WC may have only random boundaries with impurities, introduced during the fabrication process. So one should keep in mind only the marked temperature dependance of the stress exponent for the coarse-grained material.

The apparent activation energy is a supplementary parameter to be used. Figure 5 gives the results and compares the values obtained for all the materials used in this study. It is striking that all the cobalt containing materials have the same apparent activation energy ($Q_A \sim 130$ kcal/mole) independant on grain size, cobalt content and applied stress. On the other hand the pure carbides

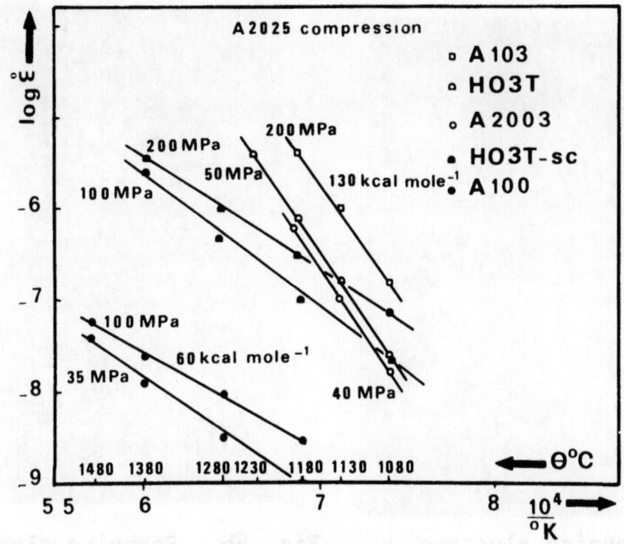

Fig. 5. Creep activation energies for the various materials. A refence is given for a A2025 ($V_V(Co)$ = 37 % ; $\bar{D}_{WC}$ = 0.7 μm) material.

have a value ($Q_A \sim$ 60 kcal/mole) which is approximately the half of the above 130 kcal/mole and depends on the applied stress. If the stress exponent were n = 1 over the whole stress range, one could conclude to a viscous creep through diffusion of iron (A100) or residual cobalt (HO3T-sc) as a second phase in the grain boundaries. At least two reasons however exclude this phenomenon. Firstly the stress exponent has also higher values and the same apparent activation energy should be measured for the cobalt containing materials and secondly there exists a stress dependance for both pure carbides.

An indication that the deformation behavior of pure carbide and carbide with cobalt should be the same is that if heavily deformed they damage in the same way. Figures 6a and 6b show the tensile surface of an A100 and an A103 specimen deformed up to 20 %. Large cavities which start by grain decohesion appear and develop. In both cases, owing to the stress exponent, to the SEM observations (9) and TEM investigations (14), the deformation of the material should occur through accomodation of carbide grain boundaries.

Microcracking phenomena due to the absence of cobalt cannot explain the lower activation energies since the creep rate of the HO3T-sc material should be higher than the one of the HO3T material. Thus at least, the creep behavior of the HO3T-sc material must be explained either by a different stress distribution in the carbide phase or by interfacial interactions between the cobalt and the car-

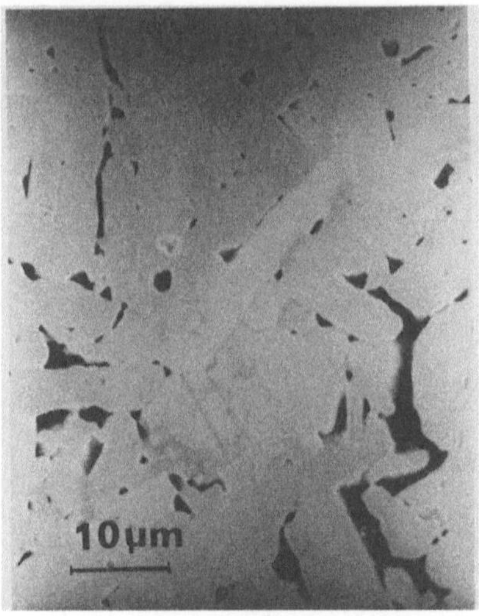

Fig. 6a. Scanning electron micrograph of heavily deformed A100 material.

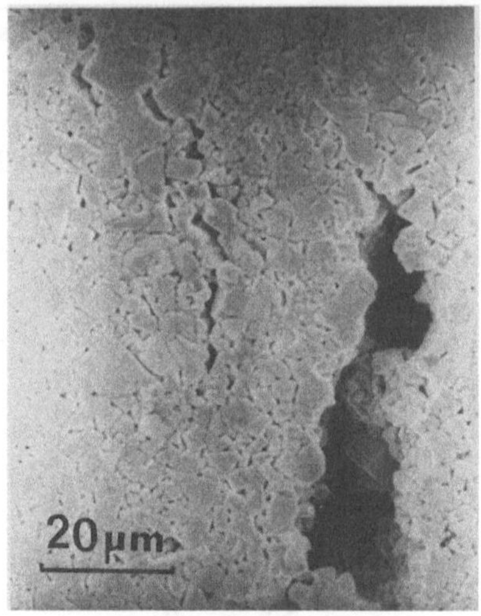

Fig. 6b. Scanning electron micrograph of heavily deformed A103 material.

bide phases. Unfortunately, current results are lacking to choose between the two possibilities.

CONCLUSION

The creep behavior of WC-Co alloys has been investigated in stage I and II of a sigmoidal log $\dot{\varepsilon}$ - log σ plot. The results show that the extrapolation which was made for the stress exponent on the basis of earlier data on high cobalt content materials is hold. However direct comparison of two series samples having the same carbide structure but the one containing no cobalt showed that the cobalt is heavily involved in the creep behavior. The precise way in which this occurs has not yet been determined.

AKNOWLEDGEMENTS

The authors thank the Metallwerk Plânsee and Dr. E. Kny for providing the H03T samples. This work has been made under contract ATP-CNRS n° 91.81.80.

REFERENCES

1. G. Altmeyer and O. Jung, Z. Metallkde. 52:576 (1961).
2. W. Dawhil and G. Altmeyer, Z. Metallkde. 54:645 (1961).
3. J.T. Smith and J.D. Wood, Acta Met. 16:1219 (1968).
4. F. Ueda, H. Doi, F. Fujiwara, H. Masatomi and Y. Oosawa, Trans. Jap. Inst. Met., 16:591 (1975).
5. J.G. Baldoni and W.S. Williams, Ceram. Bull. 57:1100 (1978).
6. S.R. Schenk, R.J. Gootschall and W.S. Williams, Mat. Sci. Eng., 32:229 (1978).
7. R.J. Gottschall, W.S. Williams and T.D. Ward, Phil. Mag. A, 41:1 (1980).
8. F. Osterstock, Thesis, University of Caen, France (1980).
9. F. Osterstock, Int. Conf. on Hard Metals, Jackson Lake Lodge, Wyoming, USA, 1981 (in press,Plenum Press, 1983).
10. J.L. Chermant, G. Leclerc and B.L. Mordike, Z. Metallkde, 71:465 (1980).
11. S. Hagège, Thesis, University of Caen, France (1979).
12. S. Hagège, G. Nouet and P. Delavignette, Phys. Stat. Sol. (a), 61:97 (1980).
13. W. Bollman, Crystal defects and crystalline interfaces, Springer Verlag, Berlin (1980).
14. S. Lay, J.L. Chermant and J. Vicens, this conference.

PLASTIC FLOW OF PLASMA SPRAYED CERAMICS

J. Padovan, G. McDonald*, R.C. Hendricks*,
R.L. Mullen+, B.T.F. Chung and M.J. Braun

University of Akron
Akron Ohio 44325

*NASA Lewis Research Center, Cleveland OH 44135

+Case-Western Reserve University, Cleveland OH 44106

ABSTRACT

Ceramics plasma sprayed onto metal substrates have a considerable potential application in gas turbine engines and other high temperature devices as thermal/mechanical insulation since plasma spraying is convenient and widely available. However, there is disturbing evidence of expulsive shedding of the ceramic after cyclic heating to approximately 1000° C. Analyses were made and experiments were designed to measure the effect of tensile and compressive strain in the ceramic when at temperature in order to determine the cause of failure. Plastic flow was predicted and inelastic behavior measured for plasma sprayed binary oxide ceramics both on metal substrates and also when stripped from the substrate before exposure to temperature. Differential expansion between the metal substrate and plasma sprayed materials ZrO_2-Y_2O_3, ZrO_2-CaO or Al_2O_3-TiO_2 results in irreversible flow. Loading of the stripped binary oxide ceramic sheet materials at temperatures up to 1200° C also produced plastic flow, but annealing before loading eliminated the inelastic behavior. The conclusion is that plasma spraying of binary oxide ceramics results in a metastable state which is inelastic at temperature but can also be stabilized or devitrified through heat treating, so as to decrease plastic properties. Mechanical properties of the as-plasma sprayed and devitrified binary oxide ceramic sheet materials have been measured. Stresses associated with sheet formation and component applications have been calculated using the finite element codes MARC and ADINA/ADINAT. These studies point to significant inelastic behavior at elevated temperatures.

INTRODUCTION

Ceramics have potential application in high performance gas turbine engines for seals as well as for thermal insulation in order to conserve fuel by permitting higher operating temperature and thereby increase thermodynamic efficiency. Use of plasma sprayed refractory ceramic is an economical, rapid, and widely available method of fabricating conformal parts.

However, there is disturbing evidence of inadequate durability of plasma sprayed ceramic under conditions of high heat transfer, (ref. 1,2,3). Previous investigations indicated (ref. 4) that plasma sprayed ZrO_2-$8Y_2O_3$ plastically deformed at elevated temperature under low stress.

Since the high temperature design application of ceramics requires accurate predicition of performance, it is necessary to determine the exact effect of temperature and stress on inelastic behavior of plasma sprayed ZrO_2-$8Y_2O_3$. Understanding the basis of such effects may permit either designing around inelastic behavior or eliminating the cause. This report discusses recent research measurements and calculations on plasma sprayed ceramics.

EXPERIMENTAL PROCEDURE

Materials and Ceramic Deposition

The ceramic specimens used to measure high temperature plastic flow were of two types, either bonded or free form. For the bonded type samples, sheet metal stainless or tungsten was first grit blasted with Al_2O_3 for cleaning and surface roughening and then was plasma sprayed in air with 0.013 cm Ni-18Cr-12Al-0.3Y followed by plasma spraying with 0.038 cm ZrO_2-$8Y_2O_3$. Free form samples of ZrO_2-$8Y_2O_3$ or Al_2O_3 were plasma sprayed on 10 micro-cm RMS stainless steel, and then removed by acid dissolution of the substrate.

Apparatus and Test Procedures

For determination of inelastic flow on a bonded type sample a 0.15 cm metal strip plasma sprayed with NiCrAlY and ZrO_2-$8Y_2O_3$ was accurately measured in length on an optical comparitor. Next the sample was heated to the desired temperature and when cool, the

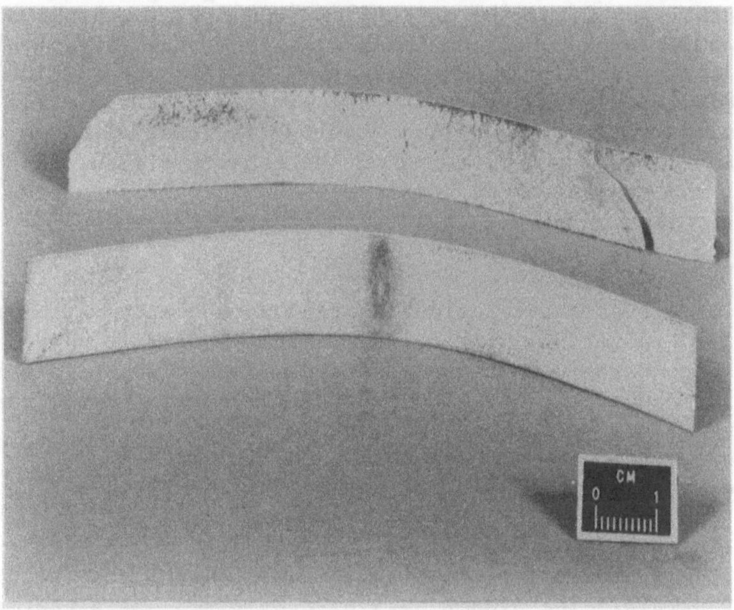

Fig. 1, Curved o.038 cm Thick ZrO_2-$8Y_2O_3$ Resulting from Bending at 1090° C.

length was accurately remeasured. (The temperature and times are recorded in the date tables). Next the metal was dissolved away and the length of the ceramic alone was measured. Samples were heated to temperature in either an argon or a hydrogen filled furnace; in one case a vacuum furnace was used.

RESULTS AND DISCUSSION

Preliminary Measurement of Inelastic Flow

The inelastic flow of plasma sprayed ZrO_2-$8Y_2O_3$ was shown experimentally by deforming at 1090° C (2000° F) to a radius as small as 30 cm a flat 0.12 cm thick stainless steel strip which had been plasma sprayed with 0.012 cm of NiCrAlY and with 0.037 cm of ZrO_2-$8Y_2O_3$ and then separating the ceramic by dissolution of the stainless (ref. 4, 5). This resulted in integral ceramic sheets; one curved by stretching on the tension side of the bent specimen, and one curved by compression on the inside of the bent substrate. The separate, curved ceramic sheets are shown in fig. 1.

The effect of these residual stresses on coating life was investigated in ref. 6. For a long substrate-ceramic coated specimen, the ceramic near the ends separated by shear failure. It was also shown that near the center, the coating could separate in bending (adhesive/cohesive) failure.

Extensive Mesurements of Plasma Sprayed Ceramic Plastic Flow

The inelastic flow of plasma sprayed ceramic was further investigated by measuring the plastic flow caused by a measured amount of strain. The plasma sprayed ZrO_2-$8Y_2O_3$ has a thermal expansion coefficient of 10^{-6} cm/cm/°C (ref. 7) and the 304 stainless steel has an expansion coefficient of 20×10^{-6} cm/cm/°C (ref. 8). If ZrO_2-$8Y_2O_3$ plasma sprayed on stainless at ambient temperature is raised in temperature, ceramic strain is produced by the difference in thermal expansion coefficient. If the metal substrate is dissolved away after returning to ambient, then any plastic flow of the ceramic will result in a difference in length between the starting ceramic-substrate specimen and the free ceramic. The amount of plastic extension of the ZrO_2-$8Y_2O_3$ varies linearly with the temperature to which the sample is heated.

The results of such tests are presented in Table I and fig. 2. If the full differences in thermal expansion produces plastic flow then the increase in length at 1205° C would be 0.012 cm/cm instead of the 0.006 cm/cm which was observed. The reason why the inelastic flow is less than the difference in

Table 1. Plastic Stretch of ZrO_2-Y_2O_3 at Various Temperatures.

Temperature, °C	As plasma sprayed	595	815	982	1205
$\Delta l / l$	0.0008	0.0022 .0021	0.0039	.0060	.0062

Table 2. Plastic Stretch of ZrO_2-Y_2O_3 After Heating 24 Hours at 1205° in Different Atmospheres.

Atmosphere	Plastic flow, cm/cm
Argon	0.0056
Hydrogen	.0046
Hydrogen	.0044
Vacuum	.0042 (4 hr only)

Table 3. Plastic Flow of Plasma Sprayed Ceramic After Heating 24 Hours.

Ceramic	Temperature °C	Plastic flow, cm/cm
ZrO_2-$8Y_2O_3$	1205	0.0053
ZrO_2-$12Y_2O_3$	1205	.0055
ZrO_2-8CaO	1000	.004
Al_2O_3-$2.5TiO_2$	1000	.004

Table 4. Plastic Flow on Heating to 1205°C ZrO_2-$8Y_2O_3$ Plasma Sprayed on Tungsten.

Sample	Plastic flow, cm/cm
1	-0.0034 cm/cm
2	-0.0027

Table 5. Plastic Flow of ZrO_2-$8Y_2O_3$ After Slow Cool.

Temperature, °C	Cooling rate,	$\Delta l / l$ cm/cm
1205	15° C/hr	0.0056

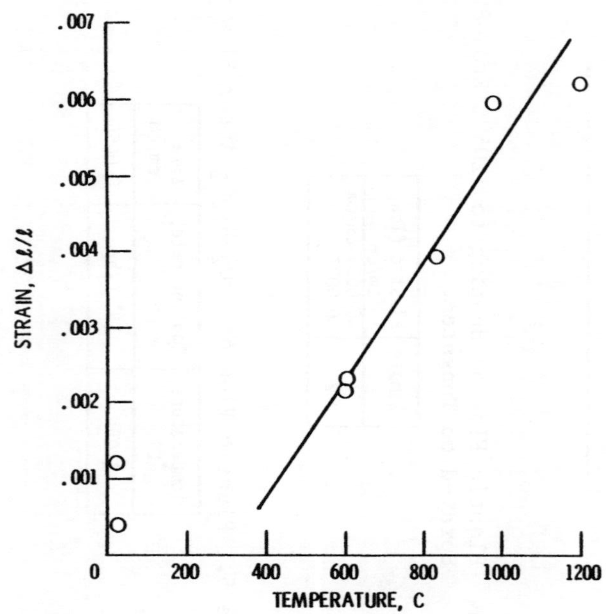

Figure 2. Inelastic Extension of ZrO_2-Y_2O_3 with Temperature.

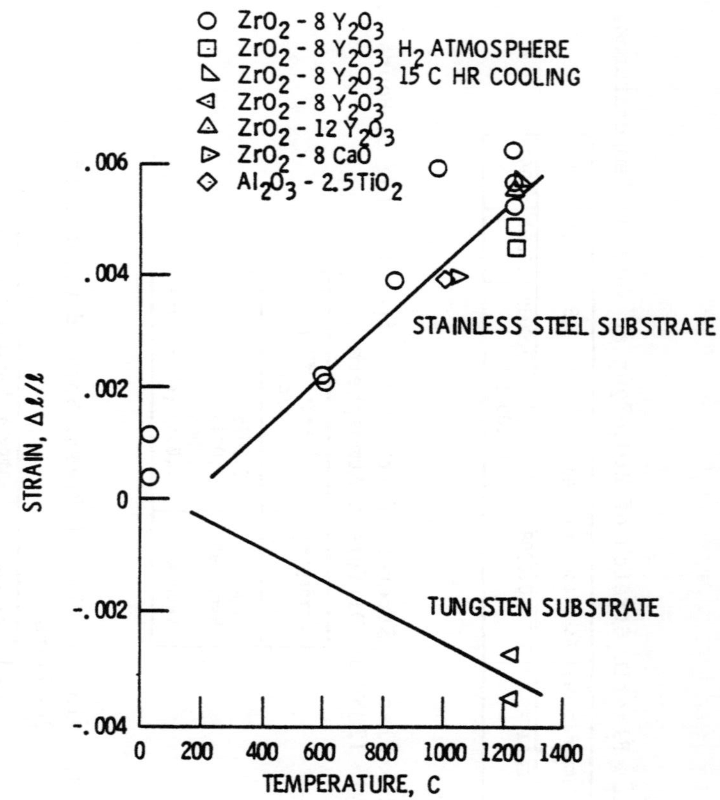

Figure 3. Inelastic Flow of Plasma Sprayed Ceramic as a Function of Temperature at which the Specimen was Treated.

TABLE VI. THERMAL AND MECHANICAL MATERIALS PROPERTIES

% YSZ/CoCrAlY:		100/0	85/15	70/30	40/60	MAR-M-50
material set #:		1	2	3	4	5
Property	Temp(°F)					
density (lb/in^3)	all	.155	.180	.205	.254	.320
poisson ratio	all	.25	.26	.27	.28	.30
thermal cond. (BTU/min-in-°F)	0	4.09	4.09	4.09	4.09	422.98
(X 10^{-4})	200	4.09	4.09	4.09	4.09	"
	500	4.09	4.66	4.98	5.38	"
	1000	4.33	6.26	7.46	8.35	"
	1500	4.74	8.51	10.60	12.20	"
	2000	5.70	11.64	14.85	17.66	"
	2500	8.11	16.30	20.95	25.52	"
specific heat (BTU/lb-°F)	0	.126	.121	.116	.107	.097
(X 10^{-2})	2500	.161	.161	.161	.158	.155
coef. of expansion (in/in-°F)	0	4.08	3.28	3.36	3.64	6.40
(X 10^{-6})	2500	4.83	7.70	8.38	9.52	12.20
Young's modulus (lb/in^2)	0	6.90	3.80	5.15	8.30	34.70
(X 10^6)	2500	2.00	2.00	8.00	17.75	15.60
tensile strength (lb/in^2)	0	4.25	6.00	8.10	33.25	40.0
(X 10^3)	2500	3.30	7.25	11.75	1.00	40.0

Approximate Conversion Factors

lb/in^3	* 2769 = kg/m^3	in/in-F	* 1.8 = m/m-C
B/min-in-F	* 1246 = W/m-K	psi	/145 = MPa
B/lb-F	* 4186 = j/kg-K		

thermal expansion is not known*. Calculating the stress on the plasma sprayed ZrO_2-$8Y_2O_3$, assuming no plastic flow from values of E = 150 GPa ($22x10^{-6}$ psi), ref. 7 p. 777, and using values of 10^{-6} cm/cm/°C and 1000° C = ΔT for the differential expansion, then the stress becomes sig. = Ee = 1.5 GPa (222 ksi). This stress from the differential expansion is much larger than the strength of the ceramic. This calculation confirms that some type of inelastic flow of the plasma sprayed ceramic is responsible for maintaining the ceramic as an 'interlocking' integral piece which can be removed from the substrate as a unit.

In the attempt to further define the cause of the extreme plastic flow of the plasma sprayed ZrO_2-$8Y_2O_3$, a series of differential expansion/plastic flow experiments were made on different materials and under different conditions to determine whether the results observed above were special to ZrO_2-$8Y_2O_3$ and stainless as they were treated under the conditions described above.

Specimens of ZrO_2-$8Y_2O_3$ were heated in hydrogen and other samples were heated in vacuum to 1205° C. The results are presented in table II. There is no significant difference in the plastic flow of the ZrO_2-$8Y_2O_3$ regardless of the atmosphere in which it was heated.

Next, the plastic flow of several different ceramics were measured to determine if the chemistry of the ceramic controlled the occurence of the plastic flow. The results are presented in Table III. There is no significant difference in the ability to undergo plastic flow of plasma sprayed ZrO_2-$8Y_2O_3$, ZrO_2-$12Y_2O_3$, ZrO_2-CaO, or Al_2O_3-$2.5TiO_2$; apparently the tendency to flow plastically originates in the plasma spraying process since it is not significantly different in the different ceramics. That is, the inelastic behavior, while much material dependent, is also dependent on how the ceramic was applied or processed.

---

* It is possible that a portion of the stress on the ceramic due to differential thermal expansion is absorbed in elastic strain. One could also consider the NiCrAlY bond coat as controlling the extension but then the tungsten results would not show compressive effects - which is incorrect. The measured extensions could also be divided into two effects, Kelvin visco-elastic type behavior and Coloumb type inelastic behavior. Differences in elastic modulus are significant, e.g.,

Material Specification	E	sig, calc
Sintered stabilized zirconia	150 GPa	1.5 GPa
Plasma sprayed yttria stabilized zirconia (YSZ)	48	.48
Free form YSZ sheet	7	.07

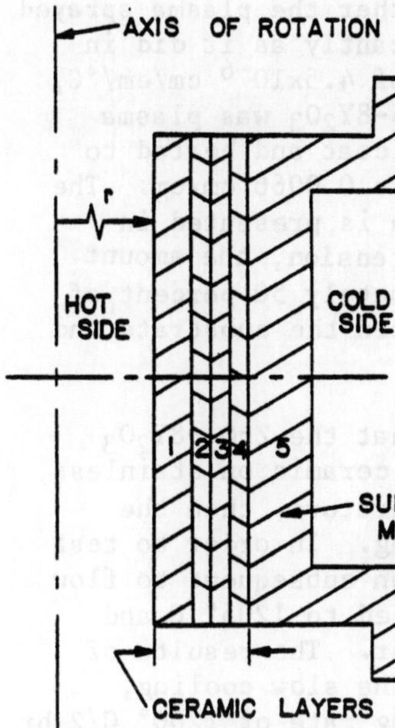

Figure 4. Geometry of Ceramic Coated Outer Gas Path Seal.

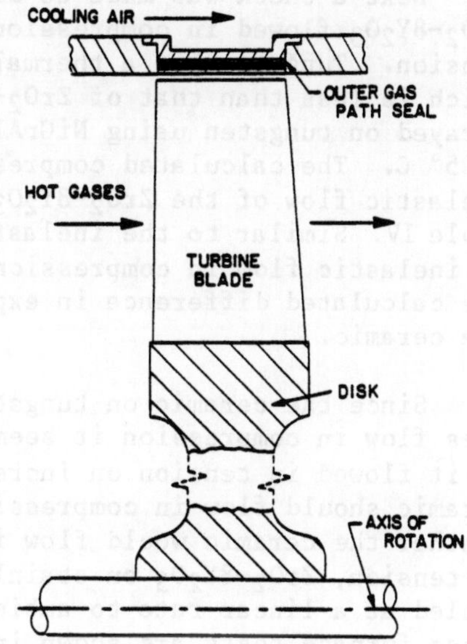

Figure 5, Placement of Outer Gas Path Seal in Turbine Section of Jet Engine.

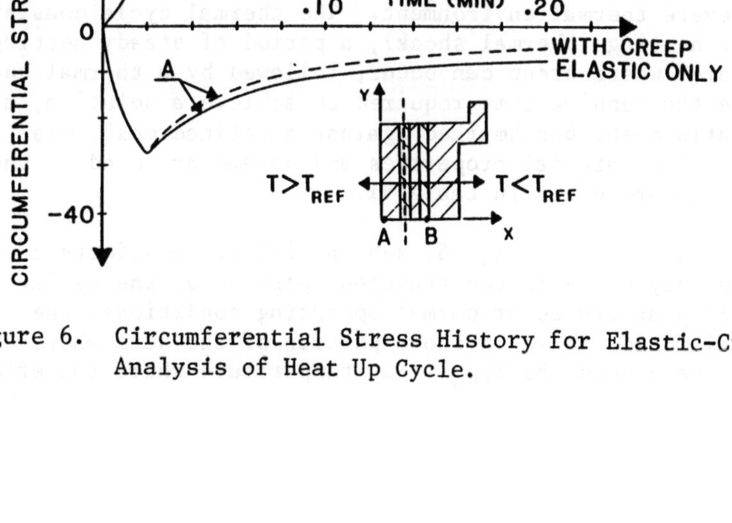

Figure 6. Circumferential Stress History for Elastic-Creep Analysis of Heat Up Cycle.

Next a check was made to determine whether the plasma sprayed ZrO_2-8Y_2O_3 flowed in compression as significantly as it did in tension. Tungsten has a thermal expansion of $4.5x10^{-6}$ cm/cm/°C, which is less than that of ZrO_2-8Y_2O_3. ZrO_2-8Y_2O_3 was plasma sprayed on tungsten using NiCrAlY as a bond coat and heated to 1205° C. The calculated compressive stain is 0.0066 cm/cm. The inelastic flow of the ZrO_2-8Y_2O_3 on tungsten is presented in table IV. Similar to the inelastic flow in tension, the amount of inelastic flow in compression is approximately 50 percent of the calculated difference in expansion between the substrate and the ceramic.

Since the ceramic on tungsten showed that the ZrO_2-8Y_2O_3 does flow in compression it seemed that the ceramic on stainless, if it flowed in tension on increase in temperature, then the ceramic should flow in compression on cooling. In order to test whether the ceramic would flow in compression subsequent to flow in tension, ZrO_2-8Y_2O_3 on stainless was heated to 1205° C and cooled at a linear rate to ambient over 15 hr. The results of these furnace tests are shown in table V. The slow cooling, 1200° C/15 hr, compared to the normal cooling rate of 1200° C/2 hr produced identical effect on plastic flow of the plasma sprayed ZrO_2-8Y_2O_3. Apparently some effect is occuring when the plasma sprayed ceramic is at temperature which decreases its capacity for inelastic flow in either compression or tension. All of the experiments on inelastic flow of plasma sprayed ceramic attached to a substrate are summarized in fig. 3.

ANALYSIS AND APPLICATION

To assess the influence of inelastic material behavior on the thermomechanical response of ceramic coated seal components, figs. 4 and 5, an analysis was undertaken which required the development of an improved finite element algorithm, ref. 9. The improved algorithm was necessary to achieve an efficient stable solution of the inherently nonlinear thermomechanical field equations coupled with a severe thermal environment. The thermal cycle consisted of transient heating (thermal shock), a period of steady heating, at temperatures where creep can occur, followed by a thermal quench. To reduce the running time required to achieve a solution, a high aspect ratio mesh, benchmarked against a refined mesh, was employed. The material properties and parameters used in the seal computations are given in table VI.

As illustrated in fig. 6, substantial creep effects are evidenced very early in the transient portion of the cycle. Specifically at cruise or normal operating conditions, the mechanical fields relax in those portion of the seal where the temperatures exceed the T_{ref} (that temperature where the effects

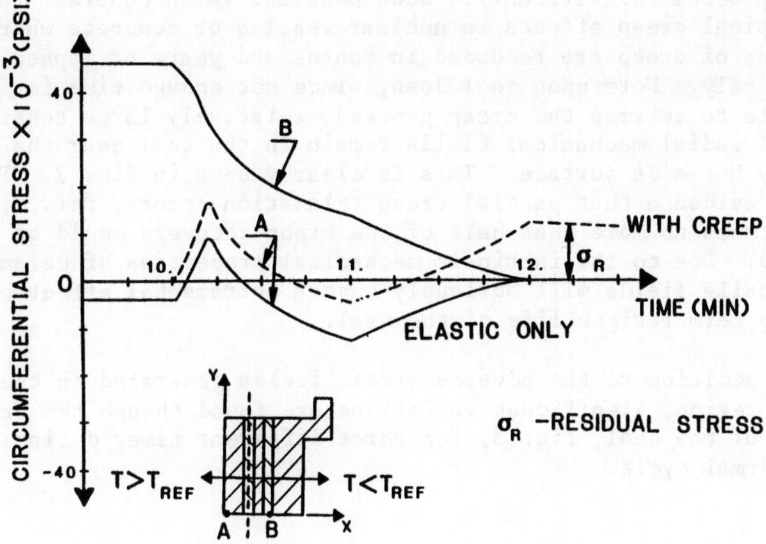

Figure 7. Circumferential Stress History for Elastic-Creep Analysis of Cool Down Cycle.

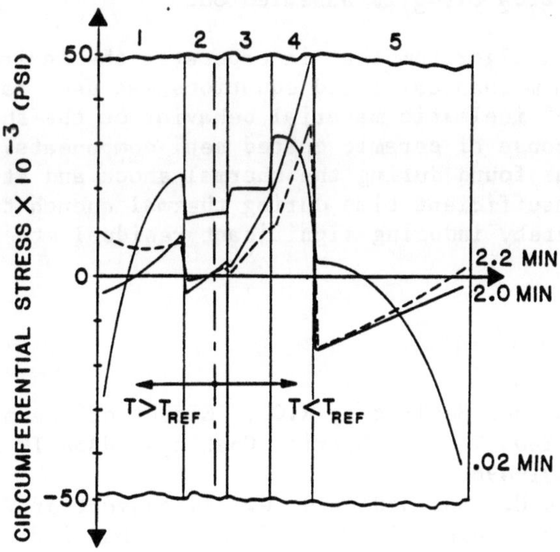

Figure 8. Variation of Circumferential Stress Through the Seal Configuration.

of creep become significant). Such behavior is in contrast to more typical creep effects in nuclear vessles or concrete wherein the rates of creep are recorded in months and years of lapsed time respectively. Note upon cool down, since not enough time is available to reverse the creep process, relatively large tensile residual radial mechanical fields remain in the seal near the normally hot side surface. This is clearly seen in fig. 7. There is also evidence that partial creep relaxation occurs, ref. 5, but in all cases no more than half of the creep recovery could be expected. Due to the intrinsic mechanical properties of ceramics, such tensile fields will obviously have a detramental effect on the long term fatigue life of the seal.

In addition to the adverse stress fields generated in the surface region, significant variations are found though the cross section of the seal, fig. 8, for three different times during the seal thermal cycle.

SUMMARY

Plastic flow of plasma sprayed ZrO_2-$8Y_2O_3$ ceramic has been measured at temperatures up to 1250° C (2200° F) and compared to the plastic flow of pressed and sintered ZrO_2-$8Y_2O_3$. The plastic flow of the plasma sprayed ZrO_2-$8Y_2O_3$ is longer than the plastic flow of sintered ZrO_2-$8Y_2O_3$ prepared from powder and is thermolabile in that at temperature the tendency to creep of the plasma sprayed ZrO_2-$8Y_2O_3$ is annealed out.

An improved algorithm that incorporates the inherently nonlinear thermomechanical field equations was used to determine the influence of inelastic material behavior on the thermomechanical response of ceramic coated seal components. Significant creep was found during the thermal shock and steady heating periods with insufficient time during thermal quench to reverse the process thereby inducing significant residual stresses into the component.

REFERENCES

1. McDonald, G. and Hendricks, R.C.: Effect of Thermal Cycling on ZrO_2-Y_2O_3 Thermal Barrier Coatings, NASA TM X-81480 (1980) 491-496.
2. Hendricks, R.C., and McDonald, G.: Assessment of Variations in Thermal Cycle Life Data of Thermal Barrier Coated Rods, NASA TM-81743 (1981). See also, Thin Solid Films 74 (1981)
3. Hendricks, R.C.; McDonald, G. and Poolos, N.J: Prolonging Thermal Barrier Coated Specimen Life by Effective Thermal Management. NASA TM-81742(1981).

4. Hendricks, R.C., McDonald, G., and Bill, R.C.: Some Inelastic Effects of Thermal Cycling on ZrO_2-Y_2O_3 Materials, Ceram. Eng. Sci. Proc. 3, 750-757 (1982).
5. Hendricks, R.C.; McDonald, G.; and Mullen, R.L.: Effect of Annealing on the Creep of Plasma Sprayed Ceramics; NASA TM 83396, 1983. See also, Ceramic Proceedings Seventh Annual Conference on Ceramics and Advanced Materials American Ceramic Society; Cocoa Beach, Fl.; Jan. 16-23, 1983, paper 54-C-83C.
6. Mullen, R.L.; McDonald, G.; Hendricks, R.C.; and Hofle, M.M.: Correlation of Compressive and Shear Stress With Spalling of Plasma Sprayed, NASA TM 63406,See also, Ceramic Proceedings Seventh Annual Conference on Ceramics and Advanced Materials American Ceramic Society; Cocoa Beach, Fl.; Jan. 16-23, 1983, paper 53-C-83C.
7. Kingery, W.D.; Bowen, H.K.; and Uhlmann, D.R. : Introduction to Ceramics, 2nd Ed., Wiley, New York ,1975, p 595.
8. Lyman, T: Metals Handbook - Properties and Selection of Metals, Vol. 1, 8th Ed., American Society of Metals, Metals Park, OH, 1961, p 442.
9. Padovan, J.; Dougherty, D.; Hendricks, R.C.; Braun, M.J.; Chung, B.T.F. : High Temperature Thermomechanical Analysis of Ceramic Coatings, J. Thermal Stress, in press, 1983.

HIGH TEMPERATURE FAILURE MECHANISMS IN CERAMIC POLYCRYSTALS*

A. G. Evans, and W. Blumenthal

Materials and Molecular Research Division,
Lawrence Berkeley Lab., Dept. of Materials Science and
Mineral Engineering, Univ. of Calif., Berkeley, CA 94720

ABSTRACT

The high temperature failure of ceramics has been observed to occur by the nucleation, growth and coalescence of cracks. Each process involves diffusive cavity nucleation and growth, either within a damage zone or at microstructural heterogeneities. The specific cavitation and cracking mechanisms pertinent to ceramics are described and available models are presented. Particular emphasis is placed on continuous crack nucleation at microstructural heterogeneities and the crack coalescence process that causes eventual failure. Failure time data are also reviewed and correlated with models.

1. INTRODUCTION

The high temperature failure of fine grained polycrystalline ceramics generally occurs by the nucleation, growth and coalescence of cracks, as illustrated by the sequence depicted in Fig. 1. The limiting stage of failure depends on the material, temperature and environment. Each stage requires thorough investigation, in order to derive a generic model of creep rupture pertinent to ceramic polycrystals. Some progress toward this objective has been realized, but a complete model is not yet available. This article describes the present status of understanding, as well as the phenomena that require further investigation before an adequate failure model can be developed.

---

*Work supported by the U.S. Department of Energy under Contract No. DE-AC03-76SF00098.

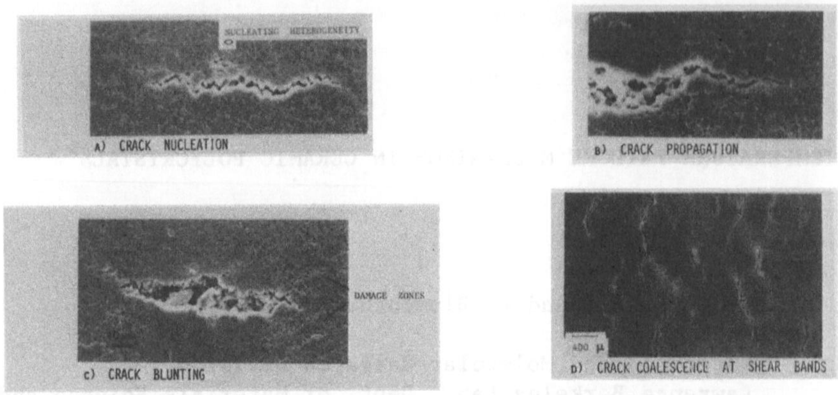

Fig. 1. Scanning electron micrograph of Al_2O_3 specimens illustrating the various stages of failure (a) crack nucleation at heterogeneities (b) crack propagation (c) crack blunting (d) crack coalescence.

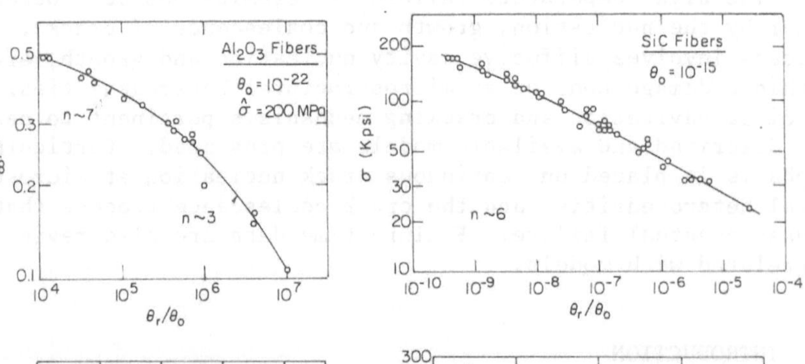

Fig. 2. Relationships between failure time and stress for ceramic polycrystals.

2. GENERAL CHARACTERISTICS

Observations performed on ceramic materials reveal[1] (Fig. 1a) that cracks typically nucleate, at microstructural heterogeneities, well before eventual failure. However, the original cracks usually blunt[2] (Fig. 1b) and new cracks nucleate continuously throughout the failure process. Creep-related failure appears to involve the coalesence of blunted cracks, along shear bands between crack tips (Fig. 1d).[3] Consequently, the processes of continuous nucleation, growth, blunting and coalescence of creep cracks all require study.

The measured relationships between failure time and stress for most ceramics are highly non-linear[3-6] (Fig. 2). For example, recent data obtained on "single phase" alumina[3] indicate that the failure time, t_f, varies with stress, σ_∞, as;

$$t_f \, \dot{\varepsilon}_\infty = (\sigma_o/\sigma_\infty)^6 \tag{1}$$

where $\dot{\varepsilon}_\infty$ is the remote strain-rate and σ_o is a constant. The specific source of the non-linearity is not yet evident. The eventual objective of research studies on high temperature failure is to provide an adequate non-linear model, based on analysis of each stage in the failure process.

3. CRACK NUCLEATION

Cracks in ceramic polycrystals nucleate preferentially at microstructural and chemical heterogeneities.[1,2] The microstructural heterogeneities include large grained regions (Fig. 3a) and large single grains or inclusions, as well as delaminations (Fig. 3b). The large grains are considered to be sites for premature crack nucleation for two reasons. They act as creep inhomogeneities, which induce local stress concentrations,[1] and they probably contain interfaces susceptible to cavity nucleation. Analysis of the stress concentrations that accrue, due to a viscosity differential, reveals that stresses up to about twice the applied stress are generated in and around the heterogeneity[1] (Fig. 4) until cracking occurs. This level of stress concentration is generally not large enough to account for cavity nucleation on interfaces free of second-phase. Consequently, the additional presence of precipitates or second phases that encourage nucleation, within the zones of stress concentration, are inferred. For example, large grained regions Al_2O_3 have been associated with the presence of TiO_2,[7] which may create a second phase amenable to cavity nucleation on sliding interfaces. Additional research is needed to establish a full understanding of these inhomogeneities.

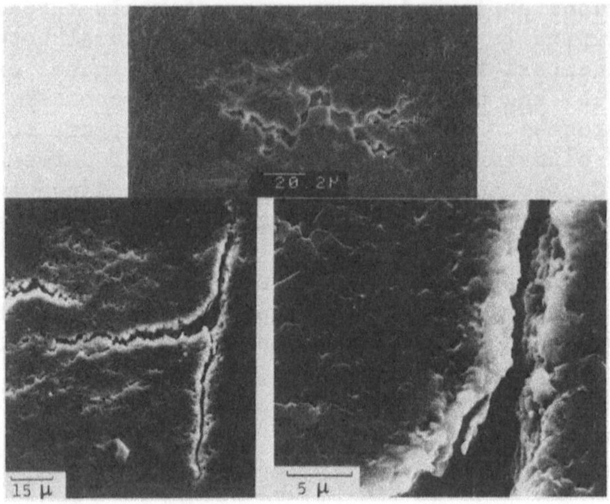

Fig. 3. Crack nucleation at microstructural heterogeneities. (a) Large grained region and (b) a delamination; note the residual fine porosity.

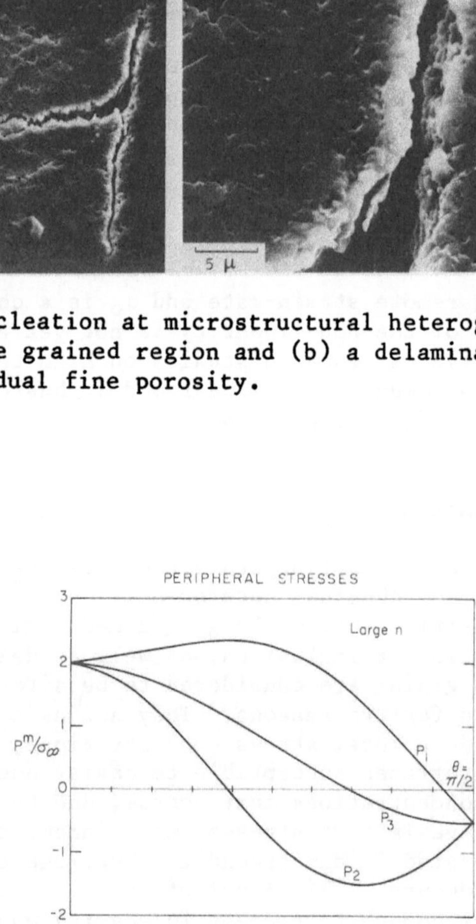

Fig. 4. The variation of stress in and around a spherical heterogeneity as a function of the relative viscosity of the matrix and the defect.

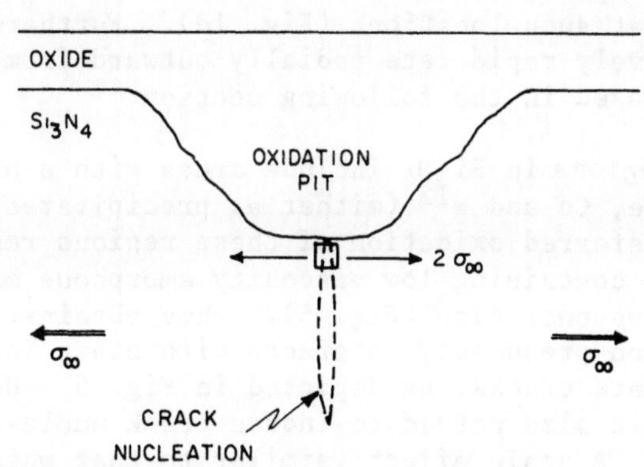

Fig. 5. A schematic illustrating the stress concentration that occurs around oxidation pits and the consequent nucleation of cracks.

The delaminations that are observed as intrinsic cracks[1] (Fig. 3b) are typical hot pressing defects. Similar crack-like entities form between agglomerates[8,9] during rapid local densification, and resist sintering. The presence of some remaining fine porosity (Fig. 3b) supports the supposition that the present defects result from incomplete local sintering.

Chemical heterogeneities often consist of second phases that oxidise preferentially during the test. Typical examples are inclusions in Si_3N_4,[10,11] and regions containing excess amorphous material. Evidence for the latter heterogeneity has been obtained in Al_2O_3, and manifests as spots that thermally etch at a more rapid rate than the surrounding microstructure[1-3] (Fig. 1a).* The presence of excess Si in these regions has been used to imply the existence of an amorphous silicate phase. Profuse crack nucleation invariably occurs at such locations (Fig. 1d). Furthermore, cracks extend at a relatively rapid rate radially outward from these regions,[2] as discussed in the following section.

Oxidizable regions in Si_3N_4 include areas with a high local concentration of Fe, Co and W[12] (either as precipitates or in solid solution). The preferred oxidation of these regions results in the formation of pits, containing low viscosity amorphous material. An attendant stress concentration (Fig. 5). thus obtains. Again the stress concentration presumably interacts with other local heterogeneities to nucleate cracks, as depicted in Fig. 5. However, the magnitude of the pit size needed to induce crack nucleation has not been investigated. A scale effect (similar to that which pertains for cracking due to thermal expansion mismatch[13]) is expected, but has not been verified.

4. CRACK GROWTH AND BLUNTING

Crack Growth

The growth characteristics of both induced surface cracks and intrinsically nucleated cracks have been subject to investigation.[2,14] The former have been most extensively studied and exhibit the characteristics depicted in Fig. 6a. Crack growth occurs between a threshold K_{th} and the critical stress intensity K_c. In Al_2O_3, for example, blunting occurs at $K \approx 0.4\ K_c$ and crack healing initiates at $K \leqslant 0.2\ K_c$. The crack growth rates are asymptotic with respect to both the threshold and critical stress intensities. The crack growth rate, $\dot{a}$, in the intermediate range, can generally be characterized by;

---

*Evidence suggests that of least some of the observed regions result from surface contamination.

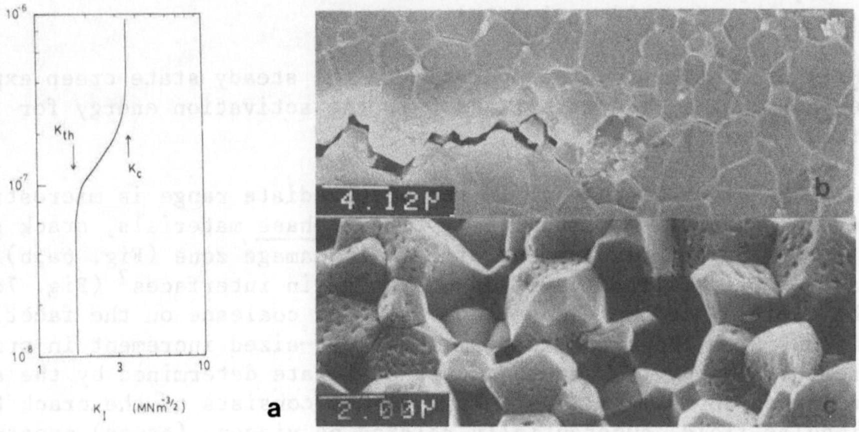

Fig. 6. a) A schematic of crack growth characteristics for extrinsic cracks indicating the threshold, K_{th}, the critical stress intensity, K_c, and the intervening region of creep crack growth.
b) A scanning electron micrograph of a growing crack profile, indicating coplanar damage.
c) A scanning electron micrograph of a crack growth surface revealed by brittle fracture at room temperature, showing cavities on two grain interfaces.

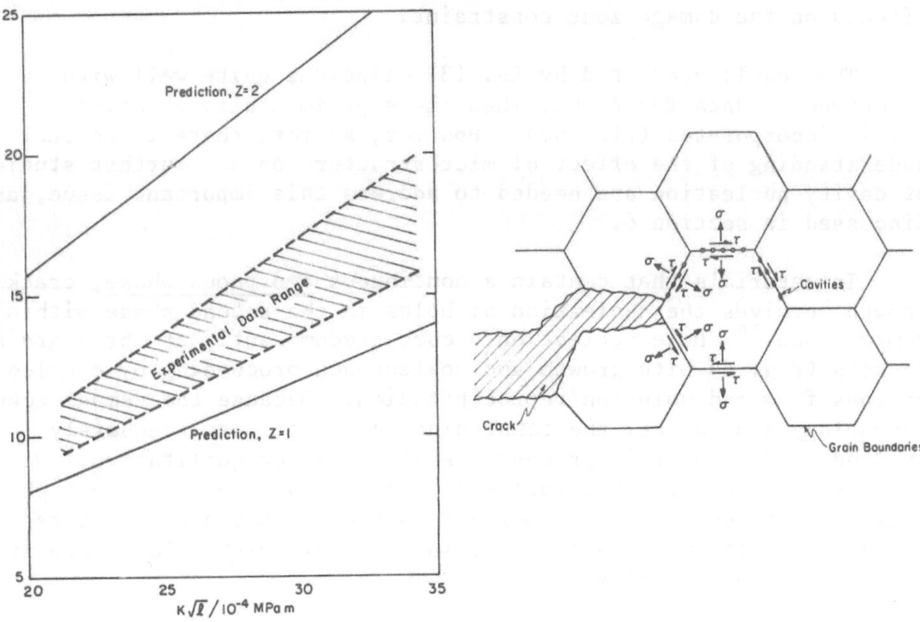

Fig. 7. The crack growth rate predicted by a coplanar damage model compared with experimental data for Al_2O_3.

$$\dot{a} = \dot{a}_o\ (K/K_c)^n \exp[-Q/RT] \quad (2)$$

where n is an exponent comparable to the steady state creep exponent ($\dot{\varepsilon} \sim \sigma^n$), $\dot{a}_o$ is a constant and Q is the activation energy for steady state creep.

The growth mechanism in the intermediate range is microstructure dependent. In essentially single phase materials, crack growth proceeds by the development of a small damage zone (Fig. 6a,b) consisting primarily of cavities on two grain interfaces[2] (Fig. 7c). The cavities in the damage zone grow and coalesce on the facet contiguous with the crack to permit a facet-sized increment in crack growth. The cavity growth occurs at a rate determined by the stress acting on the damage zone. This stress consists of the crack tip singular field, substantially relaxed by viscous (creep) constraint of the dilatation occurring in the damage zone.[15] The cavity spacing, λ, is also an important rate controlling quantity. Consequently, the crack growth rate has the form[15]

$$\dot{a}\eta = K\sqrt{l}\ F(\lambda/l,\ z,\ A_f) \quad (3)$$

where η is the viscosity of the material, l is the facet size, z is the damage zone length, A_f is the area of facet subject to cavitation at the coalescence condition and F is the function plotted in Fig. 7. The damage zone length and the coalescence area are significant determinants of the crack growth rate, by virtue of their effects on the damage zone constraint.

The result predicted by Eq. (3) coincides quite well with experimental data for Al_2O_3, when the experimentally observed λ/l is incorporated (Fig. 6c). However, as yet, there is no real understanding of the effect of microstructure on λ. Further studies of cavity nucleation are needed to address this important issue, as discussed in section 6.

In materials that contain a continuous amorphous phase, crack growth involves the nucleation of holes in the second phase within a damage zone.[16] Hole nucleation occurs predominantly at three grain pockets (Fig. 8) with growth and coalescence proceeding by coupled viscous flow and solution/reprecipitation. Because the damage zone morphology and, hence, the constraint, have not been adequately studied, models of this process are still rather qualitative. However, the parameters that influence crack growth in the single phase polycrystals must also be important in these materials, since rapid crack growth is expected to occur in materials with a low viscosity and a small hole spacing.

Intrinsically nucleated cracks are often observed to grow at more rapid rates than induced surface cracks. The growth rate differential is especially marked during initial extension of the

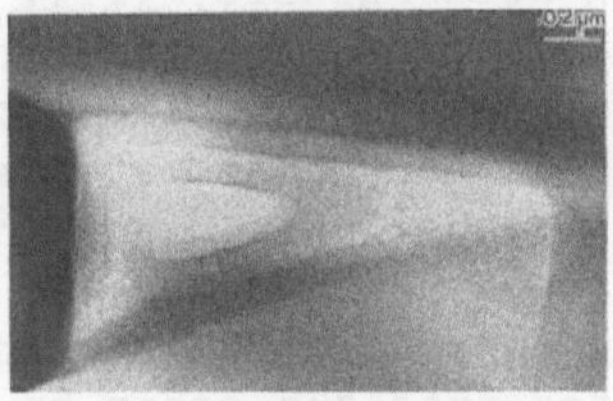

Fig. 8. Hole nucleation at an amorphous pocket in Si_3N_4.

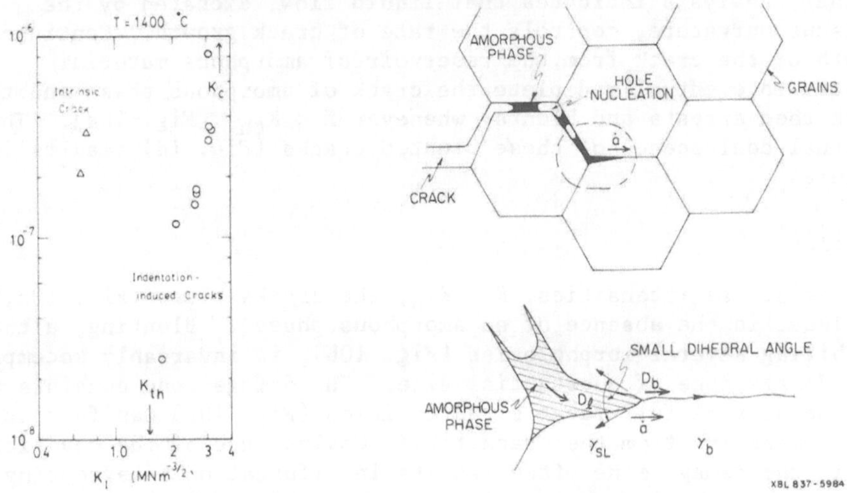

Fig. 9. The growth characteristics of intrinsically nucleated cracks. a) The growth rate for initial growth compared with that for extrinsic cracks. b) A schematic illustrating the role of an amorphous phase in the acceleration of intrinsic cracks.

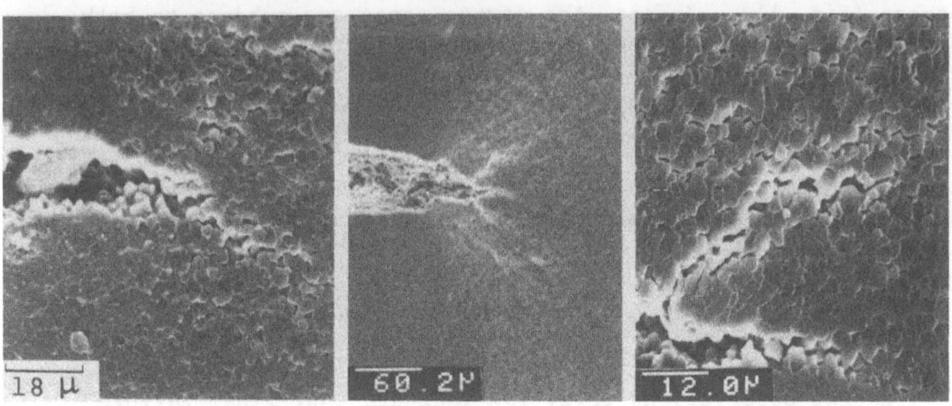

Fig. 10. a) The blunting configuration for intrinsic cracks.
b) Blunting configuration for extrinsic cracks.
c) The cavitation damage within the side lobes of blunted cracks.

crack out of the nucleating heterogeneity (Fig. 9a). The enhanced growth could be attributed, in part, to creep induced stress concentrations. However, local excesses of amorphous material are also of paramount importance. Amorphous material accelerates crack growth by the mechanism depicted in Fig. 9b. The amorphous phase penetrates the grain boundary by dissolving the major phase. The stress on the liquid then causes hole nucleation. The dissolved major phase material migrates into the grain boundary from the crack surfaces, by means of solution/precipitation process. Preliminary analysis indicates that liquid flow, dictated by the meniscus curvature, controls the rate of crack growth. Continued growth of the crack from the reservoir of amorphous material appears to eventually deplete the crack of amorphous phase and the crack then arrests and blunts, whenever $K < K_{th}$ (Fig. 10a). The eventual coalescence of these blunted cracks (Fig. 1d) results in failure.

Blunting

At stress intensities, $K < K_{th}$, the cracks blunt (Fig. 10a,b) (at least in the absence of an amorphous phase). Blunting, although exhibiting several morphologies (Fig. 10b), is invariably accompanied by a damage zone of substantial size. The damage zone consists of a high density of full-facet sized cavities (Fig. 10c) manifest as side lobes emanating from the crack tip.[2] Coalescence of the cavities within the damage zone often results in bifurcation intercepting the crack surface behind the original crack tip (Fig. 10b). The transition from frontal damage at $K > K_{th}$, to side lobe damage, at $K \approx K_{th}$, is clearly a major cause of crack blunting. However, the reasons for this transition are not well understood.

The damage envelope resembles the contours of maximum shear stress (Fig. 11). Infact, the damage zones are subject to both appreciable shear and tension.[2] However, normal stress in uniaxial tension is relatively constant ahead of the crack tip. Consequently, since cavity growth is motivated by the normal stress--and occurs on boundaries normal to the maximum tension (Fig. 12)--the role of the shear stress is presumed to reside in its effect either on cavity nucleation (e.g., due to grain boundary sliding transients)[17] or on enhanced creep accommodation of the dilatation in the developing damage zone.[18]

5. CRACK COALESCENCE

Recent observations (Fig. 13) indicate that final failure occurs by the coalescence of blunted cracks across intervening shear bands. These shear bands, which consist of material substantially damaged by full-facet cavities (see Fig. 10c), presumably support most of the creep strain in the latter stages of failure.

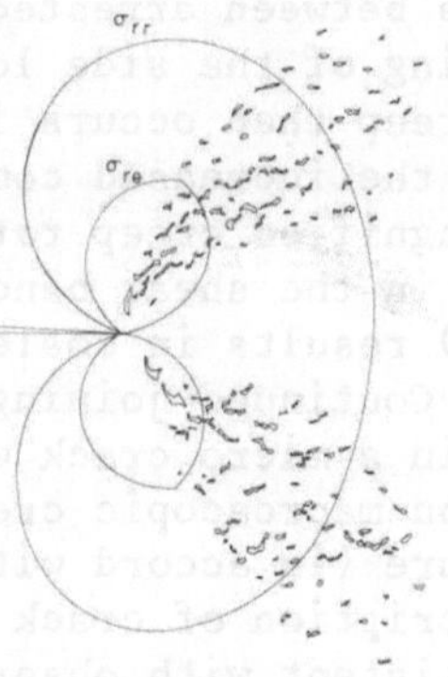

Fig. 11. Stress contours around a crack tip compared with the damage zone contours.

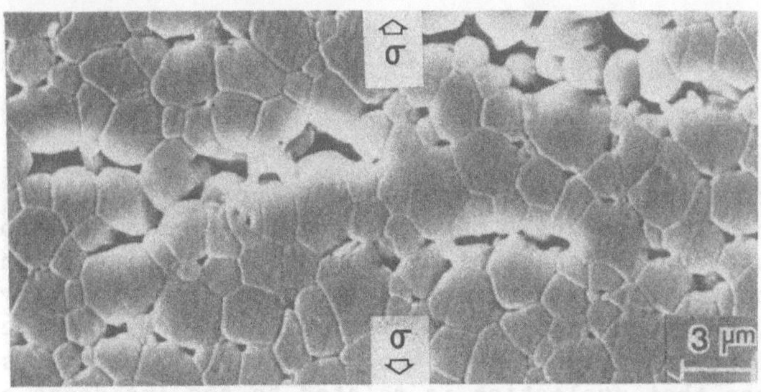

Fig. 12. Cavitation damage in a uniaxial test growing normal to the principal tension.

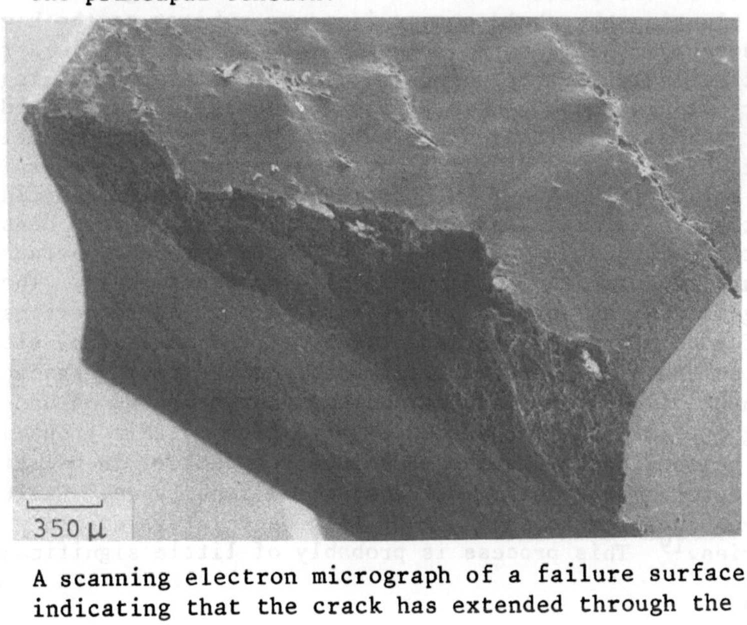

Fig. 13. A scanning electron micrograph of a failure surface indicating that the crack has extended through the shear bands.

Shear strain concentrations between arrested cracks thus ensue and undoubtedly cause the linking of the side lobe damage between neighboring cracks. Enhanced creep then occurs in the intervening damage zone, as motivated both by the increased compliance provided by the cracks (Fig. 14) and the magnified creep rate allowed by the damage. The generation of cavities in the shear band, due to the presence of the normal stress (Fig. 14) results in coalescence of the cavities and linkage of the crack. Continued joining of contiguous, blunting cracks eventually results in a macro crack with sufficient size that $K > K_{th}$ (Fig. 6a), whereupon macroscopic creep crack growth can occur and cause final failure (in accord with the process described in Section 4.1). This description of crack coalescence and failure, although qualitatively consistent with observations, has not yet been adequately modeled. Further discussion of various possibilities is presented in the following section.

6. DISCUSSION

The Cavity Spacing

The preceding description of high temperature failure has revealed that the spacing between cavities exerts an important influence on several stages of failure. This spacing is undoubtedly dependent on the microstructure and the mechanism of cavity nucleation. Yet, the quantitative association between microstructure and cavity nucleation have not yet been formulated. There is substantial evidence that nucleation is related to the incidence of grain boundary sliding transients.[17] This mode of cavity nucleation is supported by the present observation that the damage around cracks at the blunting threshold occurs in zones subject to the highest shear stresses. Sliding transients generate appreciable stress concentrations at triple points, non-deformable precipitates and boundary undulations. These stress concentrations undoubtedly account for sliding induced nucleation, perhaps in conjunction with local fluctuations in dihedral angle associated with precipitates and/or solutes. However, the precise requirements for nucleation are not known; albeit that most data suggest an average density of nucleation sites that increases with strain (probably because of the associated cumulative incidence of sliding transients). The cavity spacing is also subject to uncertainty, except in situations where non-deformable precipitates are the preferred nucleation sites, whereupon the precipitate spacing controls the cavity spacing. The wavelength of boundary undulations has been proposed as one possible spacing determinant, but adequate validation of this proposal has yet to be obtained. Studies of boundary structure in relation to cavity spacing are clearly of substantial importance. Cavities have also been observed to nucleate at slip band intersections with grain boundaries.[19] This process is probably of little significance in

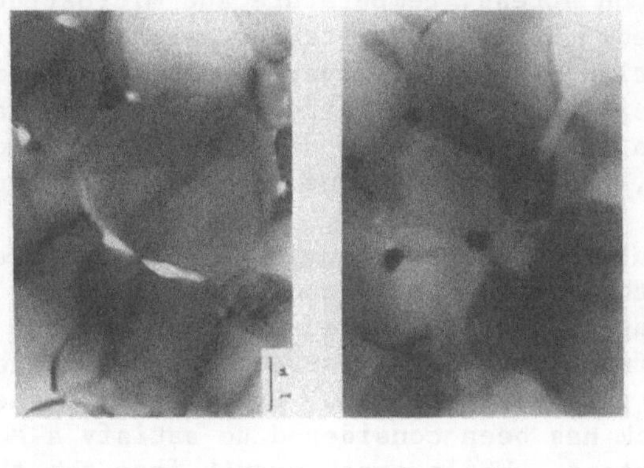

Fig. 15. A transmission electron microscope image of material in the damage zone at the blunting threshold.

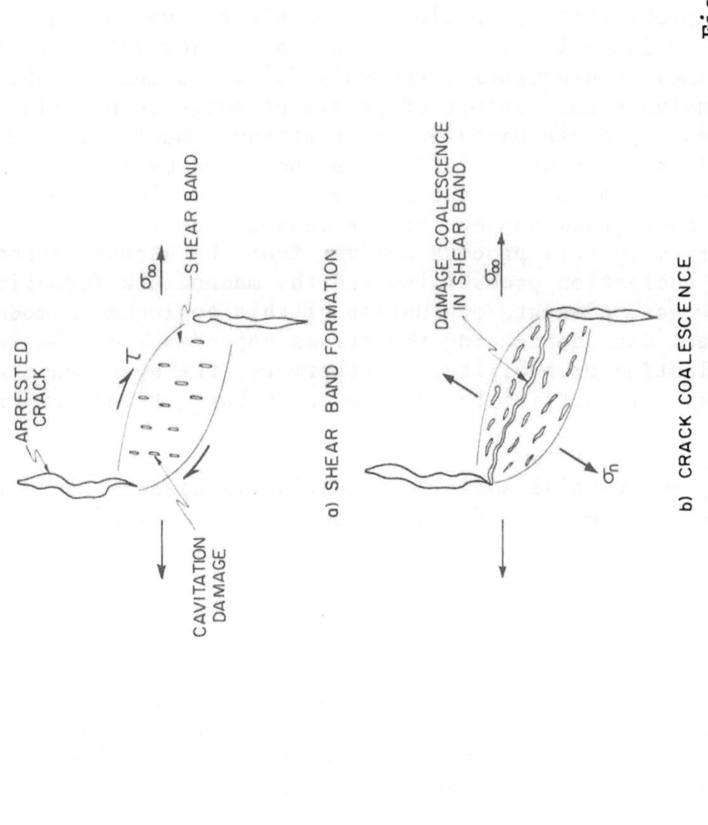

Fig. 14. A transmission indicating failure by cracking across shear bonds.

fine-grained materials which are devoid of dislocations[2] in the damage zone* (Fig. 15).

The Failure Time

The specific stage that dominates the failure time and its dependence on stress, temperature and microstructure has not been well studied. Hence, quantitative assessment of failure times is not presently justified. Nevertheless, useful insights can be gained by examining preliminary models of the various stages involved in failure. Such models are summarized in the Appendix and herein, their implications are cursorily reviewed.

Present comprehension suggests that failure is not dominated by either the initial crack nucleation stage or the final macrocrack propagation stage, because both processes exhibit small deviation from linearity in stress (at least when intrinsic crack nucleation precedes failure). Specifically, the nucleation of a single crack has been considered to satisfy a Monkman-Grant law, $t_f \dot{\varepsilon}_\infty \approx$ constant. While crack growth from the threshold, K_{th}, to the critical stress intensity, K_c, yields a failure time, $t_f \sim \sigma_\infty^{-2}$.

The paucity of data concerning the velocity of intrinsic cracks and their probability of nucleation prohibits evaluation of the growth of individual intrinsic cracks as a major determinent of the failure time. Consequently, the only failure stage amenable to present analysis and capable of providing adequate non-linearity is a coalescence process based on the continuous nucleation of intrinsic cracks. In this process, failure is dominated by the time taken to nucleate cracks on sufficient contiguous nucleation sites which then creates a macrocrack subject to a stress intensity $K > K_{th}$. The non-linearity in this process derives from the stress dependence of the crack nucleation probability and the macrocrack formation condition. However, adequate evaluation of this coalescence model awaits experimental data concerning the stress dependence of the intrinsic crack nucleation probability. Furthermore, the specimen size effect that necessarily obtains for this process has not been experimentally assessed.

Therefore, substantial additional testing is needed before the development of a generic failure model can be considered.

*The absence of dislocations is not necessarily construed to exclude the presence of some dislocation activity, but to negate the incidence of slip bands with stress concentrations of a sufficient spatial extent to induce stable cavities.

The Damage Zone Transition

Little attention has yet been devoted to the transition, at the blunting threshold, between frontal and side lobe damage. Some preliminary considerations of this problem are presented, as a basis for detailed subsequent investigation. An appropriate basis for initiating analysis is to presume that cavity nucleation during crack growth is microstructurally determined and confined to that facet contiguous with the crack tip.* Conversely, side lobe damage may be considered to initiate in response to the continuum crack tip field. Hence, a tentative criterion for damage zone initiation may be predicated on the prerequisite that the shear displacement exceed some critical value; a criterion suggested by observations of cavity nucleation during creep.[20] Analysis of the shear displacement on these notions, provides some insight concerning the transition.

The continuum displacement field in a linearly viscous solid has the form

$$\dot{w} \sim K\sqrt{r}/\eta \tag{4}$$

where $\dot{w}$ is the displacement rate and r is the distance from the crack tip. Continuum damage initiates when the sliding displacement reaches a critical value, w_c. In order to initiate side-lobe damage the sliding displacement on a facet a distance, l (facet length), beside the crack tip must satisfy,

$$\beta K\sqrt{l}\Delta t/\eta > w_c \tag{5}$$

where Δt is the waiting time between crack growth increments and β is a constant. Noting that, $\Delta t = l/\dot{a}$ and that $\dot{a}$ is given by Eq. (2), the side-lobe damage threshold requirement becomes;

$$\beta K_c^n l^{3/2} \exp(Q/RT)/w_c \dot{a}_o \eta > K^{n-1} \geq K_{th}^{n-1}$$

and the crack growth threshold condition becomes;

$$(K_{th}/K_c)^{n-1} = \beta K_c l^{3/2} \exp(Q/RT)/w_c \dot{a}_o \eta \tag{6}$$

Furthermore, since the activation energies for crack growth and viscous flow are generally similar, the relation

$$\eta = \eta_o \exp[Q/RT] \tag{7}$$

where η_o is temperature independent, can be used to simplify Eq. (7), to;

*Cavity nucleation on this facet could, for example, be initiated by a sliding transient induced during a prior crack advance increment.

$$\frac{K_{th}}{K_c} \simeq \left[\frac{\beta K_c l^{3/2}}{w_c \dot{a}_o \eta_o}\right]^{1/(n-1)} \tag{8}$$

This result suggests that K_{th}/K_c is temperature independent, consistent with preliminary measurements (Fig. 6a). Furthermore, K_{th}/K_c is anticipated to increase with an increase in grain size; however, data to assess this prediction are not yet available.

7. CONCLUDING REMARKS

The process of high temperature failure is of sufficient complexity, involving stages of crack nucleation, growth, blunting and coalescence, that simple models are deemed unlikely to be of general merit. However, progress toward definition of the important stages of failure and the development of preliminary models for each process provides encouragement that continued research is capable of generating the requisite models. The research must devote equal attention to the acquisition of data and the development of models. In addition to failure time data, measurements of the crack nucleation probability and its dependence on the distribution of heterogeneities would be most revealing. Further studies of the coalescence of blunted extrinsic cracks along shear bands are also considered to be a high priority, as well as experiments that allow interpretation of the damage zone transition at the blunting threshold.

APPENDIX

Some Models of Failure

Crack Nucleation. The nucleation of cracks generally occurs by the coalescence of cavities growing by diffusion. The cavity growth rate is usually limited by the viscous relaxation of the dilatation occurring in the damage zone. Consequently, the coalescence time depends principally on the creep rate of the material and the creep mechanism. For a power law creeping material, creep within the grain results in an elongated cavitated zone and a nucleation time,[21]

$$t_c \dot{\varepsilon}_\infty = (2/3)(\lambda/l)\operatorname{cosec}\Psi[1/(1+\cos\Psi) - (\cos\Psi)/2] \tag{A1}$$

where Ψ is the dihedral angle. In material subject to diffusional creep, and a low grain boundary viscosity, the cavitation is subject to a larger constraint and the nucleation time is given by;[22]

$$t_c \dot{\varepsilon}_\infty \approx 50\pi(\gamma_s/\sigma_\infty l)^{1/2}\sin(\Psi/4)^{3/2}(D_s\delta_s/D_b\delta_b)^{1/2} \tag{A2}$$

where γ_s is the surface energy, $D_s\delta_s$ is the surface diffusion parameter and $D_b\delta_b$ is the grain boundary diffusion parameter.

In both cases, the product $t_c \dot{\varepsilon}_\infty$ is essentially stress independent and thus at variance with the substantial stress dependence of the failure time.

Macrocrack Propagation. The growth time for macrocracks can be obtained by invoking the stress intensity relation for a semi-circular surface crack,

$$K = (2/\pi)\sigma_\infty\sqrt{a} \quad . \tag{A3}$$

Then, from Eq. (2), we obtain,

$$da/a^{n/2} = (2\sigma_\infty/\pi K_c)\, \dot{a}_o e^{-Q/RT}\, dt \tag{A4}$$

Integration of Eq. (A4) gives a macrocrack propagation time,

$$t_p = \left(\frac{\pi K_c}{2\sigma_\infty}\right)^n \frac{2\exp(Q/RT)}{(n-2)\dot{a}_o}\left[\frac{1}{a_*^{(n-2)/2}} - \frac{1}{a_c^{(n-2)/2}}\right] \tag{A5}$$

where a_* is the initial size of the macrocrack and a_c is the critical size. For failure from a pre-existing crack of initial size, a_*, Eq. (A5) clearly gives the dependence, $t_p \sim \sigma_\infty^{-n}$. Some failure data obtained at high stresses and intermediate temperatures may be explained by this relation.[6] However, at higher temperatures and longer times when macrocracks form by the coalescence of several small intrinsic cracks, it seems more pertinent to allow the macrocrack to initiate growth at K_{th}, whereupon;

$$a^* = (\pi K_{th}/2\sigma_\infty)^2 \tag{A6}$$

Then, the propagation time becomes;

$$t_p = \frac{2\exp(Q/RT)}{(n-2)\,\dot{a}_o}\left(\frac{\pi K_c}{2\sigma_\infty}\right)^2\left[\left(\frac{K_c}{K_{th}}\right)^{n-2} - 1\right] \tag{A7}$$

Consequently, the relation, $t_p \sim \sigma_\infty^{-2}$, invariably pertains. Again, this level of stress dependence is at variance with the failure time data.

Crack Coalescence. When cracks nucleate and grow to their arrest size in a fully independent manner, i.e., no interaction effects are allowed, a relatively straightforward probabilistic treatment of failure can be adopted.[23,24] In particular, a probabilistic distribution of times, t_a, required for cracks to attain their arrest dimensions may firstly be assumed,

$$p = (t_a/t_o)^k \tag{A8}$$

where p is the cumulative probability of observing a blunted intrinsic crack in unit volume of material at time t_a, k is the shape parameter and t_o is the scale parameter.* Furthermore, if it is assumed that the joining of blunted cracks across the intervening shear bands is not rate limiting, macrocrack formation depends on the probability of forming sufficient contiguous intrinsic cracks to create a macrocrack of size, a_*. Then, it can be readily shown that the failure time at the median probability level varies as;[23]

$$\ln t_f = \ln t_o - \left(\frac{4\sigma_\infty^2 c_*}{\pi k K_{th}^2} \right) \ln (4A/c_*^2) \tag{A9}$$

where c_* is the length of the blunted intrinsic cracks and A is the area of specimen subject to the stress σ_∞. The time t_o has a stress dependence characteristic of the intrinsic crack nucleation and growth process. However, even if this process is essentially linear (as would probably obtain if the arrest sequence were nucleation limited, (Eqs. (A1,A2)), then the failure time is still nonlinear, due to the effect of stress on the crack size at the threshold, K_{th} (the second term on the right hand side of Eq. (A9)).

The primary limitation of the model is the requirement that crack linkage across the shear band occur readily, and that failure is dominated by the crack formation probability. There is, as yet, no evidence to support this premise. In fact, observations (Fig. 1d) indicate that the linkage across shear bands may be the limiting process, whereupon Eq. (A9) would not apply.

REFERENCES

1. S. M. Johnson, W. Blumenthal and A. G. Evans, J. Am. Ceram. Soc., in press.
2. W. Blumenthal and A. G. Evans, J. Am. Ceram. Soc., in press.
3. B. J. Dalgleish, S. M. Johnson, W. Blumenthal and A. G. Evans, J. Am. Ceram. Soc., in press.
4. K.F.A. Walles, Proc. Brit. Ceram. Soc. 15 (1972) 157.
5. R. J. Charles, "Fracture Mechanics of Ceramics," (eds. R. C. Bradt, F. F. Lange, D. P. H. Hasselman) Plenum, N. Y. (1978), vol. 4, p. 623.

*The arrest time includes a time for nucleation, as well as a growth time for the intrinsic cracks. No assumption is required regarding the more time consuming of these two process.

6. G. G. Trantina and C. A. Johnson, J. Am. Ceram. Soc. 58 (1975) 344.
7. B. J. Dalgleish and A. G. Evans, J. Am. Ceram. Soc., in press.
8. A. G. Evans, J. Am. Ceram. Soc. 65 (1982) 497.
9. F. F. Lange, J. Am. Ceram. Soc. 66 (1983) 398.
10. J. E. Ritter, S. M. Weiderhorn, N. J. Tighe and E. R. Fuller, Ceramics for High Performance Applications (Ed. E. M. Lenoe, R. N. Katz) Plenum, N. Y., vol. 6, p. 503.
11. S. M. Wiederhorn and N. J. Tighe, J. Mat. Sci. 13 (1978) 1781.
12. F. F. Lange, J. Am. Ceram. Soc. 61 (1978) 270.
13. A. G. Evans, Acta Met. 26 (1978) 1845.
14. M. H. Lewis, B. S. B. Karunaratne, J. Meredith and C. Pickering, "Creep and Fracture of Engineering Materials and Structures" (Eds. B. Wilshire and D. R. J. Owens), Pineridge, U.K. (1981) 365.
15. M. D. Thouless, C. H. Hseuh and A. G. Evans, Acta Met., in press.
16. J. Marion, M. D. Drory, A. G. Evans, and D. R. Clarke, Acta Met., in press.
17. A. S. Argon, Scripta Met. 17 (1983) 5.
18. J. W. Hutchinson, Acta Met., in press.
19. R. Davies, this volume.
20. R. G. Fleck, D. M. R. Taplin, C. J. Beevers, Acta Met. 23 (1975) 415.
21. J. R. Rice, Acta Met. 29 (1981) 675.
22. C. H. Hseuh and A. G. Evans, Acta Met. 29 (1981) 1907.
23. A. G. Evans and A. Rana, Acta Met. 28 (1980) 129.
24. F. A. McClintoch, "Fracture Mechanics of Ceramics," (Eds. R. C. Bradt, F. F. Lange, and D. P. H. Hasselman) Plenum, N. Y. (1974), vol. 1, p. 93.

MODELS FOR CREEP CRACK GROWTH

D.S. Wilkinson

Department of Metallurgy and Materials Science
and Institute for Materials Research
McMaster University
Hamilton, Ontario, CANADA L8S 4L7

INTRODUCTION

The failure of ceramic structures at elevated temperatures involves the nucleation, growth and coalescence of cavities. The final stage of this process requires the propagation of a crack through the body. Depending on conditions, the fraction of life spent in this crack growth stage may be insignificant, or it may be life limiting. For example, a structure may contain pre-existing flaws either as cracks, or regions of high porosity which readily nucleate cracks. Damage then occurs preferentially in the region surrounding the flaw, and the life of the structure is governed by the rate of crack propagation. On the other hand, if the material is very homogeneous, then cavitation develops uniformly, and it is the nucleation and growth of individual cavity arrays, which govern the life. In real structural ceramics, an intermediate situation is likely to prevail. Large initial flaws can be eliminated by careful processing. However, materials typically contain sufficient inhomogeneity that local regions develop microcracks long before the end of life. These then grow and link until a crack large enough to propagate rapidly across the structure develops. The growth of microcracks may then be the life-limiting process.

It is therefore clear that a thorough understanding of failure in structural ceramics at elevated temperatures requires a knowledge of both cavity and crack growth, and the relationship between them. In this paper, we examine the current models for crack propagation. The effect of simplifying assumptions on the results of these models will be discussed. We will also consider the extent to which improving the level of sophistication of the models can be expected to improve the results. Finally, the extension of the models to microcracks, in which the size of the damage zone ahead of the crack tip is comparable to the crack length, will be considered.

MODELS OF CAVITY GROWTH

Cracks propagate by the nucleation and growth of cavities in the field of concentrated stress ahead of the crack tip. It is therefore appropriate that we start by considering the mechanisms of cavity growth and their associated governing equations. This subject has been extensively reviewed in recent years [1,2], and that is not the intention here. We need only build a framework with which to model the process of crack propagation.

Cavities form on grain boundaries and grow by a combination of boundary diffusion, surface diffusion and creep plasticity. In ceramic materials, at the applicable service temperature, there is little evidence of extensive creep plasticity in the form of dislocation motion. There is, however, extensive grain boundary sliding. This, especially in the presence of a liquid phase, may contribute to cavity growth.

We need to model the growth of cavities ahead of a planar crack front. This requires treating an array of cavities for which the size of individual cavities varies only with the distance from the crack tip. To a reasonable level of approximation, this may be done by replacing each row of axisymmetric cavities parallel to the crack front, by a single cylindrical cavity, an approach followed in several recent papers [3-8].

Diffusive Cavity Growth in Single Phase Materials

In ceramic materials which contain little or no glassy phase, cavities may grow by the diffusion of matter out of the cavities and down the grain boundary. In modelling this process, a variety of assumptions are generally made:

1) that grain boundary diffusion transports more material than lattice diffusion;
2) that grain boundaries and surfaces behave as perfect sources and sinks for vacancies;
3) that cavities are uniformly distributed over the grain boundary, and grow at equal rates;
4) that matter is deposited uniformly over the ligament between cavities; and
5) that local equilibrium is maintained at the interface between the cavity and the grain boundary.

From these assumptions, we can set up and solve a boundary value problem for the flux of atoms away from the cavity. But to describe the rate at which the local area fraction of cavities increases, we also need to assume something about the shape of cavities. Generally, if surface diffusion is rapid, then cavities maintain an equilibrium shape with constant curvature (Fig. 1a), while if surface diffusion is slow, they

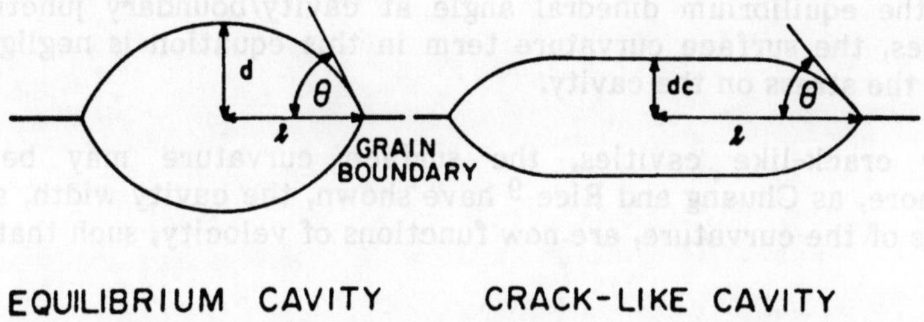

Fig. 1 Cavities may grow with either (a) an equilibrium shape, or (b) a crack-like shape, depending on whether the ratio of surface to boundary diffusivity is high or low.

become crack-like in shape (Fig. 1b). If we define the cross-sectional area of any cavity to be

$$A = 2\ell\bar{d} \tag{1}$$

where (2ℓ) is the cavity length, then $\bar{d}$ contains the shape- dependence. The general equation for cavity growth which now results is

$$\eta = \frac{\lambda}{c^3}\,\frac{12}{\frac{\bar{d}}{c}(1-\eta)^2}\left(\frac{\sigma}{G} - \frac{\gamma_s}{cG}\,\frac{1-\eta}{\frac{r}{c}}\right) \tag{2}$$

where η is the local area fraction of cavities (i.e. η = 2ℓ/c, c being the cavity spacing).

For equilibrium-shaped cavities, this reduces to

$$\dot{\eta} = \frac{\lambda}{c^3}\,\frac{24}{\eta(1-\eta)^2 g(\theta)}\left(\frac{\sigma}{G} - 4\,\frac{\gamma_s}{cG}\,\frac{1-\eta}{\eta\sin\theta}\right) \tag{3}$$

where

$$g(\theta) = \frac{\theta - \sin\theta\cdot\cos\theta}{\sin^2\theta} \tag{4}$$

θ being the equilibrium dihedral angle at cavity/boundary junction. In most cases, the surface curvature term in this equation is negligible, so that $\dot{\eta}\sim\sigma$, the stress on the cavity.

For crack-like cavities, the surface curvature may be large. Furthermore, as Chuang and Rice [9] have shown, the cavity width, and also the radius of the curvature, are now functions of velocity, such that

$$d \sim r \sim \dot{\eta}^{-1/3} \tag{5}$$

This leads to an equation for cavity growth,

$$\dot{\eta} = \frac{2\lambda}{c^3}\{\frac{3}{2}\frac{\frac{\gamma_s}{cG}}{\frac{d_o}{c}\cdot\frac{r_o}{c}(1-\eta)}[\{\frac{4}{3}\frac{d_o}{c}(\frac{r_o}{c})^2(\frac{\gamma_s}{cG})^{-2}\frac{\sigma}{G}+1\}^{1/2}-1]\}^3 \tag{6}$$

where

$$\frac{d_o}{c} = (2-2\cos\theta)^{1/2}(\frac{\delta_s D_s}{\delta_b D_b}\frac{\gamma_s}{cG})^{1/3} \tag{7}$$

and

$$r_o = d_o/(1-\cos\theta) \tag{8}$$

This equation has two limiting forms. If the surface curvature term is small, then

$$\dot{\eta} = \frac{2\lambda}{c^3}(\frac{r_o}{c}\cdot\frac{1}{1-\eta}\cdot\frac{cG}{\gamma_s}\frac{\sigma}{G})^3 \tag{9}$$

while if surface curvature is imporant (i.e., cavities are very sharp in shape), then

$$\dot{\eta} = 6\sqrt{3}\frac{\lambda}{c^3}\frac{1}{(1-\eta)^3}(\frac{c}{d_o}\frac{\sigma}{G})^{3/2} \tag{10}$$

Again, $\dot{\eta}$ depends on a power of stress, ranging between 1.5 and 3.

The transition from equilibrium to crack-like cavity shape occurs when surface diffusion becomes slower at supplying atoms to the boundary junction than boundary diffusion is at moving them down the boundary. This situation has been analyzed in detail by Chuang et al. [10], and more recently by others [3], to give

$$\frac{1-\eta}{\eta} < \frac{\sin\theta}{\theta(1-\cos\theta)^{3/2}}\,\frac{d_o}{c}\,/\left(\frac{\sigma}{G}\right)^{1/2} \tag{11}$$

as the condition for equilibrium cavities. The cavity length which can be maintained in an equilibrium shape decreases as η (and thus, σ) increases, and depends on the ratio of surface to boundary diffusivities. From several studies of the emergence of crack-like cavities [6], it appears that the transition period can be extensive. During this period, neither model adequately describes the cavity growth rate, which lies between equations (3) and (6). Several approaches to smoothing this transition are possible. The one used in the crack growth analysis, assumes that cavities become crack-like, but with a constant width, until the Chuang and Rice analysis suggests that further acceleration requires a decreasing cavity width, as given by equation (5).

The models just presented all assume that cavities are close enough that their diffusive fields overlap. This leads directly to assumption (4), stated earlier. It is possible, however, that cavities may grow in an isolated fashion, without any knowledge of the presence of neighbouring cavities. It can be shown [11], that this occurs whenever

$$4\sqrt{\frac{2\lambda}{c^3\eta}} \leq 1 \tag{12}$$

The growth of isolated cavities has been studied by Takasugi and Vitek [6] (16), who find that this situation is likely to occur only for crack-like cavities. For that case,

$$\dot{\eta} = \frac{2\lambda}{c^3}\,\{1\ 48\,k_1[\frac{k_2}{1.48}\frac{\sigma}{G} + 1 - (\frac{k_2}{1.48}\frac{\sigma}{G} + 1)^{1/2}]\}^3 \tag{13}$$

where

$$k_1 = \left(\frac{d_o}{c}\cdot\frac{c}{r_o}\cdot\frac{cG}{\gamma_s}\right)^2 \tag{14}$$

and

$$k_2 = (\frac{d_o}{c})^2 \frac{c}{r_o} \frac{\gamma_s}{cG} \tag{15}$$

This equation also has two limiting cases, depending on the important of the surface energy term. As a result, the stress dependence of cavity growth varies between σ^3 and σ^6.

Constrained Cavity Growth

Because cavity growth results in material plating on grain boundaries, it produces a creep strain, in addition to the strain produced by normal creep processes. Therefore, if cavitation is not uniform (and it rarely is), then the high local strain rates in cavitating regions eventually act to constrain further cavitation in the interest of material compatibility [1,12,13,14]. Under these conditions, cavity growth is governed by the overall creep rate in the material, such that [14]

$$\dot{\eta} \approx \eta^{-1/2} \frac{D}{c} \dot{\varepsilon} = \eta^{-1/2} \frac{D}{c} \dot{\varepsilon}_0 (\frac{\sigma}{\sigma_0})^m \tag{16}$$

where D is the grain size (i.e., D/c represents the density of cavities on the cavitated facets). Here, the creep rate has been described in terms of the applied stress σ, by the Norton power law

$$\dot{\varepsilon} = \dot{\varepsilon}_0 (\frac{\sigma}{\sigma_0})^m \tag{17}$$

For ceramics, the creep exponent m is typically [2] between 1 and 2.

Cavity Growth in Glassy Phase Materials

If a glassy phase is present at the grain boundary, then a new form of cavitation can be expected. This involves the growth of bubbles under a negative pressure, and their propagation as grain facet microcracks. This process has been modelled by Tsai and Raj [15], who found that

$$\dot{\eta} = 3\text{x}10^4 \pi \frac{\sigma^2}{EG} \eta (\frac{\delta}{c})^3 \tag{18}$$

where δ is the width of the glassy film along the grain boundary.

An Observation

A survey of these mechanistic rate equations shows that many of them (at least in their limiting forms) fit an equation of the type

$$\dot{\eta} = \phi(\eta)(\frac{\sigma}{G})^{\beta} \tag{19}$$

i.e., the cavity growth rate is proportional to stress to some power. This observation is exploited below to produce a generalized model for crack growth.

Cavity Nucleation

Despite the obvious importance of cavity nucleation, relatively few models have been produced. A variety of nucleation criteria are possible. In fine-grained ceramics, cavities generally form only at grain boundary triple junctions. In some cases, they may pre-exist as residual porosity. However, in structural ceramics cavity nucleation is generally required. This may result from decohesion due to stress concentrations produced by grain boundary sliding. Models for nucleation by decohesion have been reviewed by Goods and Brown [16]. They lead generally to a critical stress σ_c, for nucleation at a given site. Cavities may also nucleate by a local clustering of vacancies. This requires an incubation period. But as Argon et al. [17] have shown, since this mechanism operates only in regions of high stress concentration, it leads also to a critical nucleation stress. Therefore, in the models which follow, a critical stress, σ_c, is used as a cavity nucleation parameter.

CAVITATION AHEAD OF A CRACK

When a sharp crack is initially loaded by a far-field applied stress, an elastic stress field develops around the crack tip. At high temperatures, the material then begins to creep and to cavitate. If cavitation is more rapid than creep, then the dominant stress field ahead of the crack tip remains the linear elastic stress field and the normal stress in the plane of the crack is

$$\sigma = \frac{K}{\sqrt{(2\pi x)}} \tag{20}$$

where x is the distance ahead of the crack tip. If the material creeps at a significant rate, then stresses are relaxed surrounding the crack tip. For a power-law creeping solid (see equation (17)), then

$$\sigma = \sigma_o \left[\frac{C(t)}{\sigma_o \dot{\varepsilon}_o I_m x} \right]^{1/m+1} \tag{21}$$

where C(t) is the amplitude of the field, and I_m is an intensity factor dependent upon m and the degree of constraint [18]. Although C(t) is initially time-dependent and given approximately by [18]

$$C(t) = \frac{(1-v^2)K^2}{(m+1)Et} \tag{22}$$

(where E is Young's modulus), it does approach a steady-state value, C(t) = C* in time.

An important point to note is that the steady-state stress field for both elastic and creeping material under steady-state conditions can be written in the form

$$\sigma = \left(\frac{A}{x}\right)^{\alpha} \tag{23}$$

For an elastic solid $A = K^2/2\pi$, and $\alpha = 1/2$, while for a creeping solid

$$A = \frac{\sigma_o^m C^*}{\varepsilon_o I_m}, \quad \alpha = \frac{1}{m+1} \tag{24}$$

As mentioned earlier, for ceramics, $1 \le m < 2$ in general. Thus, the crak tip stress fields are changed very little in going from a linear elastic to creep-relaxed structure. It is therefore adequate for most cases to treat crack growth in ceramics as controlled by K. Nevertheless, in the discussion that follows, the generalized stress-field, given by equation (23), will be used.

<u>Model for Steady-State Crack Growth</u>

Consider an array of cavities, equally spaced ahead of a crack. The average stress on the nth cavity ahead of the crack is given by

$$\sigma = \left(\frac{A}{cn}\right)^{\alpha} \tag{25}$$

This has been used by Wilkinson and Vitek [7] to develop a general model for steady-state crack growth by cavitation. This is done by combining the general parametric equation for cavity growth as a function of local stress, equation (19), with the relationship just derived for the local stress on a cavity ahead of crack, equation (23).

We start by considering a plane strain crack loaded to tension, as shown in Fig. 2. Cavities are assumed to nucleate and grow with uniform spacing c, ahead of the crack. When the cavity nearest the crack tip reaches a critical size η_c, it links with the crack, increasing the crack length by c.

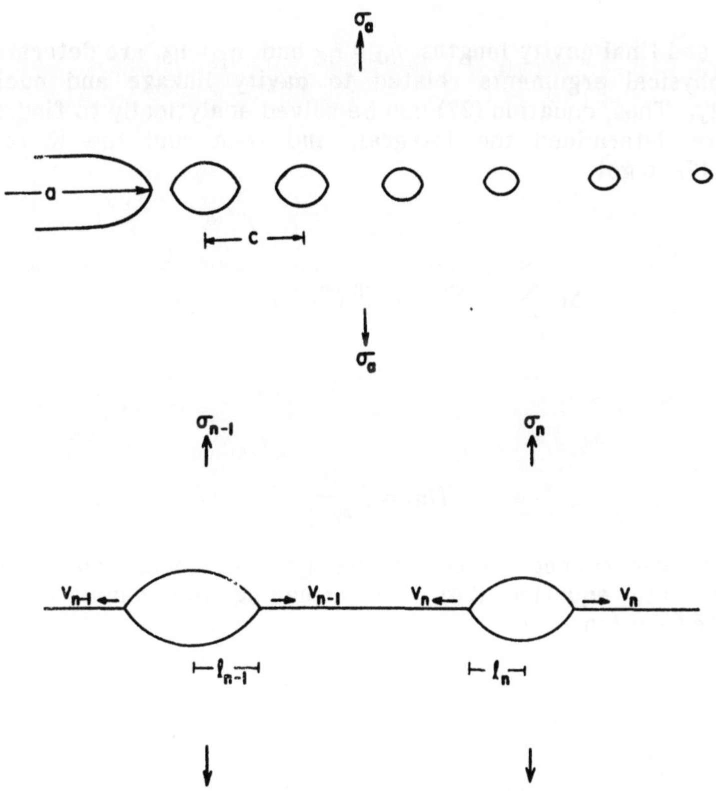

Fig. 2 A schematic diagram for uniformly spaced cavities, growing in the stress field ahead of a crack. Each cavity grows under a different local stress σ_n.

We wish to model the steady-state for which the number of cavities growing ahead of the crack N, remains constant. Thus, cavity size and the local stress loading the cavities remain constant in the frame of reference which moves with the crack tip. If the time required for each successive cavity to link with the main crack is Δt, the average crack growth rate is

$$\dot{a} = \frac{c}{\Delta t} \tag{26}$$

Because of the steady-state assumption, this time interval also sees each cavity grow by an amount equal to the difference between it and its neighbour, nearer the crack. Thus,

$$\Delta t = \int_{\eta_n}^{\eta_{n-1}} \frac{d\eta}{\dot{\eta}}, \eta = 1, \ldots, N \tag{27}$$

The initial and final cavity lengths, $\eta_o = \eta_c$ and $\eta_N = \eta_s$, are determined by separate physical arguments related to cavity linkage and nucleation respectively. Thus, equation (27) can be solved analytically to find Δt. To do this, we determined the integral, and then sum the N resulting equations (17) to get

$$\Delta t \sum_{n=1}^{N} n^{-\alpha\beta} = A^{-\alpha\beta} c^{\alpha\beta} [f(\eta_c) - f(\eta_s)] \tag{28}$$

where

$$f(\eta) = \int \frac{d\eta}{\phi(\eta)} \tag{29}$$

contains the dependence on the cavity growth mechanism through ϕ. Substituting into equation (26), and replacing the summation by an approximate function

$$\sum_{n=1}^{N} n^{-\alpha\beta} \approx h(\alpha\beta, N) = 1 + \frac{1}{2} N^{-\alpha\beta} + \frac{N^{1-\alpha\beta} - 2^{-\alpha\beta} - \frac{1}{2}}{1 - \alpha\beta} \tag{30}$$

we get an equation for crack growth:

$$\dot{a} = h(\alpha\beta, N) \frac{A^{\alpha\beta} c^{1-\alpha\beta}}{f(\eta_c) - f(\eta_s)} \tag{31}$$

This is a general equation for steady-state crack growth. It contains dependencies on the crack-tip stress field (through A and α), the mechanism of cavity growth (f and β), and the mechanism of cavity nucleation (N).

Extension to Non-Steady Crack Growth

An extension to this model by Bassani and Vitek [19] considers the effect of a time-dependent stress field on crack growth. They postulate a field of the form

$$\sigma = \left(\frac{A_t}{xt}\right)^{\alpha} \tag{32}$$

consistent with equations (21) and (22). If the stress field varies slowly with respect to the development of the damage field, then the steady-state model can be reworked for this case. They find

$$\dot{a} = h(\alpha\beta, N) \frac{A_t^{\alpha\beta} c^{1-\alpha\beta}}{f(\eta_c) - f(\eta_s)} \frac{1}{t^{\alpha\beta}} \tag{33}$$

a result very similar to that for a time-independent stress field.

MICROCRACK PROPAGATION

Most models for creep crack growth, such as those just presented, apply to macroscopic cracks only. This means that the zone of damage ahead of the crack is always assumed to lie within the singular portion of the stress field, whether it be K- or C*- controlled. However, for microcracks, the damage zone ahead of the crack may be of the same order or larger than the crack length itself. The stress field which applies in this case is not simply the singular portion, but must contain extra terms to describe the long range field. This is clearly of relevance to understanding the life of structures, which may contain microcracks.

An approximate model for microcracks can be developed from that just discussed, by replacing the singular stress field, equation (23), with

$$\sigma = \left(\frac{A}{x}\right)^{\alpha} + \sigma_a \tag{34}$$

where σ_a is the far-field applied stress. This is clearly an approximation to the true stress field. However, it is correct in the limits of x << a and x >> a. The general model for crack growth can now be reworked, and an equation similar to equation (28) results, the only difference being that $n^{-\alpha\beta}$ is replaced by

$$[n^{-\alpha} + \sigma_a (\frac{A}{c})^{-\alpha}]^{\beta} \tag{35}$$

The summation of this term for n = 1,N cannot be derived analytically. However, taking the dominant terms only (which, for $\beta > 1$ is reasonable), we get

$$\sum_{n=1}^{N} [n^{-\alpha} + \sigma_a (\frac{A}{c})^{-\alpha}]^{\beta} \approx h(\alpha\beta, N_c) + (N - N_c)\sigma_a^{\beta}(\frac{A}{c})^{-\alpha\beta} \tag{36}$$

where N_c represents the point at which the two terms are equal, i.e.

$$N_c = \sigma_a^{-1/\alpha}\frac{A}{c}, \quad 0 \leq N_c \leq N \tag{37}$$

By substituting into equation (26), we get a new solution for the crack growth rate, namely

$$\dot{a} = h(\alpha\beta, N_c)\frac{A^{\alpha\beta}c^{1-\alpha\beta}}{f(\eta_c) - f(\eta_s)} + (N - N_c)\frac{\sigma_a^{\beta}c}{f(\eta_c) - f(\eta_s)} \tag{38}$$

In the limit where the singular stress field is dominant, then N_c = N, and the solution derived here reduces to equation (31). For linear elastic conditions, equation (37) reduces to N_c = a/2c, i.e. crack-like behaviour results only if the damage zone does not extend more than a/2 from the crack tip.

DISCUSSION

The models just presented are rather simplistic in several ways. They assume a steady-state has been reached, that cavities are uniformly spaced ahead of the crack, that cavitation does not influence the crack stress field, and that cavities grow by the same mechanism from nucleation to linkage. These problems are discussed below. Nonetheless, the models provide some useful insights into the processes at work ahed of a crack. Let us look at these.

Cavity Nucleation

First consider cavity nucleation. The ease of nucleation determines the number of cavities N, which grow ahead of a crack. This affects a through the function h, plotted in Fig. 3. It is apparent that crack growth is sensitive to nucleation only if the parameter $\alpha\beta$ is small (essentially $\alpha\beta <$ 1). If is large, the size of the nucleation zone has very little effect on the rate of crack growth, and $h \approx 1$. Physically, this means that if cavity growth is sensitive to stress (β large), and stresses decay rapidly ahead of the crack tip (α large), then most of the damage occurs very close to the crack tip, even though it may be nucleated far away. For this situation, models based on even a single cavity growing ahead of a crack [3] are reasonable. Furthermore, no critical crack opening displacement is required for the crack to propagate. This situation is most often encountered in ceramics, for which m is small (i.e., α is large). From equation (31), we see that the crack growth, $\dot{a}$, is proportional to $A\alpha\beta$, i.e.,

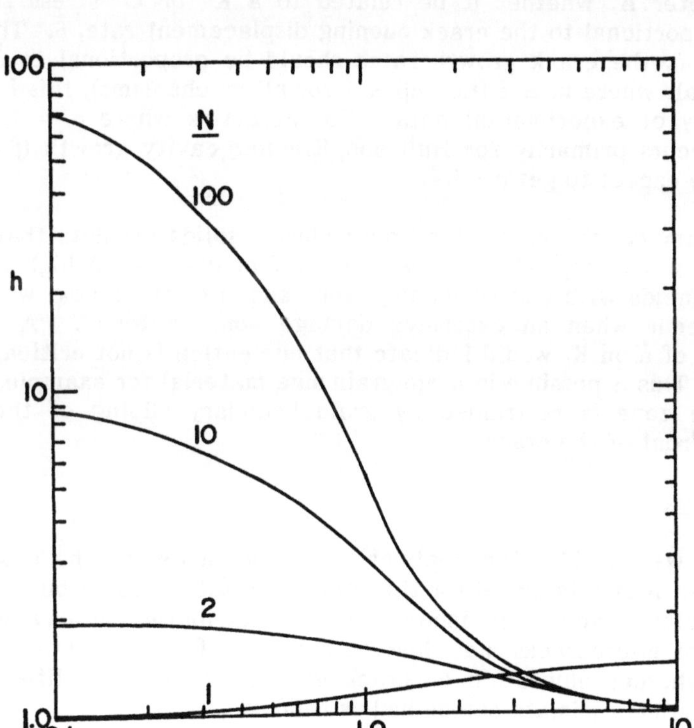

Fig. 3 The effect of cavity nucleation on crack growth is contained in the function h

$\dot{a} \sim K^{\beta}$. Thus, the K-dependence of $\dot{a}$ is mechanism-dependent and ranges between K^1 and K^6 for the mechanisms discussed earlier.

For $\alpha\beta < 1$, nucleation is important. If a critical stress, σ_c, controls nucleation, then (17)

$$N = \frac{A}{c} \sigma_c^{-1/\alpha} \tag{40}$$

and

$$\dot{a} = \frac{A}{(1-\alpha\beta)\sigma_c^{(1-\alpha\beta)/\alpha}[f(\eta_c) - f(\eta_s)]} \tag{41}$$

For this case, $\dot{a} \sim A$, independent of the mechanism of cavity growth, although the mechanism does affect the absolute rate of crack growth through the function f. This result has important physical implications. The parameter A, whether it be related to a K- or C*-stress field, is directly proportional to the crack opening displacement rate, $\dot{\delta}$. Thus, for this situation, the crack growth rates should be proportional to $\dot{\delta}$. For ductile metals where $m \geq 5$ (i.e., $\alpha\beta \leq 1$ for all mechanisms), this fits with the majority of experimental data. For ceramics where $\alpha \approx 1/2$, this situation occurs primarily for Hull and Rimmer cavity growth (β=1). In this case we expect to get $\dot{a} \sim K^2$.

In summary, the model for linear elastic solids predicts that crack growth rate should depend on K^p, where p is 2 or greater. A high value of p should coincide with a short damage zone ahead of the crack, while $p \approx 2$ should occur when an extensive damage zone is found. A weaker dependence of $\dot{a}$ on K, would indicate that nucleation is not critical stress controlled. This is possible in large grain size material for example, where the damage zone is restrained by grain boundary sliding of the grain directly in front of the crack.

Microcracking

Next, we consider the application of the model to the growth of microcracks. A simple model for this process results in equation (38). In a material which develops an inhomogeneous distribution of cavities, the nucleation of microcracks may develop early in life. It is interesting to speculate whether cavitation or crack growth is then the life-limiting process. A crude estimate can be had as follows.

Let us assume that the mechanism of cavitation is the same thorughout. The time to produce a microcrack in a region of accelerated cavitation is

$$t_u = \int_{\eta_o}^{\eta_f} \frac{d\eta}{\dot{\eta}} = \frac{f(\eta_c) - f(\eta_s)}{\sigma_a^\beta}$$

The result of this process is a microcrack of length 2a, corresponding to the width of zone of accelerated damage. This crack then propagates through the rest of the structure. Because this material has not cavitated the critical stress for cavity nucleation σ_c, must in general exceed the applied stress σ_a. Thus, comparing equations (37) and (40), then $N_c < N$, and the damage zone of the propagating microcrack is confined to a region which is less than a. Equation (31) is therefore applicable. The time required to grow this microcrack through the structure is

$$t_c = \int_{a_o}^{a_f} \frac{da}{\dot{a}}$$

The result (except for β=1), is

$$t_c = \frac{2\sqrt{2}}{2-\beta} \frac{1}{\pi^{(\beta-1)/2}} \frac{f(\eta_c) - f(\eta_s)}{h(\alpha\beta,N)\, c^{1-\beta/2}} [a_f^{1-\beta/2} - a_o^{1-\beta/2}] \frac{1}{\sigma_a^\beta} \quad (43)$$

We can now compare equations (42) and (43) which, if $\beta < 2$ and $a_f >> a_o$, yields

$$\frac{t_c}{t_u} = \frac{2\sqrt{2}}{\beta-2} \frac{1}{\pi^{(\beta-1)/2}} \left(\frac{a_o}{c}\right)^{1-\beta/2}$$

Since in general, $a_o >> c$, we can expect the time spent in crack growth to greatly exceed that spent in producing the initial microcrack.

There are two features which complicate this picture. First, the mechanism of cavity growth is not likely to remain the same throughout this process. Instead the material chooses whichever mechanism promotes the development of damage. Secondly, the material properties which impeded cavitation during the early life of the structure, will tend to slow crack growth also. Thus the function f will in general be smaller for uniform cavitation than for crack growth, thus increasing the importance of the crack growth regime.

Improvements

The models presented above make use of several assumptions. Improved models can be produced by removing some of these, but only

with a loss in the simplicity of the development. In many cases, one must resort to numerical solutions.

A major assumption in the approach taken here is that the development of the stress field ahead of a crack is not influenced by the growth of cavities. For brittle materials, the contribution of cavitation to the local strain rate can be significant. Since cavities are largest near the crack tip, the cavitation strains serve to constrain cavity growth in that region. This results in an altered stress field. The extent to which this process influences the rate of crack growth, and its dependence on K, is not clear. We are currently investigating this problem numerically.

The models also assume that cavities grow by the same mechanism throughout their life. Some recent numerical calculations show this is rarely the case. Typically cavities are rounded while small, but become crack-like as they get larger and the crack tip approaches. This does not mean however, that the crack growth rate cannot be characterized by equation (38), but only that a careful analysis is required to assess the range of dominance of each mechanism. This is also the subject of current study.

The cavity spacing may not be uniform. In this case, small microcracks develop ahead of the crack due to the coalescence of closely spaced. Vitek [20] has shown that this leads to a small but significant increase in the crack growth rate for a given average cavity spacing.

Finally, the models take a simple approach to cavity nucleation, in asusming that cavities all nucleate once they are within a certain distance from the crack tip, and that no further nucleation occurs. While this is certainly not what happens, there is insufficient data to improve on this assumption. It is clear that if the average cavity spacing decreases as the crack approaches, it is important to use a representative value for c. In most cases, the dimple spacing, measured from a fracture surface is used. Recent work has shown [21], that since this is always less than the smallest cavity spacing before fracture (due to early coalescence of some cavities), it may provide the best available estimate of cavity spacing ahead of a crack.

ACKNOWLEDGEMENTS

This work is supported by Imperial Oil Ltd., and by NSERC Canada.

REFERENCES

1. Cocks, A.C.F. and Ashby, M.F., Prog. Mat. Sci., 1982.

2. Evans, A.G., in Recent Advances in Creep and Fracture of Engineering Materials and Structures (B. Wilshire and D.R.J. Owen, eds.), Pineridge Press (1982).

3. Vitek, V. and Wilkinson, D.S., Proc. 5th Intl. Conf. on Strength of Metals and Alloys, 1979, p. 321.

4. Vitek, V., Met. Sci. J., 14, 1980, p. 403.

5. Wilkinson, D.S., Mat. Sci. Engrg., 49, 1981, p. 31.

6. Tagasuki, T. and Vitek, V., Met. Trans., 12A, 1981, p. 659.

7. Wilkinson, D.S. and Vitek, V., Acta Met., 30, 1982, p. 1723.

8. Wilkinson,D.S. in Advances in Life Prediction Methods, ASME (1983), p. 259.

9. Chuang, T-J. and Rice, J.R., Acta Met., 21, 1973, p. 1625.

10. Chuang, T-J., Kagawa, K.I., Rice, J.R. and Sills, L.B., Acta Met., 27, 1979, p. 265.

11 Wilkinson, D.S., unpublished research, 1979.

12. Dyson, B.F., Metal Sci., 10, 1976, p. 349.

13. Dyson, B.F., Can. Met. Quart., 18, 1978, p. 31.

14. Rice, J.R. Acta Met., 29, (1981), p. 675.

15. Tsai, R.L., and Raj, R. Acta Met., 30 (1982), p. 1043.

16. Goods, S.H. and Brown, L.M., Acta Met., 27 (1979), p. 1.

17. Argon, A.S., Chen, L-W and Lau, C.W., "Creep and Fatigue Fracture", ed. Pelloux, R.M.N. and Stoloff, N., A.I.M.E., 1980, New York.

18. Riedel, H. and Rice, J.R., ASTM STP 700, 1980, p. 112.

19. Bassani, J.L., Proc. 9th Natl. Cong. Appl. Mech., in press.

20. Vitek, V., private communication (1983).

EFFECT OF CRACKS ON MECHANISMS AND KINETICS OF CREEP DEFORMATION OF BRITTLE CERAMICS

D. P. H. Hasselman and A. Venkateswaran

Department of Materials Engineering
Virginia Polytechnic Institute and State University
Blacksburg, Virginia 24061 USA

ABSTRACT

Cracks can affect the creep deformation of brittle ceramics through two principle mechanisms. Time-dependent decreases in elastic moduli due to the growth of cracks leads to "elastic" creep. The transfer of stress from the crack-tip regions to the adjacent material leads to "crack-enhanced" creep, which results in an increase in the rate of creep of the underlying creep process.

Elastic and crack-enhanced creep are shown to have an effect on the shape of the creep curve and can lead to apparent non-linear creep. The presence of the cracks can lead to a time-dependent activation energy and results in a creep rate which increases with grain size, in contrast to the inverse grain size dependence of Nabarro-Herring and Coble creep. Cracks can also have a major effect on creep rate under multi-axial stress conditions. The effect of cracks on the dominant mechanism of creep is demonstrated by creep-deformation maps for a polycrystalline aluminum oxide.

INTRODUCTION

Creep deformation of polycrystalline ceramics by volume[1,2] or grain-boundary diffusion[3] or dislocation motion[4,5] is well understood and has been used extensively for the interpretation of experimental data. Diffusional creep, which depends linearly on the magnitude of stress, is dominant at relatively low values of temperatures and stress, In contrast, dislocation climb, dominant at high temperatures and stresses, exhibits a stress dependence which is highly non-linear. It may be noted also that the rate of creep by volume or grain-boundary diffusion should be an inverse function of the grain size.

Numerous experimental studies have shown the existence of non-linear creep in regimes of stress and temperature at which such behavior should not be significant [6-12]. Furthermore, observations of increasing rate of creep with increasing grain size [10], have been reported. Such effects appear to be more prevalent in materials with a large grain size or those with a residual pore phase. Extensive intergranular cracking during the deformation of polycrystalline ceramics at elevated temperatures is well known [9,12]. The existence of such cracks may be responsible for the unusual stress- and grain size-dependence. Two effects of cracks on creep can be identified. "Crack-enhanced" creep, originally analyzed by Weertman [13], represents the increase in creep over the value in the absence of cracks, due to the local stress field near the cracks and the associated transfer of stress to the material adjacent to the cracks. The second effect arises from the growth of existing or the nucleation of new cracks, which leads to "elastic creep", as identified by the present writers [14], due to the time-dependent decrease in elastic moduli. In a more recent analysis, Hasselman and Venkateswaran [15] showed that as a direct result of the stress dependence of the density of cracks, crack-enhanced creep can lead to the existence of apparent non-linear creep. Furthermore, the same effect can lead to a grain size dependence of creep rate which deviates from the values predicted for materials without cracks. Again, the existence of "elastic creep" can serve to explain unpredicted transient effects such as a decreasing creep rate and increasing activation energy with increasing duration of creep deformation. Also, crack-enhanced creep due to permanent crack-opening-displacement leads to the creation of a permanent pore-phase.

It is the purpose of this paper to review the earlier analyses of the above mechanisms and their expected effects on the creep behavior of polycrystalline ceramics.

ANALYSIS

During deformation at elevated temperatures, crack formation usually occurs by intergranular cracking along grain-boundaries oriented more or less perpendicularly to the direction of a uniaxially applied tensile stress. For this reason, appropriate mechanical models for the derivation of the required expressions for crack-enhanced and elastic creep were chosen in the form of two- and three-dimensional solids with parallel through- and penny-shaped cracks respectively, oriented perpendicularly to the stress σ applied at the surface of the cracked solid. Crack interaction is assumed to be absent. For simplicity, all cracks, will be assumed to be of equal size. However, the theory is easily extended to distributions in crack sizes. In fact, the existence of such crack size distributions serves to explain a number of observed creep phenomena. Finally, since emphasis is placed on the effect of cracks on creep, this study will not address the mechanisms of formation of such cracks nor will an a-priori choice of mechanisms of crack growth be made.

For the two-dimensional model with parallel non-interacting cracks, the rate of crack-enhanced creep ($\dot{\varepsilon}_c$) for linear creep as derived by Weertman [13] is:

$$\dot{\varepsilon}_c = \dot{\varepsilon}_o\ (1+2\pi Na^2) \qquad (1)$$

where $\dot{\varepsilon}_o$ is the rate of creep in the absence of cracks, N is the number of cracks per unit area and a is the half-length of the cracks.

For non-linear creep, the rate of crack-enhanced creep is:

$$\dot{\varepsilon}_c = \dot{\varepsilon}_o\ (1+2\pi Na^2 n^{1/2}) \qquad (2)$$

where n is the stress exponent of the basic creep mechanism.

The rate of elastic creep due to the rate of crack growth ($\dot{a}$) and rate of nucleation ($\dot{N}$) of cracks of size a can be obtained as [15]:

$$\dot{\varepsilon}_{el} = 2\pi a\sigma(2\dot{a}N + a\dot{N})/E_o \qquad (3)$$

For the three-dimensional model with parallel penny-shaped cracks, from the solution for the effective Young's modulus [16], the rate of crack-enhanced linear creep can be derived to be [15]:

$$\dot{\varepsilon}_c = \dot{\varepsilon}_o\ (1 + 4Na^3) \qquad (4)$$

where a is the crack radius and N is the number of cracks per unit volume. The corresponding equation for non-linear creep is not available from the literature and still has to be derived.

The rate of elastic creep is [15]:

$$\dot{\varepsilon}_{el} = 16(1-\nu^2)\sigma(Na^2\dot{a} + a^3\dot{N}/3)/E_o \qquad (5)$$

Crack-enhanced and elastic creep can occur concurrently. The total creep rate ($\dot{\varepsilon}_t$) is obtained by the addition of eq. 1 or 2 with 3 and eq. 4 with 5 for the two- and three-dimensional models, respectively.

DISCUSSION

A brief discussion on the nature of crack formation and growth is in order, before the effects of crack-enhanced and elastic creep can be ascertained. As indicated earlier, metallographic evidence shows that during high-temperature deformation, cracks form preferentially along grain-boundary facets. A number of different mechanisms can be postulated for the formation of such cracks. For purposes of this discussion it will be assumed that these cracks are initiated at crack-like pores located at triple-points or along grain boundaries. The size of such pores generally increases with increasing grain size. For this reason, for simplicity it will be assumed that the ratio of pore to grain size is a constant (with exceptions to be noted where appropriate). Such pores can develop into grain boundary cracks at a stress of sufficient magnitude. The total extent of crack propagation, however, is limited to the grain boundary facet. This implies that when the crack has traversed the facet completely, the rate of crack growth equals zero and crack arrest occurs. The maximum rate of crack growth will occur at some intermediate position of the crack front along the grain-boundary facet. This is shown schematically in Fig. 1 for a range of increasing stress levels. This figure also indicates qualitatively that the magnitude of stress required for the initiation of crack growth increases as the size of the crack pre-cursor decreases. For a quantitative estimate of the kinetics of crack growth along a grain boundary facet in any given material, additional theoretical analysis will be required.

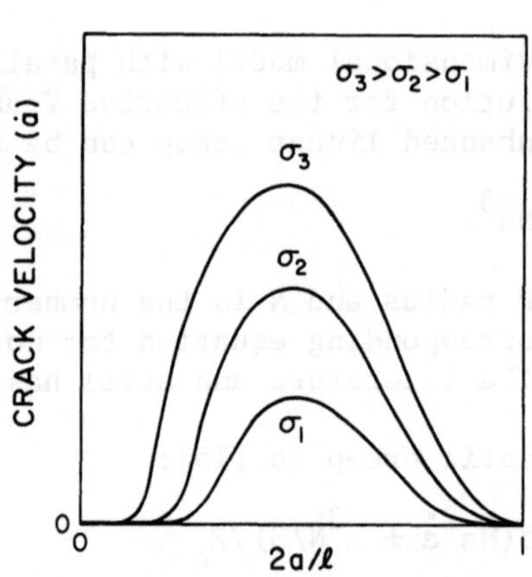

Figure 1. Schematic variation in crack velocity with relative position along grain boundary facet, for different stress values.

Figures 2a and 2b show schematically the rate of elastic creep for increasing N at constant σ and for increasing σ at constant N respectively, as inferred from the crack-growth behavior shown in Fig. 1 and the analysis presented earlier. The corresponding total elastic creep strain as a function of time is shown in Figs. 3a and 3b. The finite value of strain at t = 0 corresponds to the elastic strain which results from the imposition of the stress.

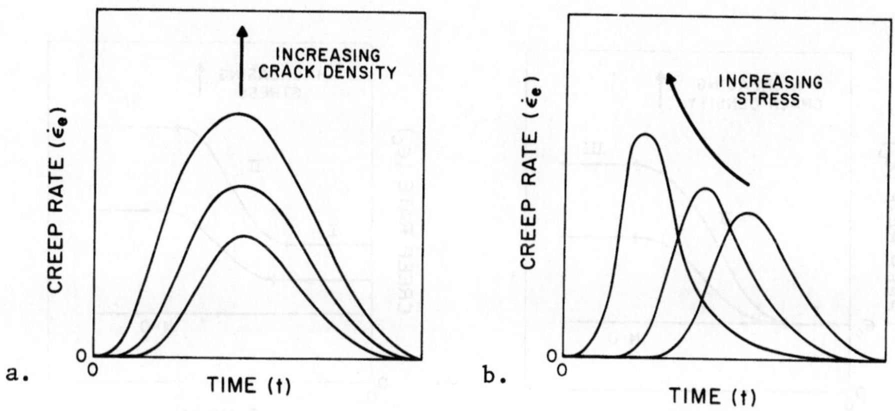

Fig. 2. Elastic or compliance creep rate as a function of time for (a) increasing crack density, constant applied stress and (b) increasing applied stress, constant crack density.

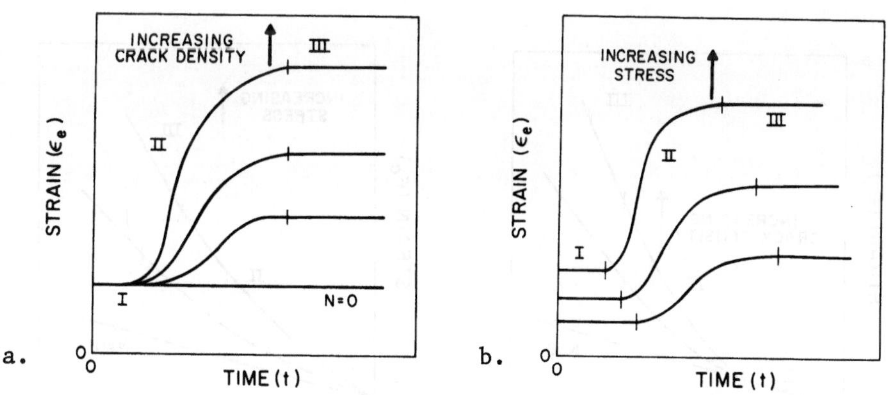

Fig. 3. Cumulative elastic creep strain due to growth of cracks along grain-boundary facets for (a) increasing crack density, constant stress and (b) increasing stress, constant crack density.

In analogy to the terminology for the multi-stage creep behavior found in metallic materials, the incubation period for crack nucleation is referred to as stage I. Elastic creep by crack growth occurs in stage II. In stage III, in which the cracks have reached their final size, crack-enhanced creep occurs, with no contribution from elastic creep.

Figures 4a and 4b show schematically, the rate of crack-enhanced creep as a function of time for increasing N and σ,

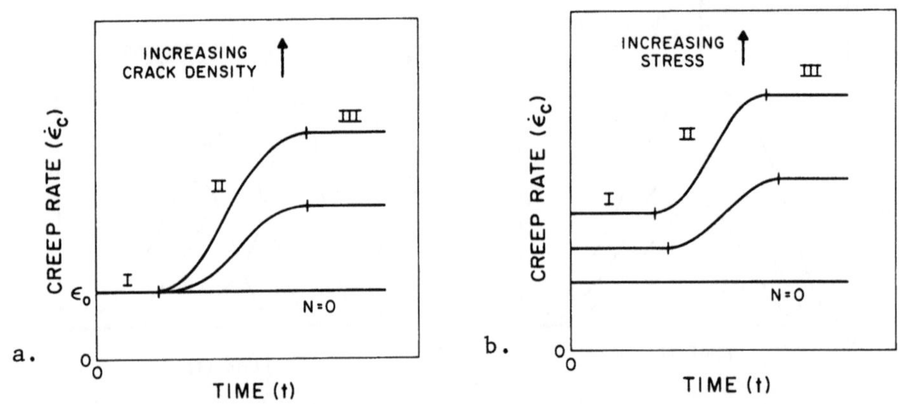

Fig. 4. Crack-enhanced creep rate as a function of test time for (a) increasing crack density, constant stress and (b) increasing stress, constant crack density.

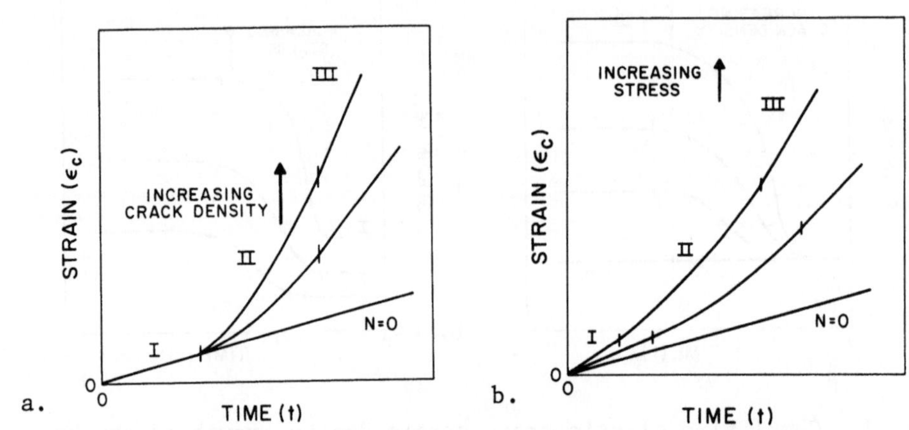

Fig. 5. Crack-enhanced creep stain as a function of time for (a) increasing crack density, constant applied stress and (b) increasing applied stress, constant crack density.

similar to Figs. 2 and 3. During stage I the creep rate corresponds to the creep rate ($\dot{\varepsilon}_o$) of the basic creep mechanism in the absence of cracks. In stage II, the creep rate becomes crack enhanced in direct relation to the instantaneous value of crack length at any value of time. In stage III, a constant rate of crack-enhanced creep is achieved since the cracks have reached their final size, a_∞. The corresponding value of total crack-enhanced creep strain as a function of time, for increasing N at constant σ and increasing σ at constant N, are shown in Figs. 5a and 5b respectively.

The total creep curve, which combines the elastic creep and the crack-enhanced creep of Figs. 3 and 5, is shown in Fig. 6. The total creep curve is drawn in a manner such that the sigmoidal character of the elastic strain contribution is retained. The actual shape in stage II will critically depend on the crack velocity as a function of position along the grain-boundary facet.

Important to note is that the contribution of elastic creep by crack growth to the total creep deformation, is a transient phenomenon which is expected to be active primarily during the beginning period of creep deformation. If crack formation occurs quite rapidly, as may well be the case at highly elevated temperatures and/or high values of stress, it is conceivable that unless care is taken, the elastic contribution to the creep deformation may not be recorded. On the other hand, at low temperatures and moderate stress levels, elastic creep due to crack formation could make a contribution over the total duration of creep deformation. If, indeed, stage II creep has the sigmoidal form shown in Fig. 6, this would manifest itself as stage III is approached as a diminishing creep rate with increasing creep duration. In fact, this effect has been observed in creep studies [17]. Possibly,

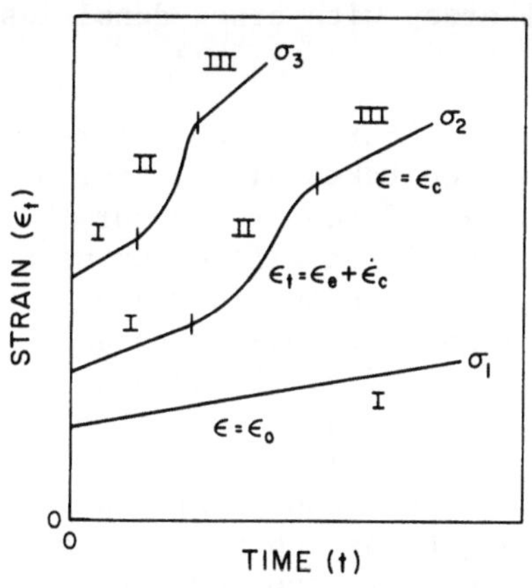

Fig. 6. Schematic creep curve for brittle polycrystalline ceramic, arising from combined contributions of elastic and crack-enhanced creep, for different stress values ($\sigma_3 > \sigma_2 > \sigma_1$).

the diminishing contribution of elastic creep with increasing time may provide a satisfactory explanation. For the same reason, a second effect should occur. The activation energy for crack growth, depending on the specific mechanism involved, may differ from the activation energy for the basic creep process. For this reason, the apparent activation energy which can be inferred from the experimental data for stage II will differ from the value for stages I and III. If the activation energy for crack growth is smaller than for the basic creep process (as may well be the case) as stage III is approached the apparent activation energy will show an increase with increasing duration of the creep deformation, which would be reflected in a creep rate that diminishes with time.

In the event that well developed, sub-facet size cracks are already present at the onset of creep, stage I and possibly part of stage II might not be observed. Instead, the creep rate would gradually decrease to the steady state value, analogous to conventional creep behavior, i.e., the primary creep observed in metals.

Possibly, the role of cracks as outlined above provides a reasonable explanation.

The stress dependence or rather the "apparent" stress dependence of the rate of creep can also be affected by the presence of the cracks. This effect arises because the density of the cracks itself can be a function of the magnitude of stress. For simplicity, the discussion of this effect will concentrate on stage I and stage III with fully developed cracks. In general, the stress exponent for the dominant mechanism of creep can be ascertained by measurement of the slope of the log-log plot of creep rate vs. stress. At values of stress which are sufficiently low that crack formation will not occur, the creep rate will correspond to stage I creep, with a stress exponent corresponding to that of the basic creep rate $\dot{\varepsilon}_o$. The same is true at high stress levels corresponding to stage III creep with crack densities and sizes that are fully developed.

A transition from stage I to stage III occurs at the intermediate stress level at which crack formation takes place. For a very homogeneous material in which the sizes of the pre-cursor cracks (grain size) are identical, this transition will occur at a single value of stress, manifested by a discontinuous increase equal to the factor $(1+2\pi Na^2)$ or $(1+4Na^3)$ for the two- or three-dimensional cracks respectively, from eqs. 1 and 4. In actual practice, however, the crack-precursors (grains) will exhibit a distribution in size, so that crack formation will occur over a range of stress. Over this range of stress the density of cracks will increase with increasing magnitude of stress, up to a value of stress at which all possible crack precursors have developed

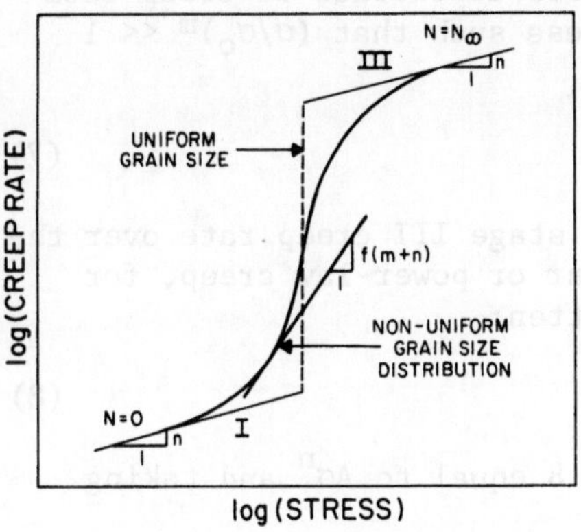

Fig. 7. Steady state creep rate in stages I and III, as a function of applied stress, for a uniform and non-uniform grain-size distribution and a constant ratio of precursor-crack to grain size.

into full-sized cracks. Above this value of stress, the crack density will remain constant at a value designated N_∞. The crack density at intermediate values of stress will be designated N_σ. Over the range of stress at which N_σ increases with increasing stress, the creep rate will exhibit a corresponding increase with stress. Critical to note is that this increase is superimposed on the normal stress dependence of the underlying creep process. The net result is that the overall stress dependence will be non-linear, regardless of the linearity or non-linearity of the basic creep mechanism.

Once all cracks have formed (i.e., the values of the quantities Na^2 or Na^3 in eqs. 1 and 4 are constant), the stress dependence of the creep rate will return to the corresponding dependence in the absence of cracks (stage I). This implies that as this condition is approached, the plot of log $\dot{\varepsilon}$ vs. log σ will exhibit a downward curvature as indicated in Fig. 7. No literature data appear to be available which confirm the existence of this prediction. This may possibly be masked by the effect of crack coalescence. Nevertheless, an experimental search for such behavior could be of interest.

An analysis of the apparent stress dependence of the rate of creep, due to a stress-dependent crack density, was presented by the present writers [15]. Briefly, it was assumed that the crack density (N) as a function of stress, in a steady-state stage III creep could be expressed by a Weibull distribution function of the form:

$$N/N_\infty = 1 - \exp\,(-\sigma/\sigma_o)^m \qquad (6)$$

where N_∞ is the maximum crack density attainable, and σ_o and m are constants. Substitution of this expression in eqs. 1 or 4 for crack-enhanced creep will result in a creep rate dependence on stress, shown by the sigmoidal curve in Fig. 7.

A simple expression for the stress dependence of creep rate can be derived for low values of stress such that $(\sigma/\sigma_o)^m << 1$ which yields:

$$N/N_\infty = (\sigma/\sigma_o)^m \tag{7}$$

The incremental increase in the stage III creep rate over the corresponding stage I value for linear or power-law creep, for the two-dimensional model can be written:

$$\dot{\varepsilon}_c - \dot{\varepsilon}_o = 2\pi\dot{\varepsilon}_o Na^2 n^{1/2} \tag{8}$$

Setting $\dot{\varepsilon}_o$ in the r.h.s. of eq. 8 equal to $A\sigma^n$ and taking logarithms yields:

$$\log(\dot{\varepsilon}_c - \dot{\varepsilon}_o) = (n+m)\log\sigma + \log(2\pi AN_\infty \sigma_o^{-m} a^2 n^{1/2}) \tag{9}$$

which indicates that the slope of the logarithm of the incremental increase in creep rate due to the cracks, against the logarithm of stress, is equal to the quantity (m+n). This implies that the stress exponent of steady-state creep of materials which undergo crack formation, will be a function of the stress exponent of the basic creep process as well as the stress dependence of the crack density. In the above analysis, the form of the dependence of crack density on stress was chosen to result in the stress exponent (n+m) of the incremental creep rate. Clearly, in practice other dependencies of crack density on stress can be found and should be established experimentally for the specific material under study. Regardless of the quantitative details, the presence of cracks can be used to explain the occurrence of non-linear creep without having to postulate the existence of dislocation-controlled creep.

The effect of grain size on the rate of creep deformation can also be a function of the presence of cracks. The rate of diffusional creep by the Nabarro-Herring or Coble mechanisms is inversely proportional to the square and the cube of the grain size respectively. The presence of cracks can have an effect on the apparent value of these grain size exponents. For simplicity, it will be assumed that the ratio of the size of the precursor crack to the grain size is a constant for both a uniform grain size and a distribution in grain sizes. Of course, for any given material, such size relationships need to be established by experimental means. Again, the discussion of the effect of grain size on creep rate will focus on stage I and III creep only, with no cracks and fully-developed cracks of size a_∞ respectively.

For grain sizes and associated crack precursors which are too small for crack formation to occur, the grain size exponent should correspond to that of the underlying creep process. For establishing the effect of cracks on creep rate it will be assumed first, that the grain size (and associated precursor size) is uniform. For any given stress level, this implies that cracks will form at and above a critical grain size. At this latter value the plot of log $\dot{\varepsilon}$ vs log d (where d represents grain size) will exhibit a discontinuous increase in creep rate corresponding to the factor $(1+2\pi N\sigma a_\infty^2)$. Such a discontinuity is shown in Fig. 8. For a distribution of grain (precursor) sizes, the creep rate will vary with grain size in a continuous manner. Included in Fig. 8 is the grain size dependence of creep for a narrow and broad grain size distribution respectively, with a value of mean grain size which equals that of the uniform grains. For the narrow grain size distribution, the creep rate increases with increasing grain size, near the value of critical grain size for the uniformly sized grains. For the broad grain size distribution, the creep rate decreases monotonically with increasing grain size. However, the slope near the critical grain size is less than the grain size exponent for the underlying mechanism of creep.

The effect of cracks on the apparent grain size exponent can be derived in a manner analogous to the derivation of the apparent stress exponent as presented earlier.

It will be assumed that at any given value of stress, the density of cracks as a function of grain size (d) can be expressed by:

$$N/N_o = 1 - \exp\{-(d/d_o)^m\} \tag{10}$$

where N_o, d_o and m are constants.

For the condition that $(d/d_o)^m \ll 1$ eq. (10) yields:

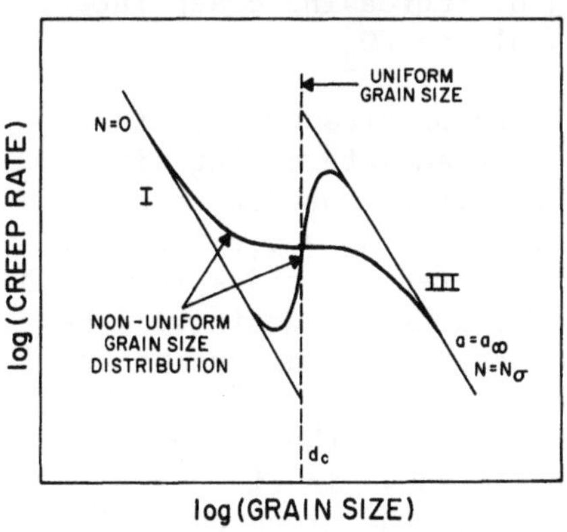

Fig. 8. Grain size dependence of stage I and III steady stage creep rate for uniform and non-uniform grain size distributions and a constant ratio of size of precursor-crack to grain size.

$$N/N_o \simeq (d/d_o)^m \tag{11}$$

The rate of crack-enhanced diffusional creep for the two-dimensional model can be written in terms of the grain size as

$$\dot{\varepsilon}_c = Cd^b\ (1 + 2\pi Na_\infty^{\ 2}) \tag{12}$$

where C is a constant which depends on stress and temperature. The constant b represents the grain size exponent and equals -2 and -3 for Nabarro-Herring and Coble creep, respectively.

The crack size can be expressed in terms of the grain size by:

$$a_\infty = d/k \tag{13}$$

where k is a constant (with k = 4 for hexagonally shaped grains and k = 2 for grains of cubic shape).

With the aid of eqs. 11 and 13, eq. 12 can be rewritten in terms of the increase in creep rate due to the cracks as

$$\dot{\varepsilon}_c - \dot{\varepsilon}_o = 2\pi CN_o d^{(b+m+2)}/k^2 d_o^{\ m} \tag{14}$$

Taking logarithms yields:

$$\log\ (\dot{\varepsilon}_c - \dot{\varepsilon}_o) = (b+m+2)\log d + \log\ (2\pi CN_o/k^2 d_o^{\ m}) \tag{15}$$

Equation 14 indicates that the grain size exponent of the incremental increase in creep equals (b+m+2). This implies that for grain size distributions with (m+2) > b, an increase in creep rate with increasing grain size can be obtained over the range of grain sizes at which grain boundary cracking occurs. This latter situation will arise in particular for grain size distributions which are relatively narrow as indicated in Fig. 8. It is suggested that the above analysis offers a reasonable explanation for the observations of Burton et al. [10] of increasing creep rate with increasing grain size in polycrystalline UO_2.

Grain size and the associated absence or presence of cracks at any given stress level should also have an interesting effect on the shape of the creep curve. Let us first consider three materials with highly uniform grain sizes $d_3 > d_2 > d_1$. Furthermore $d_1 < d_c$, the critical grain size required for crack formation. For grain sizes d_3 and d_2 the number of cracks per grain will be assumed to be a constant. This implies that when the cracks are fully developed in stage III, the quantities Na_∞^2 or Na_∞^3 will have constant values, independent of grain size. For these three materials the expected creep curves are as shown in Fig. 9a. For $d = d_1$ no crack formation occurs so that the creep curve is linear with a slope corresponding to the creep rate of the basic creep process. The materials with grain sizes d_2 and d_3 will undergo equal amounts of elastic strain and will also be subject to crack-enhanced creep. This can lead to a cross-over in the creep curve for the larger-grained materials by the corresponding curve for the fine-grained material. However, since the quantities Na^2 or Na^3 are constant, no mutual cross-over will occur for the creep curves of the materials which have undergone cracking.

A different type of cross-over in the creep curves is expected to occur for grain size distributions such that, at a particular stress level, the quantities Na^2 or Na^3 increase with increasing grain size. As indicated in Fig. 9b, the relative order of the total creep strain for grain sizes $d_3 > d_2 > d_1$ in stage I and III can be reversed.

In general, depending on the relative grain sizes and grain size distributions, other relative orders for the creep strains can be obtained.

Critical to note is that, because of the role of cracks in creep, the effects of grain size and stress no longer are independent. This is shown schematically in Fig. 10 for three materials with uniform grain size distributions and mean grain sizes $d_3 > d_2 > d_1$, such that in stage III, $N_3a_3^2 > N_2a_2^2 > N_1a_1^2 = 0$. A complete reversal of the relative order of the creep rates in stages I and III can take place as illustrated.

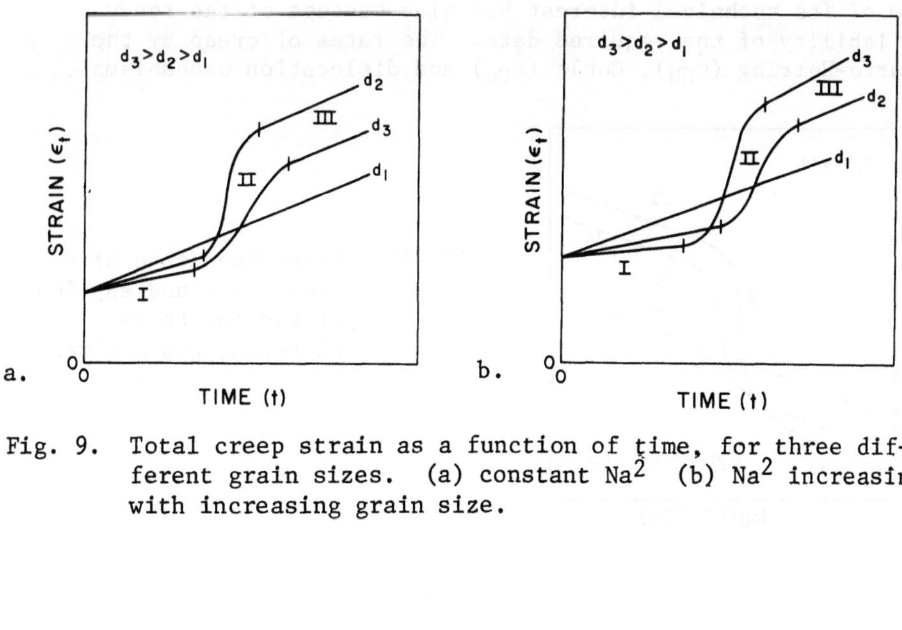

Fig. 9. Total creep strain as a function of time, for three different grain sizes. (a) constant Na^2 (b) Na^2 increasing with increasing grain size.

The interaction between grain size and stress is also indicated schematically in Fig. 11 which is a log-log plot of stage III creep rate against grain size for a series of values of stress. The critical grain size, d_c, required for crack formation decreases with increasing stress level. At any given value of stress, the number of cracks increases with increasing grain size. In general, such interaction between stress and grain size, of course is implicit in the analysis presented earlier and can be expressed quantitatively for any material for which the relationship between crack density and stress and the crack size and grain size are established.

The mechanism of creep which is dominant at any value of stress and temperature can also be affected by the presence of the cracks.

Experimental observations [18] indicate that crack growth will occur at temperatures well below those at which Nabarro-Herring, Coble or dislocation creep cause creep strains measurable within a reasonable time-span in any creep study. For this reason, elastic creep by crack growth is expected to be the dominant, if not the only, mechanism of creep at low and moderate temperatures. In fact, elastic crack by crack growth by a stress-corrosion mechanism was suggested to be responsible for observable creep in microcracked polycrystalline ceramics at room temperature [19].

The effect of cracks on the dominant mechanism of creep can be examined most effectively by the use of creep deformation maps, originally proposed by Weertman [20,21] and developed by Ashby [22]. Polycrystalline aluminum oxide was judged to be an appropriate material for the compilation of such maps, not only in view of its technical interest but also because of the ready availability of the required data. The rates of creep by the Nabarro-Herring ($\dot{\varepsilon}_{NH}$), Coble ($\dot{\varepsilon}_{Co}$) and dislocation mechanisms

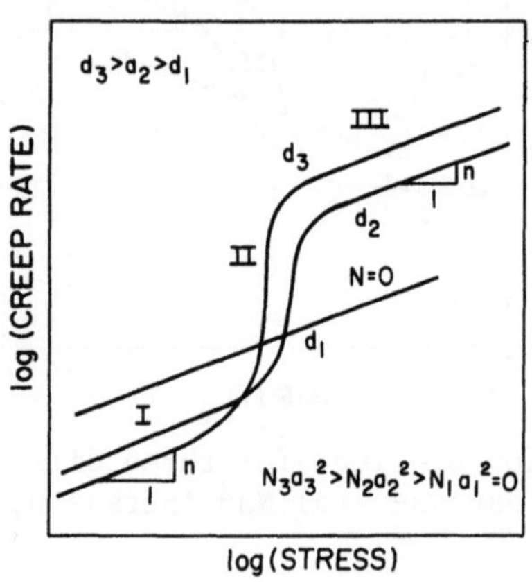

Fig. 10. Interdependence of creep rate and applied stress for three different grain sizes.

($\dot{\varepsilon}_d$) taken to be identical to those used by Mohamed and Langdon [23], are as follows:

$$\dot{\varepsilon}_{NH} = 20\Omega\sigma D_\ell^+/d^2kT \quad (16)$$

$$\dot{\varepsilon}_{Co} = 24\Omega\sigma D_{gb}^+/d^3kT \quad (17)$$

$$\dot{\varepsilon}_d = 4\vec{b}\sigma^3 D_\ell^-/kTG^2 \quad (18)$$

where Ω, the molecular volume, equals $4.2\times10^{-29}m^3$, k is the Boltzmann constant, T = absolute temperature and $\vec{b}$ the Burgers vector for basal slip = 4.75A°. Lattice diffusion coefficients for the cation and anion respectively are $D_\ell^+ = 28\times10^{-4}\exp(-478{,}000/8.31T)$ and $D_\ell^- = 1.9\times10^{-1}\exp(-637{,}000/8.31T)$. Enhanced grain boundary diffusivity for the cation, $D_{gb}^+ = 8.6\times10^{-10}\exp(-419{,}000/8.31T)$ (all in M.K.S. units) and the shear modulus is given by [24] $G = 1.71\times10^{11} - 23.4\times10^6T$ (in Nm^{-2}).

For the calculation of the rate of elastic creep by crack growth and crack-enhanced creep, the two-dimensional model was assumed to be most appropriate, in view of the known dependence of the stress-exponent on the enhancement factor (eq. 2).

The rate of crack growth ($\dot{a}$) was chosen to be of the form:

$$\dot{a} = Ak^g \exp(-Q/RT) \quad (19)$$

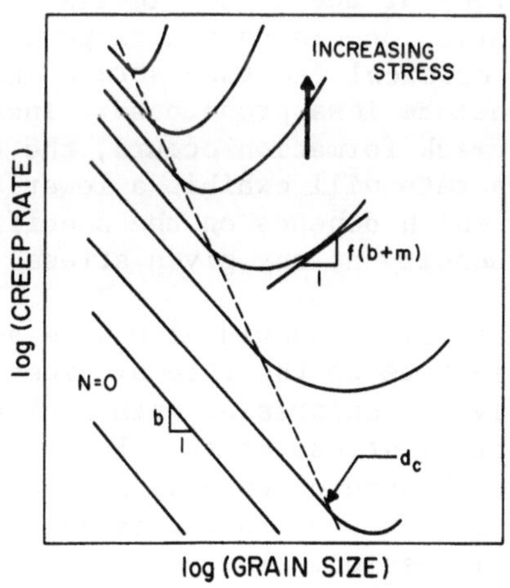

Fig. 11. Creep rate as a function of grain size, for increasing applied stress; d_c represents the critical grain size for crack formation.

where A is the pre-exponential factor, K is the stress intensity factor (with $K = \sigma\sqrt{\pi a}$) g is the exponent of the stress intensity factor and Q is the activation energy for crack growth. Values for A, n and Q were taken as $A = 10^9$ m/s, n = 10 and Q = 420 KJ/mole^{o}K. These values may possibly require modification, pending future acquisition of reliable experimental data. Grain boundary crack formation was assumed to occur at a single value of stress of 17.2 MPa. The shear modulus G was used for normalization of the stress values. For stage III creep the crack size, a_∞ was taken as, a = 10, and a crack density of $N = 5.2\text{x}10^9$ cracks was assumed. This yields a value for $Na_\infty^2 = 0.25$, corresponding to a value of the enhancement factor $(1+2\pi Na^2) = 2.57$. In stage II the crack size was taken as 6 μm.

Figures 12a, b, c, for a grain size d = 20 μm, show the creep deformation maps in stages I, II and III respectively. Comparison of stage I and III creep shows that at the stress value required for grain boundary cracking, the line for a given creep rate exhibits a discontinuous shift towards a lower value of stress or temperature, indicative of the crack-enhancement effect. Furthermore, because of the role of the stress-exponent 'n' in the crack-enhancement factor, the range of stress and temperature over which dislocation creep is dominant has grown at the expense of Nabarro-Herring and Coble creep. Clearly, since the crack size is constant, Figs. 12a and 12c do not include elastic creep by crack growth as a dominant mechanism.

This latter effect, however, as indicated in Fig. 12b, becomes dominant in stage II creep at the higher levels of stress. In general, if crack formation occurs over a range of stress, the discontinuities at the critical fracture stress σ_{th}, as evident in Figs. 12b and c, will become less pronounced. Instead, over the stress range at which crack formation occurs, the curve which indicates a given creep rate will exhibit a lower slope at the boundary, with a value which depends on the specific stress dependence of the crack density at any given stress value.

The presence of the cracks should also have an effect on the influence of the stress-state on the rate of creep deformation. Rates of creep generally are suppressed with increasing degree of uniformity of a multi-axial stress state. For the conventional creep processes (in the absence of cracks) no creep deformation can occur under conditions of a uniform triaxial stress state, either in tension or compression. However, for a uniform triaxial tensile stress state, elastic creep by crack growth becomes the only possible and therefore dominant mechanism regardless of the magnitude of stress or temperature. Elastic creep by crack growth under these latter stress conditions can be derived from the expression for the effective bulk modulus, B_{eff}, of a solid with randomly oriented penny-shaped cracks, derived by Walsh [25].

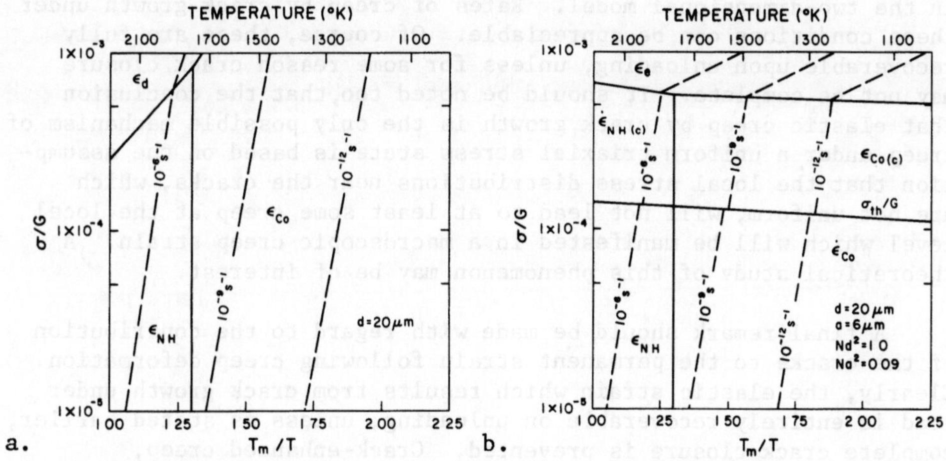

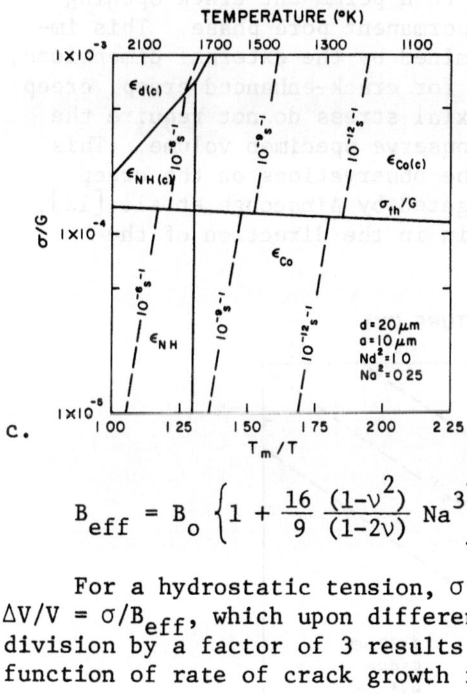

c.

Fig. 12. Deformation mechanism maps for polycrystalline aluminum oxide with grain size 20 μm. (a) no cracks (b) stage II creep and (c) stage III creep.

$$B_{eff} = B_o \left\{1 + \frac{16}{9}\,\frac{(1-\nu^2)}{(1-2\nu)}\,Na^3\right\}^{-1} \tag{20}$$

For a hydrostatic tension, σ, the relative volume change $\Delta V/V = \sigma/B_{eff}$, which upon differentation with respect to time and division by a factor of 3 results in the elastic creep rate as a function of rate of crack growth in one direction:

$$\dot{\varepsilon} = \frac{16(1-\nu^2)Na^2\dot{a}\sigma}{9(1-2\nu)B_o} \tag{21}$$

Figure 13 shows the deformation map for aluminum oxide with penny-shaped cracks of radius a = 8 μm and a crack density N = $2.5\times10^{14}m^{-3}$. The rate of crack growth in polycrystalline aluminum oxide is expressed by eq. 19 with the same values of g, A and Q as in the two-dimensional model. Rates of creep by crack growth under these conditions can be appreciable. Of course, these are fully recoverable upon unloading, unless for some reason crack closure may not be complete. It should be noted too, that the conclusion that elastic creep by crack growth is the only possible mechanism of creep under a uniform triaxial stress state is based on the assumption that the local stress distributions near the cracks, which are not uniform, will not lead to at least some creep at the local level which will be manifested in a macroscopic creep strain. A theoretical study of this phenomenon may be of interest.

A final remark should be made with regard to the contribution of the cracks to the permanent strain following creep deformation. Clearly, the elastic strain which results from crack growth under load is entirely recoverable on unloading, unless as stated earlier, complete crack closure is prevented. Crack-enhanced creep, governed by the acceleration of the creep rate of the material adjacent to the cracks, should lead to a permanent crack-opening-displacement and the creation of a permanent pore phase. This implies that specimen volume as determined by the external dimensions, is not conserved. For this reason, for crack-enhanced creep, creep strains in the direction of a uni-axial stress do not require the appropriate transverse strains to conserve specimen volume. This latter conclusion is supported by the observations on the creep behavior of Nb_2O_5-doped UO_2, investigated by Ainscough et al. [12], which indicated that the creep strain in the direction of the

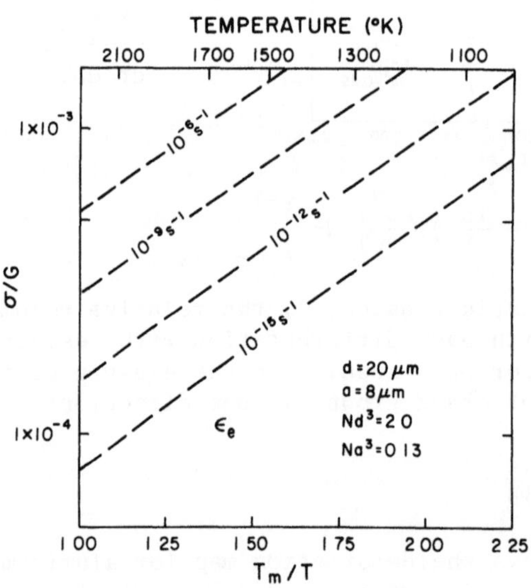

Fig. 13. Deformation mechanism map for polycrystalline aluminum oxide with grain size 20 μm, subjected to uniform triaxial tensile stress.

stress can be almost fully accounted for by the internal volume of the generated cracks. In this case, the rate of creation of porosity ($\dot{P}$) can be directly related to the rate of stage III crack-enhanced creep by [26]:

$$\dot{P} = \dot{\varepsilon}_o(1 + 2\pi Na^2) \qquad (22)$$

which in turn can be related to the rate of average crack-opening-displacement $\dot{\delta}$, for the two-dimensional crack model by:

$$\dot{\delta} = \dot{P}/2aN \qquad (23)$$

which with the aid of eq. 22 can be written

$$\dot{\delta} = \dot{\varepsilon}_o(1 + 2\pi Na^2)/2aN \qquad (24)$$

The above discussion has so far concentrated on cracks in materials subjected to tensile loading. Under these conditions, the cracks can be infinitely thin. Under conditions of compressive loading, however, infinitely thin cracks will have no effect on the creep deformation as the stresses are readily transmitted across the cracks. This latter situation, however, will not arise for cracks of a geometry such that even under conditions of zero load, a finite value of crack-opening-displacement exists. Such cracks under conditions of compressive loading should have the same effect on the rate of creep as crack-enhanced creep under conditions of tensile loading, as discussed above. An analysis of this effect, however, should include the effect of crack shape. From the solutions for the shear moduli of a two-dimensional plate with oriented elliptical shaped cracks, originally derived by Hasselman, Gebauer and Manson [27], the present writers [28] derived an expression for the rate of crack-enhanced creep valid for both tension and compression:

$$\dot{\varepsilon}_c = \dot{\varepsilon}_o\,[1 + 0.5(\alpha + 1)^2 P/\alpha] \qquad \alpha \ll 1 \qquad (25)$$

Similarly, for randomly oriented penny-shaped cracks, the enhanced creep rate from the solutions of Walsh [29] can be derived to be:

$$\dot{\varepsilon}_c = \dot{\varepsilon}_o\,[1 + 4P/5\pi\alpha] \qquad \alpha \ll 1 \qquad (26)$$

where P is the area/volume fraction of the elliptical/penny-shaped cracks and α is the ratio of the minor to major axis (b/a). The quantity P can be related to the dimensions and densities of the cracks by $P = \pi abN$ (elliptical) or $P = \frac{4}{3}\pi abcN$ (ellipsoidal). It was suggested [28] that the above equation can be used to advantage for the porosity correction required for the interpretation of rate of densification of porous compacts during pressure-sintering.

It is hoped that this study has demonstrated the important role of cracks during creep deformation of polycrystalline ceramics. Regardless of the validity of the quantitative theories presented, the possible effect of cracks must be considered in the prediction or interpretation of creep behavior. Experimental studies should be conducted in a manner which will yield the required information. Measurement of crack sizes and densities should be made as a function of stress and time. If practically possible, measurement of the elastic behavior as a function of creep duration should be made. The interpretation of the data would be aided by the development of theories or the measurement of crack growth kinetics along grain boundaries or other preferred paths of crack propagation.

ACKNOWLEDGEMENT

This study was conducted as part of a research program supported by Army Research Office under contract DAAG 29-79-C-0193. Approval from the J. Mat. Sci. for publication of some of the figures is gratefully acknowledged.

REFERENCES

1. F. R. N. Nabarro, "Report of a Conference on Strength of Solids," The Physical Society, London, p. 75 (1948).
2. C. Herring, J. Appl. Phys., 21, 437 (1950).
3. R. L. Coble, J. Appl. Phys. 34, 1679 (1963).
4. N. F. Mott, Conference on Creep and Fracture of Metals at High Temperatures, 0.21, H. M. Stationary Office, London (1956).
5. J. Weertman, J. Appl. Phys. 26, 1213 (1955); 28, 362 (1957).
6. S. I. Warshaw and F. H. Norton, J. Am. Ceram. Soc. 45, 479 (1962).
7. R. L. Coble and Y. H. Guerard, J. Am. Ceram. Soc. 46, 353 (1963).
8. J. H. Hensler and G. V. Cullen, J. Am. Ceram. Soc. 51, 557 (1973).
9. A. Crosby and P. E. Evans, J. Mat. Sci. 8, 1573 (1973).
10. B. Burton, G. L. Reynolds and J. P. Barnes, J. Mat. Sci. 8, 1690 (1973).
11. R. M. Arons and J. K. Tien, J. Mat. Sci. 15, 2046 (1980).
12. J. B. Ainscough, F. Rigby and S. A. Morrow, J. Am. Ceram. Soc. 64, 315 (1981).
13. J. Weertman, Trans. A.S.M. 62, 502 (1969).
14. A. Venkateswaran and D. P. H. Hasselman, J. Mat. Sci. 16, 1627 (1981).
15. D. P. H. Hasselman and A. Venkateswaran, J. Mat. Sci. 18, 161 (1983).
16. D. P. H. Hasselman and J. P. Singh, Bull. Am. Ceram. Soc. 58, 856 (1979).

17. R. L. Coble, p. 706 in "High Strength Materials", Ed. V. Zackay, I. M. R. D., Wiley, Inc. (1964).
18. Y. Tree, A. Venkateswaran and D. P. H. Hasselman, J. Mat. Sci. (in press).
19. D. P. H. Hasselman, J. Am. Ceram. Soc. 52, 517 (1969).
20. J. Weertman and J. R. Weertman, p. 793 in 'Physical Metallurgy', Ed. R. W. Cahn, North-Holland, Amsterdam (1965).
21. J. Weertman, Trans. A.S.M. 61, 681 (1968).
22. M. F. Ashby, Acta Met. 20, 887 (1972).
23. F. A. Mohamed and T. G. Langdon, J. Mat. Sci. 13, 473 (1978).
24. T. G. Langdon, Metals Forum, 1, 59 (1978).
25. J. B. Walsh, J. Geophys. Res. 70, 381 (1965).
26. D. P. H. Hasselman and A. Venkateswaran, Communications, J. Am. Ceram. Soc. 66, C-36 (1983).
27. D. P. H. Hasselman, J. Gebauer and J. A. Manson, J. Am. Ceram. Soc. 55, 588 (1972).
28. A. Venkateswaran and D. P. H. Hasselman, J. Mat. Sci. Letters 1, 400 (1982).
29. J. B. Walsh, J. Geophys. Res. 74, 4333 (1969).

17. R. L. Coble, p. 706 in "High Strength Materials," Ed. V. F. Zackay, J. Wiley, Inc. (1965).
18. Y. Tree, A. Venkateswaran and D. P. H. Hasselman, J. Mat. Sci. (in press).
19. D. P. H. Hasselman, J. Am. Ceram. Soc. 52, 457 (1969).
20. J. Weertman and J. R. Weertman, p. 793 in "Physical Metallurgy," Ed. R. W. Cahn, North-Holland, Amsterdam (1965).
21. [illegible] Herrmann, Trans. AIME [illegible], 561 (1966).
22. [illegible] Ashby, Acta Met. 20, [illegible] (1972).
23. [illegible] A. Mohamed and T. G. Langdon, J. Mat. Sci. [illegible] (1975).
24. [illegible] Langdon, [illegible] (1975).
25. [illegible] Walsh, J. Geophys. Res. 70, 381 (1965).
26. D. P. H. Hasselman and A. Venkateswaran, Communications J. [illegible] Am. Ceram. Soc. 66, C-36 (1983).
27. D. P. H. Hasselman, [illegible] Gebauer and J. A. Coppola, J. Am. Ceram. Soc. [illegible] (1977).
28. A. Venkateswaran and D. P. H. Hasselman, [illegible] (1983).
29. J. B. Walsh, J. Geophys. Res. 70, 5249 (1965).

THE KINETICS OF TIME DEPENDENT PLASTIC FLOW AND FRACTURE IN CERAMIC MATERIALS

K. Krausz and A.S. Krausz*

*Dept. of Mech. Engineering, University of Ottawa
Ottawa, Canada K1N 6N5

INTRODUCTION

With the increasing use of ceramic materials in the industry, special attention is being paid to their general physical and mechanical properties. In this presentation, plastic flow, and subcritical crack propagation that is assisted by the environment and leads to fracture will be discussed.

It is widely recognized that both are time dependent, rate controlled processes. There are several theories developed for the study of rate-controlled processes; they may be divided into two major approaches. One is the transition state theory (or absolute rate theory), established by Becker, Eyring, Polanyi and Wigner in the 1930's. The other, the non-transition state theories include, most notably, the Slater theory and the Kessel-Rice-Rumsperger treatment (for an extended discussion of the non-transition state theories of the rate processes see, for example, references |1-2|). The kinetics analysis presented here is based on the transition state theory; the application to plastic deformation was formulated by Krausz and Eyring |3|, and reviewed by Krausz and Faucher |4|. The further extension of the theory to environment assisted crack propagation is due to Brown |5|; Wiederhorn |6|; Krausz and co-workers |7-11|. Most recently, Krausz et al |12| pointed out the advantages of the transition state theory application to time dependent, stable crack propagation. It is the purpose of this presentation to bring attention to the advantages of the fracture and deformation kinetics analysis approach, as applied to ceramic materials.

DISCUSSION

In order to present the kinetics analysis of time dependent processes, a brief description of the physical background of thermal activation is in place. Time- and temperature-dependent, rate-type processes such as subcritical crack propagation, dislocation movement, diffusion, and other transport processes are taking place on the atomic level. Since the same basic, physical mechanisms govern all of these rate-type processes, the presentation will be formulated in terms of sub-critical crack propagation that is assisted by the environment.

By thermal activation we mean that the heat content, the thermal energy of the material and the mechanical energy (the applied stress) act together and produce the macroscopically observed effect. The thermal energy is an inherent property of all materials. The particles (Fig. 1) vibrate about their equilibrium distances, and this vibrational energy is the source of the thermal activations. Because there are a large number of atoms interconnected

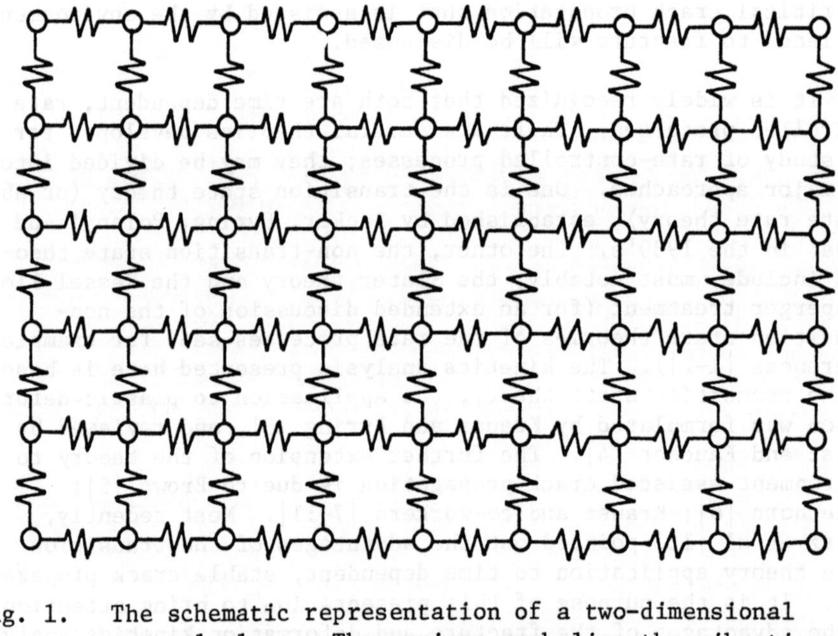

Fig. 1. The schematic representation of a two-dimensional crystal plane. The springs symbolize the vibrational motion of the atoms (open circles).

by some kind of interatomic forces, the vibrational amplitudes and the direction of the particle motion change constantly and in a random manner (Fig. 2). These changes move through the solid in waves, called phonons, and when a large amplitude reaches the crack tip where the applied stress already enlarged the equilibrium distance, the bond will break (Fig. 3). This leads to the familiar elementary rate constant of a step that is controlled by a single energy barrier |3|:

$$k = \frac{kT}{h} \exp(- \frac{\Delta G}{kT}), \qquad (1)$$

where k and h are the Boltzmann and Planck's constants, T is the absolute temperature and ΔG in the exponent is the thermal energy of activation. This is a familiar relation, in many cases it describes the process very well indeed: at high stresses, near the critical stress intensity factor K_{IC}, Region III of stress corrosion cracking can be described by the elementary rate constant |7,8|.

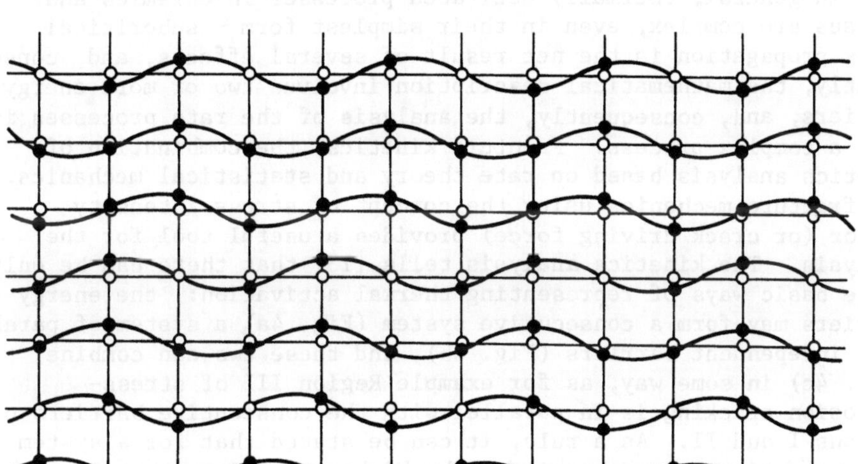

Fig. 2. The figure illustrates the propagation of the vibrational motion in a solid. The open circles represent the equilibrium position of the atoms, the solid circles show the position of the atoms during vibration. The heavy lines indicate the wave-like character of the atomic displacement, and consequently, the wave-like character of the energy content associated with the atomic vibration.

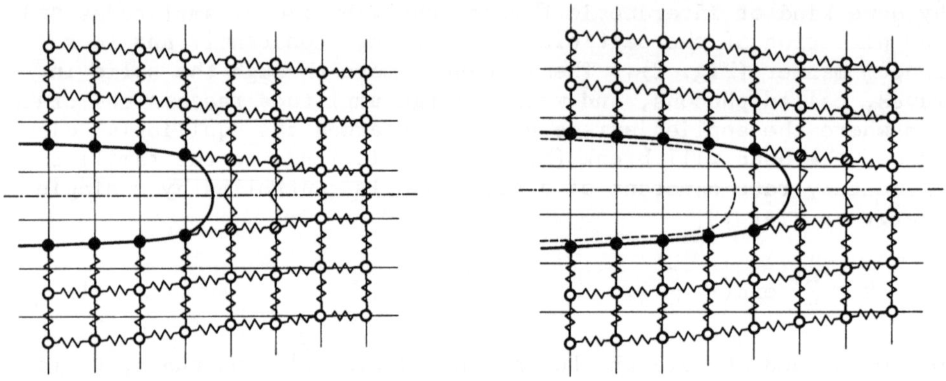

Fig. 3. In crack propagation an elementary step of thermal activation occurs when the atomic bond in front of the crack tip breaks and the crack moves from the position indicated in (a) by one atomic distance to the right as shown in (b). The intact springs intend to represent intact atomic bonds, where atomic interaction exists, and the broken bonds are shown by the broken springs.

In general, thermally activated processes in ceramics and glasses are complex, even in their simplest form — subcritical crack propagation is the net result of several effects, and, consequently, the mathematical description involves two or more energy barriers, and, consequently, the analysis of the rate processes is also a complex process. Fracture kinetics, the combination of kinetics analysis based on rate theory and statistical mechanics, and fracture mechanics using the concept of stress intensity factor (or crack driving force) provides a useful tool for the analysis. The kinetics analysis tells |13| that there can be only three basic ways of representing thermal activation: the energy barriers may form a consecutive system (Fig. 4a), a system of parallel, independent barriers (Fig. 4b), and these two can combine (Fig. 4c) in some way, as for example Region III of stress-corrosion cracking is in parallel with the consecutive barriers of Regions I and II. As a rule, it can be stated that for a system of consecutive energy barriers, the highest barrier, that is, the slowest activation ("flow") controls the rate; when the process is described by independent parallel energy barriers, then the lowest barrier is controlling the rate.

The simplest case that can be handled in a greatly simplified manner is when the process consists of two consecutive energy barriers and only forward activation takes place (Fig. 5). This has been applied to Regions I and II of the environment assisted crack propagation, and the total time of overcoming the two barriers, each representing a region, is the sum of the waiting times in front of the barriers — the waiting times between the thermal acti-

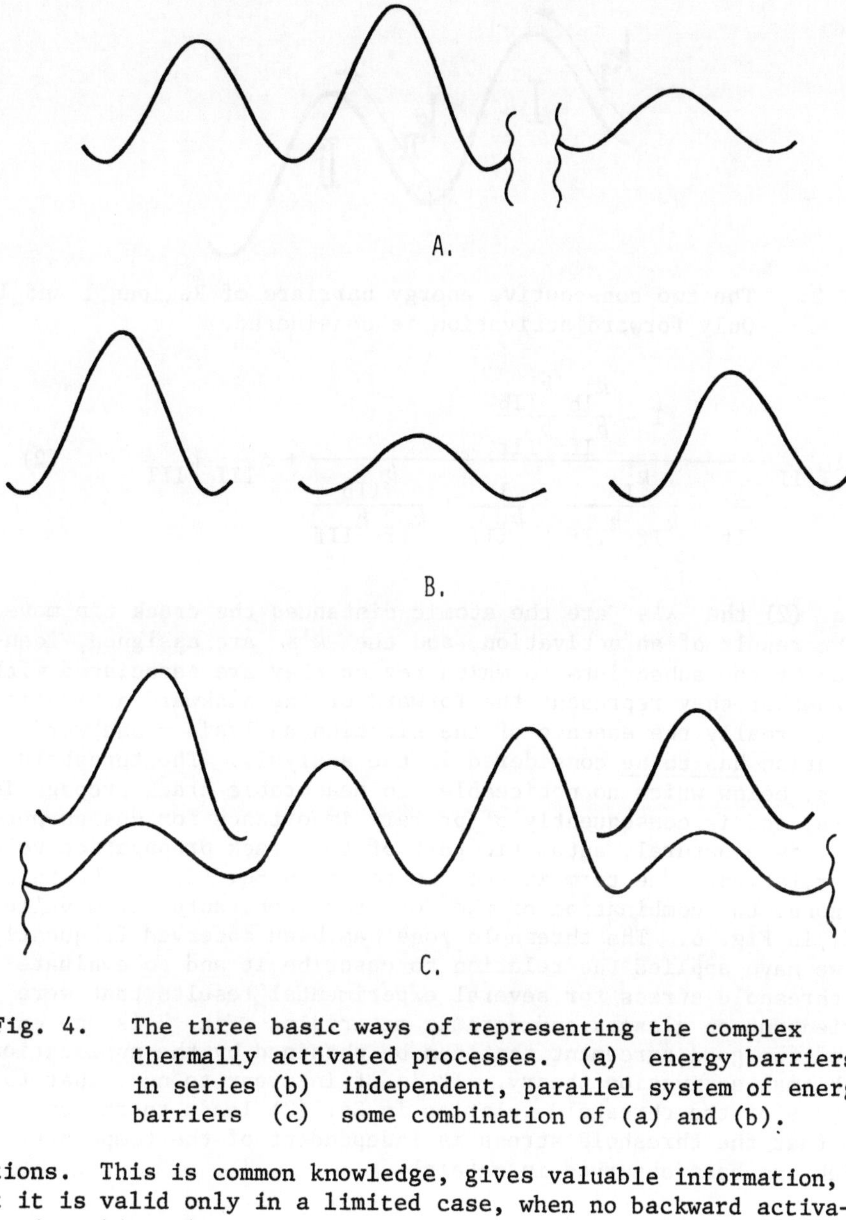

Fig. 4. The three basic ways of representing the complex thermally activated processes. (a) energy barriers in series (b) independent, parallel system of energy barriers (c) some combination of (a) and (b).

vations. This is common knowledge, gives valuable information, but it is valid only in a limited case, when no backward activation is taking place.

Since the kinetics analysis is presented here by developing it for environment assisted crack propagation, the next logical step to take in our inquiry is to consider the backward activation over the two consecutive energy barriers. In Fig. 6, the subscripts identify the sense of direction the activation takes place. It has been shown |7| that the analysis results in a velocity relation that describes the three regions and the threshold zone as well, and is of the form

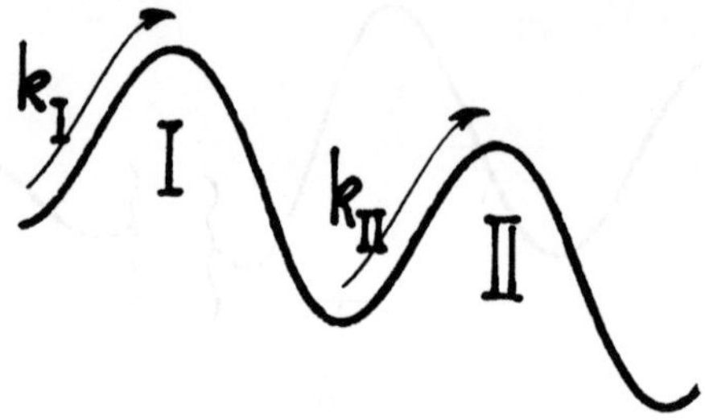

Fig. 5. The two consecutive energy barriers of Regions I and II. Only forward activation is considered.

$$V = \lambda_{I,II} \frac{1 - \dfrac{k_{Ib}\, k_{IIb}}{k_{If}\, k_{IIf}}}{\dfrac{1}{k_{If}} + \dfrac{k_{Ib}}{k_{If}\, k_{IIf}} + \dfrac{1}{k_{IIf}} + \dfrac{k_{IIb}}{k_{If}\, k_{IIf}}} + \lambda_{III}\, k_{III} \qquad (2)$$

In Eq. (2) the λ's are the atomic distances the crack tip moved as the result of an activation, and the k's are assigned, identified by the subscripts to which region they are associated with, and whether they represent the forward or the backward activation. This is really the essence of the kinetics analysis — backward activation has to be considered in the analysis. The threshold stress, below which no noticeable, no measurable crack propagation occurs, and is consequently of primary importance for design purposes, is a natural, automatic part of the crack propagation velocity relation. The term in the numerator in Eq. (2) is the threshold zone, the combination of the four rate constants, from valley I to I, in Fig. 6. The threshold zone has been observed frequently and we have applied the relation to describe it and to evaluate the threshold stress for several experimental results that were carried out on glasses and ceramic materials. Fig. 7 is one example of the good agreement that can be obtained by the application of thermal activation theory. It is of interest to note that the analysis of the threshold relation in Eq. (2) leads to the conclusion that the threshold stress is independent of the temperature (which has been observed in metals).

A further, and unexpected result of the full fracture kinetics analysis is that in the velocity relation that includes the backward activation also in Regions I and II of stress-corrosion cracking, there are four terms in the denominator. Of the four, only two are affecting the process, only two are operating — which two depends on the particular conditions: on the activation energies, on the work factor (the stress intensity factor), the temperature, and can be determined from the evaluation of the experi-

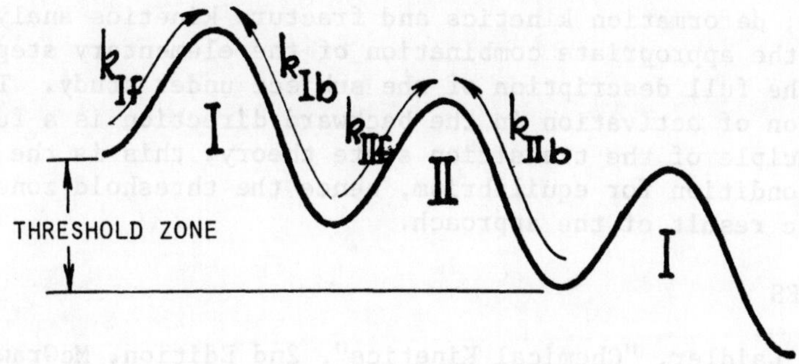

Fig. 6. The two energy barriers of Regions I and II, showing the associated rate constants in the forward and backward direction, and the threshold zone as well.

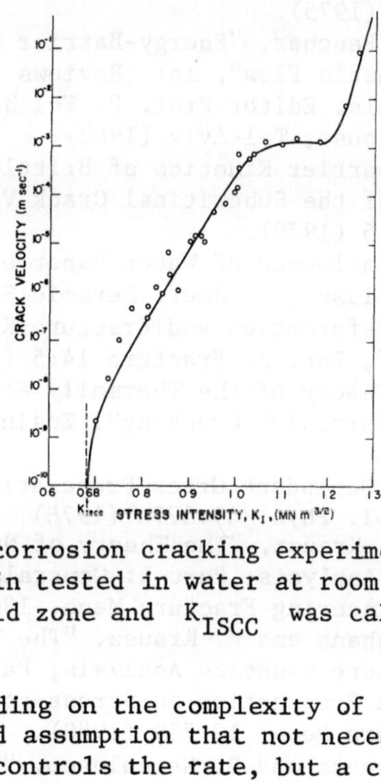

Fig. 7. Stress corrosion cracking experimental results (o); porcelain tested in water at room temperature. The threshold zone and K_{ISCC} was calculated with Eq.(2) |7|.

ments. Thus, depending on the complexity of the process under study, it is a valid assumption that not necessarily the highest energy barrier controls the rate, but a combination of the elementary rate constants is the determining factor.

SUMMARY

Absolute rate theory was initially developed for chemical reactions by kineticists — it is a powerful approach to thermally activated, time dependent processes. It applies equally well to plastic flow and fracture in all materials: ceramics; amorphous; pure metals; alloys; composites (eutectics). These processes are complex; each step is controlled by an elementary rate constant; deformation kinetics and fracture kinetics analyses determine the appropriate combination of the elementary steps, providing the full description of the subject under study. The consideration of activation in the backward direction is a fundamental principle of the transition state theory; this is the fundamental condition for equilibrium, hence the threshold zone is an automatic result of the approach.

REFERENCES

1. K.J. Laidler, "Chemical Kinetics", 2nd Edition, McGraw-Hill, N.Y. (1965).
2. S.W. Benson, "The Foundations of Chemical Kinetics", McGraw-Hill, N.Y. (1960).
3. A.S. Krausz and H. Eyring, "Deformation Kinetics", Wiley-Interscience, N.Y. (1975).
4. A.S. Krausz and B. Faucher, "Energy-Barrier Systems in Thermally Activated Plastic Flow", in: Reviews on the Deformation Behavior of Materials, Editor Prof. P. Feltham, Vol. IV, No.2, Freund Publishing House, Tel-Aviv (1982).
5. S.D. Brown, "Multibarrier Kinetics of Brittle Fracture: I, Stress Dependence of the Subcritical Crack Velocity", J. Amer. Ceramic Soc. 62: 515 (1979).
6. S.M. Wiederhorn, "Influence of Water Vapor on Crack Propagation in Soda-lime Glass", J. Amer. Ceramic Soc. 50:407 (1967).
7. A.S. Krausz, "The Deformation and Fracture Kinetics of Stress Corrosion Cracking", Int. J. Fracture 14:5 (1978).
8. A.S. Krausz, "The Theory of the Thermally Activated Processes in Brittle Stress Corrosion Cracking", Engineering Fracture Mech., 11:33 (1979).
9. A.S. Krausz, "Time Dependent Crack Propagation in Linear Elastic Solids", J. Appl. Phys. 49:3774 (1978).
10. A.S. Krausz and K. Krausz, "The Theory of Non-steady State Fracture Kinetics Analysis; Part I: General Theory of Crack Propagation", Engineering Fracture Mech. 13:751 (1980).
11. A.S. Krausz, J. Mshana and K. Krausz, "The Theory of Non-steady State Fracture Kinetics Analysis; Part II: Non-steady State Crack Propagation in Stress-corrosion Cracking", Engineering Fracture Mech. 13:759 (1980).
12. A.S. Krausz, K. Krausz and D. Necsulescu, "Reliability Theory of Stochastic Fracture Processes in Sustained Loading: Part I", Waturforsch. 38a:497 (1983).
13. T. O'D. Hanley and A.S. Krausz, "Thermally Activated Deformation, I: Method of Analysis", J. Appl. Phys. 45:2013 (1974).

CHARACTERIZATION OF CRACKS SUBJECT TO CREEP*

W. Blumenthal and A. G. Evans

Materials and Molecular Research Division
Lawrence Berkeley Laboratory and
Dept. of Materials Science and Mineral Engineering
University of California, Berkeley, CA 94720

ABSTRACT

The deformation response of small, semi-elliptical, indentation-induced surface cracks in fine-grain alumina has been monitored using crack opening and surface displacement measurements. Results indicate values of the secondary creep exponent, n, between 1.5 and 2, and a temperature dependence consistent with creep data. However at temperatures $\lesssim$1400 C, crack tip damage in the form of grain boundary cavities exert a strong influence on the displacement field as predicted by recent theories. Use of the stress intensity factor as a loading parameter, based on Newtonian-viscous behavior, does not produce adequate correlation with experimental results and is therefore unsuitable for non-linear creeping polycrystals.

INTRODUCTION

Recent studies[1-4] have indicated that high temperature failure in alumina involves the growth of cracks from microstructural heterogeneities. Similar processes accompany 'shrinkage cracking' during sintering. The comprehension and prediction of crack growth

*This work was supported by the Director, Office of Energy Research, Office of Basic Energy Sciences, Materials Sciences Division of the U.S. Department of Energy under Contract No. DE-AC03-76SF00098.

and failure is thus contingent upon an ability to characterize and model crack growth rates. The requisite characterization procedures are complicated, in the creep range, by the various modes of deformation that occur in a cracked component, in response to an applied load. The instantaneous response is elastic, but this response is followed by the development of zones of primary and secondary creep, spreading from the moving crack tip. In view of this complexity, the primary intent of the present study is to explore the response to load of the small cracks that cause creep rupture and shrinkage cracking, as a preliminary basis for devising pertinent characterization schemes. The study is conducted with particular emphasis on the crack opening displacements (COD) and surface displacements associated with small, indentation-induced surface cracks. A secondary objective of the study is to examine the suitability of the stress intensity factor, K_I, as a crack growth characterization parameter in a non-linear, creeping ceramic polycrystal.

CREEP DISPLACEMENTS

The Loading Parameter

Creep displacement and creep crack growth rates are dependent on a loading parameter which characterizes the amplitude of the singular stress field near the crack tip. When the crack tip is stationary, initial loading gives rise to the development and disappearance of various singular stress fields at the crack tip.[10,11] For example, in the presence of secondary creep, the strain rate, $\dot{\varepsilon}$, and the stress, σ, are related by;

$$\dot{\varepsilon} = \dot{\varepsilon}_o(\sigma/\sigma_o)^n \tag{1}$$

where n is the secondary creep exponent and $\dot{\varepsilon}_o$ and σ_o are constants. The associated singular stress field is

$$\sigma_{ij}/\sigma_o = \left[\frac{C_s(t)}{\sigma_o \dot{\varepsilon}_o I_n r}\right]^{\frac{1}{n+1}} \tilde{\sigma}_{ij}(\phi,n) \tag{2}$$

The quantity I_n is a non-dimensional function given by Hutchinson,[12] r is the distance from the crack tip and $C_s(t)$ is a time-dependent amplitude parameter which approaches the steady-state value,

$$C_s^* = F\,\sigma_o \dot{\varepsilon}_{so} c(\sigma_\infty/\sigma_o)^{n+1} \tag{3}$$

where c is the crack length and F is a constant dependent on the geometry and mode of loading. Note that after extended loading, when n = 1,

$$C_s^* \propto c\,\sigma_\infty^2 \quad K_I^2/\pi \tag{4}$$

Hence, K_I can be used as a loading parameter in lieu of C_s^* for Newtonian-viscous materials.

It is also important to recognize that the corresponding singular fields for moving cracks have not been ascertained. It is usually assumed that the stationary crack solutions still apply. But this premise has yet to be adequately assessed.

In the present study, the stress intensity factor K is used as a trial loading parameter, because of the near-Newtonian-viscous character of the material. An additional rationale for exploring the validity of K concerns the recognition that experimental control of the parameter C^* [13] from the specimen compliance is impractical for very small cracks.

The Crack Opening Displacement Rate

Adopting a trial, linear approximation for the material behavior, as noted above, allows the COD rate to be expressed explicitly in terms of K_I. Specifically, for an edge crack in a semi-infinite solid, the near tip displacements are given by;

$$\dot{u} = \frac{AK_I\, r}{\eta} \tag{5}$$

where A is a constant (1.2 for plane strain and 1.6 for plane stress with $\nu = 0.5$), r is the distance behind the crack tip and η is the viscosity. More remote from the tip, the displacements depend upon the crack shape. Most notably, a penny-shaped crack subject to uniform tensile loading exhibits displacements,[14]

$$\dot{u} = \frac{BK_I c^{1/2}\sqrt{1-(r/c)^2}}{\eta} \tag{6}$$

where B = 0.735 for the loading and geometry used in this study.

These simple results are modified in the presence of both nonlinearity and creep damage, such that, in general, a material with n > 1 undergoing grain boundary cavitation, exhibits a near tip COD rate given by

$$\dot{u} = \dot{\varepsilon}_o \left[\frac{C^*}{\dot{\varepsilon}_o \sigma_o I(n,\rho)} \right]^{\frac{n}{n+1}} r^{\frac{1}{n+1}} \dot{\tilde{u}}_i(\phi,n,\rho) \tag{7}$$

where $I(n,\rho)$ is a normalizing factor approximated by Hutchinson,[15] $u_i(\phi,n,\rho)$ describes the angular dependence of the displacement field and ρ is a damage parameter related to the uniaxial creep rate by[15]

$$\dot{\varepsilon} = \dot{\varepsilon}_o \left[\frac{\sigma_\infty}{\sigma_o} \right]^n (1+\rho) \tag{8}$$

In the absence of creep damage, fully plastic crack solutions for the COD can be applied to creeping materials. For example, the opening displacement rate at the center of a stationary crack is given by;

$$\dot{u}_c = c \quad \dot{\varepsilon}_\infty h_3$$

or,

$$\dot{u}_c = \dot{\varepsilon}_o \; c \; (\sigma_\infty/\sigma_o)^n \; h_3 \tag{9}$$

where h_3 is a dimensionless function of the loading, the sample geometry, and the creep exponent, n. Typical values of h_3 are summarized in Table I.

EXPERIMENTAL PROCEDURES

The displacement measurements are conducted using small surface cracks formed by indentation. Special procedures, pertinent to the effective use of these cracks for such measurement, have been devised.

Residual Stresses

The plastic deformation zone created by an indenter is constrained by the surrounding elastically deforming material and despite some relaxation due to the formation of a crack system, large residual stresses endure.[17] These residual stresses superimpose on subsequent applied stresses and must be removed to obtain accurate measurements of crack response. Residual stresses can be eliminated by either annealing or removing the plastic zone. Careful surface removal has been chosen for three reasons. First the indentation impression obscures and complicates measurement of crack opening and surface displacements associated with the crack tip strain field. Surface removal followed by polishing affords clear observation of crack response. Secondly, the material plastically

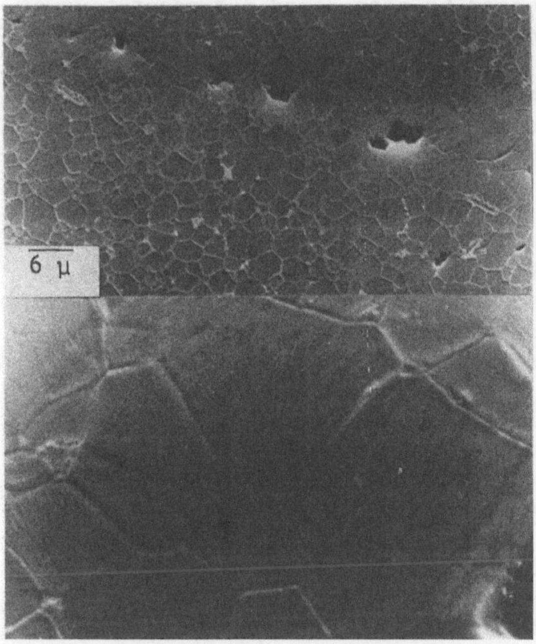

Fig. 1. (A) Missing grains indicate the path of a crack which was "healed" at high temperatures.
(B) Arrows reveal the predominantly transgranular fracture assoicated with with indentation-induced flaws in fine-grained alumina.

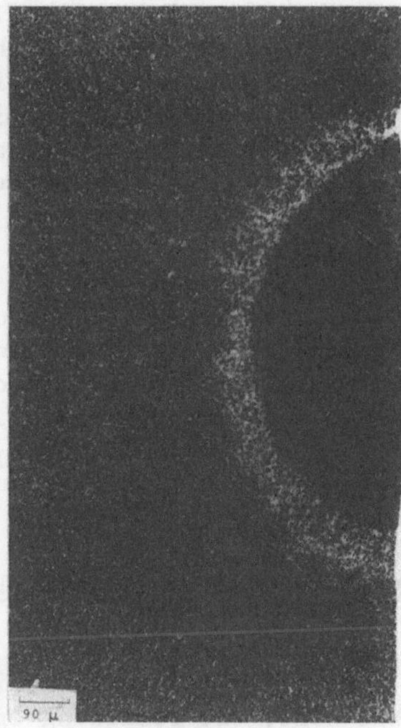

Fig. 2. Fracture from a semi-elliptical, indentation-induced surface crack (dark region) after a period of creep growth (white region). Note: creep crack growth appears to be uniform along the crack front.

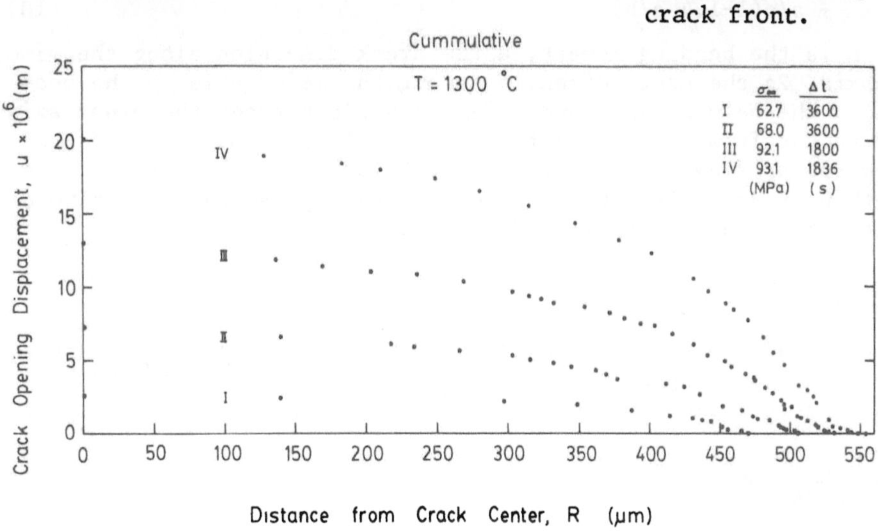

Fig. 3. Cumulative crack opening displacement (COD) as a function of the distance, R, from the center of the crack at 1300° C.

deformed by indenting can no longer be considered representative of the bulk microstructure and should be removed. Thirdly, annealing depends upon the activation of deformation mechanisms to relax the plastic zone. The specific time/temperature requirements of this method vary extensively between materials and consequently, crack healing may be difficult to avoid.

Flaw Healing

Flaw healing (Fig. 1), during the first loading iteration and during subsequent heating, can impede successful deformation measurements. This problem was overcome by sufficiently pre-loading the sample to separate the crack faces and inhibit significant sintering. Up to 1500° C, preloading at $K_I \approx K_{IC}/4$ was sufficient to prevent healing and induced only a minimal amount of crack opening (~ 2 μm at the crack center).

Determination of the Loading Parameter

A Knoop diamond indenter causes the formation of a reproducible, nearly semi-circular median crack. After removal of the plastic zone, the remaining crack is semi-elliptical (Fig. 2). In general, the stress intensity, K, varies with the mode of loading (tension or bending), the position (θ) along the crack front, the ratio of the crack depth to the specimen thickness (a/h), and the flaw shape (a/c).[18,19] In the present study, the surface crack length, 2c, and load were used to assess K_I from an expression developed by Cherapanov,[21] based on previous numerical approximations,[20]

$$K_I = 6.8Ma^{1/2}/(bh^2)[(a/c)(2\theta/\pi)^2(a/h - 0.4 + 0.6\ a/c) + a/c(1-1.4\ a/h) \qquad (10)$$

where M is the bending moment, a the crack dimension along the minor half-axis, 2c the crack dimension along the major axis, b the beam width, h the beam height, and θ is the angle between the minor axis and a vector from the origin to the crack front. Note that for a semi-circular flaw (a/c = 1), K_I is 20% larger at the surface (θ = 90°) than at depth (θ = 0°). More importantly, a constant value of K_I along the crack front is indicated when a/c approaches 2/3.

Incremental Stress Testing

Cracked samples were loaded in four-point flexure (5 mm upper span, 25 mm lower span) at temperatures between 1300°C to 1500°C. An Instron* testing machine with constant cross-head rate and load

*Canton, MA

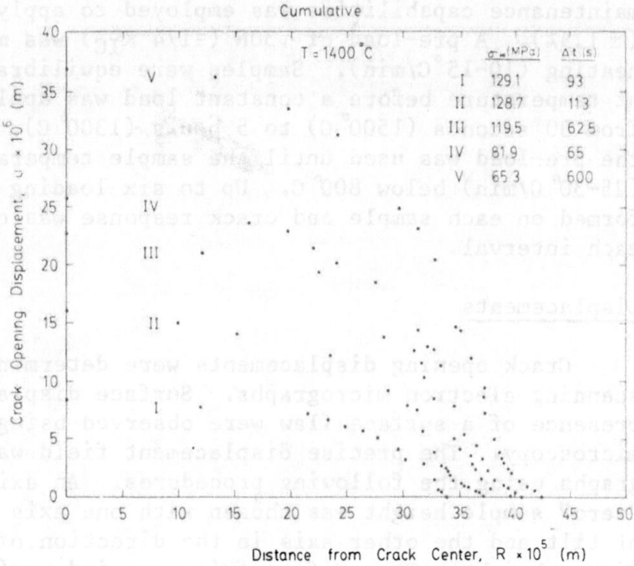

Fig. 4. Cumulative COD versus R at 1400° C

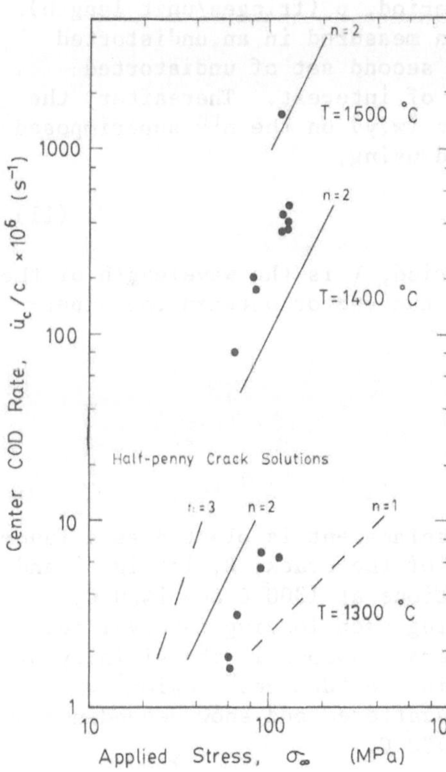

Fig. 5. Center crack opening displacement (CCOD) rate, normalized by the crack length, as a function of the applied stress.

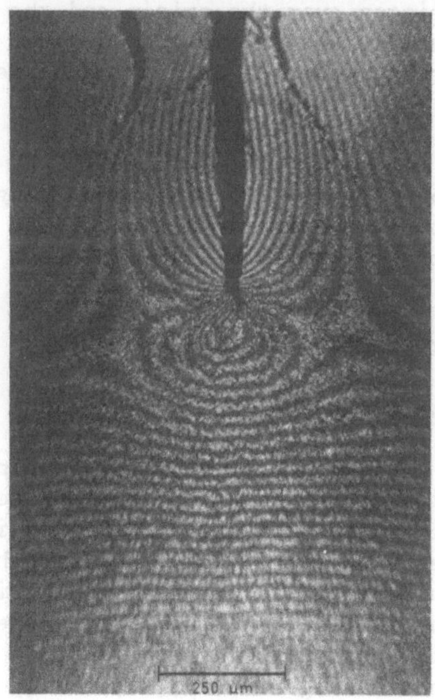

Fig. 6. Mirau interference micrograph of a crack tip region.

maintenance capabilities was employed to apply constant loads ($\pm$ 1.5%). A pre-load of $\sim$30N ($\equiv$1/4 K_{IC}) was maintained during heating (10-15°C/min). Samples were equilibrated for 5-10 minutes at temperature before a constant load was applied for times ranging from 30 seconds (1500° C) to 5 hours (1300° C). A post-load equal to the pre-load was used until the sample temperature was lowered (25-30° C/min) below 800° C. Up to six loading iterations were performed on each sample and crack response was characterized between each interval.

Displacements

Crack opening displacements were determined using optical and scanning electron micrographs. Surface displacements due to the presence of a surface flaw were observed using Mirau* interference microscopy. The precise displacement field was obtained from micrographs using the following procedures. An axis origin establishing "zero" sample height was chosen with one axis parallel to the axis of tilt and the other axis in the direction of increasing sample-to-mirror height. The uniform fringe period, p (fringes/unit length), representing the mirror tilt was then measured in an undistorted region and was used to superimpose a second set of undistorted fringes over the origin and the area of interest. Thereafter, the surface displacement at a given point (x,y) on the nth superimposed fringe from the origin may be deduced using,

$$w(x,y) = dp\lambda/2 \tag{11}$$

where p is the undistorted fringe period, λ is the wavelength of the monochromatic light source, and d is the vector between the superimposed and actual n^{th} fringe.

EXPERIMENTAL RESULTS

The Crack Opening Displacement

The cumulative crack opening displacement is plotted as a function of the distance from the center of the crack, R, in Fig. 3 and Fig. 4 for a series of loading iterations at 1300° C and 1400° C, respectively. Slow crack growth during each loading interval results in deviations from ideal elliptical crack shapes in the vicinity of the final crack tip. Displacements in the "damaged" region, where slow crack growth has occurred, are scattered and show dependence on the loading history, especially at 1400° C.

*Leitz, Germany

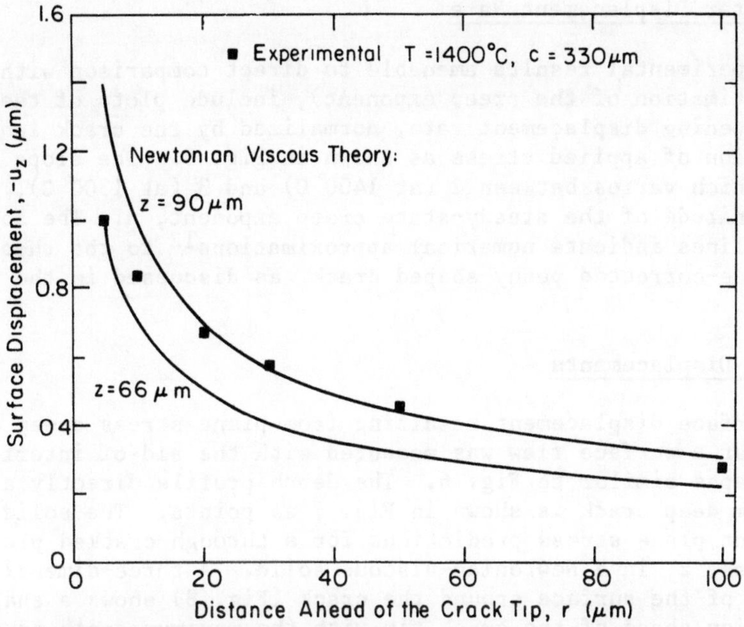

Fig. 7. Depth profile directly ahead of a 300 μm deep crack. Solid lines are Newtonian-viscous, plane stress predictions for a plate of thickness, Z.

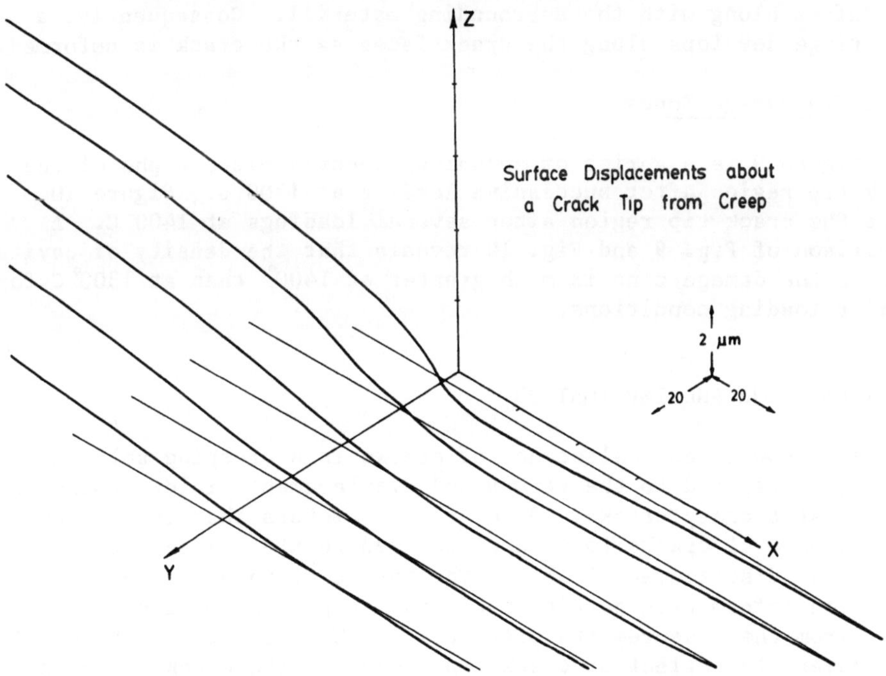

Fig. 8. A three-dimensional profile of the surface displacements about a crack tip after creep.

The Center Displacement Rate

Experimental results amenable to direct comparison with theory (and estimation of the creep exponent), include plots of the center crack opening displacement rate, normalized by the crack length, as a function of applied stress as shown in Fig. 5. The slope of this plot, which varies between 2 (at 1400° C) and 3 (at 1300° C), reflects the magnitude of the steady-state creep exponent, n. The solid and dashed lines indicate numerical approximations[12] to the theory for a surface-corrected penny-shaped crack, as discussed in the following section.

Surface Displacements

Surface displacement resulting from plane stress creep deformation near a surface flaw was measured with the aid of interference micrographs similar to Fig. 6. The depth profile directly ahead of a 300 μm deep crack is shown in Fig. 7 as points. The solid lines represent plane stress predictions for a through-cracked plate of thickness, z, in a Newtonian-viscous solid. A three-dimensional profile of the surface around the crack (Fig. 8) shows a shallow depression ahead of the crack tip with the maximum depth occurring along a vector 60° from the plane of the crack. However, these effects are over-shadowed by the effects of low stress in the region behind the crack tip. The region adjacent to the crack faces does not deform along with the surrounding material. Consequently, a high ridge develops along the crack faces as the crack is deformed.

Crack Tip Damage Zones

Figure 9 is a series of scanning electron micrographs of the crack tip region after successive loading at 1300° C. Figure 10 shows the crack tip region after several loadings at 1400° C. A comparison of Fig. 9 and Fig. 10 reveals that the density of cavitation in the damage zone is much greater at 1400° than at 1300° C for similar loading conditions.

INTERPRETATION AND IMPLICATION

The growth and coalescence of cracks in a creeping solid is intimately related to the stress and displacement fields around growing and stationary cracks. Also, the parameters that characterize the response of cracks to an applied load relate uniquely to the stress and displacement fields. The intent of this section is thus to obtain information pertinent to crack tip stresses and displacements from the measurements presented in the preceding section. In particular, the effect of crack tip damage on the crack response will be considered. In addition, an estimate of the stresses near

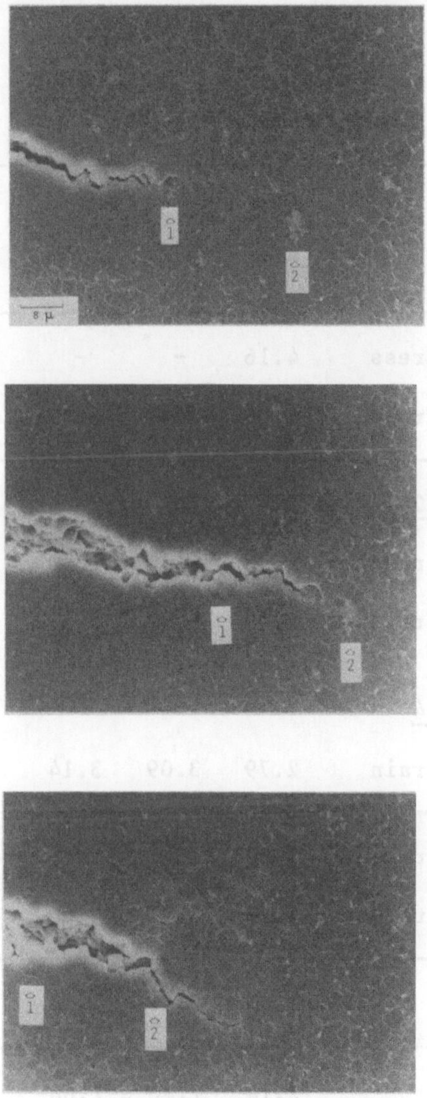

Fig. 9. Scanning electron micrographs of the crack tip after successive loading at 1300°C. Arrows denote previous crack tip positions.

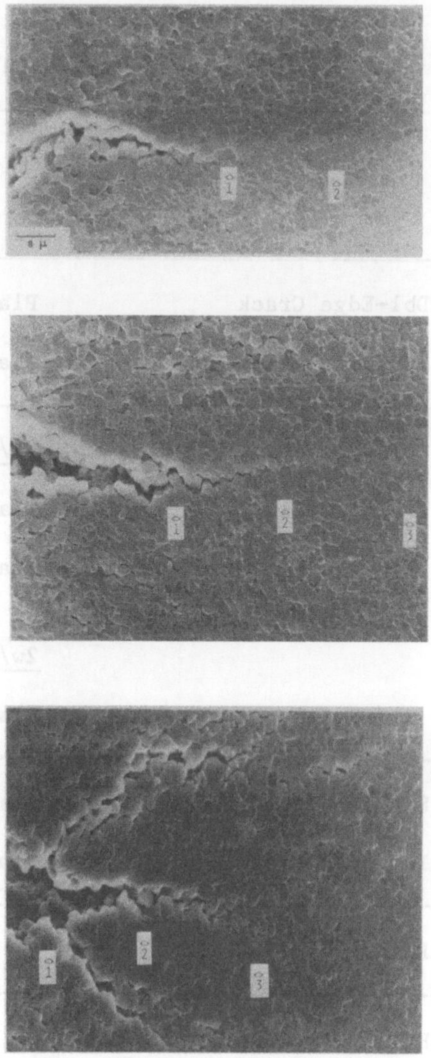

Fig. 10. SEM micrographs of the crack tip after successive loadings at 1400°C.

Table 1. Values of $h_3(n)$ for Crack Systems in Tension

Crack System		$h_3(n)$ n=1	n=2	n=3
Dbl-Edge Crack	Plane Stress	4.16	-	-
Infinite Plate	Plane Strain	3.00	3.49	3.57
Dbl-Edge Crack (2c)	2w/2c = 2			
Finite Plate Width, 2w	Plane Stress	2.38	2.00	1.70
	Plane Strain	2.09	1.99	1.61
	2w/2c = 4			
	Plane Strain	2.79	3.09	3.14
Single-edge Crack,	Plane Stress	4.66	-	-
Infinite Plate	Plane Strain	3.36	3.91	4.00
Penny-shaped Crack		1.91	2.22	2.38
Half-Penny Crack		2.14	2.48	2.66

the crack tip will be presented and the suitability of K_I as a loading parameter will be discussed.

Center Displacement Rates

The center crack opening displacement, $\dot{u}_c$, is the simplest and most accurate method of monitoring crack deformation, since the crack center is easily located and exhibits the largest displacements. Experimental measurements of the center displacement rate obtained at 1300° C (Fig. 5) agree very favorably with deformation theory (Eq. (9)) for n = 2. However, the predicted temperature dependence of $\dot{u}_c$ (with n = 2), based on steady-state creep data, underestimates experimental results at temperatures $\lesssim$ 1400° C, by a factor of 1.5 to 4. This deviation can be rationalized in terms of the effect of creep damage upon the crack tip deformation rate (without invoking mechanistic transitions in the deformation behavior), as shown in the following section.

Effects of Damage

The displacement rate, $\dot{u}$, is known to exhibit a dependence on damage (Eq. (7)) through the parameters $I(n,\rho)$ and $\dot{\tilde{u}}_i(\phi,n,\rho)$. Estimates of $I(n,\rho)$ are available; however results for the angular component, $\dot{\tilde{u}}(\phi,n,\rho)$ have not yet been obtained. Nevertheless, the dependence of the displacement rate on damage can be presumed identical to the effect of damage on the strain-rate, since both $\dot{\tilde{\varepsilon}}$ and $\dot{\tilde{u}}$ are linearly dependent on the damage parameter, ρ.[15]

The influence of damage should be characterized (Eq. (8)) by an increase in the damaged-to-undamaged strain-rate or displacement-rate by a factor of $\sim(\rho+1)$. The magnitude of the parameter ρ is estimated herein from the effect of porosity on the uniaxial creep rate of a fine-grained alumina[16] (Fig. 11) which indicates a strong influence of ρ on $\dot{\varepsilon}$ (or $\dot{u}$) in moderately porous (>5%) material.* Hence, by noting that 10-15% porosity typically develops near the crack tip at 1400° C (Fig. 10), the four-fold increase in the center displacement rate, compared with the predictions based on undamaged material, (Fig. 5) is fully consistent with the corresponding increase in damage (Fig. 11). It is concluded, therefore, that crack tip damage exerts an important influence on the displacement field at temperatures $\lesssim$ 1400° C.

Estimates of Near Tip Stresses

Stress field solutions pertinent to stationary cracks provide a useful basis for analyzing the mode of creep deformation near a

*However, the strain rate is weakly damaged dependent when the porosity < 5%.

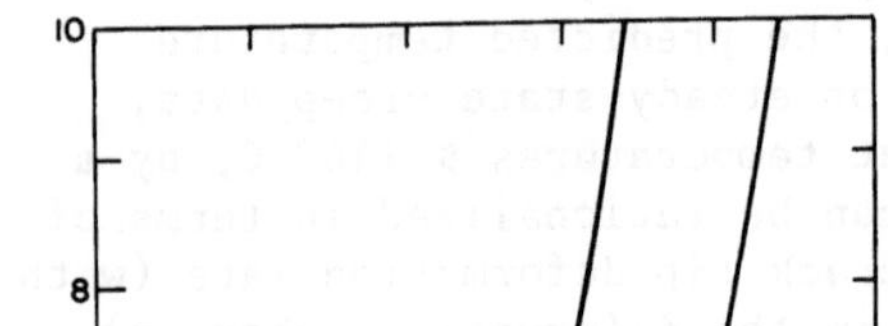

Fig. 11. Damage parameter, ρ, as a function of porosity for fine-grained alumina.

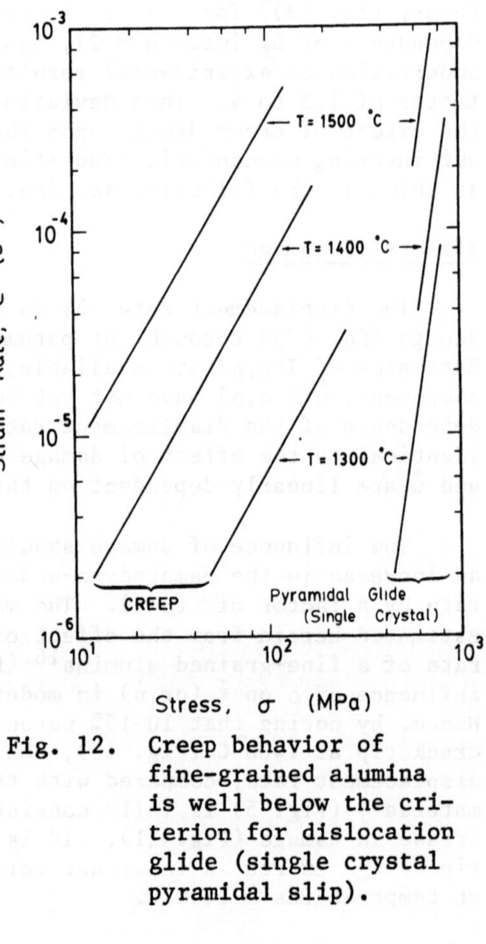

Fig. 12. Creep behavior of fine-grained alumina is well below the criterion for dislocation glide (single crystal pyramidal slip).

crack tip and hence, for modelling the crack growth process. These stresses can be obtained by substituting the generalized expression of C* for a half-penny crack of radius, c, into Eq. (2) using,

$$C^* = \sigma_\infty \dot{\varepsilon}_\infty c \, h_1 \qquad (12)$$

where h_1 is a function of the creep exponent and the damage parameter, ρ. Numerical results[12] give h_1 as 1.42 for an undamaged (ρ=0), non-linear (n=2) material.

To illustrate the largest values of stress expected near the crack tip, near-critical loading conditions at 1300 C (σ_∞ = 100 MPa, c = 550 μm, $\dot{\varepsilon}_\infty = 5.6 \times 10^{-6} s^{-1}$) are substituted into Eq. (12). The stress amplitude at the mid-point of the first grain (r = 1 μm) is ∿400 MPa, or four times the applied stress and reduces to 200 MPa at the mid-point of the fifth grain (r = 9 μm).

The amplitude of the stress field is reduced by damage. For conditions similar to the present study (n = 3, ρ = 1) the ratio of damaged-to-undamaged amplitudes is estimated as[14]

$$\sigma_{ij}(\rho = 1)/\sigma_{ij}(\rho = 0) = 0.74$$

This ratio decreases with increasing values of ρ and n, implying that the above stress calculations are upper limit estimates.

These stress levels are not sufficient to activate general dislocation motion (Fig. 12), supporting the lack of microstructural evidence for dislocation activity (Fig. 13).

K_I as a Loading Parameter

The Newtonian-viscous relationship between the opening displacement rate and K_I in Eq. (5) implies that the viscosity, η, should be single-valued at a given temperature. A plot of the normalized displacement rate, $\dot{u}/A\sqrt{r}$, as a function of the applied stress intensity factor (Eq. (10)) should then be linear with a slope, η^{-1}. The data obtained at 1300° C and 1400° C plotted according to this scheme (Fig. 14) are highly non-linear. Specifically, the inferred viscosity, at a given temperature, varies continuously over the range of K_I, by almost a factor of three. This demonstrates that the Newtonian-viscous assumption (n = 1) and the use of K_I as a loading parameter are fundamentally invalid for fine-grained alumina.

CONCLUSIONS

Monitoring the response of indentation-induced surface flaws is a unique and viable technique for studying creep and creep rupture.

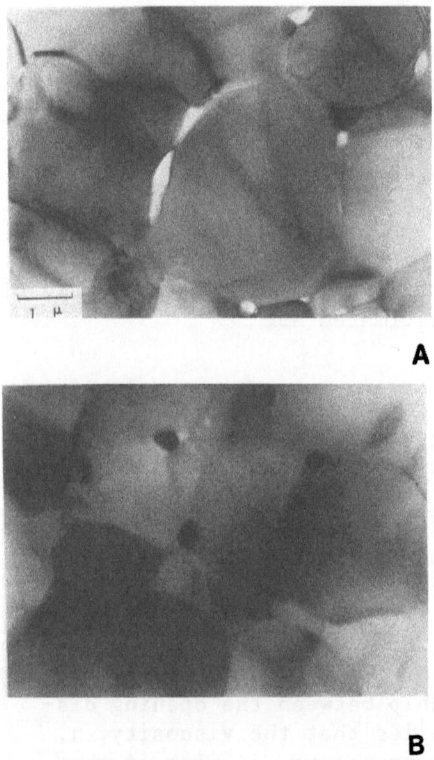

Fig. 13. (A) TEM micrograph of as-received hot-pressed alumina. (B) TEM micrograph near a creep crack tip where cavitation is prevalent. Remnants of dislocation activity are not observed.

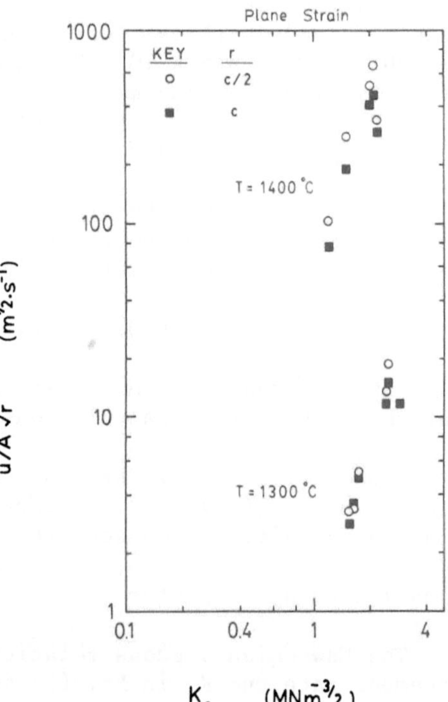

Fig. 14. COD rate versus K_I. The viscosity is not single-valued as a given temperature.

Both crack opening displacements and crack growth can be measured simultaneously using small flexure samples. The technique is suitable for low ratios of grain size to crack radius ($<10^{-2}$) and may be extended to the study of inherent or contamination flaws.

Center crack opening displacement rates are readily obtained, and indicate a creep exponent, n, between 1.5 and 2, with a temperature dependence similar to creep data.

Damage in the form of grain boundary cavities appears to increase the rate of crack opening displacement, consistent with the influence of the damage parameter, ρ, estimated from uniaxial tension data. Specifically, center crack opening displacement rates exclude a significant influence of damage at 1300° C, whereas a strong dependence on damage obtains at 1400° C, characterized by a four-fold increase in displacement rate. This behavior corresponds to microstructural observations which show a small damage zone relatively free of coalesced cavities at 1300 C but an extensive damage zone at 1400° C.

The amplitude of the crack tip stress field, conservatively estimated from a stationary crack solution, is below the stress required for dislocation glide. Hence, a transition in deformation mechanisms is not expected, nor is there microstructural evidence to support such a change under conditions of interest.

Finally, the stress intensity factor, K_I, is shown to be unsuitable for characterizing crack response due to the non-linear deformation behavior of fine-grain alumina.

REFERENCES

1. B. J. Dalgleish, A. G. Evans, "The High Temp. Failure of Polycrystalline Alumina, Part I: Comparative Microstructures," to be published.
2. S. M. Johnson, W. Blumenthal, A. G. Evans, "Part II: Crack Nucleation Sites," to be published.
3. W. Blumenthal, A. G. Evans, "Part IV: Creep Crack Growth and Blunting," to be published.
4. B. J. Dalgleish, S. M. Johnson, W. Blumenthal, A. G. Evans, "Part V: High Temperature Time-to-Failure," to be published.
5. A. G. Evans, "Fracture Toughness: The Role of Indentation Techniques," Fracture Mechanics Applied to Brittle Materials, ASTM STP678, S. W. Freiman, ed., (1979) 112.
6. B. R. Lawn, D. B. Marshall, "Hardness, Toughness and Brittleness: An Indentation Analysis," J. Am. Ceram. Soc. 62 (1979) 347.

7. D. B. Marshall, T. Noma, A. G. Evans, "A Simple Method for Determining Elastic-Modulus-to-Hardness Ratios using Knoop Indentation Measurements," J. Am. Ceram. Soc. 65, (1982) C-175.
8. M. G. Mendiratta, J. J. Petrovic, "Slow Crack Growth from Controlled Surface Flaws in Hot-pressed Si_3N_4," J. Am. Ceram. Soc. 61, (1978) 226.
9. J. J. Petrovic, M. G. Mendiratta, "Fracture from Controlled Surface Flaws," Fracture Mechanics Applied to Brittle Materials, ASTM STP678, S. W. Freiman, ed., (1979) 83.
10. H. Riedel, "Creep Deformation at Crack Tips in Elastic-viscoplastic Solids," Brown University Report MRL E-114 (1979).
11. J. L. Bassani, "Creep Crack Extension by Grain-Boundary Cavitation," Creep and Fracture of Eng. Materials and Structures; B. Wilshire, D. R. J. Owens, eds., Pineridge Press (1981) 329.
12. J. W. Hutchinson, "Singular Behavior at the End of a Tensile Crack in a Hardening Material," J. of the Mech. Phys. Solids, 16, (1968) 13.
13. J. D. Landes, J. A. Begley, "A Fracture Mechanics Approach to Creep Crack Growth," Mech. of Crack Growth, ASTM STP590 (1976) 128.
14. M. Y. He, J. W. Hutchinson, "The Penny-Shaped Crack and the Plane Strain Crack in an Infinite Body of Power-law Material," J. Appl. Mech., 48 (1981) 830.
15. J. W. Hutchinson, "Constitutive Behavior and Crack Tip Fields for Materials Undergoing Creep-Constrained Grain Boundary Cavitation," Harvard University Report #MECH-34 (1982).
16. G. Fryer, P. Thompson, "Compressive Creep of Porous Polycrystalline Alumina," Trans. J. Br. Ceram. Soc., 71 (1972) 61.
17. A. G. Evans, T. R. Wilshaw, "Quasi-Static Solid Particle Damage in Brittle Materials," Acta Metall., 24 (1976) 939.
18. I. Raju, J. Newman, "Improved Stress-Intensity Factors for Semi-elliptical Surface Cracks in Finite-Thickness Plates," NASA Report No. TMX-72815 (1977).
19. A. F. Grandt, G. M. Sinclair, "Stress Intensity Factors for Surface Cracks in Bending," ASTM STP513 (1972) 37.
20. R. C. Shah, A. S. Kobayashi, "Stress Intensity Factor for an Elliptical Crack Approaching the Surface of a Plate in Bending," ASTM STP513 (1972) 3.
21. G. P. Cherepanov, Mechanics of Brittle Fracture, McGraw-Hill (1979) 842.

KEYWORDS

Indentation-induced surface flows, fine-grained alumina, crack opening displacement rate, surface displacement, crack tip damage, grain boundary cavitation, stress intensity factor, penny-shaped crack, non-linear creep, interference microscopy.

REGIMES OF CREEP AND SLOW CRACK GROWTH IN HIGH-TEMPERATURE RUPTURE OF HOT-PRESSED SILICON NITRIDE.

G. Grathwohl

Institut für Werkstoffkunde II
Universität Karlsruhe
Postfach 6380
D-7500 Karlsruhe

INTRODUCTION

Hot-pressed silicon nitride (HPSN), sintered silicon nitride and sialons form a class of high performance ceramics in which the highest strength values of today's polycrystalline technical ceramics are found. Since their application is mostly related to high temperature engineering purposes, the material response on thermal and mechanical loading is of high practical and fundamental interest. It may especially be asked how the strength of these materials changes due to a long term application of load at high temperatures.

High temperature measurements of strength, fracture toughness, R-curve, slow crack growth, life time and creep were reported[1-6], all of these properties revealing time- or deformation-dependent effects, which are often not completely understood. In fact, the crucial point in many investigations relates to the unknown interaction of different mechanisms. E.g. it is necessary that at elevated temperatures the increasing importance of plasticity effects on slow crack growth is taken into account, while for creep measurements it has to be realized that strain or deflection phenomena can be originated by deformation and crack processes. High temperature failure can then be caused by different mechanisms of creep and slow crack growth and knowledge should be available about the regimes of the individual mechanisms in order to perform a valuable analysis of the kinetics leading to stress rupture.

OBJECTIVES

The present work is characterized by the following objectives. Stress rupture behaviour of HPSN is measured at high temperatures with simultaneous recording of the complete creep curve. This provides a comprehensive data base for the mechanical behaviour under static loading conditions at high temperatures. From this basis the evaluation of the processes governing the materials response is tried. Creep phenomena and slow crack growth of preexisting flaws are considered as competitive processes which are decisive in the high temperature mechanical properties. For a valuable analytical treatment of the kinetic data of both cases the clear distinction between the relevant regime regions is needed.

EXPERIMENTAL

The mechanical high-temperature tests were performed using a 4-point-bending fixture made from hot-pressed SiC. The specimens with the dimensions 3,5x4,5x45 mm were loaded by SiC-rollers with 20 and 40 mm as inner and outer span widths, resp. After diamond-machining the tensile surface of the specimens was carefully checked and the edges were rounded off by a polishing treatment in order to avoid any outstanding edge or surface flaws. The mid-point deflection measured at the tensile surface of the bending bars was used to calculate the deformation. Static tests were performed in a dead-load system, while for the interrupted static tests with controlled partial unloading/reloading sequences a closed-loop low cycle fatigue test machine+ was used. This unloading to e.g. 70% of the static load and reloading to 100% was automatically performed during the creep test by means of a program module in a very short time of ca 2s. From these tests the evolution of the "elastic" modulus of the creep specimens along their creep curves can be followed at high temperatures. Usually the tests were extended up to the specimen failure.

MATERIALS

Commercially available HPSN-materials are used for this study. While various HPSN's are investigated in this institute, the body of work for this study was elaborated with the particular batch II of NH 206*. Further tests were performed with NH 206/I*, NC 132** and NH 209*; the last one uses yttria as sintering additive, the previous ones contain magnesia additions. Further results of the chemical analysis are listed in Table I.

\+ Instron Ltd. High Wycombe, UK
* Annawerk GmbH, Rödental, W.Germany
** Norton Co., Worcester, MA

Table 1. Chemical analysis of 4 HPSN-materials [w/o]

	NH 206/I	NH 206/II	NC 132	NH 209
N	36.0	37.1	35.5	35.9
O	2.2	2.0	1.1	2.1
C	0.37	0.71	n.d.	0.30
Mg	0.87	1.8	0.60	0.03
Y	0.1	-	n.d.	4.6
Li	n.d.	0.005	n.d.	0.02
Ca	0.22	0.10	n.d.	0.02
Al	0.10	0.41	0.13	0.12
Fe	1.67	1.4	0.25	1.8

n.d. = not determined

The densities of NH 206/I and II were found to be 3.20 g/cm^3, the values for NC 132 and NH 209 are 3.24 and 3.30 g/cm^3, resp.

The comparison of the creep properties of these commercial HPSN-materials is demonstrated in Fig. 1 and 2. Creep rates and the rupture behaviour differ considerably not only between the various materials but also between the two batches of NH 206. An extremely high creep resistance is measured for NH 209 even at 1400°C, the temperature potential being generally more than 200°C higher compared with NH 206. In order to study the effect of high-temperature rupture under creep conditions the material NH 206/II was chosen; its behaviour is thought to be representative for other ceramics with the same type of microstructure.

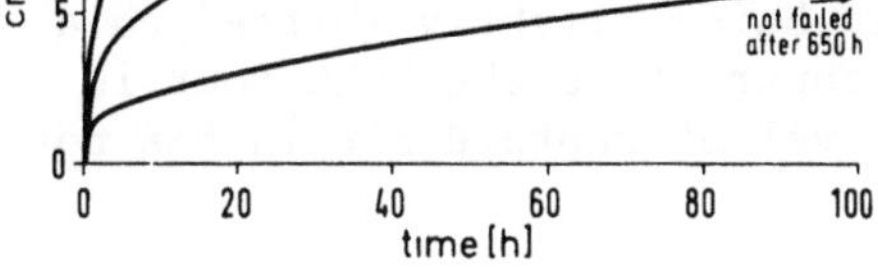

Fig. 1. Creep curves of 3 commercial HPSN-materials in air

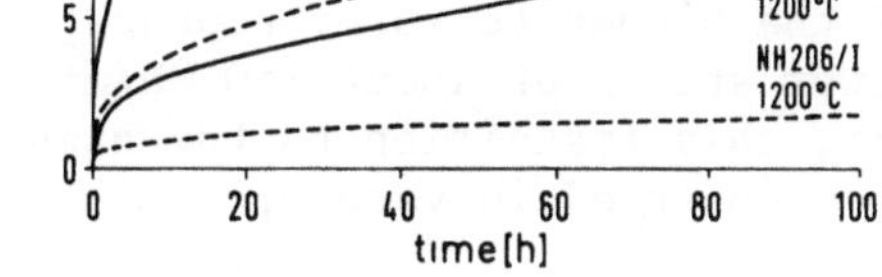

Fig. 2. Creep curves of 2 batches of NH 206 in air

The microstructure of NH 206 was recently investigated[7] extensively. As in other hot-pressed and sintered Si_3N_4-materials amorphous intergranular phases of about 10 to 20 Å width are found, while glass pockets in triple junctions contain crystalline segregations. Nevertheless, this commercial material with a relatively high amount of impurities may also contain a large fraction of "glass-free" grain boundaries[7].

RESULTS

If creep of HPSN is measured up to specimen failure, rather conventional creep curves are found as illustrated in Fig. 3 for tests at 1200°C. Very similar curves are recorded at 1300°C. Primary, secondary and tertiary creep stages are observed, the tertiary range being more pronounced at lower stresses (Fig. 3c). At an applied stress of σ = 180 MN/m^2 and higher, specimen failure occurs already in the primary creep range. While the time to failure shows a distinct dependence on stress (Fig. 3b), the rupture strain does not strongly vary with the applied stress (Fig. 3c). In Fig. 4 the influence of the applied stress on the creep rate $\dot{\varepsilon}$ is evaluated. This is done for the creep rates at constant creep strains and constant times and for the minimal creep rates $\dot{\varepsilon}_{min}$. Using Norton's law

$$\dot{\varepsilon}\ (\sigma) = \text{const} \cdot \sigma^n \qquad (1)$$

the stress exponent n reaches a value of 1.9 for the minimal creep rates. This value is approached from lower values in the constant time plot while in the constant strain plot rather high n-values are measured at small creep strains. However, these values are considered to be not valuable since based on primary creep. It is further obvious that in the secondary creep stage the fitting lines can not be extended beyond a stress of 160 MN/m^2. Thus, a value of n = 1.9 is taken as the valid exponent for the stress dependence of creep at stresses up to 160 MN/m^2.

It is illustrated in Fig. 5 that in the low stress region the rupture strain is almost constant ranging from 1.5% to 1.75% at 1200°C in air. However, the rupture strain drops sharply down to smaller values of about 0.4% at $\sigma \geq 180$ MN/m^2. The situation is rather similar at 1300°C, the critical stress for the transition of the higher to lower rupture strain being slightly shifted to a lower stress of about 150 MN/m^2. In contrast to the behaviour in air, this transition to the higher level of rupture strain has not been observed in vacuum.

The comparison of the creep curves in air and vacuum in this critical stress region is given in Fig. 6. While the creep rates are remarkably higher, the times to failure and the rupture strains are strongly decreased in vacuum compared to the tests in air.

This is also demonstrated in Fig. 7 where the stress rupture data are presented in the commonly used plot log σ vs log t_f (t_f = time to failure). It is evident that the experiments at

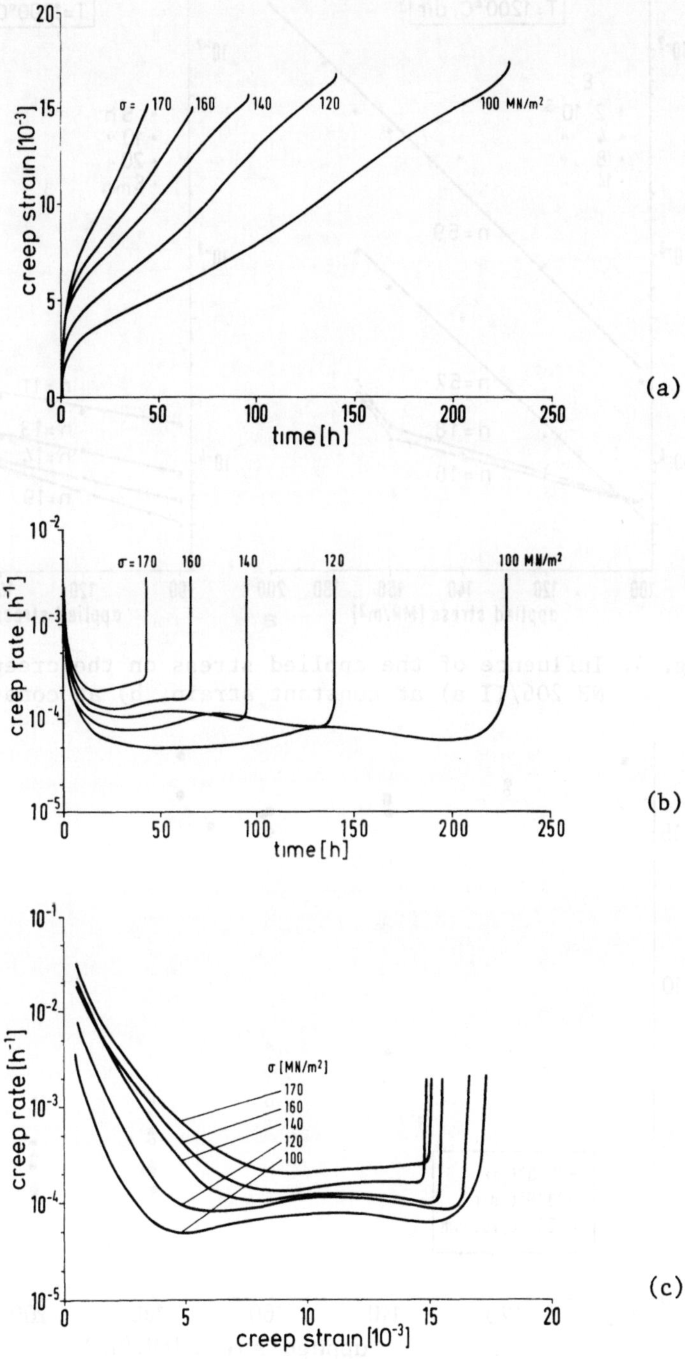

Fig. 3. Creep curves of NH 206/II at 1200°C in air

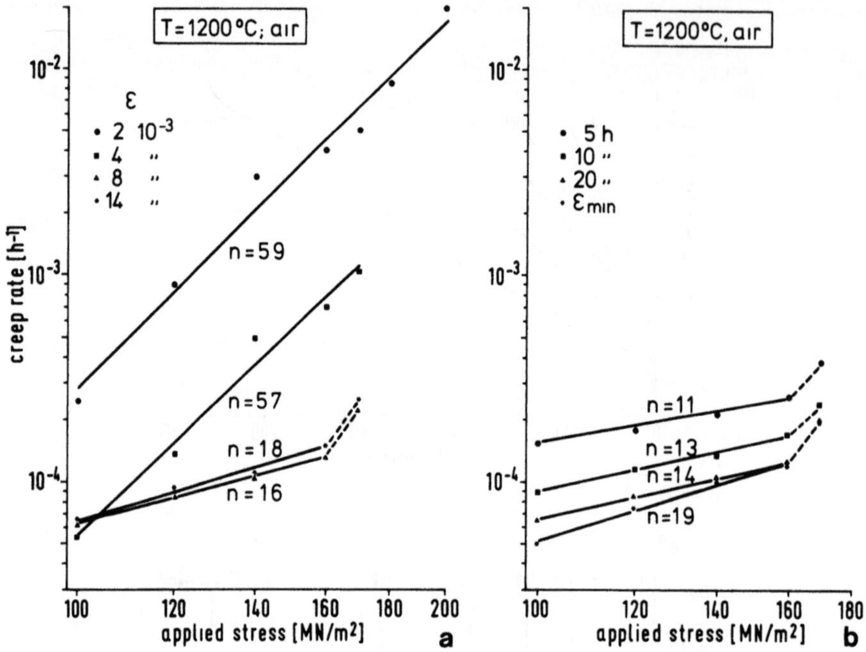

Fig. 4. Influence of the applied stress on the creep rates of NH 206/II a) at constant strain, b) at constant time

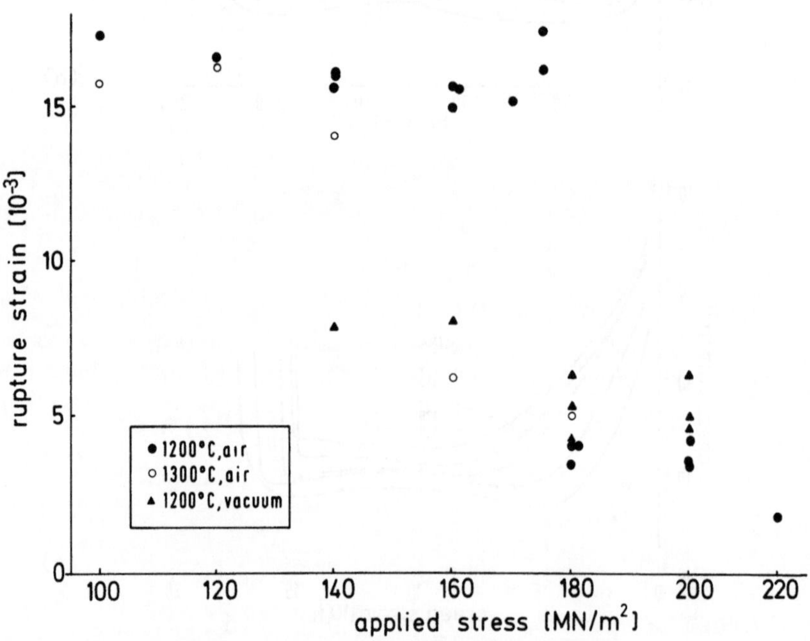

Fig. 5. Dependence of rupture strain of NH 206/II on applied stress

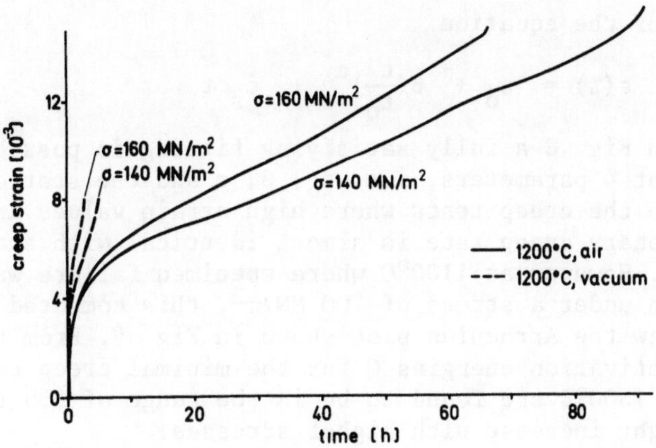

Fig. 6. Creep curves of NH 206/II in air and vacuum

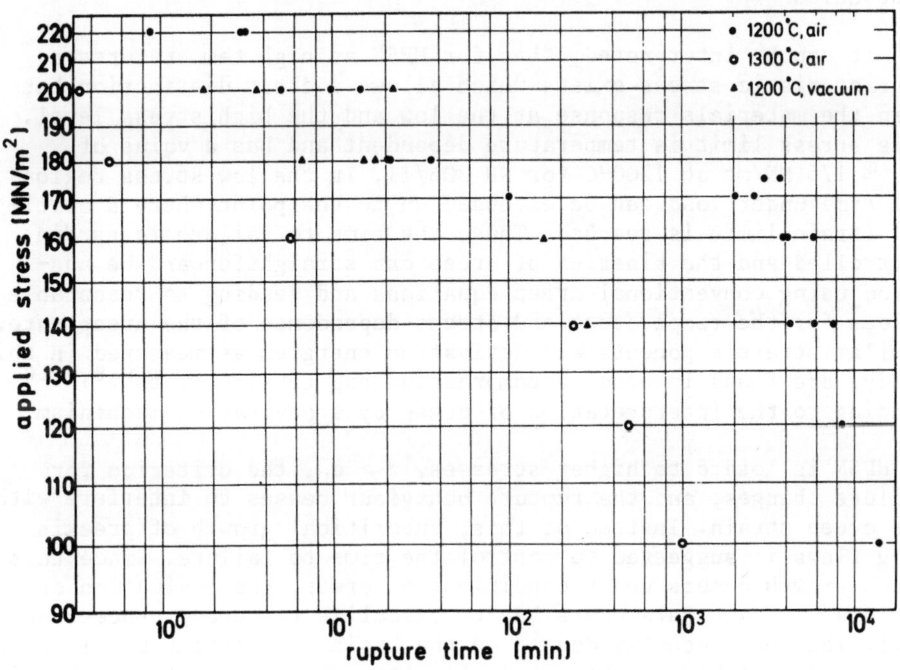

Fig. 7. Stress rupture of NH 206/II

1200°C in air reveal a transition in the behaviour at high and low stresses, both being separated by a stress of about 170 to 180MN/m^2. While this transition is missing in the vacuum test, a similar behaviour is found at 1300°C in air at a somewhat decreased stress level compared to 1200°C.

In order to extrapolate the temperature dependence of the creep rate the creep curves $\varepsilon(t)$ were analysed by a computer program on the basis of the equation

$$\varepsilon(t) = \varepsilon_o + \beta(\frac{t}{t_o})^c + \dot{\varepsilon}_s t \qquad (2)$$

As shown in Fig. 8 a fully satisfying fitting is possible taking into account 4 parameters, i.e. ε_o, β, c and the stationary creep rate $\dot{\varepsilon}_s$. In the creep tests where high strain values are reached, this stationary creep rate is almost identical with the minimal creep rate. However at 1100°C where specimen failure was not reached after 260 h under a stress of 200 MN/m^2, this computed $\dot{\varepsilon}_s$-values are used to draw the Arrhenius plot shown in Fig. 9. From this plot apparent activation energies Q for the minimal creep rate between 1100°C and 1300°C are found to be in the range of 558 to 709 kJ/mol with a slight increase with higher stresses.

DISCUSSION

It can be interpreted, that for HPSN at high temperatures a characteristic stress exists which allows a clear distinction between the materials response at the low and the high stress level. This stress limit is temperature dependent and has a value of $\sigma_o \approx 175$ MN/m^2 at 1200°C for NH 206/II. In the low stress region the time under load can be extended up to the point where a critical strain limit is reached. Thus, the time to failure is strain controlled and the kinetics of creep can straightforward be analysed using conventional creep equations and leading to reasonable values for the temperature and stress dependence of the creep rate. Similar stress exponents and activation energies as measured in this study, are found in bending, compression and tension tests[6,8,9,10] leading to the interpretation of creep by a cavitation mechanism.

If HPSN is loaded to higher stresses, $\sigma > \sigma_o$, the criterion for failure changes, and the rupture behaviour ceases to interfere with the creep strain. Instead of this, subcritical growth of preexisting flaws is suggested to control the time to failure. Since this crack growth occurs under conditions of creep, the evaluation of the crack growth parameters has to visualize the creep process as well. This was recently done[11] on the basis of constant strain rate tests and is taken in this study for the static creep tests. On the basis of the power law for the slow crack growth rate

$$\frac{da}{dt} = AK_I^N \quad (3) \qquad \text{and} \qquad K_I = \sigma Y a^{1/2} \quad (4)$$

the time to failure t_f is given by

$$B \cdot \sigma_c^{N-2} \cdot \sigma^{-N} = \int_o^{t_f} f(t)^N \, dt \qquad (5)$$

with crack length a, and A, Y and B being numerical factors. In (5) a function $\sigma(t) = \sigma \cdot f(t)$ is used, the time to failure in the

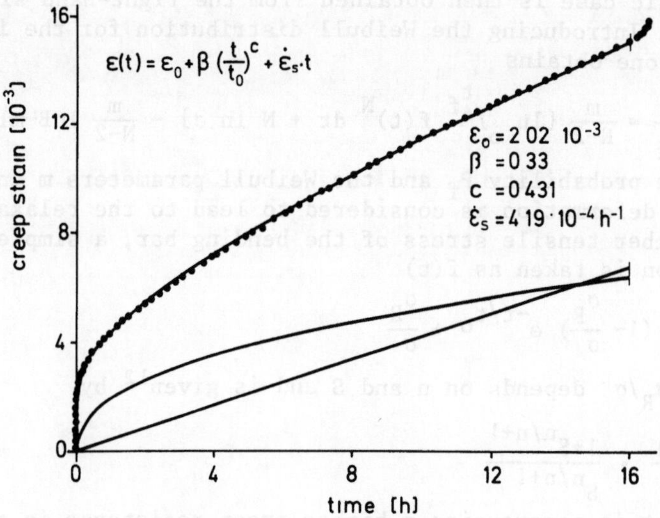

Fig. 8. Creep curve of NH 206/II at 1300°C and 100 MN/m² in air; dotted line: experimental; 3 drawn lines: computed

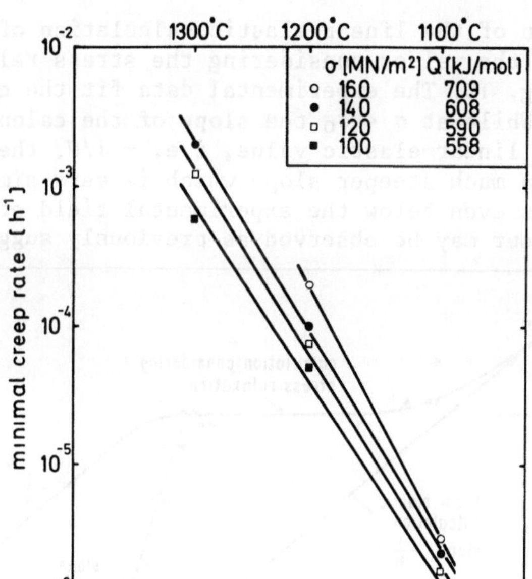

Fig. 9. Minimal creep rate of NH 206/II versus 1/T; data at 1100°C are calculated

linear-elastic case is then obtained from the right-hand side of (5) with f(t)=1. Introducing the Weibull distribution for the inert strength σ_c one obtains

$$\ln\ln \frac{1}{1-P_f} = \frac{m}{N-2}\{\ln \int_0^{t_f} f(t)^N dt + N \ln \sigma\} - \frac{m}{N-2} \ln B - m \ln \sigma_{co} \quad (6)$$

with failure probability P_f and the Weibull parameters m and σ_{co}. Since creep deformation is considered to lead to the relaxation of the outer fiber tensile stress of the bending bar, a simple relaxation function is taken as f(t)

$$f(t) = (1-\frac{\sigma_R}{\sigma}) e^{-t/t_o} + \frac{\sigma_R}{\sigma} \quad (7)$$

The ratio σ_R/σ depends on n and S and is given[12] by

$$\frac{\sigma_R}{\sigma} = \frac{2n+1}{6n} \cdot \frac{1+S^{n/n+1}}{S^{n/n+1}} \quad (8)$$

The fact that in compression a higher creep resistance is observed than in tension, can be taken into account by the factor S. A value of S = 7 was suggested[13], with n = 1.9 the final stress ratio comes out as $\sigma_R/\sigma = 0.54$. Using an iterative regression analysis[14] on the basis of (6) the values of N = 10 and m = 12.6 are obtained.

The comparison of the linear-elastic calculation of the life time diagram with the values considering the stress relaxation is illustrated in Fig. 10. The experimental data fit the calculated curve at $\sigma > \sigma_o$. While at $\sigma < \sigma_o$ the slope of the calculated curve comes back to the linear-elastic value, i.e. - 1/N, the experimental data exhibit a much steeper slope which is very similar to - 1/n. At stresses even below the experimental field of this study asymptotic behaviour may be observed as previously suggested[15].

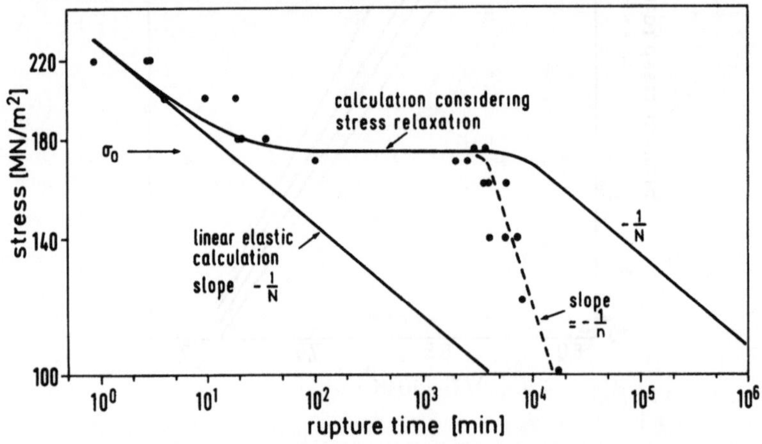

Fig. 10. Stress rupture diagram of NH 206/II at 1200°C in air.

In order to prove this approach a new technique was evolved to follow the microstructural variations during the static tests. Initiated by Hasselman's work on elastic creep[16] the idea was to investigate how the elastic modulus changes during creep due to the different damaging processes. As described before the load-deflection response ΔP/Δf was measured "in situ". Parallel to this procedure the strength was also determined on unfailed specimens after reaching particularly interesting strain values.

The load-deflection relation measured at different strain levels yielded some surprising results. The tests were performed at two characteristic stress levels, 140 and 180 MN/m^2, both being representative for the low and the high stress regimes, resp. As shown in Fig. 11 the "elastic" modulus decreases sharply in the early beginning of the creep curve at both stress levels. While the specimen at 180 MN/m^2 breaks after this sharp decrease at a strain of nearly $4 \cdot 10^{-3}$, the "elastic" modulus of the specimen at 140 MN/m^2 increases again with further strain increase and even regains its original value shortly before the tertiary creep stage begins.

The strength of specimens after creep tests which were interrupted after 90% of the expected rupture strain (i.e. 0.315% and 1.436% at 180 and 140 MN/m^2, resp.) yielded values of 42 and 88% of the original strength resp. From this and the results in Fig. 11 it is to conclude that starting the test at both stress levels the original stress intensity on preexisting flaws is high enough to start slow crack growth. Based on the calculation of the compliance

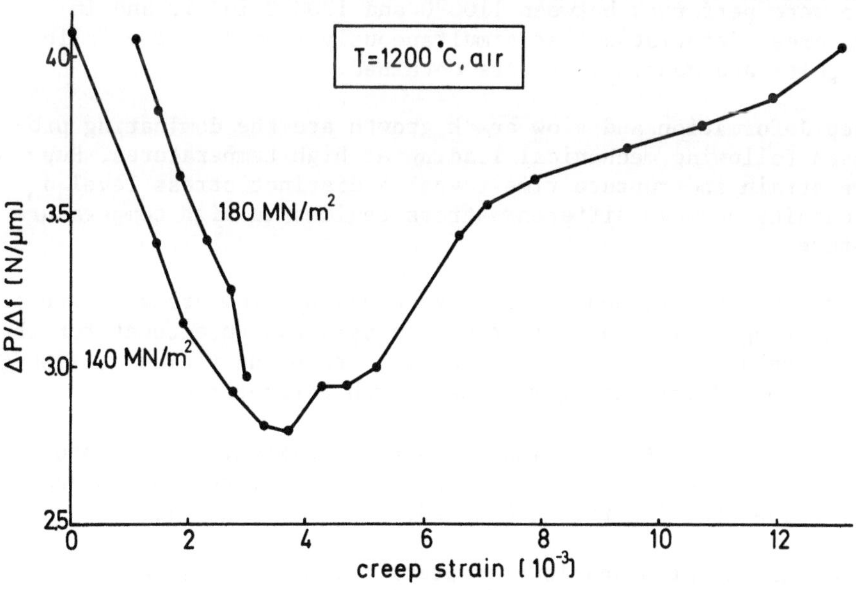

Fig. 11. Load-deflection response at various creep strains of NH-206/II bending specimens

of a bending bar with cracks growing from the outer tensile surface, the minimum in Fig. 11 can be related to crack lengths of 0.05 w (w = specimen thickness), if the growth of 10 parallel cracks is assumed. This seems to be reasonable in the light of microscopic findings.

During slow crack growth the actual stress intensity is modified by creep processes, e.g. stress relaxation and probably also by crack healing processes; however at the high stress level, these processes are not effective in preventing the slow crack growth-induced failure of the specimen; thus, the time to failure is controlled by the crack velocity.

In contrast at the lower stress level, the actual crack velocity is low enough that the specimen can reach the critical creep strain. The time to failure is then controlled by this strain and follows approximately the σ^{-n} law. Once the criterion for the effective stress relaxation and crack healing (minimum in Fig. 11) is reached, the "elastic" modulus increases with further increasing creep strain. This is in spite of damaging processes taking place during creep, e.g. cavitation[9]; but it may also be possible that processes which raise the modulus, e.g. crystallisation of amorphous phases, overcompensate these effects.

CONCLUSIONS

Flexural stress rupture experiments of hot-pressed silicon nitride were performed between 1100°C and 1300°C in air and in vacuum; creep deformation was simultaneously recorded. The following results and conclusions were obtained.

1. Creep deformation and slow crack growth are the dominating processes following mechanical loading at high temperatures. Rupture strain and rupture time reveal a distinct stress level σ_o discerning between different stress regimes in high temperature rupture.

2. For $\sigma > \sigma_o$ rupture behaviour is governed by slow crack growth of preexisting flaws; the life time analysis has to account for the stress relaxation due to simultaneous creep and yields at 1200°C a value of 10 for the slow crack growth parameter N.

3. For $\sigma < \sigma_o$ rupture behaviour is strain limited. Creep rupture is then considered as premature failure. Creep analysis results in Q values ranging from 558 to 709 kJ/mol and n = 1.9.

4. Higher creep rates and lower rupture times are measured in vacuum compared to air; a distinct σ_o limit was not yet observed underlining the importance of oxidation-related processes in crack "healing".

5. At all applied stresses the "elastic" modulus decreases during the early test period followed by a minimum and increase in the case of $\sigma < \sigma_0$; this increase may partly be explained by the crystallisation of amorphous phases, since creep-related damaging processes as e.g. cavitation are expected to lower the modulus. The case $\sigma > \sigma_0$ leads to specimen failure after the decline of the modulus due to the growth of cracks.

ACKNOWLEDGEMENT

It is a pleasure to acknowledge the cooperation and support by the group members; special thanks are due to H. Iwanek for able experimental assistance, to D. Bethge and H. Cohrt for the computer work related to the figures 8, 10 and 11 and to F. Porz for critical discussion. Financial support by the Bundesministerium für Forschung und Technologie, Bonn is gratefully acknowledged.

REFERENCES

1. R.J. Fields, T.J. Chuang, E.R. Fuller Jr. and N.J. Tighe, Effect of deformation on the fracture of Si_3N_4 and Sialon, in: "Progress in nitrogen ceramics", F.L. Riley, ed., Martinus Nijhoff Publishers, The Hague (1983) p. 507.
2. K. Kriz, The fracture behaviour of hot-pressed silicon nitride between room temperature and 1400°C, in: "Progress in nitrogen ceramics", F.L. Riley, ed., Martinus Nijhoff Publishers, The Hague (1983) p. 523.
3. F.W. Kleinlein, Langsame Rißausbreitung in spröden Werkstoffen im Biegeversuch, Dissertation, Universität Erlangen 1980.
4. D. Steinmann, Untersuchung des langsamen Rißwachstums von heißgepreßtem Siliziumnitrid bei hohen Temperaturen, KfK-Bericht Nr. 3414 (1982).
5. A.G. Evans and S.M. Wiederhorn, Crack propagation and failure prediction in silicon nitride at elevated temperatures, J. Mat. Sci. 9:270 (1974).
6. R. Kossowsky, D.G. Miller and E.S. Diaz, Tensile and creep strengths of hot-pressed Si_3N_4, J. Mat. Sci. 10:983 (1975).
7. T. Epicier and G. Orange, Electron microscopy study of the microstructure of a hot-pressed silicon nitride, Ceramics Internat. 8:154 (1982) and
G. Fantozzi, T. Epicier and G. Orange, Comportement à la rupture des nitrures de silicium, L'industrie ceramique 757:51 (1982).

8. J.M. Birch and B. Wilshire, The compression creep behaviour of silicon nitride ceramics, J. Mat. Sci. 13:2627 (1978).
9. F.F. Lange, B.J. Davis and D.R. Clarke, Compressive creep of Si_3N_4/MgO alloys, J. Mat. Sci. 15:601 (1980).
10. B.S. B. Karunaratne and M.H. Lewis, High-temperature fracture and diffusional deformation mechanisms in Si-Al-O-N ceramics, J. Mat. Sci. 15:449 (1980).
11. T. Fett and D. Munz, Lifetime prediction for hot-pressed silicon nitride at high temperatures, presented at ASTM-Symposium: "Methods for assessing the structural reliability of brittle materials", San Francisco, 13.12.1982.
12. H. Cohrt, G. Grathwohl and F. Thümmler, Transient creep of ceramics in bending tests, Res Mechanica Letters 1:159 (1981).
13. F.F. Lange, High temperature deformation and fracture phenomena of polyphase silicon nitride materials, in: "Progress in nitrogen ceramics", F.L. Riley, ed., Martinus Nijhoff Publishers, The Hague (1983) p. 467.
14. K. Jakus and J.E. Ritter Jr., Static fatigue of Si_3N_4, J. Am. Ceram. Soc. 61:274 (1978).
15. A.G. Evans and A. Rana, High temperature failure mechanisms in ceramics, Acta Met. 28:129 (1980).
16. D.P. H. Hasselman and A. Venkateswaran, Role of cracks in the creep deformation of brittle polycrystalline ceramics, J. Mat. Sci. 18:161 (1983).

KEYWORDS

Silicon nitride, creep, stress rupture under creep, stress relaxation life time prediction, elastic modules, change during creep

CREEP CAVITATION AND CRACK GROWTH IN SILICON NITRIDE

N.J. Tighe, S.M. Wiederhorn, T.-J. Chuang and
C.L. McDaniel

Inorganic Materials Division
National Bureau of Standards
Washington, DC 20234

ABSTRACT

Cavities and microcracks occur in hot-pressed MgO doped and Y_2O_3 doped Si_3N_4 during high temperature creep. Specimens selected from static load experiments at 1200°C and 1300°C were examined by transmission electron microscopy. In the MgO doped Si_3N_4 an amorphous $MgO \cdot SiO_2$ phase, which bonded the grains, softened and cavitated readily at 1200°C leading to a creep exponent of 4.2. In the Y_2O_3 doped Si_3N_4 a crystalline yttrium-silica-oxynitride phase bonded the Si_3N_4 grains. Cavitation started at the triple junctions where these phases were present. Crack-like cavities also started at triple junctions and grew between the Si_3N_4 grains. The nucleation and growth stages of creep cavitation in both materials are related to failure mechanisms.

INTRODUCTION

Hot pressed silicon nitride compositions creep at high temperature by processes which involve cavitation and crack propagation.[1-4] The creep or plastic deformation is time, temperature and stress dependent and affects lifetime predictions based on stress rupture data.[5-7] A number of papers in the current literature discuss the relationships between cavity growth and high temperature failure of structural ceramics.[8-10] The present paper addresses the systematic acquisition of microstructural evidence for cavitation that occurred in magnesia doped

(NC 132)* and yttria doped (NCX 34)* silicon nitride four point bend bars from creep tests at 1200°C and 1300°C. The test data show that these two compositions behave differently. Observations of the microstructure suggest that the creep mechanism differences can be related to microstructural differences between the materials.[6,7]

In both silicon nitride compositions the magnesia or yttria additives are present as sintering aids which bond the grains together. The magnesia addition forms an amorphous magnesium silicate phase which appears as a film (∿ 1nm) between grains and also fills the spaces between the grain boundary triple junctions. The yttria additive (8 wt %) forms discrete crystalline grains of yttrium silicon oxynitride compounds which sinter the beta silicon nitride grains together.[11-14] Creep cavities were found in magnesia doped silicon nitride at triple junctions and along grain boundaries in samples which deformed at 1200 to 1400°C. The pictures in Figure 1 show the characteristics of grain boundary cavitation that were observed in magnesia doped silicon nitride (NC 130)* double torsion plate tested at 1400°C.[1,15] In this type of test, which is used to determine the stress intensity factor, a single crack is propagated along a center groove by applying a stress across the groove at a constant rate.[16] The specimen exhibited slow crack growth and the resulting deformation was characterized by a zone of oxide lined grain boundary cracks around the primary crack. The oxide was identified as amorphous aluminum silicate and, as seen in Figure 1a the phase was pulled between the separated grains along the primary crack and along some of the smaller cracks in the zone. In a thinned foil (Figure 1b) the cracks are observed to follow grain boundaries. The small cavities visible at the grain marked W resulted from interaction with a tungsten based inclusion. Such tungsten based inclusions are distributed throughout the sample as seen in Figure 1d and they contribute to the general cavitation visible in Figure 1c.

The characterization of creep cavitation in the magnesia doped hot pressed silicon nitride is complicated because grain boundary cavitation is caused by oxidation at 1200 to 1400°C when there is no applied stress.[17] In developing the reliability map[18] (see Figure 2) as a graphical description of the time dependence of failure probability, it was found that strength degradation due to cavitation started after approximately 100 hours with an initial applied elastic stress# of 250 MPa at 1200°C. Thus both the chemical reaction due to oxidation and the strain produced by

---

*The material designations are used for identification purposes only and do not represent endorsement of the materials.

#Here and throughout the paper the expression stress is meant to be the initial applied elastic stress.

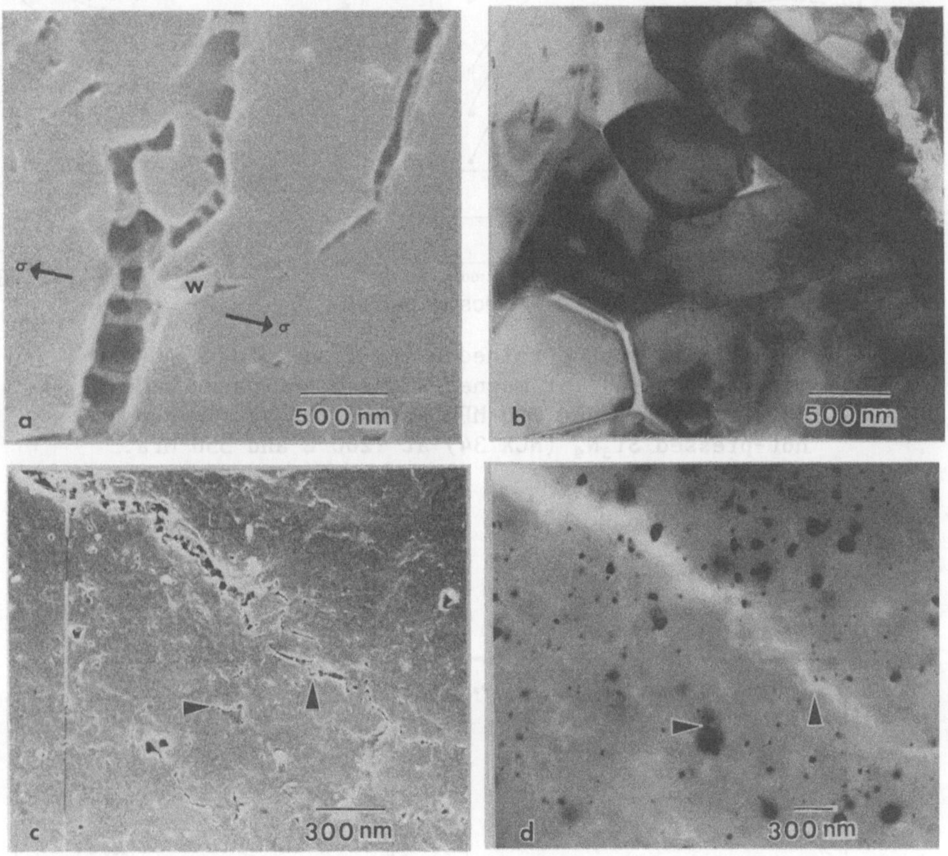

Figure 1. Microstructural characteristics of creep in a double torsion specimen tested at 1400°C in air (a,c) secondary electron image of crack tip region (b) TEM image of primary crack tip region (d) back scattered electron image of area in (c) reverse contrast shows tungsten and iron based particles as dark images (1,15).

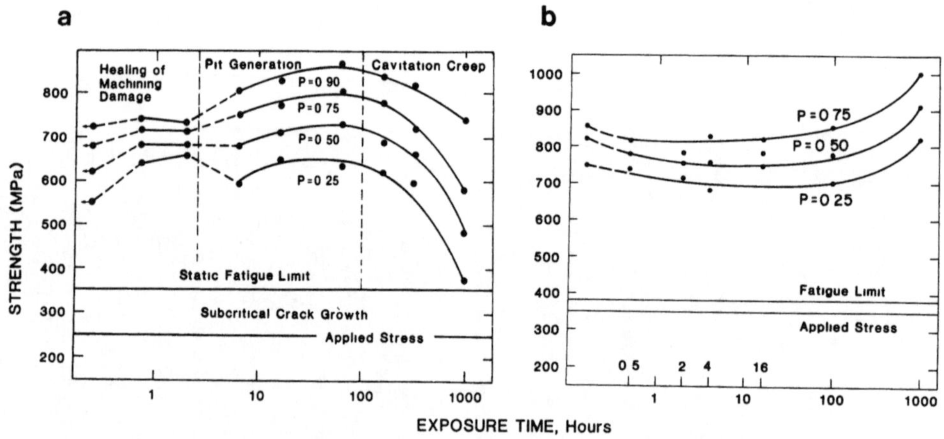

Figure 2. Reliability maps obtained using four point bend configurations (a) for magnesia doped hot-pressed Si_3N_4 (NC 132) at 1200°C and 250 MPa stress (b) yttria doped hot-pressed Si_3N_4 (NCX 34) at 1200°C and 350 MPa.

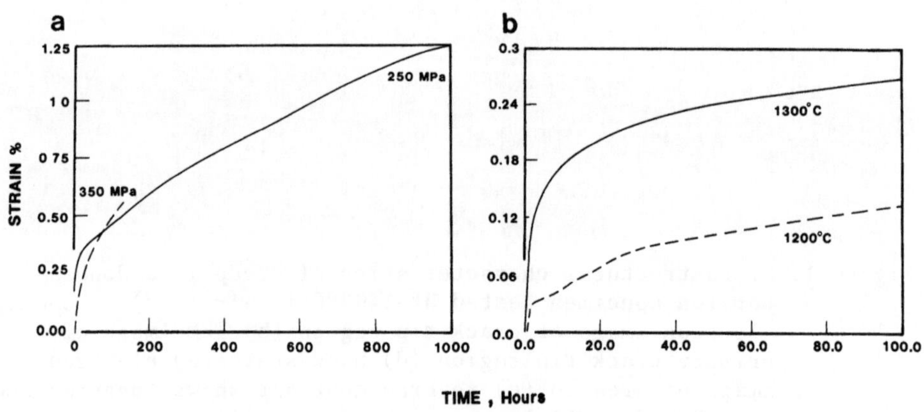

Figure 3. Static stress data for (a) magnesia doped silicon nitride at 1200°C and 250 and 350 MPa stress (b) yttria doped silicon nitride under 350 MPa stress at 1200°C and 1300°C.

the applied stress contribute to cavitation creep. The strength reliability maps for magnesia doped and yttria doped hot-pressed silicon nitride are reproduced in Figure 2. The strength distribution in the upper portion of the map was obtained by breaking specimens at temperature at the end of the test period; and, the fracture origins were related to the flaw population that was produced during the creep tests. The specimens included in the present examination were from the set of samples used to produce the reliability maps.

EXPERIMENTAL PROCEDURE

The identification and characterization of creep cavitation was carried out on specimens used in static load tests on magnesia-doped hot pressed silicon nitride (NC 132) and yttria-doped hot pressed silicon nitride (NC-X34 with 8% Y_2O_3). The specific test bars are identified in Table 1. The test bars were 50 mm long with cross-sections of 4 x 5 mm for MgO-doped and 3 x 4 mm for Y_2O_3 doped silicon nitride and were ground along their lengths. The load point deflections were measured during the static load tests and the strain vs. time curves for several specimens are shown in Figure 3. The upper and lower test spans were 10 mm and 40 mm respectively. A description of the test apparatus and test procedures was given in an earlier publication.[6]

The test bars were sectioned transverse to the bend axis along the boundaries of the upper load points. Slices parallel and transverse to these interfaces were made from these sections. Disc specimens for the electron microscope were made from the thin slices. The discs were lapped and then ion thinned to electron transparency from one or both sides. The direction of the applied stress was noted by keeping track of the as-received grinding marks or the edge of the sectioned bar. In this way it was possible to examine a specific region of interest from the test bar for example near the tensile surface, the compression surface, and the neutral axis. These regions are indicated on the figures.

Specimens were examined in a scanning transmission electron microscope operating at 120 keV. The microscope is equipped with X-ray energy dispersive spectrometer, back-scattered election detector and secondary electron detector.

EXPERIMENTAL RESULTS

Creep Cavitation at 1200°C

In 4 point bend bars of magnesia-doped, hot-pressed silicon nitride that were held for 1000 hours under a static load of

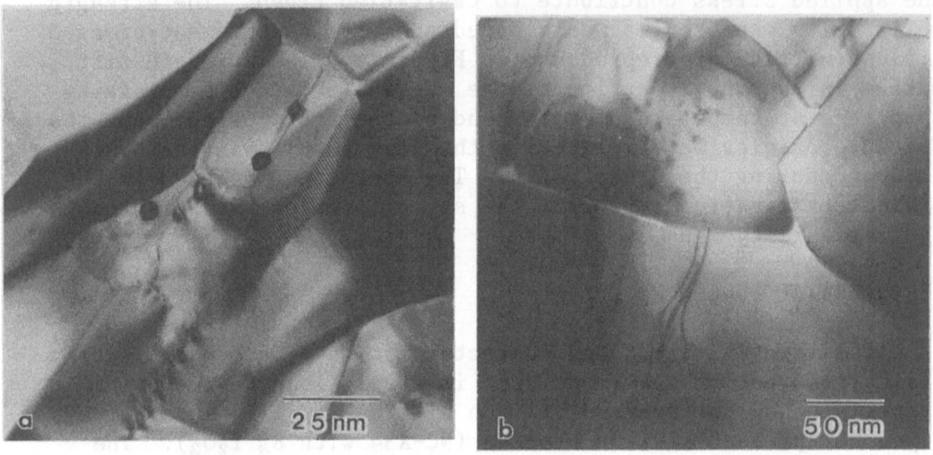

Figure 4. Creep cavitation near tensile surface in magnesia doped Si_3N_4 tested at 1200°C. (a) Nucleation of grain boundary cracks in spec #35 after 168 hours. (b) Grain boundary cracks in spec #348 after 1000 hours.

Table 1. Test conditions for the hot pressed silicon nitride

Specimen No., Dopant	Temperature °C	Initial Applied Elastic Stress (MPa)	Time h
31(MgO)	1400	*	
348(MgO)	1200	250	1000
35(MgO)	1200	2	168
44(Y_2O_3)	1200	350	100
141(Y_2O_3)	1200	350	1000
50(Y_2O_3)	1200	350	1000
28(Y_2O_3)	1300	350	117

* double torsion specimen.

250 MPa at 1200°C, strength degradation was extensive (Figure 2a) and the specimens deformed beyond the tolerable limits of practical use. The creep curves (figure 3a) for the test conditions show that the strain after 1000 hours was ∿1.2%. The cavitation microstructure in a 1000 hour specimens (No. 348) is shown in figure 4. Near the tension side of the bend bar cavities are found at all triple junctions and along many grain boundaries. Thus, in this material, both the chemical reaction of oxidation and the applied stress have caused extensive cavitation damage.

The yttria-doped silicon nitride did not exhibit significant creep at 1200°C as seen in Figure 3b. The reliability diagram in Figure 2b shows a strength increase after 1000 hours of exposure under a static load of ∿1/3 mean strength. The creep strain of the yttria doped silicon nitride (as measured at 1200°C with an initial applied stress of 350 MPa) was ∿0.12% after 100 hours and ∿0.2% after 1000 hours exposure. Specimens made from the test bars showed some cavitation at triple junctions and WC inclusions. Figure 5 shows the initial microstructure and the microstructural changes that occurred during 1200°C tests. In Figure 5a in the as-received material the yttria phases are the black grains that bond the β silicon nitride grains and their distinctive shapes are readily apparent. Figure 5b shows the $Y_2Si_2O_7$ phase in a 100 hour test specimen (No. 44) under dark field imaging conditions. The phase is a single crystal and extends over the boundaries of several silicon nitride grains. In the high magnification picture in Figure 5c the yttrium silicate phase is present between silicon nitride grains. These grains are oriented so that the $10\bar{1}0$ lattice fringe images are visible. The yttrium-silicate compound makes dihedral angle of ∿90° at the triple junction which suggests diffusional bonding. After 1000 hours there is good evidence for triple junction cavitation as seen in figure 5d. The yttria phase remains crystalline throughout and is able to accommodate the deformation by diffusion. Crack healing and compositional changes near grain boundaries resulting from such diffusion could have contributed to the observed strengthening.

Creep Cavitation at 1300°C

As seen in Figure 3b the yttria doped sample No. 28 crept at 1300°C with 350 MPa stress. When this sample was removed after 117 hours (∿70 hours of steady state creep) it had achieved about 0.27% strain and was noticeably bent. Five of six samples broke before the steady state creep region was achieved. The microstructure was examined at different distances from the tensile surface to identify nucleation and the growth stages of cavitation. At the tensile surface, extensive creep cavitation was observed as illustrated by the micrograph in Figure 6a. The specimen in Figure 6a was thinned from one side to preserve some of the surface oxide on the tensile side for examination. The surface

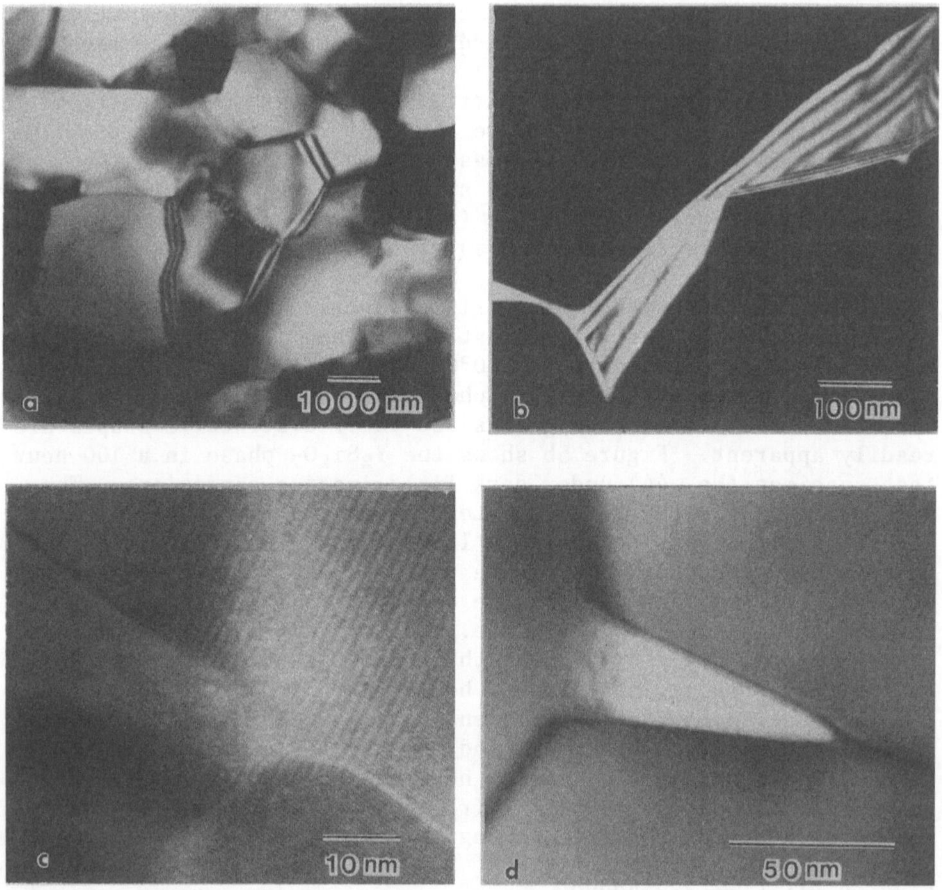

Figure 5. Microstructure of yttria-doped Si_3N_4 as received (a) and after creep test at 1200°C with 350-MPa initial applied load (b) dark field image of $Y_2Si_2O_7$ phase in spec #44 (c) grain boundary phase near tensile surface in spec #44 after 100 hours. (d) Creep cavity at triple junction in spec #141 after 1000 hours.

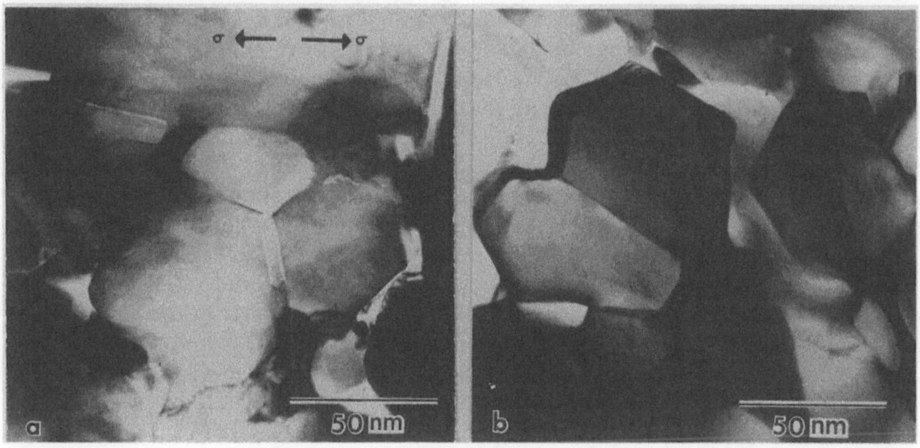

Figure 6. Microstructure near tensile (a) and compressive (b) surfaces in yttria doped Si_3N_4 after 117 hours at 1300°C under a static load of 350 MPa.

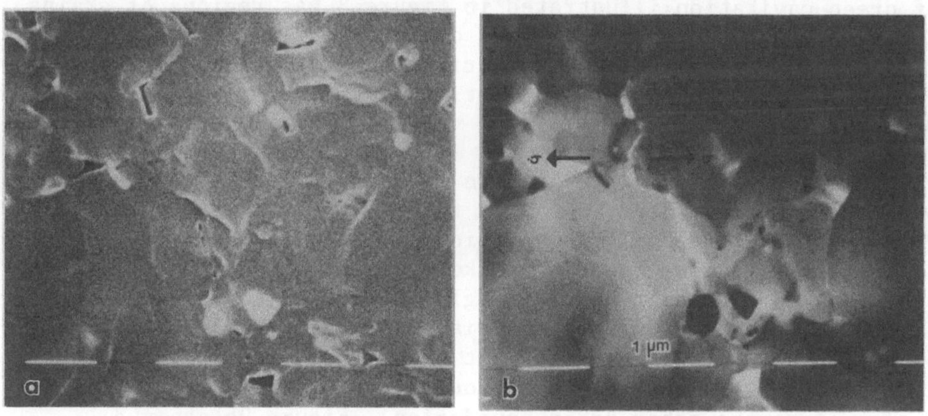

Figure 7. Microstructure near the tensile surface of spec #28 deformed at 1300°C. (a) Secondary electron image showing cavitation at grain boundaries (b) scanning transmission image showing continuous grain boundary crack and tungsten and yttria phases.

oxide adjacent to the matrix is the opaque region at (0) in the figure. This tensile surface interface is honeycombed with arrays of grain boundary cavities and microcracks such as those shown in this figure. Figure 6b shows the grain structure near the compression surface of the same test bar. The dark bands around the grains show a thickening and distortion of the yttria phase around Si_3N_4 grain boundaries. No cavitation or cracking was observed. Figure 7 shows another area of extensive cavitation which starts at triple junctions and continues along grain boundaries of the same test bar as shown in Figure 6a. From energy dispersive X ray analysis (EDS) the precipitates of WC (black rounded phases) and the yttria phase (dark grey triangular phases) along the boundary interfaces were identified. To make this thin foil the tensile surface was polished to remove the oxide layer; and, then the disc was ion thinned from the other side to retain that polished surface. The secondary electron image of the polished surface shows considerable grain boundary cavitation but does not show the complete path of the macro crack along several grain boundaries which is seen in the STEM picture in Figure 7b. This observation suggests that some caution must be exercised in interpreting cavitation and cracking from scanning electron microscope pictures only. A macro crack along grain boundaries may continue slightly below the surface and be too narrow to be resolved in an SEM picture.

The micrographs in Figures 8 and 9 show several stages in the production of grain boundary cavitation. The classical picture[10] of creep cavitation illustrated in Figure 8 has regions of grains that are within a few micrometers of the tensile interface. At this distance from the tensile surface both stress and stress corrosion from the oxidation reaction contribute to the cavitation process.

Different forms of cavitation around a β Si_3N_4 grain with two boundaries approximately parallel to the stress axis are shown in Figure 8. Two of the large cavities probably resulted from loss of the yttria phase, two crack like cavities extend from triple junctions, one boundary has a single spheroidal cavity and one triple junction cavity remains triangular. The uncavitated boundary has moiré fringe contrast which could result from rotation or from coherent precipitation of the yttria phase by diffusion along the other grain boundaries. Figure 9b shows a different region of the foil where the tensile axis is normal to the grain boundaries that are fractured. For the two cracks to connect, cavitation must occur across the yttria phase, which joins the silicon nitride grains. This connection has not occurred, possibly because the yttria phase deformed to accommodate the stress. Figure 8c shows cavities that have nearly joined to form a microcrack and a club shaped yttria phase (Y) that remains between several grains and partially covers the triple junction.

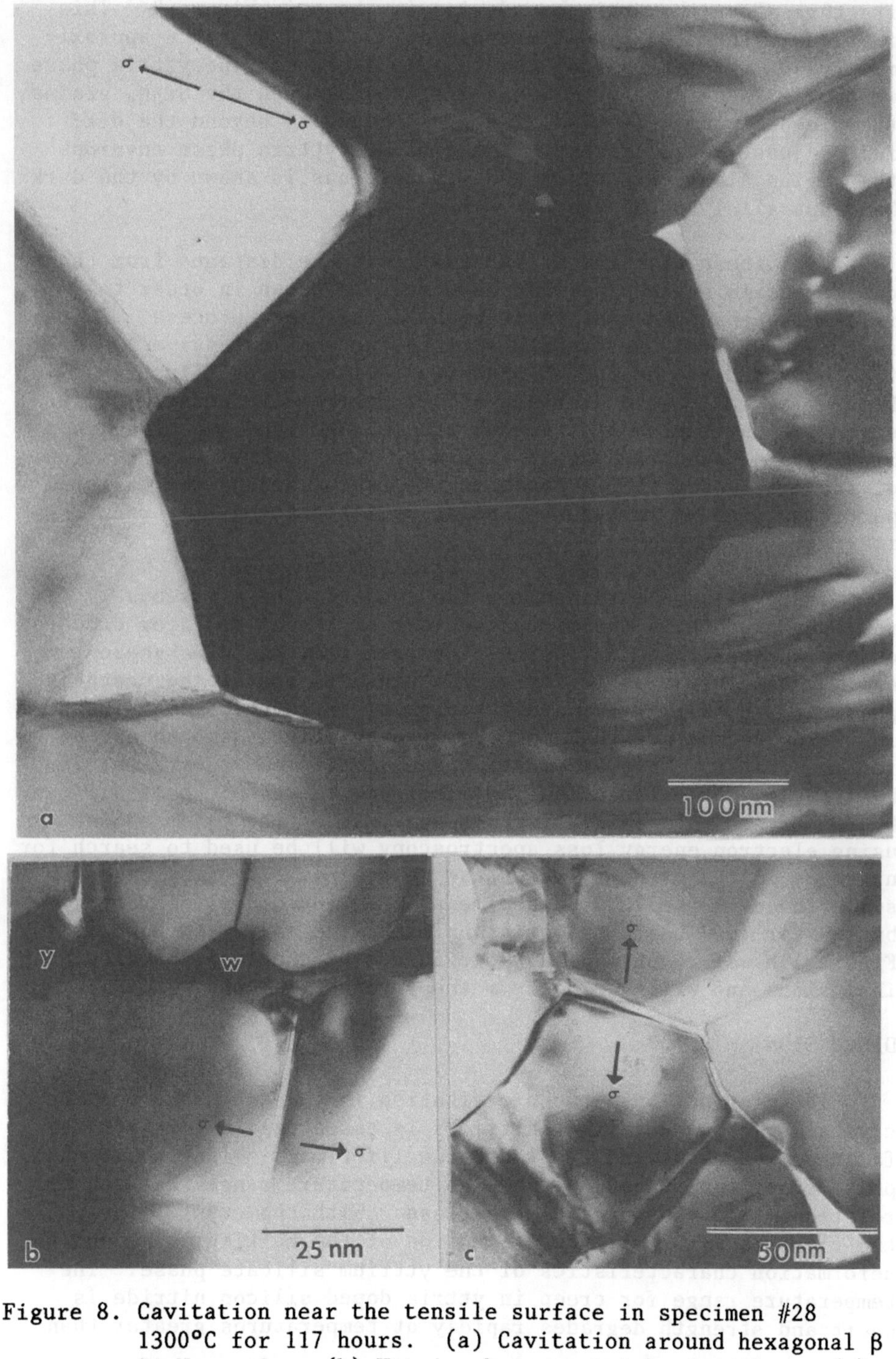

Figure 8. Cavitation near the tensile surface in specimen #28 1300°C for 117 hours. (a) Cavitation around hexagonal β Si_3N_4 grain. (b) Yttria phase at triple junction. (c) Yttria (Y) and tungsten (W) phases and grain boundary cracks.

The nucleation stage of crack-like cavitation followed by fracture along the grain boundaries is shown in Figure 9. This foil was thinned to reveal microstructure at a distance approximately 30 micrometers from the tensile surface. The yttria phase which was identified by EDS has diffused between the Si_3N_4 grains, the dihedral angle is $\sim 84°$ The grain boundary beyond the diffusion junction is cracked. The diffused yttria phase envelops the region along this crack and its terminus is shown by the dark bands parallel to the crack.

From these pictures it is clear that the distance from the outer tensile surface of bend bars must be known in order to relate the appearance of cavitation to the creep process. In the region 30 μm from the tensile surface the most extensive cavitation occurred around the WC and WSi_2 inclusions as illustrated in Figure 9b, while at a distance of 1-2 μm from the tensile surface extensive cavitation occurred at triple junctions and along the Si_3N_4 grain boundaries as in Figure 8. The spacial distribution of inclusions and yttria phase in the region around the tungsten particle is shown in Figure 9 using back scattered electron imaging.

There is some evidence for the evolution of a gaseous reaction product in the microstructure of the oxide layer produced on the surface. The diffusion of oxygen from the atmosphere can induce cavitation if the yttria phase is one of the unstable phases (J, K, H) which contain nitrogen. Figure 10a shows bubbles in the oxide surface that is otherwise relatively smooth except for the cracks. This oxide scale was identified as cristobalite by x-ray powder diffraction and was found to contain only silicon by EDS of the thin foil. Further analysis of the oxide scale using electron energy loss spectroscopy will be used to search for nitrogen in the oxide films on the test bars. The SiO_2 oxide scale found in the 1200°C specimens after 100 hours is smooth, but, after 1000 hours the cristobalite oxide scale as shown in Figure 11b was rough and showed evidence of chemical reaction with inclusions and with the Mo from the furnace elements.

DISCUSSION

The distinction between cavitation in the amorphous and the crystalline bonded silicon nitrides is essential in identifying failure mechanisms and in predicting lifetimes. With an amorphous phase creep can extend over a wide temperature range that can be related to the viscosity of the glass. With the crystalline phase the creep behavior will be a function of the melting point and the deformation characteristics of the yttrium silicate phase. The temperature range for creep in yttria doped silicon nitride is short and strength degrades rapidly at temperatures greater than

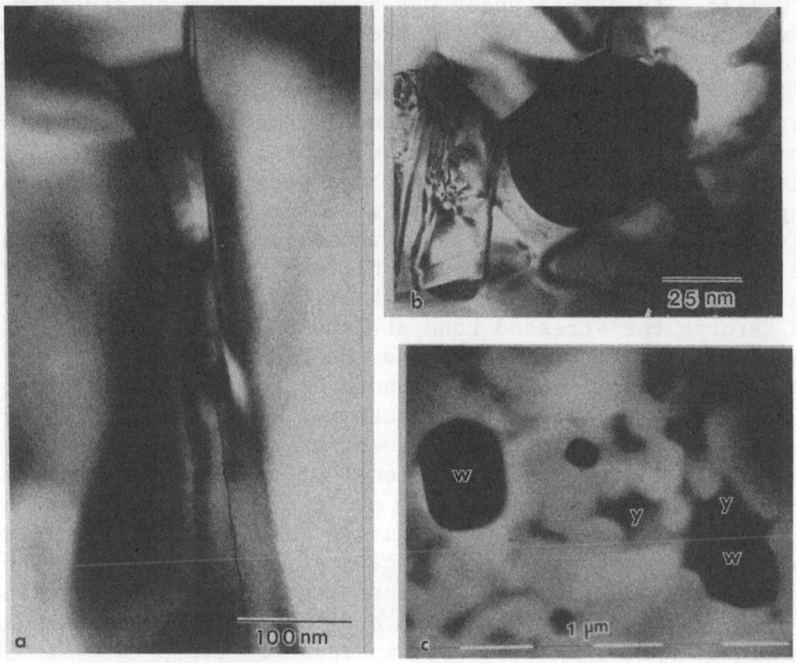

Figure 9. Cavitation ~30 μm from the tensile surface in spec #28. (a) Creep crack between Si_3N_4 grains and diffused yttria phase (b) cavity near WC inclusion, (c) spacial distribution of yttria-silicon nitride compounds (Y) and tungsten (W) particles near area of (b).

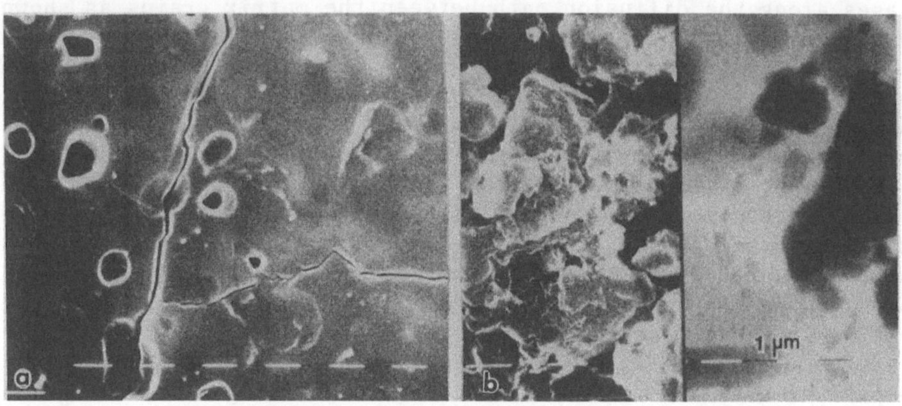

Figure 10. Oxide surface microstructure in yttria doped Si_3N_4 (a) 1300°C 117 hours. (b) 1200°C 1000 hours.

~1250°C.[7,13] The yttria phase was added to increase strength and to reduce creep,[12] as it did for the billets examined here. The width of the amorphous phase in magnesia doped silicon nitride can be as little as one nanometer between grains and as large as 1 μm when it fills the void space between silicon nitride grains that have not fully sintered. Cavitation occurs readily at triple junctions that have the amorphous phase present and grain boundary sliding with an amorphous phase between the boundaries is relatively easy. Further, in many instances there are cavities at triple junctions in as hot-pressed billets. As shown in Figure 1 the amorphous phase pulled as though molten when the creep crack passed through the stressed zone at 1400°C. At 1200°C the amorphous phase is more viscous and the steady state creep continued for a period of several hundred hours. The growth of crack like cavities into micro cracks was slower at 1200°C because of the higher viscosity and consequent greater resistance to both boundary sliding, and diffusional processes. The stress exponent obtained from a series of creep experiments on magnesia doped silicon nitride at 1200°C was 4.2 which can be associated with a diffusional grain boundary cavitation process. It is clear from figure 4b that there was extensive grain boundary cracking after 1000 hours under static load and that these grain boundary cracks linked up to cause a rapid increase in creep rate at 350 MPa stress. Thus the flaw population contributing to failure consisted of grain boundary cavities, and micro and macro cracks which were generated during the experiments as a result of both stress and oxidation corrosion.

In the yttria doped Si_3N_4 the crystalline second phase can be one of the several $Si_3N_4 + Y_2O_3$ phases. The phase most frequently identified in the present study are $Si_2Y_2O_7$ which is one of the stable phases and $Y_{10}Si_7O_{23}N_4$ which is unstable in oxygen.[13,19] If $Si_2Y_2O_7$ reacts with the Si_3N_4 grains by a diffusional process during the creep then it can form one of the unstable nitrogen phases along the diffusion path between the matrix grains as shown in Figures 5c and 9a. These phases oxidize readily and form SiO_2 and N_2 as reaction products as shown in Table 2. The bubbles seen in the oxide film in Figure 11a could result from such a process. The rapid disintegration of yttria doped silicon nitride samples which were prepared by Lange et al.[19] to include the J phase ($Y_4Si_2O_7N_2$), K phase ($YSiO_2N$) and H phase ($Y_{10}Si_7O_{23}N_4$), was attributed to the decomposition of the phases during reaction with excess oxygen. In the billets used for the present study the H phase was detected by X-ray diffraction and the $Y_2Si_2O_7$ phase was found in the thin foils by electron diffraction after exposure. Some _in situ_ experiments on Si_3N_4 + 13% Y_2O_3 in a high voltage electron microscope showed extensive grain boundary cavitation at temperatures as low as 700°C in billets containing excess WSi_2 and H phase ($Y_{10}Si_7O_{23}N_4$).[20]

Table 2. Reactions of Quaternary Phases of $Si_3N_4 + Y_2O_3$ with Oxygen (from Lange et al.[19])

Phase	Reaction		Reactants
	$Si_3Y_2O_3N_4$	$+ 3O_2$	$Y_2Si_2O_7 + SiO_2 + 2N_2$
K	$2SiYO_2N$	$+ 3O_2$	$5Y_2SiO_5 + 2SiO_2 + 2N_2$
J	$Si_2Y_4O_7N_2$	$+ 1.5O_2$	$2Y_2SiO_5 + N_2$
H	$Si_7Y_{10}O_{23}N_4$	$+ 3O_2$	$5Y_2SiO_5 + 2SiO_2 + 2N_2$

The variation in cavitation with distance from the tensile surface is a function of the diffusional process which is stress and temperature dependent. A more extensive study of the cavitation in the silicon nitride should include tensile specimens so that cavitation throughout the sample thickness can be examined. The cavitation produced in the yttria-doped silicon nitride is similar to cavitation in multiphase crystalline material where the typical cavitation process involves diffusion, grain boundary sliding and rotation. The diffusional process which produces an elongated cavity was discussed by Chuang and Rice[21] and and Chuang et al.[22] and the elongated cavities in this paper are the first examples seen in a non-oxide ceramic.

Localized cavitation and crack growth in the crystalline interface should produce a stress exponent different from that obtained for the magnesia doped silicon nitride because the grain boundary sliding and diffusion reaction between two crystalline phases is different. The weaker phase, in this case the yttria phase could deform by a dislocation mechanism of grain boundary sliding. The series of creep experiments has not been completed. At the present time little is known about the physical properties of the various yttrium silicon nitride phases although the chemistry and crystal structure have been studied.[14,19] The occurrence of the crack shaped cavities in yttria doped silicon nitride makes this material a good candidate for development of models of creep behavior in a ceramic material.

The grain size dependence of creep in silicon nitride can be assessed by considering the creep as a three dimensional problem. The beta silicon nitride grains have the morphology of hexagonal prisms about 1 μm by 5 μm and a random cross section shows an heterogeneous microstructure. As seen in the micrographs, the orientation of any group of grains with the stress axis can be random and the cavity growth will vary from ideal conditions usually drawn for grain reactions.[8,9,22] The constraints introduced by heterogeneity of the microstructure in yttria doped silicon nitride may be useful in developing failure predictions based on cavity growth as done for alumina.[23]

The tungsten based inclusions provide one of the most consistent sources of cavitation in both materials. As seen in figures 1, 4, 7 and 9 in this paper inclusions have reacted around their boundaries and cavities are usually present. The nominal content of tungsten in hot-pressed silicon nitride is approximately 3%. It would be useful to carry out stress rupture tests on the hot-pressed silicon nitride materials which do not have these inclusions in order to determine quantitatively the influence of such reactive inclusions on the creep process.

CONCLUSIONS

Microstructural examination of silicon nitride which crept at 1200, and 1300°C shows the existence of crack like cavitation along grain boundaries as a result of thermally activated chemical reaction and applied mechanical stress. The creep crack propagation rate depends on the properties of the bonding phase and on the applied stress. Amorphous magnesium silicate phase in magnesia doped hot-pressed Si_3N_4 restricts the creep dependent lifetime at 1200°C. The crystalline yttrium silicate phase restricts the creep dependent lifetime at 1300°C.

ACKNOWLEDGMENT

The work was supported in part by the Air Force Office of Scientific Research and by the Department of Energy and by the Army Materials and Mechanics Research Center. The authors thank D. Dulik for photographic assistance.

REFERENCES

1. N.J. Tighe, The structure of slow crack interfaces in silicon nitride, J. Mat. Sci. 13:1455 (1978).
2. R. Kossowsky, D.G. Miller and E.S. Diaz, Tensile and compressive creep strength of hot pressed Si_3N_4, J. Mat. Sci. 10:983 (1975).
3. J.M. Birch, and B. Wilshire, The compression creep behavior of silicon nitride ceramics, J. Mat. Sci. 10:983 (1975).
4. S.U. Din and P.S. Nicholson, Creep of hot-pressed silicon nitride, J. Mat. Sci. 10:1375 (1975).
5. S.M. Wiederhorn and N.J. Tighe, Proof testing of hot-pressed silicon nitride, J. Mat. Sci. 13:1781 (1978).

6. N.J. Tighe and S.M. Wiederhorn, Effect of oxidation on the reliability of silicon nitride, pp. 403-423 in "Fracture Mechanics of Ceramics", V. 5, R.C. Bradt, A.G. Evans, D.P.H. Hasselman and F.F. Lange, ed., Plenum Press, New York (1983).
7. S.M. Wiederhorn and N.J. Tighe, Structural reliability of yttria-doped hot pressed silicon nitride at high temperature, J. Am. Ceram. Soc. 66:884 (1983)
8. A.G. Evans and A. Rana, High temperature failure mechanisms in ceramics, Acta Met. 28:129 (1980).
9. J.R. Porter, W. Blumenthal and A.G. Evans, Creep fracture in ceramic polycrystals. I. Creep cavitation effects in polycrystalline alumina, Acta Met. 29:1899 (1981).
10. A.G. Evans and W. Blumenthal, High temperature failure in ceramics, in "Fracture Mechanics of Ceramics, V. 6, R.C. Bradt et al., Plenum Press, New York (1983).
11. D.R. Clarke and G. Thomas, Grain boundary phases in a hot pressed MgO fluxed silicon nitride, J. Am. Ceram. Soc. 60:491 (1977).
12. G.E. Gazza, Effect of yttria additions on hot-pressed Si_3N_4, J. Am. Ceram. Soc. 57:269 (1974).
13. G.Q. Weaver and J.W. Lucek, Optimization of hot pressed Si_3N_4-Y_2O_3 materials, Bull. Am. Ceram. Soc. 57:1131 (1978).
14. D.R. Clarke and G. Thomas, Microstructure of Y_2O_3 fluxed hot-pressed silicon nitride, J. Am. Ceram. Soc. 61:114 (1978).
15. N.J. Tighe, Analysis of creep microstructure in silicon nitride by AEM, Proc. 10th Intl. Congress on Electron Microscopy, Springer Verlág, p 399 (1982).
16. A.G. Evans and S.M. Wiederhorn, Crack propagation and failure prediction in silicon nitride at elevated temperature, J. Mat. Sci. 9:270 (1970).
17. N.J. Tighe, Analysis of oxide and oxide matrix interfaces in silicon nitride, in "Advances in Ceramic," V. 6, Character of Grain Boundaries, M.F. Yan and A.H. Heuer, Am. Ceram. Soc. Columbus (1983).
18. S.M. Wiederhorn, A probabilistic framework for structural design, pp. 197-226 in "Fracture Mechanics of Ceramics," 5, R.C. Bradt, A.G. Evans, D.P.H. Hasselman, and F.F. Lange, ed., Plenum Press, New York.
19. F.F. Lange, S.C. Singhal and R.C. Kuznicki, Phase relations and stability studies in the Si_3N_4-SiO_2-Y_2O_3 pseudoternary systems, J. Am. Ceram. Soc. 60;249 (1977).
20. N.J. Tighe, K. Kuroda, T.E. Mitchell and A.H. Heuer, In Situ oxidation of Y_2O_3 doped Si_3N_4, in Electron Microscopy 1980, V.4 High Voltage, North Holland, P. Prenderoo and J. Van Landuyt, ed.
21. T.-J. Chuang and J.R. Rice, The shape of intergranular creep cracks growing by surface diffusion, Acta Met. 21:1625 (1973).

22. T.-J. Chuang, K.I. Kagawa, J.R. Rice and L.B. Sills, Non-equilibrium models for diffusive cavitation of grain interfaces, Acta Met. 27:265 (1979).
23. C.H. Hsueh and A.G. Evans, Creep fracture in ceramic polycrystals. II. Effects of inhomogeneity in creep rupture, Acta Met. 29:1907 (1981).

ACKNOWLEDGMENT

The work was supported in part by the Air Force Office of Scientific Research and by the Army Materials and Mechanics Research Center under funds provided by the Department of Energy Interagency Agreement DOE/AMMRC (DOE-1A No. DE-A101-77-CS5-1017). The authors thank D. Dulik for photographic assistance.

HIGH-TEMPERATURE CREEP AND FRACTURE OF $\beta' Si_3N_4$ CERAMIC ALLOYS

M. H. Lewis, G. R. Heath, S. M. Winder and R. J. Lumby*

University of Warwick
United Kingdom
*Lucas Syalon plc
West Midlands, United Kingdom

β′ CERAMIC ALLOYS

Fabrication and Microstructure

The simultaneous substitution of Al^{3+} and O^{2-} for Si^{4+} and N^{3-}, respectively, in the $\beta' Si_3N_4$ crystal structure produces a solid solution' (β′) phase $Si_{3-x}\,Al_x\,O_x\,N_{4-x}$ which is the basis for a range of engineering ceramics. These ceramic 'alloys' may approach the mono-phase state[1] or be 'composites' of β′ with a matrix phase which is the residue of a liquid-sintering aid.[2,3] Typically, the substitution level x is less than 0.5 and there is little change in the intrinsic properties of the $\beta\ Si_3N_4$ crystal. However, it is the indirect influence of crystal substitution on the volume and composition of the residual phase which has a dominant influence on ceramic properties, especially at elevated temperatures.

The complex reaction sequence during ceramic fabrication is summarised in Fig.1a together with the process of liquid-phase sintering in which $\alpha\ Si_3N_4$ dissolves in a M-Si-Al-O-N liquid and is reprecipitated as the substituted $\beta\ Si_3N_4$. Sintering liquid compositions are based on ternary oxide compositions of the type analysed for M = Yttrium in Fig. 1b, with nitrogen substitution dependent on initial composition.

The formation of mono-phase β′ ceramics requires hot-pressing with a transient liquid phase of small volume in which components are removed by solid solution. The achievement of this state is critically dependent on the 'balance' between surface SiO_2 (on $\alpha\ Si_3N_4$) and an AlN additive.[1] The residual liquid content also depends on the metallic oxide (MO) addition since M atoms are

$$\underbrace{\alpha\text{-}Si_3N_4 + SiO_2}_{\text{SURFACE}} + \underbrace{AlN + MO}_{\text{ADDITIVES}} \rightarrow \underbrace{\beta'\text{-}Si_{3-x}Al_xO_xN_{4-x}}_{\text{MAJOR PHASE}} + \underbrace{M\text{-}Si\text{-}Al\text{-}O\text{-}N}_{\text{GLASS RESIDUE}}$$

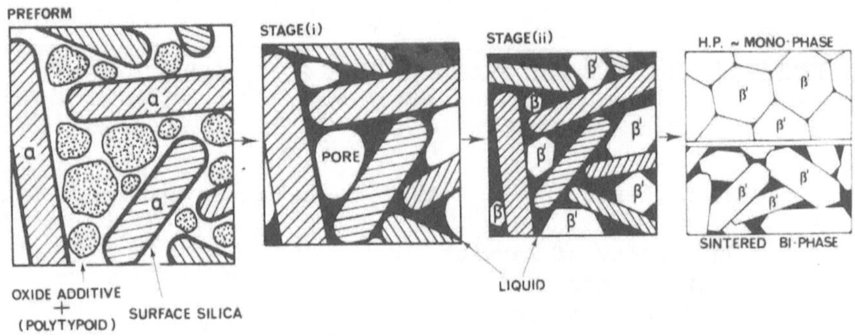

Fig. 1(a) The liquid-phase sintering process during fabrication of β′ ceramics.

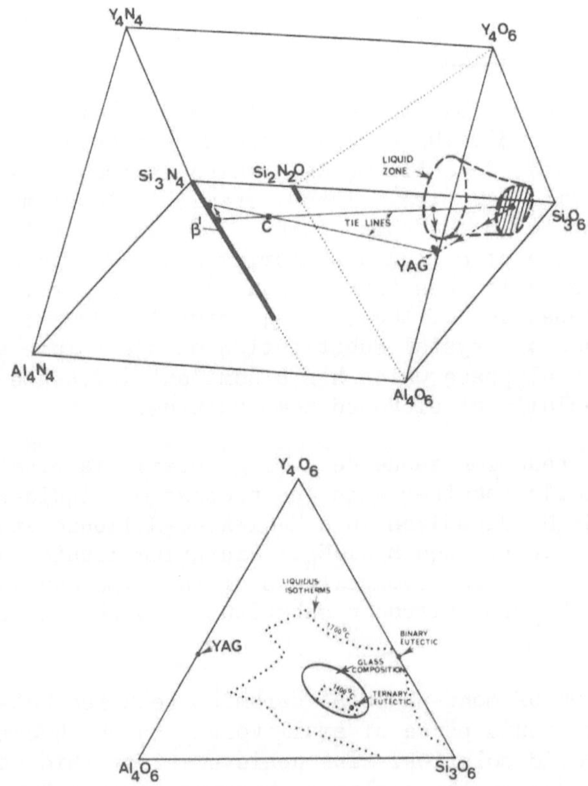

Fig. 1(b) The relation between terminal components, compositions of β′ and glass phase and the product of matrix crystallisation in high-nitrogen ceramics.

insoluble in β′. These microscopic residues are mainly in the form of nitrogen-containing silicate glasses concentrated at β′ triple-junctions.[4] A typical mono-phase polycrystalline β′ microstructure is illustrated in Fig. 2a and is distinct from that of the pressureless-sintered bi-phase ceramics which require comparatively large liquid volumes for complete densification. The latter contain faceted β′ in hexagonal prism morphology within a semi-continuous matrix of residual nitrogen containing glass (Fig. 2b). The volume fraction of glass matrix, its composition and hence its viscosity are determined by the N/O ratio for the system which is controlled by the AlN addition*. Substitution levels in β′ increase with AlN which, in a conservative system, means that the N/O ratio in the liquid must increase. This effect is amplified by the simultaneous decrease in glass volume such that a rapid increase in glass viscosity is anticipated.

Both types of β ceramics (hot pressed and, more recently, pressureless-sintered) have been the subject of a joint research programme between Warwick University and Lucas Industries over the past decade. The bi-phase pressureless sintered 'Syalon' ceramics*, fabricated with a Y_2O_3 addition, have now achieved commercial importance as engineering ceramics, especially in the metal-cutting field.

Microstructural Stability

Glass residues are metastable with respect to crystallisation of oxide or oxynitride phases. Such transformations are anticipated during creep deformation or prior heat-treatment at temperatures considered for heat-engine applications (1000-1400°C). The primary crystallisation product in pressureless-sintered ceramics varies with N/O ratio in the parent glass from β yttrium disilicate (low nitrogen) to yttrium aluminium garnet (YAG) at high nitrogen content. It has previously been shown that controlled nucleation and slow growth of YAG may occur with a diffusion of excess Si and N into with reduced substitution level thus avoiding residual glass or additional crystalline phases.[2]

An alternative form of microstructural instability results from an interaction with oxidising environments in a similar temperature interval. The origin of this process and its influence on high-temperature creep and fracture was first identified in hot-pressed monophase ceramics by Karunaratne and Lewis[4,5,6] and subsequently recognised as a general phenomenon in other Si_3N_4 based ceramics containing liquid residues.[7] A silica-rich oxidation layer acts as a 'sink' for metallic-ions contained within the residues and there is a time-dependent change in intergranular composition and structure

*In Lucas Syalon ceramics this is added as a component of a 21R polytypoid phase of composition ~ 6 AlN SiO_2. 'Syalon' is a trade name of Lucas Syalon plc.

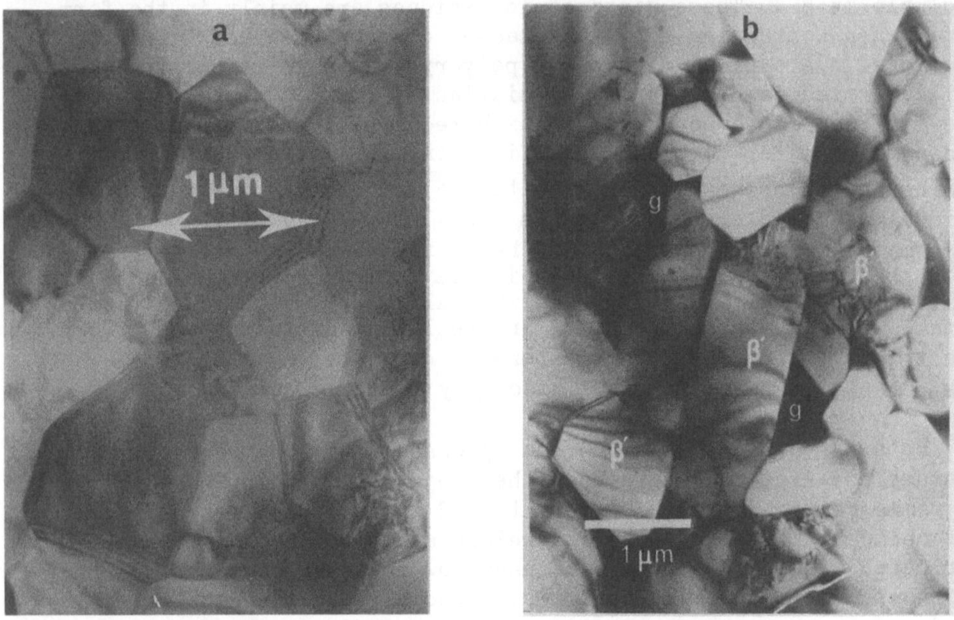

Fig. 2 Typical microstructures of hot-pressed mono-phase and pressureless-sintered bi-phase ceramics.

which results in a reduced creep rate and suppression of creep cavitation. These transient phenomena are briefly reviewed in this paper and a comparison made with the bi-phase ceramics in which there is a more pronounced change in sub-surface microstructure following a diffusional reaction with the surface oxidation layer.

A major theme of the work presented in this paper is the large variation in high-temperature creep and fracture behaviour attainable via control of initial composition, post-fabrication heat-treatment or environmental reaction.

HIGH TEMPERATURE DEFORMATION

Creep Data and Micromechanisms

A comparison of creep rates for the different ceramics in various microstructural states is given in Fig. 3 for a temperature of 1300°C. This is selected as being near to the temperature limit at which an oxidation-induced catastrophic fracture occurs in bi-phase ceramics.[8]

True steady-state creep is rare throughout the ceramic alloy series. A typical behaviour is that of a short-term primary creep of, partially recoverable, viscoelastic form followed

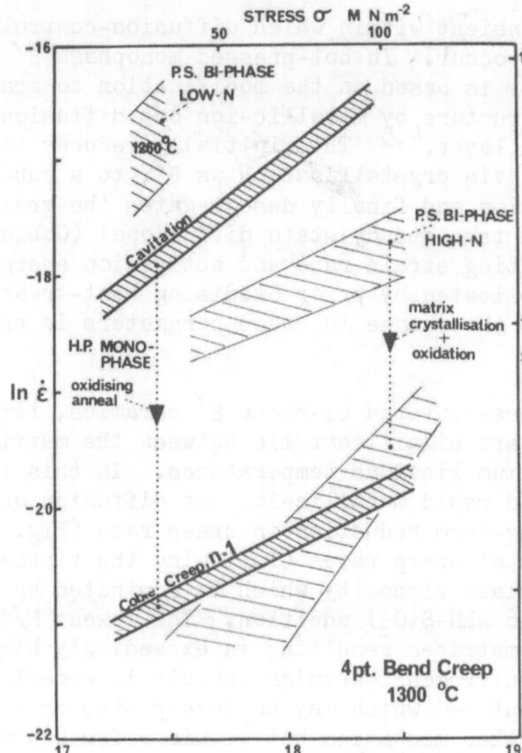

Fig. 3 A comparison of initial and final pseudo steady-state creep rates in mono and bi-phase ceramics of differing composition and microstructure.

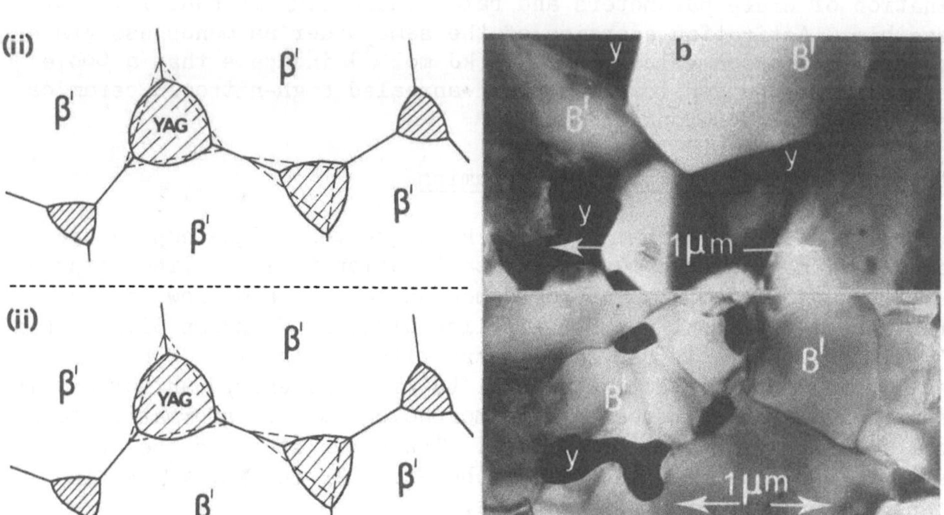

Fig. 4 The morphological change following complete matrix crystallisation in a β′ - YAG ceramic with substitution of excess Si and N in β′.

by a long-term transient within which diffusion-controlled microstructural changes occur. In hot-pressed monophase β′ ceramics the transient behaviour is based on the modification to grain-boundary composition and structure by metallic-ion out-diffusion into the silicate oxidation layer.[4,5] This initially reduces the triple junction residues, via crystallisation as β′, to a sub-critical size for cavity nucleation and finally desegregates the grain boundaries to a level where a true steady-state diffusional (Coble) creep occurs with a limiting strain rate and activation energy.[5,8] These changes may be replicated by prior oxidising heat-treatment and a typical example of the change in creep parameters is presented in Fig. 3.

In pressureless-sintered bi-phase β′ ceramics, temperatures at which creep rates are significant lie between the matrix glass transition and equilibrium liquidus temperatures. In this interval matrix crystallisation and rapid metallic-ion out-diffusion occur, giving rise to a very long-term reduction in creep rate (Fig. 3). Initial 'pseudo-steady state' creep rates (following the viscoelastic primary) are sensitive to glass viscosity which is dominated by N/O ratio and hence polytypoid (6 AlN-SiO_2) addition. The lowest N/O ratio ceramics contain fluid matrices resulting in exceedingly high creep rates at 1300°C. 'High-nitrogen' ceramics exhibit intermediate creep rates with stress exponent $n \sim 1$ which may be interpreted as a 'viscous' matrix-flow controlled mechanism but within a few hours of initial loading creep-rates are influenced by matrix crystallisation and sub-surface structural changes described in the next section. Ceramics in which matrices have been precrystallised continue to exhibit a slow reduction in creep rate such that unambiguous determination of creep parameters and rate-controlling mechanism is not feasible. Activation energies of the same order as monophase grain-boundary diffusion values (800-900 kJ mol^{-1}) indicate that a Coble mechanism may be applicable to well-annealed high-nitrogen ceramics containing a YAG second phase.

Structural changes in bi-phase ceramics

High nitrogen ceramics, which have greatest high-temperature potential, may undergo matrix crystallisation to YAG in the temperature range 1200-1400°C. A programmed nucleation and slow crystal growth heat-treatment permits a solid-solution of excess Si+N within β′ with an observable decrease in substitution level. If crystallisation is complete the interfacial (β′/matrix) energy anisotropy is reduced with the consequence of a morphological transformation to a more isolated granular form of YAG. These two steps in the crystallisation reaction are illustrated schematically in Fig. 4a and in transmission micrographs in Fig. 4b.

Within an oxidising environment either during creep or precrystallising heat-treatment there are a series of sub-surface

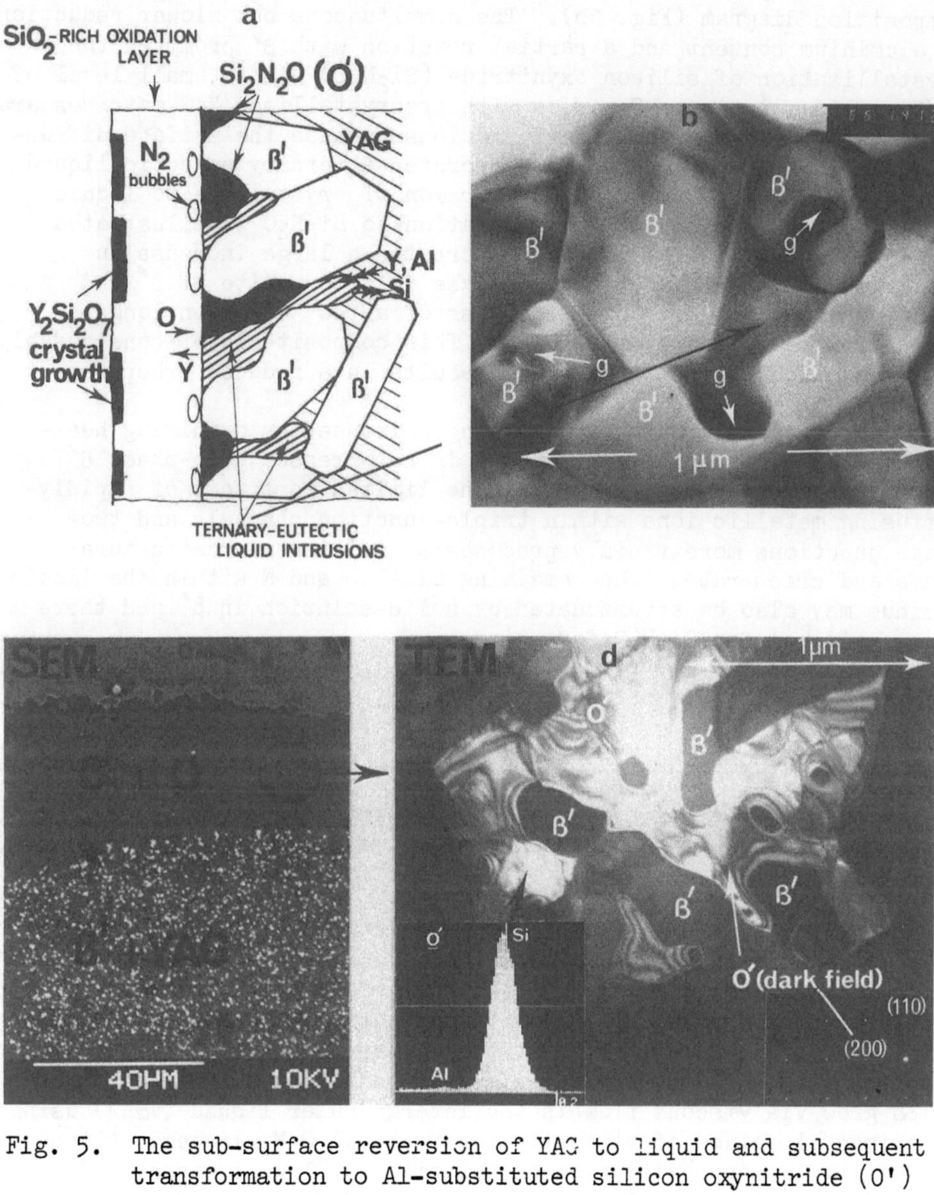

Fig. 5. The sub-surface reversion of YAG to liquid and subsequent transformation to Al-substituted silicon oxynitride (O') during the oxidising anneal of a high-nitrogen bi-phase ceramic.

transformations dominated by the relatively rapid out-diffusion of yttrium into a SiO_2-rich oxidation layer. For initially glassy matrices the removal of yttrium results in a susceptibility to alternative crystallisation products within the Si_3N_4-SiO_2-Al_2O_3-AlN composition diagram (Fig. 1b). The simultaneous but slower reduction in aluminium content and a partial reaction with β′ promotes the crystallisation of silicon oxynitride (Si_2N_2O) with a small level of Al/Si solid solution. Ceramics with precrystallised YAG matrices are also susceptible to such transformations because the surface diffusion couple between SiO_2 and YAG generates a ternary eutectic liquid above ~1300°C (Fig. 1b). This reversion of crystalline to liquid matrix and the subsequent transformation to Si_2N_2O is illustrated in Fig. 5. The liquid/β′ reaction creates a large increase in 'matrix' volume such that the subscale is a composite of β′ and Si_2N_2O which may exist to a thickness of a few 100 μm on long annealing (~1000 hours at 1300°C). This composite layer considerably enhances oxidation resistance and results in a reduced creep rate.

Subsurface microstructural changes induced by oxidising heat-treatment are comparable with those in hot-pressed mono-phase β′ ceramics. However, in the latter the limited reservoir of rapidly-diffusing metallic ions within triple-junction channels and two-grain junctions more rapidly produces a limiting microstructural state and creep rate. The remaining Si,Al,O and N within the liquid residue may also be accommodated by solid-solution in β′ and there is no evidence for Si_2N_2O formation.

CREEP CRACK GROWTH

Comparison of Mono and Bi-phase Ceramics

A review of earlier double-torsion test data for sub-critical crack growth in a range of hot-pressed β′ ceramics is presented in Fig. 6a. Ceramics containing triple-junction liquid residues above a critical size for cavity nucleation (within the stress range up to K_{1c}) exhibit relatively low exponents in the logarithmic crack velocity-stress intensity (K_1) plot. K_1 exponents of ~7 typical of commercial silicon nitrides in the 1250-1400°C range are identified with a mechanism of multiple cavity nucleation and interlinkage within a large creep damage zone at the primary crack tip.[6] This is compared with a theoretical model (Tsai-Raj[9]) for linkage of cavities which grow via viscous flow in the intergranular liquid ($V\alpha K^4$) using experimental values of viscosity for Mg-Si-Al-O-N glasses. Although predicting the approximate crack velocity for low stresses it fails to satisfactorily model the experimental values over a range of K_1. This is possibly due to the simplicity of the model in assuming the interlinkage of a linear array of microcracks rather than that distributed in a large creep-damage zone but, in β′ ceramics, the assumption of growth by viscous flow is unrealistic. A more precise description of the mechanism, based on microstructural evidence, is

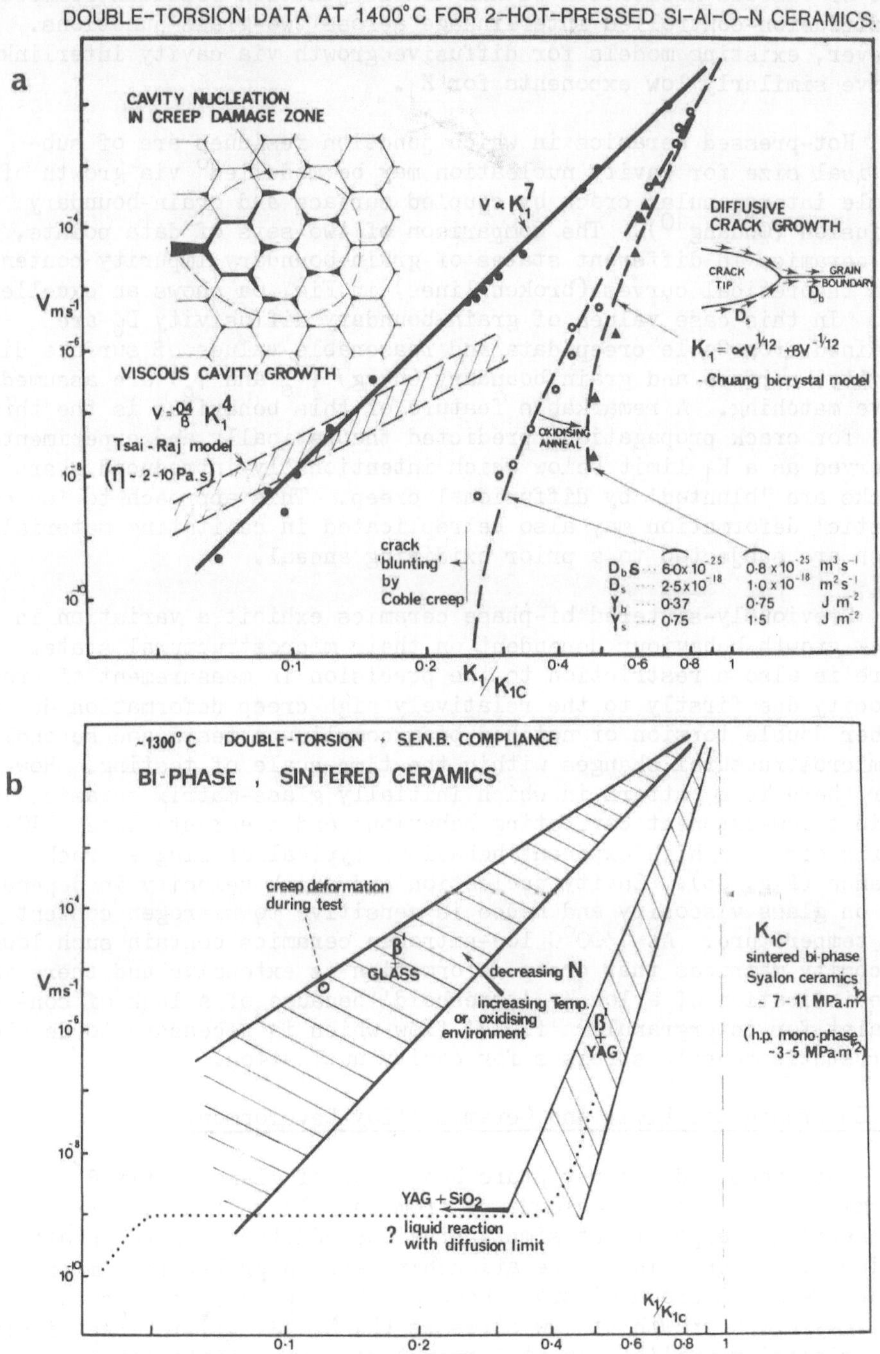

Fig. 6 The relation between sub-critical crack growth velocity (V) and normalised stress intensity K_1/K_{1c} for (a) mono and (b) bi-phase β′ ceramics in different microstructural states.

that of viscous nucleation within triple-junction residues followed by diffusion-controlled interlinkage across two-grain junctions. However, existing models for diffusive growth via cavity interlinkage derive similarly low exponents for K_1.

Hot-pressed ceramics in which junction residues are of sub-critical size for cavity nucleation may be modelled[6] via growth of a single intergranular crack by coupled surface and grain-boundary diffusion (Chuang[10]). The comparison of two sets of data points, for ceramics in different states of grain-boundary impurity content, with theoretical curves (broken lines) in Fig. 6a shows an excellent fit. In this case values of grain-boundary diffusivity D_b are obtained from Coble creep data and reasonable values of surface diffusivity, surface and grain boundary energy (γ_s and γ_b) are assumed in curve matching. A remarkable feature of this behaviour is the 'threshold' for crack propagation, predicted theoretically and experimentally observed as a K_1 limit below which intentionally-introduced sharp cracks are 'blunted' by diffusional creep. This approach to 'superplastic' deformation may also be replicated in cavitating materials which are subjected to a prior oxidising anneal.

Previously-sintered bi-phase ceramics exhibit a variation in crack growth behaviour dependent on their microstructural state. There is also a restriction to the precision in measurement of crack velocity due firstly to the relatively high creep deformation during either double torsion or notched-beam compliance tests and secondly to microstructural changes within the time-scale of testing. However there is a pattern in which initially glass-matrix ceramics exhibit low-exponent cavitating behaviour and precrystallised YAG-matrix ceramics high exponent behaviour typical of single crack advance (Fig. 6b). Cavity nucleation and crack velocity is dependent on glass viscosity and hence is sensitive to nitrogen content and temperature. At 1300°C low-nitrogen ceramics contain such low-viscosity matrices that creep deformation is extensive and there is some indication of a low K_1 'threshold' because of a lack of constraint for intergranular 'fluid'-flow which is necessary to develop hydrostatic tensile stresses for cavity nucleation.

The Environmental Limit and Ceramic Alloy Development

The creep and creep-rupture behaviour of glass-matrix β' ceramics is of little consequence from the ultra-high temperature engineering viewpoint but simply provides additional data in attempts to identify trends in the relationship between properties, microstructure and microchemistry. However the suppression of sub-critical crack growth at low stresses ($K_1/K_{1c} \gtrsim 0.3$) is a key factor in engineering application at temperatures appropriate for many heat-engine components. At such temperatures the intrinsic stability of precrystallised β′ - YAG composites and their V-K_1 characteristics are required. However, reference to Fig. 5 shows an upper limit to the stability of sub-surface microstructure in oxidising environment at $\gtrsim$ 1300°C determined by YAG-SiO_2 interdiffusion and a reversion to the eutectic liquid state with low nitrogen content. Above ~ 1350°C the liquid viscosity is low and there is a rapid advance of liquid intrusions within the matrix together with β′ solution (Fig. 5). Creep failure at these temperatures is rapid and, although not studied quantitatively, a diffusion-rate limited transformation in V-K_1 behaviour is probable at lower temperatures (indicated by the broken line in Fig. 6b).

An improvement in oxidation-induced fracture behaviour may be achieved firstly by prolonged crystallisation heat-treatment to establish the isolated YAG morphology, described above, thus inhibiting access of the SiO_2-rich oxidation layer. This is only partially effective because of the dissolution of β′ by the eutectic liquid. A more complete solution is provided by the conversion of a sub-surface layer, via programmed oxidation heat-treatment near 1300°C, to a 'composite' of β′ and substituted Si_2N_2O (Fig. 5) which exhibits a remarkable improvement in resistance to oxidation and chemically-induced creep fracture.

In summary, the development of pressureless-sintered β′ ceramic alloys for application up to temperatures of ~ 1000°C has already been achieved via control of the nitrogen content of glass-matrices. Above these temperatures even high-nitrogen glasses undergo crystallisation. A controlled post-sintering crystallisation treatment produces stable, creep and creep-rupture resistant ceramics useable to ~ 1250°C. The onset of oxidation-induced degradation at ~ 1300°C requires a controlled subsurface transformation to β′ /Si_2N_2O for application at the highest temperatures. A further study of such transformations is required with reference to the chemistry and initial microstructure of engineering component surfaces in the pressureless-sintered condition

Finally it is worth noting the impressive levels of fracture-toughness currently attainable in bi-phase Syalon ceramics. Values of 7-11 M Pa $m^{\frac{1}{2}}$ for glass-matrix ceramics compare with values of 3-5 M Pa $m^{\frac{1}{2}}$ for mono-phase hot-pressed β′ ceramics. The improvement in K_{1c} for the pressureless-sintered ceramics is believed to originate in the enhanced morphological anisotropy of β′ crystals growing with greater freedom from impingement in a liquid environment. Also although there is a greater scatter in data for high-nitrogen ceramics their mean values indicate an enhancement in the intrinsic fracture surface energy over the low-nitrogen ceramics.

REFERENCES

1. M.H. Lewis, B.D. Powell, P. Drew, R.J. Lumby, B. North and A.J. Taylor, J. Materials Sci. 12 (1977) 61.

2. M.H. Lewis, A.R. Bhatti, R.J. Lumby and B. North, J. Materials Sci. 15 (1980) 103.

3. M.H. Lewis and R.J. Lumby, 'Powder Metallurgy', June (1983)

4. B.S.B. Karunaratne and M.H. Lewis, J. Materials Sci. 15 (1980) 449.

5. B.S.B. Karunaratne and M.H. Lewis, J. Materials Sci. 15 (1980) 1781.

6. M.H. Lewis and B.S.B. Karunaratne, in 'Fracture Mechanics for Ceramics, Rocks and Concrete', eds. Freiman and Fuller, ASTM, STP745, (1981) 13.

7. F.F. Lange, B.I. Davis and D.R. Clarke, J. Materials Sci 15 (1980) 616.

8. M.H. Lewis, B.S.B. Karunaratne, J. Meredith and C. Pickering, in 'Creep and Fracture of Engineering Materials and Structures', eds. Wilshire and Owen (Pineridge Press 1981) 365.

9. R.L. Tsai and R. Raj, Acta. Met. 30 (1982) 1043.

10. T.J. Chuang, J. Amer. Ceram. Soc. 65 (1982) 93.

CREEP AND CREEP RUPTURE OF HP-SiC CONTAINING AN AMORPHOUS INTERGRANULAR PHASE

Rédouane Moussa, Jean-Louis Chermant and Frédéric Osterstock

Equipe Matériaux-Microstructure du L.A. 251
ISMRA - Université
14032 Caen Cedex, France

INTRODUCTION

Ceramic bulk materials like SiC may be processed following different ways. Boron or boron carbide may be added to obtain creep resistant materials. They contain however some amount of porosity which lowers their fracture resistance and toughness (1)(2). Full density may be performed for example by adding aluminium and subsequently pressing the powder at approximatively 2200°C. An amorphous phase is thus present in the intergranular space and densities higher than 99.5 % are achieved

This secondary phase influences also the high temperature deformation behavior. Since the carbide crystals do not deform,the intergranular phase is a controlling factor. Its scarcity makes that deformation takes place through cavitation at the interfaces and triple point boundaries. The influence of these cavitation processes on the creep parameters have been addressed theoretically but lack of experimental data is a common claim (3)(4)(5)(6)(7)(8).

The aim of this work is to measure creep rates on a HP-SiC materials which contains a small quantity of second phase mainly located at triple point boundaries and four crystals channels. The values of the resulting creep parameters (stress exponent and apparent activation energy) will be discussed in parallel with analysis of SEM investigations.

MATERIAL AND EXPERIMENTAL PROCEDURES

The material investigated in this paper is a HP- SiC provided by the Elektroschmelzwerk Company, Kempten, W-Germany. It was hot-pressed at 2200°C at a pressure of 400 MPa. To achieve maximum density, 1.5 % of aluminium was added. This produces a second phase in intergranular spaces. TEM investigations (Jeol 120 CX) show that this second phase is amorphous and is manely located in the triple point junction. These pockets become thinner and thinner as they advance in the grain boundaries :it was not possible, within the resolution of the microscope, to show that the glassy phase exists across the whole grain boundaries. We thus assume that it is confined in triple point boundaries and four point channels (9).

The creep experiments have been performed in three point bending under vacuum ($\simeq 10^{-6}$ torr). Specimen size is 20 x 4 x 2 mm^3 with a bending span of 14 mm. All the specimen were polished (diamond paste) and cautiously chanferred to avoid unexpected rupture from machining flaws. The furnace was of VMDI Sesame-type with a heating capacity up to 2000°C. Heating elements are two semi-cylindrical half-shells of tantalum. The pushrods are of tungsten 20 mm in diameter and the knives are TiC rods, 5 mm in diameter.

Constant load conditions have been achieved by using a lever system. The lower pushrod is fixed on the furnace whereas the upper one is fixed on a crosshead which is allowed to move along two columns. Transmission of load from lever system to the crosshead is done via a steel ball to avoid excentricity effects.

Displacement is measured on the external part of the pushrod. The signals of two symetrically located inductive transducer are added and plotted as a function of time on a flat-bed recorder. Two 1/1000e dial strain gages, also diametrically opposed, were used in order to check effect and to calibrate the electrical signal of the inductive transducer.

The creep temperature was between 1300°C and 1700°C and was reached within 3-4 hours. A supplementary hour was allowed for the temperature to stabilize. The loading sequence was always increasing and done at one temperature. Some experiments were run using only one or two loads to check that in the loading sequence the previous deformation had no, or only few, influence on the measured strains.

After deformation, the specimens were observed in a Jeol 20TS scanning electron microscope.

RESULTS

As these ceramics creep by cavitation processes, a pseudo steady-state was assumed, if, after the marked transient state, the

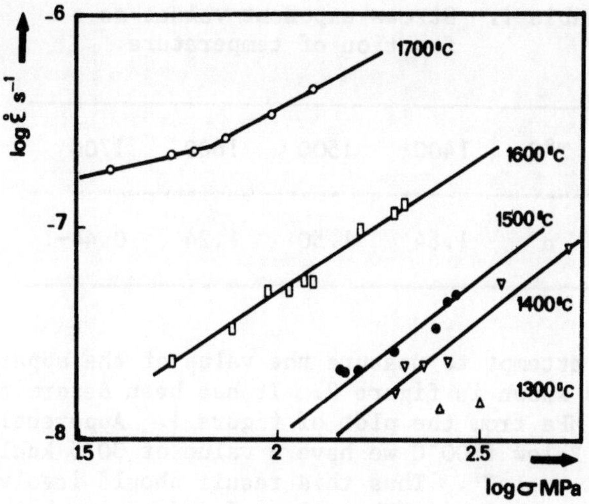

Fig. 1. Strain-rate as a function of applied stress and temperature.

creep rate revealed to be constant during some hours. We are aware that it may solely correspond to the final phase of the primary creep or to the direct transition from primary to tertiary creep. The results are shown in figure 1, as a log $\dot{\varepsilon}$ - log σ plot for all temperatures. Due to practical limitations only strain-rates higher than 10^{-8} s^{-1} could be resolved. The applied load range varies thus between 0.4 - 0.8 of the fracture loads which were previously determined (2). At 1300°C very few creep was observed and the specimen ruptured mostly rapidly. Some values, from the most favorable cases are reported as a reference. It must be recalled here that the fracture stress and toughness of these materials remain constant up to and decrease at temperatures higher than 1000°C, giving thus evidence of softening of the glassy phase.

We describe the strain-rate stress behavior as following :

$$\dot{\varepsilon} = A\,\sigma^n \exp(-Q/RT)$$

where : $\dot{\varepsilon}$ is the measured strain-rate.
σ is the applied stress.
Q is the apparent activation energy.
A is a constant which may contain microstructure parameters.
R and T have the usual meaning.

The measured values of the stress exponent, n, are reported in Table 1. In a first step viscous creep with n = 1 may be expected. It is however only observed at 1700°C. For the other temperatures, n increases with decreasing temperature.

Table 1. Stress exponent values as a function of temperature.

T°C	1400	1500	1600	1700
n	1.64	1.50	1.24	0.44-1

A first attempt to measure the value of the apparent activation energy is shown in figure 2. It has been determined at a stress of 250 MPa from the plot of figure 1. Apparently, two domains exist. Below 1500°C we have a value of 30.5 kcal. $mole^{-1}$ and above 100 kcal. $mole^{-1}$. Thus this result should involve two completly different mechanisms. The values found are much lower than those reported for Si or C self diffusion by Hong (170-190 kcal. $mole^{-1}$) (10)(11). They are also lower than the 145 kcal/mole or 170-350 kcal/mole reported for creep in the same temperature range (12)(13). It is thus concluded that diffusion phenomena in the SiC phase are not involved in the creep of our materials.

On the other hand they may be compared to the 110 kcal. $mole^{-1}$ reported for diffusion of aluminium in SiC (14), or the 30-35 kcal $mole^{-1}$ reported by Tajima and Kingery (15)(16) for the segregation

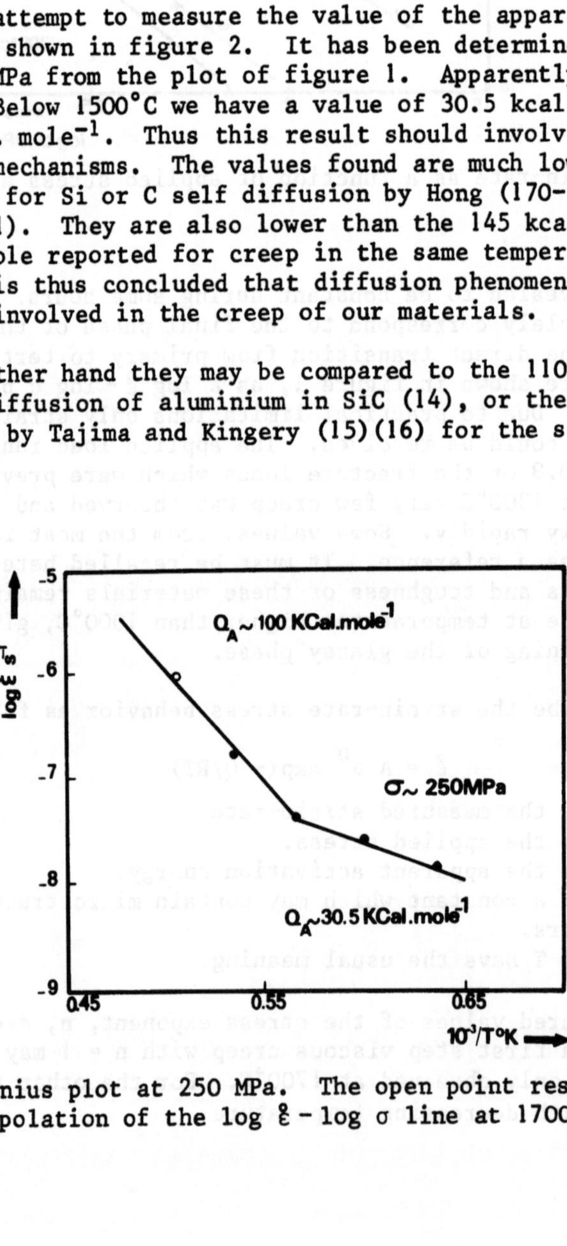

Fig. 2. Arrhenius plot at 250 MPa. The open point results from an extrapolation of the log $\dot{\varepsilon}$ - log σ line at 1700°C in Fig. 1.

of aluminium on the grain boundaries. However these mechanisms are not likely to be involved in the deformation process.

On the other hand cavitation is believed to be a major factor in the high temperature deformation of such ceramics. Rana and Evans (4) and Hasselman and Venkateswaran (5)(6) have proposed models to explain the apparent activation energy to be rather low. The next step will thus be the SEM observation of deformed and/or broken specimens in order to get qualitative features to explain the observed variation.

SCANNING ELECTRON MICROSCOPIC OBSERVATIONS

The most commonly proposed cavitation mechanism has the following sequence :

- due to grain boundary sliding high stresses arise at triple point boundaries, a bubble appears thus in the glassy phase, increases in size and propagates across the grain boundary through diffusion processes. The controlling event in this case seems to be the extension of the cavity across the grain boundary.
- if the grain boundary is subjected to tensile stresses, the grains may separate and create a penny-like crack which also may propagate. This mechanism however involves the grains to be completly encompassed by a thin layer of the amorphous phase. The claim of Raj (8) that this may be the case only for higher volumic ratios of second phase is confirmed in our case by fractographic observations of specimens fractured at room temperature (1)(2). Fracture is completely intergranular and second phase located on grain facets is scarcely observed ; it is mainly present in the triple point boundaries and the four-grain channels.

Keeping this features in mind the high temperature deformation and/or failure characteristics will be examined :

1300°C : As mentionned, at this temperature most of the specimens fractured within short times after loading. Rupture seems to be still brittle since the main fracture and secondary cracks are sharp edged (Fig. 3a). Independent cracks seem also to have propagate in competition with the main crack (Fig. 3b). If the specimen deforms a high density of grain separation occurs (Fig. 3c). Cavitation at triple point boundaries is very scarce.

1400°C : The same features as at 1300°C are observed. Damage appears to be more generalized. Figure 4 is a fractograph of a specimen which fractured during deformation. The smooth fracture surface which appears at room temperature is not longer present. The specimen had deformed before rupture.

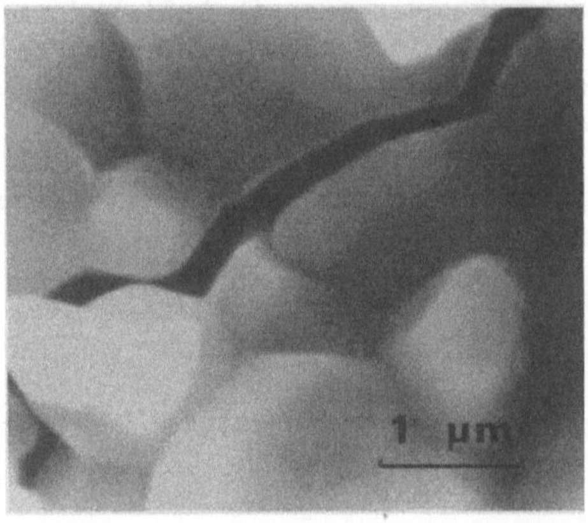

Fig. 3a. Fracture surface of a specimen subjected successively to stresses of 220, 250, 320, 340 and 430 MPa. It broke after 2 hours at 1300°C under 430 MPa.

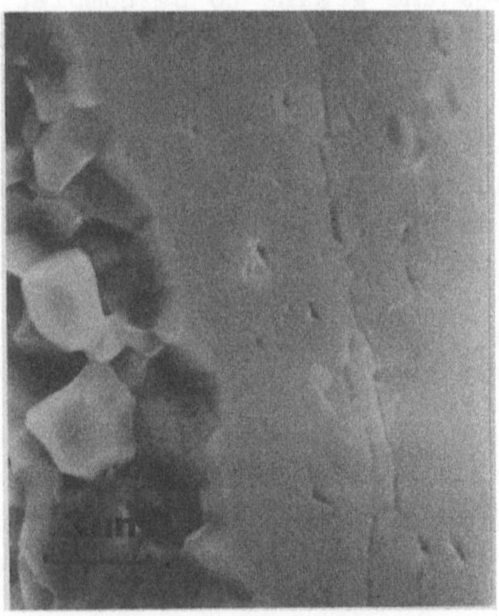

Fig. 3b. Same sample as in Fig. 3a after polishing.

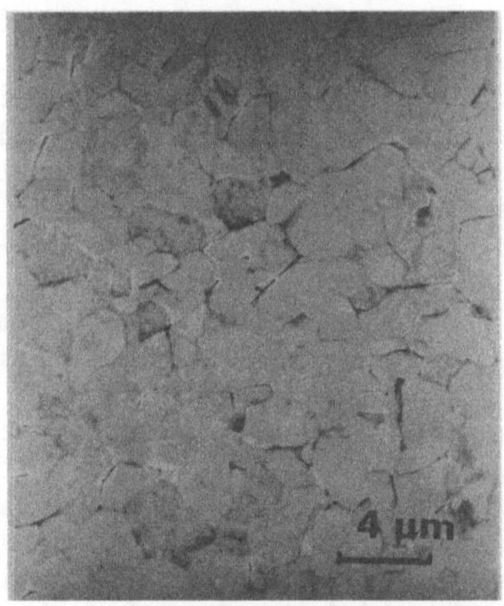

Fig. 3c. Polished and etched tensile surface of a specimen loaded successively to 200, 277 and 416 MPa at 1300°C.

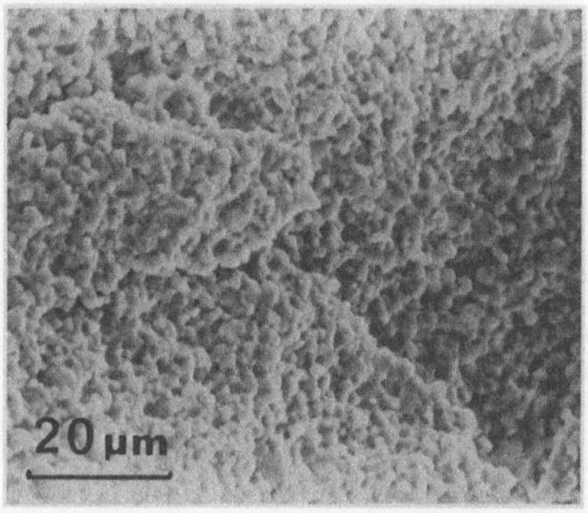

Fig. 4. Fracture surface of a specimen which broke at 1400°C while the stress was raised from 260 to 415 MPa.

1500°C : At this temperature the cavitation phenomenon becomes more pronounced (Fig. 5a). Cavitation at triple point boundaries and separation of grain facets are simultaneous (Fig. 5b). The overall cavitation is shown in figure 5c.

1600°C : Figures 6a and 6b show the influence of deformation on the polycrystalline structure. If the specimen fractures during loading, the grain boundaries are still sharp edged although nuclei of cavitation are visible. If on the other hand, the material fractures after some amount of deformation, cavitation and grain separation are fully developed. Furthermore the second phase seems to be more distributed and the grain edges rounded. Examination of each cavity shows that the starting point is a triple point boundary. The grain boundary separation which exists at 1300°C and 1400°C are no longer present. If grain facet separation occurs it has the final form shown in figure 6c. After deformation the specimen has been polished and etched (boiling Murakami solution), hence the glassy phase should have been removed and facet separation should result in an elongated hole of facet length.

1700°C : At this temperature the specimens may be heavily deformed. Figure 7a is a view of a specimen which has been deformed up to 20 %. Cavitation is mainly located on triple point boundaries. Furthermore if one considers the deformation amount, the specimen appears to be less damaged than at 1500°C (see Fig. 5c) and is mainly located at triple point boundaries. Facet separation is very scarce (Fig. 7b).

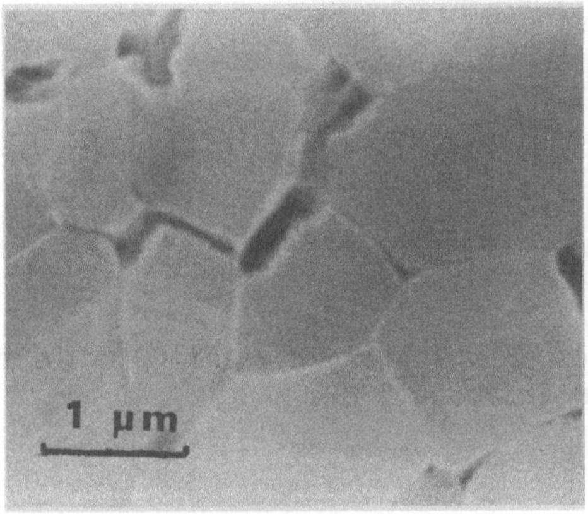

Fig. 5a. Specimen deformed at 1500°C under 150 MPa ($\dot{\varepsilon}$ = 2.10^{-8} s^{-1}) then slightly polished and etched.

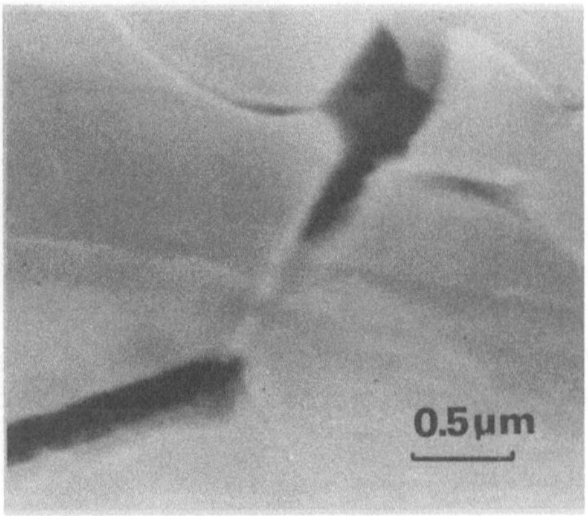

Fig. 5b. Same specimen as in figure 5a. Note simultaneous triple point cavitation and facet separation.

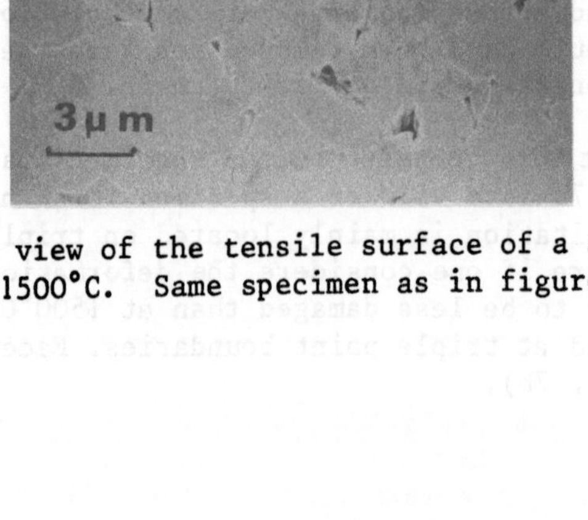

Fig. 5c. Overall view of the tensile surface of a specimen deformed at 1500°C. Same specimen as in figure 5a and 5b.

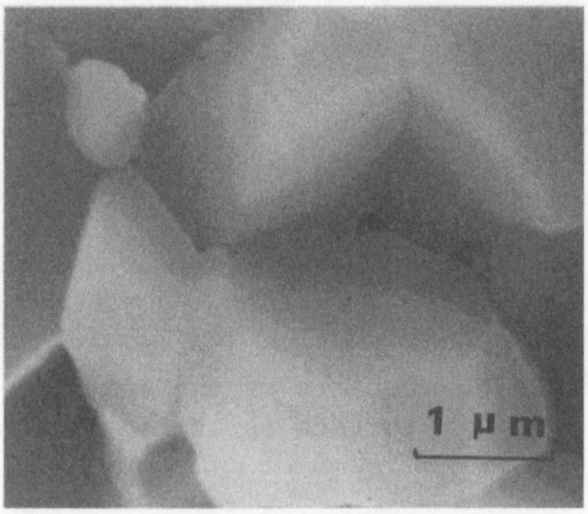

Fig. 6a. Fracture surface of a sample fractured at 1600°C without previous creep during first loading to 160 MPa.

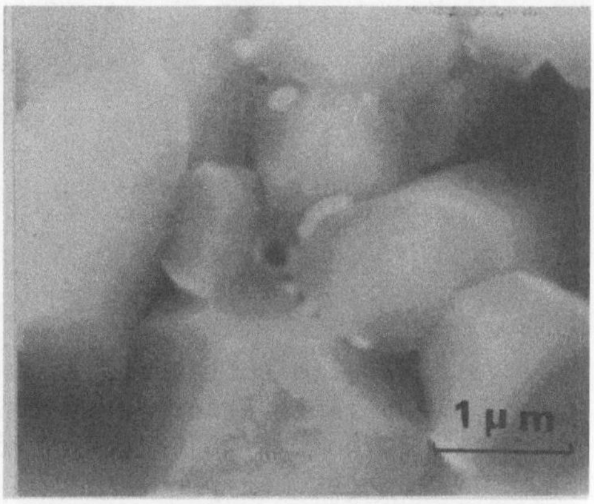

Fig. 6b. Fracture surface of a specimen fractured at 1600°C. It has been loaded to 160, 190, 210 and 260 MPa. It broke after 1 hour at 160 MPa.

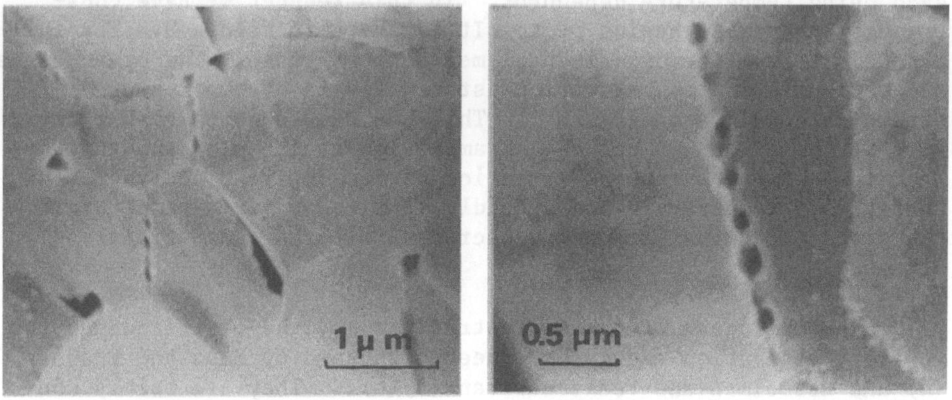

Fig. 6c. Details of the tensile surface after polishing and etching of specimen in figure 6b.

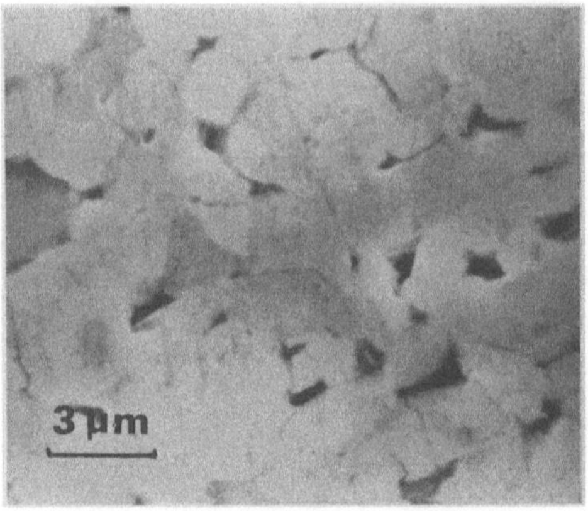

Fig. 7a. Tensile surface of a specimen deformed up to 20 % at 1700°C (polished and etched).

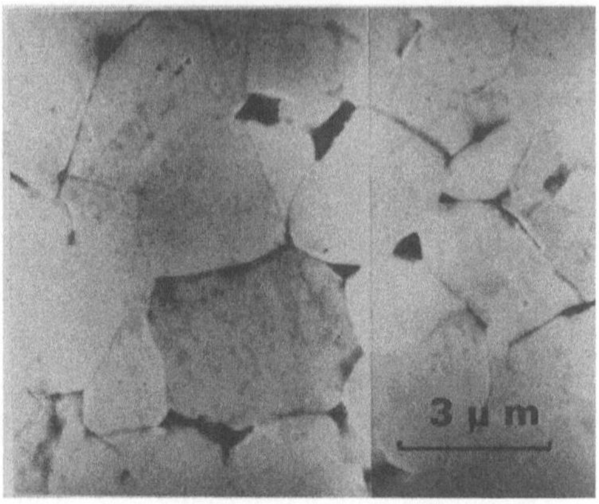

Fig. 7b. Details of the same surface as in figure 7a.

DISCUSSION

In view of the SEM observations it appears that several mechanism are activated during the deformation process, the amount of each being temperature dependent. In this respect we have constructed another Arrhenius plot. It is shown in figure 8. The full points are from the experimental measurement range. The open points are extrapolation ot the various stress levels of the measured strain-rate values in figure 1. This has been done to ensure the form of the Arrhenius curves. Examination of the plot shows first, that there is a continuous variation of the apparent activation energy with temperature and secondly that for a given temperature the apparent activation energy decreases with increasing applied stress.

The influence of the microstructure and the cavitation on the apparent activation energy have been discussed by Rana and Evans (4) and Hasselman and Venkateswaran (5)(6). They are mainly concerned with the superposition of anelastic as well as viscoelastic phenomena through diffusion processes of the second phase superposed to an elastic creep due to the existence of the cavities. The development of the cavities increases the compliance of the material and causes a supplementary amount of deformation. The resulting

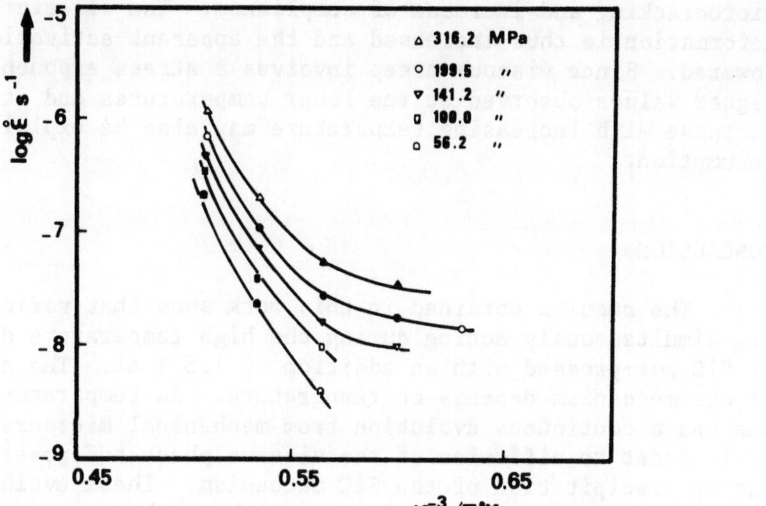

Fig. 8. Arrhenius plot at various stresses. The open point are extrapolation to the choosen stress of the plots in figure l.

apparent activation energy is thus lower than the activation energy of the diffusion processes.

Precise values of diffusion and viscosity activation energies concerning the amorphous phase are missing. The discussion may thus be only qualitative. At temperature higher than 1500°C all the mechanisms caused by the second phase are present. The influence of the applied stress is also minimized. Furthermore the stress exponent becomes closer to 1, which indicates viscous creep. At temperatures lower than 1500°C the cavitation phenomenon in the usual meaning is less pronounced. The grains are separated without cavitation at triple point boundaries, so an another mechanism than cavity formation and extension should be involved. One thinks first on the subcritical crack growth which could increase the compliance. However in the examined stress range the size of the critical crack is small owing to the measured toughness and lies between 50 and 100 μm for a semi-circular flaw. The elastic displacement which would result is very low and cannot be detected.

Owing to the presence of the observed facet rupture at the lower temperatures and the fact that the second phase is not continuous, we assume that great part of the increase in compliance is due to the mechanical fracture of grain boundaries. This mechanism is assumed to be not thermally activated and explains the very low apparent activation energies at 1300°C and 1400°C and the influence of the applied stress. Higher applied stresses mean more mechanical microcracking and increase of compliance. The elastic component of deformation is thus increased and the apparent activation energy lowered. Since viscous creep involves a stress exponent of 1 the higher values observed at the lower temperatures and its continuous decrease with increasing temperature may also be explained by this assumption.

CONCLUSION

The results obtained in this work show that various mechanisms are simultaneously acting during the high temperature deformation of SiC hot-pressed with an addition of 1.5 % Al. The predominence of one mechanism depends on temperature. As temperature increases one has a continuous evolution from mechanical microcracking of the grain facet to diffusion of the viscous phase and possibly a dissolution precipitation of the SiC mechanism. These evolutions with temperature and the extrapolation which can be done on the activation energy and the deformation damage seem to be confirmed by the work of Carry and Mocellin (17) in a higher temperature range 1850°-2100°C.

AKNOWLEDGMENTS

The authors thank the Elektroschmelzwerk Company for kindly providing the SiC material.

REFERENCES

1. R. Moussa, Thesis, University of Caen, France (1981).
2. R. Moussa, J.L. Chermant and F. Osterstock, Proc. Brit. Ceram. Soc., 32;261 (1982).
3. A.G. Evans, Acta Met., 28;1155 (1980).
4. A.G. Evans and A. Rana, Acta Met., 28;129 (1982).
5. D.P.H. Hasselman and A. Venkateswaran, J. Mat. Sci., 18;161 (1983).
6. D.P.H. Hasselman, A. Venkateswaran, J. Amer. Ceram. Soc., 66;C36 (1983).
7. R.L. Tsai, R. Raj, J. Amer. Ceram. Soc., 63;513 (1980).
8. R.L. Tsai, R. Raj. Acta Met., 30;1043 (1982).
9. G. Nouet, S. Hagège, R. Moussa, F. Osterstock, J. Vicens and J.L. Chermant, Rev. Int. Htes Temp. Réfract., 19,335 (1982).
10. J.D. Hong and R.F. Davis, J. Amer. Ceram. Soc., 63;546 (1980).
11. J.D. Hong, M.H. How and R.F. Davis, Ceram. International, 5 (1979).
12. A. Djemel, Thesis, University of Paris VI, France (1980).
13. G. Grathwohl, Th. Reets and F. Thummler, Science of Ceramics, 11;425 (1981).

14. L.B. Griffiths, J. Appl. Phys., 36;571 (1965).
15. Y. Tajima and W.D. Kingery, J. Amer. Ceram. Soc., 65;C27 (1982).
16. Y. Tajima and W.D. Kingery, J. Mat. Sci., 17;2289 (1982).
17. C. Carry and R. Mocellin, this conference.

CRACK GROWTH IN Al_2O_3 WITH 3 WT% GLASSY PHASE

H.G. Schmid, T. Haug, A. Bornhauser, V. Gerold and R.F.Pabst

Max-Planck-Institut für Metallforschung and Institut für Metallkunde, Seestraße 71, D-7000 Stuttgart 1

INTRODUCTION

Previous measurements, e.g. Kromp et al.[1], Bornhauser[2], have shown that the fracture behaviour of alumina with 3 wt% glassy phase is of non linear elastic type at temperatures above 900°C. The measured fracture toughness K_{IC} or the subcritical K_I data heavily depend on test temperature as well as on the crack velocity. Difficulties appear with the physical interpretation as the relationship between the stress intensity K_I and the crack velocity $\dot{a}$ is not clear.

This paper, however, presents an energy concept instead of a stress intensity characterisation which gives a more reasonable interpretation of the reactions. The energy data reveal that the crack resistance R depends strongly on the crack velocity and in a negligible way on the crack length.

It is the aim of this paper to discuss the following effects by means of the experimental values:

1. the kind of the micromechanism, i.e. the creep mechanism and adhesive effects between the crack surfaces as the result of the viscous nature of the glassy phase;

2. the influence of the creep mechanism on the subcritical crack growth;

3. the influence of the adhesive viscous mechanism between the crack surfaces as an explanation for the observed high crack resistance.

EXPERIMENTAL

Method of measurement

The micromechanisms responsible for the material behaviour are not clear. So we can only use a method of measurement to which the material behaviour does not have an influence. An estimation of crack resistance R with the K-concept does not work first of all, because this relation is only usable having mainly linear elastic behaviour (see list of symbols).

$$R = K^2/E = \sigma^2 Y^2 a/E \tag{1}$$

The estimation of crack resistance R with the J-integral is only possible if time independent material behaviour is assumed[2,3].

$$R = J_R = \frac{1}{b} \left| dU/da \right| \tag{2}$$

It is for this reason that we do not make use of it. If there is an appreciable time dependence of the crack propagation energy Landes and Begley[4] propose the suitability of C^* integral:

$$C^* = \frac{1}{b} \left| \partial \dot{U}/\partial a \right| \tag{3}$$

Using equation (3) we get an equation for the crack resistance R at a constant crack velocity $\dot{a}$.

$$R = \frac{1}{b} \left| dU/da \right|_{\dot{a} = \text{const}} \tag{4}$$

Comparison with equation (2) reveals that we may use normal J-evaluation methods if $\dot{a}$ is constant. In the following we will use a method described by Ernst et al.[6,2] which does not propose anything about the material behaviour and may be used for experiments with constant crack velocity.

Testing procedure

Three point bend specimens having an artificial notch of width 70 µm were tested in a displacement controlled manner at temperatures of 1000 and 1100°C in air. The displacement rate was varied between 0.3 and 30 µm/min. The specimen dimension were 35 x 7 x 2.3 mm^3 and the span width was 30 mm. The displacement of the specimens was directly measured with an inductive displacement pick-up which controls the testing machine. Some material parameters are listed in the following table:

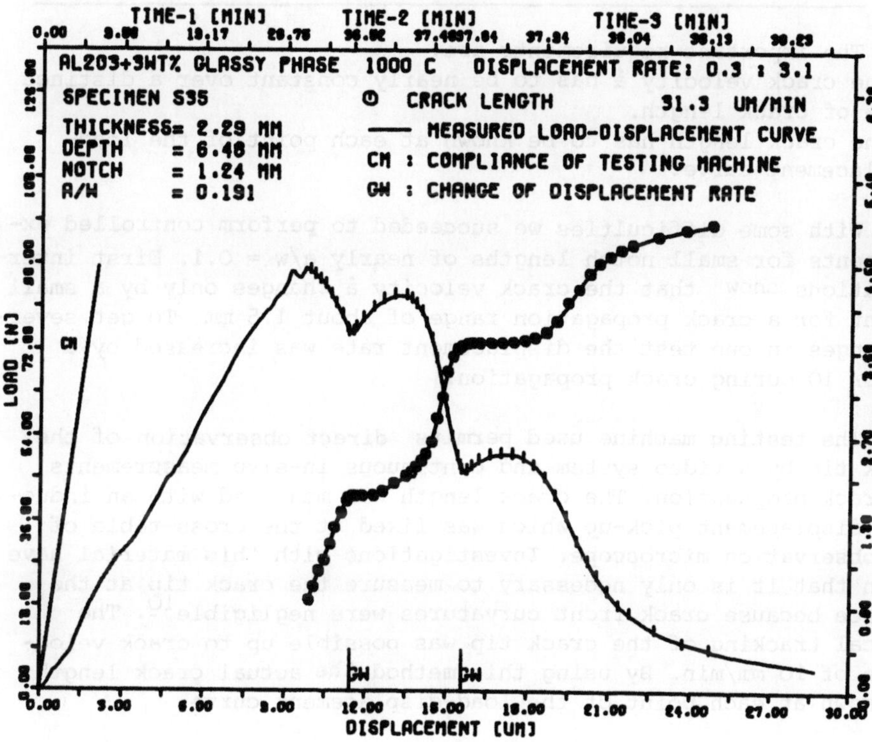

Fig. 1. Load and crack length versus displacement at 1000°C

Table 1. Material parameters

Material	grain size µm	K_{IC} (20°C) $MPam^{1/2}$	E-Modul (20°C) GPa
Al_2O_3 + 3 wt% glassy phase	10-15	4.2 +/- 0.4	365

The important requirements are:
1) The crack velocity $\dot{a}$ has to be nearly constant over a distinct range of crack length.
2) The crack length has to be known at each point of the load-displacement curve.

With some difficulties we succeeded to perform controlled experiments for small notch lengths of nearly a/w = 0.1. First interpretations show that the crack velocity $\dot{a}$ changes only by a small amount for a crack propagation range of about 1.5 mm. To get several ranges in one test the displacement rate was increased by a factor 10 during crack propagation.

The testing machine used permits direct observation of the crack tip by a video system and continuous in-situ measurements of crack propagation. The crack length was measured with an inductive displacement pick-up which was fixed at the cross-table of the observation microscope. Investigations with this material have shown that it is only necessary to measure the crack tip at the surface because crack front curvatures were negligible[10]. The optical tracking of the crack tip was possible up to crack velocities of 10 mm/min. By using this method the actual crack length is known at each point of the load-displacement curve.

Experimental results

Fig. 1 shows the load and crack length as a function of displacement at 1000°C and a displacement rate of 0.3 µm/min. During the test the displacement rate was increased twice by a factor of 10 (marked with GW in Fig. 1). After each increase the crack stops until the load has reached a higher level. Thereafter the load starts to decrease again and the crack proceeds with a higher velocity.

In Fig. 2 the crack growth energy U, evaluated from the test of Fig. 1, and the crack velocity $\dot{a}$ are plotted versus the normalized crack length. In this case the crack growth energy U is the work done irreversibly in the specimen. For the ranges with constant $\dot{a}$, U(a) is a linear function (straight lines in Fig. 2). The crack resistance R increases with the crack velocity $\dot{a}$. Furthermore, Fig. 2 shows that for each change in displacement rate the energy U increases until changing over into the next linear range. It is possible to calculate to a certain crack velocity $\dot{a}$ the corresponding crack resistance R from the slope of the linear increase. This is only meaningful for crack extensions of $a/w < 0.85$ (limited specimen dimensions). In Fig. 2 only the first two linear a/w-ranges are interpretable.

In Fig. 3 a double-logarithmic plot of R versus $\dot{a}$ is shown for several tests at 1000 and 1100°C. The values for R range between 30 and 120 J/m^2. The dependence is a linear one for both temperatures with the same slope. The correlation of the data is quite satisfactory in respect of the inhomogenious structure of the material.

DISCUSSION

Possible micromechanisms

The crack propagation experiments have shown a considerable deviation from linear elastic fracture mechanics. In addition, the crack resistance R depends considerably on the crack velocity $\dot{a}$ but not on the crack length. It reaches high values of the order of 120 J/m^2 at $\dot{a}$ = 10 mm/min (0.6 m/h) and T = 1100°C. There should exist an energy consuming mechanism which is connected with the viscous properties of the glassy phase. Two different effects are possible. At first, local creep deformation can occur due to the high stresses in front of the crack tip. Secondly, from an inspection of the crack surface a pull-out of the glassy phase has been found. This may occur only after the crack front has passed these areas. Obviously, the glassy phase links the crack surfaces and has to be torn off during further crack propagation. Both mechanisms consume energy and will be discussed in the next two chapters.

Crack growth under creep conditions

The upper part of Fig. 4 shows an idealized presentation of the grain structure and the interconnected glassy phase. An applied stress σ effects viscous flow of the glassy phase which increases the thickness of the layers normal to the stress axis at the expense of the parallel ones. This causes a net creep rate $\dot{\varepsilon}$ of the compound system. According to Drucker[7] there exists a relationship between the mean stress $\bar{\sigma}$ and the growth rate $\dot{h}$ of a normal glassy

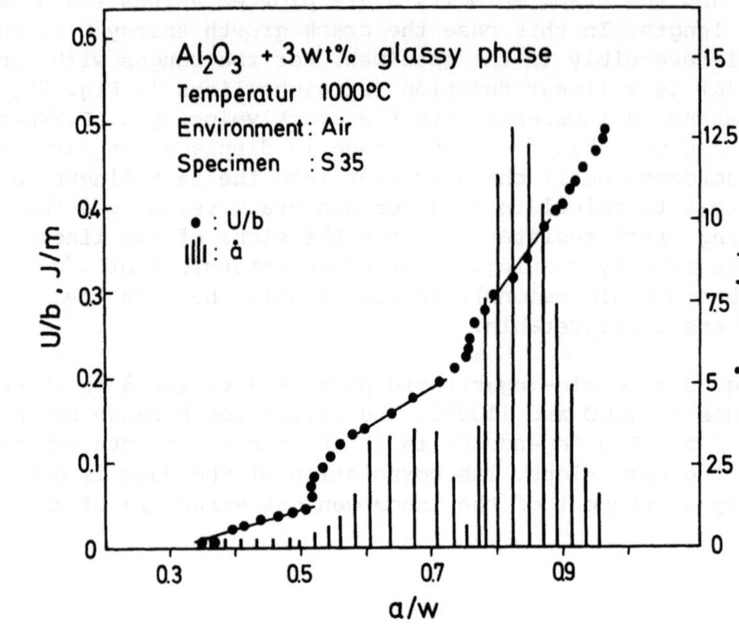

Fig. 2. Crack growth energy U, evaluated from the test of Fig. 1 and the crack velocity ȧ versus normalized crack length

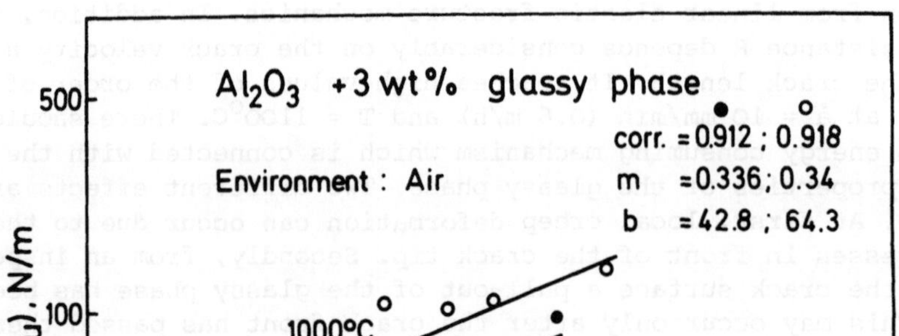

Fig. 3. Double-logarithmic plot of crack resistance R versus ȧ for several tests at 1000 and 1100°C

layer:

$$\bar{\sigma} = \alpha\eta\dot{h}d^2/h^3. \quad (5)$$

In this equation α is a numerical constant depending on geometric details of the model, η is the viscosity, d the grain size and $\dot{h}$ the original thickness of the glassy layers. From h the creep rate $\dot{\varepsilon}$ can be calculated with the Norton equation[8]

$$\dot{\varepsilon} = B\sigma^n \quad (6)$$

with

$$B = (h/d)^3 \ (1/\alpha\eta)\ ; \ n = 1.$$

Eqs. (5) and (6) show the strong influence of the microstructural parameters d and h on the stress-strain rate.

From creep experiments at 1100°C using the same material a value for B could be derived:

$$B = 1 \times 10^{-6} \ MPa^{-1}h^{-1}. \quad (7)$$

Using the values of the grain size and the volume fraction of the glassy phase the structural parameters are

$$d = 10 \ \mu m \ ; \ h = 0.1 \ \mu m. \quad (8)$$

The value for h represents an upper limit because a large amount of the glassy material is concentrated in triple points between the ceramic grains which has not been taken into account. From Eqs. (6) to (8) it follows for the viscosity

$$\eta \leq 0.5 \ MPa \ h. \quad (9)$$

This value is in the same order of magnitude as the viscosity for such glasses given by Scholze[9]. Thus the proposed viscous creep mechanism would be responsible for the creep behaviour.

To estimate the possibility of creep effects during crack propagation Riedel[5] gives a characteristic time t_R which has to be exceeded by the experiment:

$$t_R = K_I^2/((n+1) \ C^*E) \quad (10)$$

The quantity C^* is approximately K_I^2B in our case and E is the Young modulus. This leads to $t_R \simeq 1$ h. Compared to the experiments only the test with the slowest $\dot{a}$ reaches this time. All other test times are considerably smaller. Therefore, the creep mechanism seems

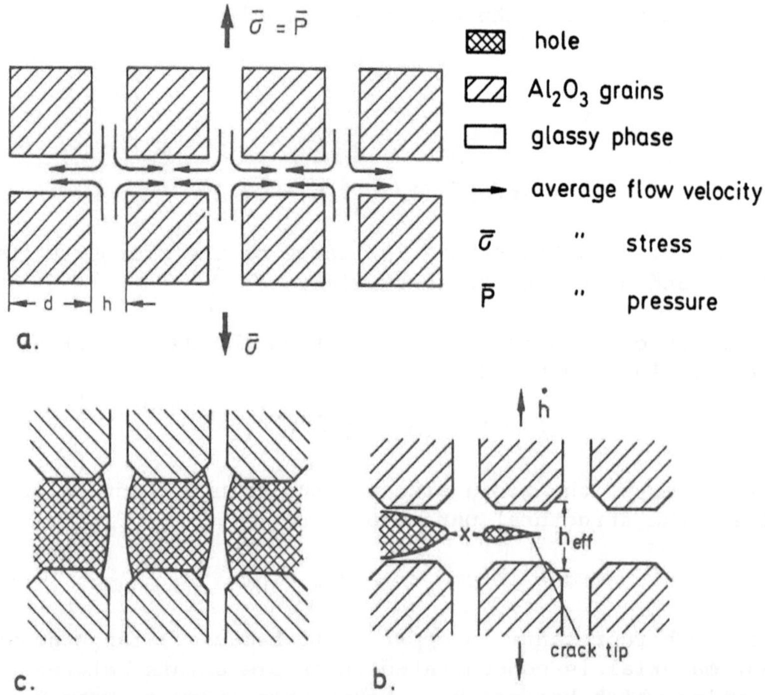

Fig. 4. Two dimensional schematic model of a ceramic material with glassy phase

a. Creep behaviour without a crack tip.
   The tensile stress $\bar{\sigma}$ induces viscous flow of the glassy phase which increases the thickness h of the horizontal layers at the expence of the vertical ones.
b. Evolution of the glassy islands at the crack tip.
   The fracture stress in the triple points is not reached and therefore remains an unbroken glassy island behind the crack tip.
c. The situation of the glassy islands far from the crack tip.

to be negligible in the present experiments.

Crack growth under influence of adhesive effects between the crack surfaces

As mentioned above the viscous phase is pulled out of the grain boundaries normal to the crack surface. This viscous flow uses up energy which will be estimated in this chapter.

At very high propagation rates $\dot{a}$ (much higher than in the present study) there is no time left for viscous flow. The local stresses at the crack tip become very large and the crack will propagate in a quasi brittle manner. When the rate $\dot{a}$ is lowered the local stress in some areas in the crack front like triple points are holes filled with glassy phase (Fig. 4) may have lowered by viscous flow such that the fracture stress is not reached in these areas. They remain as unbroken islands behind the crack front and will be pulled out until failure during further crack opening.

For the following rough estimate of the viscous flow in these islands Eq. (5) has been used. It relates the size of these islands (x_o and h_{eff}) to the crack opening rate h and the resulting average stress $\bar{\sigma}_i$ in their cross section (Fig. 4). The following equation has been taken from the two-dimensional creep model by Drucker[7] where it is assumed that only the shape factor α of Eq. (5) is altered by the totally different boundary conditions of the present problem. The following relation should be valid:

$$\bar{\sigma}_i = \alpha\eta\ (x_o^{\ 2}/h_{eff}^{\ 3})\ \dot{h} \tag{12}$$

If the rate $\dot{h}$ is assumed to be proportional to the crack velocity $\dot{a}$

$$\dot{h} = \beta\dot{a} \tag{13}$$

an expression for the crack resistance $R_{p\ell}$ can be derived:

$$R_{p\ell} = \frac{1}{4}\ \alpha\beta\eta g(\dot{a})\,\dot{a} \tag{14}$$

with

$$g(\dot{a}) = f_o x_o^{\ 2}/h_{eff}^{\ 2}.$$

It can be shown that the main contribution to $R_{p\ell}$ comes from islands immediately behind the crack front because h is increasing with the distance from the crack tip. This results in a marked decrease of the area fraction and of σ_i.

For very large crack velocities $\dot{a}$ the parameter g becomes zero since $f_o = 0$ as discussed before. With decreasing $\dot{a}$ g is increasing and finally reaches a more or less constant value. For $R_{p\ell}$ this results in a maximum value for a distinct propagation rate $\dot{a}_c$. It is supposed that our experiments have been performed at rates $\dot{a} < \dot{a}_c$. To check the order of magnitude for $R_{p\ell}$ and σ_i the following set of data have been used: $\alpha = 1$, $\beta = 10^{-3}$, $\eta = 0.5$ MPa h, $\dot{a} = 0.1$ m/h, $f_o = 6\%$, $x_o = 2$ µm, $h_{eff} = 0.3$ µm. The estimates according to Eqs. (12) and (14) are $\sigma_i = 7400$ MPa and $R_{p\ell} = 34$ J/m^2.

The values used for f_o, x_o and h_{eff} are estimates from preliminary TEM investigations. The input data as well as the estimated values for $R_{p\ell}$ and σ_i are quite reasonable. For example, the fracture stresses for glassy phases are given[9] in the order of 10000 MPa.

CONCLUSIONS

1. The subcritical crack propagation in Al_2O_3 ceramics containing 3 wt% of a glassy phase at high temperatures can be described by a crack resistance R which increases considerably with increasing crack propagation rate $\dot{a}$.

2. The value of R can be derived experimentally from a load-displacement curve of a three point bending test at a constant displacement rate. In addition, an in-situ measurement of the crack length is necessary.

3. The relatively high increase of R (30 to 120 J/m^2) cannot be explained by creep effects in front of the crack tip ($K_{IC}^2/E \simeq 30$ J/m^2)[10]. It is probably caused by the existence of many islands of the glassy phase bridging the crack immediately behind the crack tip which have to be torn off during further crack propagation.

4. A rough estimate is given for the contribution of this mechanism to the crack resistance R. The observed increase of R with increasing $\dot{a}$ can be explained with this model.

ACKNOWLEDGEMENTS

The authors would like to thank Prof. R. Coble, MIT Cambridge, U.S.A., for helpful discussions and for reading the manuscript. This work was supported by the Deutsche Forschungsgemeinschaft.

REFERENCES

1. K. Kromp, R. F. Pabst, Z. f. Materialprüf. 22, 241 (1980)
2. A. C. Bornhauser, phil. thes. Univ. Stuttgart (1983)
3. J. R. Rice, G. F. Rosengren, J. Mech. Phys. Sol. 16, 1 (1968)
4. J. D. Landes, J. A. Begley, ASTM-STP, 590, 128 (1976)
5. H. Riedel, Advanced Seminar on Fracture Mechanics, Ispra (1981)
6. H. Ernst, P. C. Paris, J. D. Landes, ASTM-National Symposion on Fracture Mechanics, Philadelphia (Penn.) USA (1980)
7. D. C. Drucker in "High Strength Materials", Ed. F. Zackay, J. Wiley & Sons New York (1980)
8. G. E. R. Schulze, "Metallphysik", Akademie-Verlag Berlin (1974)
9. H. Scholze, "Glas", Springerverlag, New York (1977)
10. H. Wieninger, phil. thes., to be published

LIST OF SYMBOLS

a	crack length
$\dot{a}$	crack velocity
α	shape factor
b	width of specimen
B	empirical constant in creep equation
β	numerical constant
C^*	C^*-integral
d	diameter of ceramic grains
E	Young's modulus
$\dot{\varepsilon}$	creep rate
η	viscosity
f_o	area fraction of uncracked islands in the crack plane
h	thickness of glassy phase
h_{eff}	effective thickness of glassy phase
$\dot{h}$	velocity of seperating grains
J	J-integral
K	stress intensity factor
$\underline{n}$	exponent of creep equation
p	average pressure
R	crack resistance
σ	applied stress
$\bar{\sigma}$	average stress in creep regions
$\bar{\sigma}_i$	average stress in islands during viscous flow
t_R	typical time for creep crack growth
U	crack growth energy
x_o	width of glassy islands
Y	correction function

CRACK RESISTANCE OF PREDEFORMED NaCl SINGLE CRYSTALS

M. Breucker and R. Steinbrech

Lehrstuhl für Werkstoffwissenschaften A
Abteilung Chemietechnik, Universität Dortmund, F.R.G.

ABSTRACT

The resistance (R) of NaCl against crack propagation on {100} planes was measured on double torsion (DT) specimens, where fracture is preceeded by large scale yielding. The influence of plastic predeformation on R was investigated in detail by producing different dislocation substructures for the propagating crack. As crack growth is observed to occur in individual steps R was determined for each step by comparing the released energy with the corresponding increase in fracture area. The measured R-values (1-20N/m) indicate that the crack resistance is determined by the distribution, density and mobility of screw dislocations produced prior to crack propagation. The results are discussed in terms of an interaction between the moving crack and these dislocations. Two mechanisms are considered as a possible cause for the energy dissipation: the cutting of screw dislocations by the crack and, more effective, the movement of dislocations in front of the crack tip.

1. INTRODUCTION

The resistance (R) of NaCl against crack propagation on {100} cleavage planes has been investigated in the past by several authors /1-3/ mainly to determine the effective surface energy ($\gamma_{eff} = R/2$). All these fracture experiments were carried out with the double cantilever beam (DCB) test geometry using precleaved specimens and the effective surface energy was calculated from the critical load at catastrophic fracture, applying formulae derived under the assumption of linear elastic fracture mechanics (LEFM). Generally the resulting values for γ_{eff} differ from the specific surface energy γ and this is usually accounted for by an additive term γ_p

$$\gamma_{eff} = \gamma + \gamma_p \quad .$$

γ_p is attributed to the influence of plastic deformation in a near crack-tip region (small scale yielding).

Depending on the relative orientation of {100} fracture plane and active <110> {110} glide systems both positive and negative values of γ_p were observed /1,3/. While $\gamma_p>0$ has been assigned to plane stress deformation, $\gamma_p<0$ was determined in relation with plane strain conditions /1/.

However, in contrast to the small scale yielding regime, little experimental attention has been paid to the influence of macroscopic plastic deformation on subsequent crack growth. Thus it was the aim of this work to get information about the post yield fracture behaviour of NaCl single crystals and thus to gain insight into the interaction mechanisms between a growing crack and obstacles as represented by the dislocation structure.

2. EXPERIMENTAL

The fracture experiments were carried out in air at room temperature using a double torsion (DT) arrangement, which for the purposes of this work offers important advantages compared to the DCB-geometry: i) pure NaCl-crystals show post yield fracture ii) the amount of plastic deformation during the DT-experiment can be influenced significantly by deformation and/or irradiation prior to testing iii) the crack advances in discrete steps parallel to the longest specimen dimension which enables multiple measurements of R within a single specimen.

The NaCl single crystals have been supplied by the "Kristall Labor SFB126" (Goettingen/Clausthal, FRG). Original crystals, grown in air from p.a. starting material by the CZOLCHRALSKI method were cut into plates with {100} surfaces with a string saw. Some of the plates were predeformed in tension up to 1% shear strain. The direction of this deformation relative to the later fracture plane and the corresponding denotations are introduced in Fig. 1a and Tab.1. In addition to predeformation, some of the specimens where irradiated by X-ray.

Prior to DT-testing a sharp notch was introduced into the specimens with a razor blade (Fig.1b). Loading of the specimens was performed at a constant crosshead speed (V_T = 0,33 x 10^{-6}m/sec).

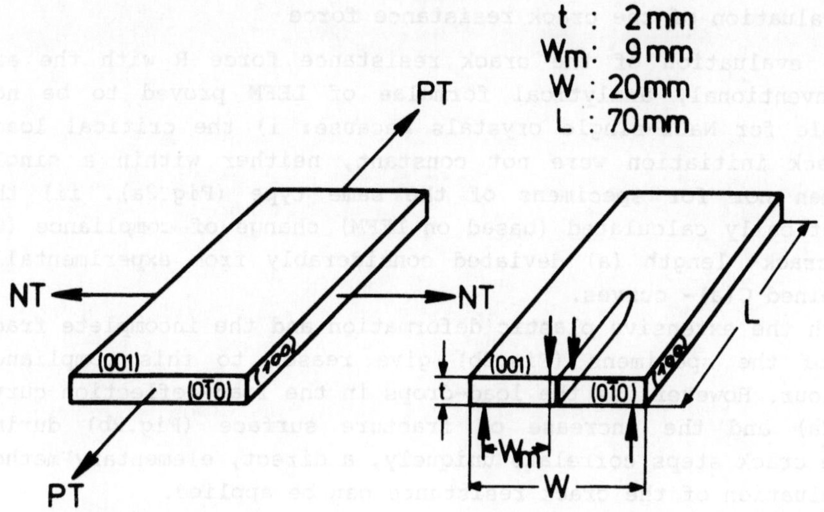

Fig.1a Predeformation under tension loading

Fig.1b Geometry of precleaved DT-specimen

Table 1. Pretreatment of DT Specimens

Type	Pretreatment
U	untreated crystal plates
PT	predeformation under tensional loading shear strain 1%; axis of tension parallel to crack plane
NT	same as PT, but axis of tension normal to crack plane
NTX	same as NT, but in addition X-ray irradiation ($Cu-K_{\alpha}$, 500 kR) before DT-test

After each of the larger crack steps, the crosshead was stopped and the specimen was partially unloaded to measure the actual compliance at the corresponding crack length.

The method applied for the evaluation of the R-values, (Sec. 3.1) required a microscopic analysis of the fracture surface. In addition the height of the "cleavage" steps was investigated using multiple beam interferometry after TOLANSKI /4/.

To analyze the glide systems, active during predeformation and/or DT-testing, the specimens were etched (etching solution after MORAN /5/). The identification of glide systems was based on a double trace analysis on two orthogonal {100} planes.

3. RESULTS

3.1 Evaluation of the crack resistance force

The evaluation of the crack resistance force R with the aid of conventional, analytical formulae of LEFM proved to be not feasible for NaCl-single crystals because: i) the critical loads at crack initiation were not constant, neither within a single specimen nor for specimens of the same type (Fig.2a). ii) the theoretically calculated (based on LEFM) change of compliance (C) with crack- length (a) deviated considerably from experimentally determined C(a)- curves.

Both the extensive plastic deformation and the incomplete fracture of the specimens (Fig.2b) give reason to this compliance behaviour. However, as the load-drops in the load-deflection curve (Fig.2a) and the increase of fracture surface (Fig.2b) during single crack steps correlate uniquely, a direct, elementary method of evaluation of the crack resistance can be applied.

The crack resistance force R was determined for individual crack steps by the ratio of dissipated elastic energy ΔU (from the load-displacement curve, Fig.2c) per fracture surface increase ΔF (from microscopic inspection)

$$R = \Delta U / \Delta F$$

The details of this method have been reported elsewhere /6,7/. It should be emphasized, that this procedure yields a mean value of R for one crack step, and characterizes the behaviour of crack propagation rather than that of crack initiation. Moreover, this method should be extended to other test-geometries providing the crack propagates in discrete steps.

At first, experiments were performed to measure such R-values for several crack steps within one specimen and thus to determine the scatter of the R-values and their significance for the material. Although ΔF varies by a factor of 4, the relation to ΔU, as plotted in Fig.3 for a single specimen, is linear with respect to the regression line and passes through the origin. The slope of this line determines the R-value within a specimen. As shown in Fig.4 there does not exist a systematic dependence of R on the deflection, i.e. the location of the crack step within the specimen length L does not influence the crack resistance. The elementary method therefore yields values of R characteristic for the material.

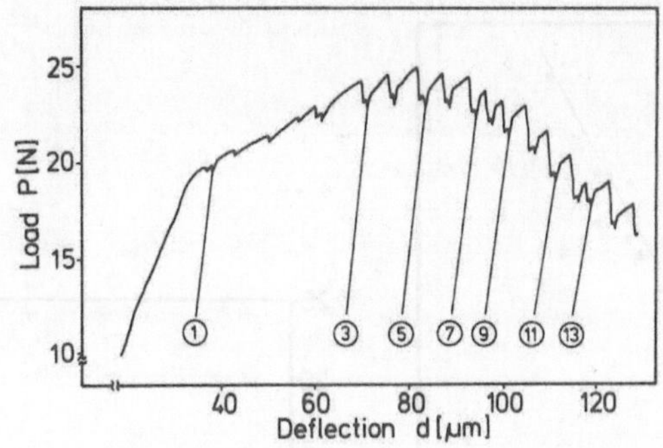

Fig.2a Load-deflection curve with stepwise crack growth. Compliance measurement by partially unloading (U-specimen).

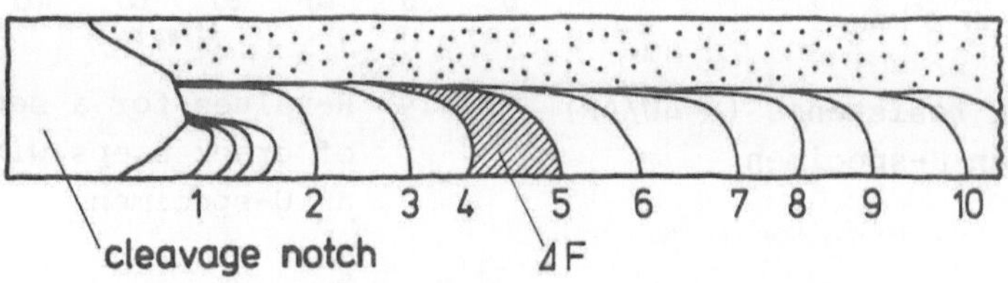

Fig.2b Fracture surface of the specimen from Fig.2a. ΔF: increase of fracture surface for a crack step.

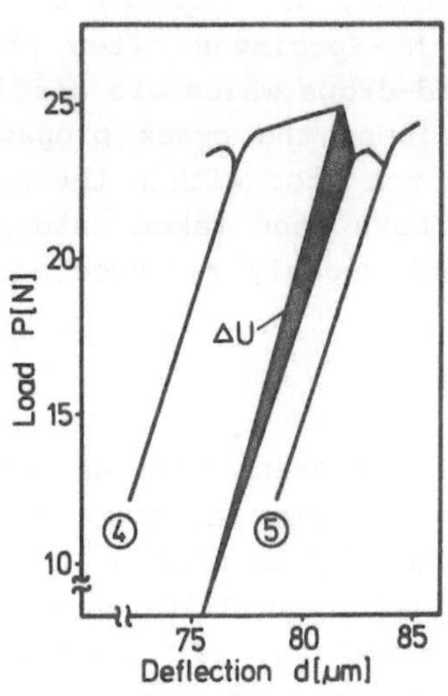

Fig.2c Determination of released energy ΔU for a crack step (enlarged section of Fig.2a)

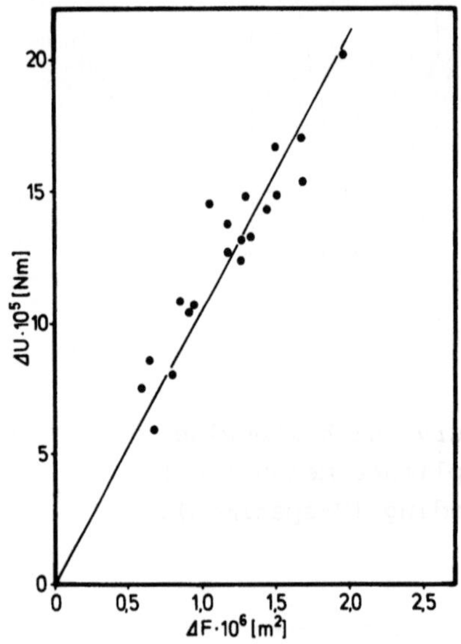

Fig.3 Crack resistance (R=ΔU/ΔF) for an U-specimen

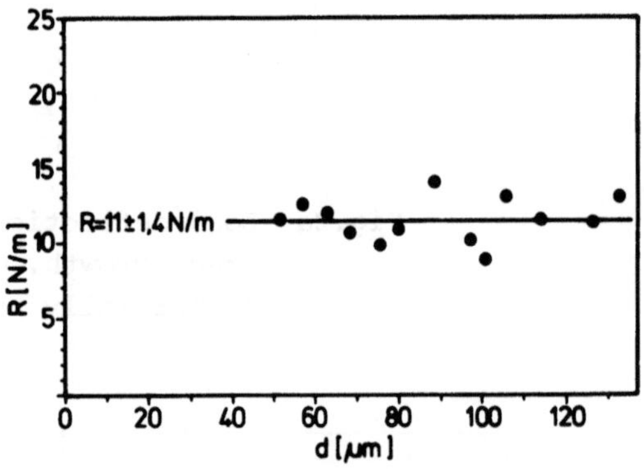

Fig.4 R-values for a sequence of crack steps within an U-specimen

Following this procedure, the R-values for differently treated specimens have been determined. However, as compared to U-specimens two major problems do arise. NT-Specimens often show very small crack steps and thus small load-drops which are difficult to resolve whereas in NTX- and PT-specimens the crack propagates in such large steps that often it does not stop within the specimen. In any case only those crack steps have been taken into consideration where ΔU and ΔF could be unambiguously measured and correlated.

3.2 Influence of predeformation.

The R-values, resulting from the different pretreatments, are listed in Table 2. All R-values exceed twice the specific surface energy ($R_o = 2\gamma = 0,6$ N/m /8/) and vary by as much as a factor of 20. The most striking fact is the significant difference resulting from the direction of predeformation (NT vs. PT) and irradiation (NT vs. NTX). However the different R-values are in accordance with the general observation, that R is mainly determined by the amount of plastic deformation during the DT-test.

Table 2. Influence of pretreatment on crack resistance

Crack steps	Type	R [N/m]
60	U	11
4 3	NT NTX	20 1,4
4	PT	1

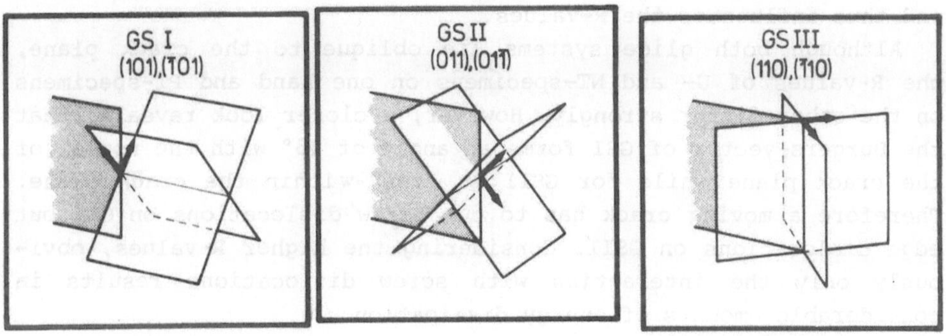

Fig.5 Orientation between (100) crack plane and <110>{110} glide systems

To rationalize this effect, the geometry of slip during predeformation and DT-testing has been analyzed. Fig.5 shows the three possible relations of a (100) crack plane with respect to the <110>{110} glide systems, any two orthogonal systems assigned as GSI, GSII and GSIII.

The etch pit study, performed for U-specimens at the beginning of deformation, with relatively low dislocation density, revealed that plastic deformation concentrates on GSI, covering the whole area between the upper loading points. According to WIEDERHORN et al./1/ this corresponds to a plane stress condition. GSIII, corresponding to plane strain conditions is thereby hardened latently.

In fact, extensive glide on GSIII could not be detected by

etchpit studies in any of the specimen-types. Since GSIII is thought to assist crack inition /3/, the fact that the R-values of this paper exceed 2γ, is in qualitative agreement with results in literature /1,3/.

The effect of predeformation is due to its influence on GSI. Because of the geometry of the thin crystal plates, PT-predeformation activates primarily GSII /9/ and hardens GSI and GSIII latently /10/ whereas NT-predeformation causes slip on GSI which is even increased during the DT-testing, probably due to a BAUSCHINGER effect.

So the influence of X-ray irradiation after predeformation (NTX) has to be seen by the impediment of slip on GSI during the DT-test.

It can be generalized as observed by the etch pit studies in the cracked specimens that either GSI (U, NT, NTX) or GSII (PT) provide the dislocation structure, that interacts with the crack and thus influences the R-values.

Although both glide systems lie oblique to the crack plane, the R-values of U- and NT-specimens on one hand and PT-specimens on the other differ strongly. However, a closer look reveals, that the Burgers-vector of GSI forms an angle of 45° with the normal of the crack plane while for GSII it lies within the crack plane. Therefore a moving crack has to cut screw dislocations on GSI but edge dislocations on GSII. Considering the higher R-values, obviously only the interaction with screw dislocations results in considerable amounts of energy dissipation.

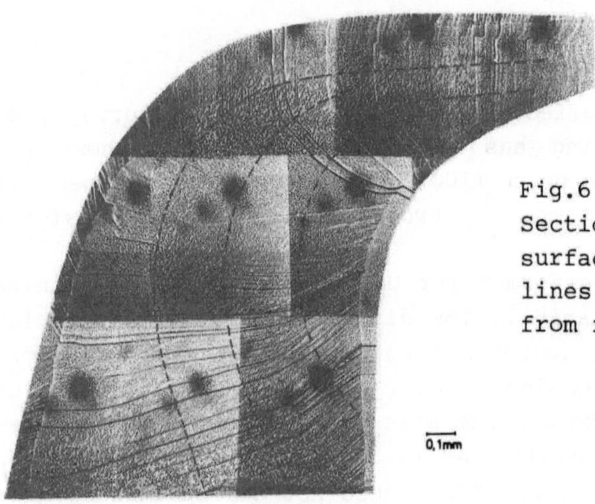

Fig.6
Section of etched fracture surface between crack arrest lines. Crack propagation from right to left

One reason for this effect could be the formation and growing of cleavage steps producing additional fracture surface. Specimens with high R-values do in fact exhibit a higher density of cleavage steps. As an illustration Fig.6 shows a section of the fracture surface (U-specimen) with such "cleavage" steps . The curved crack arrest lines mark the boundaries of one crack step. The shape of the propagating crack front can be constructed by curves normal to the "cleavage" steps. Three stages of crack growth are outlined. Multiple beam interferometry of such surfaces yielded mean "cleavage" step heights of $h \cong 0.1\mu m$ and step densities along the crack front of $\rho \cong 2\cdot 10^4\ m^{-1}$. Thus only an increase in fractured area of about 2°/o compared to the flat surface is observed.

4.DISCUSSION

Effects of crack initiation, which are supposed to be primarily caused by glide on GSIII cannot be prevailing in determining the R-values listed in table 2. This is due to the DT deformation arrangement with primary glide on GSI.

The results thus suggest that the R-values are determined by an interaction of the propagating crack with a preexisting dislocation substructure.

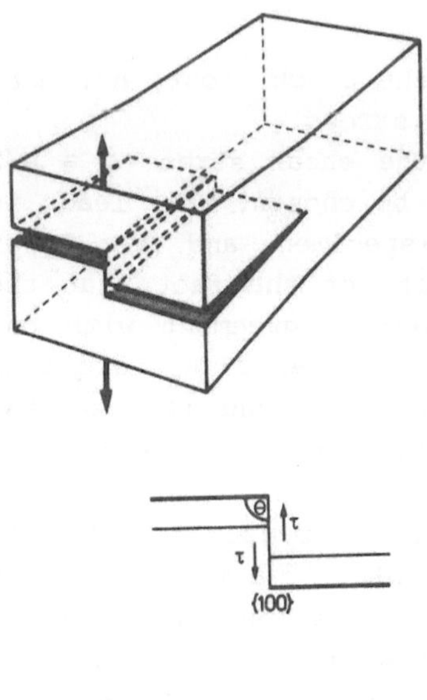

Fig.7 Shearing of cleavage steps

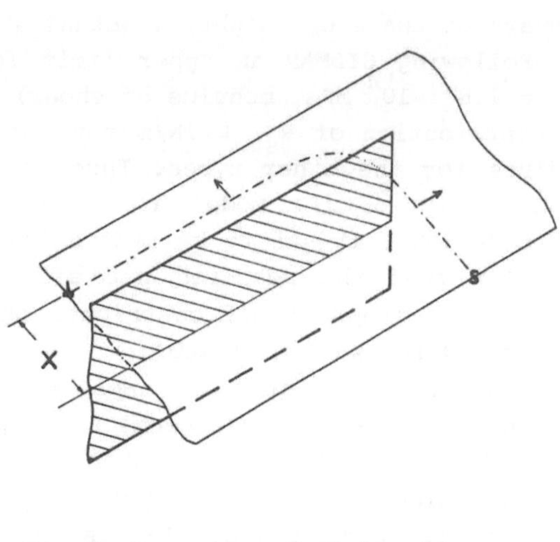

Fig.8 Dragging of dislocations (crack plane hatched)

Dislocations on GSI caused by deformation characteristic of plane stress conditions, seem to provide the most effective obstacles to crack growth. As illustrated by Fig.5 screw dislocations on this glide systems lie crosswise to the crack front at an angle of 45°.From the results of the previous section it has to be concluded that the density and mobility of these screw dislocations determines the R-value.

This statement is based on a comparison of the different types of specimens. The density of mobile screws on GSI decreases in the order NT-, U-, PT-specimens, which correspondends to a decrease in R respectively, and a diminution of their mobility (NT-vs.-NTX-specimens) lowers R too.

Thus two mechanisms of energy dissipation have to be taken into account: i) the cutting of the oblique screw dislocations by the crack and ii) the dragging of the screw dislocations in the stress field of the advancing crack.

As has already been shown, the increase in fracture area by cleavage steps can safely be ignored. On the other hand, as has been proposed by GILMAN /11/ plastic shearing can occur during the formation of the cleavage steps (Fig.7) giving rise to an energy dissipation

$$R_{st} = (\rho \cdot h^2 \cdot \tau) / 2$$

with ρ: density of cleavage steps along the crack front, h^2: mean square of the step height; τ actual shear stress.

Following GILMAN an upper limit for the shear stress $\tau \cong \mu/5$ ($\mu = 1.8 \cdot 10^4$ MPa, modulus of shear) can be chosen. This leads to a contribution of $R_{st} \cong 1$N/m for the NT-specimens and even lower values for the other types. Thus, in spite of the fact that the conclusions of the model are in qualitative agreement with our observations, it has to be stated that the values of R_{st} are much smaller than the measured net effect on R. So shearing of the "cleavage" steps should contribute only by about 1N/m or less to the crack resistance of NaCl.

However, a satisfactory explanation of the measured high differences in the R-values seems to be possible, if a model proposed by BURNS and WEBB /12/ is taken into consideration, according to which dislocations are dragged in the stress field of the advancing crack. To move a single dislocation over an area Q under an effective shear stress τ requires the work $b \cdot \tau \cdot Q$ (b: Burgers vector). The area Q is determined by the penetration depth x of the dislocation (Fig.8) and the drag length l. The contribution to crack resistance by dragging screw dislocations of density N on GSI thus becomes

$$R_V = N\cdot b\cdot\tau\cdot(x\cdot l)$$

A calculation of R_V is rendered difficult because τ, x and l depend on N and furthermore the drag area $Q = x\cdot l$ is sensitive to the dislocation mobility. A consistant set of values based on the apparent dislocation density is required, regarding that $\tau = 0.4\cdot b\cdot\mu\cdot\sqrt{N}$ and that the slip line length and thus l, as in uniaxial deformation, is a function of inverse shear stress /14/. The screw dislocation density on the fracture surfaces of U- and NT-specimens can be assumed to be in the order of $5\cdot10^{12}\ m^{-2}$, which means that shear stresses of about 6.5 MPa are necessary to move dislocations and that the drag length should be in the order of 40μm /14/. Assuming from measurements in LiF /13/ a penetration depth of about 10μm then a contribution of about $R_V \cong 14$ N/m to the crack resistance can be calculated increasing with an increasing density of mobile dislocations on GSI. This is for both order of magnitude and tendency in fair agreement with the measured R-values.

However, the range of "reasonable" values of the parameters N,x and l allows some variation in R_V. Thus the R_V-value should be considered as a rather crude estimation, showing the importance of dislocation dragging on crack resistance.

The influence of dislocation mobility (NT- vs. NTX-specimens) on such a dragging effect becomes then evident.

The measured values of the crack resistance force R can therefore be explained by three additive energy contributions: i) the specific surface energy $R_o = 2\gamma = 0{,}6$ N/m, ii) the shearing of cleavage steps $R_{st} \cong 1$N/m and iii) the dragging of dislocations (R_V about 10 N/m, U-specimens).

5. CONCLUSIONS

By different modes of predeformation with and without subsequent pinning of dislocations the crack resistance force R in post yield fracture of NaCl can be varied by a factor of 20. The analysis of crack resistance given in this paper shows, that the interaction of the propagating crack with the dislocation structure causes cutting and dragging effects.

ACKNOWLEDGEMENTS

The authors are grateful to Prof. W. Schaarwächter for many helpful discussions and to Dr. K. Rogausch for valuable comments concerning the contents of the manuscript. Parts of this work were financially supported by the DFG (Deutsche Forschungsgemeinschaft) which is gratefully appreciated.

REFERENCES

/1/ S.M. Wiederhorn, R.L. Moses and B.L. Bean; J. Am. Ceram., Soc. 53, 18 (1970)

/2/ C.N. Ahlquist; Acta Metallurgica 22, 1133 (1974)

/3/ H. Koizumi, T. Suzuki; Phys. Stat. Sol (a), 68 (1982)

/4/ S. Tolansky; "Multiple Beam Interferometry of Surfaces and Films", Oxford (1948)

/5/ P.R. Moran; J. Appl. Phys. 29, 12, 1768 (1958)

/6/ M. Breucker, R. Steinbrech in "Fracture and Fatigue", J.C. Radon (Ed.), Pergamon Press, Oxford (1980)

/7/ M. Breucker; Thesis (doctoral) Dortmund (1983)

/8/ P.W. Tasker, T.J. Bullough; Phil. Mag. A, 43, 1 (1981)

/9/ H.L. Fotedar, M. Srinivasan, D.A. Wilson, T.G. Stoebe; Mat. Sci. Eng. 7, 272 (1971)

/10/ R. Steinbrech, W. Skrotzki, P. Haasen; Phys. Stat. Sol. (a), 46, 107 (1978)

/11/ J.J. Gilman; J. Appl. Phys. 27, 11, 1262 (1956)

/12/ S.J. Burns, W.W. Webb; J. Appl. Phys. 41, 5, 2078 (1970)

/13/ S.J. Burns, W.W. Webb; Trans AIME, Met. Soc., 236, 1165 (1966)

/14/ K.H. Matucha; Phys. Stat. Sol. 26, 291 (1968)

KEYWORDS

Sodium chloride, post yield, crack resistance, crack dislocation interaction, double torsion specimen

THE ELASTIC-PLASTIC CRACK PROPAGATION IN CERAMIC MATERIALS: THEORY AND TEST

A.S. Krausz and K. Krausz*

Dept. of Mech. Engineering, University of Ottawa
Ottawa, Canada K1N 6N5
*300 Driveway, Ottawa, Canada K1S 3M6

INTRODUCTION

As the result of great efforts invested over a long period of time, linear elastic fracture mechanics (LEFM) is now the fundamental quantitative technique of fracture analysis. However, the fact that a plastic zone surrounds the crack tip is often of concern. Essentially empirical findings support the standard qualification that as long as the plastic zone is smaller than a well-defined size, its effect can be neglected; or rather, that the same consequences are involved in the measurements of the material properties as are involved in the service function of the component that was designed according to LEFM concepts. When the plastic zone is large, however, new theories and techniques had to be introduced, most notably the J-integral, the R-curve method, and the crack opening displacement technique: all three are now part of industrial test and design practices. It is noteworthy that all three are also the derivatives and extensions of linear elastic fracture mechanics into areas that are outside the scope of LEFM.

Another aspect of the LEFM-based theories is also of concern: they are continuum stress analysis concepts with the consequence that the closer is the focus of enquiry reduced to the crack tip zone, the less valid they are. There are two reasons for this. The smaller is the zone of enquiry, the more pronounced is the effect of the discreteness of the atomic structure and the less valid is the "smoothing-over" requirement of stress analysis that considers the material to be a continuum; on the atomic scale, only the discrete forces acting on the atoms and their corresponding displacements have meaning, the concept of stress and strain loses validity. Atomic forces and displacements are interrelated in a manner that is characteristic of the material. This interrelation is different from the constitutive equation of continuum theory that relates stresses and strains: the correct form is, of course, the atomic force - atomic distance relation.

In the following a synthesis of elastic and plastic fracture mechanics, and the macroscopic and microscopic approaches will be presented with particular attention to the behavior of ceramic materials.

DISCUSSION

Figure 1(a) shows the usual macroscopic representation of an elastic-plastic crack, and Figure 1(b), the corresponding atomic scale model. Notice that the microscopic scale representation does not reveal whether the deformation was elastic or plastic: the atoms are in positions that are qualitatively independent of the character of the deformation process that is associated with the crack-tip zone, only the exact quantitative location may depend on the character of the deformation.

Figure 2 illustrates one crack propagation event on the microscopic scale: because materials are composed of atoms, crack propagation in any material, and therefore, in ceramics as well, occurs by a consecutive sequence of atomic bond breaking. To break atomic bonds, an appropriate amount of energy is needed. This energy may

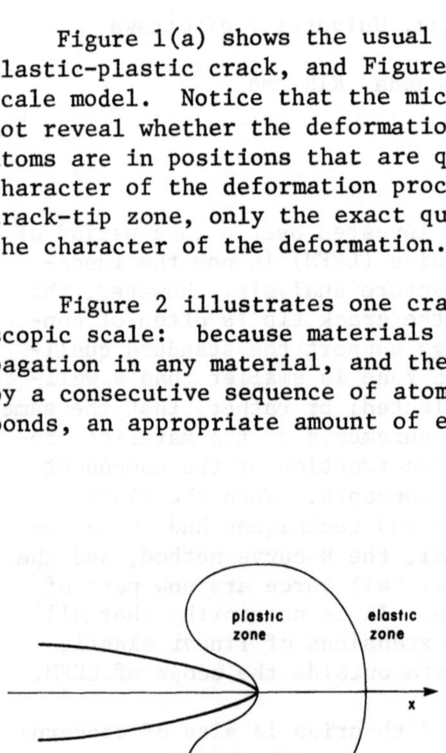

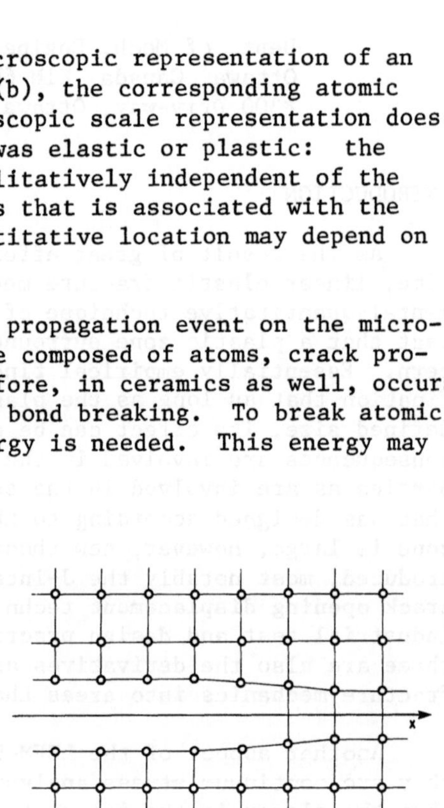

Fig. 1. (a) A macroscopic representation of the crack tip region, with a small plastic zone embedded in the surrounding elastic material. (b) The atomic scale model of the crack tip zone. The circles represent the atoms. Note that the arrangement of the atoms has not been changed by the plastic flow.

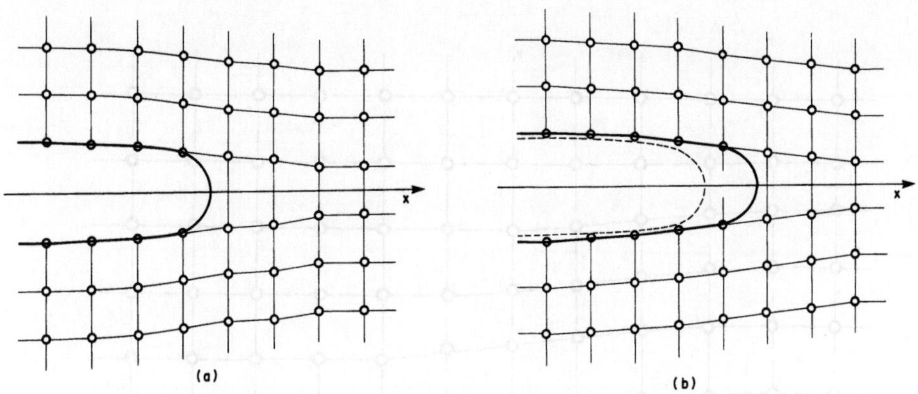

Fig. 2. (a) and (b) illustrate two consecutive steps in the crack propagation process on the atomic scale. The broken line in (b) indicates the position of the crack at the immediately preceeding step.

be fully supplied by the applied load in which case the crack propagates at the critical velocity; when the load is insufficient to supply all of the energy, the crack propagates at a subcritical velocity that depends not only on the properties of the material and the magnitude of the load, but also on the temperature as well.

In all cases, be it elastic or plastic, critical or subcritical fracture, the crack will propagate when the atoms in the crack-tip zone move sufficiently apart to bring about bond breaking, as illustrated in Figure 2; Figure 3 shows schematically the forces that act in the zone. To be quite precise, it is the energy associated with the separation, and the relative atomic distance changes in the whole body that controls bond breaking, and thus, crack propagation. The well-known interatomic force vs. interatomic distance, and interatomic energy vs. interatomic diagrams demonstrate this concept (Figure 4). Because bond breaking is really the process of increasing the interatomic distance from the equilibrium (lowest energy) value, r_o, to a distance far enough so that the force reduces to approximately zero, [1-3] it requires the separation energy as shown in Figure 4. Because the character of the force and energy diagrams is the same for any material composed of whatever atoms, and only the exact quantitative shape of the curve depends on the composition, the general validity of the approach of this study is substantiated.

Crack propagation depends, therefore, on the energy content of the atoms in the crack tip zone: if there is enough energy, and distributed appropriately so that the atomic displacements corres-

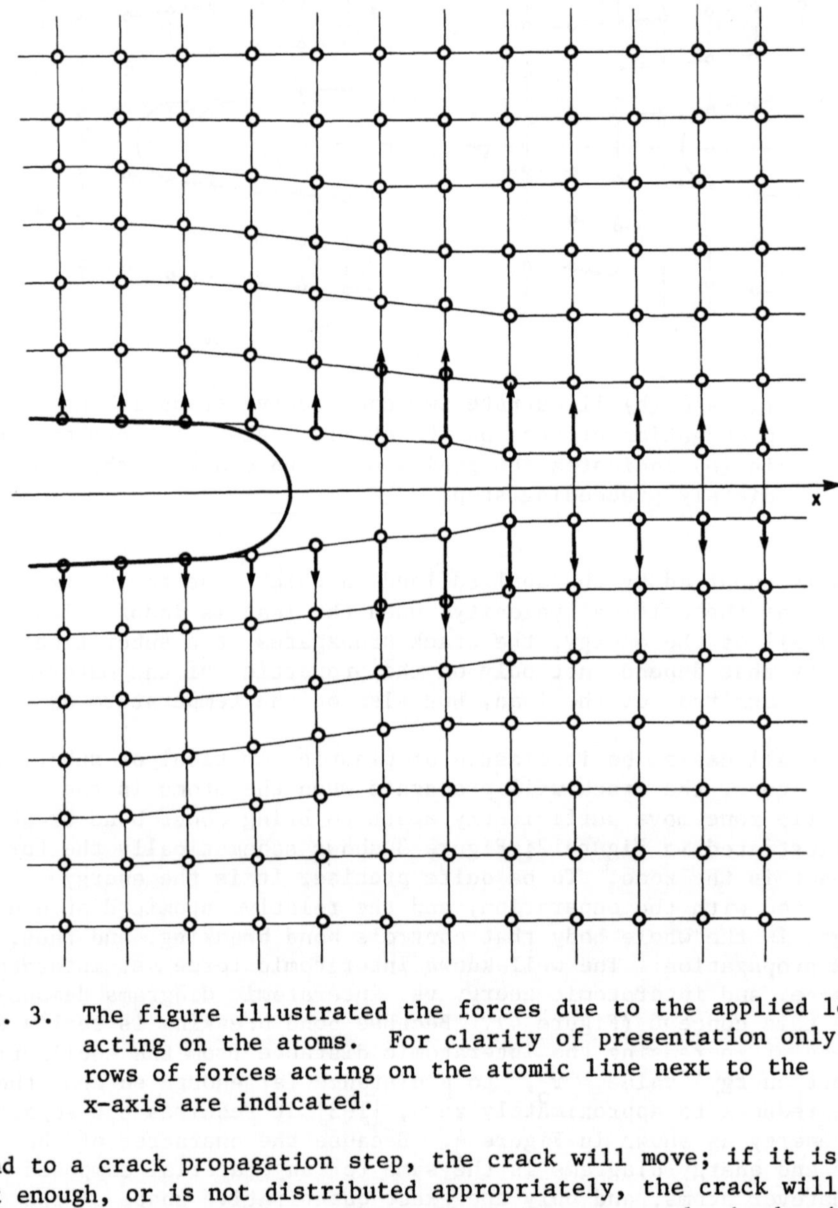

Fig. 3. The figure illustrated the forces due to the applied load, acting on the atoms. For clarity of presentation only the rows of forces acting on the atomic line next to the x-axis are indicated.

pond to a crack propagation step, the crack will move; if it is not enough, or is not distributed appropriately, the crack will not move. This is the essence of any fracture process, be it ductile or brittle, fast or slow.

There are only two sources from which energy can be drawn: the mechanical energy from the work done by the applied load, and the

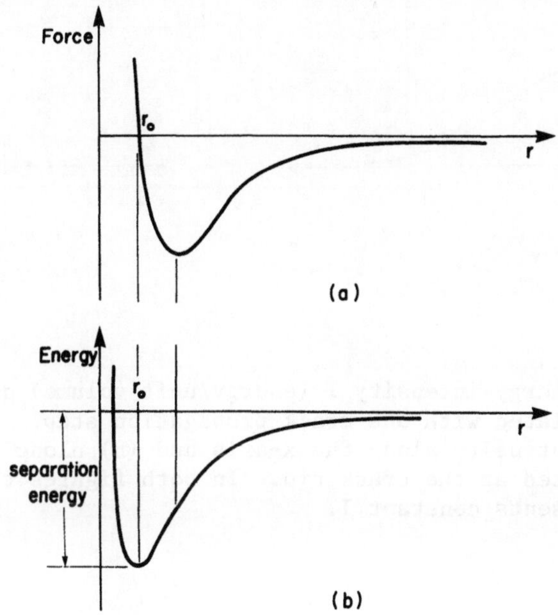

Fig. 4. The schematic representation of the typical, interatomic force vs. interatomic distance r relation (a), and the interatomic energy vs. interatomic distance r (b). At the minimum energy the atomic forces are zero: this corresponds to the equilibrium atomic distance r_o.

thermal energy from the heat stored in the body. Consequently, when the load is not sufficient to supply enough work to break the bonds, the crack cannot propagate unless energy is supplied from the heat content of the body so that

$$\Delta G^{\ddagger} = W + \Delta G \tag{1}$$

where $\Delta G^{\ddagger}$ is the total change in bond energy when the crack moves by one step, W is the work and ΔG is the thermal energy. Equation (1) is the fundamental statement of any crack propagation process; it is also a gratifying simple equation of energy "book-keeping".

The role and physical significance of the energy change, $\Delta G^{\ddagger}$, has already been discussed for the simple example shown in Figure 4: in a ceramic solid it expresses the sum of all interatomic potential energy changes. In principle, all atoms of the solid are more or less displaced when the crack takes a step, but these

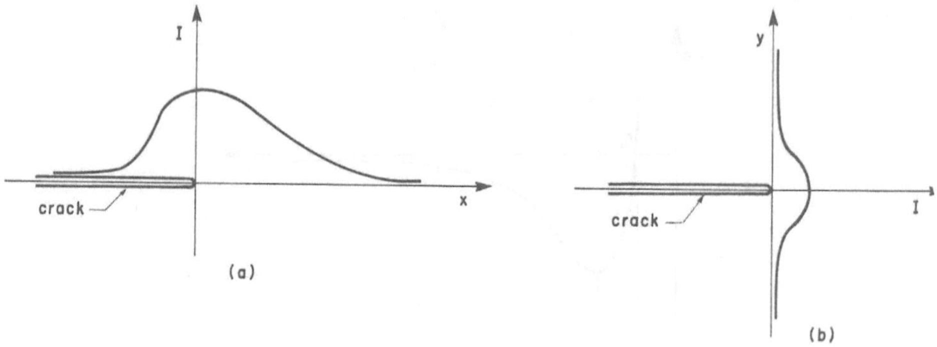

Fig. 5. The energy intensity I (energy/unit volume) distribution associated with one crack propagation step. (a) represents the intensity along the x-axis and (b) along the y-axis situated at the crack tip. In both figures the curve represents constant I.

displacements are significantly large only near the crack-tip zone and their extent reduces quickly with the distance from the tip: the corresponding energy intensity distributions are illustrated in Figure 5. The thermal energy, ΔG, supplied to the rearrangement of the atoms when the crack tip moves by one step, is tied in with the thermal vibration of the atoms: the larger the heat energy content of the solid, the larger is their average vibrational amplitude. The atoms vibrate with amplitudes and directions that change from period to period in a random manner. When the direction and amplitude are of the right orientation and magnitude to bring about the atomic distance changes needed to be superposed on the displacements already created by the mechanical forces, a crack propagation step results, accompanied by the appropriate rearrangement of the other atoms in the solid. The wave-like random to-and-fro flow of the magnitude and direction of the thermal energy (that is, the propagation of atomic vibrations) is independent whether plastic flow accompanies crack propagation or not. Therefore, two of the terms $\Delta G^{\ddagger}$ and ΔG [Eq. (1)], are independent from the elastic or plastic deformation. The mechanical work acting on the atoms need to be considered now.

At this scale, where the physical process occurs, the concept of stress has lost its meaning: the discreteness of the actual structure requires that the forces acting on the atoms must be considered. The mechanical work contributed by the load is then the sum of the work done by the forces acting on the individual atoms

$$W = \sum_i w_i = \sum_i \int \overline{F}_i \cdot \overline{dr}_i \tag{2}$$

where w_i is the work and $\overline{F}_i$ is the force acting on the i-th atom, dr_i is the displacement vector of the atom. The integral is taken over the whole displacement path from the initial state that corresponds to the unloaded state at the beginning of loading, to the activated state that will be defined later. The scalar product of the force and displacement vector is the usual expression of the work. In Equation (2) $\overline{F}$ is a function of the interatomic distance r, as defined in Figure 4 and represents, therefore, exactly the behavior: the discrepancy between the physical reality and the linear (or non-linear) elastic model of the continuum stress analysis approach, indicated in the introduction, is removed. The work expressed by Equation (2) is the contribution from the external load system. It is only noted here that the change in the stored elastic energy can be considered also: the method for this is well known.

THE EFFECTS OF PLASTIC FLOW

Consider two specimens: as long as the work W and its distribution over the atoms is the same in both, the crack is in the same state and will propagate with the same velocity (if they are of the same material and at the same temperature, so that $\Delta G^{\ddagger}$ and ΔG is identical in both). In principle, therefore, the presence or absence of plastic flow is immaterial. Plastic flow can have quantitative effects only, but not qualitative: the course and mechanism of bond breaking is not affected in its fundamental character, only in its secondary details. These can be of four kinds:

(1) The stress distribution is affected by plastic flow; consequently, the forces on the individual atoms and so the magnitude of the work w_i depends on the plastic flow. With reference to Figure 5, it is to be expected, however, that only a relatively small volume is affected by the stress distribution particular to the plastic deformation. This volume, of course, is critical because it controls the rate of crack propagation: it is clear now that crack propagation tests carried out with a specific w_i distribution define crack propagation uniquely. It remains for the designer to establish the particular plastic-zone-affected stress and strain distribution change as the crack propagates and identify this with the test results. The test and design procedure is then the same as in LEFM, or with the J-integral concept. At any rate, it is a task in stress analysis, the same as in any design problem: the difference between elastic and plastic case is

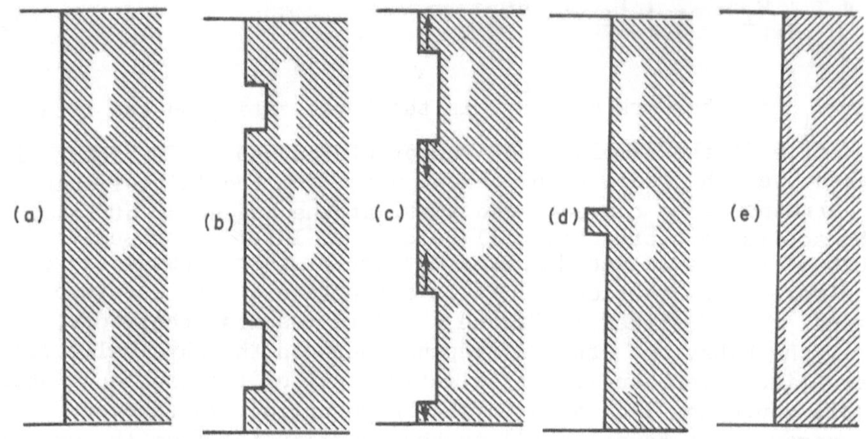

Fig. 6. The formation of double kinks along the crack front at weak regions in the material (a), the subsequent sideways spreading (b) to (d), and the merger of neighboring kinks (e).

only quantitative but not a difference in kind. Blunting of the crack tip is a natural part of this analysis.

It is self-evident that these test and design concepts are valid for steady or monotonously increasing $\overline{F}_i$ distributions.

(2) It is well established that in the plastic zone structural changes occur: micro-cracks develop at hard particles; vacancies and vacancy conglomerates form; cavities develop, grow and coalesce. This deterioration in the structural homogeneity of the material produce stress concentrations and affect the w_i distribution. The major consequence of this is that crack propagation occurs by the preferential movement of the crack front at the structural weak spots due to $\overline{F}_i$ concentrations as well as to the lesser energy needed to break bonds. Consequently, the crack front does not move as a rigid line but by first forming a double kink at the weak spots and spreading sideways afterwards, as illustrated in Figure 6. When two neighboring kinks meet they merge and the crack front is straight again. The process of kink formation, sideways spreading, merger, repeats itself periodically. This whole process is also identical to the crack propagation mechanism in LEFM, only the quantitative aspects are affected

by the plastic flow induced further deterioration of the structure.[2,4-6].

(3) Because crack propagation is a discrete process there is a waiting time between two steps. During this period the parallel process of plastic flow progresses. The dislocation movement changes the plastic strain and the stress distribution as well: by the time the next step occurs, the distribution of w_i will be different by δw_i. This effect, however, is very small and is fluctuating about w_i. It is expected that this will be only a part of the overall stochastic behavior and is, therefore, negligible.

(4) Plastic flow is a thermally activated process and is, therefore, affected by the rate and the temperature. The atomistic description made clear that crack propagation also depends on the temperature: this is also a thermally activated process. While in subcritical LEFM the bond breaking may be thermally activated, the elastic deformation is not. It follows that rate and temperature effects must be taken into consideration more critically as the plastic zone increases.

The Crack Propagation Process

It is well established that during a bond breaking event the potential energy (and so, the free energy) changes, as shown in Figure 7. In state a the bond is already stretched by the force $\overline{F}_i$; when the appropriate vibration direction with an energy content of ΔG_i reaches the atom pair its distance increases as shown at state b; the corresponding energy condition is also represented. The system then reached the top of the energy barrier indicated by the symbol ‡: this is the so-called activated state. The atom pair can lower its energy by "rolling" down into the valley c spontaneously. During this, the interatomic distance increases further and the bond breaks. [1,7,8].

As was discussed before, not only the atom pair and the cracktip is affected, but the whole system of atoms. The energy diagram of Figure 7 should be considered to be a composite of the energy change of the atom group while the crack moves by one step. As also noted, the crack front moves by a consecutive system of steps as the consequence of the double kink formation and the multistep sideways movement. Each of these steps is thermally activated and can be represented by an energy barrier. The result is a system of consecutive barriers, represented schematically in Figure 8.

Thermally activated bond breaking rate is described by the rate theory as [1,9]

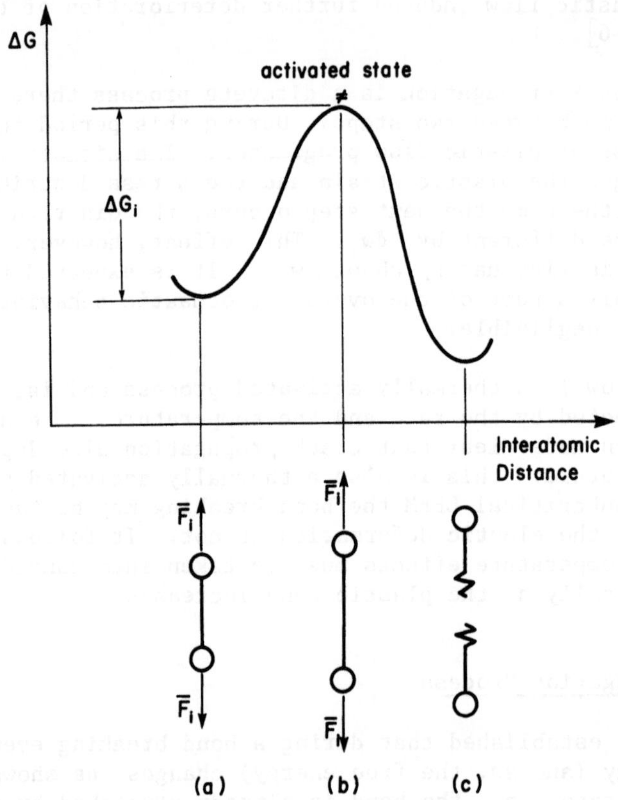

Fig. 7. The figure illustrates the relation between the state of a pair of atoms at the crack tip and the associated energy condition. The heavy line connecting the atoms indicate the bond in (a) and (b). The broken bond is shown in (c).

$$k = \frac{kT}{h} \exp\left(-\frac{\Delta G}{kT}\right) = \frac{kT}{h} \exp\left[-\frac{\Delta G^{\ddagger} - W(F)}{kT}\right] \quad (3)$$

where the rate constant k defines the number of activations over a single energy barrier in unit time, that is, the number of steps in unit time. It depends on the total energy $\Delta G^{\ddagger}$ needed to move the atomic conglomerate so that their rearrangement corresponds to the movement of the crack by one step. This movement is either the formation of the double kink or the sideways spreading step. For either, $\Delta G^{\ddagger}$ is a specific value which depends on the composition, the structure and the defect system of the material. The contribution of the mechanical work is expressed by W as the function of the forces acting on the individual atoms. This was discussed earlier; it is expected that in test and design practice

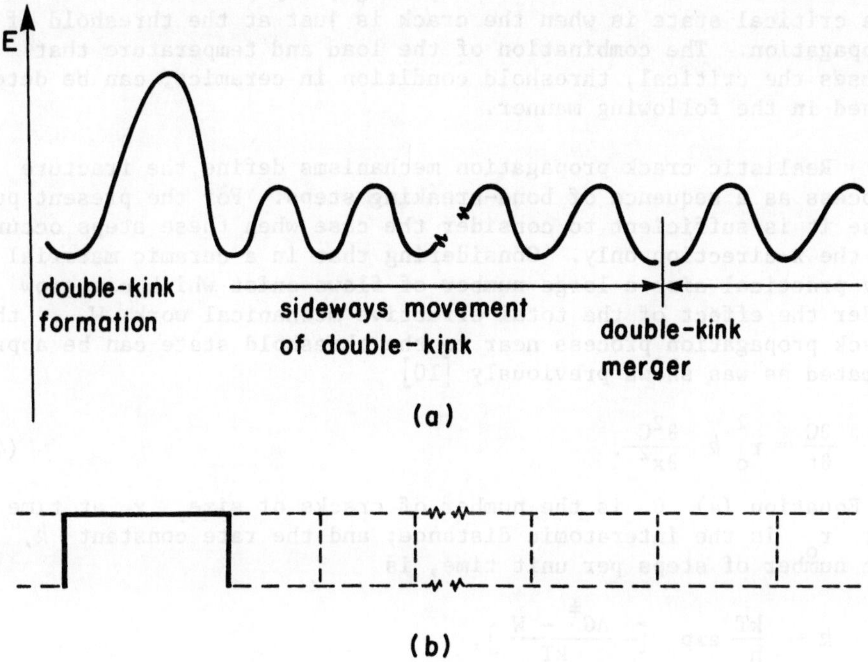

Fig. 8. The system of energy barriers associated with the formation and spreading of double kink crack fronts. (a) illustrates the energy barrier system, and (b) the corresponding shape of the crack front: the solid line shows the formation of the double kink, the dashed line indicates the subsequent spreading steps.

it will eventually be related to the stress intensity factor K or to the crack driving force G carrying with it all the advantages of the use of these parameters. In Equation (3) k is the Boltzmann constant, h is Planck's constant, and T is the temperature in degree Kelvin. [1].

The Character of Threshold State in Ceramics

The preceding analysis demonstrates that plastic-zone associated fracture occurs in ceramic materials when the combined action of the mechanical work components, w_i, and the thermal energy flow, ΔG, break the bonds at a high enough rate that enables the crack to follow the development of the plastic flow; when the

$W + \Delta G$ effect is not sufficiently large, only plastic flow occurs. The critical state is when the crack is just at the threshold of propagation. The combination of the load and temperature that causes the critical, threshold condition in ceramics, can be determined in the following manner.

Realistic crack propagation mechanisms define the fracture process as a sequence of bond-breaking steps. For the present purpose it is sufficient to consider the case when these steps occur in the x-direction only. Considering that in a ceramic material of any practical size a large number of flaws exist which can grow under the effect of the total effective mechanical work W, the crack propagation process near to the threshold state can be approximated as was shown previously [10]

$$\frac{\partial C}{\partial t} = r_o^2 \, k \, \frac{\partial^2 C}{\partial x^2}. \tag{4}$$

In Equation (4) C is the number of cracks of size x at time t ; r_o is the interatomic distance; and the rate constant k, the number of steps per unit time, is

$$k = \frac{kT}{h} \exp \left[- \frac{\Delta G^{\ddagger} - W}{kT} \right],$$

where W is the total work (the appropriate sum of w_i) contributed by the mechanical load. The previous argument indicates that the work can be expressed as the function of K, G or the J integral. For this somewhat simplified case, the solution of the differential equation is

$$C(x, t) = \frac{C_t}{(4\pi r_o^2 \, kt)^{1/2}} \exp \left(- \frac{x^2}{4r_o^2 \, kt} \right) \tag{5}$$

where C_t is the total number of cracks. Equation (5) shows that there will be a distribution of cracks of various sizes rather than a single size. The average crack will not have grown, confirming the condition that at the critical load the cracks are on the threshold of growth. However, because Equation (5) is the representation of a Gaussian probability curve, the crack size distribution is probabilistic: there are cracks that have grown although the conventional, deterministic, fracture mechanics analysis would predict no crack growth. A simple example will illustrate the significant consequences which the probabilistic physical process has on crack initiation.

In Equation (5) the term $r_o(kt)^{\frac{1}{2}}$ is the standard deviation, σ, of the Gaussian curve: it expresses, therefore, the dispersion of the cracks. For

$$\Delta G^{\ddagger} = 2 \text{ eV} \quad \text{and} \quad W = 1.52 \text{ eV},$$

after a service life of 30 years at room temperature the standard deviation is

$$\sigma = 5 \text{ mm}.$$

At a slightly higher load of $W = 1.56$ eV the standard deviation is

$$\sigma = 10 \text{ mm}.$$

and even after only two months it is almost 1 mm large. This is equivalent with the conclusion that one third of the components that were subjected to a load that may have caused plastic flow will have also cracks larger than 1 mm size, at a condition that was considered to be safe by threshold design.

SUMMARY

The argument presented here shows the interrelation between crack propagation and plastic flow in ceramic materials: it describes the physical basis for the development of an elastic-plastic fracture mechanics for the appropriate interpretation of time dependent subcritical crack propagation. Because it is an essentially energy argument, the J-integral approach and the stress intensity or crack driving force related R-curve technique should also be applicable for the analysis of the elastic-plastic fracture of ceramics, using the concepts developed in this study. The theory is based on thermal activation concepts and is eminently suitable for the description of the temperature dependent fracture of ceramic materials.

While ceramics have highly desirable qualities from engineering service point of view such as high melting points, chemical resistance, great hardness, they are brittle. [11]. The low toughness hinders many of their potential application: it is expected that the physical interpretation of the interdependence between the plastic deformation and linear elastic fracture mechanics concepts will facilitate the development of stronger ceramic materials.

REFERENCES

1. A.S. Krausz and H. Eyring, "Deformation Kinetics," Wiley-Interscience, N.Y. (1975).
2. B.R. Lawn and T.R. Wilshaw, "Fracture of Brittle Solids," Cambridge University Press, Cambridge (1975).
3. E.R. Fuller, Jr. and R.M. Thomson, Lattice theories of fracture in: "Fracture Mechanics of Ceramics, Vol. 4, R.C. Bradt, D.P.H. Hasselman and F.F. Lange, eds., Plenum Press, N.Y. (1978).
4. R.M. Thomson, C. Hsieh and V. Rana, "Lattice trapping of fracture cracks," J. Appl. Phys. 42:3154(1971).
5. A.S. Krausz, "The deformation kinetics of consecutive energy barriers," J. Phys. Chem. Solids, 40:351 (1979).
6. S.D. Brown, "A multibarrier rate process approach to subcritical crack growth," in: "Fracture Mechanics of Ceramics," Vol. 4, R.C. Bradt, D.P.H. Hasselman and F.F. Lange eds., Plenum Press, N.Y. (1978).
7. S.D. Brown, "The multibarrier kinetics of subcritical brittle crack growth," in: Proc. Sixth Canadian Cong. Appl. Mech. Vol. 1, Vancouver, B.C. (1977).
8. B.R. Lawn, "An atomistic model of kinetic crack growth in brittle solids," J. Mater. Sci. 10:469 (1975).
9. S.M. Wiederham, "Mechanisms of subcritical crack growth in glass," in: "Fracture Mechanics of Ceramics," Vol. 4, R.C. Bradt, D.P.H. Hasselman and F.F. Lange, eds., Plenum Press, N.Y. (1978).
10. A.S. Krausz, "Crack-size distribution in homogeneous solids," Int. J. Fract., 15:337(1979).
11. B.R. Lawn, "Physics of fracture," J. Amer. Ceram. Soc. 66:83(1983).

CHANGING THE SURFACE PLASTICITY AND HARDNESS OF SAPPHIRE BY ION-IMPLANTATION

P.J. Burnett and T.F. Page

Department of Metallurgy and Materials Science
University of Cambridge, Pembroke Street
Cambridge, CB2 3QZ, England

1. INTRODUCTION

Plasticity of the near-surface regions of ceramics is known to play a critical rôle in determining materials' response to contact deformation during indentation, wear and erosion[1,2]. Ion-implantation has been reported to improve the wear resistance of various components including WC-Co hard metal dies and diamond tools[3-5]. This paper presents some of the results from a study of the effects of ion-implantation on the contact deformation mechanisms occurring in a range of ceramic materials.

Previous work[6] established that the implantation of both silicon and silicon carbide with N_2^+ ions above a critical dose of $\sim 4\times10^{17}$ ions cm^{-2} resulted in a significant softening of the surface coupled with a suppression of the lateral cracking mode of indentation fracture. The effect on plasticity was attributed either to amorphisation (see section 3) or to a solid solution effect of the implanted species, while many of the changes in fracture behaviour were accounted for by implantation-induced surface stresses. Further work by us[7] has correlated the surface softening specifically with the formation and size of a radiation-produced layer of amorphised material and has shown that the critical dose corresponds to that at which a large change in amorphous layer thickness occurs. We have also reported a similar surface softening in sapphire[8]. Conversely, other workers[9,10] have reported that a number of ion species (e.g. Ti, Cr, Zr) produce a hardening of sapphire. Thus a more detailed examination of the dose and species dependence of the surface plasticity of sapphire has now been initiated including the effect of post-implantation annealing.

2. EXPERIMENTAL METHODS: IMPLANTATION, INDENTATION AND ANNEALING

High purity single crystal {10$\bar{1}$2} sapphire sections (courtesy of GEC (Wembley, UK)) were supplied with one side pre-polished to a good surface finish into which implantations with 300keV Cr^+, Ti^+, Y^+ and Zr^+ ions were carried out in the Cockcroft-Walton Facility at AERE Harwell, UK. These ion species were chosen on the criteria of ionic radius and valency as those the most likely to cause significant solid-solution hardening in sapphire[11]. The doses were determined by post-implantation Rutherford backscattering (RBS)[8] and are shown in Table 1. Given the ion species and energy, the doses were chosen to amply span that at which the onset of amorphisation was expected to occur (see section 3 and Figs. 1 and 2).

Both low-load microhardness tests and single pass scratch tests were used to determine materials' response to surface contacts. All specimens were indented on a Leitz "Miniload" using a Knoop profile indenter under ambient conditions and at constant crystallographic orientation (parallel to <$\bar{2}$021>). Six indentations of loads each of 25gf and 50gf were made on each specimen, the long diagonal size being typically 10-15μm which corresponds to a depth of penetration of ∿0.3-0.5μm; i.e. comparable with the mean ion ranges[12] (∿0.15μm for Ti and Cr; ∿0.1μm for Y and Zr).

For scratch experiments, a 90° diamond cone, mounted on a counterbalanced arm was loaded to 20gf and dragged from the unimplanted to the implanted regions of the specimen at a speed of ∿30μm sec^{-1}.

TABLE 1. Implantations and resultant surface states.

Specimen	Ion Species	Dose (ion cm^{-2})	Irradiation Damage: crystalline surface (I-II)	sub-surface amorphous (II)	amorphous surface (III)
ST1	Ti^+	$2{\cdot}5 \cdot 10^{16}$			
ST2	Ti^+	$3{\cdot}4 \cdot 10^{16}$	ECP		
ST3	Ti^+	$6{\cdot}3 \cdot 10^{16}$	ECP, RBS	RBS	
ST4	Ti^+	$1{\cdot}5 \cdot 10^{17}$	ECP, RBS	RBS	
ST5	Ti^+	$2{\cdot}3 \cdot 10^{17}$			
ST6	Ti^+	$7{\cdot}0 \cdot 10^{17}$			ECP, TEM
SC1	Cr^+	$3{\cdot}2 \cdot 10^{16}$	ECP		
SC2	Cr^+	$2{\cdot}6 \cdot 10^{17}$	RBS	RBS	
SC3	Cr^+	$6{\cdot}0 \cdot 10^{17}$			ECP
SY1	Y^+	$1{\cdot}3 \cdot 10^{16}$			
SY2	Y^+	$2{\cdot}8 \cdot 10^{16}$	ECP,TEM,RBS		
SY3	Y^+	$9{\cdot}9 \cdot 10^{16}$	RBS	RBS	ECP
SY4	Y^+	$1{\cdot}1 \cdot 10^{17}$			ECP
SY5	Y^+	$3{\cdot}8 \cdot 10^{17}$			ECP
SY6	Y^+	$8{\cdot}5 \cdot 10^{17}$			ECP,TEM,RBS
SZ1	Zr^+	$1{\cdot}5 \cdot 10^{16}$	ECP,TEM		
SZ2	Zr^+	$5{\cdot}0 \cdot 10^{16}$			ECP
SZ3	Zr^+	$1{\cdot}6 \cdot 10^{17}$			
SZ4	Zr^+	$5{\cdot}5 \cdot 10^{17}$			ECP

Annealing of specimens ST2, ST3 and SZ1 was carried out both in air using a high temperature resistance element furnace and in low vacuum using an r.f. furnace with graphite susceptor. Temperatures were chosen to be below the minimum solvus temperature suggested for the possible precipitate phase expected (~1800°C for $A\ell_2O_3$-TiO_2 and ~1900°C for $A\ell_2O_3$-ZrO_2)[13] and are detailed in Section 4.

3. STRUCTURAL CHANGES DUE TO ION-IMPLANTATION

Ion-implantation can result in massive changes in both the chemistry and structure of the near-surface region. Structural changes are principally due to displacement damage creating Frenkel pairs etc.[14]. In ceramic materials these defects are relatively immobile at the temperatures occurring during implantation (~250°C) and so metastable structures may result. At very high doses, cumulative displacement damage causes a disordered structure or amorphous layer to be formed. A simple quantitative model has been produced for the estimation of both the size and spatial position of this amorphous layer with increasing dose[7]. Typically, three damage regimes are found as shown schematically in Fig. 1. At doses below that for amorphisation, a damaged but crystalline surface results (I). At higher doses, and corresponding to the peak in the damage profile, amorphisation begins below the surface (II). Finally as the damage level rises, the amorphous layer thickens to both reach the surface and extend further inwards (III). Fig. 2 shows the predicted variation in amorphous layer thickness with dose for sapphire implanted with 300keV Ti^+, Cr^+, Y^+ and Zr^+.

Three techniques have been used to assess the microstructural state of the specimens. (i) Electron channelling patterns (ECPs) obtained in the scanning electron microscope (SEM) were used to determine the crystallographic perfection of the top ~100nm of the crystal. Both selected area channelling patterns (SACP) and wide area channelling patterns (WACP) were used (e.g. Fig. 3a). (ii) Channelled RBS yielded depth profiles of the damage distribution[14] and has been applied to several specimens using both the [0001] axial and (0001) planar channelling directions. (iii) Electron diffraction in the transmission electron microscope (TEM) was used to directly determine the extent of any amorphisation. TEM Specimens were prepared by ion-beam thinning from the unimplanted side only of 3mm discs cut from the implanted specimens. Fig. 3b shows a typical diffraction pattern obtained from a specimen with a fully amorphised surface. The results obtained from these three techniques are summarised in Table 1.

4. RESULTS

With the ion/dose combinations given in Table 1, the effects of ion-implantation on low load microhardness are shown in Fig. 4. For each ion species, there is at least one specimen implanted to a dose below that required for the onset of amorphisation (indicated by the

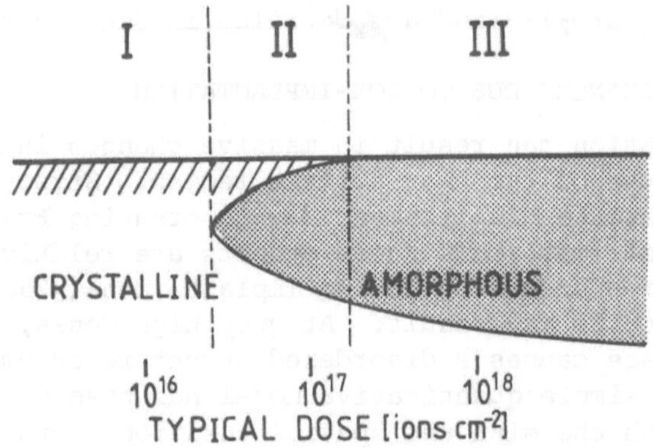

Fig.1. A schematic representation of the variation in size and position of implantation-induced amorphous zones showing:- I, a damaged crystalline region; II, a sub-surface amorphous zone; III, a surface amorphous layer.

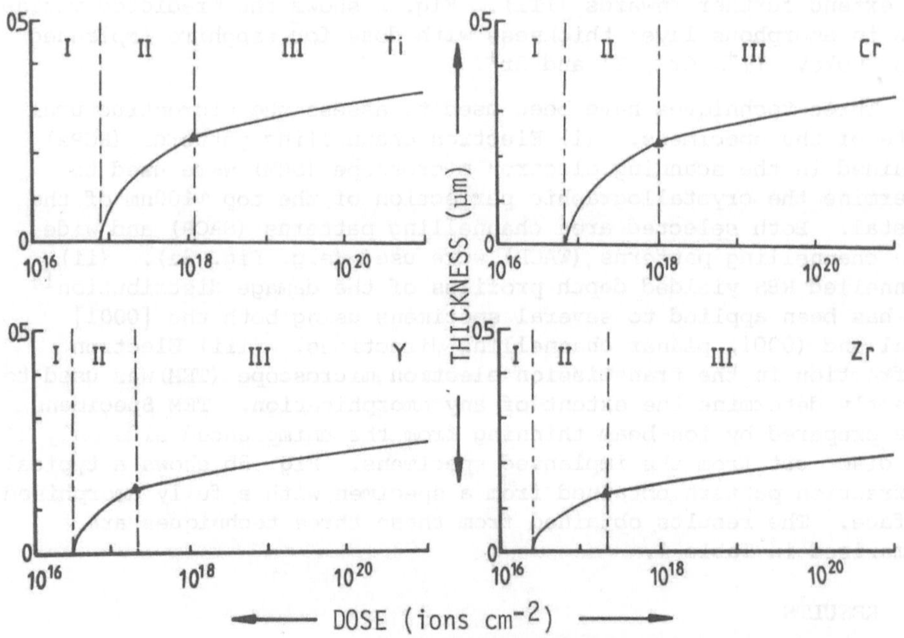

Fig.2. Predicted variation of amorphous layer size and position with dose for sapphire implanted with Ti^+, Cr^+, Y^+ and Zr^+ at 300keV. Regions, I, II and III as Fig. 1.

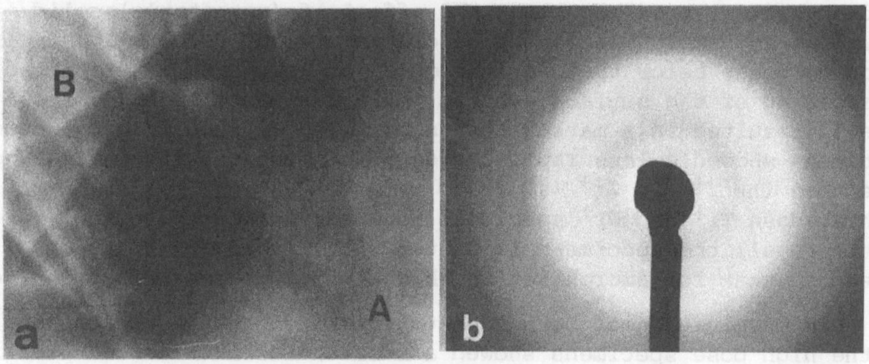

Fig.3. (a) WACP of SY2 taken across both implanted (A) and unimplanted (B) regions. (b) electron diffraction pattern from a thin foil of the surface of ST6 showing diffuse rings characteristic of amorphous material.

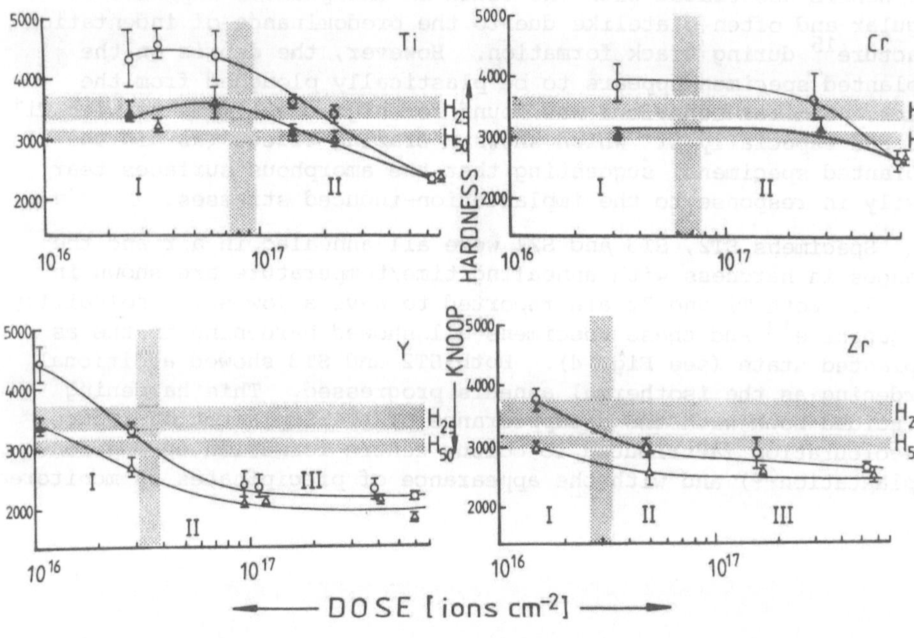

Fig.4. The experimental variation of 25gf (o) and 50gf (Δ) Knoop hardnesses with dose for sapphire implanted with Ti^+, Cr^+, Y^+ and Zr^+ at 300keV. The vertical shaded bar indicates the expected onset of amorphisation. Regions I, II and III are defined in Fig.1. The horizontal shaded bars indicate the range of 25gf and 50gf hardness values (H_{25},H_{50}) for unimplanted sapphire. Indicated errors are ±(standard deviation).

shaded vertical bars) enabling the effect of (metastable) solid solutions and amorphisation to be distinguished. Generally, implantation to doses below that for the onset of amorphisation resulted in a hardening of the surface, whilst implantation to higher doses usually resulted in a marked softening. The Ti^+ and Cr^+ implanted specimens showed a less rapid decrease in hardness with increasing dose than the Y^+ and Zr^+ implanted specimens, i.e. only the highest dose Ti^+ and Cr^+ specimens show any marked softening. By contrast, all the specimens implanted with Y^+ and Zr^+ to doses exceeding that for amorphisation show similar degrees of softening.

SEM examination of both indentations and scratch tracks made on the high dose specimens showed a substantial increase in plasticity in contrast to the more brittle behaviour of the unimplanted specimens. For example, Figs. 5a,b show Knoop indentations in unimplanted sapphire and SY5. In the latter case, a substantial amount of plastic pile-up around the indentation can be seen. Similarly, Figs. 6a,b show scratch tracks in unimplanted sapphire and SY3. The debris associated with the track in unimplanted sapphire is angular and often platelike due to the predominance of indentation fracture[15] during track formation. However, the debris on the implanted specimen appears to be plastically ploughed from the track. Similar behaviour was found for high dose implants with Ti^+ Cr^+ and especially Zr^+ which shows a crazing effect (as did the Y^+ implanted specimens) suggesting that the amorphous surfaces tear easily in response to the implantation-induced stresses.

Specimens ST2, ST3 and SZ1 were all annealed in air and the changes in hardness with annealing time/temperature are shown in Fig. 7. Both Ti and Zr are reported to have a low solid solubility in sapphire[13] and these specimens all showed hardening in the as implanted state (see Fig. 4). Both ST2 and ST3 showed additional hardening as the isothermal anneals progressed. This hardening coincided both with the disappearance of the implantation-induced discolouration (attributed to colour centre formation during ion-implantation[16]) and with the appearance of precipitates as monitored

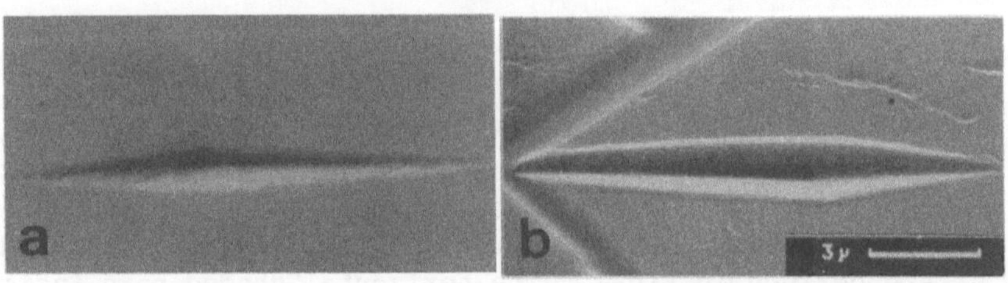

Fig.5. SEM micrographs of 25gf Knoop indentations in (a) unimplanted sapphire, (b) SY5. Note the plastic pile-up around the indentation in (b) and the stress-induced crazes which appear as tears in the surface.

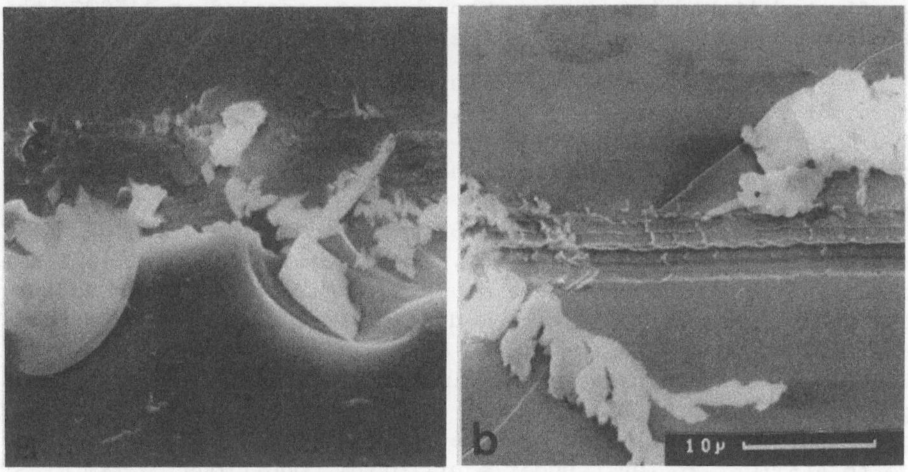

Fig.6. SEM micrographs of portions of 20gf diamond cone scratch tracks on (a) unimplanted sapphire, (b) SY3. Note the angular debris in (a) and the rounded debris in (b).

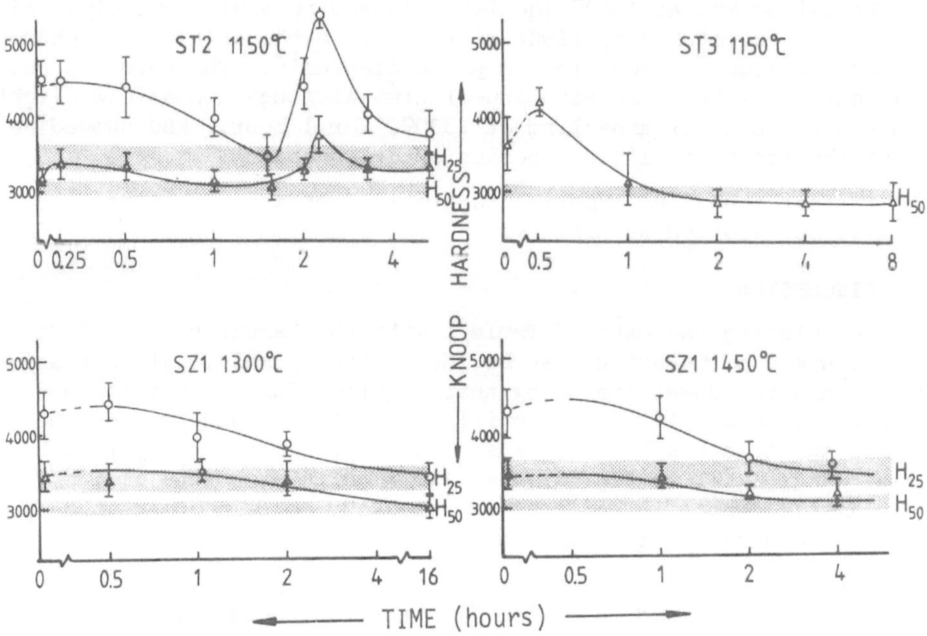

Fig.7. The observed variation of 25gf (o) and 50gf (Δ) Knoop hardness with annealing time and temperature (1150°C for ST2,3 and 1300°C, 1450°C for SZ1). The horizontal shaded bars are as in Fig. 4.

by reflected light microscopy. Though also annealed at 1150°C, this hardening occurs far earlier in the case of ST3 than ST2.

The precipitates first seen in both ST2 and ST3 were equiaxed but, after similar annealing times, those in ST2 were much finer (<1μm) than those in ST3 (∿1-3μm). For another sample of ST3, a separate anneal at 1400°C for one hour produced coarse needle-like precipitates (∿15μm in length) for which both electron diffraction patterns and lattice images (Fig. 8) were consistent with the orthorhombic phase $A\ell_2TiO_5$ (a=9.6Å, b=9.63Å, c=3.6Å) which is the expected equilibrium second phase at this temperature. Further, in situ EDX analysis in the TEM showed that these particles contained both aluminium and titanium. These observations are not consistent with the precipitate phase under similar conditions being rutile as suggested by Naramoto et al.[10]. By comparison, the anneals at 1150°C might have been expected to produce a second phase of TiO_2 in preference to $A\ell_2TiO_5$, at least under equilibrium conditions[17]. TEM of ST3 after 8 hrs. at 1150°C revealed both the equiaxed particles already mentioned together with a second population of finer tabular particles as shown in Fig. 9. However, these particles have not yet been positively identified but appear to be neither TiO_2 nor $A\ell_2TiO_5$.

Annealing SZ1 at 1300° and 1450° caused no visible precipitation, at least as detectable by light microscopy, although a second phase is expected from the equilibrium phase diagram[13]. The hardness showed a decrease with increasing anneal time although a possible slight rise was found after annealing at 1300°C for 1 hour. RBS showed no redistribution of implanted species after annealing.

Further annealing experiments are in progress on all samples and will be reported separately.

5. DISCUSSION

Correlating the data of Table 1 with the hardness data of Fig.4, three régimes of behaviour can be identified, namely: (i) hardening occurs when the doses are below that required for amorphisation (region I in Fig. 1); (ii) significant softening occurs when the dose is sufficiently high to cause a surface amorphous layer to be formed (region III in Fig. 1) and (iii) an intermediate region, where little change in hardness is observed, corresponds to the presence of a sub-surface amorphous layer (II in Fig. 1).

Since TEM of hardened specimens showed no precipitates, then the hardening observed in region I is attributed to solid solution effects (the mean concentration being typically 5 at.%). Despite reports[11] that Ti^{4+} produces greater solution strengthening in sapphire than Cr^{3+} (or Ti^{3+}), Ti and Zr implants here seemed not to be more potent as hardening agents than Cr and Y. This might indicate that, in the structure, all implanted species have a similar final charge (e.g. 3^+) with any hardening due solely to ion misfit (see Table 2). Additional evidence that the Ti implant in specimen ST3 is probably

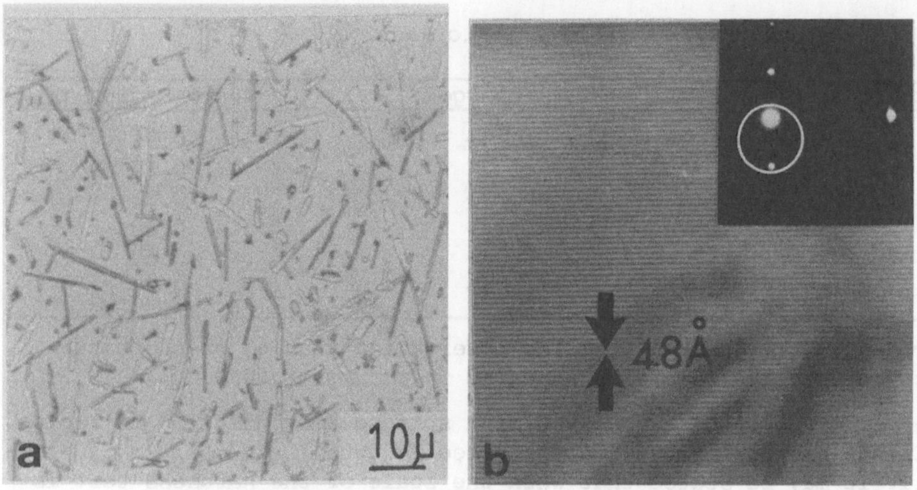

Fig.8. Precipitates formed in ST3 after annealing at 1400°C for 1 hr. (a) reflected polarized light micrograph (b) TEM lattice image and diffraction pattern from one of the Al_2TiO_5 needles seen in (a).

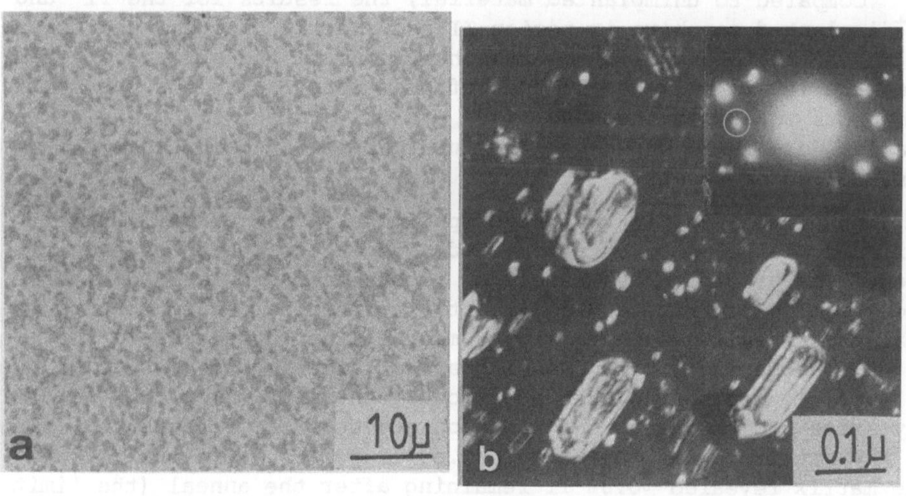

Fig.9. (a) Equiaxed precipitates seen in ST3 after annealing at 1150°C, for 8 hrs. (reflected polarized light), (b) TEM centred dark field micrograph (reflection shown) of the fine tabular precipitates found between the coarser particles seen in (a)

Table 2 Ionic Radii[18]

Atomic Species	Charge	Ionic Radius (Å)
Al	3+	0.51
Cr	3+	0.63
Ti	3+	0.76
Ti	4+	0.68
Y	3+	0.89
Zr	4+	0.79

present as Ti^{3+} stems from the observation that no precipitation occurs when the sample is annealed *in vacuum* at 1400°C.

The softening associated with surface amorphisation in region III has been described and discussed in detail elsewhere[7,8]. Generally it may be stated that when the scale of the hardness test is such that it samples both the surface layer and the substrate then the thicker the layer, the greater the decrease in hardness. Due to the tail of the implantation profile, some solid solution hardening of the material beneath the amorphous layer might be expected, but this is probably completely masked by the softening effect of the amorphous layer itself.

Compared to unimplanted material, the results for the Ti^+ and Cr^+ implanted sapphire in region II show little change in hardness. It may be that the solid solution hardening of the surface crystalline material is balanced by the softened material in the sub-surface amorphous zone. In the case of Y^+ and Zr^+, region II covers a much smaller dose range and hence the intermediate hardness régime is less marked or absent.

Annealing the Ti^+ implanted specimens *in air* showed further hardening believed to be associated with the formation of precipitates. The precipitates visible by light microscopy are considered too large to contribute significantly to this hardening but the smaller particles shown in Fig. 9 may well be responsible. The loss of this hardening with increasing anneal time is presumed to be due to particle coarsening. An additional contribution to this hardening may arise from a change in oxidation state of the Ti ions remaining in the matrix during the anneal, i.e. $Ti^{3+} \rightarrow Ti^{4+}$. EDX analysis of the matrix revealed ~0.3% Ti remaining after the anneal (the limit of solid solubility)[13].

The rapidity of nucleation and growth of the precipitate phase is dependent on the level of displacement damage present i.e. precipitation (and the subsequent rise in hardness) occurs more rapidly in the higher dose specimen, ST3, than ST2. Heterogeneous nucleation can occur as witnessed by precipitates forming preferentially around indentations made prior to annealing (Fig. 10). Since precipitation in general only seems to occur after the removal of the colour centres, it may be that nucleation is always heterogeneous occurring on

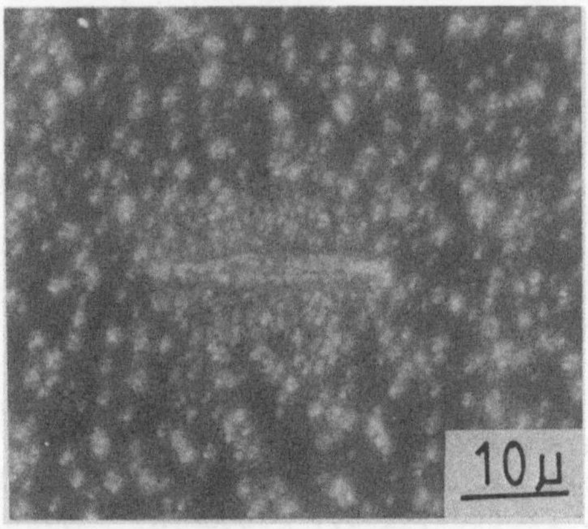

Fig.10. Heterogeneous precipitation seen around a Knoop indentation on ST3 after annealing for 1 hr. at 1150°C (Nomarski; interference contrast).

the Frank loops and small voids probably found in the early stages of annealing.

6. CONCLUSIONS

A range of near-surface softening and hardening phenomena can be induced in sapphire as follows:-

(i) High dose implantations result in a softening of the near-surface layer detectable by low-load microhardness tests. Both damage calculations and microstructural examination confirm that the effect is due to surface amorphisation.

(ii) Lower dose implantations may produce solid solution hardening dependent upon the implant species, the resultant concentration and the final charge state.

(iii) Post implantation heat treatment may produce precipitates and concommitant hardening depending on implant species, annealing temperature and atmosphere.

ACKNOWLEDGEMENTS

The authors wish to thank Prof. R.W.K. Honeycombe, FRS, for provision of laboratory facilities, SERC and AERE Harwell for support, GEC (Wembley, UK) for materials and Mrs. E. Palmer for kindly typing this manuscript.

REFERENCES

1. G.R. Sawyer, P.M. Sargent and T.F. Page, Microhardness Anisotropy of Silicon Carbide, J. Mater. Sci., 15:1001 (1980).

2. M.G.S. Naylor and T.F. Page, Microstructural studies of the temperature-dependence of deformation structures around

hardness indentations in ceramics, J. Microsc., 130:345 (1983).
3. G. Dearnaley, Ion Implantation for Improved Resistance to wear and Corrosion, Mats. in Eng. Applications. 1:28 (1978).
4. N.E.W. Hartley, Applications for Ion Implantation as a Surface Treatment Process in Production Engineering, Rad Eff., 44:19 (1979).
5. N.E.W. Hartley, Mechanical Property Improvements on Ion Implanted Diamond, in: "Metastable Materials Formation by Ion Implantation," S.T. Picraux and W.J. Choyke, Eds., Elsevier, New York, (1982).
6. S.G. Roberts and T.F. Page, The Effects of N_2^+ Ion Implantation on the Hardness and Wear Behaviour of Brittle Materials, in: "Ion Implantation into Metals," V. Ashworth, W.A. Grant and R.P.M. Procter, eds., Pergamon, Oxford, (1982).
7. P.J. Burnett and T.F. Page, Surface Softening in Silicon by Ion Implantation, J. Mater. Sci., in press.
8. P.J. Burnett and T.F. Page, Changing the Surface Mechanical Properties of Si and α-Al_2O_3 by Ion-Implantation, J. Mater.Sci., to be published.
9. C.J. McHargue, H. Naramoto, B.R. Appleton, C.W. White and J.M. Williams, Structure and Properties of Single Crystal Al_2O_3 Implanted with Chromium and Zirconium, in: "Metastable Materials Formation by Ion-Implantation," S.T. Picraux and W.J. Choyke, eds., Elsevier, New York (1982).
10. H. Naramoto, C.W. White, J.M. Williams, C.J. McHargue, O.W.Holland, M.M. Abraham and B.R. Appleton, Ion Implantation and Thermal Annealing of α-Al_2O_3 single crystals, J. Appl.Phys., 54:683 (1983).
11. T.E. Mitchell and A.H. Heuer, Solution Hardening by Aliovalent Cations in Ionic Crystals, Mats. Sci. Eng., 28:81 (1977).
12. J. Lindhard, M. Scharff and H.E. Schiøtt, Range Concepts and Heavy Ion Ranges, Mat. Fys. Medd. Dan. Vid. Selsk, 33 No.14 (1963)
13. E.M. Levin and H.E. McMurdie, "Phase Diagrams for Ceramists, 1975 Supplement," The American Ceramic Society, Columbus (1975).
14. G. Carter and W.A. Grant, "Ion Implantation into Semiconductors," Arnold, London (1976).
15. B.R. Lawn and T.R. Wilshaw, Indentation Fracture: Principles and Applications, J. Mater. Sci., 10:1049 (1975).
16. J.H. Crawford, A review of Neutron Damage on Corundum Crystals, J. Nucl. Mats., 108-9:644 (1982).
17. O.S. Phillips, A.H. Heuer and T.E. Mitchell, Precipitation in Star Sapphire, I., Phil. Mag. A., 42:385 (1980).
18. R. C. Weast, "Handbook of Chemistry and Physics," CRC Press, Ohio (1976).

KEYWORDS

Amorphisation, ion-implantation, microhardness, precipitation hardening, radiation damage by ions, sapphire, scratch test, solid solution hardening

DEFORMATION-INDUCED CRACK INITIATION BY INDENTATION OF SILICATE MATERIALS

H. Multhopp, B.R. Lawn and T.P. Dabbs

Center for Materials Science
National Bureau of Standards
Washington, DC 20234

ABSTRACT

The micromechanics of radial crack initiation produced in indentation of soda-lime glass, fused silica and quartz are discussed in terms of a two-step, nucleation and growth model. Particular attention is focussed on the strong rate effects in the presence of environmental water, as manifested by a tendency to delayed crack pop in with decreasing contact duration. Microscopic examination of the indentations indicates that deformation "shear faults", which accommodate the intense strains associated with the penetrating indenter, control the kinetics of the initiation process. The geometrical constraints which determine the stress concentrations for crack nucleation from these faults are structure-sensitive.

INTRODUCTION

Recent studies on the micromechanical response of soda-lime glass in Vickers indentation have shed new light on the mechanisms of crack initiation in brittle amorphous solids.[1-3] Essentially, the initiation process takes place within the deformation zone beneath the indenter, and thereby generates characteristic "radial" cracks from the corners of the impression.[4,5] There is an intrinsic scale effect in this process, in that the radial cracks appear only above some threshold in the loading; fracture instability may be identified with the attainment of a critical flaw size in the near-contact region.[6,7] The feature of this threshold which is arousing the greatest current interest, particularly in the context of strength in the ultra-small flaw region (e.g. optical fibers),[8,9] is a strong rate dependence in the presence of environmental moisture.[1] The rate effect is most dramatically manifested as an increasing tendency for the radial cracks to "pop in" after completion of the indentation cycle with decreasing height or width of the force-time pulse. Attempts to model the kinetic response have thus far been restricted to qualitative descriptions of the nucleation and growth of microcracks from incipient "shear faults" within the deformation zone, with due recognition of the residual contact field as a primary driving force.[1]

In this paper we expand our data base on soda-lime glass, and extend our material compass to two other silicate modifications, fused silica and quartz. Fused silica is known to be "anomalous" in its deformation behavior, in that the penetrating indenter is accommodated by a densification rather than by a cooperative shear failure of the glass structure.[10] Quartz, the "ordered" counterpart of fused silica, introduces a crystallographic element into the indentation response. By encompassing a broader range of deformation modes we may hope to gain some insight into the fundamental structural requirements for the operation of crack-initiation processes, and ultimately to develop quantitative models for characterizing flaws in the subthreshold region.

EXPERIMENTAL

A schematic of the experimental arrangement used to observe the indentations is shown in Fig. 1. The Vickers indenter arm is activated by an electromagnetic coil drive, with facility to control (via a function generator) the load-time pulse, $P(t)$. A piezoelectric cell in the arm allows for continuous monitoring of this pulse. The most commonly used pulse shapes in our experiments are square and half-sine, characterized by height P_m and base-width T. The indenter system seats on to the stage of an inverted microscope for in situ viewing of the contact area. In this way the pop-in time t_c from start of the pulse to appearance of the first radial crack can be measured directly. The specimen is housed within a simple enclosure for environmental control.

The soda-lime and fused silica glass specimens were in the form of microscope slides. Quartz specimens were cut from synthetic single crystals, parallel to the basal plane. Some of the specimens were specially prepared for post-indentation examination by scanning electron microscopy for clues as to the nature of the underlying initiation processes.

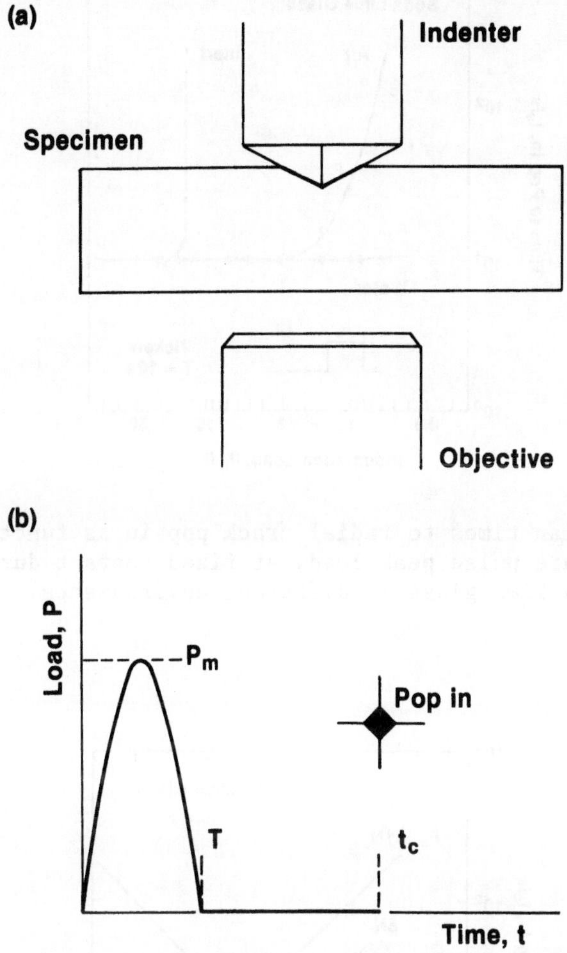

Figure 1. (a) Schematic of optical system for in situ viewing of Vickers indentations in transparent specimens. (b) Definition of load-time parameters used to characterize kinetics of radial crack pop in.

RESULTS

Radial Pop-in Kinetics

Let us now examine some of the more general features of the delayed pop-in kinetics. Figure 2 illustrates the influence of environment and indentation load on the fracture time response.

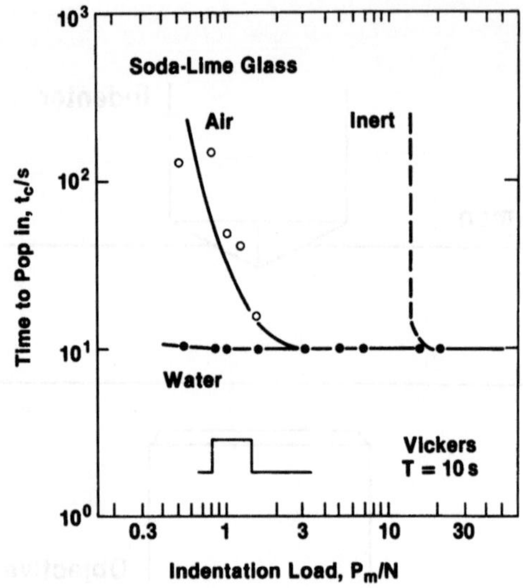

Figure 2. Median times to radial crack pop in as function of square pulse peak load, at fixed contact duration, for soda-lime glass in different environments.

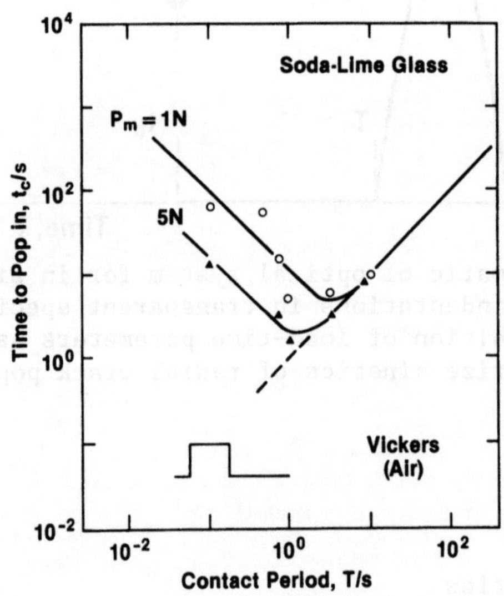

Figure 3. Median times to radial crack pop in as function of square pulse indentation hold time, at fixed peak loads, for soda-lime glass in air.

The data in this figure are for soda-lime glass using square-shaped load pulses of fixed duration 10 s. For the inert environment (dry N_2 gas) there is a well-defined cutoff load below which radial cracks cannot be made to pop in. In water, on the other hand, pop in occurs almost spontaneously as the indenter is removed, down toward the lowest attainable loads with our apparatus. An intermediate response is observed in air (55% relative humidity). Thus the delay time for radial cracking increases as the indentation load and the moisture content in the environment diminish.

The influence of contact duration is seen in Fig. 3. These data are for loads of 1 N and 5 N on soda-lime glass in air, taken under effectively the same test conditions as represented in Fig. 2. The delays to pop in increase abruptly below some critical duration, or "incubation time", this quantity diminishing with increasing load.

The use of square load pulses to obtain the data in Figs. 2 and 3 immediately provides us with useful information as to the nature of the critical condition for pop in. The fact that even under the most conducive of fracture conditions (i.e. high water contents, large loads, or long contact times) initiation occurs only on unloading the indenter means that some constraint must exist to prevent radial development while the system is maintained at maximum load. (Actually, at sufficiently high loads, well above the threshold limits defined here, cracks can generate on loading, but these are of a different kind - subsurface medians rather than surface radials.[1]) On the other hand, it is impossible with square pulses to determine exactly where on unloading this constraint is released.

It is in this context that the facility to change over to half-sine pulses proved useful. With this continuous load-time function it could be established that the constraint on crack instability disappears at $P \simeq 0.3\ P_m$, on the unload half-cycle, marking the instant at which the residual field begins to dominate. As a consequence of this capacity for initiating cracks <u>within</u> the contact cycle, the load-time configurations for which the instant of pop in just coincided with completion of the indentation cycle, defined by the condition $P_m = P_*$ and $t_c = T_*$, could be determined to relatively high accuracy. (Experimentally, these points were obtained by making series of indentations at preselected loads and systematically adjusting the pulse duration, decreasing or increasing T depending on whether pop in occurred without or within the load cycle.) Figure 4 shows data thus obtained for soda-lime glass and fused silica in air. A certain analogy may be noted between the plotting scheme in this figure and the lifetime plots used to characterize the fatigue strength of brittle materials[11]

The data in Fig. 4 suggest a strong dependence of the initiation mechanics on molecular structure. We see that the threshold load in inert environments is substantially greater for soda-lime than for fused silica glass, ≈10 N compared to ≈5 N. On the other hand, soda-lime shows a relatively larger susceptibility to hydrolytic rate effects (as measured by the inverse slope of the lines through the data points), so that at loads below ≈1 N cracking generates more easily than in fused silica. The thresholds for quartz are much lower than for their glassy counterparts, below 0.5 N under even inert conditions; those indentations which did not produce radial cracks immediately did so within minutes on exposure to air. Similar observations on quartz have been reported by Nadeau.[12] It would appear that the main influence of structure lies in the magnitude, rather than the nature, of the initiation driving force that can be generated; the same qualitative effects are seen in all three materials studied, but with substantially different sensitivities to extraneous variables.

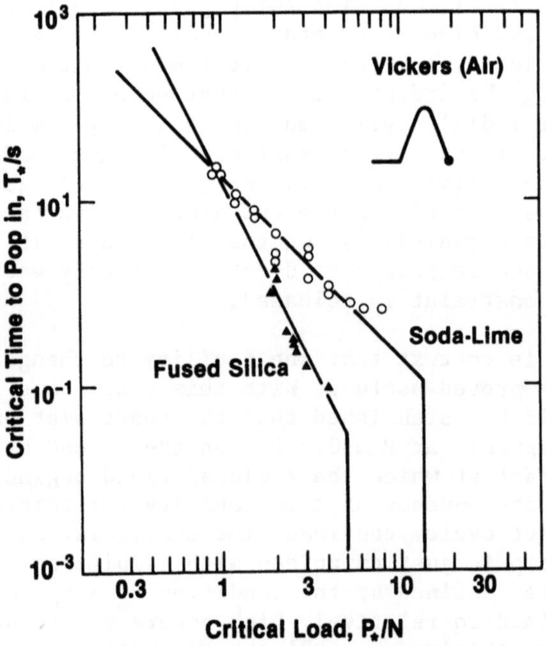

Figure 4. Plot of critical sinusoid load-time conditions to produce radial pop in at immediate unload point of contact cycle, for soda-lime and fused silica glasses in air. Vertical high-load cutoffs indicate spontaneous (time-independent) threshold limits in inert environments.

Microscopy of Indentations

Scanning electron microscopy provides some insight into the underlying mechanisms of radial crack initiation.[1-3] One of the most revealing procedures for observing the contact zone is to indent across a pre-existing hairline crack, and then to run this crack through the specimen to obtain a cross section view. Examples of Vickers indentations made in this way are shown in Fig. 5 for soda-lime and fused silica glass.

Such observations indicate that the non-recoverable component of the indentation deformation is a far from homogeneous process. Distinctive fine structure, on the micrometer scale, is apparent within the hardness impressions. This discrete detail may be interpreted in terms of an intermittent "faulting" mechanism, whereby the intense strains associated with the penetrating indenter are accommodated by local, catastrophic shear failures of the structure at the specimen surface.[1,10,13,14] The periodicity of this "punching" mode is attributable to the need to build stresses back up to the theoretical strength of the glass after each fault is formed. The shear offsets associated with the fault traces on the free surfaces are measurably large, again on the order of micrometers,[14] attesting to the catastrophic nature of the failure events. On etching in dilute hydrofluoric acid the fault traces show up much more clearly,[2] demonstrating that the slipped regions must have the character of high energy interfaces. It is clear from micrographs such as those in Fig. 5 that it is indeed these shear faults which provide the nuclei from which the radial cracks generate.

It is interesting at this point to examine the features of the fault patterns which differ in the materials studied. Some of these features may be distinguished in Fig. 5 for the two glass types. In soda-lime the faults propagate well into the subsurface region, ultimately intersecting with neighbors from other quadrants of the hardness impression. The high concentration of network modifier ions in this glass apparently provides more or less continuous paths of "easy slip" through the structure.[10,15] In fused silica the absence of modifying elements precludes such cooperative slip modes, and the subsurface material accommodates the indentation volume by network densification.[10] Accordingly, the stress intensifications associated with the surface faults are not so easily relieved in the fused silica (although in some favorable instances the constraints can be overcome by propagation into a cone crack configuration, Fig. 5), so the thresholds for radial initiation are somewhat lower than for soda-lime. On the other hand, the densification mode leaves the surrounding material in a relatively low state of residual strain, so the newly initiated radials do not propagate as far outward from the impression corners as do their counterparts in soda-lime glass.[10]

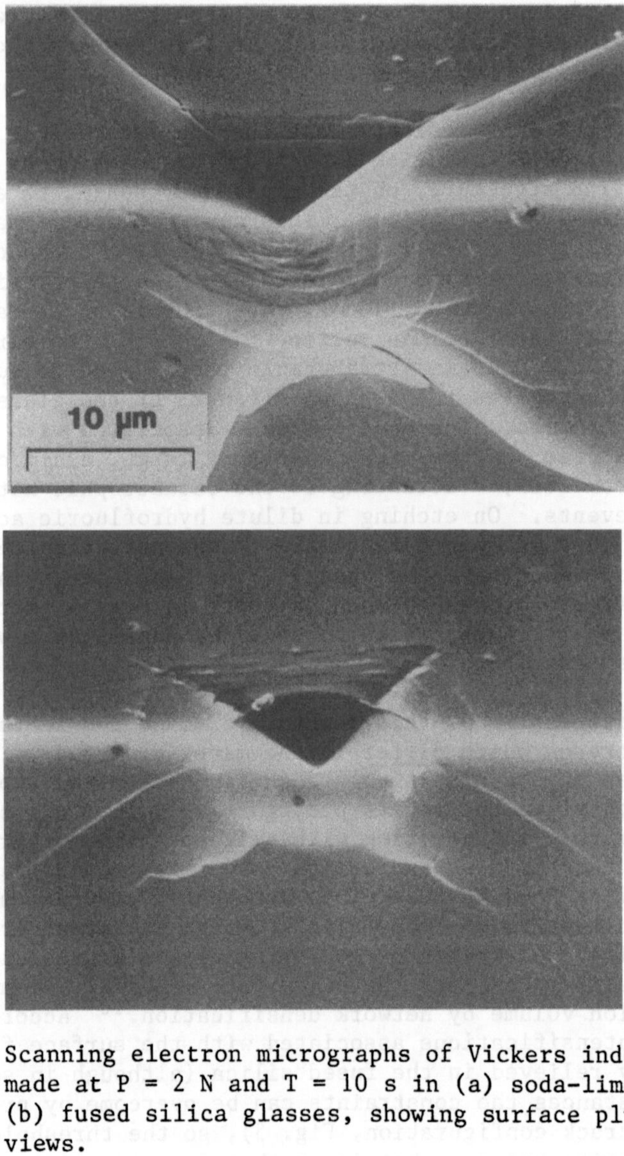

Figure 5. Scanning electron micrographs of Vickers indentations made at P = 2 N and T = 10 s in (a) soda-lime and (b) fused silica glasses, showing surface plus section views.

Similar fault markings are seen within surface indentations in quartz,[16,17] with some indication that slip now occurs preferentially on certain crystallographic planes.[16] It is conceivable that the additional constraints on the slip flexibility imposed by such crystallographic requirements could account for the ease with which cracks initiate in quartz, but our current understanding of the basic mechanisms of deformation in this material,[12] particularly in the subsurface deformation zone, is too obscure to allow for any more definitive a conclusion than this.

The question we now need to address is, do the observed shear fault patterns show any distinctive changes with contact duration? Our only attempts to answer this question thus far have been confined to observations using the optical microscope setup of Fig. 1. At the limits of resolution attainable with this setup one could just discern the incidence of individual shear events as lines of enhanced reflection within the square contact shadow of the Vickers pyramid. On reducing the contact duration the intensity of these reflections appeared to diminish. This trend correlated with an increasing "pincushion" distortion of the final impression; in soda-lime glass at P = 1 N, for instance, the ratio of "short" (edge center to opposing edge center) to "long" (corner to corner) dimensions (0.707 for ideal, square impression) dropped from 0.61 ± 0.02 at T = 10 s to 0.55 ± 0.02 at T = 0.1 s. The implication here is that the faults have less time to develop fully at the faster loading rates, indicating that the deformation process itself has an intrinsic kinetic element.

MODEL

The preceding observations lead us to a simple, two-step model for the crack initiation, depicted in Fig. 6.[1] The first step involves nucleation of the microcrack from which pop in ultimately ensues. The nucleation is expected to occur at points where the constraints on fault expansion are high, notably along the impression diagonals. Thus the major dimensions of the responsible defect configuration may remain effectively invariant during this stage of initiation; the process of stress intensification would then be more akin to flaw "sharpening" than to flaw "lengthening". The driving force for this first step comes from the shear component of the field, which reaches its maximum intensity at peak load but persists to a significant degree (as evidenced by the fact that the impression depth does not fully recover[18]) after removal of the indenter.

Step two involves growth of the newly formed microcrack to some critical instability point. The driving force now comes from the tensile component of the field which, as we have alluded, develops in the nucleation region only as the indenter is being

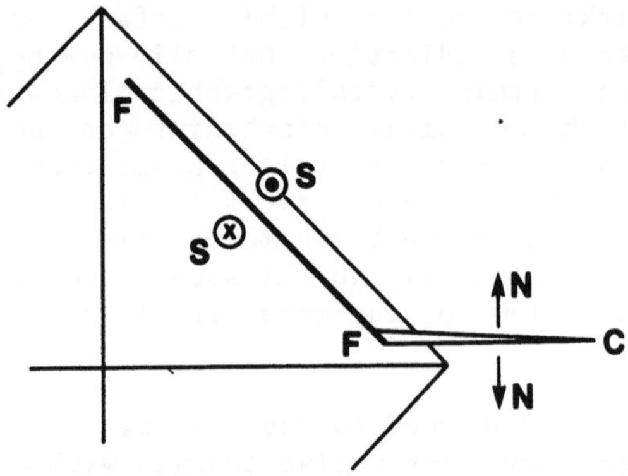

Figure 6. Schematic model of radial microcrack initiation FC from fault FF, in one quadrant of Vickers impression. Shear SS and normal NN stresses drive fault and crack respectively.

unloaded. Accordingly, this step should operate more effectively after than during the contact cycle. (This last assertion gains weight from the observation that reloading contact half-cycles cause any fully developed radial cracks to close up at the specimen surface.[1])

The fact that crack initiation occurs more easily at higher indentation loads (Fig. 2) could have its explanation in either of the two steps above. Generally, the effect of increasing the load is not to increase the intensity of the stress field (as governed by the hardness, which remains more or less invariant), but rather the spatial extent of this field.[4] Thus in the nucleation stage we would expect the size of the embryo crack to increase in some proportion to the size of the critical shear fault. In the growth stage this same crack would be subject to smaller gradients of tensile stress in the field of larger impressions.[6] At a sufficiently large indentation size the driving forces reach a level where both stages occur spontaneously, thereby defining the critical conditions for inert environments.

Insofar as rate effects are concerned, either step could exert a controlling influence. Here the paramount consideration is the role of water. This role has been explored in depth in the general deformation[19,20] and fracture[21,22] properties of amorphous and crystalline silicates. In terms of the specific model of Fig. 6, we expect the shear faults to provide preferred diffusion paths for atmospheric water; the interfaces, although in a basically compressive indentation zone, nevertheless represent planes of cohesive weakness, as evidenced by their susceptibility to attack in etch solutions. Further, hydrolytic weakening of these interfaces ("decohesion") may then redistribute the active shear stresses, so as to build up the stress concentrations at the impression corners. Once nucleation has occurred, the water is free to enter the newly formed microcrack and promote slow growth.

This leaves us with the influence of contact duration to account for. Clearly, the data in Figs. 3 and 4 are qualitatively consistent with any load-dependent rate process. Decreasing T at given P effectively reduces the time available to drive the initiation during actual contact, thereby increasing the tendency to delayed pop in under the persistent action of the residual field. Our observation that the initiation does not reach completion within the initial, loading half-cycle now takes on a special significance. It suggests strongly that of the two stress components pertinent to our model, the shear driving the fault and the tension driving the microcrack (Fig. 6), it is the former which is the more crucial. We have implied that the normal stresses at potential initiation centers must remain highly compressive up to (and beyond) the peak contact configuration; and yet rate effects are evident even in square pulse experiments where the development of the residual tension is effectively instantaneous. Moreover, if initiation does not occur within the contact cycle the delay times can be inordinately large, orders of magnitude greater than the contact duration itself (Fig. 3), consistent with the assertion that the shear driving forces remain operative, but at diminished intensity, in the unloaded state. Finally, we may recall the tendency to less pronounced development of the shear fault configurations (as reflected in the increased pincushion distortion of the overall impressions) on going from relatively long (T = 10 s) to short (T = 0.1 s) contacts, corresponding to a shift from the spontaneous to the delayed pop-in regions in Fig. 3.

CONCLUSIONS

We have proposed a two-step, nucleation and growth, model of radial crack initiation beneath sharp indenters. The evidence on silicate structures presented in support of this model leads us to conclude that it is the micromechanics of "shear fault" evolution in the presence of interactive (moist) environments which hold the key to the strong rate dependence apparent in the threshold conditions. The nature of this evolution remains somewhat obscure. The faults clearly represent catastrophic failure events, operating at or close to the theoretical strength of the silicate structures, but do not fit in with the classical description of slip by dislocation motion and regeneration, or of fracture by a shearing mode; in the first instance the process occurs in both glassy and crystalline modifications of the SiO_2 structure, and in the second the existence of remnant interfacial cohesive forces violates the requirement of stress-free boundaries. There is a need for closer investigation of these defect structures (e.g. by transmission electron microscopy) in different brittle materials, as well as of the crack nuclei that develop from these structures.

ACKNOWLEDGEMENTS

Funding for this study was provided by the U.S. Office of Naval Research, Metallurgy and Ceramics Program.

REFERENCES

1. B.R. Lawn, T.P. Dabbs and C.J. Fairbanks, J. Mater. Sci., in press.

2. T.P. Dabbs, C.J. Fairbanks and B.R. Lawn, in "Methods for Assessing the Structural Reliability of Brittle Materials," S.W. Freiman, ed., A.S.T.M. Special Technical Publication, in press.

3. B.R. Lawn, in "Strength of Glass", C.R. Kurkjian, ed., Plenum Press, in press.

4. B.R. Lawn and T.R. Wilshaw, J. Mater. Sci. 10:1049 (1975).

5. B.R. Lawn, A.G. Evans and D.B. Marshall, J. Amer. Ceram. Soc. 63:574 (1980).

6. B.R. Lawn and A.G. Evans, J. Mater. Sci. 12:2195 (1977).

7. K.E. Puttick, J. Phys. D.; Appl. Phys. 11:595 (1978).

8. T.P. Dabbs and B.R. Lawn, Phys. Chem. Glasses 23:93 (1982).

9. B.R. Lawn, D.B. Marshall and T.P. Dabbs, in "Strength of Glass," C.R. Kurkjian, ed., Plenum Press, in press.

10. A. Arora, D.B. Marshall, B.R. Lawn and M.V. Swain, J. Non-Cryst. Solids 31:915 (1979).

11. R.F. Cook and B.R. Lawn, in "Methods for Assessing the Structural Reliability of Brittle Materials," S.W. Freiman, ed., A.S.T.M. Special Technical Publication, in press.

12. J.S. Nadeau, J. Amer. Ceram. Soc. 53:568 (1970).

13. J.T. Hagan and M.V. Swain, J. Phys. D.; Appl. Phys. 11:2091 (1978).

14. J.T. Hagan, J. Mater. Sci. 15:1417 (1980).

15. F.M. Ernsberger, J. Non-Cryst. Solids 25:293 (1977).

16. W.F. Brace, J. Geology 71:581 (1963).

17. M.V. Swain, Ph.D. Thesis, University of New South Wales, Australia, 1973.

18. B.R. Lawn and V.R. Howes, J. Mater. Sci. 16:2475 (1981).

19. D.T. Griggs and J.D. Blacic, Science 147:292 (1965).

20. S.P. Gunasekera and D.G. Holloway, Phys. Chem. Glasses 14:45 (1973).

21. S.M. Wiederhorn and L.H. Bolz, J. Amer. Ceram. Soc. 53:543 (1970).

22. C.H. Sholz and R.J. Martin, J. Amer. Ceram. Soc. 54:474 (1971).

14. J. T. Regan, J. Mater. Sci. 15:1447 (1980).

15. [illegible], J. Non-Cryst. Solids 25:293 (1977).

16. W. F. Brace, J. Geology 71:581 (1963).

17. M. V. Swain, Ph.D. Thesis, University of New South Wales, Australia, 1973.

18. [illegible] and T. R. [illegible], J. Mater. Sci. 16:2475 (1981).

19. D. T. Griggs and J. D. Blacic, Science 147:292 (1965).

20. H. P. Gunasekera and D. G. Holloway, Phys. Chem. Glasses 14:45 (1973).

21. S. M. Wiederhorn and L. H. Bolz, J. Amer. Ceram. Soc. 53:543 (1970).

22. C. H. [illegible] and S. J. [illegible], J. Amer. Ceram. Soc. 56:[illegible] (1973).

INDUCED PLASTIC DEFORMATION OF ZIRCONIA

R.H.J. Hannink and M.V. Swain

CSIRO, Division of Materials Science
Advanced Materials Laboratory
P.O. Box 4331, Melbourne, Australia 3001

Indentation techniques have been used to study the low temperature deformation characteristics of zirconia based ceramics. Analysis of Knoop hardness anisotropy showed the low temperature slip systems were {100}<011>, the same as the high temperature slip system. Hardness as a function of temperature, up to 600°C revealed the possibility of a different deformation process being active above 400°C. Deformation about large load indents in Mg-PSZ were studied to examine the extent of cooperative transformation phenomena.

INTRODUCTION

Indentation hardness is the most commonly used technique for studying the low temperature deformation charcteristics of 'brittle' materials (1). By using indenters, plastic deformation may be induced at low temperatures (<0.3Tm) because the large hydrostatic stress component around the indenter inhibits brittle failure.

The technique of indentation hardness as well as being used to study the yield and flow stress criteria of normally brittle materials, has also been successfully used to determine the hardness anisotropy and slip systems of hard and/or brittle materials (2).

Zirconia based ceramics have recently become the focal point of intense scientific and technological interest because of their ability to achieve enhanced thermo-mechanical properties as a result of heat treatment (3). The improved properties are predominantly the result of a volumetric expansion associated with the tetragonal to monoclinic transformation of precipitates dispersed in a cubic stabilized matrix. These materials have found application in a diverse variety of mechanically and environmentally demanding areas

(4). To-date virtually no data is available on the low temperature plastic deformation behaviour of these materials.

We have used indentation techniques to examine the low temperature ($\leq 600^{\circ}C$) deformation characteristics of single crystal, cubic stabilized (CSZ), and polycrystalline, partially stabilized (PSZ), zirconia materials.

EXPERIMENTAL

The PSZ materials studied were prepared by standard powder and thermal treatment routes (5). Compositions were 3.3 wt% $MgO-ZrO_2$ (Mg-PSZ)+ and 3.4 wt% $CaO-ZrO_2$ (Ca-PSZ). Single crystal samples of fully stabilized zirconia were purchased* and comprised 14 mole% $CaO-ZrO_2$ (Ca-CSZ) and 14.8 mole% $Y_2O_3-ZrO_2$ (Y-CSZ).

Surfaces suitable for indentation were cut and mechanically polished to a 1μm diamond finish using a tin lap technique (6). Any residual surface damage was removed by either heating the samples for 1 hr at $1000^{\circ}C$ in air or 30 sec etch in HF. The type of surface treatment used was dependent upon the composition, prior thermal history and microstructure of the sample.

Single crystal surfaces were orientated using a Laue back reflection camera, cut and then prepared as per above.

All indentations were made with indenters shaped to a Knoop profile (1). This was done so that hardness anisotropy measurements made on single crystals were directly comparable to measurements made on polycrystalline samples. It was found, that five to ten indents were sufficient to highlight statistically significant differences. Indentations were made on a modified Shimadzu Microhardness Tester using loads of either 500 gmf or 200 gmf.

Micro hardness indentations to $600^{\circ}C$ were made using an improved version of a previously described hot stage (7). Samples were seated on a WC-Co anvil, mounted in a flat bottomed cavity on the top of a silver rod encased in a small furnace. A flatened 0.2 mm diam. R-type thermocouple was placed in contact with the sample surface. All tests were performed in air.

Large load indents were made using a diamond indenter fitted to the cross head of an Instron testing machine. For indentations made at other than room temperature a furnace was fitted around the specimen platform.

+Nilsen Sintered Products (Aust.)Pty.Ltd., 239 Separation St., Northcote, Victoria, Australia, 3070.

*W. & C. Spicer Ltd., St. Mary's Winchcomb, Gloucestershire, UK.

RESULTS AND OBSERVATIONS

1) Hardness anisotropy of single crystal CSZ

Hardness anisotropy was measured on the three major low order planes of Y-CSZ, viz, (001), (110) and (111), as a function of the long axis orientation of the Knoop indenter. Measurements were made at room temperature except for the (001) surface where the anisotropy was also measured at 600°C. The Knoop hardness (KH) measurements versus the long axis orientation on all the surfaces are presented in figs 1-2. The dashed curves through the data points are best fit curves taking into account the symmetry of the indented surfaces.

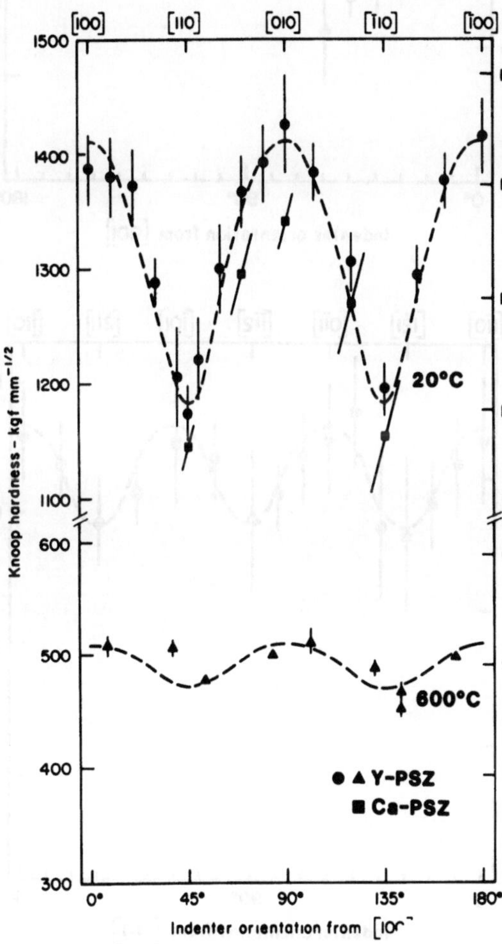

Fig. 1. Knoop hardness anisotropy of the (001) for Y-CSZ, at 20°C and 600°C. Some Ca-CSZ data included for comparison.

(a)

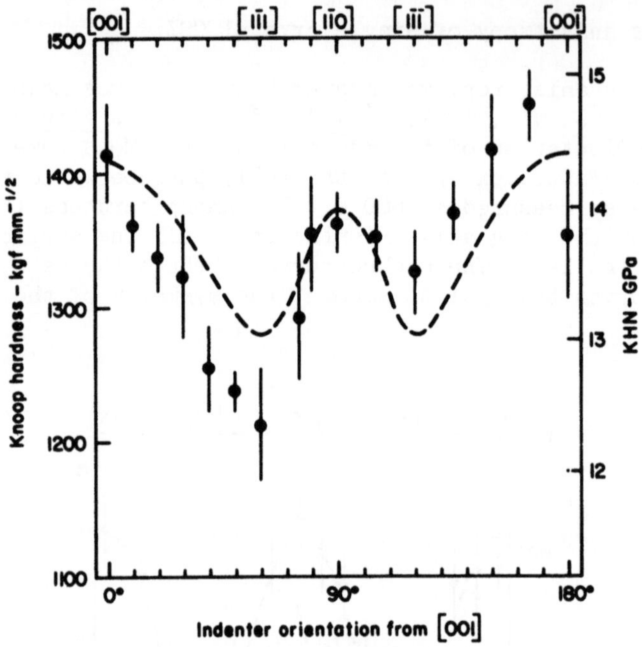

(b)

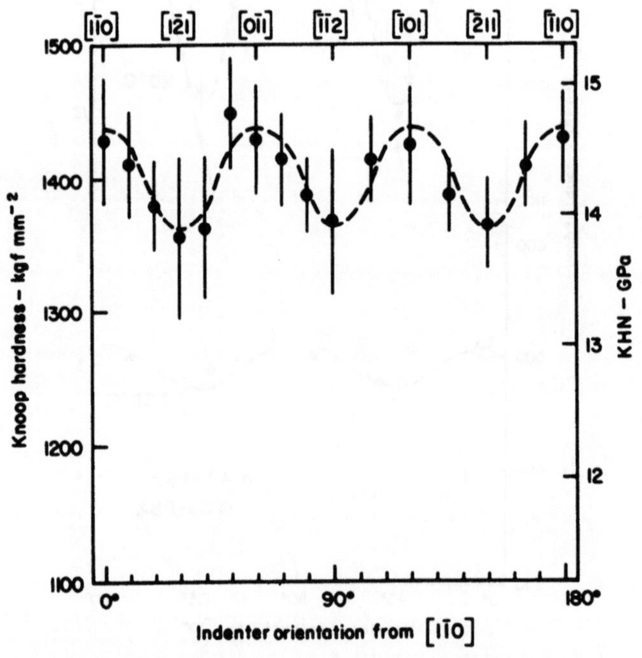

Fig. 2. Knoop hardness anisotropy of the (a) (110) and (b) (111) surfaces of Y-CSZ.

Room temperature KH was also measured on the (001) surface of single crystal Ca-CSZ. Some data points are included in Fig 1 for comparison. Since the anisotropy profile was the same as Y-CSZ, varying only slightly in intensity, and because more and larger single crystals of Y-CSZ were available the Ca-CSZ examination was not persued. It has been assumed during the course of this work that the results obtained for Y-CSZ would also pertain generally to single crystals of Ca-CSZ and Mg-CSZ. This assumption is reinforced for Ca-CSZ in the next section.

Hardness anisotropy data for indentations made at 600°C is also shown in Fig 1. The dashed line indicating a four fold symmetry of the anisotropy could equally well have been replaced by a straight horizontal line, on the basis of the data scatter, indicating virtually no hardness anisotropy at 600°C.

2) Knoop hardness versus temperature

Knoop hardness (KH) as a function of temperature, up to 600°C, was measured on a variety of PSZ materials and for comparison also on a randomly orientated Y-CSZ crystal. The materials tested were specimens of Ca-PSZ aged for various times at 1300°C and Mg-PSZ for various times either above (1420°C) or below (1100°C) the eutectoid temperature (1400°C) using loads of 500 gmf.

Hardness measurements of PSZ materials as a function of temperature were essentially the same irrespective of their thermal history. Major variations occuring only in relative starting and finishing hardness values with temperature as related to the bulk (represented by the polished surface) monoclinic contents. For this reason, only one complete set of data points has been presented. Fig 3(a) shows the KH versus temperature of Mg-PSZ material aged at 1420°C. Also shown in the figure are end point hardness results for Mg-PSZ aged at 1100°C The dashed line in Fig 3(a) shows the averaged hardness trend for the Y-CSZ, measurements being made in incremental steps of 100°C.

An Arrhenius plot showing the ln hardness versus inverse temperature for the same materials are plotted in Fig 3(b). A very marked feature of the data is the change in slope at about 350°C.

3) Deformation around indentations

Earlier work (8,9) has shown that in PSZ materials it is possible to induce the tetragonal to monoclinic transformation in a cooperative manner around indentations. In the light of these observations the nature and extent of the 'apparent' surface transformations have been made as a function of temperature. Loading has been carried out using either Vickers or Knoop profile indenters at loads up to 50 kgf.

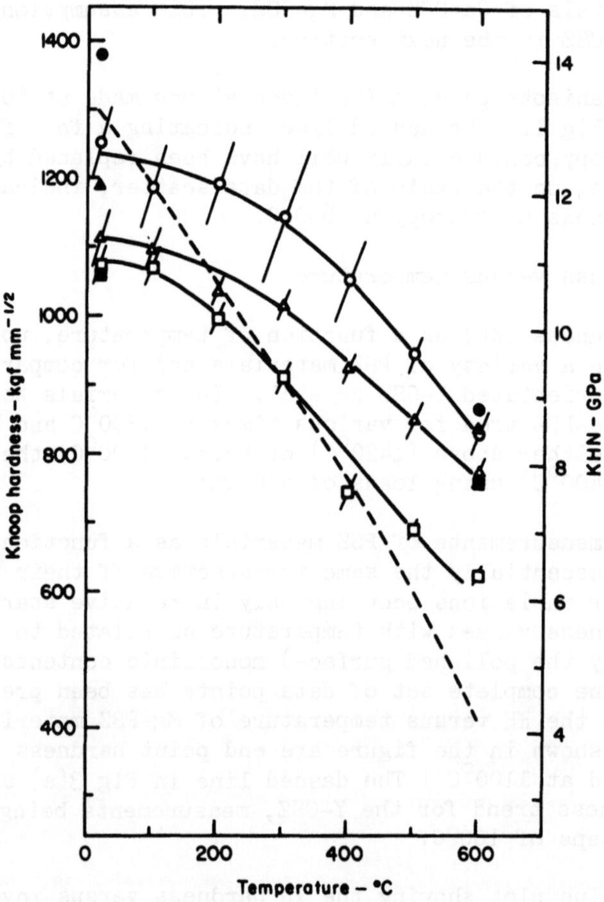

Fig. 3. (a) Knoop hardness versus temperature plot of zirconia test materials. Open symbols Mg-PSZ. as-fired (○), aged at 1400°C for 2h (△) and 8h (□). Solid symbols end point data for 1100°C aged Mg-PSZ, as fired (●), aged 2h (▲) and 8h (■). Dashed line, averaged data for Y-CSZ.

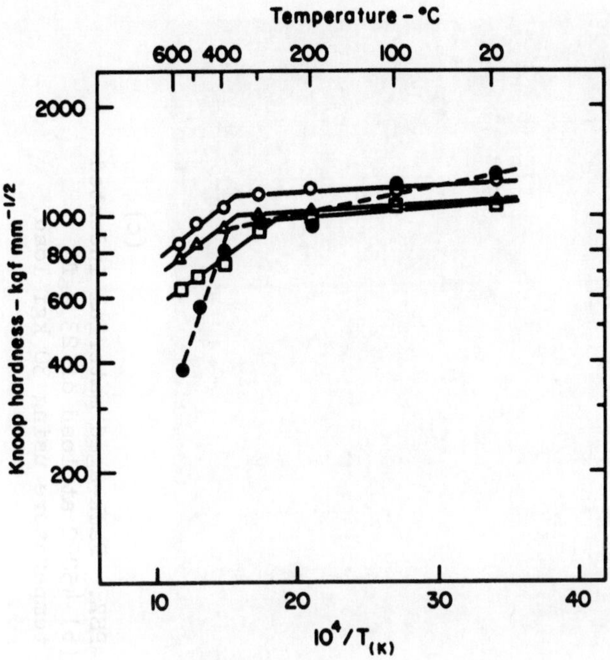

Fig. 3. (b) log knoop hardness versus inverse temperature for Mg-PSZ and Y-CSZ data data presented in (a). Open symbols as in (a), solid symbol Y-CSZ.

Figure 4(a) shows the nature of a typical deformation pattern obtained when optimally aged (peak-aged) Mg-PSZ material is indented at room temperature using a Vickers indenter and a load of 30 kgf. A feature immediately apparent is the copious amount of grain rotation and shear banding within the grains. Figure 4(b) shows the same sample indented at 450°C using a 23 kgf load. Whilst gross difference between the indentations are not immediately apparent the decrease in intensity and extent of the deformed region has occured with the 450°C deformed region being about 10% smaller.

A similar experiment on over-aged Mg-PSZ, Fig 4(c), i.e. no tetragonal phase available for transformation, shows none of the above described deformation behaviour. This material behaved in a similar manner to 'normal' ceramics, where cracking occured from the corners and faces, large lateral cracks were also present which resulted in spalling and general surface unpheaval about the indents.

Figure 5 shows the deformation about Knoop indents made in peak-aged Mg-PSZ using 30 kgf at room temperature (a) and (b) using 10 kgf at 850°C. It can be noted from these figures that the heavier load indent, made at room temperature, shows evidence for cooperative transformation around the indent and no cracking, whilst the 850°C indent has cracks eminating from the long axis corners. The cracks propagate in an intergranular manner and a very minor

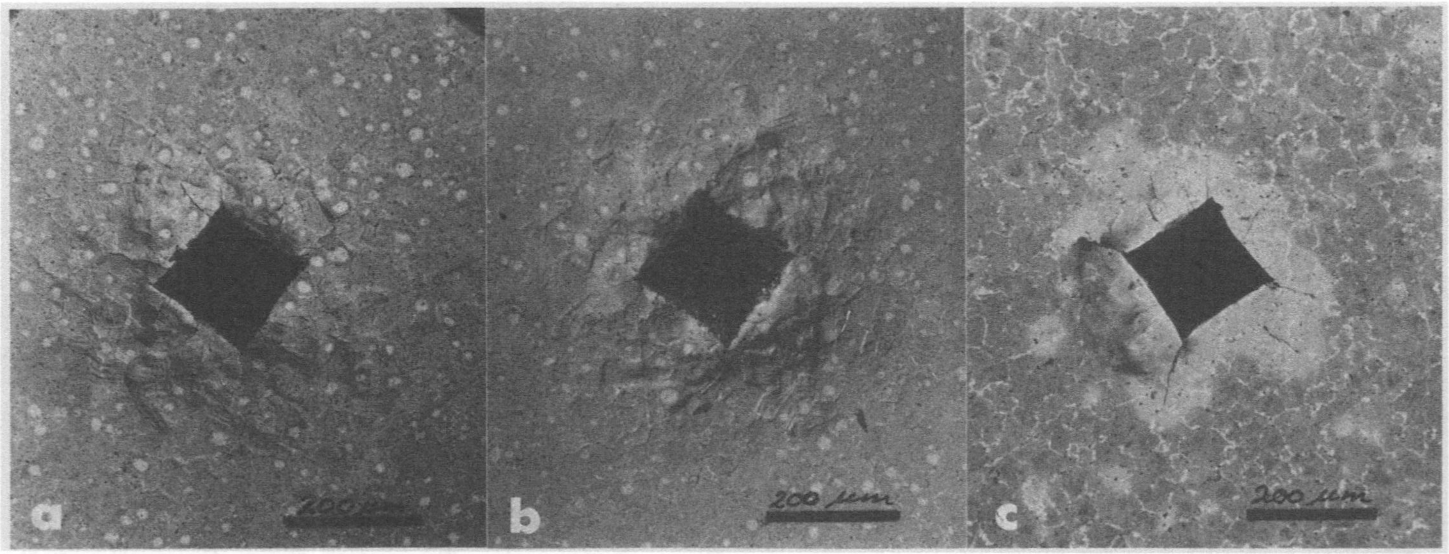

(a) (b) (c)

Fig. 4. Deformation about Vickers indents in Mg-PSZ. Pack-aged material indented at (a) room temperature using a 30 kgf (b) 450°C at load of 23 kgf (c) over-aged material indented at room temperature using 30 kgf load. Bar lengths equal 200 µm.

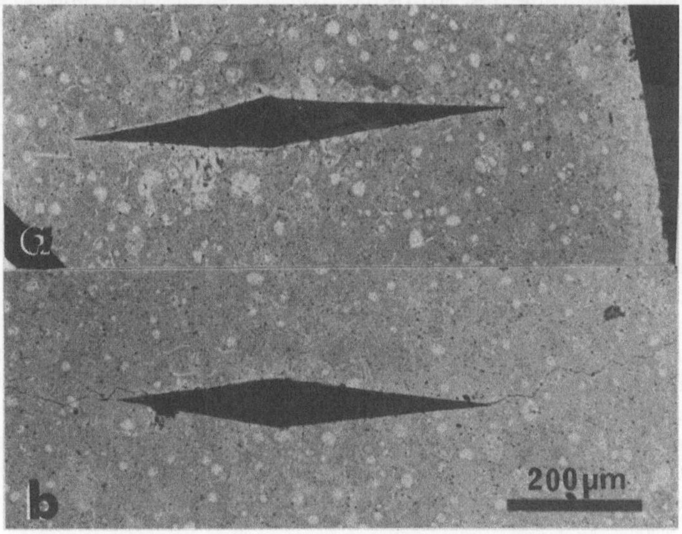

Fig. 5. Knoop indents in optimally aged Mg-PSZ, (a) room temperature at load of 50 kgf (b) 850°C at load of 30 kgf. Note cracking associated with the high temperature indent.

amount of cooperative transformation is also evident adjacent to the diagonal surface.

DISCUSSION

1) Hardness Anisotropy in Single Crystal CSZ

Analysis of the hardness anisotropy data presented in Fig 1 has been made using the effective resolved shear stress relationship (ERSS), modified by Brookes et al (2) from the original equation proposed by Daniels and Dunn (10). The premise for the relationship is the fact that the degree of hardness anisotropy is determined by the magnitude of the resolved shear stress under an indenter acting on the active slip systems of a crystal.

Whilst the relationship has a number of criticisms (1,11) it has been shown by independent means to predict correctly the operating slip systems when used to study crystals in their low homologous temperature regimes (Tm <0.3) (12,13). The analysis gives no indication of the magnitude of the hardness on the indented surface. For the present study resolution of anisotropy in terms of active slip systems is sufficient as the aim was to determine the low temperature slip systems.

In terms of the present analysis the ERSS model predicts that a crystal would appear least hard, producing a larger indent for a given load, when the indenter orientation is such that the greatest resolved shear stress acts on the crystals preferred slip system. Brookes et al (2) calculated ERSS for varying indenter orientations on (001) for cubic crystals when either {100} <011>, {110} <1$\bar{1}$0> or {111} <1$\bar{1}$0> primary slip systems are operating.

Plots of $ERSS^{-1}$ versus indenter orientation show hardness maxima, on (001) surfaces, at <100> for {100} <011> and {111} <$\bar{1}\bar{1}$0> slip system, and a minima at <100> for the {110} <1$\bar{1}$0> slip system. Now considering the room temperature data of Fig 1 we have maxima at <100> and minima at <110>. From the analysis this suggests the active slip system is either {100} <011> or {111} <1$\bar{1}$0>. Examination of isostructural CaF_2 (2) and the transmission electron microscope (TEM) analysis of high temperature compressive deformation of Ca-PSZ (14) showed the active slip system to be {100} <011>. On this basis the primary slip system for low temperature deformation in Y-CSZ and Ca-CSZ is also assumed to be {100} <011>. The primary slip system for Mg-CSZ would be expected to be similar.

A comparison of the hardness at various indenter orientations on the different surfaces can also be made, cf. Figs 1 and 2. It can be seen that the hardness values determined for the <100> directions on the (001) and (110) surfaces agree very well. However the <110> directions in the different crystal surfaces offer poor agreement, varying from values of about 1430 $kgf.mm^{-\frac{1}{2}}$ on (111) to about 1170 $kgf.mm^{-\frac{1}{2}}$ on the (001) surface. These differences highlight the variations in plastic constraint on the three surfaces.

Whilst determination of the {100} <011> slip system was not unexpected on the basis of previously reported data, it is some what unexpected that no other slip system is activated at temperature up to 600°C, or were reported in the high temperature compressive deformation study (14). As seen from Fig 1 a four fold symmetry based on {100} <011> could be drawn through the data obtained at 600°C. The von Mises criteria for slip suggests that five independent slip systems are necessary for a crystal to be able to undergo general homogenious strain by slip (see for example (15)). The lack of reported high temperature cracking and low work hardening rate in Ca-CSZ (14) is presumably due to minor cross slip and extensive dislocation climb occuring just above the brittle-ductile transition temperature. TEM examination of Knoop indentations at room temperature suggests that dislocation mobility can be induced even at these low temperatures. Using indentation techniques to induce deformation, dislocations have been observed using TEM in a variety of room temperature 'brittle' materials, eg. diamond (16), alumina (17) and niobium carbide (13).

If dislocation can be induced to propagate in the CSZ matrix as a result of shear stresses within the compressive environment generated by indents or precipitate laden matrices they could also provide the nucleating source for the tetragonal to monoclinic transformation in transformation toughened (TTZ) PSZ materials. It is now well known (3) that the tetragonal phase in suitably heat treated PSZ can undergo beneficial martensitic transformation in the presence of a crack or applied stress (18,19). However the nucleating mechanism for the martensitic transformation has remained elusive even though the problem has been examined in some detail (20). Therefore it is proposed that as a means of inducing nucleation, everything else being optimal, once the first tetragonal precipitate has transformed, placing an increased compressive shear stress into the CSZ matrix, a transformation 'cascade' of suitably sized precipitates is possible by the propagation of matrix dislocations. The first precipitates may nucleate by the passage of a moving crack (ie. stress induced transformation), microcracking at a precipitate interface or simply the spontaneous nucleation of a dislocation in the highly strained cubic matrix-tetragonal precipitate interface (strain induced transformation). Whilst strain induced transformation has been reported for PSZ (8,21), observations of dislocations have unfortunately not been reported for PSZ materials.

2) Hardness-Temperature Behaviour

The rapid decrease in hardness with temperature, Fig 3(a), is similar to the general behaviour reported for simple and complex oxides reported by Westbrook (22). In this comparison the CSZ material is likened to a simple oxides whilst the PSZ to complex oxides. The relevant feature of the data in Fig 3 is the rapid decrease in hardness for the single crystal Y-CSZ sample compared to the polycrystalline PSZ samples. This behaviour suggests that while the matrix of the two zirconias are the same, ie. cubic fluorite, the presence of the second phase reduces the rate of decrease in hardness with temperature.

The Arrhenius plots of Fig 3(b) shows that the decrease in hardness, even at low temperatures is a two stage process. For the PSZ data, stage I occurs from room temperature to about 400°C (Tm 0.1-0.23) whilst stage II is from 400°C to something in excess of 600°C (Tm 0.23->0.3). Since the hardness data is linear over the two regions it is possible to determine an activation energy from an expression of the form $H \propto \exp(Q/mRT)$, where Q is the activation energy governing plastic flow and m is defined in the steady-state creep relationship for strain rate and stress (23). Unfortunately no readily available hardness theory exists to explain activation energies obtained below 0.5Tm. For indentation processes above 0.5Tm deformation theories are based on dislocation climb mechanisms, ie. where diffusion is important (23). Using a value of m=10 (24) the activation energies for the various PSZ materials studied in the various temperature ranges was determined to be from about 1.2 kcal/mol to about 11.5 kcal/mol. The Y-CSZ material on the other hand changes from about 4 to 43 kcal/mol over the same temperature range. It is difficult to reconcile these small values with any specific structural process which may be occuring in the materials. Since creep deformation is controlled by the slowest moving species, in this case the Zr ions with Q ∿80 kcal/mol at 1300°C (25) it is not expected that creep deformation is a viable explanation. One possible explanation may be the activation of a second slip system, as has been observed in other high melting point materials (7,13).

Whilst no mechanism is readily available to explain the nature of the change in the activation energy occuring at ∿400°C (ie. Tm ∿0.25) it should be of considerable significance and reflects the CSZ matrix response to the induced plastic deformation, see for example (3,9,19).

c) Deformation About Indents

From the deformation patterns occuring about large indents it is apparent that co-operative transformation (CT) (8) is a very important mechanism enabling low temperature plastic deformation to occur in PSZ. The advent of CT is a predominantly strain induced process, because of its crystallographic aspects, where as the normal toughening process in PSZ is stress induced. Also the process predominates in the compressive region with associated shear such that the macro shape changes and grain rotations accompanying the deformation may be accommodated by micro co-operative transformation within the responding grains.

It is also noticeable, Fig 4, that the deformation zone is only marginally effected by the test temperatures used here. Whilst in tensile type test (modulus of rupture) transformation does not occur above about 500°C (19). An example once again of the two types of transformation inducing mechanisms, viz. stress and strain.

These observation show quite clearly the desirable effects that transforming precipitates can have on reducing surface damage due to particle impact (indent). The results also show the reduction in effectiveness of the stress induced transformation toughening process (at the end of the Knoop indent in Fig 5a) as temperature is increased (see (19)) or if the precipitates are allowed to transform due to thermal treatment, as in over-ageing (shown in Fig 4c).

SUMMARY AND CONCLUSIONS

The low temperature deformation of a variety of zirconia materials has been examined using indentation techniques. Hardness anisotropy measurements have shown that the {100} <011> slip system predominant at high temperatures is also operative at the low temperatures used in this study. Whilst no direct evidence for additional operating slip systems was found, on the basis of changes in deformation activation energies at about 400°C, it can be inferred that the material's response has altered, perhaps as a result of other slip systems becoming active.

Deformation patterns about large indents in PSZ materials reveal that, when suitably heat treated, plastic deformation and elimination of surface fracture is made possible by a strain induced cooperative transformation of the tetragonal precipitates within the CSZ matrix. This transformation process is thought to have relevance in the materials response to particle impact and abrasive wear situations.

ACKNOWLEDGEMENTS

The authors take pleasure in acknowledging the fruitfull discussions with Drs. D.L. Kohlstedt and S. Badwal.

REFERENCES

1. D. Tabor, "The Hardness of Solids," Rev. Phys. Tech. 1: 145 (1970).
2. C.A. Brookes, J.B. O'Neil and B.A.W. Redfern, "Anisotropy in the Hardness of Single Crystals," Proc. Roy. Soc. A332:73 (1971).
3. "Science and Technology of Zirconia," Advances in Ceramics Vol. 3, A.H. Heuer & L.W. Hobbs eds., The American Ceramic Society, Columbus, Ohio (1981).
4. R.H.J. Hannink, M.J. Murray and M. Marmach, "Magnesia-Partially Stabilized Zirconia (Mg-PSZ) as Wear Resistant Materials," Int. Conf. on Wear of Mats., Reston, Virginia (1983).
5. R.T. Pascoe, R.H.J. Hannink and R.C. Garvie, "Strengthening and Microstructural Changes in Magnesia Zirconia Alloys," Science of Ceramics Vol. 9:447 (1977).
6. A.P. Smith, "Polishing Hard Materials," Bull. Am. Ceram. Soc., TBP May (1983).
7. R.H.J. Hannink, D.L. Kohlstedt and M.J. Murray, "Slip System Determination in Cubic Carbides by Hardness Anisotropy," Proc. Roy. Soc. A326:409 (1972).
8. R.H.J. Hannink and M.V. Swain, "A Mode of Deformation in Partially Stabilized Zirconia," J. Mat. Sci. 16:1428 (1981).
9. M.V. Swain, "R-Curve Behaviour of Magnesia-Partially Stabilized Zirconia and Its Significance to Thermal Shock," Fract. Mech. of Ceramics, Vol 5&6, R.C. Bradt, A.G. Evans, D.P.H. Hasselman & F.F. Lange eds., Plenum Press, N.Y., TBP.
10. F.W. Daniels and C.G. Dunn, "The Effect of Orientation on Knoop Hardness of Single Crystals of Zinc and Silicon Ferrite," Trans. Am. Soc. Metals, 41:419 (1949).

11. R.D. Arnell, "An Investigation of Microhardness Anisotropy in Cobalt Single Crystals," J. Phys. D, 7:1225 (1974).
12. C.A. Brookes, R.P. Burnand and J.E. Morgan, "Anisotropy and Indentation Creep in Crystals with the Rocksalt Structure," J. Mat. Sci. 10:2171 (1975).
13. G. Morgan and M.H. Lewis, "Hardness Anisotropy in Niobium Carbide," J. Mat. Sci. 9:349 (1974).
14. M.L. McCartney, W.T. Donion and A.H. Heuer, "Plastic Deformation in CaO-Stabilized ZrO_2 (CSZ)," J. Mat. Sci. 15:1063 (1980).
15. W.J. McG. Tegart, "Elements of Mechanical Metallurgy," Macmillan, New York, (1966) p103.
16. P. Humble and R.H.J. Hannink, "Plastic Deformation of Diamond at Room Temperature," Nature 273:37 (1978).
17. B.J. Hockey, "Plastic Deformation of Aluminium Oxide by Indentation and Abrasion," J. Am. Ceram. Soc. 54: 223 (1971).
18. R.C. Garvie, R.H.J. Hannink and R.T. Pascoe, "Ceramic Steel?" Nature 258:703 (1975).
19. R.H.J. Hannink and M.V. Swain, "Magnesia Partially Stabilized Zirconia: The Influence of Heat Treatment on the Thermomechanical Properties," J. Australian Ceram. Soc. 18:53 (1982).
20. A.H. Heuer, N. Claussen, W. Kriven and M. Ruhle, "Stability of Tetragonal ZrO_2 Particles in Ceramic Matrices," J. Am. Ceram. Soc. 65:642 (1982).
21. J. Lankford, "Plastic Deformation of Partially Stabilized Zirconia," J. Am. Ceram. Soc., submitted.
22. J.H. Westbrook, "Temperature Dependence of Hardness of Some Common Oxides," Rev. Hautes. Temper. et Refract. 3:47 (1966).
23. A.G. Atkins and D. Tabor, "Hardness and Deformation Properties of Solids at Very High Temperatures," Proc. Roy. Soc. 292:441 (1966).
24. D.L. Kohlstedt, "The Temperature Dependence of Microhardness of the Transition Metal Carbides," J. Mat. Sci. 8:777 (1973).
25. R.H.J. Hannink, K.A. Johnston, R.T. Pascoe and R.C. Garvie, "Microstructural Changes During Isothermal Ageing of a Calcia Partially Stabilized Zirconia Alloy," in (3) p116.

AN ELASTIC FAILURE MODEL OF INDENTATION DAMAGE

B. M. Liaw, A. S. Kobayashi and A. F. Emery

University of Washington
Department of Mechanical Engineering
Seattle, Washington 98195, USA

ABSTRACT

A mechanistically consistent model for indentation damage based on elastic failure at tensile or shear overloads, is proposed. The model accommodates arbitrary crack orientation, stress relaxation, reduction and recovery of stiffnesses due to crack opening and closure, and interfacial friction due to backward sliding of closed cracks. This elastic failure model was implemented by an axisymmetric finite element program which was used to simulate progressive damage in a silicon nitride plate indented by a tungsten carbide sphere. The predicted damage patterns and the permanent impression matched those observed experimentally. The validation of this elastic failure model shows that the plastic deformation postulated by others is not necessary to replicate the indentation damage of brittle structural ceramics.

INTRODUCTION

When a glass plate is indented by a small sphere during a loading-unloading cycle, it undergoes a sequence of progressive fracture, shown schematically in Figure 1. A permanent, commonly referred to as a plastic, impression is left on the surface of glass. Marsh[1,2] attributed this impression to plastic flow due to its geometric similarity with that observed in ductile steel. The "plastic zone" in the impression occurred right below the contact area where the "yield stress" was assumed to be a function of the glass hardness. The formation of radial and lateral cracks, as shown in Figure 1, however, could not be explained by Hertzian elastic contact stress field,[3] since these

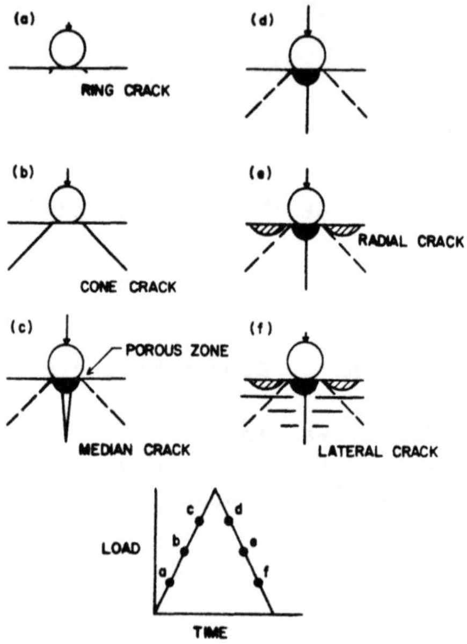

Fig. 1 Schematics of progressive fracture during indentation loading

cracks were observed to form after the "plastic" indentation. Evans and Wilshaw[4] thus suggested that during unloading, the irreversible deformation within the "plastic zone" was directly responsible for the formation of a residual tensile stress field which initiated the radial and lateral cracks. This elastic-plastic indentation model and the ensuing fracture of brittle material was also supported by Perrott,[5] Swain and Hagan,[6] and Chiang et al.,[7] although different stress fields were obtained by the different postulates used in their analyses.

The above elastic-plastic model requires the brittle material to undergo "plastic deformation" despite its observed elastic failure under uniaxial tensile loading. Such "plastic deformation" was attributed to the moderate-to-high triaxial state of stress immediately below the indenter. Plasticity theory, however, predicts no change in failure modes under a superposed hydrostatic state of stress. Furthermore, as depicted in Figure 1, previous experiments[6,8] clearly showed that median cracks and occasionally Hertzian cone cracks form during the loading half-cycle whereas radial and lateral cracks form only during the unloading half-cycle. The associated changes in stress field after the formation of median and cone cracks were not considered in the above models. Since the effect of cracks in damaged materials is to reduce the stiffness,[9] and hence to change the stress field, the onset of radial and lateral cracks could have been stimulated by the prior formation of median and cone cracks. Moreover, Lawn et al.[10] observed that minor cracking within the plastic impression was evident even though no major cracks were initiated during this process. Such microcracking could indicate that the "plastic impression" is actually due to irreversible elastic deformation which is generated by the formation of micro-cracks.

The purpose of this study is to demonstrate that radial cracks, lateral cracks, and permanent impression can be generated by elastic failure in the presence of median and cone cracks. This mechanistically-consistent elastic failure model thus replaces the popular elastic-plastic indentation model in describing the indentation damage phenomena in brittle materials.

THEORY

For convenience in numerical modeling, the crack is represented by a string of damaged elements. This approach in fracture mechanics, which is referred to as damage mechanics,[9] models crack opening and closure by the stress relaxation and the reduction/recovery of stiffness of the individual elements. Frictional resistance to relative sliding of the rough surfaces of a closed crack is generated by a high friction coefficient varying between 0.15 and 0.5 for soda-lime glass.[11] Damage mechanics has been used in numerical fracture analysis of brittle materials,[12-15] such as concrete[12], rock[13], and ceramics[15]. The basic postulates in damage mechanics are:

(1) A primary crack will form in a material point if the maximum principal tensile stress is greater than or equal to the tensile strength, T. The orientation of the crack is perpendicular to the maximum principal direction.

(2) A secondary crack will form in a cracked material once the stress component parallel to the crack is greater than or equal to T. This secondary crack is thus perpendicular to the primary crack.

(3) A set of orthogonal cracks will form in a material point if the absolute value of maximum shear stress is greater than or equal to the shear strength, S. The orientation of this set of cracks are parallel to the directions of shear stresses which always appear in pair.

(4) Zero surface traction is assumed for an open crack whereas a closed crack carries compressive load. Interfacial shear traction is governed by Coulomb's dry friction law, that is, the absolute value of shear stress cannot be greater than the absolute value of compressive stress times friction coefficient,

f. This statement can be expressed as:

$$\sigma \leq 0 \qquad \text{and} \qquad |\tau| \leq f|\sigma| \tag{1}$$

where σ and τ are the normal and shear stresses on the crack, respectively.

(5) Stiffness across the crack drops to zero once the crack opens while the stiffness along the crack remains unchanged. Hence, the finite element representing the crack becomes orthotropic in material properties. If both primary and secondary cracks open, the tensile stiffnesses along both directions are zero. Once the crack closes, the compressive stiffness simply returns to its initial value.

The axisymmetric, incremental stress-strain relation for a cracked material with only primary cracks as shown in Figure 2(a) is:

$$\begin{Bmatrix} \Delta\sigma_n \\ \Delta\sigma_t \\ \Delta\sigma_\theta \\ \Delta\tau_{nt} \end{Bmatrix} = \frac{E}{1-\nu^2} \begin{bmatrix} 0 & 0 & 0 & 0 \\ 0 & 1 & \nu & 0 \\ 0 & \nu & 1 & 0 \\ 0 & 0 & 0 & 0 \end{bmatrix} \begin{Bmatrix} \Delta\varepsilon_n \\ \Delta\varepsilon_t \\ \Delta\varepsilon_\theta \\ \Delta\gamma_{nt} \end{Bmatrix} \tag{2}$$

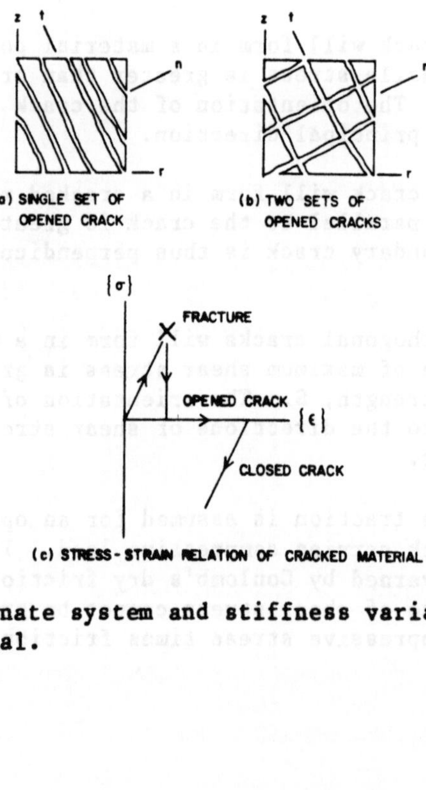

Fig. 2 Coordinate system and stiffness variations of cracked material.

When both the primary and secondary cracks as shown in Figure 2(b) exist, the incremental stress-strain relation becomes:

$$\begin{Bmatrix} \Delta\sigma_n \\ \Delta\sigma_t \\ \Delta\sigma_\theta \\ \Delta\tau_{nt} \end{Bmatrix} = \begin{bmatrix} 0 & 0 & 0 & 0 \\ 0 & 0 & 0 & 0 \\ 0 & 0 & E & 0 \\ 0 & 0 & 0 & 0 \end{bmatrix} \begin{Bmatrix} \Delta\varepsilon_n \\ \Delta\varepsilon_t \\ \Delta\varepsilon_\theta \\ \Delta\gamma_{nt} \end{Bmatrix} \qquad (3)$$

where E and ν are Young's modulus and Poisson's ratio, respectively. Note that Equation (2) is similar to the stress-strain relation for plane stress and Equation (3) resembles that for a circular bar. The subscripts n and t refer to the directions normal and tangential to the crack as shown in Figures 2(a) and 2(b) while θ stands for the circumferential direction in the cylindrical coordinate system (r, θ, z). For a closed crack, the compressive elastic modulus, as shown in Figure 2(c), is assumed to be that of the undamaged material.

Finally, indentation damage is restricted to axisymmetric cracks and thus the damage regions are likewise restricted to axisymmetric regions. Hence, the circumferential stiffnesses and stresses cannot vanish although median and radial cracks may have formed during the indentation process.

NUMERICAL MODEL OF INDENTATION DAMAGE

The stiffness of a cracked body varies with the opening or closing of a crack and thus the material[9] response of a cracked body is nonlinear and irreversible. Moreover, the formation of a crack will result in stress relaxation and redistribution which requires an iterative algorithm to assure the convergence of the redistributed stresses which must satisfy the equlibrium condition. The finite element model of indentation damage is, as shown in Figure 3, based on an incrementally linear analysis implemented with an updated Newton-Raphson iteration scheme. The first computation simulates a thick hot-pressed silicon nitride (HPSN) plate indented by a 2.4 mm diameter tungsten carbide (WC) sphere at a maximum instantaneous load of 2.4 kN. Table 1 shows the material properties of these two materials. Hertzian hemispherical contact pressure distribution was used as the prescribed[3] loading where the contact pressure, q, can be expressed as:

$$q = q_o \left[1 - \left(\frac{r}{a}\right)^2\right]^{1/2} \qquad (4)$$

q_o is the maximum contact pressure, that is,

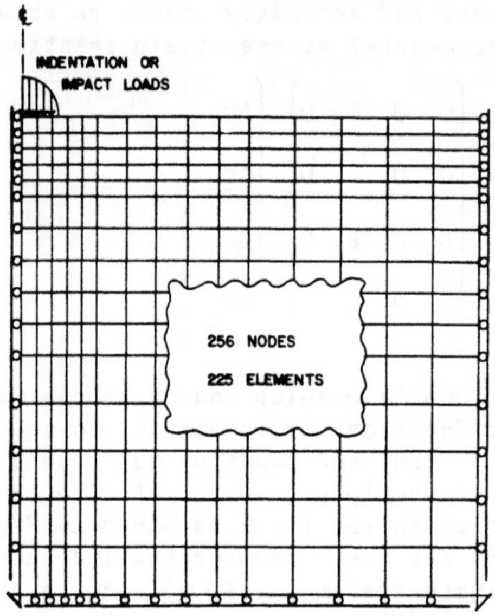

Fig. 3 Finite element breakdown of a half space subjected to indentation or impact loadings.

TABLE I. MATERIAL PROPERTIES

Material	Young's Modulus E(GPa)	Poisson's Ratio ν	Density ρ(kg/m³)	Tensile Strength T(GPa)	Shear Strength S(GPa)	Friction Coeff. f
HPSN	320	0.26	3300	1.4	14	0.15
WC	700	0.24	15000	-	-	-

$$q_o = \frac{3}{2}\frac{P}{\pi a^2} \tag{5}$$

and r is the radial distance, P is the indentation load, and a is the contact radius which can be represented as:

$$a = (\frac{3}{4} k P R)^{1/3} \tag{6}$$

Here R is the radius of the indenter and k is defined as:

$$k = \frac{1-\nu^2}{E} + \frac{1-\nu'^2}{E'} \tag{7}$$

where E, E', ν, ν' are Young's moduli and Poisson's ratios of HPSN and WC, respectively.

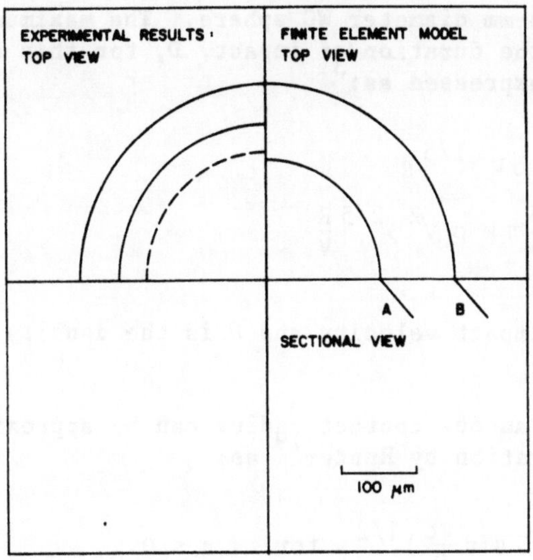

Fig. 4 Ring cracks of a HPSN plate indented up to 2.4 kN by a 2.4-mm dia. WC sphere.

EXPERIMENTAL AND NUMERICAL RESULTS

The numerical model for indentation damage was validated by replicating numerically the extent of damage observed experimentally by Dao et al.,[16] when hot-pressed silicon nitride plates were indented by a tungsten carbide ball. Figure 4 shows a comparison of the two groups of ring cracks, designated as A and B, obtained experimentally and by the finite element simulation. Although the group A crack is not clearly visible in Figure 4, photographic records[16] of the same experiment under lower indentation loads indicated the existence of this group A crack. The ring cracks are inclined 47° with respect to the surface for group A and 46° for group B. No sectional view of the tested HPSN plate was available, but the ring cracks were reported to be inclined 50° to the surface.[17] The finite element results also showed that both groups of cracks closed after complete unloading. Residual compressive stresses on both group of cracks were about 2.1 Gpa, or 12% of the peak indentation pressure indicating resistance to crack surface sliding and hence the irreversibe nature of indentation damage.

High-level indentation damage such as cone, median, radial, and lateral cracks could not be compared due to the lack of sectional views of the HPSN specimen. However, similar fracture patterns were observed by Kirchner and Larchuk[18] in an indented HPSN plate as well as an impacted HPSN plate with comparable contact radii. Such similarity allowed us to use the same elastic damage model for analyzing a thick HPSN plate impacted at

231 m/s by a 2.4-mm diameter WC sphere. The maximum contact radius, A, and the duration of impact, D, for this dynamic problem can be expressed as:[3]

$$A = \left(\frac{5}{4} \pi k \rho V \right)^{1/5} R \tag{8}$$

$$D = 2.94 \left(\frac{5}{4} \pi k \rho V^2 \right)^{2/5} \frac{R}{V} \tag{9}$$

where V is the impact velocity and ρ is the density of impact particle.

The instantaneous contact radius can be approximated by numerical integration by Hunter[19] as:

$$a = A \left(0.995 \sin \frac{\pi t}{D} \right)^{1/2} \quad \text{for} \quad t \leq D \tag{10}$$

Since Tsai[20] reported that the Hertz contact theory could be used to estimate the contact pressure during impact, Equations (4) - (10) can be used to estimate the equivalent contact pressure during impact. The same finite element model of Figure 3 can then be used to model this impact damage problem.

Figure 5 shows the the sectional view of an HPSN plate impacted by a WC sphere reported by Dao et al.[16] where qualitative agreement between numerical results generated by the finite element model and the experimental results is noted. Numerical analysis showed that most of the Hertzian cone cracks formed during the loading process while the lateral cracks only appeared during unloading and are in agreement with the experimental observations.[6,8] Finite element results also indicated that the damage-induced porous zone underneath the impact site is mainly due to shear overload, which is equivalent to uniaxial compressive crushing of ceramics,[21] and was induced by the high confining pressure in the confined contact region. This crushed damage zone was also observed in References 18 and 22.

Using Equations (6) and (8), we can compute the peak indentation load equivalence which is 33.9 kN for the given impact velocity. Figure 6 compares the fracture patterns generated by the equivalent impact and indentation loadings, where qualitative similarity is noted and is consistent with the results reported in Reference 18. Although the reduction in stiffness and stress relaxation in the circumferential direction cannot be permitted in this axisymmetric model, the potential for forming median and radial cracks can be estimated. Figure 7 shows possible regions where median and radial cracks can be formed under impact and

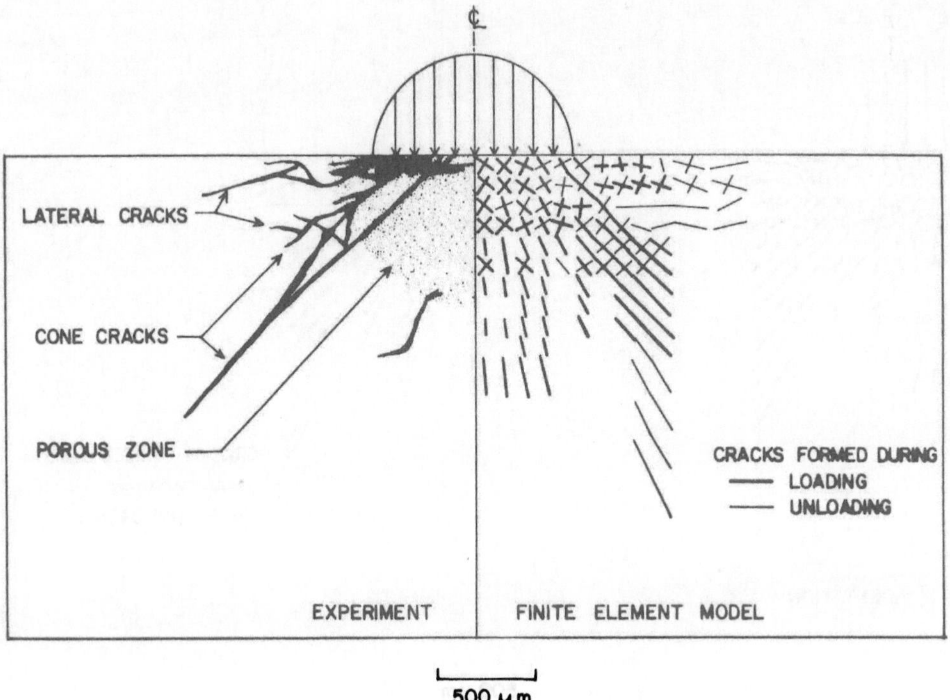

Fig. 5 Fracture pattern in a HPSN plate impacted at 231 m/s by a 2.4-mm dia. WC sphere.

indentation loadings. The finite element results also show that median crack initially extends under the impact center during the loading process and eventually breaks through the surface to form the radial crack during unloading. This again agrees with experimental observations.[6,8] The lack of stress and stiffness relief in the circumferential direction in the finite element model also results in the computed median/radial cracks to be shorter than the actual cracks.

Plastic impression can also be simulated by the numerical computation. Figure 8 shows the silhouettes of plastic impressions for the two quasi-static indentation cases where the predicted dimensions and shapes of these two plastic impressions agree with experimental results[18] within reasonable accuracy.

DISCUSSION

Despite its simplicity, the proposed elastic failure model models reproduces the complicated indentation damage patterns. With appropriate modifications, the accuracy and applicability of

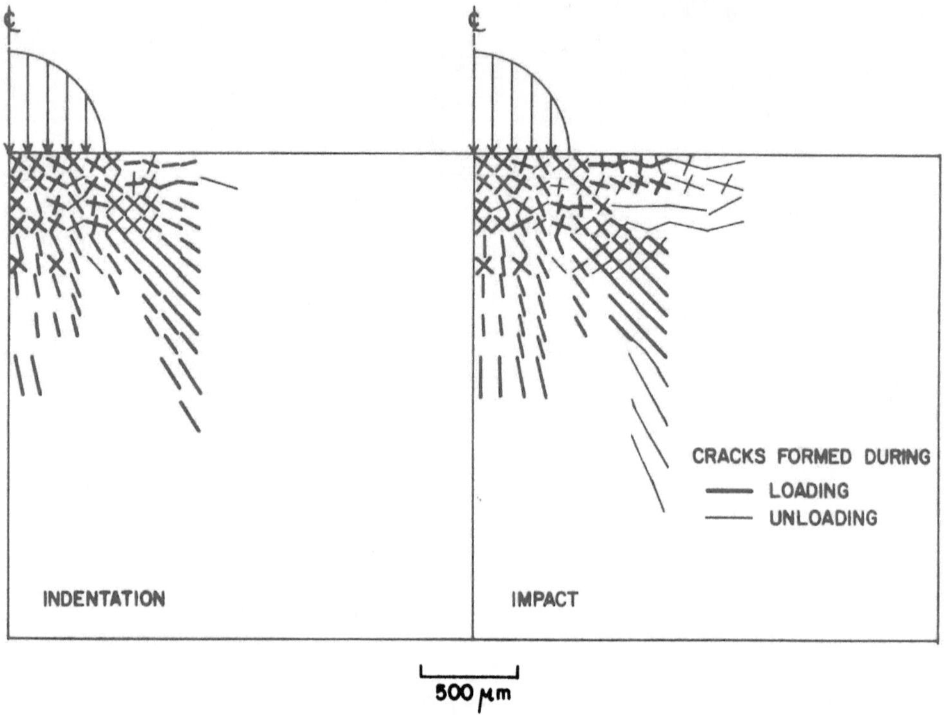

Fig. 6 Comparison of fracture patterns by indentation and impact loadings.

this model could be improved. For instance, References 23 and 24 reported that both Coulomb-Mohr and maximum strain energy theories yielded better simulations under multi-axial stress states. Stiffness reduction and hence the stress relaxation procedures of the present model are on-off type where the entire computational cell is either intact or damaged. Improvement can be made by using a progressive failure model which gradually reduces the stiffness and stress.

The values of tensile and shear strengths and friction coefficient listed in Table 1 are chosen through a sensitivity study of the present model. Dao et al.[16] reported that the equivalent fracture strength of HPSN corresponding to their indentation tests was several times higher than the handbook value which is about 0.35 GPa. This discrepancy was also observed by Rosenblatt et al.[25] in simulating impact damage in ZnSe window glass. Tensile strength of 0.35 GPa was used in this study and the resultant fracture zone was larger than the observed. This discrepancy could be due in part to natural flaws which have negligible influence on conventional strength test where the tensile

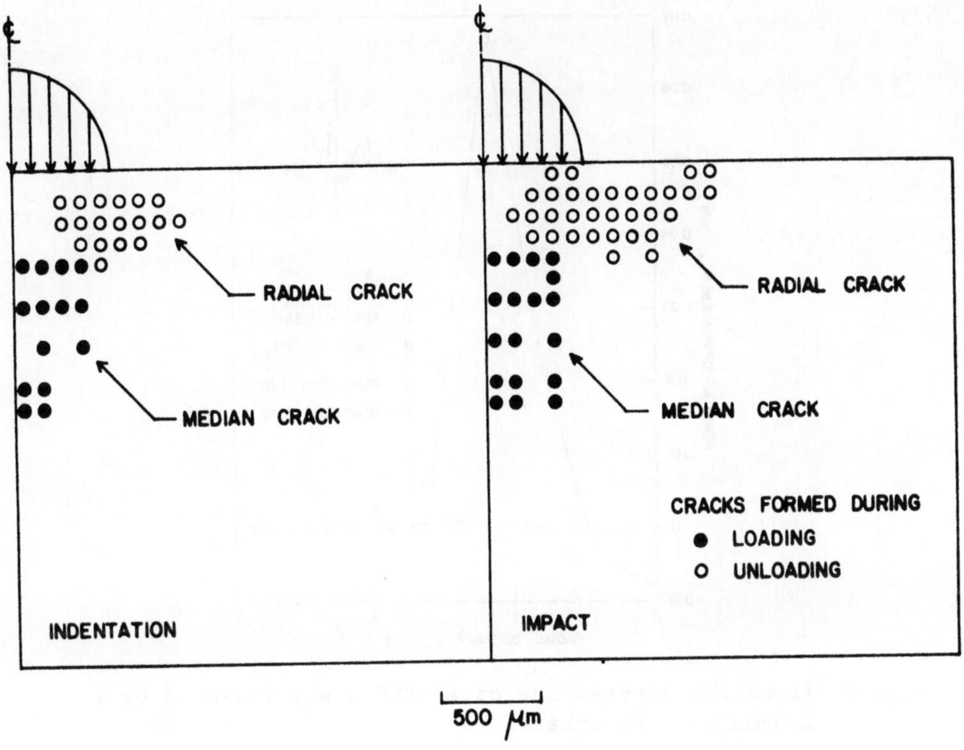

Fig. 7 Formation of median and radial cracks by indentation and impact loadings.

strength is determined by the far-field applied load. Stress concentration around natural flaws, however, requires the use of the true strength which is higher than the measured gross strength. For damage induced by a small indenter, the size of natural flaw is only one or two order less than the dimension of contact zone, implying that the computed stress in the contact region could be very close to the true strength.

CONCLUSIONS

An elastic failure model of indentation damage is presented. Not only is this model consistent with the well-known elastic behavior of brittle material, but it can also reproduce the experimentally observed permanent impression and formation of radial and lateral cracks. These deformation processes are the results of the nonlinearity and irreversibility of stiffness variations due to the presence of cone and median cracks during the loading half-cycle. The success of this model enables us to conclude that plastic deformation is not required to explain the indentation damage of the elastic, brittle materials.

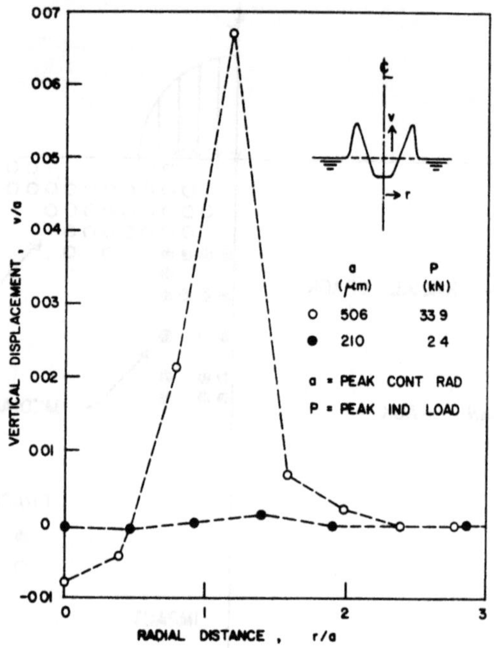

Fig. 8 Permanent impressions of a HPSN plate indented by a 2.4-mm dia. WC sphere.

ACKNOWLEDGEMENT

This investigation was supported by NASA Grant NGL 48-002-004 to the Ceramic Engineering Division, University of Washington. The authors express their gratitude to Professor J. I. Mueller, Principal Investigator of the Ceramic Material Research Program for his continuing support of this research over the past four years, and to Dr. D. A. Shockey, SRI International, for providing the original photographs of his experiments.

REFERENCES

1. Marsh, D. M., "Plastic Flow in Glass," Proc. Roy. Soc. London, Vol. A279 pp. 420-435, 1961.
2. Marsh, D. M., "Plastic Flow and Fracture of Glass," Proc. Roy. Soc. London, Vol. A282, pp. 33-43, 1964.
3. Timoshenko, S. P. and Goodier, J. N., "Theory of Elasticity," 3rd ed., McGraw-Hill, Inc., N.Y., 1970.
4. Evans, A. G. and Wilshaw, T. R., "Quasi-Static Solid Particle Damage in Brittle Solids - I. Observations, Analysis and Implications," Acta Metal., Vol. 24, pp. 939-956, 1976.
5. Perrott, C. M., "Elastic-Plastic Indentation: Hardness and Fracture," Wear, Vol. 45., pp. 293-309, 1977.

6. Swain, M. V. and Hagan, J. T., "Indentation Plasticity and the Ensuing Fracture of Glass," J. of Physics D: Applied Physics, Vol. 9, pp. 2201-2214, 1976.
7. Chiang, S. S., Marshall, D. B., and Evans, A. G., "The Response of Solids to Elastic/Plastic Indentation: I. Stresses and Residual Stresses," J. of Applied Physics, Vol. 53, pp. 298-311, 1982.
8. Ishikawa, H. and Shinka, N., "Critical Load for Median Crack Indentation in Vickers Indentation of Glasses," J. of the American Ceramic Society, Vol. 65, pp. C124-C127, 1982.
9. Krajcinovic, D. and Fonseka, G. U., "The Continuous Damage Theory of Brittle Materials, Part 1: General Theory," J. of Applied Mechanics, Vol. 48, pp. 809-815, 1981.
10. Lawn, B. R. and Howes, V. R., "Elastic Recovery at Hardness Indentations," J. of Materials Science, Vol. 16, pp. 2745-2752, 1981.
11. Chiang, S. S. and Evans, A. G., "Influence of a Tangential Force on the Fracture of Two Contacting Elastic Bodies," J. of the American Ceramic Society, Vol. 66, pp. 4-10, 1983.
12. Chen, W. F. and Suzuki, H., "Constitutive Models for Concrete," Computer and Structures, Vol. 12, pp. 23-32, 1980.
13. Kumano, A. and Goldsmith, W., "An Analytical and Experimental Investigation of the Effect of Impact on Coarse Granular Rocks," Rock Mechanics, Vol. 15, pp. 67-97, 1982.
14. Seaman, L, Curran, D. R., and Shockey, D. A., "Computational Models for Ductile and Brittle Fracture," J. of Applied Physics, Vol. 47, pp. 4814-4826, 1976.
15. Rosenblatt, M., Ito, Y. M., and Eggum, G. E., "Analysis of Brittle Target Fracture from a Subsonic Water Drop Impact," Erosion: Prevention and Useful Applications, ASTM STP 664, W. F. Adler, Ed., American Society for Testing and Materials, 1979, pp. 227-254.
16. Dao, K. C., Shockey, D. A., Seaman, L., Curran, D. R., and Rowcliffe, D. J., "Particle Impact Damage in Silicon Nitride," Annual Report, Part III, to Office of Naval Research, Arlington, VA, Contract N00014-76-057, May 1979.
17. Evans, A. G. and Wilshaw, T. R., "Dynamic Solid Particle Damage in Brittle Materials: An Appraisal," J. of Materials Science, Vol. 12, pp. 97-116, 1977.
18. Kirchner, H. P. and Larchuk, T. J., "Comparisons of Static and Impact Loading Damage in Zinc Sulfide," J. of the American Ceramic Society, Vol. 65, pp. 506-510, 1982.
19. Hunter, S. C., "Energy Absorbed by Elastic Waves During Impact," J. of the Mechanics and Physics of Solids, Vol. 5, pp. 162-171, 1957.

20. Tsai, Y. M., "Dynamic Contact Stresses Produced by the Impact of an Axisymmetrical Projectile on an Elastic Half-Space," Int. J. of Solids and Structures, Vol. 7, pp. 543-558, 1971.
21. Ryshkewitch, E., "Compression Strength of Porous Sintered Alumina and Zirconia," J. of the American Ceramic Society, Vol. 36, pp. 65-68, 1953.
22. Hagan, J. T. and Swain, M. V., "The Origin of Median and Lateral Cracks Around Plastic Indents in Brittle Materials," J. of Physics D: Applied Physics, Vol. 11, pp. 2091-2102, 1978.
23. Broutman, L. J., Krishnakumar, S. M., and Mallick, P. K., "Effects of Combined Stresses on Fracture of Alumina and Graphite," J. of the American Ceramic Society, Vol. 53, pp. 649-654, 1970.
24. Ely, R. E., "Strength of Titania and Aluminum Silicate Under Combined Stresses", J. of the American Ceramic Society, Vol. 55, pp. 347-350, 1972.
25. Rosenblatt, M., Ito, Y. M., and DeAngelo, L., "Numerical Simulation of Ceramic Target Response to Water Drop Impacts Including Effects of Surface Flaws and Pores", Proc. 5th Int. Conf. Erosion by Solid and Liquid Impact, pp. 4-1 to 4-10, Cambridge, England, 3-6 Sept., 1979.

PLASTIC DEFORMATION DURING THE ABRASIVE WEAR OF CERAMICS

S.B. Toh and R. McPherson

Department of Materials Engineering
Monash University
Clayton, Victoria, 3168, Australia

INTRODUCTION

King and Tabor[1] pointed out some time ago that, during frictional experiments, the behaviour of rock salt and other normally brittle materials is dominated by plastic and not by brittle fracture processes. They suggested that this was largely because of the high compressive component of the applied stress. X-ray diffraction studies of the structure of debris produced during abrasive wear of several ceramics have shown that they are heavily distorted and that the crystallite size and strain contributions to line width are similar to those observed in heavily cold worked metals[2-4]. The response to annealing was also similar to metals with reduction of the strain and increase in crystallite size corresponding to recovery and recrystallisation processes.

Similar studies of the abraded surfaces of ceramics, such as TiO_2 and $BaTiO_3$, in which the depth of penetration of the X-rays is limited to only a few μm, showed that the surfaces also had a distorted structure comparable to debris, and recrystallisation could be observed after annealing[3,5]. These results were interpreted as evidence that the mechanism of material removal was dominated by plastic deformation, probably assisted by transient temperature increases associated with the large plastic work done at a high strain rate. Temperature increases of up to 1200°C were estimated for Al_2O_3, however the analysis was limited by the lack of knowledge of the flow properties of Al_2O_3 at high strain rates[4].

Scanning electron microscopy (SEM) of debris, produced by light abrasion of TiO_2, revealed that segmented chips were formed which were very similar in appearance to chips produced on a larger scale during the machining of metals[6]. Transmission electron microscopy (TEM) of chips which were sufficiently thin showed that they had an extremely fine polycrystalline structure with crystallite size of ≃ 10 nm. Heating in the temperature range 500-1000°C resulted in growth in crystallite size[7].

An X-ray diffraction examination of the structure of MgO abrasive wear debris, as a function of particle size, showed that the degree of distortion increased as the particle size of the debris decreased[8]. These results suggested that the finest debris had a homogeneous structure comparable to heavily cold worked metals and were formed by plastic deformation, but that the coarser particles were more or less distorted and were probably formed by fracture of the plastically deformed surface. Such a scale effect on the mechanism of material removal during abrasive wear is supported by experiments using soda glass in which cracks, usually observed along scratches, are not formed when the scratch width falls below a critical width (≃ 1 µm). The critical width depended upon the scratching velocity[9].

The present paper reports a further TEM study of the fine scale abrasive wear of several ceramic materials.

EXPERIMENTAL PROCEDURE

Details of the materials examined are shown in Table 1.

Specimens, approximately 20 mm diameter, were polished using conventional metallographic techniques; the final polish was achieved using a tin lap impregnated with 0.5 µm diamond powder.

Table 1 Sintered Ceramics Used in Abrasion Experiments

Material	Source
Al_2O_3	Coors
MgO partially stabilised ZrO_2	CSIRO Division of Materials Science
CaO/Y_2O_3 fully stabilised ZrO_2	CSIRO Division of Materials Science
TiO_2 (rutile)	Cold compacted, sintered at 1200°C in air
Sm_2O_3	"Cerac" powder, cold compacted, sintered 1400°C in air
Y_2O_3	"Cerac" powder, cold compacted, sintered 1400°C in air

Abrasion experiments were carried out by placing the polished surface of the specimen against 1200 grade SiC abrasive paper on a glass plate under a fixed vertical load (50-100 gm) applied through a lever arm. The paper was drawn between the specimen and glass plate at a rate of approximately 10 mm sec^{-1}, and the debris, suspended in a drop of distilled water, was transferred to a carbon film on a Cu microscope grid and allowed to dry for examination in a 100 kV electron microscope. An evaporated gold film was used for electron diffraction calibration purposes.

RESULTS

All of the ceramics examined gave segmented chips when abraded on 1200 grade SiC paper under a load of 100 gm. Coarser grades of abrasive, especially under heavier loads (> 250 gm), produced fracture fragments in addition to chips.

All of the chips produced had a similar segmented structure (Figs. 1 and 2) and gave ring diffraction patterns (Fig. 3). The fine polycrystalline microstructure (Fig. 2), crystallite size < 10 nm, was also similar for all materials. Dark field examination using part of the ring pattern confirmed the polycrystalline nature of the chips (Fig. 4).

Analyses of the ring diffraction patterns showed that the TiO_2 chips consisted of rutile and brookite, and metastable phases were also detected in Al_2O_3 chips. The results are summarised in Table 2.

Table 2 Structure of Ceramic Chips

Material	Chip Width	Lamellar Spacing	Phases Detected
TiO_2	∿ 0.2 μm	30 nm	rutile, brookite
Stabilised ZrO_2	0.3 μm	-	cubic/tetragonal
Partially stabilised ZrO_2	0.1-0.4 μm	30 nm	tetragonal
Al_2O_3	0.3 μm	150 nm	α-al_2O_3, κ-al_2O_3
Sm_2O_3	0.3 μm	20 nm	cubic, monoclinic
Y_2O_3	< 0.1 μm	150 nm	cubic

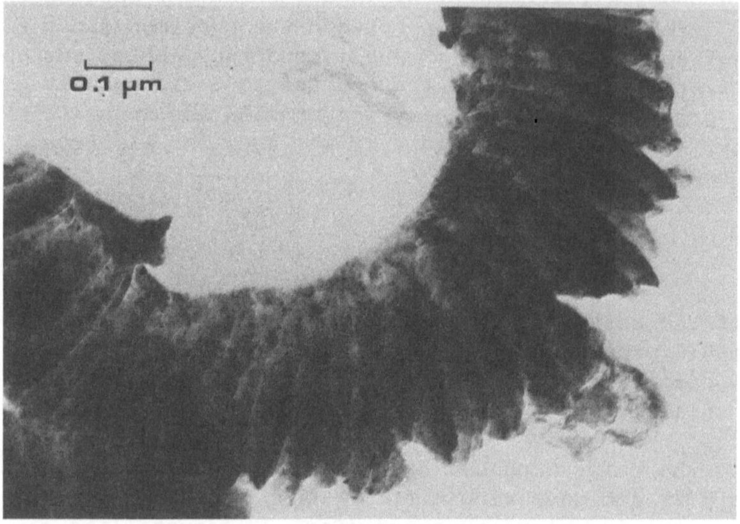

Fig. 1 Y_2O_3 chip showing serrated structure

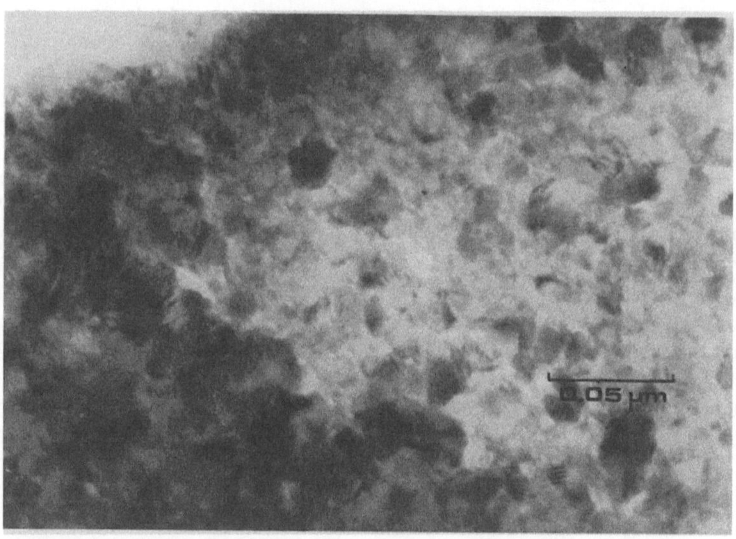

Fig. 2 Sm_2O_3 chip

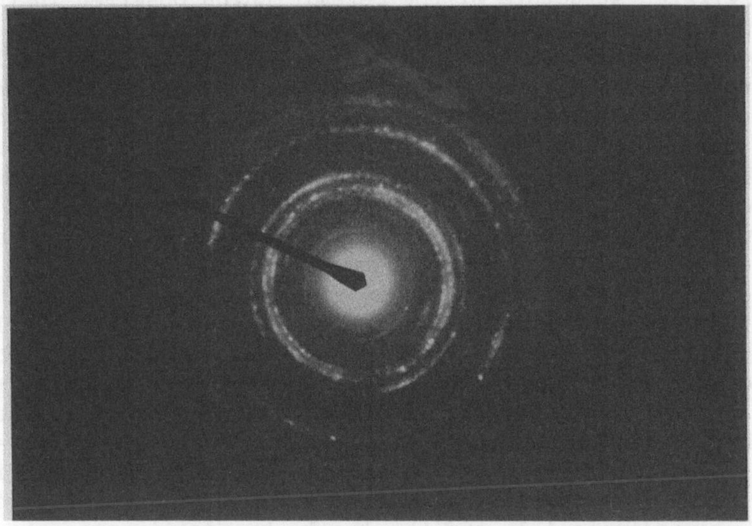

Fig. 3 Electron diffraction pattern of Y_2O_3 chip

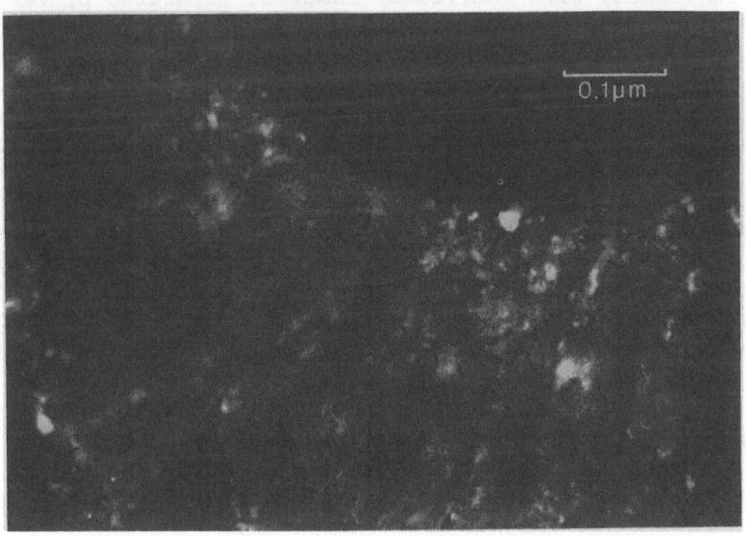

Fig. 4 Dark field image, Y_2O_3 chip

DISCUSSION

The serrated structure of the ceramic chips observed in the present work is similar to that observed in fine chips of glass[10] and metals suggesting that the process of chips formation is largely independent of the nature of the material. The major difference between the brittle materials and metals is the observation that such chips are only obtained in the former if abrasion is carried out on an extremely fine scale (chips < 0.5 μm thick) whereas this phenomenon is observed up to a macroscopic scale in the case of metals. This may be explained by a scale effect on fracture. Kendall[11] has shown, from energy considerations, that there is a lower limit to the size of a particle which will fracture. Experimentally[12] it has been observed that, as the size of hardness indentations is reduced in brittle materials, a level is reached at which cracking is no longer observed (≃ 1 μm for glass). Scratching experiments on soda glass[9] also showed that cracks were not formed along plastic grooves if they were smaller than a critical width (1-2 μm) which depended upon scratching velocity. This suggests that, in brittle ceramics, material is generally removed by fracture processes but, as the scale is reduced, cracking becomes impossible and the local stresses increase to a level at which chips can be formed by a plastic deformation process similar to that observed in metals.

The formation of serrated chips during metal cutting has been extensively studied and the formation of such structures by catastrophic shear is well established[13,14]. Thus a lamellar structure is formed, with individual lamellae separated by very thin regions of highly concentrated shear. Recht[13] has proposed that catastrophic shear will occur within a plastically deforming material when the slope of the true stress-strain function becomes zero. That is, when the local rate of change of temperature decreases strength at a rate equal to, or greater than, the rate of strain hardening.

Most studies of metal cutting have been carried out on a much larger scale than that of the present work. Black[15-17] however has performed fine scale experiments on various metals and alloys and used TEM to study the structures formed. He observed intermittent shear, giving chips consisting of well defined lamellae separated by narrow shear fronts. The separation of the shear fronts was ≃ 2 μm for f.c.c. metals and was largely independent of metal purity, cutting speed and depth of cut for depths of cut greater than 2-4 μm. The lamellar spacing was, however, reduced at smaller depths of cut. There was evidence of large scale dislocation activity, due to compressive loading, before the material entered the shear front region.

Experiments in metals[18], conducted at the same scale as the present study, have shown shear front spacings of ≃ 0.25 μm in agreement with Black's observation that the lamellar width decreases as the depth of cut is decreased below a few μm. These very small metallic chips also had a fine polycrystalline substructure similar to that observed in ceramic chips.

As the diffraction patterns and dark field examination of ceramic chips have shown, the substructure observed in between the shear bands consists of very small crystallites with large orientation differences between them and not a cellular substructure, as observed in heavily cold worked metals[19], or in hot worked metals in which dynamic recovery occurs[20]. This suggests that the structure is a result of dynamic recrystallisation arising from the temperature increase produced by the high strain rate plastic deformation.

Dynamic recovery during the hot working of metals produces a cellular substructure in which the cell size (d) is related to the strain rate ($\dot{\varepsilon}$) and temperature (T) by:

$$d \propto 1/\log\left(\dot{\varepsilon} \exp Q/RT\right)$$

Metal working experiments have been conducted at strain rates up to $\simeq 10$ $\sec^{-1}$ to give cell sizes of $\simeq 1$ μm[21]. The strain rate in chip cutting is given by:

$$\dot{\varepsilon} = V/t$$

where V is the cutting velocity and t the depth of cut. Thus, for the present experiments, $\dot{\varepsilon}$ is of the order of 10^4 $\sec^{-1}$ and subgrain diameters much less than 1 μm would not be unexpected. Presumably a stage is reached where dynamic "recrystallisation" occurs, that is, new grains form from the cellular structure by increase in the misorientation of some of the cell walls and a dynamic equilibrium structure is produced. This view is supported by observations of the substructure produced during the creep of some minerals in which smaller new grains were formed by progressive misorientation of sub-grains, an effect termed rotation recrystallisation[22]. Rotational recrystallisation occurs below a critical temperature which depends upon stress; for NaCl the critical temperature approaches 0.6 Tm at high stresses. The very small size of the chips means that the quenching rate would be very fast and grains would not grow after the material moved out of the deformation zone.

Whatever the details of the deformation mechanism and substructure produced, it is clear that the stress and strain will be extremely high and the work done at high strain rate will lead to a substantial temperature increase in the deformed zone and catastrophic shear will become possible because of thermal instability[13]. The concentration of deformation in the narrow shear band would be expected to lead to very high temperatures, perhaps approaching the melting point. Evidence of much higher temperatures in the shear band than in the bulk of a metal cutting chip is provided by Jovane et al.[23] who observed narrow bands of austenite within the shear bands of a martensitic high nickel steel in which $M_s \simeq 0°C$. This showed that the shear band temperature had risen sufficiently above the Ac_3 temperature to allow transformation to austenite in the extremely short time available (of the order of milliseconds).

The formation of metastable structures in some of the ceramics examined in the present study may be explained in a similar way. Sm_2O_3 transforms from the cubic form to monoclinic on heating above 900°C and is retained as monoclinic by rapid cooling[24]. The observation that the structure of Sm_2O_3 chips is predominantly cubic indicates that most of the chip has not been heated to a temperature sufficiently high to cause transformation (this will be much higher than the equilibrium transformation temperature because of kinetic effects). The evidence of some metastable phase however suggests that limited regions of the chip, probably the shear bands, have been heated sufficiently for a phase transformation to take place.

The case of rutile is of particular interest since there is evidence that brookite is metastable with respect to rutile at all temperatures and pressures[25]. This means that it would not be thermodynamically possible for the brookite observed in chips to have formed by direct transformation from rutile. Recent experiments have shown that brookite can crystallise from amorphous TiO_2[26] suggesting that the brookite in TiO_2 chips probably formed within the shear bands by quenching from the liquid state. Similar considerations apply to the observations of metastable Al_2O_3 in chips. The formation of δ-Al_2O_3 has also been observed during the abrasive wear testing of sintered alumina[27] in which a smeared surface layer was produced with a very fine microcrystalline structure. The authors suggested that the metastable Al_2O_3 may have been formed by intermediate reaction with water to form aluminium monohydrate which subsequently transformed to δ-Al_2O_3 by a surface heating effect. An alternative view is that it formed from material which melted and was rapidly quenched as observed in Al_2O_3 which has been thermally sprayed as fine particles or a coating[28].

Other evidence that surface melting can occur during wear of refractory materials is provided by experiments on "friction glazing" of alloys in which microstructures consistent with surface melting have been obtained[29]. Evidence of local melting has also been obtained from impact studies on glass[30] and ceramics[31].

The present study has revealed a number of interesting phenomena during the fine scale abrasion of ceramics which result from plastic deformation of ceramic materials at stresses and rates which are outside normal experimental techniques. Further work is continuing to provide data which will allow a more complete interpretation of the various effects observed.

CONCLUSIONS

Fine scale abrasion of a number of ceramics results in the formation of debris in the form of serrated chips similar to those observed in metal cutting.

Chips are formed because fracture is inhibited as the scale of individual abrasive events is reduced below a critical level.

A substructure consistent with dynamic recrystallisation is observed within regions between catastrophic shear bands.

There is evidence that the temperature may reach the melting point in the shear bands resulting in formation of metastable phases by rapid quenching of the transient liquid.

REFERENCES

1. R.F. King and D. Tabor, The strength properties and frictional behaviour of brittle solids, Proc. Roy. Soc. A223:225 (1954).
2. I.A. Cutter and R. McPherson, Examination of abraded MgO by X-ray diffraction line broadening, Phil. Mag. 20:4389 (1969).
3. I.A. Cutter and R. McPherson, Surface domain reorientation produced by abrasion and annealing of barium titanate, J. Amer. Ceram. Soc. 55:334 (1972).
4. I.A. Cutter and R. McPherson, Plastic deformation of alumina during abrasion, J. Amer. Ceram. Soc. 56: 266 (1973).
5. I.A. Cutter, Abrasion of ceramics, Ph.D. Thesis, Monash University, Australia (1970).
6. R.L. Aghan and R. McPherson, Mechanism of material removal during abrasion of rutile, J. Amer. Ceram. Soc. 56:46 (1973).
7. R.L. Aghan, Fine abrasion of ceramics, M.Eng.Sc. Thesis, Monash University, Australia (1974).
8. I.A. Cutter and R. McPherson, The structure of MgO abrasive wear debris, Wear 71:255 (1981).
9. G.S. Lomdahl and R. McPherson, A scale effect in the abrasive wear of glass, Wear 73:205 (1981).
10. J.H. Giovanola and J. Finnie, On the machining of glass, J. Mater. Sci. 15:2508 (1980).
11. K. Kendall, Complexities of compression failure, Proc. Roy. Soc. 361A:245 (1978).
12. B.R. Lawn, J. Jensen and A. Arora, Brittleness as an indentation size effect, J. Mater. Sci. 11:573 (1976).
13. R.F. Recht, Catastrophic thermoplastic shear, Trans. ASME, Series E, 31:189 (1964).
14. B.F. von Turkovich, Shear stress in metal cutting, Trans. ASME, Series B, 92:151 (1970).
15. J.T. Black, On the fundamental mechanism of large strain plastic deformation, Trans. ASME, Series B, 93:507 (1971).

16. J.T. Black, Shear front lamella structure in large strain plastic deformation processes, Trans. ASME, Series B, 94:307 (1972).
17. S. Ramalingham and J.T. Black, An electron microscopy study of chip formation, Met. Trans. 4:1103 (1973).
18. E.D. Doyle and R.L. Aghan, Mechanism of metal removal in the polishing and fine grinding of hard metals, Met. Trans. B6:143 (1975).
19. G. Langford and M. Cohen, Strain hardening of iron by severe plastic deformation, Trans. A.S.M. 62:623 (1969).
20. H. Ormerod and W.J.McG. Tegart, Substructure formation in aluminium, copper and nickel during deformation at high temperatures and high strain rates, J. Inst. Metals 92:297 (1963-64).
21. H.J. McQueen, W.A. Wong and J.J. Jonas, Deformation of aluminium at high temperatures and strain rates, Can. J. Physics 45:1225 (1967).
22. M. Guillope and J.P. Poirier, Dynamic recrystallisation during creep of single-crystalline halite: an experimental study, J. Geophys. Res. 84:5557 (1979).
23. F. Jovane, J. Carro-Cao and J. Pomey, Transformation thermo-mecanique martensite → austenite, C.R. Acad. Sc. Paris, Series C, 272:629 (1971).
24. R. McPherson, Formation of metastable monoclinic rare earth sesquioxides from the melt, J. Mater. Sci. 18:1341 (1983).
25. F. Dachille, P.Y. Simons and R. Roy, Pressure-temperature studies of anatase, brookite, rutile and TiO_2-II, Amer. Mineral 53:1929 (1968).
26. O. Yamaguchi, H. Omaki, K. Takeoka and K. Shimizu, Thermal and mechanical properties of titanium dioxide (TiO_2) prepared by alkoxy-method, J. Jap. Soc. Powder and Powder Met. 22:173 (1975).
27. J.E. Hines, R.C. Bradt and J.V. Biggers, Delta alumina formation during the abrasive wear of a polycrystalline alumina, Proc. Int. Conf. Wear of Materials, ed. K.C. Ludema, W.A. Glaeser and J.K. Rhee, ASME, New York, p.540 (1979).
28. R. McPherson, Formation of metastable phases in flame and plasma prepared alumina, J. Mater. Sci. 8:851 (1973).
29. E. Hornbogen, K.H. Klein and I. Schmidt, On friction glazing of alloys, J. Mater. Sci. Letters 1:94 (1982).
30. B.R. Lawn, B.J. Hockey and S.M. Wiederhorn, Thermal effects in sharp-particle contact, J. Amer. Ceram. Soc. 63:356 (1980).
31. C.S. Yust and R.S. Crouse, Melting at particle impact sites during erosion of ceramics, Wear 51:193 (1978).

COMPACTION OF POWDERS. II. THORIA

I. Shapiro
Universal Chemical Systems, Inc.
Los Angeles, CA 90045-0124

ABSTRACT

The ambient temperature compaction of thoria powder occurs via a fragmentation process. The progress of the compaction phenomenon has been followed by a series of photomicrographs showing relative movement of particles as a function of external pressure. There is no indication of lateral movement of particles nor is there any evidence that particles slide past one another. Displacement in the direction of applied force occurs by individual particles fragmenting with the finer fragments filling the interstitial voids between particles. The relation of apparent density and porosity as a function of pressure is described. The influence of particle size upon compaction also is presented. Density measurements indicate that the thoria powder has a grain porosity of ca 16%. Scanning electron micrographs reveal that the grains are aggregates of smaller size particles. The technique of using photomicrographs with a special die to observe the compaction process of the particles while under load is presented in some detail. This paper is part of a series dealing with a fundamental study of compaction of powders covering a wide spectrum of consolidation properties ranging from plastic deformation to brittle fragmentation.

INTRODUCTION

This paper describes the compaction behavior of thoria powder at ambient temperature. This investigation is one of a series which deals with a fundamental study of compaction of powders, covering a broad spectrum of consolidation properties ranging from plastic deformation to brittle fragmentation. As stated in a previous paper describing the compaction of zirconia powder[1]), the primary objective of this series is to develop a general compaction equation for all powders. Previous work in this area by other investigators invariably degenerated into formulation of empirical relationships of only limited scope and utility.

EXPERIMENTAL

Materials

Thoria (thorium oxide) powder was obtained from the Zirconium Corporation of America, Solon, Ohio. Optical microscopic examination of this material could be interpreted that the thoria was prepared as a gel which had then been crushed to pass a certain size. However, scanning electron microscopic (SEM) examination, which is conducted at a higher magnification than that of the optical method, reveals that even the grain particles are composed of aggregates of smaller size particles that are bonded together. This effect can be seen in Figure 1, which shows the SEM of a grain at various magnifications. The aggregates must be strongly bonded because in the preparation of the powder the particles were crushed to smaller size particles yet the aggregates retained their identity regardless of the size of the grain.

The powder in the "as received" condition was size separated by standard sieving procedures using Tyler screens; three sized fractions were prepared. The particle size of the powder was measured by use of a microscope equipped with a calibrated filar micrometer eyepiece. Two hundred particles selected randomly from each sized fraction of powder were measured according to the method of Martin[2]). The values of the individual particle size measurements were arranged into equally size-limit groups, then the arithmetic average diameter of each group was plotted against the cumulative number (percent) of particles below the maximum size of the group--the plot being made on logarithm-probability paper.

The particle size distributions of the three fractions of thoria powder are shown in Figure 2, which includes the values for the log-geometric mean diameter, d_g, the log-geometric standard deviation, σ_g, and the mean-volume diameter, d_v, for each fraction. Actually, all of the statistical diameters can be calculated conveniently by the method of Hatch and Choate[3]). The parallelism of the three lines in Figure 2 (essentially the same value for σ_g) is expected because the particles were derived from the same source and have the same general shape and degree of circularity.

The need for actually measuring particle size rather than relying solely on screen sizes is brought to light in the case of thoria. Only for spheres does the screen size have a true value of

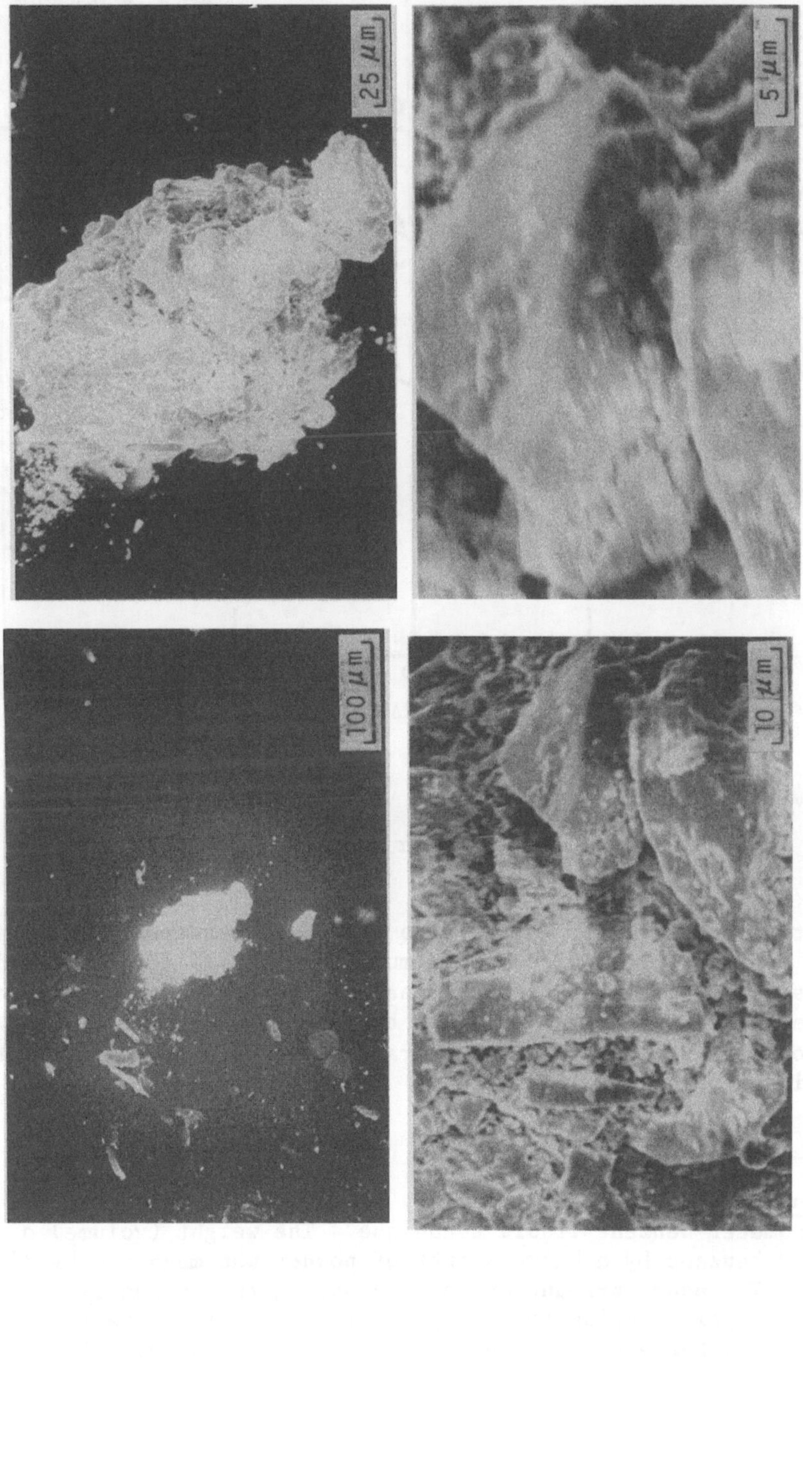

Fig. 1. Scanning electron microscopic views at various magnifications of a grain of thoria.

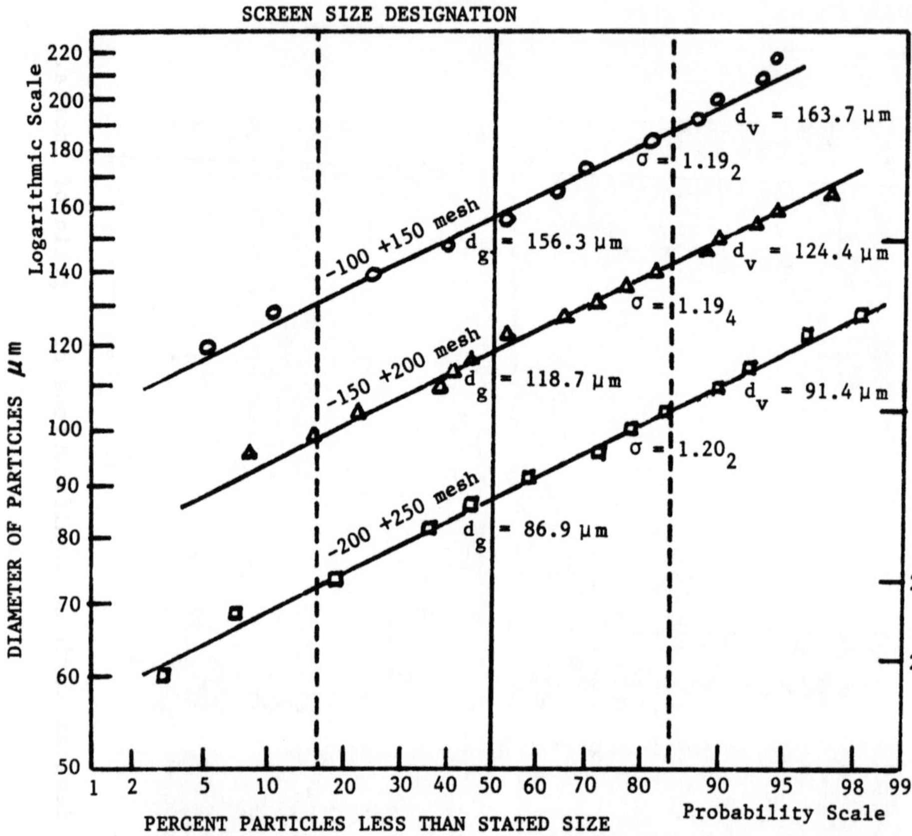

$$d_g = d_{50\%\ \text{size}}$$

$$d_v = \exp(\ln d_g + 1.5 \ln^2 \sigma_g) \qquad \sigma_g = \frac{d_{50\%\ \text{size}}}{d_{15.87\%\ \text{size}}} = \frac{d_{84.13\%\ \text{size}}}{d_{50\%\ \text{size}}}$$

Fig. 2. Particle size distribution of thoria powder.

size; it is possible for larger size irregularly shaped particles to pass a given screen size. Such must be the case for the thoria powder since in all instances the measured particle size is greater than the screen size. In each case the log-geometric mean diameter is approximately 1.25 times the average diameter opening of the two bracketing screens.

Density Measurements

True Density. The true density of the powder was measured by the pycnometer-benzene liquid technique. The weight (volume) of displaced benzene by a known weight of powder was measured by difference. The usual precautions of evacuating the pycnometer containing the powder prior to, and periodically during, the filling of the pycnometer with benzene were taken. The true density was determined to be 9.952 g/cc. The handbook density value of thoria is listed as 10.03 g/cc.

Particle Density. The particle or grain density of thoria was determined by the mercury pycnometer method. In this method the general techniques are the same as in the case of the benzene pycnometer method, except mercury is used as the liquid medium. The pycnometer containing the powder is evacuated prior to the filling of the mercury in the pycnometer. Under a differential of pressure of one atmosphere, the mercury will fill the voids between particles, but will not penetrate the intraparticle voids[4].

The particle or grain density of thoria by the mercury pycnometer method was found to be 8.341 g/cc. Hence the porosity of the individual grains is

$$[(9.952 - 8.341)/9.952]\ 100 = 16.2\%.$$

Unlike the porosity of, say, silica gel grains[5], the porosity of thoria may be due principally to the packing of the various size aggregates within the grain as indicated by the SEM photographs of Figure 1.

Bulk Density. In measuring the compaction of powders in a die, it is difficult to obtain an accurate value of the apparent density of the powder mass at "zero pressure" because even the slightest pressure on the piston during the set-up of the apparatus becomes significant. Consequently, to measure the apparent density of a powder mass at zero applied pressure, the bulk density is measured in separate experiments. Here the weight of powder in a known volume is taken after the container had been tapped repeatedly until no more settling of the powder occurs.

Bulk density measurements were made for the three particle size fractions of thoria used in the experiments reported here. The bulk density values are as follows:

Sample size Designation	Bulk Density, ρ_B g/cc
-100 +150 mesh	4.187
-150 +200 mesh	4.161
-200 +250 mesh	4.160

The overall bulk porosity for the -100 +150 mesh sample is 57.9%. Since bulk density includes both interparticle and intraparticle voids, the porosity due to particle packing (interparticle voids) for the same sample corresponds to

$$[(8.341-4.187)/8.341]\ 100 = 49.8\%.$$

Compaction Apparatus

For compaction of powder samples an evacuable three-way split die of 1.25 cm diameter bore was used. The opening of the cavity is measured by precision depth gages mounted on the exterior of the die. Actually, two precision gages are fastened on opposite sides of the piston so that any error in canting of the piston is eliminated. Pressures up to 7000 kg/cm^2 are applied to the die by means of a calibrated hydraulic press. For low pressures with the split-die an auxiliary pressure gage is mounted between the top of the piston and the movable plate of the hydraulic press. The apparatus is calibrated for the compressibility of the piston under load -- correction for depth of cavity due to compressibility of piston is critical. For visual and photographic observation of the powder under pressure, a glass-wall-lined die is used. The die body is composed of a glass sleeve encased in dental investment. An opening in the investment is made at the level of the pellet so the microscope tube and light can be inserted. The opening of the cavity between the movable and stationary pistons is measured by two precision gages arranged similarly to those for the split die. The glass sleeve must fit snugly with the pistons; the mortality rate of the glass sleeves is high. A camera attachment (Zeiss Attachment Camera) to the microscope permits taking photomicrographs of the powder through the glass sleeve.

In using the split die apparatus a known weight of powder is poured into the cavity and the apparatus assembled. A vacuum pump is connected to the outlet of the apparatus, and the two precision depth gages are read. Pressure is applied in increments, and the load is read on the auxiliary pressure gage for low-pressure region, and on the pressure gage of the hydraulic press for the high-pressure region. After each application of incremental pressure, the pressure value tends to decrease with time, the greatest rate of decrease occurring initially. The pressure is brought back up to each predetermined value—care being taken not to exceed the predetermined value at each setting. The difference in the average of the two precision depth gages with two pressure settings indicates the change in thickness of pellet after the correction for compression of the piston had been applied.

The amount of powder placed in the split-die is such that the initial height of the powder mass does not exceed 5 mm, which means the thickness-to-diameter ratio is kept below 0.4. In the case of the die used for taking photomicrographs, the amount of powder is adjusted such that the initial separation of the pistons is ca 1.5-2.0 mm, thus the entire length of the pellet will be in full view in the microscope at all times.

RESULTS

Compaction

Monolayer. Figure 3 shows a photomicrograph of a monolayer of thoria powder (-100 +150 mesh) before and after compression to 1000 kg/cm^2. It is obvious from Figure 3 that the compacting of this powder occurs via a fragmentation process rather than by plastic flow. Even after compression to 1000 kg/cm^2, some of the original thoria particles are still intact. One important observation of the fragmentation process is that a great number of minute particles are formed in addition to a rather small number of gross size fragments.

Polylayer. A study of the compaction of thoria powder was carried out in the special die so photomicrographs of the compact at various pressure stages could be taken. The photomicrograph series are shown in Figure 4. The individual photographs are arranged so that the bottom piston remains at the same level; the top piston shows the relative change in height of the filled cavity. Several colored grains act as fiducial points and their relative positions indicate there is no lateral motion of grains during compaction. As in the case of the monolayer, compaction is accompanied by fragmentation with the formation of large number of minute

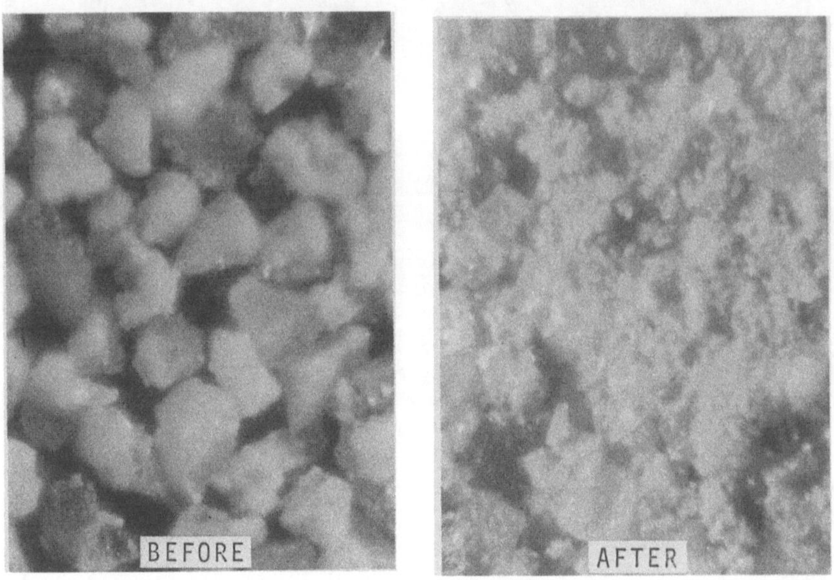

Fig. 3. Compaction of monolayer of thoria powder at 1000 kg/cm^2.

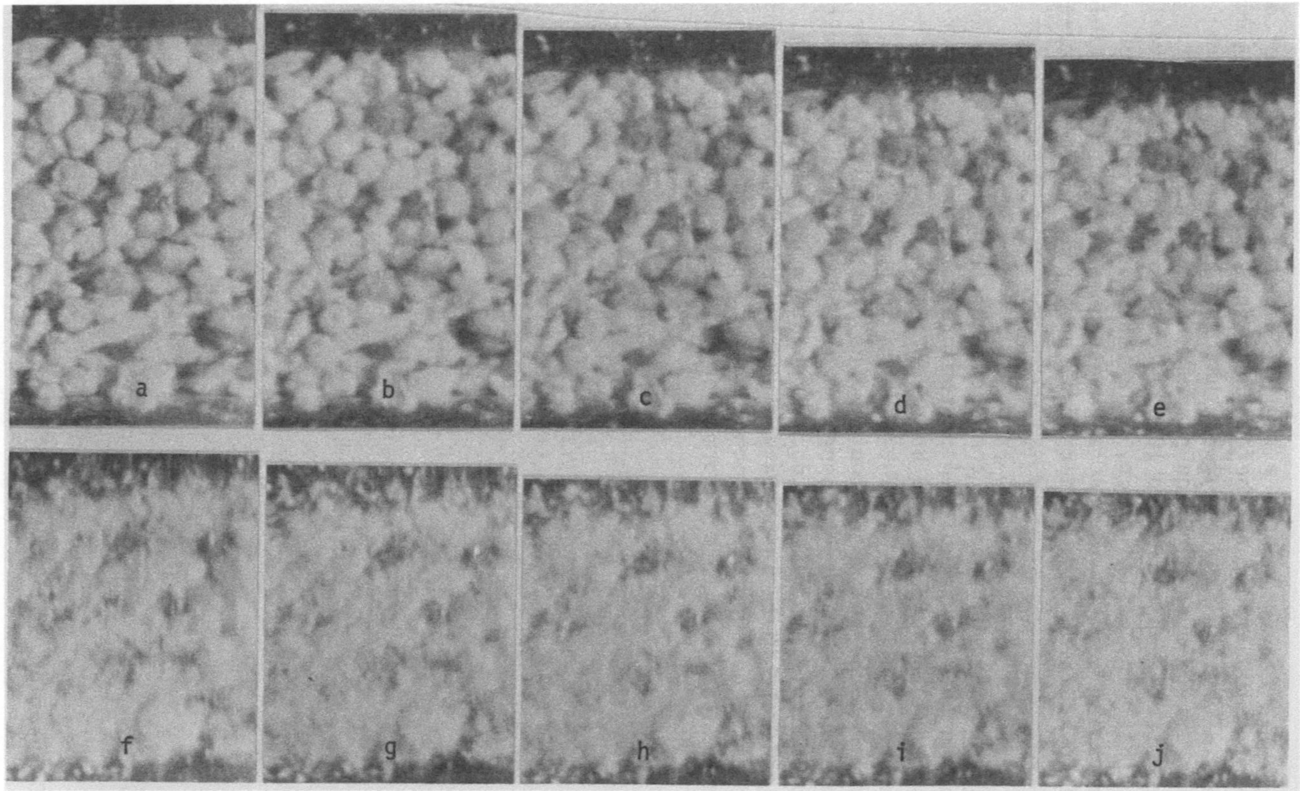

Fig. 4. Photomicrograph series showing compaction of thoria powder at various stages of pressure.

particles. Evidence of minute particle formation as a result of fragmentation is observed already at pressures in the 100 kg/cm^2 range. Presumably, at lower pressures compaction results in the fragmentation of contact points. The pressure is transmitted throughout the pellet since measurement of the relative positions of various particles in the photomicrographs show a proportional movement to the overall-piston movement.

Although considerable fragmentation has occurred at the higher pressures, a number of the original particles still are visible. The size of the fragmented particles is extremely small in comparison to that of the original particles. Many of the fine particles had fallen between the plunger and the die wall; however, it was possible to locate the surface of the plunger by scratch marks along the piston.

Apparent Density/Porosity

The apparent density and porosity values as a function of pressure for the material shown in Figure 4 are given in Table I. A comparison of the high pressure compaction of thoria powder as a function of grain size is given by the data in Table II. There is a trend that the larger size particles show a slightly greater degree of compaction at corresponding pressures; however, the actual difference in values is still very small. The observation that larger particles show a greater degree of compaction is to be expected on the basis of bearing surface exposure. For all three fractions the

Table I. Compaction of Thoria Powder at Low Pressures.

Pressure kg/cm^2	Apparent Density gm/cc	Porosity %	Ref. Fig. 4 Plate No.
0	4.18	58.6	a
42	4.32	56.8	b
87	4.56	54.4	c
134	4.80	52.0	d
178	4.99	50.1	e
657	6.00	40.0	f
982	6.38	36.3	g
1329	6.64	33.6	h
1683	6.82	31.8	i
2030	6.97	30.3	j

Particle size designation: -100 +150 mesh
Apparent density based on physical measurements of cavity in die.
(Includes porosity of individual grains)

Table II. Compaction of Thoria Powder at High Pressures.

Pressure Kg/cm^2	Percent Porosity -100 +150 mesh 1.540 gm	-100 +150 mesh 3.640 gm	-150 +200 mesh 4.662 gm	-200 +250 mesh 4.571 gm
0*	58.13*	58.13*	58.39*	58.40*
621	40.82	40.69	41.93	43.23
928	37.03	36.98	38.24	39.39
1256	34.20	34.52	35.59	36.65
1590	32.43	32.45	33.44	34.39
1918	30.78	30.83	31.85	32.75
2225	29.48	29.65	30.60	31.34
2873	27.33	27.60	28.47	29.13
3501	25.78	25.90	26.71	27.41
4156	24.55	24.75	25.51	26.09
4879	23.40	23.57	24.24	24.93
5521	22.61	22.69	23.48	24.04
6183	21.94	21.84	22.52	23.04
6811	20.99	21.21	21.82	22.42

*Bulk density values obtained from separate experiments.

general shape of the log-porosity vs pressure curves is the same, thus indicating a reliability in reproduction of data. A change in weight of a sample, corresponding to a change in a dimension of the pellet, does not alter the porosity values.

A typical apparent density plot for thoria powder as a function of pressure is shown in Figure 5. The values for ρ_o (4.187 g/cc), ρ_s (9.952 g/cc), and ρ_{grain} (8.341 g/cc) are indicated on the ordinate of the graph. Additionally, a value for ρ_∞ (9.0 g/cc) which is an extrapolated value to an "infinitely" high pressure also is included. This value represents a practical limit to which a pellet can be compacted at ambient temperature, and is determined by plotting the logarithm of the relative reduction in volume as a function of the reciprocal of the applied pressure.

DISCUSSION

The mechanism of compaction of thoria powder is one of fragmentation. For the thickness of pellets studied pressure is transmitted throughout the pellet. The observation of formation of many minute particles is understandable from the SEM photographs which show the

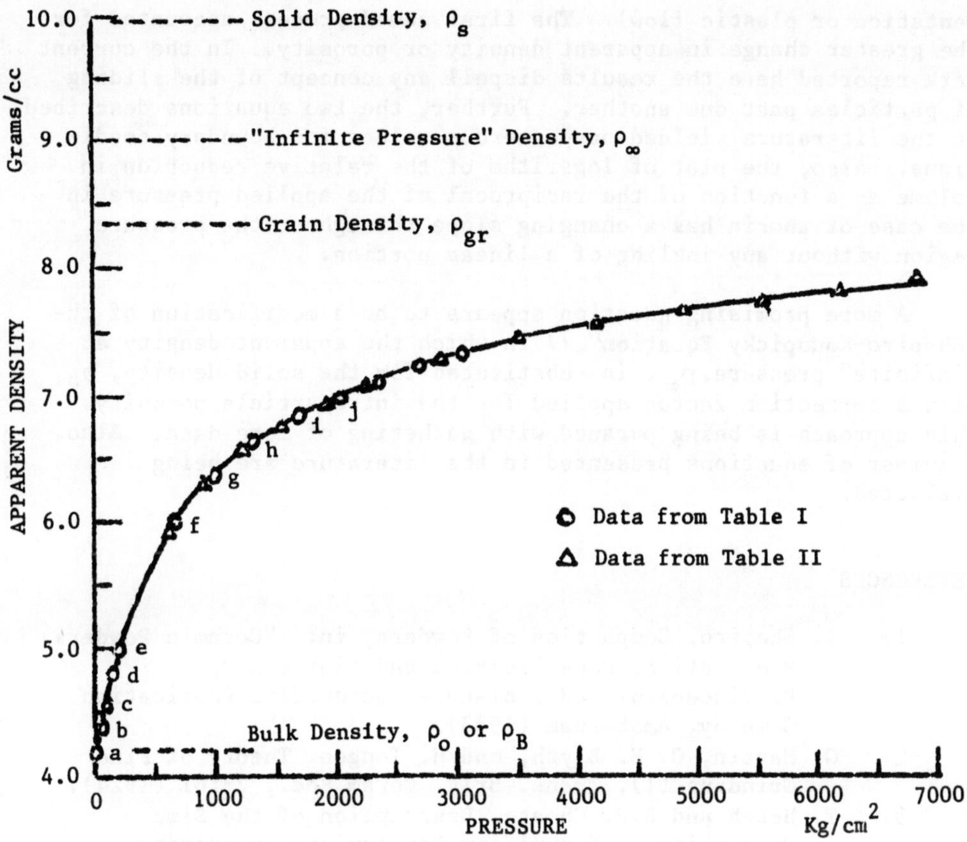

Fig. 5. Apparent density of thoria powder as a function of pressure.

grain particles really are aggregates of smaller size particles. Nevertheless, the grains have a certain rigidity as indicated by pressure transmittal.

As pressure is applied and particles fragment, the bearing surface progressively increases so the change in apparent density decreases with pressure. Since some of the original grains survive intact, the porosity of the original grains must be taken into account in any calculation of porosity.

The same plot used for calculating the practical limit of apparent density, ρ_∞, has been described in the literature[6)] to "explain" the compaction behavior of several ceramic powders. The curves were depicted as composed of two linear portions and were represented by two equations. The compaction process was based on the premise that compaction can be accounted by two largely independent processes, i.e., the filling of large holes (by particles sliding past one another), and the filling of small pores (by fragmentation or plastic flow). The first named process accounted for the greater change in apparent density or porosity. In the current work reported here the results dispell any concept of the sliding of particles past one another. Further, the two equations described in the literature yielded preposterous values for boundary conditions. Also, the plot of logarithm of the relative reduction in volume as a function of the reciprocal of the applied pressure in the case of thoria has a changing slope throughout the pressure region without any inkling of a linear portion.

A more promising equation appears to be a modification of the "Shapiro-Konopicky Equation",[7] in which the apparent density at "infinite" pressure, ρ_∞ , is substituted for the solid density, ρ_s, with a correction factor applied for the intraparticle porosity. This approach is being pursued with gathering of more data. Also, a number of equations presented in the literature are being evaluated.

REFERENCES

1. I. Shapiro, Compaction of Powders, in: "Ceramic Powders: Preparation, Consolidation and Sintering", P. Vincenzini, ed., Elsevier Scientific Publication Company, Amsterdam (1983).
2. G. Martin, C. E. Blyth, and H. Tongue, Theory of Fine Grinding (I), Trans. Brit. Ceram Soc., 23:61 (1924).
3. T. Hatch and S.P. Choate, Description of the Size Properties of Nonuniform Particulate Substances, J. Franklin Inst. 207:369 (1929).
4. H. L. Ritter and L. C. Drake, Pore-Size Distribution in Porous Materials (I), Ind. Eng. Chem., Anal. Ed. 17:782 (1945).
5. I. Shapiro and I. M. Kolthoff, Bulkiness and Porosity of Silica (Gel), J. Phys. and Colloid Chem., 52:1020 (1948).
6. A. R. Cooper, Jr., and L. E. Eaton, Compaction Behavior of Several Ceramic Powders, J. Amer. Ceram Soc. 45:97 (1962).
7. W. D. Jones, "Fundamental Principles of Powder Metallurgy", Edward Arnold (Publishers), London (1960).

CONTRIBUTORS

Co-Chairmen

R. E. Tressler, Professor and Chairman of Ceramic Science and Engineering, The Pennsylvania State University, University Park, PA

R. C. Bradt, Professor of Materials Science and Engineering, University of Washington, Seattle, WA

Conference Session Chairmen

J. R. Blachere, University of Pittsburgh, Pittsburgh, PA

R. J. Fields, National Bureau of Standards, Washington, DC

B. G. Koepke, Honeywell, Inc., New Hope, NM

J. Rabier, Labo. Metallurgie, Poitiers, France

J. L. Routbort, Argonne National Laboratory, Argonne, IL

L. Toth, National Science Foundation, Washington, DC

Conference Coordinator

Ronald Avillion, J. Orvis Keller Conference Center, The Pennsylvania State University, University Park, PA

AUTHORS

H. Alexander, Universität Köln, Köln, FRG

W. R. Blumenthal, University of California, Berkeley, CA

A. Bornhauser, Max Planck Institut für Metallforschung and Institut für Metallkunde, Stuttgart, FRG

W. J. Brindley, Michigan Technological University, Houghton, MI

S. Callahan, The Pennsylvania State University, University Park, PA

P. Carry, Lab. de ceramique, E.P.F.L., Lausanne, Switzerland

C. H. Carter, Jr., North Carolina University, Raleigh, NC

J. Castaing, Labo Physique Materiaux, Meudon-Cedex, France

J. L. Chermant, Laboratoire de Cristallographie, Chimie et Physique des Solides, Cedex, France

S. Chevacharoenkul, North Carolina State University, Raleigh, NC

P. N. Chopra, Cornell University, Ithaca, NY

T-J Chuang, National Bureau of Standards, Washington, DC

R. F. Davis, North Carolina State University, Raleigh, NC

J. C. Doukhan, Universite de Lille I, Cedex, France

R. Duclos, Universite de Lille I, Cedex, France

W. W. Durand, Honeywell, Inc., Minneapolis, MN

T. Epicier, Groupe d'Etudes de Métallurgie Physique et de Physique des Máteriaux, Cedex, France

J. D. Fitzgerald, Australian National University, Canberra, A.C.T. Australia

R. S. Gordon, Ceramatec, Inc., Salt Lake City, UT

G. Grathwohl, Universität Karlsruhe, Karlsruhe, FRG

R. H. J. Hannink, CSIRO, Melbourne Victoria, Australia

D. P. H. Hasselman, Virginia Polytechnic Institute, Blacksburg, VA

R. Hendricks, RCA Company, North Olmsted, OH

P. Hirsch, University of Oxford, Oxford, England

B. Hockey, National Bureau of Standards, Washington, DC

O. Jaoul, Universite de Paris, Cedex, France

R. S. Jupp, Michigan Technological University, Houghton, MI

A. S. Kobayashi, University of Washington, Seattle, WA

B. G. Koepke, Honeywell, Inc., New Hope, MN

D. Kohlstedt, Cornell University, Ithaca, NY

A. Krausz, University of Ottawa, Ottawa, Ontario, Canada

K. Krausz, University of Ottawa, Ottawa, Ontario, Canada

K. P. D. Lagerlof, Case Western University, Cleveland, OH

B. R. Lawn, National Bureau of Standards, Washington, DC

M. H. Lewis, University of Warwick, U.K.

R. McPherson, Monash University, Clayton, Victoria, Australia

U. Messerschmidt, Inst. of Solid State Phys. & Electron Microsc. Acad Sci, Halle/Saale, GDR

T. E. Mitchell, Case Western Reserve University, Cleveland, OH

H. J. Möller, TU Hamburg-Harburg, Halbleitertechnik, Hamburg, FRG

F. Osterstock, ISMRA - University of Caen, Caen, France

T. Page, University of Cambridge, Cambridge, England

B. Pletka, Michigan Technological University, Houghton, MI

J. Rabier, Labo. Metallurgie, Poitiers, France

R. Raj, Cornell University, Ithaca, NY

D. Rowcliffe, SRI International, Menlo Park, CA

H. Schmid, Max Planck Institut für Metallforschung, Stuttgart, FRG

W. Scott, University of Washington, Seattle, WA

I. Shapiro, Universal Chemical Systems, Inc., Los Angeles, CA

W. Skrotzki, Cornell University, Ithaca, NY

T. Soma, NGK Insulators, Ltd., Nagoya, Japan

R. W. Steinbrech, Universität Dortmund, Dortmund, FRG

N. J. Tighe, National Bureau of Standards, Washington, DC

R. E. Tressler, The Pennsylvania State University, University Park, PA

A. Venkateswaran, Virginia Polytechnic Institute, Blacksburg, VA

D. S. Wilkinson, McMaster University, Hamilton, Ontario, Canada

INDEX

MIX
Papier aus verantwortungsvollen Quellen
Paper from responsible sources
FSC® C105338

If you have any concerns about our products,
you can contact us on
ProductSafety@springernature.com

In case Publisher is established outside the EU,
the EU authorized representative is:
Springer Nature Customer Service Center GmbH
Europaplatz 3, 69115 Heidelberg, Germany

Printed by Libri Plureos GmbH
in Hamburg, Germany